SEMANTICS

An interdisciplinary reader in
philosophy, linguistics and psychology

Semantics

AN INTERDISCIPLINARY READER IN
PHILOSOPHY, LINGUISTICS
AND PSYCHOLOGY

Edited by
DANNY D. STEINBERG
Department of English as a Second Language,
University of Hawaii

& LEON A. JAKOBOVITS
Institute of Communications Research,
Center for Comparative Psycholinguistics,
University of Illinois

CAMBRIDGE
AT THE UNIVERSITY PRESS

Published by the Syndics of the Cambridge University Press
Bentley House, 200 Euston Road, London N.W.1
American Branch: 32 East 57th Street, New York, N.Y.10022

© Cambridge University Press 1971

Library of Congress Catalogue Card Number: 78–123675

ISBNs: 0 521 07822 9   hard covers

0 521 09881 5   paperback

Typeset in Great Britain
at the University Printing House, Cambridge

Printed in the United States of America

First Published 1971
Reprinted 1972
1974
1975

Contents

[v]

Preface

This anthology presents an inquiry into the nature of a variety of semantic problems. Illumination of such problems is to be found in a number of disciplines, three of which are represented in this volume. While it is known that the perspectives and aims of these disciplines differ, it is not generally recognized that once the actual prosecution of the task of studying semantics begins there is a great commonality of interest. It is hoped that the selected semantic problems considered in this volume will contribute to bringing about such a realization.

We would like to express our gratitude to John Murphy and Charles Caton of the Department of Philosophy at the University of Illinois, Kenneth Hale of the Department of Modern Languages and Linguistics at Massachusetts Institute of Technology, Oswald Werner of the Department of Anthropology at Northwestern University and to Howard Maclay of the Institute of Communications Research at the University of Illinois for their editorial assistance and many helpful suggestions. The work of the senior editor was supported (in part) by a Public Health Service Fellowship, No. 1-F2-MH 35, 206-01 (MTLH) from the National Institute of Mental Health.

D.D.S.
L.A.J.

November 1970

Acknowledgments and references to reprinted articles

Bever, Thomas G. and Rosenbaum, Peter S. Some lexical structures and their empirical validity. In R. A. Jacobs and P. S. Rosenbaum (eds.), *Readings in English Transformational Grammar*. Waltham, Mass.: Blaisdell, 1970. Reprinted by permission of the authors, editors, and publisher.

Chomsky, Noam. Deep structure, surface structure, and semantic interpretation. In Professor Shiro Hattori's 6oth birthday commemorative volume, *Studies in General and Oriental Linguistics*. Tokyo: TEC Corporation for Language and Educational Research. Reprinted by permission of the author, editors, and publisher.

Donnellan, Keith. Reference and definite descriptions. *Philosophical Review*, LXXV, 1966, 281–304. Reprinted by permission of the author and publisher.

Fodor, Jerry A. Could Meaning be an r_m? *Journal of Verbal Learning and Verbal Behavior*, IV, 1965, 73–81. Reprinted by permission of the author and publisher.

Grice, H. P. Meaning. *Philosophical Review*, LXVI, 1957, 377–88. Reprinted by permission of the author and publisher.

Harman, Gilbert H. Three levels of meaning. *The Journal of Philosophy*, LXV, 1968, 590–602. Reprinted by permission of the author and publisher.

Katz, Jerrold J. Semantic theory. From *The Philosophy of Language* by Jerrold J. Katz. Copyright © 1966 by Jerrold J. Katz. Reprinted by permission of Harper & Row, Publishers, Inc.

Kiparsky, Paul and Kiparsky, Carol. Fact. In M. Bierwisch and K. Heidolph (eds.), *Progress in Linguistics*. The Hague: Mouton, 1970. Reprinted by permission of the authors, editors, and publisher.

Lenneberg, Eric H. Author's revision of chapter 8: Language and cognition, *Biological Foundations of Language*. New York: John Wiley & Sons, 1967. Reprinted by permission of the author and publisher.

Linsky, Leonard. Reference and referents. Chapter 8 of *Referring*. New York: Humanities Press, 1967. Reprinted by permission of the author and publisher.

McCawley, James D. Where do noun phrases come from? An extended version of a paper in R. A. Jacobs and P. S. Rosenbaum (eds.), *Readings in English Transformational Grammar*. Waltham, Mass.: Blaisdell, 1970 Reprinted by permission of the author, editors, and publisher.

Miller, George A. Empirical methods in the study of semantics. A part of II: *Psycholinguistic approaches to the study of communication*. In D. L. Arm (ed.), *Journeys in Science: Small Steps – Great Strides*. Albuquerque: University of New Mexico Press, 1967. Reprinted by permission of the author and publisher.

Quine, W. V. The inscrutability of reference. A part of 'Ontological relativity', *The Journal of Philosophy*, LXV, 1968, 185–212. New York: Columbia University Press. Reprinted by permission of the author and publisher.

Searle, J. R. The problem of proper names. A section of *Speech Acts: An Essay in the Philosophy of Language*. Cambridge: Cambridge University Press, 1969. Reprinted by permission of the author and publisher.

Strawson, P. F. Identifying reference and truth-values. *Theoria*, xxx, 1964, 96–118. Reprinted by permission of the author and publisher.

Vendler, Zeno. Singular terms. Reprinted from Zeno Vendler, *Linguistics in Philosophy*. Copyright © 1967 by Cornell University. Used by permission of Cornell University Press.

Weinreich, Uriel. A part of 'Explorations in semantic theory'. In T. A. Sebeok (ed.), *Current Trends in Linguistics*, volume III. The Hague: Mouton, 1966. Reprinted by permission of the publisher.

Ziff, Paul. On H. P. Grice's account of meaning. *Analysis*, XXVIII, 1967, 1–8. Reprinted by permission of the author and publisher.

To Charles E. Osgood

PART I

PHILOSOPHY

Overview

CHARLES E. CATON

The part of philosophy known as the philosophy of language, which includes and is sometimes identified with the part known as semantics, is as diverse in its problems and viewpoints as any part of philosophy. As in the case of other philosophical fields, many of the problems of current interest have long histories. Also, as elsewhere, problems in the philosophy of language are interrelated with problems in other areas notably epistemplogy, logic, and metaphysics, but also for example the philosophy of mind and the philosophy of science. I will not try in the space of the present overview to present any serious account of the history of the subject or a general introduction to its problems.[a] I will try to give only a brief account of its recent history of and some introduction to the problems that are dealt with in the papers that are here included. First, some preliminary remarks need to be made.

The time is ripe for a collection of papers from the various disciplines that deal in their several ways with one or another aspect of meaning. I think inquirers in all these fields – in philosophy, psychology, linguistics, etc. – would each agree that there were significant questions about meaning with which they themselves didn't deal and which perhaps properly belonged to another discipline. In this overview to the philosophy section of this collection, I will try only to give some impression of the recent history of philosophical investigations concerning meaning and reference and my personal assessment of the present situation. I must emphasize the fact that this will be a *personal* assessment: in the area of philosophy of language perhaps as much as in any other philosophical field, chaos reigns, little is agreed on, new methods are to be eagerly seized upon, and we have hardly begun. Philosophy has historically both taken inspiration from and made contributions to other disciplines, ones not perhaps at the time in question thought of as 'philosophical': I think that this is very liable in the future to prove to have been true of the present time in the philosophy of language.

A preliminary statement about the provenance of the articles included should also probably be made. They all belong to that segment of philosophical opinion known

---

[a] The history of philosophy of language is comprehensively treated in Norman Kretzmann's article 'Semantics, History of' in *The Encyclopedia of Philosophy*, ed. Paul Edwards. A recent introduction is W. P. Alston's *Philosophy of Language* (Englewood Cliffs, N.J.: Prentice-Hall, 1964); see also the more treatise-like book of J. J. Katz, *The Philosophy of Language* (New York: Harper Row, 1966). There are several anthologies: Leonard Linsky's *Semantics and the Philosophy of Language* (Urbana: University of Illinois Press, 1952 – a new, expanded edition is announced), my *Philosophy and Ordinary Language* (idem, 1963), Vere Chappell's *Ordinary Language* (Englewood Cliffs, N.J.: Prentice-Hall, 1964), Katz and J. A. Fodor's *The Structure of Language* (idem, 1964), G. H. R. Parkinson's *The Theory of Meaning* (London: Oxford University Press, 1968).

as analytic philosophy.[a] Thus there are no selections included from the writings of phenomenologists or existentialists, for example, though of course philosophers of these and other viewpoints have written in recent times about language or philosophical problems concerning language.[b] Besides obvious considerations of space, in justification of leaving out certain well-known writers in the philosophy of language, it might I think be said that philosophical writings in the analytic tradition are more accessible to laymen, whether or not initially more interesting to them. And my own opinion is that, in this area at least, the writings of analytic philosophers are as liable to be of interest to workers in other fields as those written from any other current philosophical viewpoint. This is, of course, not a surprising situation, in view of the fact that the hallmark of analytic philosophy is precisely to approach philosophical problems and theories in terms of the language used to formulate them.[c]

As a final preliminary, it should be said that it has been decided that the selections to be included should fall within semantics in the narrow sense, i.e. problems of meaning and reference. Thus a number of topics in the philosophy of language which have recently been discussed extensively in the philosophical literature do not appear at all or do so only in passing. Among these topics are presupposition, contextual implication, truth, and the analytic-synthetic distinction. Especially, a large and important literature not represented here has to do with the analysis and classification of 'speech acts', things one does by or in saying something, including how such acts relate to meaning and reference.[d] But one must stop somewhere.

The division of semantics in the narrow sense into problems to do with meaning and problems to do with reference (or referring) might almost be said to be traditional in contemporary philosophy. This does not mean, of course, that there is much agreement over anything that comes under either heading or even over the distinction itself. It is better to try to convey the latter by citing examples of problems on either side, rather than by explicit distinction. Grouped under the topic of meaning are such questions as what it is for a linguistic expression to mean a particular thing or to have a particular meaning, what it is for one expression to be synonymous with or a paraphrase of another, and what it is for a sentence (or statement) to be analytic, i.e. true in virtue only of its meaning (or the meaning of a sentence which expresses it). Grouped under the topic of reference are such questions as what it is for an expression to refer to something, what it is for a person to do so, and how these two are related, what sort of meaning certain referring expressions have, and

---

[a] So does most of the literature mentioned in the preceding footnote.

[b] See the last section of Kretzmann's article mentioned in the first footnote for a general survey of recent philosophy of language.

[c] Among the accounts of the history of analytic philosophy are J. O. Urmson's *Philosophical Analysis*, its development between the two world wars (London: Oxford University Press, 1956), G. J. Warnock's *English Philosophy since 1900* (London: Oxford University Press, 1958), J. A. Passmore's *A Hundred Years of Philosophy* (London: Duckworth, 1957), and A. M. Quinton's 'Contemporary British Philosophy' in D. J. O'Connor's *A Critical History of Western Philosophy* (London: Collier-Macmillan, 1964).

[d] J. L. Austin is the prime-mover in this connection; see his *How to Do Things with Words* (Cambridge, Mass.: Harvard University Press, 1962), for example. John R. Searle has a book, *Speech Acts* (Cambridge: Cambridge University Press, 1968) (see below), as well as several important papers in this area, at least one of which, 'Austin on Locutionary and Illocutionary Acts' (*The Philosophical Review*, LXXVII (1968), pp. 405–24), is later than his book. See also W. P. Alston's 'Meaning and Use', *Philosophical Quarterly*, XIII (1963), pp. 107–24, revised in Parkinson's anthology *The Theory of Meaning*. See also Alston's *Philosophy of Language*, especially ch. 2.

whether complete sentences have references at all (candidates: facts, propositions, truth-values), whether some expressions have only references and not also or instead meanings, etc. Also, naturally, there are questions and views which involve both meaning and reference and the relations between the two: for example, one historically important type of semantical theory attempts to provide a basis for understanding language and its use entirely in terms of reference (naming, denoting, etc.), i.e. without using a distinct concept of meaning as well.[a]

Many of these inquiries might be regarded as simply the pursuit of the venerable philosophical task of clarifying a matter which is puzzling or problematical and which no one else is concerned to clarify, i.e. (if you will) as preliminary, quasi-scientific speculation concerning an essentially scientific question which is not at the moment capable of being treated as such. Philosophy has traditionally been the repository of such problems. But, from the history of semantics, it is apparent that other philosophical motives have been present from early times. For example, certain epistemological and metaphysical problems have been regarded as involving the question of what meaning is. The fact that the answers to some of these questions are not clear even now would itself be sufficient to explain why philosophers are still concerned with analyzing the concept of meaning. Consider just the ancient 'problem of universals': one aspect or version of this problem is (roughly and briefly) the question whether from the obvious fact that certain types of linguistic expressions, so-called 'general terms' (common nouns and most verbs and adjectives) occur, or occur in certain ways, in sentences used to express true or at least coherent statements about different things, it somehow follows that there are certain entities, perhaps abstract, which correspond not to the things the statements are about but rather to what is said about them. This is red, that is red, so is there redness as well? Essentially this same question has been discussed by both Plato and Quine and by both because both were pursuing ontology: a special case of what there *is* is what there has to be if language is to work in the way it does, so of course *how* it works and what there has to *be*, given that it works that way, become questions of interest to ontologists.[b] This kind of situation also arises in epistemology, logic, and other fields.

I will now sketch briefly the background in the recent history of analytic philosophy which seems to me necessary or desirable in approaching the essays in philosophical semantics included in the present section of this collection. There would probably be general agreement among philosophers that the late Ludwig Wittgenstein (died 1951) and John L. Austin (died 1960) are at least among the most important analytic philosophers in their influence on thinking about language, if indeed they aren't the two most important. I will begin by discussing their conceptions of meaning, which in the case of both involves also discussing referring (naming, etc.) in that both were opponents of what has been called the denotative or entitative theory of meaning – though it is primarily Wittgenstein whose influence was the greatest along this line.

The concept of meaning in recent philosophy is an interesting story for two reasons: first, what separates several of the recent movements in philosophy is

[a] Russell's semantics of around 1905–12, in e.g. 'On Denoting' (*Mind*, 1905, often reprinted) and *The Problems of Philosophy* (London: Oxford University Press, originally 1912), is a view of this type, with the avowed aim of doing without meanings in addition to references.

[b] See Alston, *Philosophy of Language*, Introduction, for other examples of the philosophical relevance of questions about language.

primarily differences that have to do with the concept of meaning and, second, one or two quite novel accounts of meaning have been put forward and widely accepted recently. Briefly, as to the first, the recent movements of pragmatism, logical positivism, and 'ordinary language' philosophy are separated, among other things, by different conceptions of meaning.[a] As to the second, I refer to 'one or two' conceptions of meaning because, although it is Wittgenstein's remarks on meaning that I have chiefly in mind, his writings are difficult and have been differently understood.[b] In Gilbert Ryle's well-known paper 'The Theory of Meaning'[c] one gets a pretty good idea of the background in preceding philosophy of Wittgenstein's work, what led up to his work, and some of the prevailing sorts of view – along with some indication of what he, Wittgenstein, wanted to say. In Ryle's 'Ordinary Language',[d] one gets more the form his views have taken when others elaborated them – I cite Ryle because he is a leading figure here – and I think the form in which they tend to be held by others. The latter, the popular form of Wittgenstein's ideas, can (I think not too unfairly) be regarded as a sort of rigidified, simplified form of Wittgenstein's views. I will treat of the genuine article first and then turn briefly to the more popular version.[e]

Wittgenstein's handling of the concept of meaning in the *Philosophical Investigations* and the *Blue and Brown Books* has in common with the popular version of it two things: (1) that it is essentially negative, in that it denies what is alleged by its adherents to have been a widespread, indeed practically universal tendency, in philosophical theories of meaning, and (2) that it proposes to substitute for this what it represents as the *ordinary* conception of meaning, the everyday concept in daily use, which is regarded as not suffering from the traditional sort of mistake. (It is over (2) of Wittgenstein's ideas that the genuine and popular versions primarily differ, i.e. over what the ordinary conception is regarded as being.)

The traditional mistake is described as thinking of the meaning of a word or phrase as an object or thing or entity.[f] Two questions arise here: (i) What *is* it to conceive of the meaning of a word or phrase as an object (thing, entity)? and (ii) What's wrong with thinking of the meaning of a word or phrase in this way? (i) This question arises because it is not clear exactly what Wittgenstein, Ryle, Austin, *et al.*, other philosophers, or, for that matter, the ordinary man would count as an 'object' or 'thing' or 'entity', i.e. where the dividing line is between 'objects' and non-objects, etc.; and it is clear that these words are pretty comprehensive ones. I think, actually, that this way of putting (i), though common, is not very perspicuous. What

---

[a] Of course there are differences on this matter, like others, even within a given movement; and perhaps sometimes these will be as great as differences between adherents of different schools. I am speaking broadly and of what is typical or characteristic.

[b] It is the later Wittgenstein I have in mind here, after 1930 or so. The relevant writings are his posthumously published *Philosophical Investigations* (Oxford: Blackwell, 1953) and the earlier versions of parts of that work known as *The Blue and Brown Books* (Oxford: Blackwell, 1958), originally privately circulated in the late 1930s.

[c] In *British Philosophy in the Mid-Century*, ed. C. A. Mace (London: George Allen & Unwin, 1957) originally, and several times reprinted.

[d] *The Philosophical Review*, LXII (1953), pp. 167–86. This and the preceding article are reprinted in my anthology and elsewhere.

[e] Austin's 'The Meaning of a Word', in his *Philosophical Papers* (Oxford: Oxford University Press, 1961), especially Part I, is rather like the genuine Wittgenstein article – apparently through historical coincidence.

[f] Cf. Ryle's 'The Theory of Meaning' and Austin's 'The Meaning of a Word', Part II. See also Alston's 'The Quest for Meanings', *Mind*, LXII (1963), pp. 79–87.

the point seems to come to is this: that there isn't any thing X or any such thing as the ϕ such that, for a given word or phrase W, the meaning of W is X or the ϕ. This reading seems to accord both with what Wittgenstein and the others *say* and with what they are attacking. For, as to the latter, it does seem to be a feature of many theories of meaning that had been put forward by philosophers and others before (say) 1935, that they *did* purport to provide an X or such a thing as the ϕ which allegedly answered to the description that the meaning of a given word W was X or the ϕ. And it was apparently thought to be part of being a theory of meaning to provide this thing X or the ϕ *in a general way*, so that the theory entailed or consisted in saying what kinds of things meanings were, there being perhaps several kinds according to the theory.

The alleged mistake is suggested to be that traditional theories always provided such an answer to 'What is the meaning of W?' (or 'What does W mean?') that the answer could, in principle, be put in the form 'X' or 'the ϕ', i.e. be put in the form of naming or identifying the thing X or the ϕ which *was* (the very same thing as) the meaning of W.[a] This assertion seems to me as a matter of history to be correct.

The allegation is thus that this traditional feature is an error, and in fact an error due to a (quite natural) misconception of ordinary talk about meaning. This sort of error might be called 'using a misleading model' (i.e. to understand meaning) or 'being misled by a grammatical similarity' (viz. of locutions in 'meaning', 'means', etc., to other locutions) – with the result that the concept, depth-grammar, or logic of 'meaning' or of the concept of meaning is misunderstood. That is (the suggestion is), philosophers think there must be an answer 'X' or 'the ϕ' to the question 'What is the meaning of W?' or 'What does W mean?', because they take it (perhaps unconsciously) to be like 'Who is Johnny's teacher?' or 'Who teaches Johnny?', which may have the answer 'Mary' or 'the girl in the corner'.

Thus, in belief, the error alleged is that of thinking that one can *name* and/or *identify* (the thing that is) the meaning of W, or that one can explain the meaning of W by naming or identifying the meaning of W as X or as the ϕ. It is, for example, the error of thinking that 'W means X' is like 'Mary teaches Johnny'.

The question remains: (ii) *why*, according to these philosophers, is it an error to suppose that one can name (as X) or identify (as the ϕ) the meaning of a word or phrase W? I think they would answer simply that there is no such nameable or identifiable thing (or things or sets of things) which is the meaning. This may seem just a denial of the assumption, but the support for it consists in (*a*) an account or sketch or set of reminders of how we ordinarily handle the concept of meaning (e.g. of what is involved in explaining the meaning of a word, how we explain the meaning of a word), together with (*b*) an account or sketch of what these ways of handling the concept presuppose. Thus, as to (a), Wittgenstein begins the *Blue Book* by immediately changing the subject from 'What is the meaning of a word?' to 'What is an explanation of the meaning of a word?' And similarly Austin in 'The Meaning of a Word', at the beginning of Part II. Both philosophers begin by distinguishing verbal explanantions from what Wittgenstein calls ostensive ones or what Austin describes as ones involving imagining or experiencing situations related to the meaning of the word in question. Wittgenstein examines ostensive definition at length, because verbal definitions presuppose that the pupil can already understand *some* language (viz. that in terms of which the word being explained is being explained).

[a] Ryle epitomizes this feature of these theories in his name for them, ' " Fido "-Fido theories'. The word 'Fido' means Fido, the dog, which is its meaning.

Other relations besides the one allegedly denoted by 'means' are no help to traditional theories: I mean such things as naming, standing for, referring, denoting, connoting, etc.[a] These are no help because the ordinary concepts expressed by these words won't bear the weight of a general semantical theory and because philosophers have used them as they have used the notion of meaning already discussed, i.e. in such a way that Wittgenstein's and others' considerations apply. As I have said, most or all traditional theories have taken one or more such relations as basic for semantics.

So much for the traditional error which is alleged to characterize most or all earlier theories of meaning in philosophy.

What it is proposed to substitute for the conception of meaning as some sort of entity (in the sense explained above) is what has been called the *Doctrine of Meaning as Use*, this being taken to state the ordinary conception of meaning. In the form in which it is presented in the popular version of Wittgenstein's views on this subject, this doctrine says (as in Ryle's 'Ordinary Language') that the meaning of a word or phrase is its ordinary use, i.e. its standard use (not necessarily its everyday use), in the sense of the way it is used in the language. I don't believe that anything so specific is to be found anywhere in Wittgenstein's published works, though he does repeatedly recommend, as a point of philosophical method, 'looking to' or 'asking for' the use rather than the meaning; and it is clear that the reason he does so is the tendency philosophers have to fall into the traditional error already described.[b] Also, some of Wittgenstein's remarks seem to sanction such a theory of meaning, e.g. (the one most cited):

§43. For a *large* class of cases – though not for all – in which we employ the word 'meaning' [*Bedeutung*] it can be defined [or, explained] thus: the meaning of a word is its use [*Gebrauch*] in the language.[c]

But I think probably that Wittgenstein, like Austin in 'The Meaning of a Word', was just intending to remind us of how (as he thought) one used 'meaning' (including explaining meanings), i.e. what 'meaning' and 'means' mean, just as Austin was 'explaining the syntactics' and 'demonstrating the semantics' of 'meaning', in accordance with his general account of what it is to explain the meaning of a word.

As we have just seen, denotative or entitative theories of meaning, i.e. those which attempted to provide, by a general formula, some entity or thing as the meaning of a linguistic expression (or, taking this as basic, to reduce all other varieties of meaning to it), have tended to dominate recent thinking in philosophical semantics and its history as well. Hence concepts like referring, denoting, naming, etc., on the referring side of the fence, have come in for considerable study. There is a series of articles of great philosophical interest, stretching from Frege's 'On Sense and Reference' of 1892[d] and Russell's 'On Denoting' of 1905, which is directed at Frege's views (as well as others'), to the selections in the present volume. This literature has been expounded and evaluated by Leonard Linsky in his book

---

[a] I have spoken as though meaning were all that was in question; but though this is perhaps the most favored term, various others have been used.

[b] A good bit of the *Philosophical Investigations* has to do with the ins and outs of this error, viz. of thinking of the meaning of a word or phrase as an entity.

[c] *Philosophical Investigations*, p. 20e.

[d] Translated as 'On Sense and Nominatum' by Herbert Feigl in his and Wilfred Sellars' anthology *Readings in Philosophical Analysis* (New York: Appleton-Century Crofts, 1949) and by Max Black under the above title in his and P. T. Geach's *Translations from the Philosophical Writings of Gottlob Frege* (Oxford: Blackwell, 1952; 2nd ed., 1960).

Referring,[a] which takes one up to the present articles. These earlier papers – not included here because of their easy availability elsewhere and because they can, though still well worth studying closely, reasonably be regarded as part of the background of contemporary research, rather than themselves part of it – together with Strawson's earlier 'On Referring' (1950),[b] have as their essential points the following, which I try to place in some perspective with respect to the author's own views. Frege's semantical theory was based on two relations, one of the meaning sort and one of the referring sort. He held that expressions of a natural language had the former relation, which he called 'expressing (a sense)' (*ausdrücken*) unless they were nonsense or had no 'meaning', but even so might not 'stand for' or 'refer' (*bedeuten, bezeichnen*) any object or entity, and that it was a desideratum of a scientific language that it should guarantee a reference, as well as a sense, to all of its expressions – including sentences, the reference of which (if any) he held to be a truth-value, the True or the False. Russell, though holding a similar view earlier, came in 1905, in 'On Denoting', to regard Frege's semantics as incoherent and, apart from this, involving an unnecessary postulation of types of entities, viz. the abstract entities called 'senses' which Frege held to be expressed by the meaningful expressions of any language. Russell advanced instead a semantics in which, allegedly, only a naming relation (a 'referring' concept, then) was fundamental; expressions which could not be regarded as names (e.g. because there wasn't any such thing as what they might have otherwise been regarded as naming) were *analyzed,* i.e. sentences containing them were explained to really mean[c] the same as other sentences not containing them but rather containing only Russellian names. An important or crucial tool in carrying out the analyses required to make this account plausible was his Theory of Definite Descriptions, which in his best known example required analyzing 'The present king of France is bald' as really meaning 'There is one and only one present king of France and he is bald', so that the former statement can be regarded as false (rather than, botheringly, meaningless) because there is no present king of France, as the latter version clearly states there to be. Strawson's article 'On Referring' sided with Frege in holding that meaning was not referring and that 'The present king of France is bald' (or rather the statement one can make with it, which he regarded as distinct from this sentence) was without a truth-value, neither true nor false, since there was no king of France, as it (he said) *presupposed,* rather than stated or entailed (as Russell had said).

Note that both Russell and Frege supplied entities that were the meanings of linguistic expressions in their semantical theories, though Strawson did not: Frege, certainly, since he assumed a special type of entity, his 'senses', as the meanings of expressions, as well as the 'references' of those expressions that had them; but Russell too, at least in the sense that the only semantical properties and relations he was at this period prepared to recognize involved an expression as one relatum and what was certainly an entity as the other: the nominata of Russellian names are the only candidates for 'meanings' that he supplies in this theory. It is theories of this

[a] London: Routledge & Kegan Paul, 1967. See also his article 'Referring' in Edwards' *Encyclopedia.*

[b] *Mind,* LIX (1950), pp. 320–44.

[c] It is clear here, though not often insisted upon, that a relation of synonymy between sentence (-forms) is also involved in Russell's semantics of this period, i.e. besides his naming relation. If the latter relation is really the only one, as Russell apparently thought, then he would have to hold that this synonymy relation was definable in terms of naming; but he nowhere even indicates an awareness of the problem, much less offers a solution to it.

rather general sort, which includes both Frege's and Russell's, that Wittgenstein, Austin, Ryle, *et al.* attacked.

Writings of Wittgenstein and Austin themselves are not included, since unfortunately their work ended some years ago.[a] This is not by any means to say that everything that can be got out of their writings has been got out already. But it is true that their work forms the background for much current research. Significantly, in connection with the present collection, the remaining major influence on current research in the philosophy of language is linguistics – specifically and primarily, transformational or generative grammar.[b] For this reason, several of the selections in the linguistics and psychology sections of this collection are of direct interest to the concerns of some contemporary philosophers of language.[c] It should also be mentioned that one particular recent development in the philosophy of language – or perhaps rather in analytic philosophy itself – viz. the theories of J. J. Katz, is in its very conception integrated with and based on generative grammar.[d] A somewhat different approach, which seems to rely on specific syntactical results and conjectures about words important to the problem under consideration, more than on speculations as to the form semantic theory should take, is practised by Zeno Vendler, e.g. in his essay 'Singular Terms' reprinted here.[e]

Having tried to give some indication of their background in the recent philosophy of language and some introduction to the problems which are dealt with in the selections here included, I will now say something more particularly about each of the latter. Those included under the rubric Reference constitute the leading articles from the philosophical periodical[f] literature on referring since interest was renewed in the subject by Strawson and Geach in the 1950s.[g]

[a] In the case of Wittgenstein, parts of his literary remains and notes by others of his lectures and conversations are still appearing. In the case of both men, the bulk of their published works have appeared posthumously.

[b] I am still speaking broadly. At an earlier date, behaviorist psychology was an important influence on philosophical semantics, e.g. in Charles Morris' *Signs, Language, and Behavior* (Englewood Cliffs, N.J.: Prentice-Hall, 1946). This kind of affiliation is explicit also in Quine's work, e.g. in *Word and Object* (New York: Wiley, 1960). Formal semantics, i.e. the mathematical work deriving from Tarski, has also been important. And there has also been recent work along Fregean lines.

[c] I have in mind, for example, the paper of Katz in the linguistics section.

[d] The first publication on the linguistic basis for this viewpoint was a contribution to linguisitics, Katz and Fodor's 'The Structure of a Semantic Theory', *Language*, XXXIX (1963), pp. 170–210, reprinted in their anthology *The Structure of Language*. The chief proponent of this conception of philosophical conceptual analysis is Katz; see his *The Philosophy of Language*, especially chapters 1, 2 and 5.

[e] For his methodological thinking, see the first chapter of his collection of articles, *Linguistics in Philosophy* (Ithaca, N.Y.: Cornell University Press, 1967), especially §1.5. Some earlier thoughts on meaning occur in his comment on John R. Searle's 'Meaning and Speech Acts' in *Knowledge and Experience*, ed. C. D. Rollins (Pittsburgh: University of Pittsburgh Press, n.d.). Other philosophers seem to be using this same kind of approach, e.g. Dennis W. Stampe, in his paper 'Toward a Grammar of Meaning', *The Philosophical Review*, LXXVII (1968), pp. 137–74.

[f] Linsky's book *Referring*, mentioned above, must be remembered. There are now some indications that linguists will be treating reference in a serious way in the near future (if they aren't already), the bridge between the two discussions being the syntactical and semantical problems surrounding pronominal reference, especially in connection with what philosophers know as 'intensional' contexts (i.e. ones like '*X* believes that...', 'It is necessary that...', etc.). See, e.g. George Lakoff's forthcoming 'Counterparts, or the Problem of Reference in Transformational Grammar', presented at the summer meeting of the Linguistic Society of America, July 1968, in Urbana, Illinois.

[g] The relevant writings are Strawson's 'On Referring', already discussed, his *Introduction*

Linsky's 'Reference and Referents', a chapter from his book *Referring*, makes a number of important negative points while also advancing some interesting positive theses. Among the former is his emphasis (already present in Strawson's 'On Referring') on the fact that it is primarily *speakers*, rather than linguistic expressions, that refer to things; and among the latter, positive theses is the thesis that the only ordinary sense in which *expressions* refer is derivative from the more basic sense in which speakers or groups of speakers refer to things. Strawson's 1964 *Theoria* article, 'Identifying Reference and Truth-Values', presents his view of the dialectic between his early type of view and its opponents and a new account of the existential presuppositions of a speaker's using uniquely referring phrases, which is more liberal than his earlier 1950–4 view, which however is a special case of the new account. Briefly, the new account makes these presuppositions relative to the conversational context: depending on what referring expressions would occur in stating the question at issue or topic of the conversation, a statement formulated with a sentence involving a uniquely referring expression suffering from 'radical reference-failure' (i.e. denoting a non-existent thing) may or may not be without a truth-value; it is not necessarily without one, as on the earlier account.

Donnellan's 'Reference and Definite Descriptions' pursues still further, and more positively, certain themes already broached or mooted by Strawson and Linsky, while emphasizing the difference between two uses of expressions which sometimes have a uniquely referring use, viz. between what he calls the *attributive* and the *referential* use of such expressions, e.g. between the use of 'the murderer' in 'The murderer, whoever he is, must be insane' and 'The murderer, i.e. Jones, must be insane', respectively. He attempts to sort out the wheat from the chaff in the previous literature and to distribute the wheat properly between the two uses he distinguishes. Vendler's 'Singular Terms', to which reference has already been made in connection with new methods, investigates in some detail the syntax of the type of expression Russell called definite descriptions, finding that there are certain constraints on their uniquely referring use statable in terms of the overall syntax of the conversational context or this together with certain presuppositions concerning the context of utterance. John R. Searle's 'The Problem of Proper Names', a chapter from his recent book *Speech Acts*, gives his account of the use of proper names as referring expressions, which is related to that of Wittgenstein in the *Philosophical Investigations*, and criticizes the earlier accounts of Frege and Russell. It seems to me that the cumulative effect of these papers is to confirm some contentions in the previous literature, often with a difference, and to discredit others – at least when it is the use of referring expressions in ordinary language that is in question. It is pretty clear, even after these papers, that further investigation of what is involved in referring, ordinarily so-called, or of what useful technical analogues of it might be developed, is to be desired.

The Quine selection is an excerpt from his John Dewey lectures, 'Ontological Relativity',[a] in which he continues to pursue certain topics emphasized in his book

to *Logical Theory* (London: Methuen, 1952), and his reply to Wilfred Sellars' 'Presupposing', *The Philosophical Review*, LXIII (1954), pp. 197–215, ibid. pp. 216–31. The Geach article is his 'Russell's Theory of Descriptions', *Analysis*, x (1950), pp. 84–8, in which a line much like Strawson's is advanced. Geach's later book *Reference and Generality* (Ithaca, N.Y.: Cornell University Press, 1962, rev. ed. 1968) should also be mentioned. In the statement above I am omitting literature on presupposition and contextual implication, both of which have been involved in discussions of referring in the literature.

[a] *The Journal of Philosophy*, vol. LXV, no. 7 (4 April 1968), pp. 185–212.

Word and Object and elsewhere.[a] In our selection, he contends that, besides the indeterminacy in principle of the answer to the question whether an expression of another language has been translated correctly, there is a similar indeterminacy in principle concerning the answer to the question what an expression of another language refers to – hence the 'inscrutability of reference'. Quine's penetrating examination of the epistemological foundations of our understanding another's language is pressed even farther here.

The papers by Grice and Ziff form a group. Grice's 'Meaning', originally published in 1957, distinguished a 'natural' from a 'nonnatural' sense of 'meaning' and announced a program of explaining the latter, including the meaning of a linguistic expression, in terms of the notion of the intentions of the utterer of the expression, viz. in terms of what beliefs the utterer intends to induce by the utterance in his hearer. Ziff's paper, while discussing Grice explicitly, actually offers a number of general criticisms of the *kind* of analysis of meaning put forward by Grice.[b] In using intentions to explicate meaning, Grice's approach is of course quite different from that of Quine. Gilbert Harman's paper, 'Three Levels of Meaning', distinguishes different *sorts* of theories among those that have been offered as 'theories of meaning'. Harman refers to three 'levels' of meaning:

the first would offer an account of the use of language in thinking; the second, an account of the use of language in communication; the third, an account of the use of language in certain institutions, rituals, or practices of a group of speakers.

He holds that there are certain relations of priority or presupposition among theories of these different levels and that failure to distinguish them has introduced confusion into philosophical semantics. In detailing the last, he discusses among others the views of Quine, Grice, Fodor and Katz, Alston, Davidson, and Chomsky. The Wiggins–Alston symposium, here published for the first time, is devoted to our ordinary, everyday methods of distinguishing different senses or meanings of a linguistic expression, what exactly they are, and what philosophical presuppositions and consequences they have.

I feel I really should comment on one particular feature of the current situation in philosophical research, one not unrelated to some lines of research in other disciplines. This is the apparent disregard of the Meaning-as-Use Doctrine of Wittgenstein, Austin, Ryle, *et al.* For a decade or more, this doctrine in one or another of its versions tended to dominate philosophical thinking about meaning. Yet the later, most recent developments of Grice, Ziff, Katz–Fodor, and others *seem*, at least, to ignore it and to offer accounts of meaning which would yield a statement of necessary and sufficient conditions for the application of a given term from any correct statement of what it meant. (It is, however, true that they do not purport to supply some entity which is the meaning of a term.) The relevance of this observation seems to me to be that it is, in effect, simply assumed without discussion that the rather radical positions of Wittgenstein and Austin are incorrect. And, perhaps more

[a] Many of the themes of his later work are foreshadowed or explicitly stated in earlier articles of his, a number of which are collected in his *From a Logical Point of View* (Cambridge, Mass.: Harvard University Press, 1953; 2nd ed., 1961) and in his *The Ways of Paradox* (New York: Random House, 1966).

[b] An important book in recent philosophical semantics is Ziff's *Semantic Analysis* (Ithaca, N.Y.: Cornell University Press, 1960). A number of other essays of his also deal with various questions about language. See his collection of essays, *Philosophic Turnings* (Ithaca, N.Y.: Cornell University Press, 1966).

importantly in the present state of our knowledge, the question of what sort of formal representation of meanings would be appropriate *if* Wittengenstein and Austin were right is not pursued.

Despite this last worry of my own, it is, I think, quite clear from the present selections from the recent philosophical literature and the original essays published here for the first time that interdisciplinary communication between philosophy and other fields dealing with meaning, especially the ones represented in the other sections of this collection, is already bearing fruit and promises even richer yields to come.

University of Illinois, Champaign

On sentence-sense, word-sense and difference of word-sense. Towards a philosophical theory of dictionaries[a]

DAVID WIGGINS

In the actual practice of philosophy we are constantly faced with such questions as 'Does *know*, or *believe*, or *aware*, or *deception* (in self-deception and deception of another), or *good*, or *right*, or *ought*, or *necessary*, or *if*, or *because*, or *reason*, or *cause* or...have one or more than one sense?' If we were not faced with the practical necessity to decide such questions we should probably be well advised to delay considering them, if only because, so far as I am competent to judge – which I hardly am at all, but this is my tentative opinion – the practice and whole methodology of the relevant parts of linguistics are at present in too provisional and uncertain a state of development. But we cannot always delay, and there may be something to be said for doing what I shall do in any case. This is to make one final assault on the problem of sentence-sense, word-sense and difference of word-sense, from within the traditional theory of meaning – if only to commemorate the achievements (mainly Aristotle's and Frege's) of an activity perhaps even more certainly doomed to extinction than everything else which is familiar to us. Before the existing assets of this part of philosophy are transferred and vested in a renascent science of linguistics or in a new branch of model theory, it may perhaps help to get them redeployed usefully if they are identified and accurately accounted for.

Even with the antique apparatus at the disposal of the philosophical theory of meaning it is possible, I think, to show that when people have asked 'Does word *w* have more than one sense?' at least three sorts of question have been at issue. They are:

(i) Does the word type *w* have more than one lexical content?
(ii) Does the word type *w* in different contexts represent more than one kind of *proposition-factor* or *paraphrase-component*?
(iii) Where the answer to (i) is *yes*, do we really have to suppose that *w* is a mere homonym?

For reasons which will become evident I should call the first an input question, the second an output question, and the third a question about the relationship between different inputs. All these terms will be explained in due course. I shall first try to

[a] This is a revision of the opening address given to the Oberlin Philosophy Colloquium in 1968. The commentator was William Alston and the original title was 'How does one tell if a word has one, several or many senses?' At some points the notes stray some distance from the contentions of the text and may usefully be omitted.

show that the proper concern of a dictionary is question (i), and that the notion of lexical content is the most useful interpretation of the notion of sense or *Sinn*. Question (ii) really pertains to the problem of property-identity, not the problem of difference of meaning. But before I can show anything about words I must say something about the sense or meaning of sentences.

1   Sentence-sense

Frege's notion of *Sinn* is almost always debated in the straitened context of his essay *Über Sinn und Bedeutung*, and when it is debated in that way the discussion almost always terminates in the half-hearted acceptance or inadequately founded rejection of the idea that different names and referring expressions may present one and the same object by means of different manners of presentation. The idea of *Sinn* is eked out by various metaphors about telescopes and so on, and by an intuitive feeling that something like Frege's doctrine must surely be correct for different definite descriptions of one thing – if only because such definite descriptions may plainly differ in meaning. But if discussion and interpretation of *Sinn* are confined within the context of *Über Sinn und Bedeutung*, then it is only after proper names and definite descriptions have set the scene that much notice is taken of the fact that Frege assigns sense and reference to sentences and predicates as well as to proper names. As I shall shortly show, there is an interpretation of the theory which even puts us in a position to assign a *Sinn* to morphemes. If we take things in the order in which the interpretation I am caricaturing takes them, however, then the fact that Frege allows both *Sinn* and *Bedeutung* to expressions such as sentences has to be accommodated by supposing that Frege tried to explain sentence-sense from the primitive idea of naming, mentioning or designating ordinary objects such as particular planets, men, mountains, etc. Frege's whole and complete explanation of sentence-meaning is then supposed to have been built on the idea (which is Frege's but does not, I believe, play this explanatory role) that the proposition is to an object called its truth-value, The True or The False, as the sense of the words 'planet seen as the brightest body in the evening sky' is to the planet Venus. It is then complained – and with every appearance of justice – that the analogy is quite inadequate to explain sentence-sense or elucidate its alleged equivalence with the proposition.

The dissatisfaction some have felt here is unsurprising, because neither the mode of presentation idea, nor any of the metaphors of *Über Sinn und Bedeutung*, are robust enough to sustain the weight which such an analogical extension from the idea of reference would necessarily shift onto them. Designation and reference cannot both undertake the whole explanation of what it is to *say* something and *ab initio* explain saying as referring to The True or The False. There are difficulties in any case in Frege's doctrine that sentences are names of truth-values,[a] but it makes a great deal of difference whether we are simply asked to put up with it as a consequence of a theory of meaning already clear and already explanatory, or whether we are asked to accept it from the very outset as something upon which *further*

---

[a] Some would maintain that it is true and profound that a rose is a rose is a rose. The vexatious or linguistically intolerant will maintain that this is strictly speaking an empty, meaningless, and ungrammatical assertion. Frege's theory is alone in insisting that the assertion is clear, grammatical – and false. For the truth-value True (= a rose is a rose) is *not* a rose. P. T. Geach's example. Even more serious is the fact that the equivalence relation ≡ needed to abstract truth-values as objects presupposes sentences already possessed of the established properties of meaning and truth or falsehood.

explanation can be securely based – which is what would happen on the interpreta-tion I am attacking. To treat it as the second sort of doctrine legitimates awkward questions. If the target is The True, it might be complained, and if different senses or thoughts are just different avenues to The True, then why not choose the easiest way of hitting the target and always say *snow is white*, or *the cat is on the mat*, or anything at all that is as a matter of fact true? Is there one message always and an indefinite number of media? Or if there is more than one message, is the message the medium itself? The interpretation leaves Frege's doctrine without resources to explain the evident absurdity of these suggestions. (And if one is to take fully seriously the idea that one can refer to The False while aiming at The True, there will never apparently be room for anything but a rigidly extensional theory of designation and reference. What then of sense without reference?)

It will be clear that the point of Frege's theory has got lost somewhere, if one reflects that this whole line of interpretation, which starts from the 'mode of pre-sentation' notion of sense, seems to be incompatible with Frege's insistence (in the *Grundlagen* and elsewhere) that reference itself is unintelligible outside the context of a complete sentence or thought. Frege took care to imbed the theory of sense and reference of proper names in a larger context, and it is in that context that we must seek it out. So seen, and in its proper universality, the theory is first and foremost a general theory of language, and its foundation and basis is not naming at all but the notion of sentence-sense itself. The interpretation I shall commend, which turns the other interpretation upside down and makes sentence-sense, not reference, the point of leverage, explains the production and understanding of familiar and unfamiliar utterances by an account of how the constituents of sentences can systematically contribute to the meaning of the *complete sentences* within whose structure they figure. And if sentence-sense is where the theory begins, no analogy has to bear the weight of explaining the meaning of sentences. *A fortiori* no problematic passage from designation of planets to designation of truth-values has to explain saying, saying being where the theory really started. But with a going theory of saying and sentence-sense and an account of how words contribute to it, an analogy (which is *only* an analogy) does then become possible between the way in which an arithmetic func-tion ( )² determines value 4 for an argument 2 and the way in which a predicate ( ) *is wise* determines the truth-value True for argument Socrates. But the ex-planatory dividend of this analogy primarily concerns predicates not sentences. It explains them as a species of sentence-functor. Knowing their meaning is just a matter of knowing what they do within complete sentences possessed of this or that complete meaning.[a] As for the senses of referring expressions, these are simply a special case of senses in general. It is often supposed that Frege devised the theory of sense in response to the special problem of informative identity statements. This may or may not be a historical fact. The value of the general theory of sense mani-festly transcends the grave difficulties with which e.g. genuine proper names con-front his solution to that special problem.[b]

---

[a] Frege does in fact allow predicates or concept-words a reference as well as a sense, but one who upholds this feature of Frege's theory of language, imported only by the need to quantify over properties, need not maintain that coming to understand a predicate is best or most illuminatingly described as a matter of coming to be able to identify its reference. It is a matter of coming to understand what sentences with what sense the predicate is a functor in. I dwell on this case because the word-class whose senses I shall principally be concerned with is precisely this class of sentence-functors.

[b] The difficulties are especially evident in the case of true identities of the form (proper

What then is sentence-sense? In the *Grundgesetze der Arithmetik* 1.32 Frege wrote:

But to all names properly formed from our primitive signs there belongs not only a reference (*Bedeutung*) but also a sense (*Sinn*). Every such name of a truth-value expresses a sense, a *thought* (*Gedanke*). *It is determined by what we have laid down under what conditions every such name designates The True. The sense of this name, the thought, is the sense or thought that these conditions are fulfilled.* A sentence of my concept-writing or symbolism consists then of the assertion sign and a name...of a truth-value ...By such a sentence it is affirmed that this name designates the True. Since it at the same time expresses a thought, we have in every well-formed sentence of the symbolism a judgement that a thought is true. So there simply cannot fail to be a thought [associated with such a sentence]...

The simple or composite names of which the name of a truth-value consists contribute to the expression of the thought. This contribution (*Beitrag*) of each is its sense. If a name is a part of the name of a truth-value, then the sense of the former name is a part of the thought which the latter [name of a truth-value] expresses.

Wittgenstein had no difficulty in rescuing the sentences I have italicized from the technicalities of the context (*Tractatus Logico-Philosophicus* 4.021–4.024, cp. *Philosophische Bemerkungen* IV, 43) and nor should we. The suggestion is this (cp. Dummett, P.A.S. 1958–9). If we will simply take the notion of 'true' as clear enough for the purpose – not for all purposes, but for this one – then we can say that, for arbitrary sentence *s*, to know the meaning of *s* is to know under what conditions the sentence *s* would count as true. Unlike theories constructed in terms of *belief* and *intention* after the fashion of H. P. Grice's 'Meaning' (in this volume), or theories which place trusting confidence in a notion of use which they have still to delimit in a way which does not reimport all the problems, this theory of meaning offers us not a way into the circle of semantic terms but a connexion between two of them. This may be the best theory we shall have for some time, and it is certainly some sort of basis from which to speculate about word-sense and difference of word-sense. But before I can get to the philosophical theory of dictionaries, there are some problems to clear away.

(α) The intensionality of the notion of meaning creates difficulties for the doctrine.

(β) Problems arise from the association of the definition (since the time of Frege) with various positivisitic doctrines, from the limitations of the definition, and from its concentration on the indicative mood.

(γ) There is a problem in Frege's definition arising from its apparent confusion of sentences with statements.

name *n*) + ' = ' + (proper name *m*), where *m* and *n* have to have different senses for Frege's solution to work. Unfortunately this seems to be impossible without collapsing *m* and *n* into definite descriptions, and I myself should now want to see the problem as a special case of the paradox of analysis. It is simply that paradox applied to names with *identical* senses, and the problem it raises exactly parallels the problem of the supposed difference in sense of the two sentences *oculist = df eye doctor* and *oculist = oculist*. I welcome the opportunity to disown my 'Identity-Statements' in *Analytical Philosophy* (2nd series, edited by R. J. Butler), which (amongst a hundred other faults) failed to look hard enough at the possibility of supposing (*a*) that 'if anything is Hesperus then Hesperus = Phosphorus' is an informative necessary statement, empirically discovered, and (*b*) that the paradox of analysis has and was bound to have an analogue for proper names. I now think that all the peculiarities which arise under (*a*) and (*b*) can be accepted and that they are indeed to be deduced from any correct theory of reference by proper name. (If contingent information is required to fix the sense of a proper name and this is the mark of reference to particulars – cp. Strawson, *Individuals*, Part II – then we must not be too surprised to find senses giving rise to an empirically discovered logical necessity.)

(δ) There is need for a rather careful statement of the connexion between truth-conditional definitions for sentences and meaning-definitions for the words that are constituents of sentences.

(α) The notion of meaning is an intensional one, whereas the notion of truth is not. How then can we catch the meaning of *means* by defining it in terms of *true*? Frege's proposal seems to be that s means that p if and only if (s is true if and only if p). But if we fix the meaning of a sentence s by simply saying *sentence* s *is true if and only if snow is white* then, pending an account of the *if* which does not give us back the whole problem (and others too), nothing can prevent a critic from supplanting *snow is white* in such a context by another sentence with the same truth-value – e.g. *King Charles was beheaded*. What results will still be true. Matters are quite different with *sentence* s *means that snow is white*. If that is what s means it certainly does not follow that it means that King Charles was beheaded. That is not at all what we had in mind for the truth-condition. It is not what one might call the *designated* condition.

It might go some way to meet this objection to amend the suggested schema by adding a 'necessarily' and stating the revised theory like this:

s means that p if only if (necessarily (s is true if and only if p)).

Leaving aside some perhaps surmountable difficulties about mention and use and the presumed universal contingency of the fact that any sentence (as typographically identified) is assigned the sense which it is assigned, this suggestion and any emendation of it would still have the important disadvantage of leaving all sentences expressing logically or mathematically necessary truths with the same meaning. This would have to follow unless subsidiary doctrines were somehow imported to distinguish these senses from one another by reference to the syntactic structure of the sentences and the word-senses of these sentences' constituents. (The last could be fixed via contingent sentences, for which the difficulty does not arise.) As it stands, however, the proposal certainly does not solve the original problem.

These reflections about procedures for mating sentences with the conditions under which they are true, and the reliance of such procedures on the contribution of constituents, do however serve as a reminder that the difficulty is less a practical difficulty about particular meaning definitions than a theoretical difficulty about meaning as such. We understand fewer things than we need to understand about either the theoretical constraints on, or the canonical form for, the truth-conditions which our defining procedure will mate with the English sentences whose sense they explain. But it is evident that, since there are an infinite number of English sentences to be assigned truth-conditions, such conditions can only be produced by a method of systematic decomposition of the given English sentences into the basic structures and components from which they were built up in the first place. (Such a decomposition must exist. No speaker learns to produce or understand the infinite number of sentences he can produce or understand by learning their senses one by one. And for the same reason only a finite number of basic structures and components can be available within any language.)[a] Now in practice this requirement,

[a] See Donald Davidson, 'Theories of Meaning and Learnable Languages' in *Proceedings of the 1964 International Congress for Logic, Methodology, and Philosophy of Science* (North Holland, Amsterdam, 1965, ed. Bar Hillel). Had I seen and taken proper note of this and his seminal 'Truth and Meaning' (in *Synthese*, 1967) much of the present paper could have been differently and more elegantly put by means of Davidson's illuminating allusion to Tarski's truth definition. A certain complementarity and a quantum of historical background information do however result from leaving my contentions in what is substantially their original form.

which is a substantial and non-trivial one, combines with the minimal requirement of material equivalence to make it virtually inconceivable that any satisfactory set of recursive procedures for assigning truth-conditions could serve up an irrelevant or non-designated condition for a sentence. It is of paramount importance to claim neither too little nor too much importance for this. On the one hand, it seems to me to leave the theoretical problem of meaning and the intentionality of *means* completely unaffected. We really are almost as far as ever from a definition or analysis of *means* itself.[a] On the other hand the difficulty does nothing to obscure the insight that assigning or discovering a sense for *s* at least involves assigning to it or discovering for it some correct truth-condition or other. I conclude that we have made one piece of progress, both as concerns what is theoretically involved in giving truth-conditions and as concerns the general theory of making particular word-definitions. However negligible our progress with the special case of an analysis and dictionary definition of the word *means*, we can oppose to the extreme vagueness of the requirement that the truth-condition for a sentence be a designated condition the full austerity of another requirement. The truth-condition must have been produced by the operation of a systematic, general, and uniform procedure competent to analyse any sentence in the language into semantic components drawn from a finite list of such components (i.e. a vocabulary or dictionary). And the procedure must account for the semantic structure of the sentence by showing how it could be generated by a finite number of semantically interpreted modifications or steps from one of the finite number of semantically basic sentence-forms of the language. (Whether, as I should continue to hope, the generative syntax of the school of N. Chomsky could be transposed or transformed into such a 'generative semantics' I cannot judge).

I come now to (β). As he abandoned some of the objectives of the *Tractatus Logico-Philosophicus* Wittgenstein gave a more and more operationalist or positivistic slant to the insight which he had taken from Frege's *Grundgesetze* (v. *Philosophische Bemerkungen*, passim). And the process which he began was completed by the Vienna Circle and A. J. Ayer in their doctrine *The meaning of a sentence is the method of its verification*. This formula had tendentious uses and perhaps for that reason it is now insufficiently recognized that it did represent a piece of thought, which was just as serious as it was crude maybe and unsatisfactory, about what it is to know the sense of a sentence which one uses or to know what one is saying.[b] But, genuine though this problem is, Frege's truth-conditions theory itself is neutral between, and has a value quite apart from, particular answers to it, and what concerns me are other

---

[a] And so far as this problem goes Brian Loar has persuaded me that the difficulties of keeping out non-designated conditions are intimately connected, in ways which I shall leave it to him to expound in due course, with the answer to the difficulties of amending and completing Grice's 1957 definition (to leave room for the fact that people sometimes say what they don't mean etc.). Proponents of a truth-definition of meaning have not found a way to render themselves simply exempt from all of the problems which arise here.

[b] I surely cannot say or explain what *All mimsy were the borogroves* means by saying that this sentence will be true if and only if everything satisfies the open sentence *if X is a borogrove then X is mimsy*. And it is certainly a part of what would still be lacking in this explanation that it gives no idea at all of what investigations with what outcome would count for or against the assertion. This is not however to say that discovering or fixing any sense for the sentence or its predicates would be a matter of finding or fixing anything like *the* method of verification or falsification – let alone *the* method of *conclusive* verification or falsification – for the sentence or for the satisfaction of one of these predicates.

questions. I have objected to the truth-conditions theory (see (α)) that it is in danger of letting in too much, but the standard objection to it is much more likely to be that it lets in too little. 'Even in the case where the indicative reigns, the truth-condition account leaves out almost everything of some sentences' communicative significance. And for other sentences, the theory is based on a hopelessly special case and can never come to terms with the complexity and multiplicity of other grammatical moods and functions than the indicative.'

This charge may be expected to come with accusations of committing what Austin called *the descriptive fallacy*, but caution would be advisable in invoking his authority at this point. For so put the charge is dubiously consistent with something else which Austin also insisted upon, at a slightly later stage of his thought, – his distinction between the meaning of a sentence and its *force*. To draw this distinction he relied upon the notion of a *locution* (which he opposed to an *illocution* and a *perlocution*); and *locution* he glossed in the Fregean terms of *sense* and *reference*. If Austin knew anything about Frege, whose *Grundlagen* he had after all translated, he must have been aware that at least for indicative sentences (Austin might have said 'at least for straightforwardly indicative sentences') the meaning-force distinction, if glossed in this way, would tend to limit their proper meaning to all and only that which bears upon their truth-conditions. And Austin was surely aware that Frege himself had laid great emphasis upon precisely this kind of distinction when he insisted that logic could not concern itself with the colouring or illumination of language, which was something Frege considered quite irrelevant to sense, or with anything at all that did not bear directly upon truth.[a] Grice has looked for similar distinctions (of greater generality than Frege's)[b] and he distinguishes between the *implications* of a sentence *s*, what follows from it if it is true, and the *implicatures* of it which are distinct from its implications – the situational import of *a speaker's saying* s. (This import typically arises not from the meaning of *s* but from the standard working of the perfectly general conventions of helpfulness, sincerity, etc., which Grice argues govern the conduct of speech-exchanges.)

There is a clear point in attempting to make such distinctions as these. The total communicative content of an utterance, construing 'content' comprehensively, is something too complex to be accounted for by any one-level theory. An utterance or speech-act considered as a whole may be horizontally complex, e.g. in respect of the syntactic structure of the sentence used in the making of it, and it may also be vertically complex. In saying that p a man may warn someone that q. In warning him that q he may threaten that r, and in threatening that r he may insinuate, what is even worse perhaps for the listener, that s. And in doing all this he may bring off what Austin called perlocutionary effects. He may alter the listener's behaviour in some way he precisely intended to alter it. If we persist in lumping all these things together in an undifferentiated notion of meaning it seems hopeless to look for a systematic theory to account for such 'meaning'. We must rather unpack the speech-act layer by layer. And at the first layer we must, I think, try to isolate all and only

---

[a] Frege's unconcern with the illumination (*Erleuchtung*) or colouring of language is so total and he is so uncurious about what does not straightforwardly pertain to truth he prevents himself from examining what lies on the other side of his distinction or drawing it in terms which match the generality of Austin's or Grice's. But I should say that his view of the *sense* side is for the most part congruent with theirs. There are complications Frege would have to take account of, however. See note *a*, p. 21, below.

[b] In 'The Causal Theory of Perception', P.A.S.S. (1962) and in subsequent work as yet, unfortunately, unpublished.

what is strictly *said* before we can explain how circumstances, conventions, and whatever else add implicatures, forces, or illocutions; and how these in their turn secure perlocutionary effects. The one element with a claim to be really fundamental or central to linguistic communication is this element consisting of *what is strictly said*. It is this which we must begin by refining, and the suggestion implicit in Frege's theory of sense, and in Grice's and Austin's theories, is that there relates to the strict meaning of a sentence *s* all and only that which bears logically upon the *truth* of *s*. From which we recover an intelligible and satisfactory-sounding doctrine – that truth must be the central notion of *semantics*, and that the boundary between what does and what does not bear logically on the truth of what is strictly said must be the boundary between the science of *semantics* and the science of the further effects obtaining in a speech-exchange.[a] There is a precedent for calling the latter *pragmatics*, but obviously this name is still little more than a catch-all for what does not bear on strict meaning.

Even if this defence of Frege's theory of sense were accepted by the objector he might still claim that an enormous number of apparently indicative utterances were not really indicative in meaning – insisting still on the heinousness of what Austin stigmatized as the descriptive fallacy – and he might claim that an even larger body of utterances were not even apparently indicative. A full reply to this would take the form of (i) a thorough scrutiny of the class of so-called performative utterances,[b] and

---

[a] For some more of what needs to be taken account of here see the discussion below of the definition of *snub*.

It seems much more important to settle a hierarchy or ordering of the considerations which need to be taken into account in working out the total communicative content than to establish a firm frontier here. For example if the minimal connectedness of p with q were somehow shown to be a universal condition of uttering *if p then q* (where p and q are in the subjunctive) – *if* so much were shown – then it might come to seem quite unimportant whether this was due to the meaning of subjunctive *if* (semantics) or due to the operation of a convention for making a speech act (pragmatics) which was too universal or too deeply entrenched to admit of waiver.

[b] There are two aspects of this problem. Some still hold, it seems, that it is a part of the meaning of 'There is a bull in this field', said (under suitable circumstances) as a warning, that the utterance is a warning and equivalent to the 'explicit form' (so-called) 'I warn you there is a bull in this field'. Austin himself would have preferred to hold that warning was only a part of the illocutionary *force* and not any part of the meaning of 'There is a bull in this field'. But he would certainly have held that its having this force was determined by *convention*. Both Austin's convention view and the other view of 'There is a bull in this field', like any alleged equivalence between 'explicit' and 'non-explicit' forms, seem to be undermined by the obvious reflection that what makes 'There is a bull in this field' into a warning is a starkly extra-linguistic matter which a hearer would have to know already, and have to combine with a pre-existing understanding of what is already *said* by 'There is a bull in this field', in order to understand it as a warning. I should make a similar claim about 'I will be there' construed or misconstrued by the hearer as an illocutionary act of promising. This cannot be equivalent to 'I promise to be there'.

The other aspect of the problem of performatives is the status of the 'explicit' forms, e.g. 'I warn you there is a bull in this field' and 'I promise to be there'. I persist in taking these as straightforward statements.

The least unimpressive objection to construing 'I promise...' as a statement is that it could hardly be tantamount to the self-descriptive running commentary 'I am promising'. But 'I love you' cannot be converted into 'I am loving you' either. What this shows is something about what sort of verb 'love' is. It doesn't show one can state or declare that one loves someone. The performative theorist may retort that what he is arguing about are cases like promising something where something does go on and a man does do an action, and what the man says doesn't appear to describe that action. It makes all the difference, he will say, that the

(ii) an appraisal of the claims (which still seem to me to be passably good) of the indicative mood's claims to priority over other moods.[a] However many concessions the objector succeeded in wresting from a defender of the Fregean doctrine it is difficult to believe that they could imperil the following minimal contention – That any satisfactory theory of meaning (whether or not exempt from the allegedly crippling defects of the descriptive fallacy) must entail the following proposition:

To know the sense of an indicative sentence s *it is necessary to know some condition p which is true if and only if* s *is true and which is the designated condition for s.*

But from this somewhat exiguous condition of adequacy I think it is possible to squeeze enough to get started on word-sense. Not even one who believes there is a descriptive fallacy waiting to be committed (and certainly then committed by this writer) will want to suppose that sentence-constituents, i.e. words, generally have different senses according to whether they occur in indicative or 'performative' or imperative or optative sentences.

But two more difficulties remain, (γ) and (δ), before we can exploit the power of this Fregean doctrine.

(γ) As it stands in Frege and at many points in my exposition, the theory seems to confuse sentence with statement or proposition. If the *s* which figures in the theory that to know the meaning of *s* is to know under what conditions *s* is true is a statement then it is the right sort of thing to be true or false, but we have not yet been given a route back to sentences and their constituent words. If *s* is a sentence then there is no problem of this kind, but it may be said that a sentence by itself is not true or false. An amended theory would then have to rule that, for arbitrary sentence *s*, to know the meaning of sentence *s* is to know the truth-conditions of the *statement* which *s* makes, which apparently makes the theory presuppose a whole prior account

continuous present 'I am promising' is *available*, and that it still does not give an equivalent of 'I promise'.

I think that the reply to this is that 'I promise' does describe an act but describes it in a way not equivalent to the continuous form 'I am promising'. It describes the act in the grammatically *perfective* aspect instead of the grammatically *imperfective* aspect. It is the completable act, not the activity of completing it, which is of interest when a man avails himself of the 'I promise' formula, and it is this which makes the instantaneous or aoristic form 'I promise' appropriate and the continuous form inappropriate.

[a] To understand, as opposed to merely reacting satisfactorily to, the command 'shut the door' I must know what it would be for such a command actually to have been executed, i.e. what it is for it to be true that the door has been shut. (Perhaps I must also know what it is for the door to have been shut by the person commanded *because of* the command to shut the door.) But to understand what it is for it to be true that the door is or will be shut I need not understand anything about commands or the imperative mood.

If this be thought an important asymmetry it may incline us to allow both the indicative and the notion of *saying that* a kind of priority. And it may also incline us to suspicion of the whole idea of a propositional content (as it figures in Frege's and subsequent theories of the assertion sign, '⊢', conceived by Frege as an assertive vertical ' | ' prefixed to a horizontal stroke '—' which heralds a content) which is neutral between and embeddable as a common element in different acts of *saying, commanding, questioning, wishing*. At very best this common content can only be an artificiality. For if there is no strict parallellism between the indicative and the imperative, optative, and interrogative moods, then the asymmetry thesis combine with the other arguments against ' ⊢ ' to suggest that the common propositional content ⌜that p⌝ must really be got by *subtraction from the assertion* of ⌜p⌝, rather than the latter by *addition* of ' ⊢ ' to ⌜p⌝. One might maintain that ⌜p⌝ automatically says that p unless you obstruct it from doing so. (One way of so obstructing it is to embed it?)

of stating. But if we have that, it may be said, then all the hard work is over already. Nor can the dilemma be laughed off by those who say they cannot countenance propositions unless 'proposition' is a gratuitous synonym for sentence. If truth is accounted a predicate of sentences then, if context-dependent utterance is to be explained, it has to be defined for sentence-*tokens*. The dilemma is then that a theory of meaning needs to be or include a theory of sentence-*type* meaning.

It looks like the beginning of an answer to state the theory thus:

> To know the meaning of a sentence-type *s* is to know under what conditions the utterance of a token *t^s* of the type *s* is factually licensed [token-true].

s confers an intial purport on *t^s* which it enjoys in abstraction from all the circumstances of its utterance, but in working out its meaning there is also the context in which it is lodged to take into account. Only this can determine what the notion of factual licence or token-truth necessarily imports, e.g. the references of such referring devices as demonstratives. To understand *s*, then, is, amongst other things, precisely to understand how such extralinguistic factors are to determine the truth conditions for *t^s*. (They must of course be all and only the extralinguistic factors which bear on any *truth*-condition for *t^s*.) The speaker knows the truth-conditions for *t^s*, and he identifies a particular *statement*, only if he knows how to work out the contribution of, for example, these demonstrative or identificatory factors. So the truth-conditions theory does not provide any effective retrieval of all and only the semantic purport of *t^s* until the nature of these contextual inputs is fully described and we are in a position to add to what we already have such additional theses as this:

> To know the meaning of a sentence type *s* is (*a*) to know how in principle to work out the demonstrative purport of its indexical expressions, in context, and (*b*) to know under what conditions the utterance of a token *t^s* with contextual determination *d* is factually licensed [token-true].

I recognize that to make this retrieval effective we should need actually to possess a complete theory of demonstrative reference. And for a completed theory we should have to specify *exactly* what contextual determination amounted to. I expect that this would be tedious and at some points very difficult but I believe that we have enough feeling for what it would be like to make it unnecessary to say more here.[a]

---

[a] It might be objected to the words *factually licensed* which figure in the revised theory that this technical locution can only be elucidated in terms of a token's *being used to make a true proposition or statement*, which reimports both *stating* and *statement*. But I am encouraged to think that a reply could be found to this. It must surely be possible to teach someone a convention of doing something or not doing something (or permitting someone else to do something or not permitting them) according to whether or not a certain condition C obtains. The notion of convention itself is a wider notion than that of saying or stating, and is certainly not automatically or for all purposes ruled out from employment in the elucidation of saying or stating. Now if the idea of convention is carefully handled I believe we can elucidate and designate a class of performances within the convention of saying as in a special way *acceptable* performances of saying. This will be the class of performances of *saying truly*. (We need to ascribe structure to *saying truly* but, so far, the theory need not ascribe more structure than *verb + adverb*. It is only at a later stage, and for different and slightly more dubious purposes, quantification over propositions, that we need to ascribe to it the structure *verb + object*.) I say that the idea of convention has to be carefully handled because the notion of true saying will only be effectively fixed here if advantage be taken of the possibility of giving the elucidation in the presence of *shared correct* belief about whether or not C. Otherwise there will be no effective distinction between the false statement that C obtains made by one who understands

The following diagram may serve to summarize some of the conclusions of this section and the answer to (β).

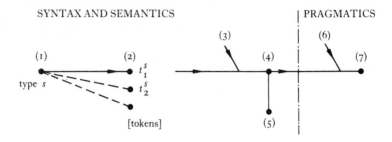

SYNTAX AND SEMANTICS | PRAGMATICS

Key

(1) Sentence type s with assigned and determinate grammatical structure and generic sense S.

(2) Tokens t_1^s, t_2^s...produced in particular speech episodes E(t_1^s), E(t_2^s)...

(3) Demonstrative (and any other) inputs to (4), these being determined by the context of utterance-episode E(t_1^s). The demonstrative purport of any referring phrases in t_1^s (plus any other semantically relevant purport as yet unaccounted for).

(4) The statement made or proposition propounded by the speaker in E(t_1^s) – what he *says*, this being determined by what has to be the case for the speaker to count as saying truly (i.e. saying something true).

(5) Truth-value of (4).

(6) Situational factors bearing on (7).

(7) What the speaker means in or by saying (4).

(δ) It has been maintained by Ryle and others that words have sense in an only derivative manner, that they are *abstractions* rather than *extractions* from sentence-sense. There is something we must acknowledge and something we must reject in this doctrine. What we must concede is that when we specify the contribution of words we specify what they contribute as verbs or predicates or names or whatever, i.e. as sentence-parts, to a whole sentence-sense. Neither their status as this or that part of speech nor the very idea of words having a sense can exist in isolation from the possibility of words' occurrence in sentences. But this is not yet to accept that words do not have sense as it were *autonomously*. And they must. If our *entire* understanding of word-sense were derived by abstraction from the senses of sentences and if (as is obviously the case) we could only get to know a finite number of sentence senses *directly*, there would be an infinite number of different ways of extrapolating to the sense of sentences whose meanings we have to work out. But we

what he is saying and a man's total failure to understand the linguistic purport of the token whose sense is being fixed.

It seems obvious that any thorough discussion would have to touch on or work its way back to Grice's project. See answer to (α) above.

The effect of the tentative suggestions I have made to answer problem (γ) is to distinguish the *sense* of a sentence (something produced in utterance as a token ·of its type) from its output, the *proposition* which the token expresses. Since Frege himself identifies *Sinn* and *Gedanke*, modifications become necessary to his theory of indirect *Sinn* and oratio obliqua when *Sinn* and *Gedanke* are distinguished. It looks as if oblique occurrences of an expression E will have to stand for the output (the proposition or proposition factor) not the input, or strict *Sinn*. But these are complications I shall not enter into at this stage.

do in fact have an agreed way of working them out. This is because word-senses are autonomous items, for which we can write dictionary entries. Quine has put the point in dispute so elegantly and concisely that it is enough to quote him:

> The unit of communication is the sentence and not the word. This point of semantical theory was long obscured by the undeniable primacy, in one respect, of words. Sentences being limitless in number and words limited, we necessarily understand most sentences by construction from antecedently familiar words. Actually there is no conflict here. We can allow the sentences a full monopoly of 'meaning' in some sense, without denying that the meaning must be worked out. Then we can say that knowing words is knowing how to work out the meanings of sentences containing them. Dictionary definitions are mere clauses in a recursive definition of the meanings of sentences.[a]

It is easy to see that by the account towards which we are working, the sense of a word being the precise contribution (*Beitrag*) which its presence makes to the true utterance conditions of the complete sentences in which it figures, a word will be ambiguous if and only if the dictionary which defines it requires more than one entry for the word. It will require this in order that the procedure of recursive definition (of which the dictionary is one part and the syntax of the language is another) should account correctly for the meaning or meanings of the sentences in which it occurs. Sometimes we can see straight off that a word has more than one definition (I say some more below about certain hazards associated with 'just seeing'), but the view I am advancing or commending makes much more hang on our being able to recognize that a *sentence* may be read or heard in one way so as to be true and read or heard in another way so as to be false. If the syntax of the language can find only one grammatical analysis of the sentence then we must try to account for the different readings by postulating a word ambiguity and try to write a multiple dictionary entry in order to account for the distinct truth-conditions for the sentence.[b]

2 Word-sense

In spite of the inconclusiveness of some of the answers to points (α), (β), (γ), and (δ) I now make a modest start with ambiguity and the theory of dictionaries. Its modesty must be emphasized. What exactly is involved in a satisfactory explanation or specification of the intended truth-condition for a sentence s? Satisfactoriness in an explanation is relative to an interest. What is the best formulation of the theoretical interest we *should* have? Till more thought has been trained on this question we have no very clear idea (cp. note b, p. 19). And what are we to make of the fact that most words don't really have dictionary definitions (certainly not intersubstitutable equivalents)? As Putnam has recently stressed, most natural-kind words do not. Webster and Larousse have large numbers of pictures, especially for such natural

---

[a] 'Russell's Ontological Development', p. 306 in *Bertrand Russell, Philosopher of the Century*, ed. R. Schoenman (London, 1967).

[b] Once there are two dictionary entries both classifying some single word as the same part of speech, this may create a much larger number of theoretical alternative readings of some sentences than it will normally occur to us to suppose they have. For some of the theoretically possible readings will be too absurd for them to occur to us as likely or credible. My own view, for what it is worth, would be that this exclusion and this absurdity almost always results from the operation of *pragmatic* factors, matters of fact, and contextual knowledge rather than from anything lying within the purview of *semantics*, for which all the absurd alternatives must count as theoretically possible.

kinds as plants, animals, birds, etc. In the *Politicus* Plato approved such explanations, and its serves a clear purpose which nothing else would serve to start off by fixing the extension of a term in this way – 'whatever belongs to the same kind (e.g. animal species) as this...' [here a picture or display is supplied].[a] But is this the sort of thing our dictionary is meant to do, or not? Having raised this problem I abandon it and try to draw out the consequences of such clarity as we do have about the minimal requirements on a dictionary.

A dictionary entry will presumably specify the part of a word and thereby determine its grammatical compatibilities in the terms of the best grammar for the language. (If a word can figure as different kinds of speech this will immediately give rise to more than one entry for it, and by our proposals for sense counting it will then rate as ambiguous from the outset.) Next, the dictionary will or may specify a *definiens* or lexical citation for the meaning of the word. The first point to be made about this we owe to Aristotle (*Metaphysics* Z).

Consider the word 'snub' in 'Kallias is snub-nosed'. If we wrote the entry for 'snub' as '(adjective) *concave*' then we might seem to have made it possible for 'this mirror is snub' to be true. But if we write '(adjective) *concave-nosed*' then it would look as if when we go back to our original sentence and write its truth-condition it must be 'Kallias is concave nosed-nosed'.[b] Nor can the trouble be cured by defining 'snub-nosed' in one piece. It would not cope with 'She had a charming nose, almost snub' or 'A retroussé nose is something like a snub nose'. Consideration of Aristotle's example suggests that at least some dictionary entries will have to carry another indication like this: 'Snub: (adjective) [of a nose], *concave*'. We may call this extra component [......], which imports what is really syntactic as well as semantic information, a *directive*, and call the other component the *analysis* or *citation* for the word. It is blatantly obvious that we are not yet in a position to explain *directive* otherwise than as 'what cannot find a place in the analysis itself'. And it is most unlikely that the need for the [.......] device always arises in the same way. I expect that the point about *snub* may seem rather insignificant. It is not, however, insignificant and it bears on a number of disputes which would have benefited by exposure to it.[c]

[a] In general it will be very difficult to determine in advance what will be relevant and what irrelevant to the collection of outlying members of the class or to testing the claims of founder members. Consider the criterion now employed in zoology to determine the bounds of the *Carnivores*. Fusion of the scaphoid and lunare is hardly part of the sense, perhaps, or anyway the intension of 'carnivore'. *Kind* as it figures in this sort of definition schema is theory-laden. Or rather it waits on whatever may be the best theory of the relevant kind of kind.

[b] It is sometimes objected to Aristotle that the objection to 'concave nosed-nosed' is (at very best) grammatical. Consider then *fetlock* = *horse's ankle* and then use this paraphrase to rewrite 'the stone hit the horse's fetlock' as the 'the stone hit the horse's horse's ankle'. If this were grammatical it would itself be ambiguous! It does not rule out that mad reading of the genitive which suggests that one horse owned another horse which owned an ankle. Paraphrase is meant to eliminate, not to introduce, ambiguity.

[c] Indeed it is part of what is a much larger flaw in a programme of philosophical analysis, the unremitting search for analysis by intersubstitutable equivalents, which has taxed the ingenuity of a whole tradition, Plato (*Theaetetus*), Aristotle, Leibniz (*Characteristica Universalis*), Frege, Moore, Wittgenstein (*Tractatus*), early and middle Carnap and even some Oxford philosophy. When Goodman, Quine and Carnap lower their demands to some looser equivalence I think that this smacks more of disappointment (and of Goodman's and Quine's dissatisfaction with the notion of *a priori* necessity) than of any fundamental reappraisal of what is involved in giving an adequate explanation of sense. Once we revise our ideas about what this really involves it is possible again to see Goodman's merely extensional requirement

Consider *and* and *but*,[a] or *aliquis* (someone) and *quidam* (a certain person, the speaker not letting on who), or even perhaps *gaggle* and *flock*. If we try to account for the differences in each pair by making different *analyses* or *citations* do the explanatory work, then we reach some absurd positions. *But* and *and* differ in meaning. Suppose we account for this by a difference in citation. Then the sentence *His speech was long but impressive* will count as false even if the speech was both long and impressive if there exists no such contrast as the one which the different citations for *and* and *but* require. Finding this absurd, we may feel obliged to accept one and the same simple truth-functional citation for both words. And we may wrongly reject the idea that *but* and *and* have any difference at all in strict *meaning*. Note, however, that we shall be doing this purely and simply because their lexical citations are identical. Frege boldly took this course but it is equally counter-intuitive. Aristotle's point shows that this reason is not a good reason to identify their meanings and that there is another option. The generalization of the distinction between a directive and a citation makes it possible for us to see how *and* and *but* have different lexicon entries and differ in meaning *without* differing in citation. The standard (sometimes called rhetorical) difference between them comes in their respective directives. But this leaves the difference within the realm of *semantics*, and we must adjust the statement of the Fregean semantic programme to find room for such effects of directives.

The point may still seem a trivial one. Perhaps I can make it seem more interesting by suggesting that it also solves one small problem about conditionals.[b] Philosophers have been puzzled about the part played by *if* with the subjunctive mood. Subjunctive conditionals seem to be strongly connected with contrary-to-factness. Moreover they seem to be connected by something semantic – or if not something semantic then something very general. On the other hand it is noticed that the truth of the antecedent *p* in *if p were true then q would be true* could hardly be held to *falsify* the conditional. A part of the answer to this conundrum is surely that a dictionary might well abandon any attempt to build counterfactuality into the *citation* for subjunctive *if* and still account for what needs to be explained by writing an entry on these lines: *if* [used with subjunctive when the speaker is either speaking as one believing the

as much too permissive. And a number of other problems are transformed too. We can take it in our stride, for instance, that different definite descriptions can fix the sense of a proper name and all in their different ways fix one and the same sense.

I hold no brief for Austin or his champions such as Stanley Cavell on the meaning or appropriateness of 'Did you tie your shoes voluntarily?' (see Austin's 'Excuses' in *Philosophical Papers* (Oxford, 1961)), but parties to the dispute ought, I think, to reflect whether any inappropriateness there may be in the question derives from an implicature of reluctance, or from a directive for 'voluntary' such as: [used when the question of willingness arises].

[a] *But* and *and* as used internally within whole sentences. Their use as particles to link different sentences, paragraphs, etc. is not in consideration here. The preface paradox ('I believe every proposition in this book but I am confident that a mistake will be found somewhere' – which rules out the possibility of seeing the book even notionally as a single conjunctive assertion) makes me doubt that the two uses are interreducible.

[b] Nor can the difference between citation and directive be irrelevant to the many disputes there have been between cognitivists, emotivists, prescriptivists and others in ethics. If the differences between 'Limey' and 'Englishman' or 'nigger' and 'negro' are differences in something like directive then even the best putative examples of 'emotive meaning' show much less than anti-cognitivists have wanted them to show against the naturalists' theories of meaning. But I must remind the reader at this point that *directive* was defined negatively. Work would need to be done, and the idea would need to be broken down, before more use could be made of it in ethics.

untruth of the antecedent or as one not believing the truth or untruth of the antecedent]...If parties to disputes about subjunctive *if* had noted the rather uninteresting case of *snub* and seen that not all meaning specification can be by provision of an intersubstituend many hours might certainly have been saved.

Even with so small a fund of information as we have about the theoretical requirements on a good dictionary, there are one or two further conclusions which follow with something near certainty.

Sometimes it will happen that, because one sort of occurrence of a word has suggested one definition and another sort of occurrence has suggested another definition, we have a *prima facie* case for ambiguity of the word. But even if the two lexicon entries for the word seem to be sound and they yield correct readings for the respective types of occurrence with which they were intended to deal, this cannot prove that one general account could not have been given to cover both simultaneously. One might have been able to do better. So ambiguity is often difficult to prove. This is as it should be.

This will raise the question – 'What will you do if someone takes an obviously ambiguous word and simply disjoins the two entries for it? Can the lexicon approach show there is anything wrong with such a concoction? And what is its test for detecting specious unity?' Now certainly *bank* means either *river verge* or *money depository*. But this does not imply that it means *either a river verge or a money depository*. It is simply a brute fact which a theory of meaning has to explain, and not explain away, that the sentence *I went to the bank this morning and wept many tears* might on many occasions be entirely and completely ambiguous between two utterly distinct conditions. It is of its nature to prompt the question *which do you mean, a river bank or money bank?* Till we know which is meant we do not know what was said. Nor is it difficult to see that the disjunctive account would result in the conversion of actually true statements into false, e.g. *All banks in the U.S.A. are now guaranteed by the Central Reserve Bank.*

But of course there are unitary accounts of problematic words which are not put forward in a captious spirit and which are much more difficult to adjudicate. And notoriously there is the word 'good', for which such unitary definitions have been advanced as *satisfies the criteria (or comes up to the standards) prevailing for the evaluation of items of its kind* or *answers to the relevant interest*. Of course these definitions have been offered in the pursuit of philosophical enterprises with which I am in general sympathy. For this very reason I am anxious to show that they go too far and that the Fregean lexicon approach is stringent enough to be committed to their actual inadequacy. The example of 'good' has the advantage of raising a large number of general problems about word-sense, so there is some point in a rather extended discussion. Since the arguments against both proposals are broadly similar, I shall choose only one of them, the second having the attraction of being interestingly defended by Paul Ziff.[a]

In the first instance Ziff's proposal might be put in something like this form – *good*: (adj.), [used in context of evaluation with some determinate evaluative interest. Let the interest in the case be interest *i*,] *answers to the interest i*. Could anything on these lines possibly be adequate? There is certainly something which the proposal explains. What it explains, however, is what I originally labelled a type (iii) question, not a type (i) question. I do not think it fully explains the utterance conditions of all, or indeed any, sentences containing the word *good*.

[a] *Semantic Analysis* (last chapter).

Consider the sentence *She has good legs.*[a] It has a number of interesting properties, of which the most striking is that it is a straightforward counter-example to the notion which has first occurred to many philosophers who have reflected on good-ness – that to any determinate thing of a determinate kind there is annexed (anyway for a *determinate speaker or hearer*) a determinate criterion of goodness for the good-ness of that kind of thing. For the trouble with legs is very roughly this – that they can be assessed aesthetically as good = beautiful, they can be assessed functionally or 'instrumentally' as good = strong or well suited to this or that activity, and they can be assessed medically as good = not maimed, healthy. The undoubted con-nexion (to which I shall come in a moment – this is the type (iii) question) between these dimensions of assessment does nothing to establish their identity. There are other dimensions of assessment for legs too, but it is interesting (or anyway it is to be noted) that the verdict of goodness in these other dimensions has then to contain express reference to that special dimension. *She has legs to show to Dr X* (who would be interested in such and such an anatomical feature). *She has good legs for doing such and such a test on. Her legs would be good to eat.*

How, if at all, does this threaten the unified account of *good*? The unified account could be correct if *good* combined with *legs* to produce one set of conditions; or perhaps it might be correct if there were an indefinite or potentially infinite number of such conditions corresponding to a supposably indefinite number of logically possible dimensions of assessment for legs. (Goodness would then be a very in-determinate property.) Neither situation obtains. The number of interpretations sticks obstinately at three or four and I submit that we know what they are in ad-vance of any context of utterance. It may be said that it is in the nature of legs to fix and limit to three or four these three or four interpretations for good legs. This is true but not to the point, if there really is a genuine indeterminacy between three and only three genuinely distinct genuine interpretations of the sentence *She has good legs*. It is pertinent to add that when Von Wright wrote a book called *The Varieties of Goodness* he found that the pattern *good + noun* (with 'good' unqualified) was as-sociated with some six or seven categories of assessment. Not many more. Not many less. This did not result from his restricting his choice of values for the noun-place. What he discovered does no doubt result from the fact that as a fact of natural his-tory some six or seven dimensions of assessment, instrumental goodness, medical goodness, the beneficial, the useful, hedonic goodness, etc., dominate all our stan-dard evaluations. And no doubt there is something one can say to set them in a pat-tern. But if I am right this fact of natural history has infiltrated the semantics of the *good*. When *good* and some noun with which it is joined coalesce to produce their joint output, *good* carries something more distinctive to the compound than the colourless idea of evaluation *per se*. What it carries depends on the noun, but for some nouns there is more than one such distinctive meaning imported by the word *good*. It is this which embarrasses the unitary account, which might otherwise have been able to plead that it was fair enough to produce a unitary account by classifying nouns into some seven categories and giving a compound instruction for the pro-duction of truth-conditions according to the category of the noun combined with *good*.

It would be empty, I think, for the unity theorist to appeal to the possibility that contextual factors would always decide. In fact they do not. The remark *She has good legs* really is found to be indeterminate (and I should say ambiguous) in some

[a] An example I have several times discussed with Michael Woods, to whom I am greatly indebted.

contexts. And there are some where it would surely *have* to be found ambiguous. It is certainly vacuous to count the larger conversational context *including* elucidations of the remark itself as part of the context. By this proposal all words indiscriminately would admit of one account.

If I am right my contention can be reinforced by testing the *answers to the relevant interests* account the other way round. Suppose that I have been engaged to find a woman with legs about a foot shorter than the statistical average for women of her height. Suppose that Mr *Y*, an imaginative and powerfully backed film director, badly needs an extra with this physical peculiarity. If I find a woman with this characteristic then her legs answer the relevant interest. (Does she not herself answer it? – which doesn't make her a good woman.) She has good legs *to show to Mr Y* or good legs *for my purpose* or good legs *for the part of so and so in such and such a film*. But none of these qualifications is detachable, and this is a linguistic fact which cries out for explanation. This awkward fact would also obstruct the attempt to give a unitary account of the word *good* which started from the schema *good* (noun) *to/for* (. . . . .) with two slots and then explained *good* (noun) by the operation of some kind of ellipsis. The explanation of the admissibility of some ellipses and the inadmissibility of others would most probably involve a virtual acknowledgement of the ambiguity – theorist's account of the word *good*. (In this connexion it is also worth remarking that the deleted 'to . . .' or 'for . . .' is extremely difficult to specify properly, with the required degree of vagueness, etc. Are good (= beautiful) legs precisely legs which are good legs to *look at*? And what is a good man a good man to or for?)

These difficulties in the unitary account of 'good' lead to another difficulty in Ziff's view. In the case where we have a substance of kind *X* called a good *X* and there is one and only one interest or set of standards which is agreed to be relevant for being a good *X simpliciter*, there is little or no trouble in meeting the ancient difficulty which troubled Stevenson and Ayer about evaluative disagreement. If disputants understand what an *X* is then in *this* sort of case that does simply fix a common interest to decide the question whether *X* is good or not. But where there is no clear and agreed interest, or where the whole substance of dispute is 'which interest is the relevant interest?' matters are not so simple. Examples tend to bunch in areas of some importance in moral philosophy, e.g. 'good man', 'good plan', or 'good thing to do'. A disagreement about what to do need not always, even between partners with a common interest such as a man and his wife, be a disagreement about means to ends. A disagreement may be a disagreement about which interest or end *is* the relevant interest or end (all relevant questions of means being relatively easy in the case, let us suppose). This is quite compatible with its being a disagreement about whether *X* is or is not a good plan, or a good thing to do. What seems to be wanted is to transfer the word 'relevant' from the directive to the citation or analysis, thus transforming the analysis itself into *answers to the relevant interest*. But this proposal can only aggravate all the difficulties of unifying the sense of *good*. The dictionary has still to say more than it can by means of one entry about what interests are the ones that qualify, and about which interest is the relevant one in a given case. Suppose scaring birds is the relevant interest. Then in dressing a scarecrow a used and worthless hat will come to count as a good hat. A good hat *simpliciter*, not just a good hat to give to a scarecrow. But don't we want to hold onto the idea that a bad hat is a good hat to give to a scarecrow? (And see the remarks above about the elliptical analysis. It is possible, I suppose, that we could altogether drop the

that we may try to account *focally* for the particular array of interests represented in the separate entries for *good X*, but by an anthropological explanation which finds a pattern in the interests which have found their way into the several senses of *good* kept strictly separate from one another by the lexicon. What *partly* organizes the pattern of interests picked out by the users of the language is what the users of the language conceive that they want and need in the general framework of the life they desire for themselves (*eudaimonia*). If it fits anywhere it is here that the piece of information that 'good' is the most general word of commendation in English comes in. (But neither this piece of information nor Ziff's account is a recipe for constructing a single lexical analysis of *good*.)

This is very tentative and nothing hangs on the details. The generalizable point is that when we distinguish senses we do not necessarily condemn a word to *pun* status, and that it is a part of the equipment of a mature speaker of a language that he possesses some principles (e.g. extension by analogy, extension of senses from a focus) which enable him to invent or 'cotton onto' new uses of a word and see their rationale. The new uses which do get into the dictionary should be kept apart there from one another and apart from the original use. For this very reason it is not necessary, and at any moment of time it will never be desirable or possible to put all of the new uses in the dictionary. (Consider, every week somebody may invent yet another use, technical, erotic, penal, topological or far-fetched, for, say, the word 'inside'.) But if and when uses do get sufficiently standardized to merit a place in the dictionary, they will give rise to new and distinct entries. New uses get into the dictionary when they settle down into distinct and distinctive senses which are worth noticing in the dictionary, and they only arrive there because they make some recurrent difference to truth-conditions. It is difference to truth-conditions of sentences, not type (iii) connexions, which must primarily concern the dictionary.

The account I have offered of ambiguity is in some ways rather remote from traditional discussions, which concentrated on paraphrases for particular words. Paraphrase is the method employed by Aristotle in his, in my opinion, defective proofs of the non-univocality of *good*. (See *Nicomachean Ethics* I. 6, the earlier *Eudemian Ethics* I. 8 and his fullest early discussion of the matter, *Topics* 107.) Traditional discussions find it enough to do what I think is not enough for a type (i) question. They point to the fact that a word like *good* can be paraphrased in completely different ways in its occurrences in such contexts as *good food* and *good knife* and *good lyre*. At least in the last two, however, it seems obvious that the word *good* itself makes exactly the same input to the sense of the sentence. But it is also obvious that the question of sense is not necessarily the question to which the traditional discussions were directed. It takes no great charity to suggest that they are really interested not in the input but in the *output* of a word, the part of the paraphrase of a whole sentence for whose presence some particular word in the sentence is responsible. In fact they are interested in what one might call *proposition-factors*. There is no difficulty in constructing a diagram analogous to the p. 24 diagram where (1) assumes as a new value a word type with a sense, e.g. *good*, (2) has as its new value a word-token of that word type, e.g. *good* as it occurs in *this is a good knife* and (4) takes the value *has a fine edge suitable for cutting*, this being one component of the paraphrase *This knife has a fine edge suitable for cutting*, which paraphrases *This is a good knife*.

It may be asked, why should anyone interest himself in these output questions. I think the answer is that type (ii) questions sometimes constitute interesting problems about *property*-identity. Surely, it has been thought, goodness in a knife is not the same property as goodness in food? And not even the same as goodness in a lyre? And the inferential consequences of the possession of one sort of goodness will be very different from those of another. (I can imagine that similar questions about the relation of input and output might matter for the discussion of some implications, e.g. '*ought* implies *can*'.)[a]

The sign that this may be the best view of the apparent conflict between the approaches, and that more clarification is required than resolution, is provided by the fact that Aristotle's alarm in *Physics*, Book VII about definitions themselves turning out to be ambiguous looks most comical when he suggests that *double* or *much* may be ambiguous. (The worry is that *much air* is not comparable with *much water*.) The reason why this looks like a senseless anxiety is that there is neither an input question nor an interesting output question here. *Much* is not a first order property at all and *a fortiori* it is not a first order property about which there can be questions like questions about goodness in lyres and goodness in knives or food. There is no interesting or anyway straightforward type (ii) question to be asked about it.

Harvard University and University of London (Bedford College)

---

[a] This difference between input and output questions may be one of the things at issue in one so far fruitless dispute between Fred Sommers and his critics (conveniently summarized in Jonathan Bennett, *Journal of Symbolic Logic*, September 1967, p. 407).

How does one tell whether a word has one, several or many senses?

WILLIAM P. ALSTON

1   Mr Wiggins is certainly correct in supposing that the answer to the title question should be based on an account of what it is for a word to have a certain meaning, and the associated account of how specifications of word-meaning are to be tested. If we take a Lockean view according to which having a meaning is a matter of being regularly used as the sign of a certain (Lockean) idea, then the job of showing that 'necessary' has several different meanings is quite a different affair from what it would be if we accept the view that for a word to have a certain meaning is for utterances of that word to stand in certain causal relations with physical stimuli and/or overt responses. For my part I share with Mr Wiggins, and many other theorists, the conviction that for a word to have a certain meaning is for it to make a certain contribution to some appropriate semantic property of sentences in which it occurs, a property that we may dub 'sentence-meaning'. I say 'dub' because I doubt that there is enough talk about the meaning of *sentences* (unlike the situation vis-a-vis words), to yield any substantial pre-theoretical concept properly so called. The kind of sentence property that I believe both Mr Wiggins and I (along with others) have in mind is most per-spicuously indicated in ordinary language by some such term as 'what the sentence can be used to say' (in a sense of say in which 'what he said' is not synonymous with 'what sentence he uttered'), or more barbarously, the sentence's 'saying potential'. Mr Wiggins wishes to explicate this notion in terms of the conditions under which an indicative sentence can be used to make a true statement (or, in his lingo, the conditions under which an utterance of the sentence will be factually licensed). He correctly anticipates that some will find this approach intolerably restricted, and I count myself among that number. I think it not too vaulting an ambition to search for a more general account of 'saying potentials' of sentences, such that the poten-tiality of sentences of a certain sub-class for being used to make statements would be a special case of this more general notion. I have elsewhere provided the beginnings of such an account, using the term 'illocutionary act potential'.[a] (This is *not* Austin's concept of an illocutionary act, whatever that is, though I did filch the term from him.) Roughly, for a sentence to have a certain illocutionary act potential is for it to be subject to a rule that enjoins members of the language community from uttering the sentence, in certain kinds of contexts, unless certain specified conditions hold. Thus the central illocutionary act potential of 'Please pass the salt' would be specified by making it explicit that a rule is in force in the English language com-munity that could be roughly formulated as follows:

[a] *Philosophy of Language* (Englewood Cliffs, N.J.: Prentice-Hall, 1964), pp. 34 ff.

One is not to utter 'Please pass the salt' in a normal[a] context unless the following conditions hold:

1. Someone, H, is being addressed.
2. It is possible for H to pass the speaker some salt.
3. The speaker has some interest in getting H to pass him some salt.[b]

For the special case where the illocutionary act, a potential for which is in question, is something statemental or assertive, something that is straightforwardly true or false, the conditions imbedded in the rule would coincide with the truth conditions for the statement. Thus the most basic illocutionary act potential for 'My uncle sold his wireless' could be specified by making it explicit that the following rule is in force in the English language community:

One is not to utter 'My uncle sold his wireless' in a normal context unless the following conditions hold:

1. Some particular uncle, x, of the speaker is contextually indicated.
2. At some time prior to the time of utterance, x sold a wireless that had belonged to him.

Thus the filling for the specification of illocutionary act potential for this sentence would coincide with what Wiggins calls conditions for the utterance being factually licensed. Any doubt as to whether a certain set of conditions, e.g. those listed above, do constitute the truth conditions for the statement (or conditions for an utterance of the sentence being factually licensed), would equally be a doubt as to whether the sentence is subject to a rule of the above sort that requires just those conditions to hold.

Instead of continuing the exposition of my views on sentence-meaning, I shall address myself in this paper to problems that have specifically to do with distinguishing senses of words. Let us take it as agreed between Wiggins and myself that we have some appropriate sense of 'sentence-meaning', and that we are thinking of a meaning of a word as a constant contribution it makes to the meaning of any sentence in which it occurs with that meaning. Then how are we to conceive the job of determining how many senses a word has? Again I find myself in sympathy with the general thrust of Wiggins' account. A word is properly assigned as many senses as is necessary to account for the facts about the meanings of sentences in which it occurs. I should like to put more *stress* than Wiggins does on the ineradicably systemic character of the evaluation of particular semantic hypotheses, given this approach to meaning. (Though I have no reason to think he would disagree with what I am about to say.) If dictionary entries are to be evaluated in terms of the 'readings' or 'interpretations' of sentences they yield, any given entry will have to be evaluated in conjunction with a number of other semantic hypotheses. Suppose that I have two p oposed entries for 'run' (as a transitive verb), very roughly as follows: (a) *operate*, (b) *force*, together in each case with what Wiggins calls 'directives' and what Katz calls 'selection restrictions'.[c] Obviously we can't use these entries by themselves to derive any readings for any sentences, e.g., 'How long have you been running the engine?' or 'You are going to run me into debt'. To derive readings for these sentences we need not only entries for 'run', but also entries for the other words, specifications of the syntactical structure of the sentences and constituent complex

---

[a] This qualifier is meant to rule out such contexts as those in which the speaker is acting in a play or giving an example of a request.
[b] This list of conditions is intended to be illustrative rather than exhaustive.
[c] *Philosophy of Language* (New York: Harper and Row, 1966), pp. 154 ff.

expressions, and what Katz calls 'projection rules',[a] which for a given kind of complex expression tell us how to go from facts like the above to a reading for the whole expression. Thus on this approach dictionary entries cannot be tested in isolation, one-by-one, any more than can any other constituents of the semantic description of a language. Just as a particular general hypothesis in science can be empirically tested only in the context of some system of principles, so with particular semantic and syntactical hypotheses about a particular language.

This means that our title problem merges into the very large methodological problem: How can we formulate and evaluate a semantic description of a language? For the above considerations show that my justification for supposing that, e.g., 'run' has at least two meanings, *operate* and *force*, can be no greater than my justification for supposing that an adequate lexicon for English would include two such entries for 'run'. The most fundamental justification of this latter claim would be, of course, the demonstration that a given lexicon containing theses entries is an adequate one, and, as indicated above, we could not show that without showing that an adequate semantic description of the language could be constructed using this lexicon.[b] Thus an ideally thorough treatment of this particular methodological problem would include consideration of all the methodological problems involved in the formulation and evaluation of the semantic description of a language and the components thereof. This includes such problems as:

1. What form should dictionary entries take?
2. What form should the specifications of sentence-meaning take?
3. What facts about sentence-meaning can be used as data for the testing of semantic systems, and how can these facts be gleaned from the behavior of language users?
4. All the methodological problems involved in developing the syntactical description of a language.

These problems do not come piecemeal. Language itself is so systematic that the investigation of language fails to be equally systematic only at its peril.

2 Rather than emit *obiter dicta* on such large issues as these, I shall turn to the consideration of a more modest methodological problem, one small enough to be illuminated in the time at my disposal. Suppose that we find ourselves required to decide whether a given word is used in the same or different senses in two contexts,[c] but we do not have the time, resources, or ingenuity to consider this problem in the light of some proposal for a complete semantic description of the language. This is a situation in which lexicographers and philosophers often find themselves. For the traditional lexicographer such problems come up in the course of deciding on a set of entries for a word in a dictionary of the usual sort, one with much more modest pretensions than the ideal lexicon about which we have been talking. For the philosopher, on the other hand, they most usually arise in the course of giving analyses of

[a] Ibid.

[b] This not to say that I cannot be justified in semantic claims without having established them in this full-dress fashion. If one could not know, or be justified in believing, many things without being able to give ideally complete justifications of them, we would be in a pretty mess in many areas of thought.

[c] This is a less ambitious task than that of determining *what* sense the word has in each of these contexts, although the lexicographer always, and the philosopher sometimes, asks a question like this in the course of trying to answer the more ambitious question.

the concepts expressed by certain terms. If a philosopher is concerned to analyze the concept expressed by 'remember' over a certain stretch of contexts, he will have to determine how many senses the word has over that stretch, so that he will know how many concepts to analyze. Again, if a philosopher A claims that 'true' as it occurs in

⟨1⟩ What you say is true.

means *corresponds to the facts*, and then B attacks this by pointing out that one can be a true friend without corresponding to any facts, then if A is to defend his analysis he will have to either show that B's claim is mistaken, or show that in

⟨2⟩ He is a true friend.

'true' is used in a sense different from the one in which he is interested. My relatively modest methodological question is: what devices, short of a consideration of schemes for a complete description of the language, can we deploy for the resolution of such questions? And what difficulties will we encounter in this enterprise?

It may be useful to sharpen our intuitions and initially formulate our principles in the context of some pedestrian examples far removed from difficult and philosophically exciting terms like 'true' and 'remember'. For this purpose, consider the occurrences of the humble term 'run' in

⟨3⟩ Harold Stassen is still running.
⟨4⟩ The boundary ran from this tree to that tree.
⟨5⟩ John ran from this tree to that tree.
⟨6⟩ The engine is still running.
⟨7⟩ Is the vacuum cleaner still running?

We may pose as our initial sample problems: how can we show that 'run' has at least two senses in ⟨3⟩, but only one sense in ⟨4⟩? And how can we show that 'run' has different senses in ⟨4⟩ and ⟨5⟩, but the same sense in ⟨6⟩ and ⟨7⟩?

If these are to be precise questions, we shall have to make explicit what resources we are allowing our inquirer. I suggest that we regard him as capable of making reliable judgments of (at least approximate) sameness and difference of sentence meaning. Thus he can determine that ⟨6⟩ and ⟨7⟩ have different meanings, that ⟨4⟩ and

⟨8⟩ The boundary extended from this tree to that tree.

have the same meaning, and that ⟨3⟩ has at least two different meanings. Moreover we shall regard him as capable of determining what entailment relations hold between sentences, e.g., that ⟨6⟩ entails 'The engine is in working condition'. That is, we are going to trust his 'intuitions' about the semantic relations of sentences taken as (semantically) unanalyzed units, even though, lacking a fine grained semantic description of the language, or even a proposal for such, he is unable at this stage to analyze sentence-meanings into the components contributed by the words that make up the sentence. In granting him even these relatively crude capacities we are making some large, and recently highly controversial, assumptions about the possibility of drawing a line between what one means by a sentence and what happens to be true of the subject matter, between the analytic and the synthetic, and so on. But for purposes of this paper we can do no more than note these assumptions in passing.

Proceeding, armed with these tools, to our sample problems, the first thing to note is that the tools do not suffice in and of themselves for the resolution of the problems. It does not follow just from the fact that ⟨4⟩ and ⟨5⟩ have different meanings that 'run' is used in a different sense in them. To establish that we have to rule out the possibility that other features of this sentence pairing, e.g., the interchange of 'The boundary' and 'John', are solely responsible for the difference in sentence-meaning. After all, ⟨6⟩ and ⟨7⟩ differ in meaning too; but here we are not inclined to attribute this difference, even in part, to a difference in the sense of 'run'. Again, admitting that ⟨3⟩ can be used to say two quite different things, why attribute this difference to differences in senses of 'run'? The mere fact that there are two different meanings of this sentence is compatible with a number of alternative hypotheses, e.g., that one of the other words, e.g., 'still', has two different senses in this context, or that the sentence can be grammatically construed in two different ways. And yet we are strongly inclined to attribute the plurality of sentence-meanings to a plurality in the senses of 'run'. How can these inclinations be justified?

If we had adequate analyses of the meanings of each of these sentences, we could read off from those analyses what each sentence-constituent contributes to the whole. Hence we could say *what* the contribution of 'run' is to the meaning of ⟨4⟩ and to the meaning of ⟨5⟩, thereby determining whether it is the same contribution in each case. But this is just what we do not have, short of an adequate systematic semantics of English. What we are looking for is the closest approximation to this procedure that is possible within the specified limitations. A technique that looks promising is that of partial substitution. Although we cannot say *what* 'run' contributes to the meaning of ⟨4⟩ and of ⟨5⟩, we can try various substitutions for 'run' in each of the two sentences, noting which ones do and which ones do not preserve sentence-meaning. Thus we can preserve (approximately) the meaning of ⟨4⟩ while substituting 'extend' for 'run', and we can preserve (approximately) the meaning of ⟨5⟩ while substituting 'locomote springily'[a] for 'run', but opposite substitutions will produce a marked change in (or destroy) sentence-meaning. Similarly, if we consider two contexts in which ⟨3⟩ would be used to say different things, in one of these contexts sentence-meaning can be preserved by a substitution of 'actively seek public office' for 'run', in the other context sentence-meaning can be preserved by a substitution of 'locomote springily' for 'run', but again not vice versa. These results support the claim that at least part of the difference in sentence-meaning is due to differences in the senses with which 'run' is used. What this technique gives us is an indirect, symptomatic approach as a substitute for an unavailable direct account of the underlying structure. This technique is related to the ideal demonstration as a symptomatic diagnosis of a disease is related to a pathological description. The fact that non-convertible replacements for 'run' in ⟨4⟩ and ⟨5⟩ will preserve sentence-meaning is taken as a symptom of a change of sense in 'run' between ⟨4⟩ and ⟨5⟩, just as an increase in body temperature is taken as a symptom of an increase in infection somewhere in the body. Let us consider the assumptions we make in performing this symptomatic inference. I will begin with the same-sentence case, since the other case involves all the assumptions present there, in addition to some others.

Let us consider two situations, S_1 and S_2, in which ⟨3⟩ is being used to say two different things. In S_1 we can say the same thing by

⟨9⟩ Harold Stassen is still locomoting springily.

In S_2 we can say the same thing by

⟨10⟩ Harold Stassen is still actively seeking public office.

but not vice versa. This would seem to show conclusively that in S_1 'run' is being

---

[a] I am using this phrase as an abbreviation for 'move rapidly by springing steps so that there is an instant in each step when neither foot touches the ground'.

used to mean the same thing as 'locomote springily' but not the same thing as 'actively seek public office' and in S_2 vice versa. Since 'locomote springily' and 'actively seek public office' are not interchangeable in this sentential context, they do not mean the same thing in this context. Therefore, since 'run' is synonymous with the one in S_1 and with the other in S_2, it must have different meanings in S_1 and S_2.

However, without falling into the last extremities of Cartesian scepticism we can note the following possibility, which, if realized, would throw off this inference. Suppose that some other word or words in the sentence frame 'Harold Stassen is still — ' shifts its meaning in the course of one of these substitutions, e.g., the substitution of 'actively seeking public office' for 'running'. In that case the fact that in S_2 ⟨10⟩ has the same meaning as ⟨3⟩ would not show that in S_2 'run' has the same meaning as 'actively seek public office'. It would show the opposite. For if some part of the frame changes its meaning, *and* the substitute for 'run' is a synonym, then the resulting sentence would have a *different* meaning from ⟨3⟩. Under those conditions, ⟨10⟩ would have the same meaning as ⟨3⟩ only if 'actively seeking public office' were to differ in meaning from 'running' in such a way as to neutralize the other differences in the sentences. By the same argument we can show that if some part of the frame were to shift its meaning through the substitution, the fact that ⟨3⟩ (in S_2) and ⟨9⟩ differ in meaning does not show that 'run' and 'locomote springily' differ in meaning in this context. For even if they have the same meaning in this context, ⟨3⟩ and ⟨9⟩ would differ in meaning because of the semantic shift elsewhere in the sentence. These considerations show that in using the substitution test we are assuming that the rest of the sentence holds fast semantically. If we had to justify such an assumption we would have to carry out substitutions with respect to each of the other constitutents of the sentence. We would have to show, e.g., that 'still' can be replaced by 'yet', *salva meaning* in both ⟨3⟩ and ⟨9⟩. But this argument would be subject to analogous assumptions concerning the semantic fixity of the rest of these sentential contexts, as well as the assumptions that 'yet' has the same meaning in the two sentences. Hence the attempt to justify every such asumption one-by-one would lead to an infinite regress. It would seem that at some point we are forced to invoke a principle of simplicity, according to which terms are held to retain the same meaning over two contexts unless we are forced to recognize a difference.

Consider now the case where we make non-convertible substitutions in different sentences, e.g. ⟨4⟩ and ⟨5⟩. ⟨5⟩ has the same meaning as

⟨11⟩ John locomoted springily from this tree to that tree.

but not the same meaning as

⟨12⟩ John extended from this tree to that tree.

whereas in ⟨4⟩ 'ran' is replaceable by 'extended' but not by 'locomoted springily'. From this we infer that 'ran' has different meanings in ⟨4⟩ and ⟨5⟩. Here the reasoning is subject to the same assumption as in the same-sentence case, viz., that in each substitution the rest of the sentence remains semantically fixed; but there is an additional problem that stems from the difference in sentence-contexts. For suppose that 'extended' shifts its meaning from ⟨8⟩ to ⟨12⟩. In that case the fact that ⟨4⟩ preserves the same meaning under a substitution of 'extended' for 'ran' while ⟨5⟩ does not, does not show that 'ran' does not have the same meaning in ⟨4⟩ and ⟨5⟩. For even if 'ran' has the same meaning in both contexts, we would not expect that meaning to be preserved in both contexts under a substitution of 'extended' for 'ran', if 'extended' has a different meaning in the two contexts. We would expect just the reverse. Thus the substitution test employed here assumes that our substituends, 'locomoted springily' and 'extended', do *not* have different meanings in the two contexts. And again any attempt to justify this assumption would lead us to make further analogous assumptions.

I uncover these assumptions not in order to show that we are never justified in concluding that a word has two different senses in two contexts, a conclusion I would not embrace, but rather to re-emphasize the systemic character of investigations in this area, as in other areas of language. If we had a workable semantic description of the whole language, each component would receive its justification from its presence in the system, which in turn would be justified by the fact that it would do what we expect a semantic description of a language to do.[a] But so long as we are confined to piecemeal inquiries we will correspondingly be forced, at any stage, to rely on assumptions of sameness and difference of meaning that have received no justification. While this is our condition we are undoubtedly well justified here, as elsewhere, in proceeding on the basis of principles of simplicity like the one cited two paragraphs back, and in these terms we are often justified in drawing conclusions as to sameness of difference of sense. But ideally such conclusions are provisional, pending a systematic development of a total semantics of the language.

It is not in every case of a suspected difference in meaning that we are able to carry out substitutions just for the word in question. Sometimes the sentence meaning cannot be so neatly dissected, and we have to get along with still more indirect indications. Consider

⟨13⟩ I ran him a close second.

and

⟨14⟩ He always runs everything together.

I daresay we will not find any substitution for 'run' in either of these sentences which, while leaving the rest of the sentence unchanged, will preserve sentence meaning. We can find near-equivalents of the two sentences as wholes. Thus ⟨13⟩ has about the same meaning as

⟨15⟩ I placed second, close behind him.

and ⟨14⟩ has about the same meaning as

⟨16⟩ He fails to distinguish things sufficiently.

⟨15⟩ and ⟨16⟩ are so different from each other that it encourages us to say that 'run' must have some different meaning in ⟨13⟩ and ⟨14⟩. Moreover the fact that ⟨15⟩ is concerned with the speaker's rank in the results of some contest, while ⟨16⟩ is concerned with failing to make distinctions, encourages us to say that in ⟨13⟩ 'run' means something like *be ranked as*, while in ⟨14⟩ 'run' means something like *confuse*. But only *something* like. After all, what we have in ⟨15⟩ and ⟨16⟩ are paraphrases of whole sentences in which several constituents are replaced and the structure changed. Hence this technique fails to pinpoint the semantic contribution of 'run'.

The same disability attaches to the demonstration of mutually non-substitutable entailments. ⟨13⟩ but not ⟨14⟩ entails that the speaker was in some contest, while ⟨14⟩ but not ⟨13⟩ entails that the person referred to has been talking about several

---

[a] This is not to deny that any judgments we make about the adequacy of a total system are highly fallible, nor that given any such system we are likely to construct, it is highly likely that a better one could be constructed.

distinguishable topics. But again it is the whole sentence that has this implication in each case, and there remains a question as to what this shows us about the specific semantic contribution of 'run'.

In these cases our conclusions do not have even the kind of provisional validity enjoyed by those based on the word-substitution test. Here our grounds are shakier because less explicit. In the case of the former we could identify assumptions such that if these assumptions were correct the conclusions would be established. Furthermore the assumptions were of the same type as the conclusion in question, and so any one of them could itself be tested, subject, of course, to the same necessity for dependence on other like assumptions. Nothing like that is possible here. Since we have no resources, analogous to the word-substitution test, for prying the sentence apart semantically, we are unable to identify any testable assumptions on which our conclusion depends.[a] Such basis as we have is intuitive rather than discursive. It seems to us, as we mouth the sentences, that the word 'run' is 'doing something different' in the two cases. But of course such impressions are notoriously fallible. That is not to say that they are worthless. We frequently have to make do with this sort of thing in an undeveloped stage of a discipline, and the impressions of sensitive trained observers are by no means to be taken lightly. Nevertheless it is salutary to realize just what status our conclusions have.

3   I should now like to apply the results of the last section to a consideration of some cases in which philosophers find themselves called on to decide questions of sameness or difference of meaning. In many such cases the philosopher is able to make use of the pinpointed substitution test. Thus we can show that in ⟨1⟩ but not ⟨2⟩ 'true' can be replaced by 'correct', while in ⟨2⟩, but not in ⟨1⟩, 'true' can be replaced by 'real'. Again, a philosopher doing philosophical psychology may want to distinguish the senses of 'want' in the most common uses of

⟨17⟩ I want an ice cream soda.

and

⟨18⟩ That child wants a good spanking.

This he can do by pointing out that in ⟨17⟩ but not ⟨18⟩ 'want' can be replaced by 'have a desire for', while in ⟨18⟩ but not ⟨17⟩ 'want' can be replaced by 'need'. These conclusions will have the status we earlier saw such conclusions to have. That is, they can be considered justified, given certain plausible assumptions that can either be accepted on a simplicity principle or investigated in the same way as the conclusions in question.

Wiggins' contentions about 'good' employs this technique. I take him to be claiming that

⟨19⟩ She has good legs.

has several meanings such that in one and only one of these meanings 'good' can be replaced by 'beautiful' *salva* sentence-meaning, in one and only one meaning of the sentence 'good' can be replaced by 'healthy' and so on. Let me just note in passing that the only serious opposition to Wiggins' thesis will come at the level of sentence-meaning. One who wishes to maintain the univocity of 'good' will (be well advised to) claim that ⟨19⟩ is used with the same meaning (is used to make the same

---

[a] Of course one might say that the conclusion, e.g., that 'run' has different senses in ⟨13⟩ and ⟨14⟩ depends on the assumption that 'run' contributes at least part of the difference in the meanings of the two sentences. But this is just our conclusion over again.

assertion) in all the contexts of which Wiggins is thinking, and that the differences to which he alludes are differences in the considerations that are relevant to the evaluation of this one and the same assertion in one or another context. In other words, the defender of univocity would challenge Wiggins' claim that ⟨19⟩ sometimes means the same as 'She has beautiful legs', sometimes the same as 'She has healthy legs', etc. He would maintain that each of those latter sentences means more than ⟨19⟩ ever *means*, for it adds to the claim made by ⟨19⟩, a claim about the proper criteria of evaluation for that context. Thus the main issue here has to do with the relation of meaning and criteria for evaluative terms (or sentences). I do think that if Wiggins' claims on the sentence-meaning level are granted, there can be no serious doubt that he is correct about the word 'good'.

However the philosopher is not always in a position to carry out the word-substitution test. Suppose that a philosopher has proposed an analysis of 'see' such that one of the defining conditions consists of the actual existence of the object of sight in the physical environment of the perceiver. Now he is confronted with locutions like

⟨20⟩ I see Mt Rainier before my mind's eye.

where there is obviously no implication that Mt Rainier is actually in the physical environment. He replies, of course, that this is a different sense; he means to be elucidating the concept of seeing as the exercise of a sense organ, not the concept of the exercise of visual imagination. But how can he show that the word 'see' is used in different senses in ⟨20⟩ and

⟨21⟩ The clouds have lifted; I now see Mt Rainier.

He can, of course, find equivalents to ⟨20⟩ of a sort not available for ⟨21⟩. Thus ⟨20⟩ is roughly equivalent to 'I have a mental image of Mt Rainier' and to 'I am visualizing Mt Rainier', while ⟨21⟩ is paraphrasable in no such ways. But does this show that 'see' is used in different senses in ⟨20⟩ and ⟨21⟩? A partisan of the single sense view will (be well advised to) concede that these results show that 'see x in the mind's eye' has a different meaning from 'see x'. For the former phrase, but not the latter, can be replaced by 'have a mental image of x' or 'am visualizing x'. But this is not at issue. The question is as to whether the word 'see' itself makes a different contribution to the meaning of the two phrases. One who maintains that it does not can hold that the difference in meaning in ⟨20⟩ and ⟨21⟩ comes just from the presence of 'in the mind's eye' in ⟨20⟩ and not in ⟨21⟩. And unless we can pry apart these phrases by making substitutions just for 'see', we will have no *argument* against him.

Again suppose that H. P. Grice is attempting to separate out and analyze a specially 'semantic' sense of 'mean' as used of speakers, a sense of 'mean' in 'What S meant (by what he said) was —', such that from a statement of this form we can derive a specification of the sense in which he was using whatever sentence he uttered, or a specification of what the sentence meant as he was using it on that occasion.[a] In order to focus on this we shall have to filter out other senses of 'mean' as used of speakers, e.g., in

⟨22⟩ What do you mean?

together with answers thereto. Sometimes ⟨22⟩ is a request for a justification of what was said ('I'm not going to the party', 'What do you mean, you're not going'?). Sometimes ⟨22⟩ is a request to be more specific (Doctor: 'You'll be out of the

[a] See his article 'Meaning', reprinted in this volume, pp. 53 ff.

hospital soon.' 'What do you mean, "soon"?'). We shall want to distinguish these senses of ⟨22⟩ and correlated senses of 'I mean — ' from the case in which ⟨22⟩ is a request for a more intelligible paraphrase of the sentence uttered. But can we show that 'mean' has different meanings, in ⟨22⟩, in cases of these three sorts? Let us agree that the equivalents of the whole sentence are different in the three cases. In one it is 'What is your justification for saying that?', in another it is 'Be more specific', in the third, 'Give me a paraphrase of your sentence that I can understand'. But how do we show that these differences in sentence-meaning are due to differences in the sense of 'mean' rather than, or as well as, differences in the meaning of 'what' or 'do', or differences in grammatical structure? It certainly seems intuitively plausible to suppose that differences in the sense of 'mean' must be at least partially responsible, but how to *show* this?

Again, suppose that in the course of trying to understand emotion-concepts, I feel the need to distinguish 'occurrent' from 'dispositional' uses of emotion terms. That is, I want to show, e.g., a difference in the senses of 'afraid' in

⟨23⟩ I have always been afraid of snakes.

and

⟨24⟩ When he started toward us I became very afraid of him.

The search for non-convertible replacements for 'afraid' alone is doomed to failure because any substitution that will work in one, e.g., 'frightened', will work in the other as well. If 'afraid' has different senses in these contexts, so does 'frightened'. We may then be led to pointing out differences in entailment patterns. Thus ⟨23⟩ entails 'Usually when I see snakes I get frightened', but there is no comparable entailment for ⟨24⟩, e.g., 'Usually when he starts toward us, I get frightened'. But of course this is a difference between ⟨23⟩ as a whole and ⟨24⟩ as a whole. We still have not *shown* that this difference is due, at least in part, to a difference in senses of 'afraid'.

In these three cases our conclusions have the purely intuitive character we have seen earlier to attach to such conclusions. Again this is not to claim that the conclusions should be abandoned, but only to point out where we are at this stage of the game.

I have been surveying cases in which the philosopher is in a relatively good position to establish difference of meaning, and less favorable cases in which, nevertheless, it seems plausible to suppose that there is a difference of meaning which would be established with more adequate devices. Now I want to call attention to some cases in which philosophers make blatantly unjustified claims of multivocality through neglect of some of the points set out above.

Often when I feel inclined to cavil at multivocality claims the controversy is properly located at the sentence-meaning level. Thus B. F. McGuiness maintains that 'want' must be used in a different sense when talking of conscious and of unconscious wants.[a] He bases this claim on the claim that 'P wants S' entails 'P knows that he wants S' when we are using 'want' in an ordinary sense of the term, but not when we are talking about repressed wants. (Hence a sentence of the form 'P wants S' means something different in these two contexts.) If he is right about this difference in sentence-meaning, his multivocality claim is surely justified, but I see no reason to accept the entailment claim. Again Norman Malcolm notoriously maintains that 'He had a dream last night' has a meaning different from its ordinary

---

[a] 'I Know What I Want', *Proc. Arist. Soc.*, N.S. LVII, 1956–7.

meaning if it is based on REM evidence rather than on the subject's reports on awakening.[a] For the sentence will be logically related to different 'criterial' statements in the two cases. Again if this difference of sentence-meaning can be established, it will be at least very plausible to suppose that the word 'dream' has different meaning in the two contexts. But again I do not feel constrained to recognize that what I am prepared to take as decisive evidence enters into meaning in the way Malcolm supposes.

From the point of view of this paper the more interesting cases of philosophical folly are those in which the putative differences in sentence-meaning are uncontroversial, but where word-multivocality is inferred from this without heeding the cautions insisted on earlier. The most flagrant cases are those in which a multiplicity of senses is ascribed to some term on the ground that different sentences in which it occurs are tested or verified or established differently. Thus 'there is' is said to have different meanings in

⟨25⟩ There is a fireplace in my study.

and

⟨26⟩ There is a prime number between 6 and 10.

on the grounds that the justifications of these two statements are widely different. There is no doubt that the justifications are widely different and that ⟨25⟩ and ⟨26⟩ have different meanings (not that we needed an appeal to verifiability to show *that*). However this is radically insufficient to show that 'there is' has different meanings in the two sentences. It would seem that the difference in meaning between 'fireplace in my study' and 'prime number between 6 and 10' is quite sufficient to yield the difference in meaning between ⟨25⟩ and ⟨26⟩ that is reflected in the different justifications. Analogous remarks are to be made about the similar argument that 'true' has different meanings in

⟨27⟩ It is true that there is a fireplace in my study.

and

⟨28⟩ It is true that there is a prime number between 6 and 10.

on the grounds that ⟨27⟩ and ⟨28⟩ are tested in radically different ways. Once more the difference in what follows 'It is true that' in the two cases is quite sufficient to account for that difference.

Again philosophers will argue for multivocality from differences in patterns of entailment, heedless of the possibility that these differences may be explained by differences in other constitutents of the sentence. Thus 'know' is said to have different senses in

⟨29⟩ I know that I feel disturbed.

and

⟨30⟩ I know that my car is in the garage.

on the grounds that ⟨29⟩ is entailed by 'I feel disturbed',[b] while ⟨30⟩ is not entailed by 'My car is in the garage'. But it still remains to be shown that this difference in entailment patterns reflects a difference in the meaning of 'know', rather than just a difference between 'I feel disturbed' and 'My car is in the garage'. Again, it is common for philosophers to hold that even when we restrict ourselves to specifica-

[a] *Dreaming* (London: Routledge and Kegan Paul, 1959), p. 80.
[b] This entailment is itself highly disputable, but I shall accept it for the purposes of this illustration.

tions of linguistic meaning 'mean' has a different meaning depending on whether we are talking of a 'categorematic' word like 'thermometer' or a 'syncatetorematic' word like 'if'. This claim is rarely given any justification other than a question-begging one to the effect that a favored analysis of the concept of meaning (usually of a referential sort) applies to the former but not to the latter. But one might try to justify the claim by pointing out that '"Thermometer" means *instrument for measuring temperature*' entails that 'thermometer' denotes instruments for measuring temperature, whereas '"If" means *provided that*' does not entail that 'if' denotes 'provided that'. However, even if we grant this entailment claim, it will not follow that 'mean' has different senses in the two contexts. For the difference in entailments may be adequately accounted for by the differences between 'thermometer' and 'instrument for measuring temperature' on the one hand and 'if' and 'provided that' on the other. In other words, it may be that these expressions just have different *kinds* of meanings, not meanings in different senses of 'mean'.

4   We have seen that the attempt to decide questions of multivocality piecemeal is beset with serious difficulties. In the most favorable cases we have to depend on assumptions that receive such justification as they have from considerations of simplicity. In less favorable cases our conclusions rest on unadulterated intuition, pending more systematic constructions. Now it *may* be that these difficulties are not basic problems for the philosopher just because, contrary to first impressions, the philosopher, unlike the lexicographer, does not have to settle questions of word-multivocality. It is true that most philosophers who have thought of conceptual analysis as a linguistic enterprise have thought of it as essentially concerned with the meanings of words. Thus much meta-ethics has been concentrated on questions about the meaning(s) of 'good', there has been much talk in epistemology about the meaning(s) of 'know' and so on. But it may not be necessary for the analytical philosopher to couch his problems in this way. If one is trying to understand value-judgments, or moral judgments, it is not essential for the accomplishment of that purpose to provide a set of dictionary entries for 'good' that would be satisfactory as a part of an adequate semantic description of English. To meet *that* requirement the entries would have to be such as to yield, together with the appropriate other parts of the description, acceptable interpretations of any sentence, declarative or otherwise, in which 'good' occurs. However it seems that one could arrive at a philosophically illuminating understanding of value-judgments just by giving patterns of interpretation for sentence-types, like 'x is a good ψ', without attempting to spell out what is contributed to the interpretation of the whole sentence by each of its meaningful components. (Many philosophers would prefer to call such results 'contextual definitions' of 'good', for various contexts). Again if I am interested in understanding linguistic meaning, then I am interested in giving interpretations of sentences of the form 'x means y', where x is a variable ranging over linguistic expressions, sentences of the form 'P knows what x means', etc. It is not essential for this purpose that I determine whether or not 'mean' has the same sense in these contexts as it does in various contexts where it is used in conjunction with something other than designations of linguistic expressions, e.g., 'I mean Susie', or 'That look on his face means trouble'. Here too we can give interpretation of the *sentence-types* with which we are concerned without going into their fine semantic structure. Again, philosophers who make claims about 'there is' and 'true' of the sort we criticized a few pages back may not have to make such claims in order to accomplish their main

purposes. It may be that they too can restrict themselves to the sentence-level, e.g., by just pointing out the differences in what we are saying in uttering sentences like ⟨25⟩ and ⟨26⟩, without having to trace these differences to differences in the meanings of 'there is' in these sentences.

I am not claiming that it is never important for the philosopher, *qua* philosopher, to provide adequate dictionary entries, or to establish word-multivocality. The boundaries of what philosophers do *qua* philosophers are too fuzzy to permit us to establish such a claim. I *have* tried to suggest, via a few examples, that often when philosophers think they are essentially concerned with word-meaning, they are really concerned with the meaning or interpretation of sentence-types, and that they can formulate their semantic hypotheses in terms of sentence-sized units, thereby avoiding the special difficulties one encounters in analyzing sentence-meaning into the semantic contributions made by the various meaningful constituents of the sentence. This suggests, in turn, one reason for not assimilating the job of the analytical philosopher to that of the lexicographer. It would still remain true that the question of this paper is crucial to the methodology of linguistics, even if not to the methodology of analytical philosophy.

University of Michigan

A reply to Mr Alston

DAVID WIGGINS

Nothing that has happened since J. L. Austin's 1950s lectures 'Words and Deeds' or their publication[a] seems to me to have undermined the approach or made obsolete the kind of semantical theory typified by Frege or Russell or, in our own times, Carnap. In the broadest possible terms, however, to look at what a speaker does in saying what he says (a formula vague and inclusive enough to comprehend the very different researches both of Grice and of Austin) may enable us to place the philosophy of language in its principal proper context, the philosophy of mind; it may fill certain gaps and hiatuses left by the more formalistic approach which I follow these authors in espousing[b]; and it may put to rest metaphysical perplexities which maybe nothing else can relieve.[c] It is a very pleasant discovery that, coming on the problem of ambiguity from opposite directions, Mr Alston and I can concur in the conviction 'that for a certain word to have a certain meaning is for it to make a certain contribution to some appropriate semantic property of sentences in which it occurs' and that ' a word is properly assigned as many senses as is necessary to account for the facts about the meanings of sentences in which it occurs'. I shall call this doctrine *A*. Our agreement in *A* helps me to believe that complementarity and mutual consistency, not the annihilation or unconditional surrender of the other side, should be the proper aim of both the factions to this dispute.[d] There are three or four points in Mr Alston's reply which I should like to take up, however.

(*a*) Of the semantical property of sentences to which we agree word-senses contribute Mr Alston says that the present writer wishes to explicate it 'in terms of the conditions under which an indicative sentence...will be true or factually licensed. [Wiggins] correctly anticipates that some will find this approach intolerably restrictive', and Alston says that he counts himself amongst their number, seeking as he does a theory according to which the 'saying potential' of statements would be merely 'a special case' of a 'more general notion' of saying and 'saying potential'.

Evidently the most that Alston could make peace with me on is my 'minimal contention' of page 22 of my paper, but I still believe that the indicative utterance is much more than a special case, and that the autonomy and syntactical and semantical

---

[a] Under the title *How to do Things with Words* (Oxford, 1962).

[b] See my paper 'On sentence-sense' in this volume, pp. 18, 19, 23, 24.

[c] Consider negation. If anything can do what the traditional *otherness* explanations were meant to do to make us comfortable with *not*, I suppose it is a theory of what a speaker does when he negates a sentence (proposition?). See my 'Negation, Falsity and Plato's Problem of Non-Being' in *Plato* I (ed. Vlastos, New York, 1970), vii–viii. I still think that is more important, however, to formulate what the problem of negation precisely is than to rush to answer it.

[d] A thought implicit in note *a*, p. 19 and now argued in another way by P. F. Strawson in *Meaning and Truth* (An Inaugural Lecture, Oxford, 1970).

completeness of the indicative mood, as opposed to the subjunctive, optative, and imperative, simply must confer theoretical primacy upon it. (The questioning mood, if mood it is, does match the indicative in completeness, but hardly in autonomy. There could be statements without provision for the possibility of questions perhaps. Could there be questions without provision for the possibility of answers to them?)

Leaving aside the power and theoretical simplicity of an indicative-oriented theory, if Alston still holds that performative utterances are an important obstacle to the construction of one then I must briefly resume the problem. The distinction I appealed to in footnote *b* on p. 21 between tense and aspect was not forced upon me by any sort of desperation. It is a commonplace of Russian grammar, and is in fact a pervasive feature of both English and Greek. But the self-fulfilling properties of performatives do require a little more comment.

Take 'It's yours' said by Y to X to make X a present. Encouraged by the idea that the appropriate dimension of assessment for the utterance has less obviously to do with truth than with the efficacity of the uttering, we may try to spell out conditions for this in terms of Y's seriousness, Y's commitment, the thing's being Y's to give X in the first place, etc. Note that the prudence or imprudence or motivation or whatever else of the act will have no bearing at all on the particular 'felicity' or aptness which is in question. But once we have worked out the conditions for the efficacity which is our only concern here they may surprise the opponent of a contative analysis. The conditions are nothing other than the *truth*-conditions of the statement that the object is indeed X's! If one wants to know why saying 'It's yours' verifies 'it's yours', however, he must look at the social and legal conventions governing the making of gifts. There is no further semantic problem.

I would add one question. If the sentence 'It's yours' is not in the indicative mood then what mood is it in? The donatory mood? Are there then as many myriad moods as there are myriads of illocutions? This makes nonsense of the idea that a language is a system of communication with determinate semantics (see p. 18), and as Davidson would point out it completely confuses *mood* with *mode*.[a] It is simply not worth denying that 'It's yours' is in the indicative mood.

It is in virtue, then, of the conventions of ownership and transfer that to say 'It's yours' makes it yours. But consider explicit performatives. Can there be a convention which decrees that to say that one ϕs *is* to ϕ? How could a mere convention make it true that you ϕ if you don't in fact ϕ. Consider the two following conversations:[b]

1.
    Y: 'I assert that the persistence of the Conservative party is a monument to the impossibility of intellectual progress.'
    X: 'Go on then, assert it.'
    Y: 'I just have.'
    X: 'No, all you've done so far is assert that you assert it.'

2.
    Y: 'I promise I will be there.'
    X: 'Go on then. Promise!'
    Y: 'I already have.'
    X: 'You haven't. All you've done is assert that you do.'

[a] See page 171 of Donald Davidson 'On Saying That' in *Words and Objections*, ed. D. Davidson and J. Hintikka (Dordrecht, 1969).

[b] Which I have had occasion to discuss with my student, Mr Dudley Knowles of Bedford College.

> *Y*: 'All right. I will be there.'
>
> *X*: 'But I want you to *promise* to be there, not just say that you will be!'

Mad though all this is it will steer the second conversation back to a haven of sanity to extend it.

> *Y*: 'All right. I will be there. That's a promise. Is that enough? I promise this. I will be there.'

for which *Y* might have substituted:

> *Y*: 'All right. I promise that I will be there. Which is what I said in the first place!'

The one missing piece has now come to light in the shape of the paratactic analysis advanced by Davidson, and once anticipated by Meinong,[a] of ' *Y* ϕs that.....' We analyse it into two utterances of which the first, ' *Y* ϕs this   :', makes a reference to the second which is an autonomous but subordinate utterance of the sentence '.........'.[b] It is not convention but the logical form of 'I promise that I will be there' or 'I state that the persistence of...' which makes each of them do the complex of distinct things it does do.

(*b*) I do not dissent from Alston's general insistence on the 'ineradicably systemic character' of semantic hypotheses in general. And I am gratified that we can concur in many of the consequences which Alston ends his reply by deducing from *A*. But Alston sometimes talks and he often proceeds as if *A* were equivalent to what really is a quite different doctrine, which I shall call *B*. 'My justification for supposing that e.g. "run" has at least two meanings, *operate* and *force*, can be no greater than my justification for supposing that an adequate lexicon for English would include two such entries for "run".' Now the view I took and still take is that this is at once too much and too little to ask. It is the wrong test in other words. It is too little to ask because even this justification would not show that no *other* adequate lexicon and grammar could make do with a single and more abstract entry for 'run' (see pp. 37 and 28). It is too much to ask because one may sometimes be able to show directly that no correct lexicon could get by without treating some word as ambiguous. (Consider 'bank' again and my sentence 'All banks in the U.S.A. are now guaranteed by the Central Reserve Bank', p. 28. And I suggested other methods of bringing these matters to decision.) I think it is because Alston confuses *B* and *A* that, suspicious though he is, he is less suspicious than I should counsel of the bad old traditional method for suggesting differences of sense in a word *w* in contexts '——*w*——' and '.....*w*.....'. To find a reading of *w* which makes the first sentence true and the second false if the reading is substituted for *w* in each is not enough. Nor is Alston correct in suggesting – if he does suggest it – that the validity of the old substitution method is anywhere involved in my demonstration that all unitary analyses of 'good' so far offered are defective. The method of substitution was used there to disprove an analysis, not to prove an ambiguity. My claim that we can understand or 'hear' the sentence 'she has good legs' in three different ways is itself independent of any substitution test.

(*c*) Alston identifies what I called *directives* with what Katz and Chomsky have called *selection restrictions*. This identification I very strongly deprecate. Katz and

---

[a] See *Über Annahmen* III, 9. I owe the reference to another student, Miss Glen Crowther.

[b] See again Davidson, p. 171.

Chomsky's notion of a selection restriction comes with a notion of a semantic theory, and a whole apparatus of 'semantic markers' and what not, which is expressly designed to do all sorts of things I am inclined to believe that no semantic theory should attempt to do: for instance rule out certain weird readings of sentences which are impossible, but which must in my opinion count as theoretically possible so far as *semantics* are concerned (see p. 25, n. *b*). The notion of a directive, on the other hand, was introduced in connexion with a phenomenon which is perfectly incontrovertible from whatever point of view. It was then generalized by me from the case of *snub*, quite vaguely and probably quite recklessly. Aristotle's insight, for whatever it is worth, is not the creature of any particular linguistic methodology. For the time being I should wish to leave matters like that – in the air. And I should leave the notion of 'directive' tentative and imperfectly defined.

(*d*) Similarly, Alston uses the notion of a *projection rule* in the exposition of the theoretical framework within which our common doctrine *A* is to be applied. This again for my part I deprecate, since it would seem to commit both Alston and myself to a particular view, and one I myself reject, of what a semantic interpretation or reading is –, viz. a transcription into a universal conceptual notation or *characteristica universalis*. So far as I can see, projection rules have been intended at once

(i) to give the dictionary expansions by which some words need to be supplanted in order to display the deep structures, or (as I should prefer to have it) the logical form of the sentences where they occur,

(ii) to mark out and characterize homonymous words, and

(iii) to make translations into the *characteristica universalis*.

(iii) is an independent enterprise, whatever we think of it. And (i) and (ii) are separate again. My original paper would have done better to distinguish them. (i) is an important and special task. I give here one example of it.

Consider a term occurring in the manner which the scholastics called *in sensu diviso*, e.g. 'a sloop' as it occurs in 'I want a sloop', the sentence being so read that it does *not* imply 'there is a sloop such that I want it'. If you see no hope, as I see no hope, of explaining the difference between the cases where such sentences do and do not imply their existential generalizations in terms of a difference of style of occurrence of 'a sloop' – any old sloop versus a particular one, and the rest of that bad old business – then you must unfold the *want* we have here into *want that one have*[a] and distinguish

$(\exists x)$ (x is a sloop and I want that I have x)

from

I want $(\exists x)$ (x is a sloop and I have x).

To persons as puritanical as myself about referential matters the possibility of a verb's generating the *in sensu diviso/in sensu composito* ambiguity must always spell the necessity for lexical intervention. The thrust of this suggestion will escape nobody who has tried to characterize the logical form of intentional verbs which appear to take an accusative or direct object. (Nor will the difficulties of *finding* accurate two-part definitions which will bring out these referential ambiguities.)

[a] Cp. W. V. Quine, 'Quantifiers and Propositional Attitudes' in *Ways of Paradox* (New York, 1966).

This and like business under (i) may serve to divorce some homonyms, but not of course all. The ambiguity of *bank*, for instance, does not impinge on logical form. Such matters must wait on (ii). But (ii) again, like (i), though it is a new use for the dictionary, need never involve more than assigning English sentences English sentences as their readings. For the enterprise to which my paper was devoted everything that is special and technically distinctive about the notion of a projection rule is superfluous.

University of London

Meaning*a*

H. P. GRICE

Consider the following sentences:

> Those spots mean (meant) measles.
> Those spots didn't mean anything to me, but to the doctor they meant measles.
> The recent budget means that we shall have a hard year.

(1) I cannot say, 'Those spots mean measles, but he hasn't got measles', and I cannot say, 'The recent budget means that we shall have a hard year, but we shan't have'. That is to say, in cases like the above, *x meant that p* and *x means that p* entail *p*.

(2) I cannot argue from 'Those spots mean (meant) measles' to any conclusion about 'what is (was) meant by those spots'; for example, I am not entitled to say, 'What was meant by those spots was that he had measles'. Equally I cannot draw from the statement about the recent budget the conclusion 'What is meant by the recent budget is that we shall have a hard year'.

(3) I cannot argue from 'Those spots meant measles' to any conclusion to the effect that somebody or other meant by those spots so-and-so. *Mutatis mutandis*, the same is true of the sentence about the recent budget.

(4) For none of the above examples can a restatement be found in which the verb 'mean' is followed by a sentence or phrase in inverted commas. Thus 'Those spots meant measles' cannot be reformulated as 'Those spots meant "measles"' or as 'Those spots meant "he has measles"'.

(5) On the other hand, for all these examples an approximate restatement can be found beginning with the phrase 'The fact that...'; for example, 'The fact that he had those spots meant that he had measles' and 'The fact that the recent budget was as it was means that we shall have a hard year.'

Now contrast the above sentences with the following:

> Those three rings on the bell (of the bus) means that the 'bus is full'.
> That remark, 'Smith couldn't get on without his trouble and strife', meant that Smith found his wife indispensable.

(1) I can use the first of these and go on to say, 'But it isn't in fact full – the conductor has made a mistake'; and I can use the second and go on, 'But in fact Smith deserted her seven years ago'. That is to say, here *x means that p* and *x meant that p* do not entail *p*.

(2) I can argue from the first to some statement about 'what is (was) meant' by the rings on the bell and from the second to some statement about 'what is (was) meant' by the quoted remark.

a This paper is reprinted from *Philosophical Review*, LXVI, 1957, 377–88.

(3) I can argue from the first sentence to the conclusion that somebody (viz., the conductor) meant, or at any rate should have meant, by the rings that the bus is full, and I can argue analogously for the second sentence.

(4) The first sentence can be restated in a form in which the verb 'mean' is followed by a phrase in inverted commas, that is, 'Those three rings on the bell mean "the bus is full"'. So also can the second sentence.

(5) Such a sentence as 'The fact that the bell has been rung three times means that the bus is full' is not a restatement of the meaning of the first sentence. Both may be true, but they do not have, even approximately, the same meaning.

When the expressions 'means', 'means something', 'means that' are used in the kind of way in which they are used in the first set of sentences, I shall speak of the sense, or senses, in which they are used, as the *natural* sense, or senses, of the expressions in question. When the expressions are used in the kind of way in which they are used in the second set of sentences, I shall speak of the sense, or senses, in which they are used, as the *nonnatural* sense, or senses, of the expressions in question. I shall use the abbreviation 'means$_{NN}$' to distinguish the nonnatural sense or senses.

I propose, for convenience, also to include under the head of natural senses of 'mean' such senses of 'mean' as may be exemplified in sentences of the pattern 'A means (meant) *to do* so-and-so (by x)', where A is a human agent. By contrast, as the previous examples show, I include under the head of nonnatural senses of 'mean' any senses of 'mean' found in sentences of the patterns 'A means (meant) something by x' or 'A means (meant) by x that...' (This is overrigid; but it will serve as an indication.)

I do not want to maintain that *all* our uses of 'mean' fall easily, obviously, and tidily into one of the two groups I have distinguished; but I think that in most cases we should be at least fairly strongly inclined to assimilate a use of 'mean' to one group rather than to the other. The question which now arises is this: 'What more can be said about the distinction between the cases where we should say that the word is applied in a natural sense and the cases where we should that the word is applied in a nonnatural sense?' Asking this question will not of course prohibit us from trying to give an explanation of 'meaning$_{NN}$' in terms of one or another natural sense of 'mean'.

This question about the distinction between natural and non-natural meaning is, I think, what people are getting at when they display an interest in a distinction between 'natural' and 'conventional' signs. But I think my formulation is better. For some things which can mean$_{NN}$ something are not signs (e.g., words are not), and some are not conventional in any ordinary sense (e.g., certain gestures); while some things which mean naturally are not signs of what they mean (cf. the recent budget example).

I want first to consider briefly, and reject, what I might term a causal type of answer to the question, 'What is meaning$_{NN}$?' We might try to say, for instance, more or less with C. L. Stevenson,[a] that for x to mean$_{NN}$ something, x must have (roughly) a tendency to produce in an audience some attitude (cognitive or otherwise) and a tendency, in the case of a speaker, to *be* produced *by* that attitude, these tendencies being dependent on 'an elaborate process of conditioning attending the use of the sign in communication'.[b] This clearly will not do.

[a] *Ethics and Language* (New Haven, 1944), ch. 3.
[b] Ibid, p. 57.

(1) Let us consider a case where an utterance, if it qualifies at all as meaning$_{NN}$ something, will be of a descriptive or informative kind and the relevant attitude, therefore, will be a cognitive one, for example, a brief. (I use 'utterance' as a neutral word to apply to any candidate for meaning$_{NN}$; it has a convenient act-object ambiguity.) It is no doubt the case that many people have a tendency to put on a tail coat when they think they are about to go to a dance, and it is no doubt also the case that many people, on seeing someone put on a tail coat, would conclude that the person in question was about to go to a dance. Does this satisfy us that putting on a tail coats means$_{NN}$ that one is about to go to a dance (or indeed means$_{NN}$ anything at all)? Obviously not. It is no help to refer to the qualifying phrases 'dependent on an elaborate process of conditioning. . .' For if all this means is that the response to the sight of a tail coat being put on is in some way learned or acquired, it will not exclude the present case from being one of meaning$_{NN}$. But if we have to take seriously the second part of the qualifying phrase ('attending the use of the sign in communication'), then the account of meaning$_{NN}$ is obviously circular. We might just as well say, 'X has meaning$_{NN}$ if it is used in communication', which, though true, is not helpful.

(2) If this is not enough, there is a difficulty – really the same difficulty, I think which Stevenson recognizes: how we are to avoid saying, for example, that 'Jones is tall' is part of what is meant by 'Jones is an athlete', since to tell someone that Jones is an athlete would tend to make him believe that Jones is tall. Stevenson here resorts to invoking linguistic rules, namely, a permissive rule of language that 'athletes may be nontall'. This amounts to saying that we are not prohibited by rule from speaking of 'nontall athletes'. But why are we not prohibited? Not because it is not bad grammar, or is not impolite, and so on, but presumably because it is not meaningless (or, if this is too strong, does not in any way violate the rules of meaning for the expressions concerned). But this seems to involve us in another circle. Moreover, one wants to ask why, if it is legitimate to appeal here to rules to distinguish what is meant from what is suggested, this appeal was not made earlier, in the case of groans, for example, to deal with which Stevenson originally introduced the qualifying phrase about dependence on conditioning.

A further deficiency in a causal theory of the type just expounded seems to be that, even if we accept it as it stands, we are furnished with an analysis only of statements about the *standard* meaning, or the meaning in general, of a 'sign'. No provision is made for dealing with statements about what a particular speaker or writer means by a sign on a particular occasion (which may well diverge from the standard meaning of the sign); nor is it obvious how the theory could be adapted to make such provision. One might even go further in criticism and maintain that the causal theory ignores the fact that the meaning (in general) of a sign needs to be explained in terms of what users of the sign do (or should) mean by it on particular occasions; and so the latter notion, which is unexplained by the causal theory, is in fact the fundamental one. I am sympathetic to this more radical criticism, though I am aware that the point is controversial.

I do not propose to consider any further theories of the 'causal-tendency' type. I suspect no such theory could avoid difficulties analogous to those I have outlined without utterly losing its claim to rank as a theory of this type.

I will now try a different and, I hope, more promising line. If we can elucidate the meaning of

x meant$_{NN}$ something (on a particular occasion)   and
x meant$_{NN}$ that so-and-so (on a particular occasion)

and of

> A meant$_{NN}$ something by x (on a particular occasion)   and
>
> A meant$_{NN}$ by x that so-and-so (on a particular occasion),

this might reasonably be expected to help us with

> x means$_{NN}$ (timeless) something (that so-and-so),
>
> A means$_{NN}$ (timeless) by x something (that so-and-so),

and with the explication of 'means the same as', 'understands', 'entails', and so on. Let us for the moment pretend that we have to deal only with utterances which might be informative or descriptive.

A first shot would be to suggest that 'x meant$_{NN}$ something' would be true if x was intended by its utterer to induce a belief in some 'audience' and that to say what the belief was would be to say what x meant$_{NN}$. This will not do. I might leave B's handkerchief near the scene of a murder in order to induce the detective to believe that B was the murderer; but we should not want to say that the handkerchief (or my leaving it there) meant$_{NN}$ anything or that I had meant$_{NN}$ by leaving it that B was the murderer. Clearly we must at least add that, for x to have meant$_{NN}$ anything, not merely must it have been 'uttered' with the intention of inducing a certain belief but also the utterer must have intended an 'audience' to recognize the intention behind the utterance.

This, though perhaps better, is not good enough. Consider the following cases:

⟨1⟩ Herod presents Salome with the head of St John the Baptist on a charger.

⟨2⟩ Feeling faint, a child lets its mother see how pale it is (hoping that she may draw her own conclusions and help).

⟨3⟩ I leave the china my daughter has broken lying around for my wife to see.

Here we seem to have cases which satisfy the conditions so far given for meaning$_{NN}$. For example, Herod intended to make Salome believe that St John the Baptist was dead and no doubt also intended Salome to recognize that he intended her to believe that St John the Baptist was dead. Similarly for the other cases. Yet I certainly do not think that we should want to say that we have here cases of meaning$_{NN}$.

What we want to find is the difference between, for example, 'deliberately and openly letting someone know' and 'telling' and between 'getting someone to think' and 'telling'.

The way out is perhaps as follows. Compare the following two cases:

⟨1⟩ I show Mr X a photograph of Mr Y displaying undue familiarity to Mrs X.

⟨2⟩ I draw a picture of Mr Y behaving in this manner and show it to Mr X.

I find that I want to deny that in ⟨1⟩ the photograph (or my showing it to Mr X) meant$_{NN}$ anything at all; while I want to assert that in ⟨2⟩ the picture (or my drawing and showing it) meant$_{NN}$ something (that Mr Y had been unduly familiar), or at least that I had meant$_{NN}$ by it that Mr Y had been unduly familiar. What is the difference between the two cases? Surely that in case ⟨1⟩ Mr X's recognition of my intention to make him believe that there is something between Mr Y and Mrs X is (more or less) irrelevant to the production of this effect by the photograph. Mr X would be led by the photograph at least to suspect Mrs X even if instead of showing it to him I had left it in his room by accident; and I (the photograph shower) would not be unaware of this. But it will make a difference to the effect of my picture on Mr X whether or not he takes me to be intending to inform him (make him believe something) about Mrs X, and not to be just doodling or trying to produce a work of art.

But now we seem to be landed in a further difficulty if we accept this account. For consider now, say, frowning. If I frown spontaneously, in the ordinary course of events, someone looking at me may well treat the frown as a natural sign of displeasure. But if I frown deliberately (to convey my displeasure), an onlooker may be expected, provided he recognizes my intention, *still* to conclude that I am displeased. Ought we not then to say, since it could not be expected to make any difference to the onlooker's reaction whether he regards my frown as spontaneous or as intended to be informative, that my frown (deliberate) does *not* mean$_{NN}$ anything? I think this difficulty can be met; for though in general a deliberate frown may have the same effect (as regards inducing belief in my displeasure) as a spontaneous frown, it can be expected to have the same effect only *provided* the audience takes it as intended to convey displeasure. That is, if we take away the recognition of intention, leaving the other circumstances (including the recognition of the frown as deliberate), the belief-producing tendency of the frown must be regarded as being impaired or destroyed.

Perhaps we may sum up what is necessary for A to mean something by x as follows. A must intend to induce by x a belief in an audience, and he must also intend his utterance to be recognized as so intended. But these intentions are not independent; the recognition is intended by A to play its part in inducing the belief, and if it does not do so something will have gone wrong with the fulfillment of A's intentions. Moreover, A's intending that the recognition should play this part implies, I think, that he assumes that there is some chance that it will in fact play this part, that he does not regard it was a foregone conclusion that the belief will be induced in the audience whether or not the intention behind the utterance is recognized. Shortly, perhaps, we may say that 'A meant$_{NN}$ something by x' is roughly equivalent to 'A uttered x with the intention of inducing a belief by means of the recognition of this intention'. (This seems to involve a reflexive paradox, but it does not really do so.)

Now perhaps it is time to drop the pretense that we have to deal only with 'informative' cases. Let us start with some examples of imperatives or quasi-imperatives. I have a very avaricious man in my room, and I want him to go; so I throw a pound note out of the window. Is there here any utterance with a meaning$_{NN}$? No, because in behaving as I did, I did not intend his recognition of my purpose to be in any way effective in getting him to go. This is parallel to the photograph case. If on the other hand I had pointed to the door or given him a little push, then my behavior might well be held to constitute a meaningful$_{NN}$ utterance, just because the recognition of my intention would be intended by me to be effective in speeding his departure. Another pair of cases would be (1) a policeman who stops a car by standing in its way and (2) a policeman who stops a car by waving.

Or, to turn briefly to another type of case, if as an examiner I fail a man, I may well cause him distress or indignation or humiliation; and if I am vindictive, I may intend this effect and even intend him to recognize my intention. But I should not be inclined to say that my failing him meant$_{NN}$ anything. On the other hand, if I cut someone in the street I do feel inclined to assimilate this to the cases of meaning$_{NN}$, and this inclination seems to me dependent on the fact that I could not reasonably expect him to be distressed (indignant, humiliated) unless he recognized my intention to affect him in this way. (Cf., if my college stopped my salary altogether I should accuse them of ruining me; if they cut it by 2s. 6d. I might accuse them of insulting me; with some intermediate amounts I might not know quite what to say.)

Perhaps then we may make the following generalizations.

(1) 'A meant$_{NN}$ something by x' is (roughly) equivalent to 'A intended the utterance of x to produce some effect in an audience by means of the recognition of this intention'; and we may add that to ask what A meant is to ask for a specification of the intended effect (though, of course, it may not always be possible to get a straight answer involving a 'that' clause, for example, 'a belief that...').

(2) 'x meant something' is (roughly) equivalent to 'Somebody meant$_{NN}$ something by x.' Here again there will be cases where this will not quite work. I feel inclined to say that (as regards traffic lights) the change to red meant$_{NN}$ that the traffic was to stop; but it would be very unnatural to say, 'Somebody (e.g., the Corporation) meant$_{NN}$ by the red-light change that the traffic was to stop.' Nevertheless, there seems to be *some* sort of reference to somebody's intentions.

(3) 'x means$_{NN}$ (timeless) that so-and-so' might as a first shot be equated with some statement or disjunction of statements about what 'people' (vague) intend (with qualifications about 'recognition') to effect by x. I shall have a word to say about this.

Will any kind of intended effect do, or may there be cases where an effect is intended (with the required qualifications) and yet we should not want to talk of meaning$_{NN}$? Suppose I discovered some person so constituted that, when I told him that whenever I grunted in a special way I wanted him to blush or to incur some physical malady, thereafter whenever he recognized the grunt (and with it my intention), he did blush or incur the malady. Should we then want to say that the grunt meant$_{NN}$ something? I do not think so. This points to the fact that for x to have meaning$_{NN}$, the intended effect must be something which in some sense is within the control of the audience, or that in some sense of 'reason' the recognition of the intention behind x is for the audience a reason and not merely a cause. It might look as if there is a sort of pun here ('reason for believing' and 'reason for doing'), but I do not think this is serious. For though no doubt from one point of view questions about reasons for believing are questions about evidence and so quite different from questions about reasons for doing, nevertheless to recognize an utterer's intention in uttering x (descriptive utterance), to have a reason for believing that so-and-so, is at least quite like 'having a motive for' accepting so-and-so. Decisions 'that' seem to involve decisions 'to' (and this is why we can 'refuse to believe' and also be 'compelled to believe'). (The 'cutting' case needs slightly different treatment, for one cannot in any straightforward sense 'decide' to be offended; but one can refuse to be offended.) It looks then as if the intended effect must be something within the control of the audience, or at least the *sort* of thing which is within its control.

One point before passing to an objection or two. I think it follows that from what I have said about the connection between meaning$_{NN}$ and recognition of intention that (insofar as I am right) only what I may call the primary intention of an utterer is relevant to the meaning$_{NN}$ of an utterance. For if I utter x, intending (with the aid of the recognition of this intention) to induce an effect E, and intend this effect E to lead to a further effect F, then insofar as the occurrence of F is thought to be dependent solely on E, I cannot regard F as in the least dependent on recognition of my intention to induce E. That is, if (say) I intend to get a man to do something by giving him some information, it cannot be regarded as relevant to the meaning$_{NN}$ of my utterance to describe what I intend him to do.

Now some question may be raised about my use, fairly free, of such words as 'intention' and 'recognition'. I must disclaim any intention of peopling all our talking life with armies of complicated psychological occurrences. I do not hope to solve any philosophical puzzles about intending, but I do want briefly to argue that no special difficulties are raised by my use of the word 'intention' in connection with meaning. First, there will be cases where an utterance is accompanied or preceded by a conscious 'plan', or explicit formulation of intention (e.g., I declare how I am going to use *x*, or ask myself how to 'get something across'). The presence of such an explicit 'plan' obviously counts fairly heavily in favor of the utterer's intention (meaning) being as 'planned'; though it is not, I think, conclusive; for example, a speaker who has declared an intention to use a familiar expression in an unfamiliar way may slip into the familiar use. Similarly in non-linguistic cases: if we are asking about an agent's intention, a previous expression counts heavily; nevertheless, a man might plan to throw a letter in the dustbin and yet take it to the post; when lifting his hand he might 'come to' and say *either* 'I didn't intend to do this at all' *or* 'I suppose I must have been intending to put it in'.

Explicitly formulated linguistic (or quasi-linguistic) intentions are no doubt comparatively rare. In their absence we would seem to rely on very much the same kinds of criteria as we do in the case of nonlinguistic intentions where there is a general usage. An utterer is held to intend to convey what is normally conveyed (or normally intended to be conveyed), and we require a good reason for accepting that a particular use diverges from the general usage (e.g., he never knew or had forgotten the general usage). Similarly in nonlinguistic cases: we are presumed to intend the normal consequences of our actions.

Again, in cases where there is doubt, say, about which of two or more things an utterer intends to convey, we tend to refer to the context (linguistic or otherwise) of the utterance and ask which of the alternatives would be relevant to other things he is saying or doing, or which intention in a particular situation would fit in with some purpose he obviously has (e.g., a man who calls for a 'pump' at a fire would not want a bicycle pump). Nonlinguistic parallels are obvious: context is a criterion in settling the question of why a man who has just put a cigarette in his mouth has put his hand in his pocket; relevance to an obvious end is a criterion in settling why a man is running away from a bull.

In certain linguistic cases we ask the utterer afterward about his intention, and in a few of these cases (the very difficult ones, like a philosopher asked to explain the meaning of an unclear passage in one of his works), the answer is not based on what he remembers but is more like a decision, a decision about how what he said is to be taken. I cannot find a nonlinguistic parallel here; but the case is so special as not to seem to contribute a vital difference.

All this is very obvious; but surely to show that the criteria for judging linguistic intentions are very like the criteria for judging nonlinguistic intentions is to show that linguistic intentions are very like nonlinguistic intentions.

University of Oxford

On H. P. Grice's account of meaning[a]

PAUL ZIFF

Because I believe the coin is counterfeit, because it seems to be gaining currency, I mean to examine and attempt to discredit an account of meaning circulated some time ago by H. P. Grice.[b]

Those among us concerned with problems of semantics are much concerned with the sense(s) or meaning(s) of the morpheme 'mean' in ⟨1⟩ and ⟨2⟩:

⟨1⟩ The sentence 'Snow is white' means snow is white.

⟨2⟩ The adjective 'ungulate' means having hoofs.

Other senses or meanings of 'mean' are of interest in semantics primarily only in so far as they have some bearing on 'mean' in either ⟨1⟩ or ⟨2⟩.

Grice's paper is entitled 'Meaning'. It appears to be an account of meaning that is supposed to have some bearing on the senses of 'mean' in ⟨1⟩ and ⟨2⟩.

Grice apparently says that 'mean' in ⟨1⟩ and ⟨2⟩ is used in what he calls 'non-natural' senses of the verb. He uses the abbreviation 'mean-nn' to mark the 'non-natural' senses of 'mean'. He offers something of an analysis of 'mean-nn' (and of the morphological variants, 'meant-nn', 'means-nn', and so forth).

Does 'mean' in ⟨1⟩ or ⟨2⟩ have the sense(s) indicated by Grice's analysis?

It will simplify matters to adopt the following convention: when the expression 'mean-nn' (or any of its variants) is used here, that expression is to have the sense(s) indicated by and in conformance with Grice's analysis. The problem of this paper can then be stated in a simple way.

Consider ⟨1nn⟩ and ⟨2nn⟩:

⟨1nn⟩ The sentence 'Snow is white' means-nn snow is white.

⟨2nn⟩ The adjective 'ungulate' means-nn having hoofs.

Is ⟨1nn⟩ simply a restatement of ⟨1⟩, ⟨2nn⟩ of ⟨2⟩?

After an ingenious intricate discussion, Grice arrives at the following 'generalizations' (p. 58):

(i) 'A meant-nn something by x' is (roughly) equivalent to 'A intended the utterance of x to produce some effect in an audience by means of the recognition of this intention'.

(ii) 'x meant-nn something' is (roughly) equivalent to 'Somebody meant-nn something by x'.

(iii) 'x means-nn (timeless) that so-and-so' might as a first shot be equated with some statement or disjunction of statements about what 'people' (vague) intend (with qualifications about 'recognition') to effect by x.

---

[a] This paper is reprinted from *Analysis*, XXVIII, 1967, 1–8.

[b] 'Meaning', *The Philosophical Review*, LXVI, 3 (1957). Reprinted in this volume, pp. 53–9.

It is indicated in the discussion that the letter '*A*' is supposed to be replaceable by the name of a person, that the letter '*x*' may be but need not be replaced by a sentence.

Thus it is evident that Grice's account is supposed to apply to sentence ⟨1⟩ and so to ⟨1nn⟩. Although he mentions words in the course of his discussion none of the 'generalizations' appear to apply to words. Thus apparently he is not concerned to supply, in the paper in question, an account of the sense of 'means' in ⟨2⟩. Consequently, I mean to forget about ⟨2⟩ and ⟨2nn⟩ and be concerned here only with ⟨1⟩ and ⟨1nn⟩.

Of the three 'generalizations' stated, (iii) is the only one that directly applies to ⟨1nn⟩. Unfortunately, (iii) is not particularly pellucid. (Even so, what emerges from the fog will be sufficient to establish certain points.) Let us begin by examining (i) and (ii) in the hope of gaining insight into (iii).

On being inducted into the army, George is compelled to take a test designed to establish sanity. George is known to be an irritable academic. The test he is being given would be appropriate for morons. One of the questions asked is: 'What would you say if you were asked to identify yourself?' George replied to the officer asking the question by uttering ⟨3⟩:

⟨3⟩ Ugh ugh blugh blugh ugh blug blug.

According to the dictum of (i), George meant-nn something by ⟨3⟩: he intended the utterance of ⟨3⟩ to produce an effect in his audience by means of the recognition of his intention. The effect he intended was that of offending his audience. The accomplishment of this effect depended on the recognition of his intention. (The case in question is also in accordance with the various caveats noted by Grice in the course of his discussion: the officer testing George could 'refuse to be offended' (p. 58), thus the intended effect was in some sense within the control of the audience; George's intention to offend was his 'primary' intention (p. 58); and so forth.) Consequently, as far as one can tell, ⟨3⟩ fills Grice's bill.

But even though it is clear that George meant-nn something by ⟨3⟩, it is equally clear that George did not mean anything by ⟨3⟩. Grice seems to have conflated and confused '*A* meant something by uttering *x*', which is true in a case like ⟨3⟩, with the quite different '*A* meant something by *x*', which is untrue in a case like ⟨3⟩.

The malady just noted in connection with (i) of course at once infects (ii). For even though it is clear that George meant-nn something by ⟨3⟩ and hence ⟨3⟩ meant-nn something, it is equally clear that ⟨3⟩ did not mean anything. Indeed, had ⟨3⟩ meant anything, that would have defeated George's purpose in uttering ⟨3⟩.

The preceding case admits of the following variation. On being given the test over again by another officer, instead of uttering ⟨3⟩, George uttered ⟨4⟩:

⟨4⟩ pi.hi.y pi.hi.y

Again in accordance with (i) we can say that George meant-nn something by ⟨4⟩ and what he meant-nn was precisely what he meant-nn by ⟨3⟩. (For we may suppose that he had the same intention in each case, expected the same reaction, and so forth.)

But in this case, not only did George mean something by uttering ⟨4⟩, he also meant something by ⟨4⟩: even though he rightly expected his utterance to be treated as though it were mere noise, what he meant by ⟨4⟩, and what he said, was that he didn't know. George was perversely speaking in Hopi.[a] Here one need not confuse

---

[a] See B. Whorf, *Language, Thought, and Reality* (Cambridge: The Technology Press, 1956), p. 114, for the phonetic significance of ⟨4⟩.

'George did not mean what he said', which is true, for he did know the answer to the question, with the quite different 'George did not mean anything by ⟨4⟩', which is untrue.

That George meant-nn something by ⟨4⟩ is wholly irrelevant to the question whether George meant something by ⟨4⟩. And the fact that ⟨4⟩ meant-nn something (in virtue of (ii)) is wholy irrelevant to the question whether ⟨4⟩ meant anything. This should be obvious from the fact that what ⟨4⟩ meant had nothing whatever to do with what George intended to effect by uttering the utterance and hence had nothing whatever to do with what ⟨4⟩ meant-nn.

The curious character of (ii) can be further displayed by the following sort of cases. Consider ⟨5⟩:

⟨5⟩ Claudius murdered my father.

and let us conjure up three contexts of utterance: (a) George uttered a sentence token of type ⟨5⟩, thus he uttered ⟨5a⟩, in the course of a morning soliloquy; (b) George uttered another such token, ⟨5b⟩, in the afternoon in the course of a discussion with Josef; and (c) George uttered another such token, ⟨5c⟩ in the evening while delerious with fever. Now consider ⟨6⟩ and ⟨6nn⟩:

⟨6⟩ ⟨5a⟩ meant the same as ⟨5b⟩ which meant the same as ⟨5c⟩ which meant the same as ⟨5a⟩.

where ⟨6nn⟩ is the same as ⟨6⟩ save that for each occurrence of 'meant' in ⟨6⟩, ⟨6nn⟩ has an occurrence of 'meant-nn'. Thus ⟨6⟩ says that the three tokens in question all had the same meaning; ⟨6nn⟩ says that they all had the same meaning-nn.

Although ⟨6⟩ is true, according to Grice's account ⟨6nn⟩ must be false (here taking ⟨6⟩ and ⟨6nn⟩ to stand for statements). That this is so can be seen as follows.

According to (ii), a sentence S meant-nn something (roughly) if and only if somebody meant-nn something by it. Thus ⟨5a⟩, ⟨5b⟩, and ⟨5c⟩ meant-nn something (roughly) if and only if somebody meant-nn something by them.

Did anyone mean-nn anything by ⟨5c⟩? Evidently not. For ⟨5c⟩ was uttered while George was delerious with fever, unaware of any audience. Hence ⟨5c⟩ was not intended to produce any effect in an audience. (Here one need not confuse 'What George said meant nothing', which may be true in one sense of 'what George said', with 'The expression which George uttered meant nothing', which is untrue.)

Did anyone mean-nn anything by ⟨5b⟩? Presumably so. Since ⟨5b⟩ was uttered by George in the course of a discussion with Josef, if anything fits Grice's account, ⟨5b⟩ does.

Did anyone mean-nn anything by ⟨5a⟩? It would seem not, for since George uttered ⟨5a⟩ in the course of a soliloquy, it could hardly have been intended to produce an effect in an audience. (But perhaps Grice would wish to maintain that, in so far as George was speaking to himself, he was his own audience. But then could he intend to produce an effect in himself by means of a recognition on his own part of his own intention? These are mysteries we may cheerfully bequeath to Mr. Grice.)

Evidently ⟨6nn⟩ is untrue even though ⟨6⟩ is true.

Sentence token ⟨5c⟩ exemplifies a case in which even though it was not true that the speaker meant anything by the token, the token nonetheless meant something. One can also produce cases in which a speaker did mean something by an utterance and yet the utterance itself did not mean anything.

George has had his head tampered with: electrodes have been inserted, plates mounted, and so forth. The effect was curious: when asked how he felt, George replied by uttering ⟨7⟩:

⟨7⟩ Glyting elly beleg.

What he meant by ⟨7⟩, he later informed us, was that he felt fine. He said that, at the time, he had somehow believed that ⟨7⟩ was synonymous with 'I feel fine' and that everyone knew this.

According to (i), George meant-nn something by ⟨7⟩, and according to (ii), ⟨7⟩ must have meant-nn something. But ⟨7⟩ did not mean anything at all.

The preceding examples should suffice to indicate that Grice's equivalences (i) and (ii) are untenable. But their extraordinary character can be made even plainer by the following sort of case.

A man suddenly cried out 'Gleeg gleeg gleeg!', intending thereby to produce a certain effect in an audience by means of the recognition of his intention. He wished to make his audience believe that it was snowing in Tibet. Of course he did not produce the effect he was after since no one recognized what his intention was. Nonetheless that he had such an intention became clear. Being deemed mad, he was turned over to a psychiatrist. He complained to the psychiatrist that when he cried 'Gleeg gleeg gleeg!' he had such an intention but no one recognized his intention and were they not mad not to do so.

According to Grice's equivalence (i), the madman meant-nn something by 'Gleeg gleeg gleeg!' and so, according to (ii), the madman's cry must have meant-nn something, presumably that it was snowing in Tibet. But the madman's cry did not mean anything at all; it certainly did not mean it was snowing in Tibet. Had it meant that, there would have been less reason to turn him over to a psychiatrist.

On Grice's account, good intentions suffice to convert nonsense to sense: the road to Babble is paved with such intentions.

It is time to turn to Grice's suggestion of an equivalence, (iii). Consider sentences ⟨8⟩, ⟨9⟩, and ⟨9nn⟩:

⟨8⟩ He's a son of a stickleback fish.
⟨9⟩ Sentence ⟨8⟩ means the male referred to is a son of a small scaleless fish (family Gasterosteidae) having two or more free spines in front of the dorsal fin.

where ⟨9nn⟩ is the same as ⟨9⟩ save that (as in ⟨6nn⟩) 'means' has given way to 'means-nn'.

I take it that there is no reason whatever to suppose that the sense of 'means' in ⟨9⟩ differs in any way from the sense of 'means' in ⟨1⟩. Both ⟨9⟩ and ⟨1⟩ are simply of the form: sentence *S* means *m*. But if 'means' in ⟨9⟩ has precisely the same sense as 'means' in ⟨1⟩, it follows that ⟨1nn⟩ is not simply a restatement of ⟨1⟩. For ⟨1⟩ and ⟨1nn⟩ differ as ⟨9⟩ and ⟨9nn⟩ differ, and according to Grice's account, ⟨9⟩ and ⟨9nn⟩ are radically different. That this is so can be seen as follows.

I am inclined to suppose that ⟨8⟩ has been uttered only rarely. Nonetheless (taking ⟨9⟩ and ⟨9nn⟩ to stand for statements), I am reasonably certain that ⟨9⟩ is a reasonably correct statement of the meaning of ⟨8⟩ and I am being reasonable in being so certain.

If it is a correct statement of the meaning of ⟨8⟩, and if ⟨9nn⟩ is simply a restatement of ⟨9⟩, since ⟨9nn⟩ is, ⟨9⟩ must be equivalent to some statement or disjunction of statements about what 'people' intend to effect by ⟨8⟩. So Grice evidently maintains; for despite its vagueness, that is what is indicated by (iii).

But the question 'What do people intend to effect by ⟨8⟩?' would not be a sensible question. Since hardly anyone has ever uttered ⟨8⟩ before, or so I suppose, one can hardly ask what people intend to effect by it.

'Then what would people intend to effect by ⟨8⟩?': the question is somewhat idle. What people would intend to effect by ⟨8⟩ is a matter about which one can only speculate, vaguely. However, since the obvious emendation of Grice's account invites such speculation, let us speculate.

What would people intend to effect by uttering ⟨8⟩? Given the acoustic similarity between ⟨8⟩ and a familiar form of expression, given that sticklebacks are known to be tough fish, given that the sex of a fish is not readily determined by the uninitiated, most likely by uttering ⟨8⟩ people would thereby intend to denigrate a contextually indicated male person.

What people would intend to effect by ⟨8⟩ is a subject for profitless speculation. But if one must say something about the matter, I am inclined to suppose that ⟨9nn⟩ does not convey a correct account of what people would intend to effect by uttering ⟨8⟩. Thus ⟨9nn⟩ is presumably untrue.

If ⟨9nn⟩ were simply a restatement of ⟨9⟩, only a fool would profess to being even reasonably certain that ⟨9⟩ is a correct statement and it would be unreasonable of him to be so certain. But I am reasonably certain that ⟨9⟩ is a correct statement of the meaning of ⟨8⟩ and I am not being unreasonable in being so certain. Therefore ⟨9nn⟩ cannot be simply a restatement of ⟨9⟩ and neither can ⟨1nn⟩ be simply a restatement of ⟨1⟩.

Before allowing Grice's analysis to rest in peace, the moral of its passing should be emphasized.

His suggestion is stated in terms of what people 'intend', not in terms of what they 'would intend'. As such it obviously occasions difficulty with novel utterances. But matters are not at all improved by switching to what people 'would intend'.

For first, if a sentence is such that people in general simply would not utter it, then if they were to utter it, what they would intend to effect by uttering it might very well have nothing to do with the meaning of the sentence. What would a person intend to effect by uttering the sentence 'Snow is white and snow is white and snow is white and snow is white and snow is white'? I conjecture that a person uttering such a sentence would be either a philosopher or a linguist or an avant-garde novelist or a child at play or a Chinese torturer. What people would intend to effect by uttering such a sentence would most likely have nothing whatever to do with the meaning of the sentence.

Secondly, the switch to 'would' would be of help only if there were a constructive method of determining what people would intend to effect by uttering an utterance. There is no such method. There is not likely to be any (at least in our lifetime).

Ignoring the futility of talking about what people 'would intend' to effect by uttering an utterance, one need not ignore the fact that what people generally in fact mean may be altogether irrelevant to a meaning of an utterance.

By the spoken utterance 'HE GAVE HIM HELL' people generally mean what is meant by the written utterance 'He gave him hell' and not what is meant by the written utterance 'He gave him Hell'. Quite possibly no one has ever said 'I saw the children shooting' meaning by that he saws children while he is shooting. That is nonetheless one of the meanings of that remarkably ambiguous sentence. Indefinitely many such examples could be supplied.

To be concerned with what people intend (or would intend) to effect by uttering an expression is to be concerned with the use of the expression. As I have elsewhere pointed out and argued at length, the use of an expression is determined by many factors, many of which have nothing (or have nothing directly) to do with its meaning: acoustic shape is one such factor, length another.[a]

Grice's analysis rings untrue. It was bound to; his alloy lacks the basic ingredient of meaning: a set of projective devices. The syntactic and semantic structure of any natural language is essentially recursive in character. What any given sentence means depends on what (various) other sentences in the language mean.

That people generally intend (or would intend) this or that by uttering an utterance has, at best, as much significance as a statement to the effect that when 'Pass the salt!' is uttered, generally people are eating, thus what I have elsewhere called the statement of a 'regularity'.[b] Not all regularities are semantically relevant: a regularity couched in terms of people's 'intentions' is not likely to be.

But even if such regularities were somehow relevant, that would not matter much. A regularity is no more than a ladder which one climbs and then kicks away. An account of meaning constituted by (i), (ii), and (iii) never gets off the ground. There is no reason to suppose it can.[c]

University of Illinois

[a] See my *Semantic Analysis* (Ithaca: Cornell University Press, 1960).
[b] Ibid.
[c] I am indebted to D. Stampe for useful criticisms of various points.

Three levels of meaning[a]

GILBERT H. HARMAN

Philosophers approach the theory of meaning in three different ways. (1) Carnap, Ayer, Lewis, Firth, Hempel, Sellars, Quine, etc. take meaning to be connected with evidence and inference, a function of the place an expression has in one's 'conceptual scheme' or of its role in some inferential 'language game'. (2) Morris, Stevenson, Grice, Katz, etc. take meaning to be a matter of the idea, thought, feeling, or motion that an expression can be used to communicate. (3) Wittgenstein (?), Austin, Hare, Nowell-Smith, Searle, Alston, etc. take meaning to have something to do with the speech acts the expression can be used to perform.

Familiar objections to each type of theory

1   Theories of the first sort, which take meaning to be specified by inferential and observational evidential considerations, are accused of ignoring the social aspect of language. Such theories, it is said, admit the possibility of a private language in which one might express thoughts without being able to communicate them to another; and this possibility is held to be absurd. More generally it can be argued that, even if meaning depends on considerations of evidential connection, the relevant notion of evidence involves intersubjective objectivity, which requires the possibility of communication among several people. Therefore it can be argued that one could not account for meaning via the notion of evidence without also discussion of meaning in communication.

Furthermore, there are many uses of language to which the notion of evidence has no application. If one asks a question or gives an order, it is not appropriate to look for the evidence for what has been said. But if there can be no evidence for a question, in the way that there can be evidence of a conclusion, differences in meaning of different questions cannot be explicated by means of differences in what evidence can be relevant to such questions. So theories of the first sort seem vulnerable in several respects.

2   On the other hand, theories of the second sort seem threatened by circularity from at least two directions. According to Katz, one understands the words someone

[a] This paper was presented in an APA symposium on Levels of Meaning, 28 December 1968. Commentators were Keith Donnellan and John Searle. The paper is reprinted from *The Journal of Philosophy*, LXV (1968): 590–602.

Work on this paper was supported in part by the National Endowment in the Humanities (grant No. H-67-0-28) and in part by the National Science Foundation (grant No. NSF-GS-2210).

else says by decoding them into the corresponding thought or idea.[a] But a person ordinarily thinks in words, often the same words he communicates with and the same words others use when they communicate with him. Surely the words mean the same thing when used in these different ways; but to apply Katz's account of meaning to the words one thinks with would seem at best to take us in a circle.

Similarly, consider Grice's theory of meaning. According to Grice, one means that p by one's words (in communication) if and only if one uses them with the intention of getting one's listener to think one thinks that p.[b] But what is it to think that p? On one plausible view it is to think certain words (or some other representations) by which one means that p. If so, Grice's analysis would seem to be circular: one means that p by one's words if and only if one uses them with the intention of getting one's listener to think one has done something by which one means that p.

Circularity and worse also threatens from another side, if the second type of approach is intended to explain what it is to promise to do something or if it is supposed to be adequate to exhibit the difference between asking someone to do something and telling him to do it, etc. The fact that saying something in a particular context constitutes one or another speech act cannot be represented simply as the speaker's communicating certain thoughts. For example, promising to do something is not simply communicating that you intend to do it, nor is asking (or telling) someone to do something simply a matter of communicating your desire that he do it. At the very least, to perform one or another speech act, one must communicate that one is intending to be performing that act; so at the very least, to treat all speech acts as cases of communication would involve the same sort of circularity already mentioned. Furthermore, communication of one's intention to be performing a given speech act is not in general sufficient for success. The speaker may not be in a position to promise or to tell someone to do something, no matter what his intentions and desires.

3 On the other hand, theories of the third sort, which treat meaning as speech-act potential, are also subject to familiar objections. For example, Chomsky[c] (following Humboldt) argues that this third approach (and probably the second as well) ignores the 'creative aspect of language use'. Language exists primarily for the free expression of thought. Communication and other social uses of language are, according to Chomsky, of only secondary importance. Proponents of the first approach will surely agree with Chomsky on this point.

In line with this, it can be argued that one of the most important characteristics of human language is its unbounded character. Almost anything that one says has never been said by anyone before. Surely this unboundedness reflects the unbounded

---

[a] 'Roughly, linguistic communication consists in the production of some external, publicly observable, acoustic phenomenon whose phonetic and syntactic structure encodes a speaker's inner private thoughts or ideas and the decoding of the phonetic and syntactic structure exhibited in such a physical phenomenon by other speakers in the form of an inner private experience of the same thoughts or ideas.' J. J. Katz, *The Philosophy of Language* (New York: Harper & Row, 1966), p. 98.

[b] H. P. Grice, 'Meaning', *Philosophical Review*, LXVI, 3 (July 1957), reprinted in this volume, pp. 53–9, and 'Utterer's Meaning and Intentions', *Philosophical Review*, LXXVIII, 2 (April 1969): 147–77.

[c] For example, Noam Chomsky, 'Current Issues in Linguistic Theory', in Fodor and Katz, eds., *The Structure of Language* (Englewood Cliffs, N.J.: Prentice-Hall, 1964), pp. 57–61. See also *Cartesian Linguistics* (New York: Harper & Row, 1966).

creative character of thought and is not simply a reflection of the more or less practical uses to which language can be put in a social context.

Furthermore, approaches of the third sort seem to be at least as afflicted with circularity as are approaches of the second sort. For example, Alston suggests defining sameness of meaning as sameness of illocutionary-act potential, where illocutionary acts are the relevant subclass of speech acts. He claims that two expressions have the same meaning if and only if they can be used to perform the same illocutionary acts.[a] Now, suppose we ask whether the expressions 'water' and 'H_2O' have the same meaning. They do only if, e.g., in saying 'Please pass the water' one performs the same illocutionary act as one does in saying 'Please pass the H_2O'. But it can be argued that we are able to decide whether these acts are the same only by first deciding whether the expressions 'water' and 'H_2O' have the same meaning. If so, Alston's proposal is circular.

Three levels of meaning

Each of the preceding objections is based on the assumption that the three approaches to the theory of meaning are approaches to the same thing. I suggest that this assumption is false. Theories of meaning may attempt to do any of three different things. One theory might attempt to explain what it is for a thought to be the thought that so-and-so, etc. Another might attempt to explain what it takes to communicate certain information. A third might offer an account of speech acts. As theories of language, the first would offer an account of the use of language in thinking; the second, an account of the use of language in communication; the third, an account of the use of language in certain institutions, rituals, or practices of a group of speakers.

I shall refer to theories of meaning of level 1, of level 2, and of level 3, respectively. I believe that there is a sense in which later levels presuppose earlier ones. Thus a theory of level 2, i.e., a theory of communication (of thoughts), presupposes a theory of level 1 that would say what various thoughts are. Similarly, a theory of level 3 (e.g., an account of promising) must almost always presuppose a theory of level 2 (since in promising one must communicate what it is one has promised to do).

The objections I have just discussed show only that a theory of one level does not provide a good theory of another level. A theory of the meaning of thoughts does not provide a good account of communication. A theory of meaning in communication does not provide a good account of speech acts. And so forth. On the other hand, I do not want to deny that proponents of the various theories have occasionally been confused about their objectives. In the third section of this paper (p. 72 ff.) I shall argue that such confusion has led to mistakes in all three types of theory.

But first, from the point of view of the suggested distinctions between such levels of meaning, I shall briefly review the three approaches to the theory of meaning sketched at the beginning of this paper.

1   A theory of level 1 attempts to explain what it is to think that p, what it is to believe that p, to desire that p, etc. Let us suppose we are concerned only with thinking done in language. Such a supposition will not affect the argument so long as thinking makes use of some system of representation, whether or not the system is properly part of any natural language.

[a] William P. Alston, *The Philosophy of Language* (Englewood Cliffs, N.J.: Prentice-Hall, 1964), pp. 36–7.

Even if we do not know what the various expressions of a subject's language mean, we can still describe him as thinking some sentence of his language, believing true some sentence, desiring true some sentence, etc. It seems reasonable to assume that the subject has the thought that p if and only if he thinks certain words (or other representations) by which he means that p; that he believes that p if and only if he accepts as true some sentence by which he means that p; that he desires that p if and only if he desires true some sentence by which he means that p, etc. The problem of saying what it is to think, believe, desire, etc. that p can be reduced to the problem of saying what it is to mean that p by certain words used in thinking.

Another way to put the same point is this. A theory of the nature of thought, belief, desire, and other psychological attitudes can appear in the guise of the theory of meaning. That is the best way to interpret the first sort of theory discussed at the beginning of this paper. Extreme positivists claim that what a *thought* means, i.e., what thought it is, is determined by its conditions of verification and refutation. Its meaning or content is determined by the observational conditions under which the subject would acquire the corresponding belief plus those conditions under which he would acquire the corresponding disbelief. Other empiricists argue that what a thought is or means is determined by its position in a whole structure of thoughts and other psychological attitudes, i.e., its place in a subject's conceptual scheme, including not only relations to experience but also relations to other things in that same scheme.

Several philosophers have argued a similar thesis that makes no explicit reference to meaning. Fodor,[a] Putnam,[b] and Scriven[c] have each taken psychological states to be 'functional states' of the human organism. What is important about such states is not how they are realized; for my psychological states may well be realized in a different neurophysiological way from yours. What is important is that there is a certain relationship among the various states a person can be in, between such states and observational 'input', and between such states and action 'output'. In this regard persons are sometimes compared with nondeterministic automata.[d] Just as a particular program or flow chart may be instantiated by various automata made from quite different materials, so too the 'same' person (a person with the same psychological characteristics and dispositions) might be instantiated by different neurophysiological set-ups and perhaps even by some robot made of semi-conductors, printed circuits, etc. For a person to be in a particular psychological state is like the automaton's being at a certain point in its program or flow chart rather than like something's happening at one or another transistor.

If we conceive the automaton's operation to consist largely in the formation, transfer, and 'storage' of certain representations, the analogy is even better. To say that such an automaton is at a certain point in a particular program is to say, first, that the automaton has various possible states related to one another and to input and output in such a way that it instantiates a particular program and, second, that it is in a particular one of the states or collections so indicated. For the automaton in question, the same point can be made by first specifying the role of various represen-

[a] Jerry Fodor, 'Explanations in Psychology', in Max Black, ed., *Philosophy in America* (Ithaca, N.Y.: Cornell, 1965).

[b] Hilary Putnam, 'Minds and Machines', in Sidney Hook, ed., *Dimensions of Mind* (New York: NYU Press, 1960).

[c] Michael Scriven, *Primary Philosophy* (New York: McGraw-Hill, 1966), pp. 181–97.

[d] G. A. Miller, E. Galanter, and K. H. Pribram, *Plans and the Structure of Behavior* (New York: Holt, Rhinehart & Winston, 1960).

tations it uses in its internal operation, its reaction to input, and its influence on output. Second, one may describe the present state of the computer by indicating what representations are where.[a]

It is obvious how such an account may offer a functional account of psychological states via a person's use of language. Thus, according to Sellars, the meanings of one's words are determined by the *role* of the words in the evidence-inference-action game, which includes the influence of observation on thought, the influence of thought on thought in inference, and the influence of thought on action via decision and intention.[b] Sellars is simply offering a functional account of psychological states in the guise of a theory of meaning. (I do not mean to suggest that Sellars is at all unaware of what he is doing.)

It is important of course that the analogy be with nondeterministic automata. According to Sellars, the meaning of an expression is given by its role in the evidence-inference-action game, where this role is not causal but rather defined in terms of possible (i.e., more or less legitimate) moves that can be made. A similar point would have to be accepted by anyone who would identify psychological states with functional states.

Quine's thesis of indeterminacy says that, functionally defined, the meaning of a thought is not uniquely determined. The thesis ought to be expressible directly as the following claim about instantiations of nondeterministic automata: When a set of possible states of some device can be interpreted in a particular way as instantiations of a given nondeterministic automaton, that interpretation will not in general be the only way to interpret those physical states as instantiations of the given automaton.[c]

I hope I have said enough to show how theories of the first sort may be treated as theories about the nature of meaning of thoughts and other psychological ('intentional') states.

2   A theory of level 2 attempts to say what communication is and what is involved in a message's having a particular meaning. Communication is communication of thoughts and ideas; and Katz's description of it is perfectly acceptable provided that his talk about 'decoding' is not taken too literally. It is true that Katz's description of communication would have us explain meaning in terms of meaning; but the two sorts of meaning are different. Katz would have us explain the meaning of a message in terms of the meaning of a thought, which is to explain meaning of level 2 in terms of meaning of level 1. And there is nothing wrong with that. (On the other hand I do not mean to suggest that the Katz–Fodor theory of meaning is not involved in serious confusion. On the contrary. But I shall delay discussion of that point until the final section of this paper.) Grice's theory of meaning seems to provide what Katz is looking for. And it avoids the charge of circularity by explaining the meaning of a message (what the speaker means) in terms of the meaning of the thought communicated (which the speaker intends the hearer to think the speaker has).

---

[a] Cf. my 'Psychological Aspects of the Theory of Syntax', *The Journal of Philosophy*, LXIV, 2 (February 1967): 75–87; and 'Knowledge, Inference, and Explanation', *American Philosophical Quarterly*, V, 3 (July 1968): 164–5.

[b] Wilfrid Sellars, 'Some Reflections on Language Games', in *Science, Perception and Reality* (London: Routledge & Kegan Paul, 1963).

[c] Cf. my 'Quine on Meaning and Existence I', *Review of Metaphysics*, XXXI, 1 (September 1967): 124–51.

Communication need not involve use of language. When it does, the language used need not be one either speaker or hearer is able to think in. And even when the language used is one both participants think in, it may (for the purposes of certain communications) be used arbitrarily as a code. But ordinary communication makes use of a language which both participants think in and which is not being used arbitrarily as a code. In such a case the hearer typically assigns, as his interpretation of what the speaker says, either (*a*) a thought that the hearer expresses using the same words the speaker has used (with possible minor modification, e.g., for first and second person in pronouns) or (*b*) a thought that is some simple function of a thought in those words, where the function is determined by context (irony, e.g.). Similarly, the speaker standardly uses in communication (almost) the same words he uses in expressing to himself the thought he intends to communicate. This is no accident, and one will fail to understand the nature of linguistic communication unless one grasps this point. It is obscured when linguistic communication is described as if it involved processes of coding and decoding. We would not be able to use language in communication as we do if communication really involved coding and decoding. (As I shall argue below, Katz and Fodor have gone wrong at exactly this point.)

Similarly, it would be a mistake to treat learning one's first language as simply a matter of learning how to communicate one's thoughts to others and to understand others when they attempt to communicate. When a child is exposed to language he acquires two things. First he acquires a new system of representation for use in thinking and in the formation of various psychological attitudes. This is the primary thing he acquires. Second he acquires the ability, alluded to above, to communicate with and understand other speakers of the language. This ability relies heavily on the fact that the language has been acquired as an instrument of thought. No very complicated principles of interpretation need to be learned to support this ability. All the child needs to do, at first, is to assume that other speakers express by their words thoughts the child would think using those same words. More complicated principles of interpretation are learned later to allow for lying, irony, metaphor, etc. But it would surely be a mistake to think of the child as having an ability to perform a certain sort of complicated decoding.

Aside from that point, I hope it is now clear how, e.g., Grice's theory may be treated as a promising attempt at a level 2 theory of meaning; and I hope it is clear why it should not be criticized for failing to do what can be done only by a theory of meaning of another level.

3   A theory of level 3 would be a theory of social institutions, games, practices, etc. The theory would explain how the existence of such things can make certain acts possible, e.g., how the existence of a game of football can make possible scoring a touchdown or how the existence of an institution of banking, etc. can make possible writing a check. In a sense such a theory is a theory of meaning. The game or institution confers meaning on an act like carrying a ball to a certain place or writing one's name on a piece of paper.

Some institutions, games, practices, etc. involve the use of language and can therefore confer meaning (significance) on such uses of language. But this is a different sort of meaning than that involved in levels 1 and 2. And typically, use of certain words within an institution, practice, or game presupposes that the words have meaning as a message (which standardly presupposes that they have meaning when used to express one's thoughts). In a sense (which does not destroy the priority of level 1) meaning on levels 1 and 2 can sometimes presuppose meaning on level 3; but this is only because one can think and communicate about practices, games, and institutions.

Applications

Distinguishing between the three levels of meaning can clarify many issues in philosophy and linguistics. In this final section of my paper I shall briefly give some examples.

1   The distinction of levels tends to dissolve as verbal certain philosophical worries about what has to be true before someone can be said to use a language. One may use a system of representation in thinking, without being able to use it in communication or speech acts. Children and animals presumably do so, and perhaps some computers may also be said to do so. Similarly one may use a system of representation in thought and communication without being able to engage in more sophisticated speech acts. (Compare computers that 'communicate' with the programmer.) Whether communication or more sophisticated speech acts must be possible before one's system of representation counts as a language can only be a purely verbal issue.

A special case of this issue would be the philosophical question whether there can be a private language. For the issue is simply whether there could be a language used to think in but not to communicate with. There can be a system of representation with such properties; whether it should count as a language is a purely verbal issue. On the other hand, Wittgenstein's private-language argument may be directed against a conception of language learning and of the use of language in communication similar to that put forward in transformational linguistics by Chomsky, Katz, and Fodor, among others. I shall argue below that this conception is based on failure to distinguish levels of meaning.

2   The distinction can be used to help clarify various philosophical accounts of meaning. For example, in chapter 2 of *Word and Object* Quine presents considerations mainly relevant to level 1 theories of meaning. But by describing language as a set of dispositions to verbal behavior, he suggest wrongly that he is concerned with communication or more sophisticated speech acts. And this occasionally leads him wrong. He describes the thesis of indeterminacy as the view that a speaker's sentences might be mapped onto themselves in various ways without affecting his dispositions to 'verbal behavior'. So stated the thesis would be obviously wrong. A conversation containing one sentence would be mapped onto one containing another. Dispositions to verbal behavior would therefore change under the mappings in question. Actually Quine is interested in only one particular sort of verbal behavior: assent or dissent to a sentence. And his position would be even clearer if he had entirely avoided the behavioristic formulation and spoken instead about a speaker's accepting as true (or accepting as false) various sentences.[a]

In his papers on meaning, Paul Grice presents a level 2 theory of meaning. But in a recent paper[b] he is troubled by (among other things) (a) difficulties he has in

---

[a] Cf. my 'An Introduction to Translation and Meaning': Chapter Two of *Word and Object*, *Synthese*, XIX, 1–2 (December 1968): 14–26.
[b] 'Utterer's Meaning and Intentions', *Philosophical Review*, LXXVIII, 2 (April 1969): 147–77.

accounting for the difference between telling someone one wishes him to do something and ordering him to do it, and (*b*) difficulties in accounting for meaning something by one's words in silent thought. But (*a*) can be handled only within a level 3 theory, and (*b*) can be handled only within a level 1 theory. The former point is somewhat obscured by Grice's formulation of the notion to be analyzed: '*U* meant *x* by uttering *y*.' The locution is at least three ways ambiguous. It may mean (i) that *x* is the message conveyed by *U*'s uttering *y*, (ii) that *U* intended to say *x* when he said *y*, or (iii) that *U* really meant it when he said *x*; i.e., he uttered *y* with no fingers crossed, not ironically, not in jest, etc. Grice does not make clear exactly which of these interpretations we are to assign to the locution he is analyzing. A theory of communication results if the interpretation is (i). If (ii) were the correct interpretation, Grice's analysis of meaning in terms of the speaker's intentions would be trivialized. And (iii) involves appeal to speech acts. I urge Grice to accept (i), but I fear that he may opt for (iii). That would blur the line between levels 2 and 3 and would, I think, make it much more difficult for Grice to succeed in giving any sort of general account of meaning.

Alston presents a level 3 theory of meaning. But he believes that such a theory must account for sameness of meaning of linguistic expressions. I have argued above that this cannot be done. We cannot define sameness of meaning of expressions as sameness of illocutionary-act potential. Sameness of meaning is to be accounted for, if at all, within a level 1 theory. Given a theory of level 1, we might hope to define sameness of meaning (i.e., significance) of illocutionary acts via sameness of meaning of linguistic expressions. None of this shows that meaning cannot be approached via speech acts, as long as it is understood what sort of theory of meaning a theory of speech acts is.

3   The distinction between levels of meaning can be used to show what is really wrong with Katz and Fodor's[a] semantic theory. They claim that an adequate semantic theory must show how the meaning of a sentence is determined by its grammatical structure and the meaning of its lexical items. They say that such a theory must specify the form of dictionary entries for lexical items and must say how such entries are combined, on the basis of grammatical structure, in order to give readings of sentences.

These claims are the direct result of failure to distinguish a theory of the meaning of language as it is used in thinking from a theory of the meaning of a message, plus a failure to remember that in the standard case one communicates with a language one thinks in. Thus at first Katz and Fodor purport to be describing the structure of a theory of linguistic communication. They are impressed by the fact that a speaker has the ability to produce and understand sentences he has never previously encountered. As a result they treat communication as involving a complex process of coding and decoding, where readings are assigned to sentences on the basis of grammatical structure and dictionary entries. That this is a mistake has already been noted above. In normal linguistic communication a message is interpreted as expressing the thought (or some simple function of that thought) that is expressed by the same words the message is in.

The fact that a speaker can produce and understand novel sentences is a direct consequence of the facts that he can think novel thoughts and that he thinks in the

---

[a] Katz and Fodor, 'The Structure of a Semantic Theory', in Fodor and Katz, eds., *The Structure of Language*.

same language he communicates in. One gives an account of the meaning of words as they are used in thinking by giving an account of their use in the evidence-inference-action game. For a speaker to understand certain words, phrases, and sentences of his language is for him to be able to use them in thinking, etc. It is not at all a matter of his assigning *readings* to the words, for to assign a reading to an expression is simply to correlate words with words.

Katz and Fodor do have some sense of the distinction between levels 1 and 2. Although their theory is put forward as if it were an account of communication, they describe it as a theory of meanings a sentence has when taken in isolation from its possible settings in linguistic discourse. They do this in order to avoid having to take into account special 'readings' due to codes, figurative uses of language, etc. Their theory of meaning is restricted to giving an account of the meaning of a message for that case in which the message communicates the thought that is expressed in the same words as those in which the message is expressed. They recognize that another theory would have to account for the interpretation that is assigned when a sentence occurs in a particular context.

In a way this amounts to distinguishing my levels 1 and 2. And in a sense Katz and Fodor attempt to provide a theory of level 1. More accurately, their theory falls between levels 1 and 2. It cannot provide a level 1 theory, since a speaker does not understand the words he uses in thinking by assigning readings to them. It cannot provide a level 2 theory, since it treats a very simple problem of interpretation as if it were quite complicated.

I think that perfectly analogous complaints can be raised against Paul Ziff's[a] views about meaning and against theories like Davidson's[b] that attempt to account for meaning in terms of truth conditions.

4   The foregoing points may shed some light on some of the puzzling things Chomsky says about linguistic competence and language learning.[c] He says that anyone who knows a language has (unconscious) knowledge of the grammatical rules of the language. This is puzzling, since we would ordinarily ascribe knowledge of the rules of grammar (unconscious or conscious) to a linguist or grammarian. We would not ordinarily ascribe such knowledge to a typical speaker of the language. What is more puzzling is that Chomsky does not argue for this claim but takes it to be obvious. In the same vein, Chomsky treats language learning as a special case of theory construction. The child learning a language must infer a theory of the language that is spoken by those around him, i.e., he must infer a grammar of that language. Again this is puzzling because it treats the child as if he were a linguist investigating some hitherto untranslated language.

It is easy to point out the counterintuitive nature of Chomsky's proposals and the difficulties involved if one takes them seriously. It is less easy to say what led Chomsky to make such proposals and why he continues to accept them in the face of heavy criticism. But seeing what is wrong with Katz and Fodor's theory of semantics suggests what may be wrong with Chomsky's views on linguistic competence and language learning.

[a] Paul Ziff, *Semantic Analysis* (Ithaca, N.Y.: Cornell, 1960).

[b] Donald Davidson, 'Truth and Meaning', *Synthese*, XVII, 3 (September 1967): 304–23.

[c] E.g., in *Aspects of the Theory of Syntax* (New York: Harper & Row, 1965). Cf. my 'Psychological Aspects' cited in n. *a*, p. 70, above and the discussion in the NYU Institute of Philosophy *Proceedings* for 1968, in Sydney Hook, ed., *Language and Philosophy* (New York: New York University Press, 1969).

'I am referring to you', as well as a non-continuous present form, 'I refer to Adlai Stevenson'.

What these grammatical considerations show is that referring to someone is an action; meaning someone is not an action. As an action it can be right or wrong for one to perform. Thus it can be wrong of you to refer to someone; but not wrong of you to mean someone. It can be important or necessary that you refer to someone, but not important or necessary that you mean someone. One can intend to refer to someone, but not intend to mean him.

3 In discussions of statements such as 'Edward VII is the king of England' it is sometimes said that in making them one is referring to the same person twice. Frege says that the person is referred to in different ways each time. This way of looking at them leads to their interpretation as identities. But consider the following conversation to see how odd it is to talk of referring twice to the same person in such contexts:

A He is the king of England.
B To whom are you referring?
A That man behind the flag.
B How many times did you refer to him?

Referring to some one several times during the course of a speech would be a rather different sort of thing. If I mention a man's name I would not ordinarily be said to have referred to him in so doing. Using a man's name is in some ways opposed to referring to him rather than an instance of it.

If we assume that whenever in an assertion something is mentioned by name by a speaker he is referring to that thing certain very paradoxical conclusions can be deduced. It would follow that when I write in my paper 'I am not, of course, referring to Ludwig Wittgenstein' I would be referring to Ludwig Wittgenstein. But if someone were asked to show where in my paper I had referred to Ludwig Wittgenstein it would be absurd for him to point to the statement in which I say, 'I am not referring to Ludwig Wittgenstein.' The same would be true of the statement in which I say, 'I am referring to Ludwig Wittgenstein'. In both cases I would have used Wittgenstein's name. Therefore, to mention someone by name is not necessarily to refer to him. And consider this example. Suppose the porter at Magdalen College asks me whom I am looking for. I answer, 'Gilbert Ryle'. Would anyone say I had referred to Gilbert Ryle? But if I say, in the course of a talk, 'I am not referring to the most important of present-day philosophers', I would then and there be referring to Ludwig Wittgenstein; though in saying as I just did, 'I would then and there be referring to Ludwig Wittgenstein', I could not be said to have referred to Ludwig Wittgenstein. And this is so notwithstanding the fact that Ludwig Wittgenstein is the most important of present-day philosophers. This, then, is the paradox of reference. In saying 'I am referring to Ludwig Wittgenstein' I am not referring to Ludwig Wittgenstein.[a]

Strawson has argued correctly that if I claim, e.g., 'The king of France is Charles de Gaulle', what I have said is neither true nor false. The reason for this is not, as he says, that I have failed to refer in saying, 'The king of France...' The reason is that France is not a monarchy and there is no king of France. Just so, and said of a

---

[a] Philosophical tradition sanctions the production of such paradoxes. I am thinking of Meinong's paradox about objects, of which it is true to say that no such objects exist; and Frege's paradox that the concept horse is not a concept.

spinster, 'Her husband is kind to her' is neither true nor false. But a speaker might very well be referring to someone in using these words, for he may think that someone is the husband of the lady (who in fact is a spinster). Still, the statement is neither true nor false, for it presupposes that the lady has a husband, which she has not. This last refutes Strawson's thesis that if the presupposition of existence is not satisfied the speaker has failed to refer. For here that presupposition is false, but still the speaker has referred to someone, namely, the man mistakenly taken to be her husband.

Of course a man may 'fail to refer', but not as Strawson uses this expression. For example, in your article you may fail to refer to my article.

4 Referring does not have the omnipresence accorded to it in the philosophical literature. It sounds odd to say that when I say, 'Santa Claus lives at the North Pole' I am referring to Santa Claus, or that when I say 'The round square does not exist' I am referring to the round square. Must I be referring to something? Philosophers ask, 'How is it possible to refer to something which does not exist?' But often the examples produced in which we are supposed to do this ('Hamlet was a prince of Denmark', 'Pegasus was captured by Bellerophon', 'The golden mountain does not exist') are such that the question 'To whom (what) are you referring?' simply cannot sensibly arise in connection with them. In these cases, anyway, there is nothing to be explained.

How is it possible to make a true statement about a non-existent object? For if a statement is to be about something that thing must exist, otherwise how could the statement mention *it*, or refer to *it*? One cannot refer to, or mention nothing; and if a statement cannot be about nothing it must always be about something. Hence, this ancient line of reasoning concludes, it is not possible to say anything true or false about a non-existent object. It is not even possible to say that it does not exist.

It is this hoary line of argument which, beginning with Plato, has made the topic of referring a problem for philosophers. Still, ancient or not, the reasoning is outrageously bad. Surely here is a case where philosophers really have been seduced and led astray by misleading analogies. I cannot hang a non-existent man. I can only hang a man. To hang a non-existent man is not to do any hanging at all. So by parity of reasoning, to refer to a non-existent man is not to refer at all. Hence, I cannot say anything about a non-existent man. One might as well argue that I cannot hunt for deer in a forest where there are no deer, for that would be to hunt for *nothing*.

It must have been philosophical reflections of this *genre* which prompted Wittgenstein to say in his *Remarks*,

We pay attention to the expressions we use concerning these things; we do not understand them, however, but misinterpret them. When we do philosophy we are like savages, primitive people, who hear the expressions of civilized men, put a false interpretation on them, and then draw queer conclusions from it.[a]

Let us look a bit closer at what it is to talk about things which do not exist. Of course there are a variety of different cases here. If we stick to the kind of case which has figured prominently in philosophy, however, this variety can be reduced. What we now have to consider are characters in fiction such as Mr Pickwick; mythological figures such as Pegasus; legendary figures such as Paul Bunyan, make-believe figures like Santa Claus and fairy-tale figures like Snow White. (And why not add comic-

[a] *Remarks on the Foundations of Mathematics*, Oxford, 1956, p. 39.

strip figures like Pogo?) And do not these characters really exist? Mr Pickwick really is a character in fiction, Professor Ryle is not. There really is a figure in Greek mythology whose name is 'Pegasus', but none whose name is 'Socrates'; and there really is a comic-strip character named 'Pogo'. In talking about these characters I may say things which are true and I may also say things which are not. If I say, for example, that Pogo is a talking elephant that is just not true. Neither is Pegasus a duck. In talking about these things there is this matter of getting the facts straight. This is a problem for me; it is not a problem for Dickens or for Walt Kelly. What Dickens says about Mr Pickwick in *The Pickwick Papers* cannot be false, though it can be not true to character; and in the comic strip Walt Kelly does not say anything about his possum Pogo, for Pogo talks for himself. Still, Pogo could say something about Walt Kelly (or Charles de Gaulle), and that might not be true.

There is, however, another group of cases discussed by philosophers, and this group has the important characteristic that in talking about its members there is no such thing as getting the facts straight. Here we find Russell's famous example, the present king of France; and Meinong's equally famous example of the golden mountain. What are they supposed to be examples of? Well, just things that do not exist. But in saying this we must keep in mind how different they are from Mr Pickwick, Santa Claus, Snow White, etc. Keeping this difference in mind, we can see that though it makes perfectly good sense to ask whether Mr Pickwick ran a book-store or whether Santa Claus lives at the North Pole; it makes no sense whatever to ask whether the golden mountain is in California. Similarly, though we can ask whether Mr Pickwick was married or not, *we* cannot sensibly ask whether the present king of France is bald or not.

If the question is 'How can we talk about objects which do not exist?', then it is wrong to use the examples of the golden mountain and the present king of France. These famous philosophical examples, the round square, the golden mountain, are just things we do not talk about (except in telling a story or a fairy tale or something of the kind). Meinong, Russell, and Ryle all puzzle over sentences such as 'The gold mountain is in California', as though one just had to make up one's mind whether to put it in the box with all the other true propositions or into the box with the other false propositions. They fail to see that one would only utter it in the course of telling a story or the like. It does not occur in isolation from some such larger context. If it did so occur, if someone were just to come up to us and say, 'The gold mountain is in California', we would not concern ourselves with truth or falsity, but with this man. What is wrong with him? When the sentence occurs in a fairy tale it would never occur to us to raise the question of its truth. And when we are asked to consider whether it is true or false outside of such a context we can only say that it does not so occur, we just do not say it.

Of course, we may sometimes in error, or by mistake, talk about non-existent things, e.g., Hemingway's autobiography. So here is *one* way in which it can occur that we speak of non-existent objects. As a result of a mistake!

5   It is difficult to read the philosophical literature on these topics (the existence of chimeras, the round square, imaginary objects) without arriving at the conclusion that much of the difficulty surrounding them stems from an insistence that complex questions be given simple, unqualified, and categorical answers. One asks, 'Does Santa Claus live at the North Pole?' The philosopher asks it with the assumption that there must be just one correct answer, for by the law of the excluded middle,

either Santa Claus does or he does not live at the North Pole. Russell says that the correct answer to the question is that it is false that Santa Claus lives at the North Pole, for since he does not exist he does not live anywhere. Meinong would say, I suppose, that it is true that Santa Claus lives at the North Pole, for his so-being (*So-sein*) is independent of his being (*Sein*).

But it seems to me obvious that there is no single, categorical, and unqualified answer to our question. Why must there be one? Why are we not allowed first to consider what the person who asked the question wants to know? Surely it is obvious that there are *different* things which may be at issue. Maybe we are dealing with a child who has just heard the Santa Claus story. Perhaps the child believes in Santa Claus and wants more details about him. Then if we say, 'Yes, he lives at the North Pole' we encourage the child in his mistaken belief, so from this point of view it is the wrong thing to say. But if the child knows already that Santa Claus is just make-believe, and wants more details about the Santa Claus story, surely it is perfectly correct to say, 'Yes, he lives at the North Pole', for that is the way the story goes. A third case is one in which we do not know whether the child believes in Santa Claus and we do not want to encourage him in his belief, if he does believe in Santa Claus. Here we might say, 'No, he does not *really* live at the North Pole. It is just make-believe.'

Now in view of the obvious complexity of the situation, why should philosophers insist that there be single correct answers here? Not only are there different things one may want to know in asking 'Does Santa Claus live at the North Pole?', there are different things that may be at issue when one asks, 'Does Santa Claus exist?', 'Is Mr Pickwick an imaginary man?' Here are some of the things that one may want to know in asking this last question. One may want to know if Pickwick was a real person or only a character in one of Dickens' novels. One may know that Pickwick was not a real person, and one may know that he is a character in a novel, but wonder whether he is a real person in the novel or (perhaps) the imaginary friend of a child in the novel. Clearly what is correct to say (or better, what is the least misleading thing to say) depends upon the background in which our question arises. Further, there is nothing, in logic, which requires us to give our answer in one short sentence. Why should we not say something as complicated as this: 'Pickwick is a real person in the novel, he is not the creation of the imagination of anyone in the novel. Of course he is a creation of the imagination of Dickens. He is not a real person at all.' We may *have* to say at least this much in order not to mislead our audience.

In speaking about movies, plays, novels, dreams, legends, superstitions, make-believe, etc., our words may be thought of as occurring within the scope of special 'operators'. Let me explain. Watching the western, I say, 'I thought the sheriff would hang the hero, but he didn't'. The context in which these words are said makes it clear that they are occurring under the 'in-the-movie' operator. I am telling you of my expectations concerning the course of the movie. I thought that a hanging would take place in the movie, not in the cinema. Similarly, I might say, 'Leopold Bloom lived in Dublin'. Is this true or false? Obviously it depends upon whether my words are or are not within the scope of the 'in-the-novel' operator. It depends whether or not I am talking about (say) James Joyce's *Ulysses*. If I am talking about the chief character in Joyce's *Ulysses* and if I mean to tell you that according to Joyce this character lived in Dublin, what I say is true. But if I mean to say that Joyce's fictional character is modelled on a real Dubliner it may not be true. I do not know.

But surely what is being said, and what is being asked, is generally apparent in context, and if it is not a few questions can usually make it clear.

There is no doubt that behind these discussions of Pegasus, chimeras, the round square, there is a kind of ontological anxiety produced by certain pictures. One comes across talk of realms of being, of supersensible worlds occupied by shadowy entities not occupying space and time but nevertheless real. Certain images come to mind. Images of things with shapes and sizes, yet not tangible. We think of them as being 'out there' and 'up there', yet nowhere. (It is curious that the picture requires numbers, chimeras, and Santa Claus to be 'up there', not 'down' or even 'level' with us. We picture the numbers standing next to each other like clothes pins on a line. But why should they be 'up there' rather than below our feet?)

It is the influence of such pictures, I believe, which Russell felt when he argued that talk of Hamlet is 'really' talk about Shakespeare. It is difficult even dimly to comprehend what the dispute about whether Pegasus has being, in some sense, can amount to unless one sees the operation of these pictures. One can almost feel the anxiety in this passage from Russell. 'It is argued, e.g., by Meinong, that we can speak about "the golden mountain", "the round square", and so on; we can make true propositions of which these are the subjects; hence they must have some kind of logical being, since otherwise the propositions in which they occur would be meaningless. In such theories, it seems to me, there is a failure of that feeling for reality which ought to be preserved even in the most abstract studies. Logic, I should maintain, must no more admit a unicorn than zoology can; for logic is concerned with the real world just as truly as zoology, though with its more abstract and general features. To say that unicorns have an existence in heraldry, or in literature, or in imagination, is a most pitiful and paltry evasion.'[a]

One wonders just what the issue is between Russell and those against whom these remarks are directed, for surely no one is arguing that Santa Claus is not just make-believe and that Hamlet is not just the creature of Shakespeare's imagination. How, then, does one's 'feeling for reality' enter here at all? I am unable to explain what is at issue except in terms of the 'pictures' I have discussed.

6 It is said to be an astronomical fact of some importance that

⟨1⟩ The morning star = the evening star.

This was not always known, but the identification was early made by the Greeks. Frege said that it was because the two expressions 'the morning star' and 'the evening star' had the same reference that ⟨1⟩ was true, and because these two had different senses that ⟨1⟩ was not a trivial thing to say.

Frege's way of putting the matter seems to invite the objection that the two expressions 'the morning star' and 'the evening star' do not refer to the same thing. For the first refers to the planet Venus when seen in the morning before sunrise. The second phrase refers to the same planet when it appears in the heavens after sunset. Do they refer, then, to the same 'thing'? Is it, as Carnap says,[b] a matter of 'astronomical fact' that they do? One wants to protest that it is a matter of 'linguistic fact' that they do not.

Perhaps Frege's view is better put if we think of the two expressions as names, that is, 'The Morning Star' and 'The Evening Star'. Thus Quine,[c] in repeating

---

[a] *Introduction to Mathematical Philosophy*, Allen and Unwin, London, 1920, p. 169.
[b] *Meaning and Necessity*, Chicago, 1947, p. 119.
[c] *From a Logical Point of View*, Cambridge, Mass., 1953, p. 21.

Frege's example but adding capital letters, speaks of the expressions 'Evening Star' and 'Morning Star' as names. Quine would say that what the astronomers had discovered was that

⟨2⟩ The Morning Star = The Evening Star.

This is better, for ⟨1⟩ implies (or presupposes) what ⟨2⟩ does not, that there is only one star in the sky both in the morning and in the evening. Also, a purist might object that it cannot be taken as ground for ⟨1⟩ that Venus is both the morning star and the evening star. Venus is not a star but a planet. It would be wrong to say that what the astronomers discovered was that the morning planet is the evening planet.

⟨2⟩ is free from these criticisms, but still the same protest is in order as was made against ⟨1⟩. The name 'The Morning Star' does not refer *simpliciter* to the planet Venus. It does not refer to the planet in the way in which the demonstrative 'that' might be used to refer to the planet on some occasion. The names 'The Morning Star' and 'The Evening Star' are not that sort of 'referring expression'.

It would be incorrect for me to say to my son as he awakens, 'Look to the place where the sun is rising and you will see The Evening Star', for that is not what the star is called when seen in the east before sunrise. Again, the proposal that we stay up until we see The Evening Star is quite a different proposal from the proposal that we stay up until we see The Morning Star. In dealing with failures of substitutivity in some ways like these, Frege developed his concept of 'oblique' (*ungerade*) discourse, and Quine has talked about 'referential opacity'. Names in oblique contexts, according to Frege, do not have their 'ordinary' referents but an oblique referent which is the same as their ordinary sense. But it would be absurd to suggest that when I tell my boy that if he looks to the East on arising he will see The Evening Star I am not referring to a planet but to a 'sense', whatever that might be. Using Quine's notion of referential opacity, one might suggest that the reason why the proposal to wait up until we see The Evening Star is a different proposal from the proposal to wait up until we see The Morning Star is that here the context is referentially opaque, so that the two names in these contexts do not refer to anything at all. But surely this result is too paradoxical to be taken seriously, and in any case no one has yet told us how to understand the view that a proposal can be referentially opaque.

Under the entry for 'Venus' in the *Encyclopaedia Britannica* we are given the following information: 'When seen in the western sky in the evenings, i.e., at its eastern elongations, it was called by the ancients "Hesperus", and when visible in the mornings, i.e., at its western elongations, "Phosphorus".' Did the astronomers then discover that

⟨3⟩ Hesperus = Phosphorus?

In the entry under 'Hesperus' in Smith's *Smaller Classical Dictionary* we read, 'Hesperus, the evening star, son of Astraeus and Eos, of Cephalus and Eos, or of Atlas'. From this, together with ⟨3⟩, we are able to get by Leibniz's Law

⟨4⟩ Phosphorus is the evening star.

And avoiding unnecessary complications, let us interpret this as meaning

⟨5⟩ Phosphorus is The Evening Star.

Any competent classicist knows that this is not true.

Under the entry on 'Phosphorus' in Smith's *Smaller Classical Dictionary* we find: 'Lucifer or Phosphorus ("bringer of light"), is the name of the planet Venus, when seen in the morning before sunrise. The same planet was called Hesperus, Vespergo, Vesper, Noctifer, or Nocturnus, when it appeared in the heavens after sunset. Lucifer as a personification is called a son of Astraeus and Aurora or Eos; of Cephalus and Aurora, or of Atlas.' So the stars were personified, and it seems to be a matter of mythology that

⟨6⟩ Hesperus is not Phosphorus.

Then did the astronomers discover that the mythologists were wrong?

Of course ⟨3⟩ is false, and no astronomical research could have established it. What could we make of the contention that the Greeks mistakenly believed that Hesperus was not Phosphorus? According to the *Encyclopaedia Britannica* (s.v., 'Hesperus'): 'the two stars were early identified by the Greeks.' But once the identification was made, what was left to be mistaken about here?

Could not one mistake The Evening Star for The Morning Star? Certainly one could. This would involve mistaking evening for morning. One could do this. In the morning it is just getting light, and in the evening it is just getting dark. Imagine someone awaking from a sleep induced by a soporific. 'But there aren't *two* stars, so how *could* one be mistaken for the other?'

Hence, though it is sometimes made to look as though the Greeks were victims of a mistaken astronomical belief, this is not so. And Quine suggests that the true situation was 'probably first established by some observant Babylonian'. If that is the case, a knowing Greek would not have said

⟨7⟩ The Morning Star is not The Evening Star.

unless, of course, he were in the process of teaching his child the *use* of these words. And, drawing on his unwillingness to say ⟨7⟩ (except in special circumstances when he might want to say just that), we might push him into saying that The Morning Star is The Evening Star, and even that Hesperus is Phosphorus, though now he would begin to feel that these sayings were queer.

The moral is that if we allow ourselves no more apparatus than the apparatus of proper names and descriptions, sense and reference, and the expression '$x = y$' we just cannot give an undistorted account of what the astronomers discovered, or about Hesperus and Phosphorus. Only the logician's interest in formulas of the kind '$x = y$' could lead him to construct such sentences as 'The Morning Star = The Evening Star' or 'Hesperus = Phosphorus'. Astronomers and mythologists don't put it that way.

University of Chicago

Identifying reference and truth-values[a]

P. F. STRAWSON

The materials for this paper are: one familiar and fundamental speech-function; one controversy in philosophical logic; and two or three platitudes.

We are to be concerned with statements in which, at least ostensibly, some particular historical fact or event or state of affairs, past or present, notable or trivial, is reported: as that the emperor has lost a battle or the baby has lost its rattle or the emperor is dying or the baby is crying. More exactly, we are to be concerned with an important subclass of such statements, viz. those in which the task of specifying just the historical state of affairs which is being reported includes, as an essential part, the sub-task of designating some particular historical item or items which the state of affairs involves. Not all performances of the reporting task include the performance of this sub-task – the task, I shall call it, of identifying reference to a particular item. Thus, the report that it is raining now, or the report that it was raining here an hour ago, do not. But the statement that Caesar is dying, besides specifying the historical fact or situation which it is the function of the statement as a whole to report, has, as a part of this function, the sub-function of designating a particular historical item, viz. Caesar, which that situation essentially involves. And this part of the function of the whole statement is the whole of the function of part of the statement, viz. of the name 'Caesar'.

The speech-function we are to be concerned with, then, is the function of *identifying reference* to a particular historical item, when such reference occurs as a sub-function of statement. We are to be concerned with it in relation to a particular point of philosophical controversy, viz. the question whether a radical failure in the performance of this function results in a special case of falsity in statement or, rather, in what Quine calls a truth-value gap. The hope is not to show that one party to this dispute is quite right and the other quite wrong. The hope is to exhibit speech-function, controversy and one or two platitudes in a mutually illuminating way.

I introduce now my first pair, a complementary pair, of platitudes. One, perhaps the primary, but not of course the only, purpose of assertive discourse is to give information to an audience of some kind, viz. one's listener or listeners or reader or readers. Since there is no point in, or perhaps one should say, no possibility of, informing somebody of something of which he is already apprised, the making of an assertive utterance or statement – where such an utterance has in view this primary purpose of assertion – implies a presumption (on the part of the speaker) of ignorance (on the part of the audience) of some point to be imparted in the utterance. This platitude might be called the Principle of the Presumption of Ignorance. It is honoured to excess in some philosophical proposals for analysis or reconstruction of

[a] This paper is reprinted from *Theoria*, xxx, 1964, pp. 96–118.

ordinary language, proposals which might appear to be based on the different and mistaken Principle of the Presumption of Total Ignorance. To guard against such excess, we need to emphasise a platitude complementary to the first. It might be called the Principle of the Presumption of Knowledge. The substance of this complementary platitude, loosely expressed, is that when an empirically assertive utterance is made with an informative intention, there is usually or at least often a presumption (on the part of the speaker) of knowledge (in the possession of the audience) of empirical facts relevant to the particular point to be imparted in the utterance. This is *too* loosely expressed. The connexion between the presumption of knowledge and the intention to impart just such-and-such a particular point of information may be closer than that of customary association; the connexion between the *identity* of the particular point it is intended to impart and the kind of knowledge presumed may be closer than that of relevance. Just as we might say that it could not be true of a speaker that he intended to *inform* an audience of some particular point unless he presumed their ignorance of that point, so we might often say that it could not be true of a speaker that he intended to inform the audience of just *that* particular point unless he presumed in his audience certain empirical knowledge. So the second principle, in which I am mainly interested, is truly complementary to the first.

Now this may sound a little mysterious. But at least there will be no difficulty felt in conceding the general and vague point that we do constantly presume knowledge as well as ignorance on the part of those who are the audiences of our assertive utterances, and that the first kind of presumption, as well as the second, bears importantly on our choice of what we say. The particular application that I want to make of this general point is to the case of identifying reference. To make it, I must introduce the not very abstruse notion of identifying knowledge of particulars.

Everyone has knowledge of the existence of various particular things each of which he is able, in one sense or another, though not necessarily in every sense, to distinguish from all other things. Thus a person may be able to pick a thing out in his current field of perception. Or he may know there is a thing (not in his current field of perception) to which a certain description applies which applies to no other thing: such a description I shall call an *identifying description*. Or he may know the name of a thing and be able to recognise it when he encounters it, even if he can normally give no identifying description of it other than one which incorporates its own name. If a man satisfies any one of these conditions in respect of a certain particular, I shall say he has identifying knowledge of that particular. One is bound to define such a notion in terms of its outlying cases, cases, here, of minimal and relatively isolated identifying knowledge. So it is worth emphasising that, in contrast with cases of minimal and relatively isolated identifying knowledge, there are hosts of cases of very rich and full identifying knowledge, and that, in general, our identifying knowledge of particulars forms an immensely complex web of connexions and relations – the web, one might say, of our historical and geographical knowledge in general, granted that these adjectives are not to be construed as qualifying academic subject alone, but also knowledge of the most unpretentious kind about the particular things and people which enter into our minute-to-minute or day-to-day transactions with the world.

The notion of identifying reference is to be understood in close relation to the notion of identifying knowledge. When people talk to each other they commonly and rightly assume a large community of identifying knowledge of particular items. Very often a speaker knows or assumes that a thing of which he has such knowledge is also a thing of which his audience has such knowledge. Knowing or assuming this, he may wish to state some particular fact regarding such a thing, e.g. that it is thus-and-so; and he will then normally include in this utterance an expression which he regards as adequate, in the circumstances of utterance, to indicate to the audience *which* thing it is, of all the things in the scope of the audience's identifying knowledge, that he is declaring to be thus-and-so. The language contains expressions of several celebrated kinds which are peculiarly well adapted, in different ways, for use with this purpose. These kinds include proper names, definite and possessive and demonstrative descriptions, demonstrative and personal pronouns. I do not say that *all* expression of these kinds are well adapted for use with this purpose; nor do I say, of those that are, that they are not regularly used in other ways, with other purposes.

When an expression of one of these classes *is* used in this way, I shall say that it is used to *invoke* identifying knowledge known or presumed to be in possession of an audience. It would now be easy to define identifying reference so that only when an expression is used to invoke identifying knowledge is it used to perform the function of identifying reference. But though it would simplify exposition thus to restrict attention to what we shall in any case count as the central cases of identifying reference, it would not be wholly desirable. For there are cases which cannot exactly be described as cases of invoking identifying knowledge, but which are nevertheless sufficiently like cases which *can* be so described to be worth classifying with them as cases of identifying reference. For instance, there may be within a man's current field of possible perception something which he has not noticed and cannot be said actually to have discriminated there, but to which his attention may be intentionally drawn simply by the use, on the part of a speaker, of an expression of one of the kinds I mentioned, as part of a statement of some fact regarding the particular item in question. In so far as the speaker's intention, in using the expression in question, is not so much to *inform* the audience of the existence of some particular item unique in a certain respect as to bring it about that the audience *sees for itself* that there is such an item, we may think this case worth classifying with the central cases of identifying reference. Again, there are cases in which an audience cannot exactly be credited with *knowledge* of the existence of a certain item unique in a certain respect, but can be credited with a strong *presumption* to this effect, can be credited, we might say, with *identifying presumption* rather than identifying knowledge. Such presumed presumption can be invoked in the same style as such knowledge can be invoked.

So we may allow the notion of identifying reference to a particular item to extend beyond the cases of invoking identifying knowledge. We must then face the un-surprising consequence that if, as we do, we wish to contrast cases in which a speaker uses an expression to perform the function of identifying reference with cases in which the intention and effect of a speaker's use of an expression is to inform the audience of the existence of a particular item unique in a certain respect, then we shall encounter some cases which do not *clearly* belong to either of these con-trasting classes, which seem more or less dubious candidates for both. But this is not a situation which should cause us embarrassment, in philosophy; and, having made the point, I shall for simplicity's sake speak in what follows as if all cases of identify-ing reference were, at least in intention, cases of invoking identifying knowledge.

What I have said so far, in describing the function of identifying reference, is I think, uncontroversial in the sense that the description has proceeded without my having to take up a position on any well-worn point of controversy. It has a con-sequence, just alluded to, which should be equally uncontroversial, and which I shall labour a little now, partly in order to distinguish it from any *prise de position* on a matter which is undoubtedly one of controversy.

I have explained identifying reference – or the central case of identifying reference – as essentially involving a presumption, on the speaker's part, of the possession by the audience of identifying knowledge of a particular item. Identifying knowledge is knowledge of the existence of a particular item distinguished, in one or another sense, by the audience from any other. The appropriate stretch of identifying knowledge is to be invoked by the use of an expression deemed adequate by the speaker, in the total circumstances of utterance, to indicate to the audience which, of all the items within the scope of the audience's identifying knowledge, is being declared, in the utterance as a whole, to be thus-and-so. Depending upon the nature of the item and the situation of utterance, the expression used may be a name or a pronoun or a definite or demonstrative description; and it is of course not necessary that either name or description should in general be *uniquely* applicable to the item in question, so long as its choice, in the total circumstances of utterance, is deemed adequate to indicate to the audience which, of all the particular items within the scope of his identifying knowledge, is being declared, in the utterance as a whole, to be thus-and-so.

Now one thing that is absolutely clear is that it can be no part of the speaker's intention in the case of such utterances to *inform* the audience of the *existence* of a particular item bearing the name or answering to the description and distinguished by that fact, or by that fact plus something else known to the audience, from any other. On the contrary, the very task of identifying reference, as described, can be undertaken only by a speaker who knows or presumes his audience to be already in possession of such knowledge of existence and uniqueness as this. The task of identifying reference is *defined* in terms of a type of speaker-intention which *rules out* ascription to the speaker of the intention to impart the existence-and-uniqueness information in question. All this can be put, perfectly naturally, in other ways. Thus, that there exists a particular item to which the name or description is applicable and which, if not unique in this respect, satisfies some uniqueness-condition known to the hearer (*and* satisfies some uniqueness-condition known to the speaker) is no part of what the speaker *asserts* in an utterance in which the name or description is used to perform the function of identifying reference; it is, rather, a *presupposition* of his asserting what he asserts.

This way of putting it is still uncontroversial. For it is a natural way of putting what is itself uncontroversial. But it introduces a contrast, between the *asserted* and the *presupposed*, in words which are associated with an issue of controversy.

We can come at this issue by considering some of the ways in which an attempt to perform the function of identifying reference can either fail altogether or at least fall short of complete success and satisfactoriness. There are several ways in which such an attempt can fail or be flawed. For instance, it may be that, though the speaker possesses, the audience does not possess, identifying knowledge of the particular historical item to which the speaker intends to make an identifying reference; that the speaker credits the audience with identifying knowledge the audience does not possess. It may be that though the audience possesses identifying knowledge of the particular item in question, the expression chosen by the speaker fails to invoke the appropriate stretch of identifying knowledge and leaves the audience uncertain, or even misleads the audience, as to who or what is meant. Failures of this kind may be, though they need not be, due to flaws of another kind. For it may be that the speaker's choice of expression reflects mistakes of fact or language on his part; and such mistake reflecting choices are still flaws, even where they do not mislead, as, for example, references to Great Britain as 'England' or to President Kennedy as 'the U.S. Premier' are not likely to mislead.

Though these are all cases of flawed or failed reference, they are not cases of the most radical possible kind of failure. For my descriptions of these cases imply that at least one fundamental condition of success is fulfilled, even if others are not fulfilled. They imply at least that there *is* a particular historical item within the scope of the speaker's identifying knowledge – even if not all his beliefs about it are true – such that he intends, by suitable choice of expression, to invoke identifying knowledge, presumed by him to be in possession of the audience, of that item. But this condition might fail too; and that in various ways. It might be that there just is no such particular item at all as the speaker takes himself to be referring to, that what he, and perhaps the audience too, take to be identifying knowledge of a particular item is not knowledge at all, but completely false belief. This is but one case of what might uncontroversially be called radical failure of the existence presupposition of identifying reference. It involves no moral turpitude on the part of the speaker. Different would be the case in which the speaker uses an expression, by way of apparently intended identifying reference, to invoke what he knows or thinks the audience thinks to be identifying knowledge, though he, the speaker, knows it to be false belief. The speaker in this case can have no intention actually to refer to a particular historical item, and so cannot strictly fail to carry out *that* intention. He can have the intention to be *taken* to have the former intention; and in *this* intention he may succeed. A full treatment of the subject would call for careful consideration of such differences. For simplicity's sake, I shall ignore the case of pretence, and concentrate on that case of radical reference-failure in which the failure is not a moral one.

Our point of controversy concerns the following question: given an utterance which suffers from radical reference-failure, are we to say that what we have here is just one special case of false statement or are we to say that our statement suffers from a deficiency so radical as to deprive it of the chance of being either true or false? Of philosophers who have discussed this question in recent years some have plumped uncompromisingly for the first answer, some uncompromisingly for the second; some have been eclectic about it, choosing the first answer for some cases and the second for others; and some have simply contented themselves with sniping at any doctrine that offered, while wisely refraining from exposing any target themselves. In virtue of his Theory of Descriptions and his views on ordinary proper names as being condensed descriptions, Russell might be said to be the patron of the first party – the 'special case of false statement' party. One recent explicit adherent of that party is Mr Dummett in his interesting article on *Truth* (P.A.S. 58–9). The second party – the 'neither true nor false' party – might be said, with some reservations, to have included Quine, Austin and myself. Quine invented the excellent phrase 'truth-value gap' to characterise what we have in these cases (see *Word and Object*). Austin (see *Performative Utterances* and *How to Do Things with Words*) contrasts this sort of deficiency or, as he calls it, 'infelicity' in statement with straightforward falsity and prefers to say that a statement suffering from this sort of deficiency is void – 'void for lack of reference'.

Let us ignore the eclectics and the snipers and confine our attention, at least for the moment, to the two uncompromising parties. I do not think there is any question of demonstrating that one party is quite right and the other quite wrong. What we have here is the familiar philosophical situation of one party being attracted by one simplified, theoretical – or 'straightened out – concept of truth and falsity, and the other by another. It might be asked: How does ordinary usage speak on the point? And this, as always, is a question which it is instructive to ask. But ordinary usage does not deliver a clear verdict for one party or the other. Why should it? The interests which ordinary usage reflects are too complicated and various for it to provide overwhelming support for either way of *simplifying* the picture. The fact that ordinary usage does not deliver a clear verdict does not mean, of course, that there can be no other way of demonstrating, at least, that one view is quite wrong. It might be shown, for example, to be inconsistent, or incoherent in some other way. But this is not the case with either of these views. Each would have a certain amount of explaining and adjusting to do, but each could perfectly consistently be carried through. More important, each is *reasonable*. Instead of trying to demonstrate that one is quite right and the other quite wrong, it is more instructive to see how both are reasonable, how both represent different ways of being impressed by the facts.

As a point of departure here, it is reasonable to take related cases of what are indisputably false statements and then set beside them the disputed case, so that we can see how one party is more impressed by the resemblances, the other by the differences, between the disputed case and the undisputed cases. The relevant undisputed cases are obviously of two kinds. One is that of an utterance in which an identifying reference is successfully made, and all the conditions of a satisfactory or allround successful act of empirical assertion are fulfilled except that the particular item identifying referred to and declared to be thus-and-so is, as a matter of fact, *not* thus-and-so. It is said of Mr Smith, the new tenant of the Grange, that he is single, when he is in fact married: a statement satisfactory in all respects except that it is, indisputably, false. The other relevant case is that in which an explicit assertion of existence and uniqueness is made. It is said, say, that there is one and not more than one island in Beatitude Bay. And this is false because there is none at all or because there are several.

Now, it might be said on the one side, how vastly different from the ways in which things go wrong in either of these undisputed cases of falsity is the way in which things go wrong in the case of radical reference-failure. A judgment as to truth or falsity is a judgment on what the speaker asserts. But we have already noted the uncontroversial point that the existence-condition which fails in the case of radical reference-failure is not something asserted, but something presupposed, by the speaker's utterance. So his statement cannot be judged as a false existential assertion. Nor, evidently, can it be judged as an assertion false in the same way as the first undisputed example, i.e. false as being a *mis*characterisation of the particular item referred to. For there is no such item for it to be a mischaracterisation of. In general, where there *is* such an item as the speaker refers to, and the speaker asserts, with regard to that item, that it is thus-and-so, his assertion is rightly assessed as true if the item is thus-and-so, as false if it is not. In the case of radical reference failure, the speaker, speaking in good faith, *means* his statement to be up for assessment in this way; he takes himself similarly to be asserting with regard to a particular item, that it is thus-and-so. But in fact the conditions of his making an assertion such as he takes himself to be making are not fulfilled. We can acknowledge the character of his intentions and the nature of his speech-performance by saying that he makes a statement; but we must acknowledge, too, the radical character of the way in which his intentions are frustrated by saying that his statement does not qualify as such an assertion as he takes it to be, and hence does not qualify for assessment as such an assertion. But then it does not qualify for truth-or-falsity assessment at all. The whole assertive enterprise is wrecked by the failure of its presupposition.

But now, on the other side, it could be said that what the disputed case has in common with the undisputed cases of falsity is far more important than the differences between them. In all the cases alike, we may take it a genuine empirical statement is made; a form of words is used such that if there were as a matter of fact in the world (in Space and Time) a certain item, or certain items, with certain characteristics – if, to put it differently, certain complex circumstances did as a matter of fact obtain in the world (in Space and Time) – then the statement would be true. The important distinction is between the case in which those complex circumstances do obtain and the case in which they don't. This distinction is the distinction we *should* use the words 'true' and 'false' (of statements) to mark, even if we do not consistently do so. And this distinction can be drawn equally in the disputed and the undisputed cases. A false empirical statement is simply any empirical statement whatever which fails for *factual* reasons, i.e. on account of circumstances in the world being as they are and not otherwise, to be a true one. Cases of radical reference-failure are simply one class of false statements.

It no longer seems to me important to come down on one side or the other in this dispute. Both conceptions are tailored, in the ways I have just indicated, to emphasise different kinds of interest in statement; and each has its own merits. My motives in bringing up the issue are three, two of which have already partially shown themselves. First, I want to disentangle this issue of controversy from other questions with which it is sometimes confused. Second, I want to dispel the illusion that the issue of controversy can be speedily settled, one way or the other, by a brisk little formal argument. Third, I want to indicate one way – no doubt there are more – in which, without positive commitment to either rival theory, we may find the issues they raise worth pursuing and refining. I shall say something on all three points, though most on the third.

First, then, the issue between the truth-value gap theory and the falsity theory, which has loomed so large in this whole area of discussion, has done so in a way which might be misleading, which might give a false impression. The impression might be given that the issue between these two theoretical accounts was the *crucial* issue in the whole area – the key, as it were, to all positions. Thus it might be supposed that anyone who rejected the view that the Theory of Descriptions gives an adequate general analysis, or account of the functioning, of definite descriptions was committed, by that rejection, to uncompromising adherence to the truth-value gap theory and uncompromising rejection of the falsity theory for the case of radical reference-failure. But this is not so at all. The distinction between identifying reference and uniquely existential assertion is something quite undeniable. The sense in which the existence of something answering to a definite description used for the purpose of identifying reference, and its distinguishability by an audience from anything else, is presupposed and not asserted in an utterance containing such an expression, so used, stands absolutely firm, whether or not one opts for the view that radical failure of the presupposition would deprive the statement of a truth-value. It remains a decisive objection to the Theory of Descriptions, regarded as embodying a generally correct analysis of statements containing definite descriptions, that, so regarded, it amounts to a denial of these undeniable distinctions. I feel bound to labour the point a little, since I may be partly responsible for the confusion of these two issues by making the word 'presupposition' carry simultaneously the burden both of the functional distinction and of the truth-value gap theory. Only, at most, partly responsible; for the line-up is natural enough, though not inevitable; and though there is no logical compulsion one way, there is logical compulsion the other. One who accepts the Theory of Descriptions as a correct analysis is bound to accept the falsity theory for certain cases and reject the truth-value gap theory. One who accepts the truth-value gap theory is bound to reject the Theory of Descriptions as a generally correct analysis. But it is perfectly consistent to reject the view that the Theory of Descriptions is a generally correct analysis, on the grounds I have indicated, and also to withhold assent to the truth-value gap theory.

Now for my second point. I have denied that either of the two theories can be decisively refuted by short arguments, and I shall support this by citing and commenting on some specimen arguments which are sometimes thought to have this power. First, some arguments for the truth-value gap theory and against the falsity theory:

A (1) Let *Fa* represent a statement of the kind in question. If the falsity theory is correct, then the contradictory of *Fa* is not -*Fa*, but the disjunction of -*Fa* with a negative existential statement. But the contradictory of *Fa is* -*Fa*. Therefore the falsity theory is false.

(2) If 'false' is used normally, then from *It is false that S is P* it is correct to infer *S is not P*. But it is agreed on both theories that *S is not P* is true only if there is such a thing as *S*. Hence, if 'false' is used normally, *It is false that S is P* is true only if there is such a thing as *S*. Hence, if 'false' is used normally in the statement of the falsity theory, that theory is false.

(3) The question *Is S P?* and the command *Bring it about that S is P* may suffer from exactly the same radical reference-failure as the statement *S is P*. But if an utterance which suffers from this radical reference-failure is thereby rendered false, the question and command must be said to be false. But this is absurd. So the falsity theory is false.

Now arguments on the other side:

B (1) Let *Fa* represent a statement of the kind in question (e.g. *The king of France is bald*). Then there may be an equivalent statement, *Gb*, (e.g. *France has a bald king*) which is obviously false if there is no such thing as *a*. But the two statements are equivalent. So *Fa* is false if there is no such thing as *a*. So the truth-value gap theory is false.

(2) Let *P* be a statement which, on the truth-value gap theory, is neither true nor false. Then the statement that *P* is true is itself false. But if it is false that *P* is true, then P is false. In the same way we can derive from the hypothesis the conclusion that *P* is true, hence the conclusion that *P* is both true and false. This is self-contradictory, hence the original hypothesis is so too.

The defender of either view will have little difficulty in countering these arguments against it. Thus to B2 the reply is that if a statement lacks a truth-value, any statement assessing it as true *simpliciter* or as false *simpliciter* similarly lacks a truth-value. So no contradiction is derivable. To B1 the reply is that the argument is either inconclusive or question-begging. If 'equivalent' means simply 'such that if either is true, then necessarily the other is true', it is inconclusive. If it also means 'such that if either is false, then necessarily the other is false', it is question-begging. To A3 the reply is that there is no reason why what holds for statements should hold also for questions and commands. To A2 the reply is that the inference is not strictly correct, though it is perfectly natural that we should normally make it. To A1 the reply is that it is question-begging, though again it is perfectly intelligible that we should be prone to think of contradictories in this way.

It is just an illusion to think that either side's position can be carried by such swift little sallies as these. What we have, in the enthusiastic defence of one theory or the other, is a symptom of difference of direction of interest. One who has an interest in actual speech-situations, in the part that stating plays in communication between human beings, will tend to find the simpler falsity theory inadequate and feel sympathy with – though, as I say, he is under no compulsion, exclusively or at all, to embrace – its rival. One who takes a more impersonal view of statement, who has a picture in which the actual needs, purposes and presumptions of speakers and hearers are of slight significance – in which, as it were, there are just statements on the one side and, on the other, the world they should reflect – he will naturally tend to brush aside the truth-value gap theory and embrace *its* simpler rival. For him, one might say, the subject of every statement is just the world in general. For his opponent, it is now this item, now that; and perhaps sometimes – rarely and disconcertingly enough – nothing at all.

And now for the third matter, which we shall find not unconnected with this last thought. It seems to be a fact which advocates of either, or of neither, theory can equally safely acknowledge, that the intuitive appeal, or prima facie plausibility, of the truth-value gap theory is not constant for all example-cases of radical reference-failure which can be produced or imagined. We can, without commitment to either theory, set ourselves to explain this variation in the intuitive appeal of one of them – which is also an inverse variation in the intuitive appeal of the other. The attempt to explain this fact may bring into prominence other facts which bear in interesting ways upon speech-situations in general, and those involving identifying reference in particular. I shall draw attention to but one factor – no doubt there are more – which may contribute to the explanation of this fact in some cases. In doing so, I shall invoke another platitude to set beside, and connect with, the platitudes we already have.

First, we may note that the truth-value gap theory can be expressed, in terms of the familiar idea of predication, in such a way as to secure for it a certain flexibility in application. Let us call an expression *as and when used in a statement with the role of identifying reference* – whether or not it suffers in that use from radical reference-failure – a *referring expression*. Then any statement containing a referring expression, E, can be regarded as consisting of two expression-parts, one the expression E itself, to be called the subject-expression or subject-term, and the other the remainder of the statement, to be called the predicate-expression or predicate-term. In the case of a statement containing more than one, say two, referring expressions, it is to be open to us to cast one of these for the role of subject-expression, while the other is regarded as absorbed into the predicate-term which is attached to the subject-term to yield the statement as a whole. The adherent of the truth-value gap view can then state his view as follows.[a] The statement or predication as a whole is true just in the case in which the predicate-term does in fact apply to (is in fact 'true of') the object which the subject-term (identifyingly) refers to. The statement or predication as a whole is false just in the case where the negation of the predicate-term applies to that

[a] This way of stating it is in fact implicit in the fundamental definition of predication which Quine gives on p. 96 of *Word and Object*.

object, i.e. the case where the predicate-term can be truthfully denied of that object. The case of radical reference-failure on the part of the subject-term is of neither of these two kinds. It is the case of the truth-value gap.

Now consider a statement consisting of two referring expressions one of which is guilty of reference-failure while the other is not. Then it is open to us to carve up the statement in two different ways; and different decisions as to carving-up procedure may be allowed to result in different assessments of the statement for truth-value. Thus (1) we can see the guilty referring expression as absorbed into a predicate-term which is attached to the innocent referring expression to make up the statement as a whole; or (2) we can see the innocent referring expression as absorbed into a predicate-expression which is attached to the guilty referring expression to make up the statement as a whole. Now if we carve up the statement in the second way, we must say – according to our current statement of the truth-value gap theory – that the statement lacks a truth-value. But if we carve it up in the first way, we *may* say that it is false (or, sometimes – when negative in form – that it is true). For to carve it up in the first way is to think of the statement as made up of the satisfactory or innocent referring expression together with one general term or predicate into which the guilty referring expression has been absorbed. The question whether that predicate does or does not apply to the object referred to by the satisfactory referring expression remains a perfectly answerable question; and the fact that the predicate has absorbed a guilty referring expression will, for most predicates affirmative in form, merely have the consequence that the right answer is 'No'. Thus, if we look at such a statement in this way, we can naturally enough declare it false or untrue, and naturally enough affirm its *negation* as true, on the strength of the reference-failure of the guilty referring expression.

In this way, it might seem, the truth-value gap theory can readily modify itself to take account of certain examples which may seem intuitively unfavourable to it. For example, there is no king of France; and there is, let us say, no swimming-pool locally. But there is, let us say, an Exhibition in town; and there is, let us say, no doubt of Jones' existence. If we consider the statements

⟨1⟩ that Jones spent the morning at the local swimming-pool

and

⟨2⟩ that the Exhibition was visited yesterday by the king of France

it may seem natural enough to say (1) that it is quite untrue, or is false, that Jones spent the morning at the local swimming-pool, since there isn't one; that, however Jones spent the morning, he did *not* spend it at the local swimming-pool, since there's no such place; and similarly (2) that it is quite untrue, or is false, that the Exhibition was visited yesterday by the king of France; that, whoever the Exhibition was visited by yesterday, it was *not* visited by the king of France, since there is no such person. And the modified truth-value gap theory accommodates these intuitions by allowing the guilty referring expressions, 'the local swimming-pool', 'the king of France', to be absorbed into the predicate in each case.

This modification to the truth-value gap theory, though easy and graceful, will scarcely seem adequate. For one thing, it will not be available for all intuitively unfavourable examples, but only for those which contain more than one referring expression. For another, it will remain incomplete inside its own domain unless some *principle of choice* between alternative ways of carving up a statement is supplied. The theory might resolve the latter question self-sacrificially, by declaring that the carving-up operation was always to be so conducted as to permit the assignment of a truth-value whenever possible. But this move might be too self-sacrificial, turning friends into enemies, turning intuitively favourable cases into unfavourable ones.

So let us consider further. Confronted with the classical example, 'The king of France is bald', we may well feel it natural to say, straight off, that the question whether the statement is true or false doesn't arise because there is no king of France. But suppose the statement occurring in the context of a set of answers to the question: 'What examples, if any, are there of famous contemporary figures who are *bald*?' Or think of someone compiling a list in answer to the question, 'Who has died recently?' and including in it the term 'the king of France'. Or think of someone including the statement 'The king of France married again' in a set of statements compiled in reply to the question: 'What outstanding events, if any, have occurred recently in the social and political fields?' In the first two cases the king of France appears to be cited as an *instance* or example of an *antecedently introduced class*. In the last case the statement as a whole claims to report an event as an instance of an *antecedently introduced class*. The question in each case represents the antecedent centre of interest as a certain class – the class of bald notables, the class of recently deceased notables, the class of notable recent events in a certain field – and the question is as to what items, if any, the classes include. Since it is certainly false that the classes, in each case, include any such items as our answers claim they do, those answers can, without too much squeamishness, be simply marked as wrong answers. So to mark them is not to reject them as answers to questions which don't arise, but to reject them as wrong answers to questions which do arise. Yet the answers need included only *one* referring expression for a particular item, viz. the one guilty of reference failure; and the questions need not contain any at all.

This suggests a direction in which we might look for the missing principle of choice in the case of our previous examples, those about the swimming pool and the Exhibition, which contained two referring expressions. The point was not, or was not solely, that each contained an extra and satisfactory referring expression. It was rather that we could easily see the centre of interest in each case as being the question, e.g. *how Jones spent the morning* or *what notable visitors the Exhibition has had*, or *how the Exhibition is getting on*. And the naturalness of taking them in this way was increased by the device of putting the satisfactory referring expression *first*, as grammatical subject of the sentence, and the unsatisfactory one last. We might, for example, have felt a shade more squeamish if we had written 'The king of France visited the Exhibition yesterday' instead of 'The Exhibition was visited yesterday by the king of France'. We feel very squeamish indeed about 'The king of France is bald' presented abruptly, out of context, just because we don't naturally and immediately think of a context in which interest is centred, say, on the question *What bald notables are there?* rather than on the question *What is the king of France like?* or *Is the king of France bald?* Of course, to either of *these* two questions the statement would not be just an incorrect answer. *These* questions have no correct answer and hence, in a sense, no incorrect answer either. They *are* questions which do not arise. This does not mean there is no correct *reply* to them. The correct reply is: 'There is no king of France.' But this reply is not an answer to, but a rejection of, the question. The question about bald notables, on the other hand, *can* be answered, rightly or wrongly. Any answer which purports to mention someone included in the class, and fails to do so, is wrong; and it is still wrong even if there is no such person at all as it purports to mention.

I should like to state the considerations I have been hinting at a little more generally, and with less dependence upon the notion of a question. Summarily my suggestions are as follows.

(1) First comes the additional platitude I promised. Statements, or the pieces of discourse to which they belong, have subjects, not only in the relatively precise senses of logic and grammar, but in a vaguer sense with which I shall associate the words 'topic' and 'about'. Just now I used the hypothesis of a *question* to bring out, with somewhat unnatural sharpness, the idea of the topic or centre of interest of a statement, the idea of what a statement could be said, in this sense, to be about. But even where there is no actual first-order question to pinpoint for us with this degree of sharpness the answer to the higher-order question, 'What is the statement, in this sense, about?', it may nevertheless often be possible to give a fairly definite answer to this question. For stating is not a gratuitous and random human activity. We do not, except in social desperation, direct isolated and unconnected pieces of information at each other, but on the contrary intend in general to give or add information about what is a matter of standing or current interest or concern. There is a great variety of possible types of answer to the question what the topic of a statement is, what a statement is 'about', – about baldness, about what great men are bald, about which countries have bald rulers, about France, about the king, etc. – and not every such answer excludes every other in a given case. This platitude we might dignify with the title, the Principle of Relevance.

(2) It comes to stand beside that other general platitude which I announced earlier under the title, the Principle of the Presumption of Knowledge. This principle, it will be remembered, is that statements, in respect of their informativeness, are not generally self-sufficient units, free of any reliance upon what the audience is assumed to know or to assume already, but commonly depend for their effect upon knowledge assumed to be already in the audience's possession. The particular application I made of this principle was to the case of identifying reference, in so far as the performance of this function rests on the presumption of identifying knowledge in the possession of the audience. When I say that the new platitude comes to stand beside the old one, I mean that the spheres of (*a*) what a statement addressed to an audience is *about* and (*b*) what, in the making of that statement, the audience is assumed to have some knowledge of already, are spheres that will often, and naturally, overlap.

But (3) they need not be co-extensive. Thus, given a statement which contains a referring expression, the specification of that statement's topic, what it is about, would very often involve mentioning, or seeming to mention, the object which that expression was intended to refer to; but sometimes the topic of a statement containing such an expression could be specified without mentioning such an object. Let us call the first type of case Type 1 and the second type of case Type 2. (Evidently a statement could be of Type 1 relative to one referring expression it contained and of Type 2 relative to another.)

Now (4) assessments of statements as true or untrue are commonly, though not only, topic-centred in the same way as the statements assessed; and when, as commonly, this is so, we may say that the statement is assessed *as* putative information *about its topic*.

Hence (5), given a case of radical reference-failure on the part of a referring expression, the truth-value gap account of the consequences of this failure will seem more naturally applicable if the statement in question is of Type 1 (relative to that referring expression) than if it is of Type 2. For if it is of Type 2, the failure of reference does not affect the topic of the statement, it merely affects what purports to be information *about its topic*. We may still judge the statement as putative information *about its topic* and say, perhaps, that the failure of reference has the consequence that it is *mis*informative *about its topic*. But we cannot say this if it is a case of Type 1. If it is a case of Type 1, the failure of reference affects the topic itself and not merely the putative information about the topic. If we know of the reference-failure, we know that the statement cannot really have the topic it is intended to have and hence cannot be assessed as putative information about that topic. It can be seen neither as correct, nor as incorrect, information *about its topic*.

But, it might be said, this account is self-contradictory. For it implies that in a Type 1 case of radical reference-failure the statement does not really *have* the topic which by hypothesis it does have; it implies that a statement which, by hypothesis, is *about* something is really about nothing. To this objection we must reply with a distinction. If I believe that the legend of King Arthur is historical truth, when there was in fact no such person, I may in one sense make statements *about* King Arthur, *describe* King Arthur and make King Arthur my *topic*. But there is another sense in which I cannot make statements about King Arthur, describe him or make him my topic. This second sense is stronger than the first. I may suppose myself to be making statements about him in the second, stronger sense; but I am really only making statements about him in the first and weaker sense. If, however, my belief in King Arthur were true and I really was making statements about him in the second sense, it would still be true that I was making statements about him in the first sense. This is why the first is a weaker (i.e. more comprehensive) sense than the second and not merely different from it.

Bearing this distinction of sense in mind, we can now frame a recipe for distinguishing those cases of reference-failure which are relatively favourable to the truth-value gap theory from those cases which are relatively unfavourable to it. The recipe is as follows. Consider in its context the statement suffering from reference-failure and frame a certain kind of description of the speech-episode of making it. The description is to begin with some phrase like 'He (i.e. the speaker) was saying (describing)...' and is to continue with an interrogative pronoun, adjective or adverb, introducing a dependent clause. The clause, with its introductory conjunction, specifies the topic of the statement, what it can be said (at least in the weaker, and, if there is no reference-failure, also in the stronger, sense) to be *about*; while *what is said about its topic* is eliminated from the description in favour of the interrogative expression. Examples of such descriptions based on cases already mentioned would be:

> He was describing *how Jones spent the morning*
> He was saying *which notable contemporaries are bald*
> He was saying *what the king of France is like.*

If the peccant referring expression survives in the clause introduced by the interrogative, the clause which specifies what the original statement was about, then we have a case relatively favourable to the truth-value gap theory. If the peccant referring expression is eliminated, and thus belongs to what purports to be information about the topic of the original statement, then we have a case relatively unfavourable to the truth-value gap theory. There can be no true or false, right or wrong, descriptions-of-what-the-king-of-France-is-like, because there is no king of France. But there can be right or wrong descriptions-of-how-Jones-spent-the-morning, and the description of him as having spent it at the local swimming-pool is wrong because there is no such place.

It is easy to see why the relevance of these factors should have been overlooked by those philosophers, including myself, who, considering a few example sentences in isolation from possible contexts of their use, have been tempted to embrace, and to generalise, the truth-value gap theory. For, first, it often is the case that the topic of a statement is, or includes, something referred to by a referring expression; for such an expression invokes the knowledge or current perceptions of an audience, and what is of concern to an audience is often what it already knows something about or is currently perceiving. And, second, it often is the case that the placing of an expression at the beginning of a sentence, in the position of grammatical subject, serves, as it were, to announce the statement's topic. The philosopher, thinking about reference-failure in terms of one or two short and isolated example sentences beginning with referring expressions, will tend to be influenced by these facts without noticing *all* of what is influencing him. So he will tend to attribute his sense of something more radically wrong than falsity to the presence alone of what is alone obvious, viz. a referring expression which fails of reference; and thus will overlook altogether these considerations about aboutness or topic which I have been discussing.

Let me remark that I do not claim to have done more than mention one factor which may sometimes bear on the fact that a truth-value gap theory for the case of radical reference-failure is apt to seem more intuitively attractive in some instances than it does in others.

University of Oxford

Reference and definite descriptions[a]

KEITH DONNELLAN

1   Definite descriptions, I shall argue, have two possible functions. They are used to refer to what a speaker wishes to talk about, but they are also used quite differently. Moreover, a definite description occurring in one and the same sentence may, on different occasions of its use, function in either way. The failure to deal with this duality of function obscures the genuine referring use of definite descriptions. The best-known theories of definite descriptions, those of Russell and Strawson, I shall suggest, are both guilty of this. Before discussing this distinction in use, I will mention some features of these theories to which it is especially relevant.

On Russell's view a definite description may denote an entity: 'if "*C*" is a denoting phrase [as definite descriptions are by definition], it may happen that there is one entity *x* (there cannot be more than one) for which the proposition "*x* is identical with *C*" is true...We may say that the entity *x* is the denotation of the phrase "*C*".'[b] In using a definite description, then, a speaker may use an expression which denotes some entity, but this is the only relationship between that entity and the use of the definite description recognized by Russell. I shall argue, however, that there are two uses of definite descriptions. The definition of denotation given by Russell is applicable to both, but in one of these the definite description serves to do something more. I shall say that in this use the speaker uses the definite description to *refer* to something, and call this use the 'referential use' of a definite description. Thus, if I am right, referring is not the same as denoting and the referential use of definite descriptions is not recognized on Russell's view.

Furthermore, on Russell's view the type of expression that comes closest to performing the function of the referential use of definite descriptions turns out, as one might suspect, to be a proper name (in 'the narrow logical sense'). Many of the things said about proper names by Russell can, I think, be said about the referential use of definite descriptions without straining senses unduly. Thus the gulf Russell thought he saw between names and definite descriptions is narrower than he thought.

Strawson, on the other hand, certainly does recognize a referential use of definite descriptions. But what I think he did not see is that a definite description may have a quite different role – may be used nonreferentially, even as it occurs in one and the

[a] This paper is reprinted from *Philosophical Review*, LXXV (1966), 281–304.

I should like to thank my colleagues, John Canfield, Sydney Shoemaker, and Timothy Smiley, who read an earlier draft and gave me helpful suggestions. I also had the benefit of the valuable and detailed comments of the referee for the paper, to whom I wish to express my gratitude.
[b] 'On Denoting', reprinted in *Logic and Knowledge*, ed. Robert C. Marsh (London, 1956), p. 51.

same sentence. Strawson, it is true, points out nonreferential uses of definite descriptions,[a] but which use a definite description has seems to be for him a function of the kind of sentence in which it occurs; whereas, if I am right, there can be two possible uses of a definite description in the same sentence. Thus, in 'On Referring', he says, speaking of expressions used to refer, 'Any expression of any of these classes [one being that of definite descriptions] can occur as the subject of what would traditionally be regarded as a singular subject-predicate sentence; and would, so occurring, exemplify the use I wish to discuss.'[b] So the definite description in, say, the sentence 'The Republican candidate for president in 1968 will be a conservative' presumably exemplifies the referential use. But if I am right, we could not say this of the sentence in isolation for some particular occasion on which it is used to state something; and then it might not turn out that the definite description has a referential use.

Strawson and Russell seem to me to make a common assumption here about the question of how definite descriptions function: that we can ask how a definite description functions in some sentence independently of a particular occasion upon which it is used. This assumption is not really rejected in Strawson's arguments against Russell. Although he can sum up his position by saying, '"Mentioning" or "referring" is not something an expression does; it is something that someone can use an expression to do',[c] he means by this to deny the radical view that a 'genuine' referring expression *has* a referent, functions to refer, independent of the context of some use of the expression. The denial of this view, however, does not entail that definite descriptions cannot be identified as referring expressions in a sentence unless the sentence is being used. Just as we can speak of a function of a tool that is not at the moment performing its function, Strawson's view, I believe, allows us to speak of the referential function of a definite description in a sentence even when it is not being used. This, I hope to show, is a mistake.

A second assumption shared by Russell's and Strawson's account of definite descriptions is this. In many cases a person who uses a definite description can be said (in some sense) to presuppose or imply that something fits the description.[d] If I state that the king is on his throne, I presuppose or imply that there is a king. (At any rate, this would be a natural thing to say for anyone who doubted that there is a king.) Both Russell and Strawson assume that where the presupposition or implication is false, the truth value of what the speaker says is affected. For Russell the statement made is false; for Strawson it has no truth value. Now if there are two uses of definite descriptions, it may be that the truth value is affected differently in each case by the falsity of the presupposition or implication. This is what I shall in fact argue. It will turn out, I believe, that one or the other of the two views, Russell's or Strawson's, may be correct about the nonreferential use of definite descriptions, but neither fits the referential use. This is not so surprising about Russell's view, since he did not recognize this use in any case, but it is surprising about Strawson's since the referential use is what he tries to explain and defend. Furthermore, on Strawson's

---

[a] 'On Referring', reprinted in *Philosophy and Ordinary Language*, ed. Charles C. Caton (Urbana, 1963), pp. 162–3.
[b] Ibid. p. 162.   [c] Ibid. p. 170.
[d] Here and elsewhere I use the disjunction 'presuppose or imply' to avoid taking a stand that would side me with Russell or Strawson on the issue of what the relationship involved is. To take a stand here would be beside my main point as well as being misleading, since later on I shall argue that the presupposition or implication arises in a different way depending upon the use to which the definite description is put. This last also accounts for my use of the vagueness indicator, 'in some sense'.

account, the result of there being nothing which fits the description is a failure of reference.[a] This too, I believe, turns out not to be true about the referential use of definite descriptions.

2   There are some uses of definite descriptions which carry neither any hint of a referential use nor any presupposition or implication that something fits the description. In general, it seems, these are recognizable from the sentence frame in which the description occurs. These uses will not interest us, but it is necessary to point them out if only to set them aside.

An obvious example would be the sentence 'The present king of France does not exist', used, say, to correct someone's mistaken impression that de Gaulle is the king of France.

A more interesting example is this. Suppose someone were to ask, 'Is de Gaulle the king of France?' This is the natural form of words for a person to use who is in doubt as to whether de Gaulle is king or president of France. Given this background to the question, there seems to be no presupposition or implication that someone is the king of France. Nor is the person attempting to refer to someone by using the definite description. On the other hand, reverse the name and description in the question and the speaker probably would be thought to presuppose or imply this. 'Is the king of France de Gaulle?' is the natural question for one to ask who wonders whether it is de Gaulle rather than someone else who occupies the throne of France.[b]

Many times, however, the use of a definite description does carry a presupposition or implication that something fits the description. If definite descriptions do have a referring role, it will be here. But it is a mistake, I think, to try, as I believe both Russell and Strawson do, to settle this matter without further ado. What is needed, I believe, is the distinction I will now discuss.

3   I will call the two uses of definite descriptions I have in mind the attributive use and the referential use. A speaker who uses a definite description attributively in an assertion states something about whoever or whatever is the so-and-so. A speaker who uses a definite description referentially in an assertion, on the other hand, uses the description to enable his audience to pick out whom or what he is talking about and states something about that person or thing. In the first case the definite description might be said to occur essentially, for the speaker wishes to assert something about whatever or whoever fits that description; but in the referential use the definite description is merely one tool for doing a certain job – calling attention to a person or thing – and in general any other device for doing the same job, another description or a name, would do as well. In the attributive use, the attribute of being the so-and-so is all important, while it is not in the referential use.

To illustrate this distinction, in the case of a single sentence, consider the sentence, 'Smith's murderer is insane'. Suppose first that we come upon poor Smith foully murdered. From the brutal manner of the killing and the fact that Smith was

---

[a] In a footnote added to the original version of 'On Referring' (op. cit., p. 181) Strawson seems to imply that where the presupposition is false, we still succeed in referring in a 'secondary' way, which seems to mean 'as we could be said to refer to fictional or make-believe things'. But his view is still that we cannot refer in such a case in the 'primary' way. This is, I believe, wrong. For a discussion of this modification of Strawson's view see Charles C. Caton, 'Strawson on Referring', *Mind*, LXVIII (1959), 539–44.

[b] This is an adaptation of an example (used for a somewhat different purpose) given by Leonard Linsky in 'Reference and Referents', reprinted in this volume, pp. 76–85.

the most lovable person in the world, we might exclaim, 'Smith's murderer is insane'. I will assume, to make it a simpler case, that in a quite ordinary sense we do not know who murdered Smith (though this is not in the end essential to the case). This, I shall say, is an attributive use of the definite description.

The contrast with such a use of the sentence is one of those situations in which we expect and intend our audience to realize whom we have in mind when we speak of Smith's murderer and, most importantly, to know that it is this person about whom we are going to say something.

For example, suppose that Jones has been charged with Smith's murder and has been placed on trial. Imagine that there is a discussion of Jones's odd behavior at his trial. We might sum up our impression of his behavior by saying, 'Smith's murderer is insane'. If someone asks to whom we are referring, by using this description, the answer here is 'Jones'. This, I shall say, is a referential use of the definite description.

That these two uses of the definite description in the same sentence are really quite different can perhaps best be brought out by considering the consequences of the assumption that Smith had no murderer (for example, he in fact committed suicide). In both situations, in using the definite description 'Smith's murderer', the speaker in some sense presupposes or implies that there is a murderer. But when we hypothesize that the presupposition or implication is false, there are different results for the two uses. In both cases we have used the predicate 'is insane', but in the first case, if there is no murderer, there is no person of whom it could be correctly said that we attributed insanity to him. Such a person could be identified (correctly) only in case someone fitted the description used. But in the second case, where the definite description is simply a means of identifying the person we want to talk about, it is quite possible for the correct identification to be made even though no one fits the description we used.[a] We were speaking about Jones even though he is not in fact Smith's murderer and, in the circumstances imagined, it was his behavior we were commenting upon. Jones might, for example, accuse us of saying false things of him in calling him insane and it would be no defense, I should think, that our description, 'the murderer of Smith', failed to fit him.

It is, moreover, perfectly possible for our audience to know to whom we refer, in the second situation, even though they do not share our presupposition. A person hearing our comment in the context imagined might know we are talking about Jones even though he does not think Jones guilty.

Generalizing from this case, we can say, I think, that there are two uses of sentences of the form, 'The ϕ is ψ'. In the first, if nothing is the ϕ then nothing has been said to be ψ. In the second, the fact that nothing is the ϕ does not have this consequence.

With suitable changes the same difference in use can be formulated for uses of language other than assertions. Suppose one is at a party and, seeing an interesting-looking person holding a martini glass, one asks, 'Who is the man drinking a martini?' If it should turn out that there is only water in the glass, one has nevertheless asked a question about a particular person, a question that it is possible for someone to answer. Contrast this with the use of the same question by the chairman of the

---

[a] In 'Reference and Referents' (pp. 76–85), Linsky correctly points out that one does not fail to refer simply because the description used does not in fact fit anything (or fits more than one thing). Thus he pinpoints one of the difficulties in Strawson's view. Here, however, I use this fact about referring to make a distinction I believe he does not draw, between two uses of definite descriptions. I later discuss the second passage from Linsky's paper.

local Teetotalers Union. He has just been informed that a man is drinking a
martini at their annual party. He responds by asking his informant, 'Who is the man
drinking a martini?' In asking the question the chairman does not have some particular person in mind about whom he asks the question; if no one is drinking a
martini, if the information is wrong, no person can be singled out as the person about
whom the question was asked. Unlike the first case, the attribute of being the man
drinking a martini is all-important, because if it is the attribute of no one, the chairman's question has no straightforward answer.

This illustrates also another difference between the referential and the attributive use of definite descriptions. In the one case we have asked a question about a
particular person or thing even though nothing fits the description we used; in the
other this is not so. But also in the one case our question can be answered; in the
other it cannot be. In the referential use of a definite description we may succeed in
picking out a person or thing to ask a question about even though he or it does not
really fit the description; but in the attributive use if nothing fits the description, no
straightforward answer to the question can be given.

This further difference is also illustrated by commands or orders containing
definite descriptions. Consider the order, 'Bring me the book on the table.' If 'the
book on the table' is being used referentially, it is possible to fulfill the order even
though there is no book on the table. If, for example, there is a book *beside* the table,
though there is none *on* it, one might bring that book back and ask the issuer of the
order whether this is 'the book you meant'. And it may be. But imagine we are told
that someone has laid a book on our prize antique table, where nothing should be put.
The order, 'Bring me the book on the table' cannot now be obeyed unless there is a
book that has been placed on the table. There is no possibility of bringing back a book
which was never on the table and having it be the one that was meant, because there is
no book that in that sense was 'meant'. In the one case the definite description was a
device for getting the other person to pick the right book; if he is able to pick the
right book even though it does not satisfy the description, one still succeeds in his
purpose. In the other case, there is, antecedently, no 'right book' except one which
fits the description; the attribute of being the book on the table is essential. Not only
is there no book about which an order was issued, if there is no book on the table,
but the order itself cannot be obeyed. When a definite description is used attributively in a command or question and nothing fits the description, the command
cannot be obeyed and the question cannot be answered. This suggests some analogous consequence for assertions containing definite descriptions used attributively.
Perhaps the analogous result is that the assertion is neither true nor false: this is
Strawson's view of what happens when the presupposition of the use of a definite
description is false. But if so, Strawson's view works not for definite descriptions
used referentially, but for the quite different use, which I have called the attributive use.

I have tried to bring out the two uses of definite descriptions by pointing out the
different consequences of supposing that nothing fits the description used. There
are still other differences. One is this: when a definite description is used referentially, not only is there in some sense a presupposition or implication that someone
or something fits the description, as there is also in the attributive use, but there is a
quite different presupposition; the speaker presupposes of some *particular* someone
or something that he or it fits the description. In asking, for example, 'Who is the
man drinking a martini?' where we mean to ask a question about that man over there,
we are presupposing that that man over there is drinking a martini – not just that

someone is a man drinking a martini. When we say, in a context where it is clear we are referring to Jones, 'Smith's murderer is insane', we are presupposing that Jones is Smith's murderer. No such presupposition is present in the attributive use of definite descriptions. There is, of course, the presupposition that someone *or other* did the murder, but the speaker does not presuppose of someone in particular – Jones or Robinson, say – that he did it. What I mean by this second kind of presupposition that someone or something in particular fits the description – which is present in a referential use but not in an attributive use – can perhaps be seen more clearly by considering a member of the speaker's audience who believes that Smith was not murdered at all. Now in the case of the referential use of the description, 'Smith's murderer', he could accuse the speaker of mistakenly presupposing both that some-one or other is the murderer and that also Jones is the murderer, for even though he believes Jones not to have done the deed, he knows that the speaker was referring to Jones. But in the case of the attributive use, he can accuse the speaker of having only the first, less specific presupposition; he cannot pick out some person and claim that the speaker is presupposing that that person is Smith's murderer. Now the more particular presuppositions that we find present in referential uses are clearly not ones we can assign to a definite description in some particular sentence in isolation from a context of use. In order to know that a person presupposes that Jones is Smith's murderer in using the sentence 'Smith's murderer is insane', we have to know that he is using the description referentially and also to whom he is referring. The sentence by itself does not tell us any of this.

4   From the way in which I set up each of the previous examples it might be sup-posed that the important difference between the referential and the attributive use lies in the beliefs of the speaker. Does he believe of some particular person or thing that he or it fits the description used? In the Smith murder example, for instance, there was in the one case no belief as to who did the deed, whereas in the contrasting case it was believed that Jones did it. But this is, in fact, not an essential difference. It is possible for a definite description to be used attributively even though the speaker (and his audience) believes that a certain person or thing fits the description. And it is possible for a definite description to be used referentially where the speaker believes that nothing fits the description. It is true – and this is why, for simplicity, I set up the examples the way I did – that if a speaker does not believe that anything fits the description or does not believe that he is in a position to pick out what does fit the description, it is likely that he is not using it referentially. It is also true that if he and his audience would pick out some particular thing or person as fitting the description, then a use of the definite description is very likely referential. But these are only presumptions and not entailments.

   To use the Smith murder case again, suppose that Jones is on trial for the murder and I and everyone else believe him guilty. Suppose that I comment that the mur-derer of Smith is insane, but instead of backing this up, as in the example previously used, by citing Jones' behavior in the dock, I go on to outline reasons for thinking that *anyone* who murdered poor Smith in that particularly horrible way must be insane. If now it turns out that Jones was not the murderer after all, but someone else was, I think I can claim to have been right if the true murderer is after all in-sane. Here, I think, I would be using the definite description attributively, even though I believe that a particular person fits the description.

   It is also possible to think of cases in which the speaker does not believe that what he means to refer to by using the definite description fits the description, or to imagine cases in which the definite description is used referentially even though the speaker believes *nothing* fits the description. Admittedly, these cases may be parasitic on a more normal use; nevertheless, they are sufficient to show that such beliefs of the speaker are not decisive as to which use is made of a definite description.

Suppose the throne is occupied by a man I firmly believe to be not the king, but a usurper. Imagine also that his followers as firmly believe that he is the king. Suppose I wish to see this man. I might say to his minions, 'Is the king in his counting-house?' I succeed in referring to the man I wish to refer to without myself believing that he fits the description. It is not even necessary, moreover, to suppose that his followers believe him to be the king. If they are cynical about the whole thing, know he is not the king, I may still succeed in referring to the man I wish to refer to. Similarly, neither I nor the people I speak to may suppose that *anyone* is the king and, finally, each party may know that the other does not so suppose and yet the reference may go through.

5   Both the attributive and the referential use of definite descriptions seem to carry a presupposition or implication that there is something which fits the description. But the reasons for the existence of the presupposition or implication are different in the two cases.

There is a presumption that a person who uses a definite description referentially believes that what he wishes to refer to fits the description. Because the purpose of using the description is to get the audience to pick out or think of the right thing or person, one would normally choose a description that he believes the thing or person fits. Normally a misdescription of that to which one wants to refer would mislead the audience. Hence, there is a presumption that the speaker believes *something* fits the description – namely, that to which he refers.

When a definite description is used attributively, however, there is not the same possibility of misdescription. In the example of 'Smith's murderer', used attributively, there was not the possibility of misdescribing Jones or anyone else; we were not referring to Jones nor to anyone else by using the description. The presumption that the speaker believes *someone* is Smith's murderer does not arise here from a more specific presumption that he believes Jones or Robinson or someone else whom he can name or identify is Smith's murderer.

The presupposition or implication is borne by a definite description used attributively because if nothing fits the description the linguistic purpose of the speech act will be thwarted. That is, the speaker will not succeed in saying something true, if he makes an assertion; he will not succeed in asking a question that can be answered, if he has asked a question; he will not succeed in issuing an order that can be obeyed, if he has issued an order. If one states that Smith's murderer is insane, when Smith has no murderer, and uses the definite description nonreferentially, then one fails to say anything *true*. If one issues the order 'Bring me Smith's murderer' under similar circumstances, the order cannot be obeyed; nothing would count as obeying it.

When the definite description is used referentially, on the other hand, the presupposition or implication stems simply from the fact that normally a person tries to describe correctly what he wants to refer to because normally this is the best way to get his audience to recognize what he is referring to. As we have seen, it is possible for the linguistic purpose of the speech act to be accomplished in such a case even though nothing fits the description; it is possible to say something true or to ask a question that gets answered or to issue a command that gets obeyed. For when the definite description is used referentially, one's audience may succeed in seeing to what one refers even though neither it nor anything else fits the description.

6 The result of the last section shows something to be wrong with the theories of both Russell and Strawson; for though they give differing accounts of the implication or presupposition involved, each gives only one. Yet, as I have argued, the presupposition or implication is present for a quite different reason, depending upon whether the definite description is used attributively or referentially, and exactly what presuppositions or implications are involved is also different. Moreover, neither theory seems a correct characterization of the referential use. On Russell's there is a logical entailment: 'The ϕ is ψ' entails 'There exists one and only one ϕ.' Whether or not this is so for the attributive use, it does not seem true of the referential use of the definite description. The 'implication' that something is the ϕ, as I have argued, does not amount to an entailment; it is more like a presumption based on what is *usually* true of the use of a definite description to refer. In any case, of course, Russell's theory does not show – what is true of the referential use – that the implication that *something* is the ϕ comes from the more specific implication that *what is being referred to* is the ϕ. Hence, as a theory of definite descriptions, Russell's view seems to apply, if at all, to the attributive use only.

Russell's definition of denoting (a definite description denotes an entity if that entity fits the description uniquely) is clearly applicable to either use of definite descriptions. Thus whether or not a definite description is used referentially or attributively, it may have a denotation. Hence, denoting and referring, as I have explicated the latter notion, are distinct and Russell's view recognizes only the former. It seems to me, moreover, that this is a welcome result, that denoting and referring should not be confused. If one tried to maintain that they are the same notion, one result would be that a speaker might be referring to something without knowing it. If someone said, for example, in 1960 before he had any idea that Mr Goldwater would be the Republican nominee in 1964, 'The Republican candidate for president in 1964 will be a conservative' (perhaps on the basis of an analysis of the views of party leaders), the definite description here would *denote* Mr Goldwater. But would we wish to say that the speaker had referred to, mentioned, or talked about Mr Goldwater? I feel these terms would be out of place. Yet if we identify referring and denoting, it ought to be possible for it to turn out (after the Republican Convention) that the speaker had, unknown to himself, referred in 1960 to Mr Goldwater. On my view, however, while the definite description used did *denote* Mr Goldwater (using Russell's definition), the speaker used it *attributively* and did not *refer* to Mr Goldwater.

Turning to Strawson's theory, it was supposed to demonstrate how definite descriptions are referential. But it goes too far in this direction. For there are non-referential uses of definite descriptions also, even as they occur in one and the same sentence. I believe that Strawson's theory involves the following propositions:

(1) If someone asserts that the ϕ is ψ he has not made a true or false statement if if there is no ϕ.[a]

[a] In 'A Reply to Mr Sellars', *Philosophical Review*, LXIII (1954), 216–31, Strawson admits that we do not always refuse to ascribe truth to what a person says when the definite description he uses fails to fit anything (or fits more than one thing). To cite one of his examples, a

(2) If there is no ϕ then the speaker has failed to refer to anything.[a]

(3) The reason he has said nothing true or false is that he has failed to refer.

Each of these propositions is either false or, at best, applies to only one of the two uses of definite descriptions.

Proposition (1) is possibly true of the attributive use. In the example in which 'Smith's murderer is insane' was said when Smith's body was first discovered, an attributive use of the definite description, there was no person to whom the speaker referred. If Smith has no murderer, nothing true was said. It is quite tempting to conclude, following Strawson, that nothing true *or* false was said. But where the definite description is used referentially, something true may well have been said. It is possible that something true was said of the person or thing referred to.[b]

Proposition (2) is, as we have seen, simply false. Where a definite description is used referentially it is perfectly possible to refer to something though nothing fits the description used.

The situation with proposition (3) is a bit more complicated. It ties together, on Strawson's view, the two strands given in (1) and (2). As an account of why, when the presupposition is false, nothing true or false has been stated, it clearly cannot work for the attributive use of definite descriptions, for the reason it supplies is that reference has failed. It does not then give the reason why, if indeed this is so, a speaker using a definite description attributively fails to say anything true or false if

person who said, 'The United States Chamber of Deputies contains representatives of two major parties' would be allowed to have said something true even though he had used the wrong title. Strawson thinks this does not constitute a genuine problem for his view. He thinks that what we do in such cases, 'where the speaker's intended reference is pretty clear, is simply to amend his statement in accordance with his guessed intentions and assess the amended statement for truth or falsity; we are not awarding a truth value at all to the original statement' (p. 230).

The notion of an 'amended statement', however, will not do. We may note, first of all, that the sort of case Strawson has in mind could arise only when a definite description is used referentially. For the 'amendment' is made by seeing the speaker's intended reference. But this could happen only if the speaker had an intended reference, a particular person or thing in mind, independent of the description he used. The cases Strawson has in mind are presumably not cases of slips of the tongue or the like; presumably they are cases in which a definite description is used because the speaker believes, though he is mistaken, that he is describing correctly what he wants to refer to. We supposedly amend the statement by knowing to what he intends to refer. But what description is to be used in the amended statement? In the example, perhaps, we could use 'the United States Congress'. But this description might be one the speaker would not even accept as correctly describing what he wants to refer to, because he is misinformed about the correct title. Hence, this is not a case of deciding what the speaker meant to say as opposed to what he in fact said, for the speaker did not mean to say 'the United States Congress'. If this is so, then there is no bar to the 'amended' statement containing any description that does correctly pick out what the speaker intended to refer to. It could be, e.g., 'The lower house of the United States Congress'. But this means that there is no one unique 'amended' statement to be assessed for truth value. And, in fact, it should now be clear that the notion of the amended statement really plays no role anyway. For if we can arrive at the amended statement only by first knowing to what the speaker intended to refer, we can assess the truth of what he said simply by deciding whether what he intended to refer to has the properties he ascribed to it.

[a] As noted earlier (p. 102, n. *a*), Strawson may allow that one has possibly referred in a 'secondary' way, but, if I am right, the fact that there is no ϕ does not preclude one from having referred in the same way one does if there is a ϕ.

[b] For a further discussion of the notion of saying something true *of* someone or something, see section 8.

nothing fits the description. It does, however, raise a question about the referential use. Can reference fail when a definite description is used referentially?

I do not fail to refer merely because my audience does not correctly pick out what I am referring to. I can be referring to a particular man when I use the description 'the man drinking a martini', even though the people to whom I speak fail to pick out the right person or any person at all. Nor, as we have stressed, do I fail to refer when nothing fits the description. But perhaps I fail to refer in some extreme circumstances, when there is nothing that *I* am willing to pick out as that to which I referred.

Suppose that I think I see at some distance a man walking and ask, 'Is the man carrying a walking stick the professor of history?' We should perhaps distinguish four cases at this point. (*a*) There is a man carrying a walking stick; I have then referred to a person and asked a question about him that can be answered if my audience has the information. (*b*) The man over there is not carrying a walking stick, but an umbrella; I have still referred to someone and asked a question that can be answered, though if my audience sees that it is an umbrella and not a walking stick, they may also correct my apparently mistaken impression. (*c*) It is not a man at all, but a rock that looks like one; in this case, I think I still have referred to something, to the thing over there that happens to be a rock but that I took to be a man. But in this case it is not clear that my question can be answered correctly. This, I think, is not because I have failed to refer, but rather because, given the true nature of what I referred to, my question is not appropriate. A simple 'No, that is not the professor of history' is at least a bit misleading if said by someone who realizes that I mistook a rock for a person. It may, therefore, be plausible to conclude that in such a case I have not asked a question to which there is a straightforwardly correct answer. But if this is true, it is not because nothing fits the description I used, but rather because what I referred to is a rock and my question has no correct answer when asked of a rock. (*d*) There is finally the case in which there is nothing at all where I thought there was a man with a walking stick; and perhaps here we have a genuine failure to refer at all, even though the description was used for the purpose of referring. There is no rock, nor anything else, to which I meant to refer; it was, perhaps, a trick of light that made me think there was a man there. I cannot say of anything, 'That is what I was referring to, though I now see that it's not a man carrying a walking stick'. This failure of reference, however, requires circumstances much more radical than the mere nonexistence of anything fitting the description used. It requires that there be nothing of which it can be said, 'That is what he was referring to'. Now perhaps also in such cases, if the speaker has asserted something, he fails to state anything true or false if there is nothing that can be identified as that to which he referred. But if so, the failure of reference and truth value does not come about merely because nothing fits the description he used. So (3) may be true of some cases of the referential use of definite descriptions; it may be true that a failure of reference results in a lack of truth value. But these cases are of a much more extreme sort than Strawson's theory implies.

I conclude, then, that neither Russell's nor Strawson's theory represents a correct account of the use of definite descriptions – Russell's because it ignores altogether the referential use, Strawson's because it fails to make the distinction between the referential and the attributive and mixes together truths about each (together with some things that are false).

7   It does not seem possible to say categorically of a definite description in a particular sentence that it is a referring expression (of course, one could say this if he meant that it *might* be used to refer). In general, whether or not a definite description is used referentially or attributively is a function of the speaker's intentions in a particular case. 'The murderer of Smith' may be used either way in the sentence 'The murderer of Smith is insane.' It does not appear plausible to account for this, either, as an ambiguity in the sentence. The grammatical structure of the sentence seems to me to be the same whether the description is used referentially or attributively: that is, it is not syntactically ambiguous. Nor does it seem at all attractive to suppose an ambiguity in the meaning of the words; it does not appear to be semantically ambiguous. (Perhaps we could say that the sentence is pragmatically ambiguous: the distinction between roles that the description plays is a function of the speaker's intentions.) These, of course, are intuitions; I do not have an argument for these conclusions. Nevertheless, the burden of proof is surely on the other side.

This, I think, means that the view, for example, that sentences can be divided up into predicates, logical operators, and referring expressions is not generally true. In the case of definite descriptions one cannot always assign the referential function in isolation from a particular occasion on which it is used.

There may be sentences in which a definite description can be used only attributively or only referentially. A sentence in which it seems that the definite description could be used only attributively would be 'Point out the man who is drinking my martini', I am not so certain that any can be found in which the definite description can be used only referentially. Even if there are such sentences, it does not spoil the point that there are many sentences, apparently not ambiguous either syntactically or semantically, containing definite descriptions that can be used either way.

If it could be shown that the dual use of definite descriptions can be accounted for by the presence of an ambiguity, there is still a point to be made against the theories of Strawson and Russell. For neither, so far as I can see, has anything to say about the possibility of such an ambiguity and, in fact, neither seems compatible with such a possibility. Russell's does not recognize the possibility of the referring use, and Strawson's, as I have tried to show in the last section, combines elements from each use into one unitary account. Thus the view that there is an ambiguity in such sentences does not seem any more attractive to these positions.

8   Using a definite description referentially, a speaker may say something true even though the description correctly applies to nothing. The sense in which he may say something true is the sense in which he may say something true about someone or something. This sense is, I think, an interesting one that needs investigation. Isolating it is one of the by-products of the distinction between the attributive and referential uses of definite descriptions.

For one thing, it raises questions about the notion of a statement. This is brought out by considering a passage in a paper by Leonard Linsky in which he rightly makes the point that one can refer to someone although the definite description used does not correctly describe the person:

said of a spinster, 'Her husband is kind to her' is neither true nor false. But a speaker might very well be referring to someone in using these words, for he may think that someone is the husband of the lady (who in fact is a spinster). Still, the statement is neither true nor false, for it presupposes that the lady has a husband, which she has not.

This last refutes Strawson's thesis that if the presupposition of existence is not satisfied the speaker has failed to refer.[a]

There is much that is right in this passage. But because Linsky does not make the distinction between the referential and the attributive uses of definite descriptions, it does not represent a wholly adequate account of the situation. A perhaps minor point about this passage is that Linsky apparently thinks it sufficient to establish that the speaker in his example is referring to someone by using the definite description 'her husband', that he *believe* that someone is her husband. This will only approximate the truth provided that the 'someone' in the description of the belief means 'someone in particular' and is not merely the existential quantifier, 'there is someone or other'. For in both the attributive and the referential use the belief that someone *or other* is the husband of the lady is very likely to be present. If, for example, the speaker has just met the lady and, noticing her cheerfulness and radiant good health, makes his remark from his conviction that these attributes are always the result of having good husbands, he would be using the definite description attributively. Since she has no husband, there is no one to pick out as the person to whom he was referring. Nevertheless, the speaker believed that *someone or other* was her husband. On the other hand, if the use of 'her husband' was simply a way of referring to a man the speaker has just met whom he assumed to be the lady's husband, he would have referred to that man even though neither he nor anyone else fits the description. I think it is likely that in this passage Linsky did mean by 'someone', in his description of the belief, 'someone in particular'. But even then, as we have seen, we have neither a sufficient nor a necessary condition for a referential use of the definite description. A definite description can be used attributively even when the speaker believes that some particular thing or person fits the description, and it can be used referentially in the absence of this belief.

My main point, here, however, has to do with Linsky's view that because the preposition is not satisfied, the *statement* is neither true nor false. This seems to me possibly correct *if* the definite description is thought of as being used attributively (depending upon whether we go with Strawson or Russell). But when we consider it as used referentially, this categorical assertion is no longer clearly correct. For the man the speaker referred to may indeed be kind to the spinster; the speaker may have said something true about that man. Now the difficulty is in the notion of 'the statement'. Suppose that we know that the lady is a spinster, but nevertheless know that the man referred to by the speaker is kind to her. It seems to me that we shall, on the one hand, want to hold that the speaker said something true, but be reluctant to express this by 'It is true that her husband is kind to her.'

This shows, I think, a difficulty in speaking simply about 'the statement' when definite descriptions are used referentially. For the speaker stated something, in this example, about a particular person, and his statement, we may suppose, was true. Nevertheless, we should not like to agree with his statement by using the sentence he used; we should not like to identify the true statement via the speaker's words.

[a] 'Reference and Referents', p. 81. It should be clear that I agree with Linsky in holding that a speaker may refer even though the 'presupposition of existence' is not satisfied. And I agree in thinking this an objection to Strawson's view. I think, however, that this point, among others, can be used to define two distinct uses of definite descriptions which, in turn, yields a more general criticism of Strawson. So, while I develop here a point of difference, which grows out of the distinction I want to make, I find myself in agreement with much of Linsky's article.

The reason for this is not so hard to find. If we say, in this example, 'It is true that her husband is kind to her', *we* are now using the definite description either attributively or referentially. But we should not be subscribing to what the original speaker truly said if we use the description attributively, for it was only in its function as referring to a particular person that the definite description yields the possibility of saying something true (since the lady has no husband). Our reluctance, however, to endorse the original speaker's statement by using the definite description referentially to refer to the same person stems from a quite different consideration. For if we too were laboring under the mistaken belief that this man was the lady's husband, we could agree with the original speaker using his exact words. (Moreover, it is possible, as we have seen, deliberately to use a definite description to refer to someone we believe not to fit the description.) Hence, our reluctance to use the original speaker's words does not arise from the fact that if we did we should not succeed in stating anything true or false. It rather stems from the fact that when a definite description is used referentially there is a presumption that the speaker believes that what he refers to fits the description. Since we, who know the lady to be a spinster, would not normally want to give the impression that we believe otherwise, we would not like to use the original speaker's way of referring to the man in question.

How then would we express agreement with the original speaker without involving ourselves in unwanted impressions about our beliefs? The answer shows another difference between the referential and attributive uses of definite descriptions and brings out an important point about genuine referring.

When a speaker says, 'The ϕ is ψ', where 'the ϕ' is used attributively, if there is no ϕ, we cannot correctly report the speaker as having said *of* this or that person or thing that it is ψ. But if the definite description is used referentially we can report the speaker as having attributed ψ to something. And *we* may refer to what the speaker referred to, using whatever description or name suits our purpose. Thus, if a speaker says, 'Her husband is kind to her', referring to the man he was just talking to, and if that man is Jones, we may report him as having said *of Jones* that he is kind to her. If Jones is also the president of the college, we may report the speaker as having said *of the president of the college* that he is kind to her. And finally, if we are talking to Jones, we may say, referring to the original speaker, 'He said of you that *you* are kind to her'. It does not matter here whether or not the woman has a husband or whether, if she does, Jones is her husband. If the original speaker referred to Jones, he said of him that he is kind to her. Thus where the definite description is used referentially, but does not fit what was referred to, we can report what a speaker said and agree with him by using a description or name which does fit. In doing so we need not, it is important to note, choose a description or name which the original speaker would agree fits what he was referring to. That is, we can report the speaker in the above case to have said truly of Jones that he is kind to her even if the original speaker did not know that the man he was referring to is named Jones or even if he thinks he is not named Jones.

Returning to what Linsky said in the passage quoted, he claimed that, were someone to say 'Her husband is kind to her', when she has no husband, *the statement* would be neither true nor false. As I have said, this is a likely view to hold if the definite description is being used attributively. But if it is being used referentially it is not clear what is meant by 'the statement'. If we think about what the speaker said about the person he referred to, then there is no reason to suppose he has not said something true or false about him, even though he is not the lady's husband. And

Linsky's claim would be wrong. On the other hand, if we do not identify the statement in this way, what is the statement that the speaker made? To say that the statement he made was that her husband is kind to her lands us in difficulties. For we have to decide whether in using the definite description here in the identification of the statement, we are using it attributively or referentially. If the former, then we misrepresent the linguistic performance of the speaker; if the latter, then we are ourselves referring to someone and reporting the speaker to have said something of that person, in which case we are back to the possibility that he did say something true or false of that person.

I am thus drawn to the conclusion that when a speaker uses a definite description referentially he may have stated something true or false even if nothing fits the description, and that there is not a clear sense in which he has made a statement which is neither true nor false.

9 I want to end by a brief examination of a picture of what a genuine referring expression is that one might derive from Russell's views. I want to suggest that this picture is not so far wrong as one might suppose and that strange as this may seem, some of the things we have said about the referential use of definite descriptions are not foreign to this picture.

Genuine proper names, in Russell's sense, would refer to something without ascribing any properties to it. They would, one might say, refer to the thing itself, not simply the thing in so far as it falls under a certain description.[a] Now this would seem to Russell something a definite description could not do, for he assumed that if definite descriptions were capable of referring at all, they would refer to something only in so far as that thing satisfied the description. Not only have we seen this assumption to be false, however, but in the last section we saw something more. We saw that when a definite description is used referentially, a speaker can be reported as having said something *of* something. And in reporting what it was of which he said something we are not restricted to the description he used, or synonyms of it; we may ourselves refer to it using any descriptions, names, and so forth, that will do the job. Now this seems to give a sense in which we are concerned with the thing itself and not just the thing under a certain description, when we report the linguistic act of a speaker using a definite description referentially. That is, such a definite description comes closer to performing the function of Russell's proper names than certainly he supposed.

Secondly, Russell thought, I believe, that whenever we use descriptions, as opposed to proper names, we introduce an element of generality which ought to be absent if what we are doing is referring to some particular thing. This is clear from his analysis of sentences containing definite descriptions. One of the conclusions we are supposed to draw from that analysis is that such sentences express what are in reality completely general propositions: there is a ϕ and only one such and any ϕ is ψ. We might put this in a slightly different way. If there is anything which might be identified as reference here, it is reference in a very weak sense – namely, reference to *whatever* is the one and only one ϕ, if there is any such. Now this is something we might well say about the attributive use of definite descriptions, as should be evident from the previous discussion. But this lack of particularity is absent from the referential use of definite descriptions precisely because the description is here

---

[a] Cf. 'The Philosophy of Logical Atomism', reprinted in *Logic and Knowledge*, p. 200.

merely a device for getting one's audience to pick out or think of the thing to be spoken about, a device which may serve its function even if the description is incorrect. More importantly perhaps, in the referential use as opposed to the attributive, there is a *right* thing to be picked out by the audience and its being the right thing is not simply a function of its fitting the description.

Cornell University

Singular terms*a*

ZENO VENDLER

1 The attempt to understand the nature of singular terms has been one of the permanent preoccupations of analytic philosophy, and the theory of descriptions is often mentioned as perhaps the most obvious triumph of that philosophy. As we read the many pages that Russell, Quine, Geach, Strawson, and others have devoted to this topic, and as we follow them in tracing the problems it raises, we cannot but agree with this concern. Perhaps the most important use of language is the stating of facts, and in order to understand this role one has to know how proper names function and what constitutes a definite description, one has to be clear about what we do when we refer to something, in particular whether in doing so we assert or only presuppose the existence of a thing, and, finally, one has to know what kind of existence is involved in the various situations.

As I have just implied, the collective effort of the philosophers in this case has been successful. In spite of some disagreements the results fundamentally converge and give us a fairly illuminating picture of the linguistic make-up and logical status of singular terms. This is a surprising fact. My expression of surprise, however, is intended as a tribute rather than an insult: I am amazed at how much these authors have got out of the precious little at their disposal. A few and often incorrect linguistic data obscured by an archaic grammar were more often than not all they had to start with. Yet their conclusions, as we shall see, anticipate in substance the findings of the advanced grammatical theory of today. Of course, they had their intuitions and the apparatus of formal logic. But the former often mislead and the latter tends to oversimplify. In this case the combination produced happy results, many of which will be confirmed in this paper on the basis of strictly linguistic considerations.

2 I intend to proceed in an expository rather than polemical fashion. To begin with, I shall try to indicate the importance of singular terms for logical theory; then I shall outline the linguistic features marking such terms; and finally I shall use these results to assess the validity of certain philosophical claims.

Some philosophers regard terms as purely linguistic entities – that is, as parts of sentences or logical formulae – while others consider them as elements of certain nonlinguistic entities called propositions.*b* Since my concern, at least at the beginning, is primarily linguistic, I shall use the word *term* in accordance with the first

---

a This paper is reprinted from Vendler, *Linguistics in Philosophy*, Ithaca, Cornell University Press, 1967, pp. 33–69.

b 'Quine uses the expression "term" in application to linguistic items only, whereas I apply it to non-linguistic items' (P. F. Strawson, *Individuals*, p. 154 n).

alternative, that is, to denote a string of words of a certain type or its equivalent in logical notation. This procedure, however, is not intended to prejudice the issue. We shall be led, in fact, by the natural course of our investigations, to a view some-what different from the first alternative.

3 The word *term* belongs to the logician's and not to the linguist's vocabulary. Although the use of *term* is not quite uniform, most logicians would agree with the following approximation. The result of the logical analysis of a proposition consists of the logical form and of the terms that fit into this form. These latter have no structure of their own; they are 'atomic' elements, being, as it were, the parameters in the logical equation. But this simplicity is relative: it may happen that a term left intact at a certain level of analysis will require further resolution at a more advanced level. Russell's analysis of definite descriptions and Quine's elimination of singular terms can serve as classic examples of such a move.[a] To give a simpler illustration, while in the argument

>   All philistines hated Socrates
>   Some Athenians were philistines
> ∴ Some Athenians hated Socrates

the expression *hated Socrates* need not be analysed, that is, may be regarded as one term for the purposes of syllogistic logic, in the equally valid argument

>   All philistines hated Socrates
>   Socrates was an Athenian
> ∴ Some Athenian was hated by all philistines

the expression *hated Socrates* has to be split to show validity by means of the theory of quantification.

The logical forms available to simple syllogistic logic treat all terms in a uniform fashion: any term can have universal or particular 'quantity' depending upon the quantifier (*all, some*), the 'quality' of the proposition (affirmative or negative), and the position of the term (subject or predicate). It is in the theory of quantification that the distinction between singular and general terms becomes explicit. For one thing, the schemata themselves may provide for such a distinction. Consider the second argument given above. It can be represented as follows:

$$(x) (Px \supset Hxs)$$
$$As$$
$$\therefore (\exists x) [Ax.(y) (Py \supset Hyx)]$$

Notice that the argument will not work if *Socrates* is treated like the other terms (*philistine, hated, Athenian*). Such a treatment might amount to the following:

$$(x) [Px \supset (\exists y) (Sy.Hxy)]$$
$$(\exists x) (Sx.Ax)$$
$$\therefore (\exists x) [Ax. (y) (Py \supset Hyx)]$$

This argument, of course, is not valid. Nevertheless, as Quine stipulated, *Socrates* may be represented as a term on par with the rest, provided a uniqueness clause is added to the premises, that is

$$(x) (y) (Sx.Sy. \supset x = y)$$

[a] See, for example, B. Russell, 'Descriptions', ch. 16 in *Introduction to Mathematical Philosophy*, pp. 167–80; W. V. Quine, *Methods of Logic*, pp. 220 ff.

Quine's proposal thus restores the homogeneity of terms characteristic of syllogistic logic: singularity or generality becomes a function of the logical form alone. Yet, in any case, whether the logician is inclined to follow Quine or not, he at least has to realize the difference between terms like *Socrates* and terms like *philistine* or *Athenian*, and must either represent the former by an individual constant or, if he prefers homogeneity and treats it as a predicate, then add the uniqueness clause. Then the question arises how to recognize terms that require such a special consideration, in a word, how to recognize singular terms. The possibility of an 'ideal' language without such terms will not excuse the logician from facing this problem if he intends to use his system to interpret propositions formulated in a natural language.

The linguistic considerations relevant to the solution of this problem are by no means restricted to the morphology of the term in question; often the whole sentence, together with its transforms or even its textual and pragmatic environment must also be considered. Granted, a logician who is a fluent speaker of the language is usually able to make a decision without explicit knowledge of the relevant factors. Such an intuition, however, cannot be used to support philosophical claims about singular terms with any authority. To provide such support and to make our intuitions explicit one has to review the 'natural history' of singular terms in English, to which task I shall address myself in the following sections.

4 It is not an accident that in giving an example of a singular term I selected a proper name, *Socrates*; proper names are traditionally regarded as paradigms of singular terms. Owing to a fortunate convention of modern English spelling, proper names, when written, wear their credentials on their sleeves. This, however, is hardly a criterion. Many adjectives, like *English*, have to be capitalized too. Moreover, while this convention might aid the reader, it certainly does not help, in the absence of a capitalization morpheme, the listener or the writer. Thus we had better remind ourselves of the linguists' *dictum* that language is the spoken language, and look for some real marks.

First we might fall back on the intuition that proper names have no meaning (in the sense of 'sense' and not of 'reference'), which is borne out by the fact that they do not require translation into another language. *Vienna* is the English version and not the English translation of the German name *Wien*. Accordingly, dictionaries do not list proper names; knowledge of proper names does not belong to the knowledge of a language. In linguistic terms this intuition amounts to the following: proper names have no specific co-occurrence restrictions.[a] A simple example will illustrate this.

⟨1⟩ I visited Providence

is a correct sentence, but

⟨2⟩ *I visited providence

is not (here I make use of the above-mentioned convention). The word *providence* has fairly strict co-occurrence restrictions, which exclude contexts like ⟨2⟩. The morphologically identical name in ⟨1⟩, however, waives these restrictions and permits to co-occurrence with *I visited*...Of course, our knowledge that Providence is, in fact, a city will impose other restrictions. This piece of knowledge, however,

---

[a] On the notion of co-occurrence see Z. S. Harris, 'Co-occurrence and Transformation in Linguistic Structure', *Language*, xxxiii (1957), 283–340.

belongs to geography and not linguistics. That is to say, while it belongs to the understanding of the word *providence* that it cannot occur in sentence like ⟨2⟩, it is not the understanding of the name *Providence* that permits ⟨1⟩, but the knowledge that it happens to be the name of a city. From a linguistic point of view, proper names have no restrictions of occurrence beyond the broad grammatical constraints governing noun phrases in general. Indeed, only some proper names show a morphological identity with significant words; and this coincidence is of a mere historical interest: *Providence*, as a name, is no more significant than *Pawtucket*. For these reasons some linguists regard all proper names as a single morpheme. The naming of cats may be a difficult matter, but it does not enrich the language.

A little reflection will show that the very incomprehensibility of the proper names that do not coincide morphologically with significant words, and the absence of specific co-occurrence restrictions with those that do, form a valuable clue in recognizing proper names in spoken discourse. But this mark applies to proper names only and casts little light in general on the nature of singular terms, most of which are not proper names. There are, however, other characteristics marking the occurrence of proper names that will lead us to the very essence of singular terms.

5   Names, as I implied above, fit into noun-phrase slots. And most of them can occur there without any additional apparatus, unlike the vast majority of common nouns, which, at least in the singular, require an article or its equivalent. The sentence

*I visited city

lacks an article, but ⟨1⟩ above does not. Some common nouns, too, can occur without an article. This is true of the so-called 'mass' nouns and 'abstract' nouns. For instance:

I drink water.
Love is a many-splendored thing.

Yet these nouns, too, can take the definite article, at least when accompanied by certain 'adjuncts' (italicized) in the same noun phrase:[a]

I see the water *in the glass*.
The love *she felt for him* was great.[b]

Later on I shall elaborate on the role of adjuncts like *in the glass* and *she felt for him*. For the time being I merely express the intuition that these adjuncts, in some sense or other, restrict the application of the nouns in question; *in the glass* indicates a definite bulk of water, *she felt for him* individuates love.

This intuition gains in force as we note that such adjuncts and the definite article are repugnant to proper names, or, if we force the issue, they destroy the very nature of such names. First of all, there is something unusual about noun phrases like

⟨3⟩ the Joe in our house
⟨4⟩ the Margaret you see.

---

[a]   The technical notion, 'the phrase *x* is an adjunct of the phrase *y*', roughly corresponds to the intuitive notion of one phrase 'modifying' another. See Z. S. Harris, *String Analysis of Sentence Structure*, pp. 9 ff.

[b]   Mass nouns can take the indefinite article only in explicit or implicit combination with 'measure' nouns: *a pound of meat*, *a cup of coffee*; phrases like *a coffee*, are products of an obvious deletion: *a [cup of] coffee*.

And, notice, this oddity is not due to co-occurrence restrictions:

⟨5⟩ Joe is in our house
⟨6⟩ You see Margaret

are perfectly natural sentences. The point is that while sentences like

I see a man
Water is in the glass
He feels hatred

yield noun phrases like

the man I see
the water in the glass
the hatred he feels

sentences like ⟨5⟩ and ⟨6⟩ only reluctantly yield phrases like ⟨3⟩ and ⟨4⟩. Nevertheless such phrases do occur and we understand them. It is clear, however, that such a context is fatal to the name as a proper name, at least for the discourse in which it occurs. The full context, explicit or implicit, will be of the following sort:

The Joe in our house is not the one you are talking about.
The Margaret you see is a guest, the Margaret I mentioned is my sister.

As the noun replacer, *one*, in the first sentence makes abundantly clear, the names here simulate the status of a count noun: there are two Joes and two Margarets presupposed in the discourse, and this is, of course, inconsistent with the idea of a logically proper name. *Joe* and *Margaret* are here really equivalent to something like *person called Joe* or *person called Margaret*, and because these phrases fit many individuals, they should be treated as general terms by the logician.

Certain names, moreover, can be used to function as count nouns in a less trivial sense:

Joe is not *a* Shakespeare.
Amsterdam is *the* Venice of the North.
These little Napoleon*s* caused the trouble in Paraguay.

Here again we can rely upon the grammatical setting to recognize them as count nouns, albeit of a peculiar ancestry.

It is harder to deal with another case of proper names with restrictive adjunct and article. I do not want to claim that the names in sentences like

The Providence you know is no more
You will see a revived Boston
He prefers the early Mozart

have ceased to be proper names. Still less would I cast doubt on the credentials of proper names that seem to require the definite article, like *the Hudson*, *the Bronx*, *the Cambrian*, and so forth. The difficulties posed by these two kinds require more advanced linguistic considerations, so I shall deal with them at a later stage.

Disregarding such peripheral exceptions, we may conclude that proper names are like mass nouns in refusing the indefinite article, but are unlike them in refusing the definite article as well. And the reason seems to be that while even mass nouns or abstract nouns can take *the* when accompanied by certain restrictive adjuncts, proper names cannot take *the* since such adjuncts themselves are incompatible with proper names. Clearly, then, the intuitive notion that a proper name, as such, uniquely refers to one and only one individual has the impossibility of restrictive adjuncts as a linguistic counterpart. To put it bluntly, what is restricted to one cannot be further restricted. A proper name, therefore, is a noun which has no specific co-occurrence restrictions and which precludes restrictive adjuncts and, consequently, articles of any kind in the same noun phrase.

6 This latter point receives a beautiful confirmation as we turn our attention to a small class of other nouns that are also taken to be uniquely referring. These are the pronouns, *I, you, he, she,* and *it.*[a] The impossibility of adding restrictive adjuncts and the definite article is even more marked here than in the case of proper names. Yet, once more, this is not due to co-occurrence restrictions; there is nothing wrong with sentences such as

I am in the room.
I see you.

But they will not yield the noun phrases

*(the) I in the room
*(the) you I see

which they would yield were the pronouns replaced by common nouns like *a man* or *water*. There is an even more striking point. Neither these pronouns nor proper names can ordinarily take prenominal adjectives. From the sentences

He is bald
She is dirty

we cannot get

*bald he
*dirty she.

And even from

Joe is bald
Margaret is dirty

we need poetic licence to obtain

bald Joe
dirty Margaret.

True, we use 'Homeric' epithets, like

lightfooted Achilles
tiny Alice

and, in an emotive tone, we say things like

poor Joe

or even

poor she
miserable you

[a] *We, you,* and *they* are used to refer to unique *groups* of individuals. Here, as in the sequel, I restrict myself to the discussion of definite noun phrases in the singular. It is clear, however, that our findings will apply *mutatis mutandis* to definite noun phrases in the plural as well: *those houses, our dogs, the children you see,* and so forth. From a logical point of view these phrases show a greater affinity to singular than to general terms. See P. F. Strawson, 'On Referring', *Mind,* LIX (1950), 343–4.

but such a pattern is neither common nor universally productive. These facts seem to suggest that prenominal adjectives are also restrictive adjuncts. Later on we shall be able to confirm this impression.

7 'A grammar book of a language is, in part, a treatise on the different styles of introduction of terms into remarks by means of expressions of that language.'[a] Adopting for a moment Strawson's terminology, we can say that proper names and singular pronouns introduce singular terms by themselves without any specific style or additional linguistic apparatus. These nouns are, in fact, allergic to the restrictive apparatus which other nouns need to introduce singular terms, or, reverting to our own way of talking, the restrictive apparatus which other nouns need to become singular terms. In this section I shall take up the task of the grammar book and investigate in detail the natural history of singular terms formed out of common nouns. My paradigms will be count-nouns, simply because they show the full scope of the restrictive apparatus of the language.

It does not require much grammatical sophistication to detect the main categories of singular terms formed out of common nouns. They will begin with a demonstrative pronoun, possessive pronoun, or the definite article – for instance, *this table, your house, the dog*. The first two kinds may be identifying by themselves, but not the third. This can be shown in a simple example. Someone says,

A house has burned down.

We ask,

Which house?

The answers

That house
Your house

may be sufficient in a given situation. The simple answer

The house

is not. *The* alone is not enough. We have to add an adjunct that lends identifying force – for instance:

The house you sold yesterday
The house in which we lived last year.

Nevertheless, in certain contexts *the* alone seems to identify. Consider the following sequence:

I saw a man. The man wore a hat.

Obviously, the man *I saw* wore a hat. *The*, here, indicates a deleted but recoverable restrictive adjunct based upon a previous occurrence of the same noun in an identifying context. This possibility, following upon our previous findings concerning *the*, suggests a hypothesis of fundamental importance: the definite article in front of a noun is always and infallibly the sign of a restrictive adjunct, present or recoverable, attached to the noun. The proof of this hypothesis will require a somewhat technical discussion of restrictive adjuncts. But *the*, according to Russell, is 'a word of very great importance', worth investigating even in prison or dead from the waist down.[b]

[a] P. F. Strawson, *Individuals*, p. 147.
[b] Russell, 'Descriptions', p. 167.

8 My first task, then, is to give a precise equivalent for the intuitive notion of a restrictive adjunct. I claim that all such adjuncts can be reduced to what the grammarians call the restrictive relative clause. With respect to many of the examples used thus far the reconstruction of the relative clause is a simple matter indeed. All we have to know is that the relative pronoun – *which*, *who*, *that*, and so on – can be omitted between two noun phrases, and that the relative pronoun plus the copula can be omitted between a noun phrase and a string consisting of a preposition and a noun. Thus we can complete the full relative clauses in our familiar examples:

I see the water (which is) in the glass
The love (which) she felt for him was great
The man (whom) I saw wore a hat
The house (which) you sold yesterday has burned down

and so forth. If the conditions just given are not satisfied, *wh*. . .or *wh*. . .*is* cannot be omitted:

The man *who* came in is my brother.
The house *which is* burning is yours.

The reduction of prenominal adjectives to relative clauses is a less simple matter. In most cases, however, the following transformation is sufficient to achieve this:

⟨7⟩ AN – N wh. . .is A[a]

as in

bald man – man who is bald
dirty water – water that is dirty

and so on. Later on we shall be able to show the correctness of ⟨7⟩.

In order to arrive at a precise notion of a restrictive relative clause, I have to say a few words about the other class of relative clauses, which are called appositive relative clauses. Some examples:

⟨8⟩ You, who are rich, can afford two cars.
⟨9⟩ Mary, whom you met, is my sister.
⟨10⟩ Vipers, which are poisonous, should be avoided.

Our intuition tells us that the clauses here have no restrictive effect on the noun to which they are attached. *You* and *Mary*, as we recall, cannot be further restricted, and the range of *vipers* is not restricted either, since all vipers are poisonous. Indeed, ⟨8⟩–⟨10⟩ easily split into the following conjunctions:

⟨11⟩ *You* are rich. *You* can afford two cars.
⟨12⟩ You met *Mary*. *Mary* is my sister.
⟨13⟩ *Vipers* are poisonous. *Vipers* should be avoided.

Thus we see that the appositive clause is nothing but a device for joining two sentences that share a noun phrase. One occurrence of the shared noun phrase gets replaced by the appropriate *wh*. . .and the resulting phrase (after some rearrangement of the word order when necessary) gets inserted into the other sentence following the occurrence of the shared noun phrase there. It is important to realize that this move does not alter the structure of the shared noun phrase in either of the ingredient sentences: the *wh*. . .replaces that noun phrase 'as is' in the enclosed sentence, and

---

[a] *Wh*. . .stands for the appropriate relative pronoun.

the clause gets attached to that noun phrase 'as is' in the enclosing sentence.[a] It is not surprising, therefore, that the whole move leaves the truth-value of the ingredient sentences intact: ⟨8⟩–⟨10⟩ are true, if and only if the conjunctions in ⟨11⟩–⟨13⟩ are true.

This is not so with restrictive clauses. Compare ⟨10⟩ with

⟨14⟩ Snakes which are poisonous should be avoided.

If we try to split ⟨14⟩ into two ingredients we get

⟨15⟩ Snakes are poisonous. Snakes should be avoided.

Clearly, the conjunction in ⟨15⟩ is false, but ⟨14⟩ is true. And the reason for this fact is equally obvious. The clause *which are poisonous* is an integral part of the subject of ⟨14⟩; the predicate *should be avoided* is not ascribed to *snakes* but to *snakes which are poisonous*, that is, by virtue of ⟨7⟩, to *poisonous snakes*. It appears, therefore, that while the insertion of an appositive clause merely joins two complete sentences, the insertion of a restrictive clause alters the very structure of the enclosing sentence by completing one of its noun phrases. Consequently a mere conjunction of the ingredient sentences is bound to fall short of the information content embodied in the sentence containing the restrictive clause.

There are a few more or less reliable morphological clues that may help us in distinguishing these two kinds. First, appositive clauses, but not restrictive ones, are usually separated by a pause, or in writing by a comma, from the enclosing sentence. Second, *which* or *who* may be replaced by *that* in restrictive clauses, but hardly in appositive ones:

Snakes that are poisonous should be avoided

versus

Vipers, which are poisonous, should be avoided.

Finally, the omission of *wh...* or *wh...is* mentioned above works only in restrictive clauses:

The man you met is here

versus

*Mary, you met, is here.

9 I claim that the insertion of a restrictive clause after a noun is a necessary condition of its acquiring the definite article. Therefore the definite article does not belong either to the enclosing or to the enclosed sentence prior to the formation of the clause. Consider the sentence

⟨16⟩ I know the man who killed Kennedy.

If we take *the man* to be the shared noun phrase, we get the ingredients

I know the man. The man killed Kennedy.

Here *the man* suggests some other identifying device, different from the one in ⟨16⟩, namely *who killed Kennedy*. In the case of a proper name this line of analysis leads to outright ungrammaticality. Consider, for example,

The Providence you know is no more.

---

[a] The shared noun phrase need not have an identical form in the original sentences. From *I bought a house, which has two stories* we recover *I bought a house. The house (I bought) has two stories.* These two sentences are continuous with respect to the noun *house*. This notion of continuity will be explained later.

Taking *the Providence* as the shared noun phrase we get the unacceptable

*You know the Providence. *The Providence is no more.

Thus we have to conclude that the ingredient sentences do not contain the definite article; it first enters the construction after the fusion of the two ingredients. Accordingly ⟨16⟩ is to be resolved into the following two sentences:

I know a man. A man killed Kennedy.

The shared noun phrase is *a man*. By replacing its second occurrence with *who* we obtain the clause *who killed Kennedy*. This gets inserted into the first sentence yielding

I know a man who killed Kennedy.

Since the verb *kill* suggests a unique agent, the definite article replaces the indefinite one, and we get ⟨16⟩. If the relevant verb has no connotation of uniqueness, no such replacement need take place; for instance,

I know a man who fought in Korea.

Of course we can say, in the plural,

⟨17⟩ I know the men who fought in Korea.

In this case I imply that, in some sense or other, I know all those men. If I just say

I know men who fought in Korea

no completeness is implied; it is enough if I know some such men.

It transpires, then, that the definite article marks the speaker's intention to exhaust the range determined by the restrictive clause. If that range is already restricted to one, the speaker's hand is forced: *the* becomes obligatory; a sentence like

God spoke to a man who begot Isaac

is odd for this reason. In this case the semantics of *beget* already decides the issue. In other cases the option remains:

I see a tree in our garden

is as good as

I see the tree in our garden.

This latter remark, however, would be misplaced if, in fact, more than one tree is in our garden: the speaker promises uniqueness, which, in the given situation, the clause cannot deliver.

The way of producing a singular term out of a common noun is as follows: attach a restrictive clause to the noun in the singular and prefix the definite article. It may happen that the clause is not restrictive enough; its domain, in a given speech situation, may include more than one individual. This trouble is similar to the one created by saying

Joe is hungry

when more than one person is called Joe in the house. In either case there are several possibilities: the speaker may lack some information, may be just careless, may be intentionally misleading, or some such thing. Yet *Joe* or *the tree in our garden* remain singular terms. The fact that a tool can be misused does not alter the function of the tool. Later on I shall return to infelicities of this kind.

10 But this is only half the story. I mentioned above that in many cases the addition of the definite article alone seems to suffice to create a singular term out of a common noun:

⟨18⟩ I see a man. The man wears a hat.

Obviously, we added, the man I see wears a hat. What happened is that the clause *whom I see* got deleted after *the man*, in view of the redundancy in the full sequence

I see a man. The man I see wears a hat.

The in ⟨18⟩, then, is nothing but a reminder of a deleted but recoverable restrictive clause. It is, as it were, a connecting device, which makes the discourse continuous with respect to a given noun. Indeed, if *the* is omitted, the two sentences become discontinuous:

I see a man. A man wears a hat.

Hence an important conclusion: *the* in front of a noun not actually followed by a restrictive clause is the sign of a deleted clause to be formed from a previous sentence in the same discourse containing the same noun. This rule explains the continuity in a discourse like

I have a dog and a cat. The dog has a ball to play with. Often the cat plays with the ball too

and the felt discontinuity in a text like

I have a dog and a cat. A dog has the ball.

If our conclusions are correct, then a noun in the singular already equipped with the definite article cannot take another restrictive clause, since such a noun phrase is a singular term as much as a proper name or a singular pronoun. Compare the two sequences:

⟨19⟩ I see a man. The man wears a hat.
⟨20⟩ I see a man. The man you know wears a hat.

⟨19⟩ is continuous. *The* is the sign of the deleted clause *(whom) I see*. In ⟨20⟩ the possibility of this clause is precluded by the presence of the actual clause *(whom) you know*. *The* in ⟨20⟩ belongs to this clause and any further restrictive clauses are excluded. Consequently there is no reason to think that the man you know is the same as the man I see. Not so with appositive clauses. The sequence

I see a man. The man, whom you know, wears a hat

is perfectly continuous. *The man*, in the second sentence, has the deleted restrictive clause *(whom) I see*, plus the appositive clause *whom you know*. Now consider the following pair:

⟨21⟩ I see a rose. The rose is lovely.
⟨22⟩ I see a rose. The red rose is lovely.

⟨21⟩ is continuous, ⟨22⟩ is not. This fact can be explained by assuming ⟨7⟩, that is, by deriving the prenominal adjective from a restrictive clause, which clause then precludes the acquisition of additional restrictive clauses. The assumption of ⟨7⟩, as we recall, also explains the difficulties encountered in trying to attach prenominal adjectives to proper names and personal pronouns.

11 The story, alas, is still not complete. Think of the ambiguity in a sentence like

⟨23⟩ The man she loves must be generous.

This either means that there is a man whom she loves and who must be generous, or that any man she loves must be generous. Examples of this kind can be multiplied. In some of these the generic interpretation is the obvious one. For instance,

Happy is the man whose heart is pure.

It would be odd to continue:

I met him yesterday.

The natural sequel is rather:

I met one yesterday.[a]

In other cases the individual interpretation prevails:

The man she loved committed suicide.

Yet, with some imagination, even such a sentence can be taken in the generic sense.

How do we decide which interpretation is right in a given case? In order to arrive at an answer, imagine three discourses beginning as follows:

⟨24⟩ Mary is a demanding girl. The man she loves must be generous.
⟨25⟩ Mary loves a man. The man she loves must be generous.
⟨26⟩ Mary loves a man. The man must be generous.

In ⟨26⟩ there is no ambiguity: *the man* is a singular term; in ⟨25⟩ it is likely to be a singular term; in ⟨24⟩ it is likely to be a general one. Why is this so? In ⟨26⟩ the deleted clause must be derived from the previous sentence, since, as we recall, the point of deletion is to remove redundancy. In ⟨25⟩ the clause is most likely a derivative of the previous sentence. If so, *the man* is a singular term. It remains possible, however, to imagine a break in the discourse between the two sentences: after stating a specific fact about her the speaker begins to talk about Mary in general terms. In ⟨24⟩ the reverse holds: the clause cannot be derived from a previous sentence, since there is no such sentence containing the noun, *man*. Consequently *the man* will be generic, unless a statement to the effect that Mary, in fact, loves a man is presupposed. Thus the moral of these examples emerges: a phrase of the type *the N* is a singular term if its occurrence is preceded by an actual or presupposed sentence of a certain kind in which *N* occurs, in the same discourse (the qualification, 'of a certain kind', will be explained soon). Accordingly, to take an occurrence of a *the N*-phrase to be a singular term is to assume the existence of such a sentence.

12 Since *the* always indicates a restrictive clause and since the only reason for deleting such a clause thus far mentioned is redundancy, that is, the presence of the sentence from which the clause is generated, one might conclude that no *the N*-phrase without a clause can occur if no such sentence can be found in the previous portion of the discourse. Yet this is not so. Some clause-less *the N*-phrases can occur at the very outset of a discourse. These counterexamples fall into three categories.

The first class comprises utterances of the following type:

The castle is burning.
The president is ill.

In these cases the clauses (*in our town, of our country*) are omitted simply because they are superfluous in the given situation. Such *the N*-phrases, in fact, approximate the

---

[a] It is interesting to realize that a personal pronoun, like *he*, also can occur in a generic sense – e.g., *He who asks shall be given.*

status of proper names: they tend to identify by themselves. It is not surprising, therefore, that they are often spelled with a capital letter: *the President, the Castle.* To a small circle of speakers even more common nouns can acquire this status:

Did you feed the dog?

The second category amounts to a literary device. One can begin a novel as follows:

The boy left the house.

Such a beginning suggests familiarity: the reader is invited to put himself into the picture: he is 'there', he sees the boy, he knows the house.

13 The third kind is entirely different. It involves a generic *the* without an actual clause. Examples abound:

⟨27⟩ The mouse is a rodent
⟨28⟩ The tiger lives in the jungle
⟨29⟩ The Incas did not use the wheel

and so forth. It is obvious that no clause restricting *mouse, tiger,* or *wheel* is to be resurrected here: the ranges of these nouns remain unrestricted. Shall we, then, abandon our claim that the definite article always presupposes a restrictive clause? We need not and must not. In order to see this, consider the saying:

None but the brave deserves the fair.

The obvious paraphrase is

None but the [man who is] brave deserves the [woman who is] fair.

This suggests the following deletion pattern:

the N wh...is A → the A

Then it is easy to see that sentences like

This book is written for the mathematician
Only the expert can give an answer

contain products of a similar pattern, to wit:

⟨30⟩ the N_i wh...is an N_j → the N_j

Thus *the mathematician* and *the expert* come from *the [person who is a] mathematician* and *the [person who is an] expert.* And similarly, for ⟨27⟩–⟨29⟩ the sources are:

the [animal that is a] mouse (tiger)
the [instrument that is a] wheel.

We have seen above that a redundant clause can be omitted. In ⟨30⟩ a redundant noun, N_i, is omitted and *the* is transferred to N_j. N_i is redundant because it is nothing but N_j's genus, and as such easily recoverable. This suggests that nouns that are themselves too generic to fall under a superior genus are not subject to ⟨30⟩. This is indeed so. While

Tigers live in the jungle
The Incas did not use wheels

do have ⟨28⟩ and ⟨29⟩ as paraphrases, sentences like

Objects are in space
Monkeys do not use instruments

are not paraphraseable into

> The object is in space.
> Monkeys do not use the instrument.[a]

In these sentences the *the N*-phrases have to be singular terms, consequently we are looking for the sentences from which the identifying clauses belonging to *the object* and *the instrument* are to be derived: what object (instrument) are we talking about?

This last point may serve as an indirect proof of ⟨30⟩. A more direct proof is forthcoming from the following example:

> There are two kinds of large cat living in Paraguay, the jaguar and the puma.

Obviously, *the jaguar and the puma* is derived from

> the [(kind of) large cat that is a] jaguar and the [(kind of) large cat that is a] puma.

In this case, unlike some of the previous examples, neither *a jaguar and a puma* nor *iaguars and pumas* will do to replace the generic *the jaguar and the puma*. Thus the generic *the* is not a mere variant of other generic forms. It has an origin of its own. Another illustration:

> Euclid described the parabola.

The parabola here is inadequately paraphrased by *a parabola*, *parabolas* or *all parabolas*. The given solution works again:

> Euclid described the [(kind of) curve that is a] parabola.

Incidentally, although we might say

> Euclid described curves

we cannot express this by saying

> Euclid described the curve.

Curve is too generic.[b]

14 The possibility of transferring *the* from an earlier noun, exemplified in ⟨30⟩, indicates a solution for noun phrases containing a proper name with the definite article. *The Hudson, the early Mozart, the Providence you know* are most likely derived from

> the [river called] Hudson
> the early [period/works of] Mozart
> the [aspect/appearance of] Providence you know.

Indeed, it can be shown that the clause *you know*, for instance, does not belong to *Providence* directly. For if it did, the following sequence would be acceptable:

> You know Providence.* The Providence is no more

on the analogy of, say,

> You had a house. The house is no more.

In this case the first sentence yields the clause *which you had*, which clause justifies *the* before *house* in the second sentence. In the previous example, however, the

---

[a] The existence or nonexistence of a higher genus may be a function of the discourse. In philosophical writing, for instance, one might find a generic sentence like *The idea is more perfect than the object*, which presupposes a common genus for ideas and objects.

[b] As man is an exceptional animal, *man* is an exceptional noun. It has a generic sense in the singular without any article: *Man, but not the ape, uses instruments*.

first sentence refuses to yield the clause *which you know*, precisely because *Providence* is a proper name. Thus *Providence* in the second sentence has no clause to justify *the*. Consequently *the Providence you know* does not come directly from

You know Providence.

15 Owing to the inductive nature of our investigations up to this point, our conclusions concerning the formation of singular terms out of common nouns had to be presented in a provisional manner leaving room, as it were, for the variety of facts still to be accounted for. Now, in retrospect, we are able to give a more coherent picture, at least in its main lines, for many details of this very complex affair must be left to further studies. The basic rules seem to be the following:

(*a*) The definite article is a function of a restrictive clause attached to the noun.

(*b*) This article indicates that the scope of the so restricted noun is to be taken exhaustively, extending to any and all objects falling under it.

(*c*) If the restriction is to one individual the definite article is obligatory and marks a singular term. Otherwise the term is general and the definite article remains optional.

(*d*) The clause is restrictive to one individual if and only if it is derived from a sentence either actually occurring in the previous part of the same discourse, or presupposed by the same discourse, and in which sentence N has an identifying occurrence. This last notion remains to be explained.

(*e*) Redundant clauses can be omitted.

(*f*) A clause is redundant if it is derived from a sentence actually occurring in the previous part of the discourse, or if the information content of a sentence in which N has an identifying occurrence is generally known to the participants of the discourse.

(*g*) Redundant genus nouns can be omitted according to ⟨30⟩.

16 These rules give us the following recognition-procedure with respect to any *the N*-phrase, where N is a common noun.

(i) If the phrase is followed by a clause look for the mother sentence of the clause.

(ii) If it is found, and if it identifies N, the phrase is a singular term. If it fails to identify, the phrase is a general term.

(iii) If no mother sentence can be found, ask whether the circumstances of the discourse warrant the assumption of an identifying sentence corresponding to the clause.

(iv) If the answer is in the affirmative, we have a singular term, otherwise a general one.

(v) If the phrase is not followed by a clause, look for a previous sentence in which N occurs without the definite article.

(vi) If there is such an occurrence the deleted clause after the phrase is to be recovered from that sentence.

(vii) If that occurrence is identifying we have a singular term, otherwise a general one.

(viii) If there is no such occurrence, ask whether the circumstances of the discourse warrant the assumption of a sentence that would identify N to the participants of the discourse.

(ix) If the answer is in the affirmative we have a singular term and the clause is to be recovered from that sentence.

(x) If the answer is in the negative *the N* is a general term with a missing genus to be recovered following ⟨30⟩.

In order to have an example illustrating the various possibilities for a singular term of the *the N* type, consider the following. My friend returns from a hunt and begins:

> Imagine, I shot a bear and an elk. The bear I shot nearly got away, but the elk dropped dead on the spot. Incidentally, the gun worked beautifully, but the map you gave me was all wrong.

The bear I shot is a singular term by (ii). *The elk* is singular by (vii) with the clause *I shot* according to (vi). *The gun* is singular by (ix) with a clause something like *I had with me; the map you gave me* is singular by (iv).

The appeal to the circumstances of the discourse found in (iii) and (viii) is admittedly a cover for an almost inexhaustible variety of relevant considerations. Some of these are linguistic, others pragmatic. Tensed verbs suggest singular terms, modal contexts general ones. But think of the man Mary loves, who must be generous, and of the dinosaur, which roamed over Jurassic plains. In practice it may be impossible to arrive at a verdict in many situations. You may have only one cat, yet your wife's remark

> The cat is a clever beast

may remain ambiguous. What is important to us is rather the universal presupposition of all singular terms of the *the N*-type; the actual or implied existence of an identifying sentence. This notion still remains to be explained.

17   Once more I shall proceed in an inductive manner. First I shall enumerate the main types of identifying sentences and then try to find some common characteristics.

First of all, a sentence identifies *N* if it connects *N* with a definite noun phrase in a noncopulative and nonmodal fashion. The class of definite noun phrases comprises all singular terms and their plural equivalents like *we, you, they, these boys, my daughters, the dogs*, and so forth. Consequently the following sentences are identifying ones:

⟨31⟩ I see a house. The house...
⟨32⟩ They dug a hole. The hole...
⟨33⟩ The dogs found a bone. The bone...

The order of the noun phrases does not matter:

⟨34⟩ A snake bit me. The snake...

PN adjuncts also connect both ways:

⟨35⟩ They dug a hole with a stick. The stick...
⟨36⟩ A boy had dinner with me. The boy...

and so forth.

It follows that definite nouns of the *the N*-type can form a chain of identification. For instance:

> I see a man. The man wears a hat. The hat has a feather on it. The feather is green.

But, of course, all chains must begin somewhere. This means that at the beginning of most discourses containing definite nouns, there must occur a 'basic' definite noun: a personal pronoun, a proper name, or a noun phrase beginning with a demonstrative or possessive pronoun. By these terms, as it were, the whole discourse is anchored in the world of individuals.

Copulative verbs like *be* and *become* do not connect. The following sequences remain discontinuous:

⟨37⟩ He is a teacher. The teacher is lazy.
⟨38⟩ Joe became a salesman. The salesman is well paid.

We know, of course, that these two verbs are peculiar in other respects too. Their verb object does not take the accusative and the sentences formed with them reject the passive transformation. What is more relevant for us, however, is the fact that these same verbs resist the formation of a relative clause:

*the teacher who he is
*the salesman he became.

This feature, of course, provides an unexpected confirmation for our theory about the definite article: ⟨37⟩ and ⟨38⟩ are discontinuous because the starting sentences cannot provide the clause for the subsequent *the* N-phrase.

Verbs accompanied by modal auxiliaries may or may not connect:

⟨39⟩ Joe can lift a bear
⟨40⟩ He could have married a rich girl
⟨41⟩ You must buy a house
⟨42⟩ I should have seen a play

remain ambiguous between generality and individuality concerning the second noun phrase.

In some cases nouns are identified by the mere presence of a verb in the past tense:

A man caught a shark in a lake. The shark was a fully developed specimen.

18 Finally, there is the least specific way of introducing a singular term:

Once upon a time there was a king who had seven daughters. The king...

This pattern of 'existential extraction' has great importance for us. It appears that it can be used as a criterion of identifying occurrence: a sentence is identifying with respect to an N if and only if the transform

There is an N wh......

is acceptable as a paraphrase. Thus the identifying sentences in ⟨31⟩–⟨36⟩ yield:

There is a house I see.
There is a hole they dug.
There is a bone the dogs found.
There is a snake that bit me.
There is a stick with which they dug a hole.
There is a boy who had dinner with me.

Nonidentifying sentences, like

A cat is an animal
A tiger eats meat

or the ones like ⟨37⟩–⟨38⟩, reject this form:

*There is an animal that is a cat.
*There is meat a tiger eats.
*There is a teacher he is.
*There is a salesman Joe became.

As for the modal sentences ⟨39⟩–⟨42⟩, it is obvious that the possibility of existential extraction is the sign of their being taken in the identifying sense:

There is a bear Joe can lift
There is a rich girl you could have married

and so on. We may conclude, then, that given any *the N wh... ...*-phrase, it is a singular term if and only if the sentence *There is an N wh... ...* is entailed by the discourse.

19 This conclusion should fill the hearts of all analytic philosophers with the glow of familiarity. Hence it may be worth while to review our conclusions from the point of view of recent controversies on the topic.

First of all we have found that the question whether or not a given *the N*-phrase is a singular term cannot be decided by considering only the sentence in which it occurs. Strawson's suggestion that it is the use of the sentence, or the expression, that is relevant is certainly true, but it falls short of telling us what it means to use a sentence to make a statement, or to use a certain phrase to refer to something. Our results indicate a way of being more explicit and precise. At least with respect to *the N*-phrases, their being uniquely referring expressions, that is, singular terms, is conditioned by their occurrence in a discourse of a certain type. Such a discourse has to contain a previous sentence which identifies N, and, as we remember, such a sentence is always paraphraseable by the existential extraction, *There is an N wh....* ...Therefore, although Russell's claim, according to which sentences containing *the N*-type singular terms entail an assertion of existence, is too strong, Strawson's claim, that no such assertion is entailed by the referential occurrence of such a phrase is too weak. True, it is not the sentence containing the referential *the N* that entails the existential assertion, but another sentence, the occurrence of which, however, is a *conditio sine qua non* of a referential *the N*.

But, you object, the identifying sentence need not actually occur. In many cases it is merely assumed or presupposed. My answer is that this is philosophically irrelevant. The omission of the identifying sentence is a device of economy: we do not bother to state the obvious. What matters is the essential structure of the discourse. In giving a mathematical proof we often omit steps that are obvious to the audience, yet those steps remain part of the proof. The omission of the identifying sentence, like the omission of certain steps in a given proof, depends upon what the speaker deems to be obvious to the audience in question. And this has no philosophical significance.

Our conclusion is in accordance with common sense. If a child tells me

⟨43⟩ The bear I shot yesterday was huge

I will answer

⟨44⟩ But you did not shoot any bear.

⟨44⟩ does not contradict ⟨43⟩. It contradicts, however, the sentence

⟨45⟩ I shot a bear yesterday

which the child presupposed, but wisely omitted, in trying to get me to take *the bear* ...as a referring expression. Is, then, ⟨43⟩ true or false? In itself it is neither, since the *the N*-phrase in it can achieve reference only if the preceding identifying sentence, ⟨45⟩, is true. Since this is not the case, *the bear*...fails to refer to anything in spite of the fact that it satisfies the conditions for a singular term.

Of course the logician, who abhors truth-gaps as nature does the void will be justified in trying to unmask the impotence of such singular terms by insisting upon the inclusion of a version of the relevant identifying sentence (*There is an N wh... ...*) into the analysis of sentences containing singular *the N* terms. In view of our results, this move is far less artificial than some authors have claimed.

20 The triumph of the partisans of the philosophical theory of descriptions will, however, be somewhat damped when we point out that sentences of the type

There is an *N* wh... ...

do not necessarily assert 'real' existence, let alone spatio-temporal existence. Take the following discourse:

I dreamt about a ship. The ship...

The identifying sentence easily yields the transform

There is a ship I dreamt about.

This may be true, yet it does not mean that there is such a ship in reality. If some-body suggests that that ship has a dream existence, or that the house I just imagined has an imaginary existence, or that the king with the seven daughters has a fairy-tale existence, I cannot but agree. I only add that it would be desirable to be able to characterize the various types of discourse appropriate to these kinds of 'existence'. Particularly, of course, we are interested in discourses pertaining to 'real' existence. I give a hint. I remarked above that at the beginning of almost every discourse con-taining singular terms there must be a 'basic' singular term (or its plural equivalent). Now if we find such a basic term denoting a real entity, like *I, Lyndon B. Johnson,* or *Uganda,* then we should trace the connection of other singular terms to these. As long as the links are formed by 'reality-preserving' verbs like *push, kick,* and *eat,* we remain in spatio-temporal reality. Verbs like *dream, imagine, need, want, look for,* and *plan* should caution us: the link may be broken, although it need not be. Reality may enter by another path. If I only say

I dreamt about a house. The house...

one has no reason to think that the house I dreamt about is to be found in the world. If, however, I report

I dreamt about the house in which I was born. The house...

the house I talk about is a real house, but not by virtue of *dream* but by virtue of *being born in.* It is this latter and not the former verb that preserves reality. Of course, if the 'basic' singular term is something like *Zeus,* or *the king who lived once upon a time,* the situation is clear. The development of this hint would require much fascinating detail.

For the time being, we have to be satisfied with the conclusion that the discourse in which a referential *the N*-phrase occurs entails a *There is an N...* assertion. But we should add the *caveat*: there are things that do not really exist.[a]

University of Calgary

[a] The author wishes to express his indebtedness to the work of Dr Beverly L. Robbins later published in *The Definite Article in English Transformations.* The Hague, Mouton, 1968.

The problem of proper names[a]

JOHN R. SEARLE

At first sight nothing seems easier to understand in the philosophy of language than our use of proper names: here is the name, that is the object. The name stands for the object.

Although this account is obviously true, it explains nothing. What is meant by 'stands for'? And how is the relation indicated by 'stands for' ever set up in the first place? Do proper names 'stand for' in the same way that definite descriptions 'stand for'? These and other questions which I wish to attack in this section can be summed up in the question, 'Do proper names have senses?' What this question asks, as a start, is what, if any, similarity is there between the way a definite description picks out its referent and the way a proper name picks out its referent. Is a proper name really a shorthand description? We shall see that the two opposing answers given to this question arise from the tension between, on the one hand, the almost exclusive use of proper names to perform the speech act of reference, and, on the other hand, the means and preconditions for performing this speech act especially the condition expressed in the principle of identification. The principle of identification may be stated as follows:

> A necessary condition for the successful performance of a definite reference in the utterance of an expression is that either the expression must be an identifying description or the speaker must be able to produce an identifying description on demand.

For an explanation of, arguments for, and qualifications to this principle, see: John A. Searle, *Speech Acts*, Cambridge, 1969, pp. 77 ff.

The first answer goes something like this: proper names do not have senses, they are meaningless marks; they have denotation but no connotation (Mill).[b] The argument for this view is that whereas a definite description refers to an object only in virtue of the fact that it describes some *aspect* of that object, a proper name does not *describe* the object at all. To know that a definite description fits an object is to know a fact about that object, but to know its name is not so far to know any facts about it. This difference between proper names and definite descriptions is further illustrated by the fact that we can often turn a definite description (a referring expression) into an ordinary predicative expression by simply substituting an indefinite article for the definite, e.g., 'a man' for 'the man'. No such shift is in general possible with proper names. When we do put the indefinite article in front of a proper name it is either a shorthand way of expressing well-known characteristics of the

---

[a] This paper is reprinted from J. R. Searle, *Speech Acts: An Essay in the Philosophy of Language* (Cambridge, 1969), pp. 162–74.

[b] J. S. Mill, *A System of Logic* (London and Colchester, 1949), book 1, chapter 2, para. 5.

bearer of the name (e.g., 'He is a Napoleon' means 'He is like Napoleon in many respects'), or it is a shorthand form of a formal-mode expression about the name itself (e.g., 'He is a Robert' means 'He is named Robert'). In short we use a proper name to refer and not to describe; a proper name predicates nothing and consequently does not have a sense.

Our robust common sense leads us to think that this answer must be right, but though it has enormous plausibility, we shall see that it cannot be right, at least not as it stands, for too many facts militate against it. First, let us look at some of the metaphysical traps that an uncritical acceptance of such a view is likely to lead us into. The proper name, we are inclined to say, is not connected with any *aspects* of the object as descriptions are, it is tied to the object itself. Descriptions stand for aspects or properties of an object, proper names for the real thing. This is the first step on the road that leads to substance, for it fastens on to what is supposed to be a basic metaphysical distinction between objects and properties or aspects of objects, and it derives this distinction from an alleged difference between proper names and definite descriptions. Such a muddle is to be found in the *Tractatus*, 'The name means the object. The object is its meaning' (3.203).[a] But notice to what interesting paradoxes this leads immediately: the meaning of words, it seems, cannot depend on any contingent facts in the world, for we can still describe the world even if the facts alter. Yet the existence of ordinary objects – people, cities, etc. – is contingent, and hence the existence of any meaning for their names is contingent. So their names are not the real names at all! There must exist a class of objects whose existence is not a contingent fact, and it is their names which are the real names.[b] And what does this mean? Here we see another good illustration of the original sin of all metaphysics, the attempt to read real or alleged features of language into the world.

The usual rejoinder to the thesis that there is a basic metaphysical distinction between objects and properties is that objects are just collections of properties.[c] The first thesis is derived from the distinction between referring and predicating, the second thesis is derived from the tautology that everything that can be said about an object can be said in descriptions of that object. But both theses are equally nonsensical. It is nonsense to suppose that an object is a combination of its propertyless self and its properties, and it is nonsense to suppose that an object is a heap or collection of properties. Again, both views have a common origin in the metaphysical mistake of deriving ontological conclusions from linguistic theses.

There are three objections to the view that proper names do not have senses:

1. We use proper names in existential propositions, e.g., 'there is such a place as Africa', 'Cerberus does not exist'. Here proper names cannot be said to refer, for no such subject of an existential statement can refer. If it did, the precondition of its having a truth value would guarantee its truth, if it were in the affirmative, and its falsity, if it were in the negative. (This is just another way of saying that 'exists' is not a predicate.) Every existential statement states that a certain predicate is instantiated. (As Frege put it, existence is a second order concept.)[d] An existential statement does not refer to an object and state that it exists, rather it expresses a concept and states that that concept is instantiated. Thus, if a proper name occurs in an

---

[a] Mill's: proper names have no meaning, might appear to be inconsistent with Wittgenstein's: objects are their meanings. But they are not inconsistent. (Ambiguity of 'mean' and 'bedeuten'.) Both say, proper names have referents but not senses.

[b] Cf. also Plato, *Theaetetus*.

[c] E.g., Russell, *An Inquiry into Meaning and Truth* (London, 1940), p. 97.

[d] *Grundgesetze der Arithmetik* (Jena, 1893), vol. I, section 21.

existential statement it must have some conceptual or descriptive content. Attempts such as Russell's[a] to evade this point have taken the form of saying that such expressions are not *really* proper names, a desperate maneuver which shows that something must be wrong with the assumptions which drive one to it.

2. Sentences containing proper names can be used to make identity statements which convey factual and not merely linguistic information. Thus the sentence, 'Everest is Chomolungma' can be used to make an assertion which has geographical and not merely lexicographical import. Yet if proper names were without senses, then the assertion could convey no more information than does an assertion made with the sentence 'Everest is Everest'. Thus it seems that proper names must have descriptive content, they must have a sense. This is substantially Frege's argument that proper names have senses.[b]

3. The principle of identification requires that an utterance of a proper name must convey a description just as the utterance of a definite description must if the reference is to be consummated. And from this it seems to follow that a proper name is a kind of shorthand description.

All three objections point to the same conclusion, namely, that proper names are shorthand definite descriptions.

But it seems that this conclusion cannot be right, for, aside from its grotesque unplausibility, it is inconsistent with too many obvious truths. First, if it were the case that a proper name is a shorthand description, then descriptions should be available as definitional equivalents for proper names: but we do not, in general, have definitions of proper names. In so called dictionaries of proper names, one finds descriptions of the bearers of the names, but in most cases these descriptions are not definitional equivalents for the names, since they are only contingently true of the bearers.

Not only do we not have definitional equivalents, but it is not clear how we could go about getting them to substitute in all cases for proper names. If we try to present a complete description of the object as the sense of the name, odd consequences would ensue, e.g., any true statement about the object using the name as subject would be analytic, any false one self-contradictory, the meaning of the name (and perhaps the identity of the object) would change every time there was any change at all in the object, the name would have different meanings for different people, etc. So it seems that the view that proper names are descriptions cannot be true either.

Here we have a beautiful example of a philosophical problem: on the one hand common sense drives us to the conclusion that a proper name is not a species of description, that it is *sui generis*, but against this a series of theoretical considerations drive us to the conclusion that it must be a shorthand definite description. But against this too we can adduce serious arguments. This antinomy admits of a solution toward which I shall now argue.

We might rephrase our original question, 'Do proper names have senses?' as 'Do referring uses of proper names entail any descriptive predicates?' or simply 'Are any propositions where the subject is a proper name and the predicate a descriptive

[a] 'The Philosophy of Logical Atomism', R. Marsh (ed.), *Logic and Knowledge* (London, 1956), pp. 200 ff.

[b] Though, with a characteristic perversity, he did not see that this account of identity statements provides an explanation of the use of proper names in existential statements. He thought it was nonsense to use proper names in existential statements. 'Über die Grundlagen der Geometrie II', *Jahresbericht der Deutschen Mathematiker-Vereinigung* (1903), p. 373.

expression analytic?'[a] But this question has a weaker and a stronger form: (a) the weaker: 'Are any such statements at all analytic?' and (b) the stronger: 'Are any statements where the subject is a proper name and the predicate an identifying description analytic?'

Consider the first question. It is characteristic of a proper name that it is used to refer to the *same* object on *different* occasions. The use of the same name at different times in the history of the object presupposes that the object is the same; a necessary condition of identity of reference is identity of the object referred to. But to presuppose that the object is the same in turn presupposes a criterion of identity: that is, it presupposes an ability on the part of the speaker to answer the question, 'In virtue of what is the object at time *t*. 1, referred by name *N*, identical with the object at time *t*. 2, referred to by the same name?' or, put more simply, 'The object at time *t*. 1 is the same *what* as the object at time *t*. 2?' and the gap indicated by 'what' is to be filled by a descriptive general term; it is the same mountain, the same person, the same river, the general term providing in each case a temporal criterion of identity. This gives us an affirmative answer to the weaker question. Some general term is analytically tied to any proper name: Everest is a mountain, the Mississippi is a river, de Gaulle is a person. Anything which was not a mountain could not be Everest, etc., for to secure continuity of reference we need a criterion of identity, and the general term associated with the name provides the criterion. Even for those people who would want to assert that de Gaulle could turn into a tree or horse and still be de Gaulle, there must be some identity criterion. De Gaulle could not turn into anything whatever, e.g., a prime number, and still remain de Gaulle, and to say this is to say that some term or range of terms is analytically tied to the name 'de Gaulle'.

To forestall an objection: one temptation is to say that if we continue to call an object 'Everest', the property of being called 'Everest' is sufficient to guarantee that it is the same. But the point of the above analysis is that we are only justified in calling it 'Everest' if we can give a reason for supposing it to be identical with what we used to call 'Everest' and to give as the reason that it is called 'Everest' would be circular. In this sense at least, proper names do have 'connotations'.

But the answer 'yes' to the weaker question does not entail the same answer to the stronger one, and it is the stronger form which is crucial for deciding whether or not a proper name has a sense, as Frege and I use the word. For according to Frege the sense of a proper name contains the 'mode of presentation' which identifies the referent, and of course a single descriptive predicate does not provide us with a mode of presentation; it does not provide an identifying description. That Socrates is a man may be analytically true, but the predicate 'man' is not an identifying description of Socrates.

So let us consider the stronger formulation of our question in the light of the principle of identification. According to this principle, anyone who uses a proper name must be prepared to substitute an identifying description (remembering that identifying descriptions include ostensive presentations) of the object referred to by a proper name. If he were unable to do this, we should say that he did not know whom or what he was talking about, and it is this consideration which inclines us,

[a] Of course, in one sense of 'analytic', no such subject–predicate proposition can be analytic, since it is in general a contingent fact that the subject expression has a referent at all and hence contingent that the proposition has a truth-value. To meet this objection we can either redefine 'analytic' as: 'p is analytic $= df.$ if p has a truth-value, it is true by definition' or we can rephrase the original question as, 'Is any proposition of the form "if anything is S it is P" analytic, where "S" is replaced by a proper name and "P" by a descriptive predicate?'

and which among other things inclined Frege, to say that a proper name must have a sense, and that the identifying description constitutes that sense. Think what it is to learn a proper name. Suppose you say to me: 'Consider Thaklakes, tell me what you think of Thaklakes.' If I have never heard that name before I can only reply, 'Who is he?' or 'What is it?' And does not your next move – which according to the principle of identification consists in giving me an ostensive presentation or a set of descriptions – does this not give me the sense of the name, just as you might give me the sense of a general term? Is this not a definition of the name?

We have discussed several objections to this view already; a further one is that the description one man is prepared to substitute for the name may not be the same as the one someone else is prepared to substitute. Are we to say that what is definitionally true for one is only contingent for another? Notice what maneuvers Frege is forced to here:

Suppose further that Herbert Garner knows that Dr Gustav Lauben was born on 13 September 1875, in N. H. and this is not true of anyone else; against this suppose that he does not know where Dr Lauben now lives or indeed anything about him. On the other hand, suppose Leo Peter does not know that Dr Lauben was born on 13 September 1875, in N. H. Then as far as the proper name 'Dr Gustav Lauben' is concerned, Herbert Garner and Leo Peter do not speak the same language, since, although they do in fact refer to the same man with this name, they do not know that they do so.[a]

Thus according to Frege, unless our descriptive backing for the name is the same, we are not even speaking the same language. But, against this, notice that we seldom consider a proper name as part of *one* language as opposed to another at all.

Furthermore, I might discover that my identifying description was not true of the object in question and still not abandon his name. I may learn the use of 'Aristotle' by being told that it is the name of the Greek philosopher born in Stagira, but if later scholars assure me that Aristotle was not born in Stagira at all but in Thebes, I will not accuse them of self-contradiction. But let us scrutinize this more closely: scholars might discover that a *particular* belief commonly held about Aristotle was false. But does it make sense to suppose that everything anyone has ever believed to be true of Aristotle was in fact not true of the real Aristotle? Clearly not, and this will provide us with the germ of an answer to our question.

Suppose we ask the users of the name 'Aristotle' to state what they regard as certain essential and established facts about him. Their answers would constitute a set of identifying descriptions, and I wish to argue that though no single one of them is analytically true of Aristotle, their disjunction is. Put it this way: suppose we have independent means of identifying an object, what then are the conditions under which I could say of the object, 'This is Aristotle?' I wish to claim that the condition, the descriptive power of the statement, is that a sufficient but so far unspecified number of these statements (or descriptions) are true of the object. In short, if none of the identifying descriptions believed to be true of some object by the users of the name of that object proved to be true of some independently located object, then that object could not be identical with the bearer of the name. It is a necessary condition for an object to be Aristotle that it satisfy at least some of these descriptions. This is another way of saying that the disjunction of these descriptions is analytically tied to the name 'Aristotle' – which is a quasi-affirmative answer to the question, 'Do proper names have senses?' in its stronger formulation.

[a] 'The Thought: a logical inquiry', trans. A. and M. Quinton, *Mind* (1956), p. 297.

My answer, then, to the question, 'Do proper names have senses?' – if this asks whether or not proper names are used to describe or specify characteristics of objects – is 'No'. But if it asks whether or not proper names are logically connected with characteristics of the object to which they refer, the answer is 'Yes, in a loose sort of way'.

Some philosophers suppose that it is an objection to this sort of account that the same word is sometimes used as a name for more than one object. But this is a totally irrelevant fact and not an objection to my account at all. That different objects are named 'John Smith' is no more relevant to the question 'Do proper names have senses?' than the fact that both riversides and finance houses are called 'banks' is relevant to the question, 'Do general terms have senses?' Both 'bank' and 'John Smith' suffer from kinds of homonymy, but one does not prove a word meaningless by pointing out that it has several meanings. I should have considered this point too obvious to need stating, were it not for the fact that almost every philosopher to whom I have presented this account makes this objection.

What I have said is a sort of compromise between Mill and Frege. Mill was right in thinking that proper names do not entail any particular description, that they do not have definitions, but Frege was correct in assuming that any singular term must have a mode of presentation and hence, in a way, a sense. His mistake was in taking the identifying description which we can substitute for the name as a definition.

I should point out, parenthetically, that of course the description, 'The man called X' will not do, or at any rate will not do by itself, as a satisfaction of the principle of identification. For if you ask me, 'Whom do you mean by X?' and I answer, 'The man called X', even if it were true that there is only one man who is called X, I am simply saying that he is the man whom other people refer to by the name 'X'. But if they refer to him by the name 'X' then they must also be prepared to substitute an identifying description for 'X' and if they in their turn substitute 'the man called X', the question is only carried a stage further and cannot go on indefinitely without circularity or infinite regress. My reference to an individual may be parasitic on someone else's but this parasitism cannot be carried on indefinitely if there is to be any reference at all.

For this reason it is no answer at all to the question of what if anything is the sense of a proper name 'X' to say its sense or part of its sense is 'called X'. One might as well say that part of the meaning of 'horse' is 'called a horse'. It is really quite amazing how often this mistake is made.[a]

My analysis of proper names enables us to account for all the apparently inconsistent views at the beginning of this section. How is it possible that a proper name can occur in an existential statement? A statement such as 'Aristotle never existed' states that a sufficient, but so far unspecified, number of the descriptive backings of 'Aristotle' are false. Which one of these is asserted to be false is not yet clear, for the descriptive backing of 'Aristotle' is not yet precise. Suppose that of the propositions believed to be true of Aristotle half were true of one man and half of another, would we say that Aristotle never existed? The question is not decided for us in advance.

Similarly it is easy to explain identity statements using proper names. 'Everest is Chomolungma' states that the descriptive backing of both names is true of the same object. If the descriptive backing of the two names, for the person making the assertion, is the same, or if one contains the other, the statement is analytic, if not,

[a] E.g., A. Church, *Introduction to Mathematical Logic* (Princeton, 1956), p. 5.

synthetic. Frege's instinct was sound in inferring from the fact that we do make factually informative identity statements using proper names that they must have a sense, but he was wrong in supposing that this sense is as straightforward as in a definite description. His famous 'Morning Star–Evening Star' example led him astray here, for though the sense of these names is fairly straightforward, these expressions are not paradigm proper names, but are on the boundary line between definite descriptions and proper names.

Furthermore, we now see how an utterance of a proper name satisfies the principle of identification: if both the speaker and the hearer associate some identifying description with the name, then the utterance of the name is sufficient to satisfy the principle of identification, for both the speaker and the hearer are able to substitute an identifying description. The utterance of the name communicates a proposition to the hearer. It is not necessary that both should supply the same identifying description, provided only that their descriptions are in fact true of the same object.

We have seen that insofar as proper names can be said to have a sense, it is an imprecise one. We must now explore the reasons for this imprecision. Is the imprecision as to what characteristics exactly constitute the necessary and sufficient conditions for applying a proper name a mere accident, a product of linguistic slovenliness? Or does it derive from the functions which proper names perform for us? To ask for the criteria for applying the name 'Aristotle' is to ask in the formal mode what Aristotle is; it is to ask for a set of identity criteria for the object Aristotle. 'What is Aristotle?' and 'What are the criteria for applying the name "Aristotle"?' ask the same question, the former in the material mode, and the latter in the formal mode of speech. So if, prior to using the name, we came to an agreement on the precise characteristics which constituted the identity of Aristotle, our rules for using the name would be precise. But this precision would be achieved only at the cost of entailing some *specific* descriptions by any use of the name. Indeed, the name itself would become logically equivalent to this set of descriptions. But if this were the case we would be in the position of being able to refer to an object solely by, in effect, describing it. Whereas in fact this is just what the institution of proper names enables us to avoid and what distinguishes proper names from definite descriptions. If the criteria for proper names were in all cases quite rigid and specific, then a proper name would be nothing more than a shorthand for these criteria, it would function exactly like an elaborate definite description. But the uniqueness and immense pragmatic convenience of proper names in our language lies precisely in the fact that they enable us to refer publicly to objects without being forced to raise issues and come to an agreement as to which descriptive characteristics exactly constitute the identity of the object. They function not as descriptions, but as pegs on which to hang descriptions. Thus the looseness of the criteria for proper names is a necessary condition for isolating the referring function from the describing function of language.

To put the same point differently, suppose we ask, 'Why do we have proper names at all?' Obviously, to refer to individuals. 'Yes, but descriptions could do that for us.' But only at the cost of specifying identity conditions every time reference is made: suppose we agree to drop 'Aristotle' and use, say, 'the teacher of Alexander', then it is an analytic truth that the man referred to is Alexander's teacher – but it is a contingent fact that Aristotle ever went into pedagogy. (Though it is, as I have said, a necessary truth that Aristotle has the logical sum [inclusive disjunction] of the properties commonly attributed to him.)[a]

[a] Ignoring contradictory properties, $p \ v \ {\sim}p$ would render the logical sum trivially true.

It should not be thought that the only sort of looseness of identity criteria for individuals is that which I have described as peculiar to proper names. Identity problems of quite different sorts may arise, for instance, from referring uses of definite descriptions. 'This is the man who taught Alexander' may be said to entail, e.g., that this object is spatio-temporally continuous with the man teaching Alexander at another point in space-time; but someone might also argue that this man's spatio-temporal continuity is a contingent characteristic and not an identity criterion. And the logical nature of the connection of such characteristics with the man's identity may again be loose and undecided in advance of dispute. But this is quite another dimension of looseness from that which I cited as the looseness of the criteria for applying proper names, and does not affect the distinction in function between definite descriptions and proper names, viz., that definite descriptions refer only in virtue of the fact that the criteria are not loose in the original sense, for they refer by providing an explicit description of the object. But proper names refer without providing such a description.

We might clarify some of the points made in this chapter by comparing paradigm proper names with degenerate proper names like 'the Bank of England'. For these limiting cases of proper names, it seems the sense is given as straightforwardly as in a definite description; the presuppositions, as it were, rise to the surface. And a proper name may acquire a rigid use without having the verbal form of a description: God is just, omnipotent, omniscient, etc., *by definition* for believers. To us, 'Homer' just means 'the author of the *Iliad* and the *Odyssey*'. The form may often mislead us: the Holy Roman Empire was neither holy nor Roman, etc., but it was, nonetheless, the Holy Roman Empire. Again, it may be conventional to name only girls 'Martha', but if I name my son 'Martha', I may mislead, but I do not lie. And of course not all paradigm proper names are alike with respect to the nature of their 'descriptive content'. There will, e.g., be a difference between the names of living people, where the capacity of the user of the name to recognize the person may be an important 'identifying description', and the names of historical figures. But the essential fact to keep in mind when dealing with these problems is that we have the institution of proper names to perform the speech act of identifying reference. The existence of these expressions derives from our need to separate the referring from the predicating functions of language. But we never get referring completely isolated from predication, for to do so would be to violate the principle of identification, without conformity to which we cannot refer at all.

University of California

The inscrutability of reference[a]

W. V. QUINE

1 I listened to Dewey on Art as Experience when I was a graduate student in the spring of 1931. Dewey was then at Harvard as the first William James Lecturer. I am proud now to be at Columbia as the first John Dewey Lecturer.

Philosophically I am bound to Dewey by the naturalism that dominated his last three decades. With Dewey I hold that knowledge, mind, and meaning are part of the same world that they have to do with, and that they are to be studied in the same empirical spirit that animates natural science. There is no place for a prior philosophy.

When a naturalistic philosopher addresses himself to the philosophy of mind, he is apt to talk of language. Meanings are, first and foremost, meanings of language. Language is a social art which we all acquire on the evidence solely of other people's overt behavior under publicly recognizable circumstances. Meanings, therefore, those very models of mental entities, end up as grist for the behaviorist's mill. Dewey was explicit on the point: 'Meaning...is not a psychic existence; it is primarily a property of behavior.'[b]

Once we appreciate the institution of language in these terms, we see that there cannot be, in any useful sense, a private language. This point was stressed by Dewey in the twenties. 'Soliloquy', he wrote, 'is the product and reflex of converse with others' (170). Farther along he expanded the point thus: 'Language is specifically a mode of interaction of at least two beings, a speaker and a hearer; it presupposes an organized group to which these creatures belong, and from whom they have acquired their habits of speech. It is therefore a relationship' (185). Years later, Wittgenstein likewise rejected private language. When Dewey was writing in this naturalistic vein, Wittgenstein still held his copy theory of language.

The copy theory in its various forms stands closer to the main philosophical tradition, and to the attitude of common sense today. Uncritical semantics is the myth of a museum in which the exhibits are meanings and the words are labels. To switch languages is to change the labels. Now the naturalist's primary objection to this view is not an objection to meanings on account of their being mental entities, though that could be objection enough. The primary objection persists even if we take the labeled exhibits not as mental ideas but as Platonic ideas or even as the denoted concrete objects. Semantics is vitiated by a pernicious mentalism as long as we regard a man's semantics as somehow determinate in his mind beyond what

[a] This is part of the first of two Dewey lectures delivered at Columbia University on 26 March 1968 and published under its original title, 'Ontological Relativity' in *The Journal of Philosophy*, volume LXV, no. 7, 4 April 1968.

[b] *Experience and Nature* (La Salle, Ill.: Open Court, 1925, 1958), p. 179.

might be implicit in his dispositions to overt behavior. It is the very facts about meaning, not the entities meant, that must be construed in terms of behavior.

There are two parts to knowing a word. One part is being familiar with the sound of it and being able to reproduce it. This part, the phonetic part, is achieved by observing and imitating other people's behavior, and there are no important illusions about the process. The other part, the semantic part, is knowing how to use the word. This part, even in the paradigm case, is more complex than the phonetic part. The word refers, in the paradigm case, to some visible object. The learner has now not only to learn the word phonetically, by hearing it from another speaker; he also has to see the object; and in addition to this, in order to capture the relevance of the object to the word, he has to see that the speaker also sees the object. Dewey summed up the point thus: 'The characteristic theory about B's understanding of A's sounds is that he responds to the thing from the standpoint of A' (178). Each of us, as he learns his language, is a student of his neighbor's behavior; and conversely, insofar as his tries are approved or corrected, he is a subject of his neighbor's behavioral study.

The semantic part of learning a word is more complex than the phonetic part, therefore, even in simple cases: we have to see what is stimulating the other speaker. In the case of words not directly ascribing observable traits to things, the learning process is increasingly complex and obscure; and obscurity is the breeding place of mentalistic semantics. What the naturalist insists on is that, even in the complex and obscure parts of language learning, the learner has no data to work with but the overt behavior of other speakers.

When with Dewey we turn thus toward a naturalistic view of language and a behavioral view of meaning, what we give up is not just the museum figure of speech. We give up an assurance of determinacy. Seen according to the museum myth, the words and sentences of a language have their determinate meanings. To discover the meanings of the native's words we may have to observe his behavior, but still the meanings of the words are supposed to be determinate in the native's *mind*, his mental museum, even in cases where behavioral criteria are powerless to discover them for us. When on the other hand we recognize with Dewey that 'meaning...is primarily a property of behavior', we recognize that there are no meanings, nor likenesses nor distinctions of meaning, beyond what are implicit in people's dispositions to overt behavior. For naturalism the question whether two expressions are alike or unlike in meaning has no determinate answer, known or unknown, except insofar as the answer is settled in principle by people's speech dispositions, known or unknown. If by these standards there are indeterminate cases, so much the worse for the terminology of meaning and likeness of meaning.

To see what such indeterminacy would be like, suppose there were an expression in a remote language that could be translated into English equally defensibly in either of two ways, unlike in meaning in English. I am not speaking of ambiguity within the native language. I am supposing that one and the same native use of the expression can be given either of the English translations, each being accommodated by compensating adjustments in the translation of other words. Suppose both translations, along with these accommodations in each case, accord equally well with all observable behavior on the part of speakers of the remote language and speakers of English. Suppose they accord perfectly not only with behavior actually observed, but with all dispositions to behavior on the part of all the speakers concerned. On these assumptions it would be forever impossible to know of one of these translations that it was the right one, and the other wrong. Still, if the museum myth were true, there would be a right and wrong of the matter; it is just that we would never know, not having access to the museum. See language naturalistically, on the other hand, and you have to see the notion of likeness of meaning in such a case simply as nonsense.

I have been keeping to the hypothetical. Turning now to examples, let me begin with a disappointing one and work up. In the French construction 'ne...rien' you can translate 'rien' into English as 'anything' or as 'nothing' at will, and then accommodate your choice by translating 'ne' as 'not' or by construing it as pleonastic. This example is disappointing because you can object that I have merely cut the French units too small. You can believe the mentalistic myth of the meaning museum and still grant that 'rien' of itself has no meaning, being no whole label; it is part of 'ne...rien', which has its meaning as a whole.

I began with this disappointing example because I think its conspicuous trait – its dependence on cutting language into segments too short to carry meanings – is the secret of the more serious cases as well. What makes other cases more serious is that the segments they involve are seriously long: long enough to be predicates and to be true of things and hence, you would think, to carry meanings.

An artificial example which I have used elsewhere[a] depends on the fact that a whole rabbit is present when and only when an undetached part of a rabbit is present; also when and only when a temporal stage of a rabbit is present. If we are wondering whether to translate a native expression 'gavagai' as 'rabbit' or as 'undetached rabbit part' or as 'rabbit stage', we can never settle the matter simply by ostension – that is, simply by repeatedly querying the expression 'gavagai' for the native's assent or dissent in the presence of assorted stimulations.

Before going on to urge that we cannot settle the matter by nonostensive means either, let me belabor this ostensive predicament a bit. I am not worrying, as Wittgenstein did, about simple cases of ostension. The color word 'sepia', to take one of his examples,[b] can certainly be learned by an ordinary process of conditioning, or induction. One need not even be told that sepia is a color and not a shape or a material or an article. True, barring such hints, many lessons may be needed, so as to eliminate wrong generalizations based on shape, material, etc., rather than color, and so as to eliminate wrong notions as to the intended boundary of an indicated example, and so as to delimit the admissible variations of color itself. Like all conditioning, or induction, the process will depend ultimately also on one's own inborn propensity to find one stimulation qualitatively more akin to a second stimulation than to a third; otherwise there can never be any selective reinforcement and extinction of responses.[c] Still, in principle nothing more is needed in learning 'sepia' than in any conditioning or induction.

But the big difference between 'rabbit' and 'sepia' is that whereas 'sepia' is a mass term like 'water', 'rabbit' is a term of divided reference. As such it cannot be mastered without mastering its principle of individuation: where one rabbit leaves off and another begins. And this cannot be mastered by pure ostension, however persistent.

Such is the quandary over 'gavagai': where one gavagai leaves off and another begins. The only difference between rabbits, undetached rabbit parts, and rabbit

[a] *Word and Object* (Cambridge, Mass.: MIT Press, 1960), §12.
[b] *Philosophical Investigations* (New York: Macmillan, 1953), p. 14.
[c] Cf. *Word and Object*, §17.

stages is in their individuation. If you take the total scattered portion of the spatio-temporal world that is made up of rabbits, and that which is made up of undetached rabbit parts, and that which is made up of rabbit stages, you come out with the same scattered portion of the world each of the three times. The only difference is in how you slice it. And how to slice it is what ostension or simple conditioning, however persistently repeated, cannot teach.

Thus consider specifically the problem of deciding between 'rabbit' and 'undetached rabbit part' as translation of 'gavagai'. No word of the native language is known, except that we have settled on some working hypothesis as to what native words or gestures to construe as assent and dissent in response to our pointings and queryings. Now the trouble is that whenever we point to different parts of the rabbit, even sometimes screening the rest of the rabbit, we are pointing also each time to the rabbit. When, conversely, we indicate the whole rabbit with a sweeping gesture, we are still pointing to a multitude of rabbit parts. And note that we do not have even a native analogue of our plural ending to exploit, in asking 'gavagai?'. It seems clear that no even tentative decision between 'rabbit' and 'undetached rabbit part' is to be sought at this level.

How would we finally decide? My passing mention of plural endings is part of the answer. Our individuating of terms of divided reference, in English, is bound up with a cluster of interrelated grammatical particles and constructions: plural endings, pronouns, numerals, the 'is' of identity, and its adaptations 'same' and 'other'. It is the cluster of interrelated devices in which quantification becomes central when the regimentation of symbolic logic is imposed. If in his language we could ask the native 'Is this *gavagai* the same as that one?' while making appropriate multiple ostensions, then indeed we would be well on our way to deciding between 'rabbit', 'undetached rabbit part', and 'rabbit stage'. And of course the linguist does at length reach the point where he can ask what purports to be that question. He develops a system for translating our pluralizations, pronouns, numerals, identity, and related devices contextually into the native idiom. He develops such a system by abstraction and hypothesis. He abstracts native particles and constructions from observed native sentences, and tries associating these variously with English particles and constructions. Insofar as the native sentences and the thus associated English ones seem to match up in respect of appropriate occasions of use, the linguist feels confirmed in these hypotheses of translation – what I call *analytical hypotheses.*[a]

But it seems that this method, though laudable in practice and the best we can hope for, does not in principle settle the indeterminacy between 'rabbit', 'undetached rabbit part', and 'rabbit stage'. For if one workable over-all system of analytical hypotheses provides for translating a given native expression into 'is the same as', perhaps another equally workable but systematically different system would translate that native expression rather into something like 'belongs with'. Then when in the native language we try to ask 'Is this *gavagai* the same as that?', we could as well be asking 'Does this *gavagai* belong with that?'. Insofar, the native's assent is no objective evidence for translating 'gavagai' as 'rabbit' rather than 'undetached rabbit part' or 'rabbit stage'.

[a] *Word and Object*, §15. For a summary of the general point of view see also §1 of my 'Speaking of Objects', *Proceedings and Addresses of American Philosophical Association*, XXXI (1958): 5 ff; reprinted in Y. Krikorian and A. Edel, eds., *Contemporary Philosophical Problems* (New York: Macmillan, 1959), and in J. Fodor and J. Katz, eds., *The Structure of Language* (Englewood Cliffs, N.J.: Prentice-Hall, 1964), and in P. Kurtz, ed., *American Philosophy in the Twentieth Century* (New York: Macmillan, 1966).

This artificial example shares the structure of the trivial earlier example 'ne...
rien'. We were able to translate 'rien' as 'anything' or as 'nothing', thanks to a
compensatory adjustment in the handling of 'ne'. And I suggest that we can translate 'gavagai' as 'rabbit' or 'undetached rabbit part' or 'rabbit stage', thanks to
compensatory adjustments in the translation of accompanying native locutions.
Other adjustments still might accommodate translation of 'gavagai' as 'rabbithood',
or in further ways. I find this plausible because of the broadly structural and contextual character of any considerations that could guide us to native translations of
the English cluster of interrelated devices of individuation. There seem bound to be
systematically very different choices, all of which do justice to all dispositions to
verbal behavior on the part of all concerned.

An actual field linguist would of course be sensible enough to equate 'gavagai'
with 'rabbit', dismissing such perverse alternatives as 'undetached rabbit part' and
'rabbit stage' out of hand. This sensible choice and others like it would help in turn
to determine his subsequent hypotheses as to what native locutions should answer to
the English apparatus of individuation, and thus everything would come out all
right. The implicit maxim guiding his choice of 'rabbit', and similar choices for
other native words, is that an enduring and relatively homogeneous object, moving as
a whole against a contrasting background is a likely reference for a short expression.
If he were to become conscious of this maxim, he might celebrate it as one of the
linguistic universals, or traits of all languages, and he would have no trouble pointing
out its psychological plausibility. But he would be wrong; the maxim is his own
imposition, toward settling what is objectively indeterminate. It is a very sensible
imposition, and I would recommend no other. But I am making a philosophical
point.

It is philosophically interesting, moreover, that what is indeterminate in this
artificial example is not just meaning, but extension; reference. My remarks on
indeterminacy began as a challenge to likeness of meaning. I had us imagining 'an
expression that could be translated into English equally defensibly in either of two
ways, unlike in meaning in English'. Certainly likeness of meaning is a dim notion,
repeatedly challenged. Of two predicates which are alike in extension, it has never
been clear when to say that they are alike in meaning and when not; it is the old
matter of featherless bipeds and rational animals, or of equiangular and equilateral
triangles. Reference, extension, has been the firm thing; meaning, intension, the
infirm. The indeterminacy of translation now confronting us, however, cuts across
extension and intension alike. The terms 'rabbit', 'undetached rabbit part', and
'rabbit stage' differ not only in meaning; they are true of different things. Reference
itself proves behaviorally inscrutable.

Within the parochial limits of our own language, we can continue as always to find
extensional talk clearer than intensional. For the indeterminacy between 'rabbit',
'rabbit stage', and the rest depended only on a correlative indeterminacy of translation of the English apparatus of individuation – the apparatus of pronouns,
pluralization, identity, numerals, and so on. No such indeterminacy obtrudes so
long as we think of this apparatus as given and fixed. Given this apparatus, there is
no mystery about extension; terms have the same extension when true of the same
things. At the level of radical translation, on the other hand, extension itself goes
inscrutable.

My example of rabbits and their parts and stages is a contrived example and a
perverse one, with which, as I said, the practicing linguist would have no patience.

But there are also cases, less bizarre ones, that obtrude in practice. In Japanese there are certain particles, called 'classifiers', which may be explained in either of two ways. Commonly they are explained as attaching to numerals, to form compound numerals of distinctive styles. Thus take the numeral for 5. If you attach one classifier to it you get a style of '5' suitable for counting animals; if you attach a different classifier, you get a style of '5' suitable for counting slim things like pencils and chopsticks; and so on. But another way of viewing classifiers is to view them not as constituting part of the numeral, but as constituting part of the term – the term for 'chopsticks' or 'oxen' or whatever. On this view the classifier does the individuative job that is done in English by 'sticks of' as applied to the mass term 'wood', or 'head of' as applied to the mass term 'cattle'.

What we have on either view is a Japanese phrase tantamount say to 'five oxen', but consisting of three words;[a] the first is in effect the neutral numeral '5', the second is a classifier of the animal kind, and the last corresponds in some fashion to 'ox'. On one view the neutral numeral and the classifier go together to constitute a declined numeral in the 'animal gender', which then modifies 'ox' to give, in effect, 'five oxen'. On the other view the third Japanese word answers not to the individuative term 'ox' but to the mass term 'cattle'; the classifier applies to this mass term to produce a composite individuative term, in effect 'head of cattle'; and the neutral numeral applies directly to all this without benefit of gender, giving 'five head of cattle', hence again in effect 'five oxen'.

If so simple an example is to serve its expository purpose, its needs your connivance. You have to understand 'cattle' as a mass term covering only bovines, and 'ox' as applying to all bovines. That these usages are not the invariable usages is beside the point. The point is that the Japanese phrase comes out as 'five bovines', as desired, when parsed in either of two ways. The one way treats the third Japanese word as an individuative term true of each bovine, and the other way treats that word rather as a mass term covering the unindividuated totality of beef on the hoof. These are two very different ways of treating the third Japanese word; and the three-word phrase as a whole turns out all right in both cases only because of compensatory differences in our account of the second word, the classifier.

This example is reminiscent in a way of our trivial initial example, 'ne...rien'. We were able to represent 'rien' as 'anything' or as 'nothing', by compensatorily taking 'ne' as negative or as vacuous. We are able now to represent a Japanese word either as an individuative term for bovines or as a mass term for live beef, by compensatorily taking the classifier as declining the numeral or as individuating the mass term. However, the triviality of the one example does not quite carry over to the other. The early example was dismissed on the ground that we had cut too small: 'rien' was too short for significant translation on its own, and 'ne...rien' was the significant unit. But you cannot dismiss the Japanese example by saying that the third word was too short for significant translation on its own and that only the whole three-word phrase, tantamount to 'five oxen', was the significant unit. You cannot take this line unless you are prepared to call a word too short for significant translation even when it is long enough to be a term and carry denotation. For the third Japanese word is, on either approach, a term; on one approach a term of divided reference, and on the other a mass term. If you are indeed prepared thus to call a word too short for significant translation even when it is a denoting term, then in a

[a] To keep my account graphic I am counting a certain postpositive particle as a suffix rather than a word.

back-handed way you are granting what I wanted to prove: the inscrutability of reference.

Between the two accounts of Japanese classifiers there is no question of right and wrong. The one account makes for more efficient translation into idiomatic English; the other makes for more of a feeling for the Japanese idiom. Both fit all verbal behavior equally well. All whole sentences, and even component phrases like 'five oxen', admit of the same net over-all English translations on either account. This much is invariant. But what is philosophically interesting is that the reference or extension of shorter terms can fail to be invariant. Whether that third Japanese word is itself true of each ox, or whether on the other hand it is a mass term which needs to be adjoined to the classifier to make a term which is true of each ox – here is a question that remains undecided by the totality of human dispositions to verbal behavior. It is indeterminate in principle; there is no fact of the matter. Either answer can be accommodated by an account of the classifier. Here again, then, is the inscrutability of reference – illustrated this time by a humdrum point of practical translation.

The inscrutability of reference can be brought closer to home by considering the word 'alpha', or again the word 'green'. In our use of these words and others like them there is a systematic ambiguity. Sometimes we use such words as concrete general terms, as when we say the grass is green, or that some inscription begins with an alpha. Sometimes on the other hand we use them as abstract singular terms, as when we say that green is a color and alpha is a letter. Such ambiguity is encouraged by the fact that there is nothing in ostension to distinguish the two uses. The pointing that would be done in teaching the concrete general term 'green', or 'alpha', differs none from the pointing that would be done in teaching the abstract singular term 'green' or 'alpha'. Yet the objects referred to by the word are very different under the two uses: under the one use the word is true of many concrete objects, and under the other use it names a single abstract object.

We can of course tell the two uses apart by seeing how the word turns up in sentences: whether it takes an indefinite article, whether it takes a plural ending, whether it stands as singular subject, whether it stands as modifier, as predicate complement, and so on. But these criteria appeal to our special English grammatical constructions and particles, our special English apparatus of individuation, which, I already urged, is itself subject to indeterminacy of translation. So, from the point of view of translation into a remote language, the distinction between a concrete general and an abstract singular term is in the same predicament as the distinction between 'rabbit', 'rabbit part', and 'rabbit stage'. Here then is another example of the inscrutability of reference, since the difference between the concrete general and the abstract singular is a difference in the objects referred to.

Incidentally we can concede this much indeterminacy also to the 'sepia' example, after all. But this move is not evidently what was worrying Wittgenstein.

The ostensive indistinguishability of the abstract singular from the concrete general turns upon what may be called 'deferred ostension', as opposed to direct ostension. First let me define direct ostension. The *ostended point*, as I shall call it, is the point where the line of the pointing finger first meets an opaque surface. What characterizes *direct ostension*, then, is that the term which is being ostensively explained is true of something that contains the ostended point. Even such direct ostension has its uncertainties, of course, and these are familiar. There is the question how wide an environment of the ostended point is meant to be covered by the term that is being ostensively explained. There is the question how considerably an absent thing or substance might be allowed to differ from what is now ostended, and still be covered by the term that is now being ostensively explained. Both of these questions can in principle be settled as well as need be by induction from multiple ostensions. Also, if the term is a term of divided reference like 'apple', there is the question of individuation: the question where one of its objects leaves off and another begins. This can be settled by induction from multiple ostensions of a more elaborate kind, accompanied by expressions like 'same apple' and 'another', if an equivalent of this English apparatus of individuation has been settled on; otherwise the indeterminacy persists that was illustrated by 'rabbit', 'undetached rabbit part', and 'rabbit stage'.

Such, then, is the way of direct ostension. Other ostension I call *deferred*. It occurs when we point at the gauge, and not the gasoline, to show that there is gasoline. Also it occurs when we explain the abstract singular term 'green' or 'alpha' by pointing at grass or a Greek inscription. Such pointing is direct ostension when used to explain the concrete general term 'green' or 'alpha', but it is deferred ostension when used to explain the abstract singular terms; for the abstract object which is the color green or the letter alpha does not contain the ostended point, nor any point.

Deferred ostension occurs very naturally when, as in the case of the gasoline gauge, we have a correspondence in mind. Another such example is afforded by the Gödel numbering of expressions. Thus if 7 has been assigned as Gödel number of the letter alpha, a man conscious of the Gödel numbering would not hesitate to say 'Seven' on pointing to an inscription of the Greek letter in question. This is, on the face of it, a doubly deferred ostension: one step of deferment carries us from the inscription to the letter as abstract object, and a second step carries us thence to the number.

By appeal to our apparatus of individuation, if it is available, we can distinguish between the concrete general and the abstract singular use of the word 'alpha'; this we saw. By appeal again to that apparatus, and in particular to identity, we can evidently settle also whether the word 'alpha' in its abstract singular use is being used really to name the letter or whether, perversely, it is being used to name the Gödel number of the letter. At any rate we can distinguish these alternatives if also we have located the speaker's equivalent of the numeral '7' to our satisfaction; for we can ask him whether alpha *is* 7.

These considerations suggest that deferred ostension adds no essential problem to those presented by direct ostension. Once we have settled upon analytical hypotheses of translation covering identity and the other English particles relating to individuation, we can resolve not only the indecision between 'rabbit' and 'rabbit stage' and the rest, which came of direct ostension, but also any indecision between concrete general and abstract singular, and any indecision between expression and Gödel number, which come of deferred ostension.

However, this conclusion is too sanguine. The inscrutability of reference runs deep, and it persists in a subtle form even if we accept identity and the rest of the apparatus of individuation as fixed and settled; even, indeed, if we forsake radical translation and think only of English.

Consider the case of a thoughtful protosyntactician. He has a formalized system of first-order proof theory, or protosyntax, whose universe comprises just expressions, that is, strings of signs of a specified alphabet. Now just what sorts of things, more specifically, are these expressions? They are types, not tokens. So, one might suppose, each of them is the set of all its tokens. That is, each expression is a set of inscriptions which are variously situated in space-time but are classed together by virtue of a certain similarity in shape. The concatenate xy of two expressions x and y, in a given order, will be the set of all inscriptions each of which has two parts which are tokens respectively of x and y and follow one upon the other in that order. But xy may then be the null set, though x and y are not null; for it may be that inscriptions belonging to x and y happen to turn up head to tail nowhere, in the past, present, or future. This danger increases with the lengths of x and y. But it is easily seen to violate a law of protosyntax which says that $x = z$ whenever $xy = zy$.

Thus it is that our thoughtful protosyntactician will not construe the things in his universe as sets of inscriptions. He can still take his atoms, the single signs, as sets of inscriptions, for there is no risk of nullity in these cases. And then, instead of taking his strings of signs as sets of inscriptions, he can invoke the mathematical notion of sequence and take them as sequences of signs. A familiar way of taking sequences, in turn, is as a mapping of things on numbers. On this approach an expression or string of signs becomes a finite set of pairs, each of which is the pair of a sign and a number.

This account of expressions is more artificial and more complex than one is apt to expect who simply says he is letting his variables range over the strings of such and such signs. Moreover, it is not the inevitable choice; the considerations that motivated it can be met also by alternative constructions. One of these constructions is Gödel numbering itself, and it is temptingly simple. It uses just natural numbers, whereas the foregoing construction used sets of one-letter inscriptions and also natural numbers and sets of pairs of these. How clear is it that at just *this* point we have dropped expressions in favor of numbers? What is clearer is merely that in both constructions we were artificially devising models to satisfy laws that expressions in an unexplicated sense had been meant to satisfy.

So much for expressions. Consider now the arithmetician himself, with his elementary number theory. His universe comprises the natural numbers outright. Is it clearer than the protosyntactician's? What, after all, is a natural number? There are Frege's version, Zermelo's, and von Neumann's, and countless further alternatives, all mutually incompatible and equally correct. What we are doing in any one of these explications of natural number is to devise set-theoretic models to satisfy laws which the natural numbers in an unexplicated sense had been meant to satisfy. The case is quite like that of protosyntax.

It will perhaps be felt that any set-theoretic explication of natural number is at best a case of *obscurum per obscurius*; that all explications must assume something, and the natural numbers themselves are an admirable assumption to start with. I must agree that a construction of sets and set theory from natural numbers and arithmetic would be far more desirable than the familiar opposite. On the other hand our impression of the charity even of the notion of natural number itself has suffered somewhat from Gödel's proof of the impossibility of a complete proof procedure for elementary number theory, or, for that matter, from Skolem's and Henkin's observations that all laws of natural numbers admit nonstandard models.[a]

We are finding no clear difference between *specifying* a universe of discourse – the range of the variables of quantification – and *reducing* that universe to some other. We saw no significant difference between clarifying the notion of expression and supplanting it by that of number. And now to say more particularly what numbers

[a] See Leon Henkin, 'Completeness in the Theory of Types', *Journal of Symbolic Logic*, xv, 2 (June 1950): 81–91, and references therein.

themselves are is in no evident way different from just dropping numbers and assigning to arithmetic one or another new model, say in set theory.

Expressions are known only by their laws, the laws of concatenation theory, so that any constructs obeying those laws – Gödel numbers, for instance – are *ipso facto* eligible as explications of expression. Numbers in turn are known only by their laws, the laws of arithmetic, so that any constructs obeying those laws – certain sets, for instance – are eligible in turn as explications of number. Sets in turn are known only by their laws, the laws of set theory.

Russell pressed a contrary thesis, long ago. Writing of numbers, he argued that for an understanding of number the laws of arithmetic are not enough; we must know the applications, we must understand numerical discourse embedded in discourse of other matters. In applying number the key notion, he urged, is *Anzahl*: there are *n* so-and-sos. However, Russell can be answered. First take, specifically, *Anzahl*. We can define 'there are *n* so-and-sos' without ever deciding what numbers are, apart from their fulfillment of arithmetic. That there are *n* so-and-sos can be explained simply as meaning that the so-and-sos are in one-to-one correspondence with the numbers up to *n*.[a]

Russell's more general point about application can be answered too. Always, if the structure is there, the applications will fall into place. As paradigm it is perhaps sufficient to recall again this reflection on expressions and Gödel numbers: that even the pointing out of an inscription is no final evidence that our talk is of expressions and not of Gödel numbers. We can always plead deferred ostension.

It is in this sense true to say, as mathematicians often do, that arithmetic is all there is to number. But it would be a confusion to express this point by saying, as is sometimes said, that numbers are any things fulfilling arithmetic. This formulation is wrong because distinct domains of objects yield distinct models of arithmetic. Any progression can be made to serve; and to identify all progressions with one another, e.g., to identify the progression of odd numbers with the progression of evens, would contradict arithmetic after all.

So, though Russell was wrong in suggesting that numbers need more than their arithmetical properties, he was right in objecting to the definition of numbers as any things fulfilling arithmetic. The subtle point is that any progression will serve as a version of number so long and only so long as we stick to one and the same progression. Arithmetic is, in this sense, all there is to number: there is no saying absolutely what the numbers are; there is only arithmetic.[b]

2 I first urged the inscrutability of reference with the help of examples like the one about rabbits and rabbit parts. These used direct ostension, and the inscrutability of reference hinged on the indeterminacy of translation of identity and other individuative apparatus. The setting of these examples, accordingly, was radical translation: translation from a remote language on behavioral evidence, unaided by prior dictionaries. Moving then to deferred ostension and abstract objects, we found a certain dimness of reference pervading the home language itself.

Now it should be noted that even for the earlier examples the resort to a remote

---

[a] For more on this theme see my *Set Theory and Its Logic* (Cambridge, Mass.: Harvard 1963, 1968), §11.

[b] Paul Benacerraf, 'What Numbers Cannot Be', *Philosophical Review*, LXXIV, 1 (January 1965): 47–73, develops this point. His conclusions differ in some ways from those I shall come to.

language was not really essential. On deeper reflection, radical translation begins at home. Must we equate our neighbor's English words with the same strings of phonemes in our own mouths? Certainly not; for sometimes we do not thus equate them. Sometimes we find it to be in the interests of communication to recognize that our neighbor's use of some word, such as 'cool' or 'square' or 'hopefully', differs from ours, and so we translate that word of his into a different string of phonemes in our idiolect. Our usual domestic rule of translation is indeed the homophonic one, which simply carries each string of phonemes into itself; but still we are always prepared to temper homophony with what Neil Wilson has called the 'principle of charity'.[a] We will construe a neighbor's word heterophonically now and again if thereby we see our way to making his message less absurd.

The homophonic rule is a handy one on the whole. That it works so well is no accident, since imitation and feedback are what propagate a language. We acquired a great fund of basic words and phrases in this way, imitating our elders and encouraged by our elders amid external circumstances to which the phrases suitably apply. Homophonic translation is implicit in this social method of learning. Departure from homophonic translation in this quarter would only hinder communication. Then there are the relatively rare instances of opposite kind, due to divergence in dialect or confusion in an individual, where homophonic translation incurs negative feedback. But what tends to escape notice is that there is also a vast mid-region where the homophonic method is indifferent. Here, gratuitously, we can systematically reconstrue our neighbor's apparent references to rabbits as really references to rabbit stages, and his apparent references to formulas as really references to Gödel numbers and vice versa. We can reconcile all this with our neighbor's verbal behavior, by cunningly readjusting our translations of his various connecting predicates so as to compensate for the switch of ontology. In short, we can reproduce the inscrutability of reference at home. It is of no avail to check on this fanciful version of our neighbor's meanings by asking him, say, whether he really means at a certain point to refer to formulas or to their Gödel numbers; for our question and his answer – 'By all means, the numbers' – have lost their title to homophonic translation. The problem at home differs none from radical translation ordinarily so called except in the willfulness of this suspension of homophonic translation.

I have urged in defense of the behavioral philosophy of language, Dewey's, that the inscrutability of reference is not the inscrutability of a fact; there is no fact of the matter. But if there is really no fact of the matter, then the inscrutability of reference can be brought even closer to home than the neighbor's case; we can apply it to ourselves. If it is to make sense to say even of oneself that one is referring to rabbits and formulas and not to rabbit stages and Gödel numbers, then it should make sense equally to say it of someone else. After all, as Dewey stressed, there is no private language.

We seem to be maneuvering ourselves into the absurd position that there is no difference on any terms, interlinguistic or intralinguistic, objective or subjective, between referring to rabbits and referring to rabbit parts or stages; or between referring to formulas and referring to their Gödel numbers. Surely this is absurd, for it would imply that there is no difference between the rabbit and each of its parts or stages, and no difference between a formula and its Gödel number. Reference would seem now to become nonsense not just in radical translation, but at home.

[a] N. L. Wilson, 'Substances without Substrata', *Review of Metaphysics*, XII, 4 (June 1959): 521–39, p 532.

Toward resolving this quandary, begin by picturing us at home in our language, with all its predicates and auxiliary devices. This vocabulary includes 'rabbit', 'rabbit part', 'rabbit stage', 'formula', 'number', 'ox', 'cattle'; also the two-place predicates of identity and difference, and other logical particles. In these terms we can say in so many words that this is a formula and that a number, this a rabbit and that a rabbit part, this and that the same rabbit, and this and that different parts. *In just those words.* This network of terms and predicates and auxiliary devices is, in relativity jargon, our frame of reference, or coordinate system. Relative to *it* we can and do talk meaningfully and distinctively of rabbits and parts, numbers and formulas. Next, as in recent paragraphs, we contemplate alternative denotations for our familiar terms. We begin to appreciate that a grand and ingenious permutation of these denotations, along with compensatory adjustments in the interpretations of the auxiliary particles, might still accommodate all existing speech dispositions. This was the inscrutability of reference, applied to ourselves; and it made nonsense of reference. Fair enough; reference *is* nonsense except relative to a coordinate system. In this principle of relativity lies the resolution of our quandary.

It is meaningless to ask whether, in general, our terms 'rabbit', 'rabbit part', 'number', etc., really refer respectively to rabbits, rabbit parts, numbers, etc., rather than to some ingeniously permuted denotations. It is meaningless to ask this absolutely; we can meaningfully ask it only relative to some background language. When we ask, 'Does "rabbit" really refer to rabbits?' someone can counter with the question: 'Refer to rabbits in what sense of "rabbits"?' thus launching a regress; and we need the background language to regress into. The background language gives the query sense, if only relative sense; sense relative in turn to it, this background language. Querying reference in any more absolute way would be like asking absolute position, or absolute velocity, rather than position or velocity relative to a given frame of reference. Also it is very much like asking whether our neighbor may not systematically see everything upside down, or in complementary color, forever undetectably.

We need a background language, I said, to regress into. Are we involved now in an infinite regress? If questions of reference of the sort we are considering make sense only relative to a background language, then evidently questions of reference of the background language make sense in turn only relative to a further background language. In these terms the situation sounds desperate, but in fact it is little different from questions of position and velocity. When we are given position and velocity relative to a given coordinate system, we can always ask in turn about the placing of origin and orientation of axes of that system of coordinates; and there is no end to the succession of further coordinate systems that could be adduced in answering the successive questions thus generated.

In practice of course we end the regress of coordinate systems by something like pointing. And in practice we end the regress of background languages, in discussions of reference, by acquiescing in our mother tongue and taking its words at face value.

Very well; in the case of position and velocity, in practice, pointing breaks the regress. But what of position and velocity apart from practice? what of the regress then? The answer, of course, is the relational doctrine of space; there is no absolute position or velocity; there are just the relations of coordinate systems to one another, and ultimately of things to one another. And I think that the parallel question regarding denotation calls for a parallel answer, a relational theory of what the objects of theories are. What makes sense is to say not what the objects of a theory are, absolutely speaking, but how one theory of objects is interpretable or reinter-pretable in another.

The point is not that bare matter is inscrutable: that things are indistinguishable except by their properties. That point does not need making. The present point is reflected better in the riddle about seeing things upside down, or in complementary colors; for it is that things can be inscrutably switched even while carrying their properties with them. Rabbits differ from rabbit parts and rabbit stages not just as bare matter, after all, but in respect of properties; and formulas differ from numbers in respect of properties.

Harvard University

PART II

LINGUISTICS

Overview

HOWARD MACLAY

It must seem perfectly obvious to the ordinary speaker of the language that meaning is a central and crucial element in his linguistic activity and that no account of his language which ignores this vital factor can possibly be adequate. While most scientific linguists also have acknowledged in passing the general importance of the semantic aspect of language, meaning has come to be widely regarded as a legitimate object of systematic linguistic interest only within the past decade. A collection of papers such as this one would have hardly been conceivable in the mid 1950s. In fact, most linguists of that period tended to regard a concern with meaning as evidence of a certain soft-headedness and lack of genuine scientific commitment. That this anthology is now available signals a significant shift in the prevailing conception of the goals and content of a valid linguistic description. In order to evaluate the essays which follow it is necessary to consider some of the attitudes toward meaning which have characterized the two dominant American schools of descriptive linguistics.[a]

Although all of the papers in this section either proceed directly from the assumptions of generative-transformational grammar or have been markedly influenced by this approach, they represent different tendencies within this framework, and they also differ in some important respects both from earlier transformational work and from studies based on a structuralist approach. Table 1 is an attempt to summarize, along a number of relevant dimensions, the central tendencies with regard to semantics characteristic of American structural linguistics and of four varieties of generative-transformational linguists. Figures 1–5 present diagrammatic representations of the interrelations among the components of a linguistic description as defined by each of these five approaches. These should be regarded as visual aids for the discussion to follow rather than as any sort of definitive account of the views of these schools.[b]

I will first describe the general organization of the analysis and then consider in more detail the content of each position. The discussion will attempt to present a simplified description of each position, omitting many details. Readers may consult the references listed in each section for a more complete and adequate account.

---

[a] There seems little point in merely presenting a concise abstract of each paper in an introductory essay of this sort. Rather, an attempt will be made to provide a relevant context within which readers may arrive at their own evaluations.

[b] Readers should be particularly wary of the directionality of the arrows in these diagrams as this is an issue of much dispute in linguistics at the present time. See the papers by Chomsky and Lakoff in this volume for further discussion. I believe the representations given here reflect accurately the practice of linguists at each period regardless of the ultimate logical relationship among the components of a grammar. In any case, most of them are based on published diagrams.

The entries of Table 1 are organized in chronological order with the numbers following each of the transformational positions indicating temporal order within this school (i.e. GT-3A and GT-3B are contemporaneous). A number of representative works are listed under each category and their dates provide a more precise index of the chronology involved. The fact that several of the items at the GT-3 level are included in this volume shows the very recent occurrence of this split in transformational theory.

TABLE **I**

	I Semantics part of a linguistic description?	II Form (syntax) independent of meaning?	III Semantics interpretive?	IV An independent level of syntactic deep structure?	V Semantic interpretations based exclusively on deep structure?
SL Bloomfield (1933) Harris (1951)	No	Yes	—	—	—
GT-1 Chomsky (1957) Lees (1960)	No	Yes	—	—	—
GT-2 Chomsky (1965) Katz and Postal (1964)	Yes	Yes	Yes	Yes	Yes
GT-3A Chomsky (this volume)	Yes	Yes	Yes	Yes	No
GT-3B Lakoff and McCawley (this volume)	Yes	No	No	No	—

The dimensions are presented in order of importance from left to right. They are at least partially independent in that a score on any one of them does not necessarily predict with complete accuracy the score on any other. Indeed, it is the various dependencies of fact and principle among dimensions that are most relevant in understanding these approaches to semantics. *Dimension I* simply asks whether or not meaning is to be regarded as a proper object of systematic linguistic analysis. If this is denied, on whatever grounds, then the score on *Dimension II* (Is syntax/form independent of semantics?) is determined, since a grammar must be seen as confined to formal syntax (and, perhaps, phonology) which is, necessarily, independent of semantic considerations. If, however, it is argued that meaning is a proper part of a linguistic description, it is possible either to affirm or deny the independence of syntax. *Dimension III* (Is semantics interpretive?) is relevant only if a positive response has been given on Dimension 1. Given that meaning must be included and that syntax is independent of it, one can argue that semantics is interpretive (i.e. that the semantic component of a grammar must have syntactic information as input as in GT-2 and GT-3A) or, conversely, that the reverse condition exists. If the relevance of meaning is affirmed and the autonomy of syntax denied (GT-3B) then there is no reasonable sense in which one of the factors can be said to dominate the other. While the generative semanticists have surely moved semantics into a central role in linguistics this has involved the denial of a clear syntax/semantics distinction rather than an inversion of the relationship. *Dimension IV* (Is there an independent level of syntactic deep structure?) applies directly only to GT-2, GT-3A and GT-3B. It is factually redundant with respect to Dimension III for these theories, but there seems no necessary reason why one could not propose that the independent syntactic component of grammar be organized without this distinction. That the linguists of GT-3B would deny this division follows from their rejection of independent syntax and interpretive semantics. The final dimension (V) functions to distinguish between GT-2 and GT-3A and involves the question of which part or parts of the syntactic component provide the information necessary for semantic interpretation.

Structural linguistics

The views of American structureal linguists during the period in which this school was dominant in descriptive linguistics are well represented in Bloomfield (1933) and Harris (1951). As the diagram in Figure 1 suggests, structuralists proposed a model of grammar consisting of several different levels of analysis; the importance of keeping the levels separate being particularly emphasized. This device takes as initial input a body of observable linguistic data consisting of phonetically transcribed utterances along with judgments by native speaker informants as to the sameness or difference, primarily of pairs of words, but sometimes of longer phrases and sentences. It was argued that this primary data could be processed by explicit methods of analysis so as to produce an identification and classification of higher categories such as phonemes and morphemes. The model involves a strong linear directionality away from the primary data. This means that the input to each level must come entirely from the preceding level. One cannot, for example, use morphological information in the identification of phonemes. The approach is operational in that the higher abstract categories of the grammar must be clearly connected to observable data by a series of analytic procedures applied to that data. Thus a *phoneme*, while it might be roughly defined as a functionally significant class of sounds or phones which does not bear meaning, is, strictly speaking, no more than the result of a set of operations performed on the primary data. The goals of this analysis are essentially classificatory, thus the term *taxonomic* is often applied to this school of linguistics. All of the operations are based on the notion of *formal distribution* which is for any element, the list of immediate environments defined by its co-occurrence with other elements of the same type. Two ordered operations are used in the analysis of distributions which we may label *contrast* and *substitution*. The functionally significant units of each component (phonemes and morphemes) are identified by *contrast*. Given distributions for the set of elements at Level N (phones, morphs), the elements of Level $N+1$ (phonemes, morphemes) are formed in the following way. If two elements of N occur in one or more of the same environments (i.e. if their distributions overlap), they are in *contrastive distribution* and must become members of different elements at Level $N+1$. If, on the other hand, their distributions do not overlap, they are in *complementary distribution*. Under this condition, the elements are eligible for membership in the same element of N + 1. Whether or not they are placed in the same element of N + 1 depends, in practice, on such non-distributional considerations as phonetic similarity (for phonemes), meaning (for morphemes) and perhaps some notion of symmetry. The operations are then repeated in the same way for the next highest component which takes as input a transcription provided by the lower component.

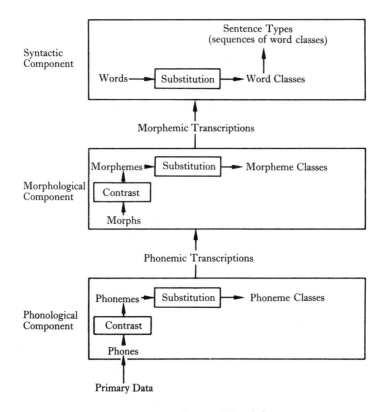

FIGURE I. *Structural linguistics*

Given, now, a set of elements on N + 1, new distributions for these are calculated and the elements classified on this level by *substitution*. If two or more elements occur in many of the same environments (i.e. their distributions are similar), they are placed in the same class. It should be noted that both operations can apply to the same data. *Contrast* is involved in the transition from level to level within components while *substitution* operates only within a given level.

An example from English may be helpful. Consider the following voiceless stops where [p] is bilabial, [t] is alveolar, [k] is velar and [h] indicates aspiration. Suppose that the distribution of each is represented accurately below where the environments are given in English orthography and the element under consideration is under-lined.

[pʰ]	[p]	[tʰ]	[t]	[kʰ]	[k]
pat	spat	tat		cat	scat
pan	span	tan	Stan	can	scan
pin	spin	tin		kin	skin
pub		tub	stub	cub	
	spoon	tune		coon	

Applying *contrast* to these distributions we may note that [p], [t] and [k] all occur only in the general environment # s–V (where # is a word boundary and V is any vowel), while [pʰ], [tʰ], and [kʰ] occur only in the environment # –V. Thus, [pʰ], for example, is in contrast with [tʰ] and [kʰ] and in complementary distribution with [p], [t] and [k]. On the basis of this evidence, along with considerations of phonetic similarity, the phonemic solution would be:

/p/ : [pʰ], [p]

/t/ : [tʰ], [t]

/k/ : [kʰ], [k]

(where / / indicates a phoneme and [ ] a phone)

Substitution is then applied to classify /p/, /t/, /k/ together on the grounds that they occur in many of the same environments when new distributions are calculated for the phonemic level. If substitution is seriously applied to *all* environments on this level the result is normally a set of intersecting classes and, given the interest of linguists in discrete categories, this may account, in part, for the relative neglect of this operation in phonology in favor of a classification based on phonetic similarity.

These operations are repeated on the level of morphology producing morphs and morphemes. Thus, the phonemically distinct English plural morphs /-s/, /-z/ and /-əz/ are classified together in the same morpheme on the basis of their complementary distribution in such words as cat*s*, dog*s*, and fox*es*.

Perhaps because syntax is so far removed from the basic data, its position in a structural analysis is rather insecure.[a] A structuralist asked to describe his methodology is likely to follow the course adopted here and concentrate on phonemics. A representative study of English syntax is that of Fries (1952) in which he ignores the problem of identifying syntactic elements (words) and uses *substitution* to categorize words into syntactic classes. His noun-like Class 1, for example, is defined on the occurrence of words in such environments as The _ was good. Sentences are then represented as sequences of word classes.

This description of structural methodology is offered to demonstrate that the role of semantics in this view of language is marginal at best. In the case of Bloomfield this was not a result of any lack of interest. In *Language*, for example, two full chapters are given to this topic and the remainder of the book contains many references to various semantic issues in historical and descriptive linguistics. There is,

---

[a] Harris (1951) includes a large part of what is normally regarded as syntactic analysis within his level of morphology and has no syntactic level as such. Figure 1 represents a guess as to what a separate syntactic component might contain. Words are classified by substitution and sentences are represented as sequences of word classes.

in fact, a good deal more emphasis on meaning in Bloomfield's work than in the writings of later structuralists or in the work of Chomsky. Consider one of the comments in *Language* on the general goals of linguistic analysis:

Man utters many kinds of vocal noise and makes use of the variety: under certain types of stimuli he produces certain vocal sounds, and his fellows, hearing these same sounds, make the appropriate response. To put it briefly, in human speech, different sounds have different meanings. To study this co-ordination of certain sounds with certain meanings is to study language. (*Language*, 27)

This is not unlike the characterization of the goal of a linguistic description given by many of the writers in this present volume. The parallelism between various formal categories (phoneme, morpheme, etc.) and their semantic counterparts (sememe, episememe, etc.) is emphasized. Why, given this clear interest in semantic questions, does Bloomfield not incorporate meaning into the structural description of language? The answer seems to lie in his definition of meaning.

We have defined the *meaning* of a linguistic form as the situation in which the speaker utters it and the response which it calls forth in the hearer. The speaker's situation and the hearer's response are closely co-ordinated, thanks to the circumstance that every one of us learns to act indifferently as a speaker or as a hearer. In the causal sequence

speaker's situation ⟶ speech ⟶ hearer's response,

the speaker's situation, as the earlier term, will usually present a simpler aspect than the hearer's response; therefore we usually discuss and define meanings in terms of a speaker's stimulus.

The situations which prompt people to utter speech, include every object and happening in their universe. In order to give a scientifically accurate definition of meaning for every form of a language, we should have to have a scientifically accurate knowledge of everything in the speaker's world. (*Language*, 139)

Given such a global definition, one may rightly be pessimistic about the prospects of describing meaning in any systematic fashion. In more recent terminology, Bloomfield is equating meaning to the encyclopedia rather than to the dictionary. It is interesting that he does distinguish different types of meaning along these lines though he does not take the narrower definition to be the domain of linguistic theories as Katz and Fodor were later to do.

People very often utter a word like *apple* when no apple at all is present. We may call this *displaced speech*...A starving beggar at the door says *I'm hungry* and the housewife gives him food: this incident, we say, embodies the *primary* or *dictionary meaning* of the speech-form *I'm hungry*. A petulant child, at bed-time, says *I'm hungry*, and his mother, who is up to his tricks, answers by packing him off to bed. This is an example of displaced speech. It is a remarkable fact that if a foreign observer asked for the meaning of the form *I'm hungry*, both mother and child would still, in most instances, define it for him in terms of the dictionary meaning...The remarkable thing about these variant meanings is our assurance and our agreement in viewing one of the meanings as *normal* (or *central*) and the others as *marginal* (*metaphoric* or *transferred*). (*Language*, 141–2)

Differential meaning does function for Bloomfield as an essential part of the methodology of linguistics since the identification of phonemes is based on differences in meaning as determined by informant judgments.

These views, while they do anticipate many future developments in semantic analysis, signal the exclusion of semantics from linguistics and represent an important attempt to constrain the scope of linguistic theories in a way that will permit linguists to reach definite, though limited, goals. This rejection of meaning became even more complete in later theoretical works by structuralists. Zelig Harris, who is a significant transitional figure between structural and generative-transformational linguistics (see Harris, 1957), attempted, in *Methods of Structural Linguistics* (1951), to specify the exact operations that were necessary to derive a linguistic description, operationally, from basic linguistic data. He accepted Bloomfield's general position on the definition of meaning and the relations between form and meaning but denied that meaning could be used as anything more than an heuristic device in exact linguistic methodology. The rejection of meaning is seen in its most extreme form in Bloch (1948) where the human informant-judge is eliminated, and it is proposed that the input to a linguistic analysis need consist of nothing more than accurate recordings of utterances.

A consequence of restricting linguistics to purely formal matters was an extreme narrowness of focus on the utterances of a language independent of *any* properties of human language users. The external and internal stimuli acting on a speaker are placed outside of linguistics and all other, perhaps more interesting, aspects of human speakers are excluded as well. Further, the results of a linguistic analysis are not taken to be relevant to an understanding of the capacities and fundamental characteristics of human beings. The independence and methodological priority of form over meaning is clearly affirmed. This assumption, that form is independent, may be regarded as one of the central conceptions of modern linguistic theory, and, though it has come under increasingly severe attack in recent years, it continues to be vigorously defended by such scholars as Chomsky in this volume.

GT-1

The extraordinary and traumatic impact of the publication of *Syntactic Structures* by Noam Chomsky in 1957 can hardly be appreciated by one who did not live through this upheaval.[a] Chomsky denies the fundamental assumption of structuralism by arguing that an adequate linguistic description of grammar cannot be derived by applying sets of operations to primary data but rather must be viewed as a formal deductive theory whose object is to separate the grammatical sentences of a language from the ungrammatical ones and to provide a systematic account of the structure of grammatical sentences. This position involves a complete reversal of the relations among the parts of a linguistic description (as symbolized by the direction of arrows in Figure 2 as opposed to Figure 1) as well as defining the primary object of linguistic theory to be the principles underlying the construction of sentences rather than the identification of minimal signaling units such as phonemes.

Chomsky's work has led to a genuine scientific revolution in that his approach has redefined the goals and methods of linguistics and thereby delineated a set of relevant problems with which linguists may be properly concerned. Sklar's (1968) account of Chomsky's efforts to develop a rigorous analysis of syntax based on structuralist principles and his conclusion that a multitude of stubborn linguistic facts simply could not be handled by operational procedures is an illustration of Kuhn's (1962) description of the way in which a paradigm, during a period of normal problem solving science, accumulates implicit counterexamples which are, at some

[a] An especially good account of Chomsky's views is given in Lees (1957) while Lees (1960) represents the first major application of the theory of *Syntactic Structures* to a natural language.

point, made explicit leading to the development of a new paradigm. Once one becomes seriously interested in syntax (a natural consequence of being interested in language), procedures which seem to work effectively on phonological and morphological problems are no longer satisfactory. Since the lexicon of a language must be finite, it may seem at least possible in principle to enumerate all of the environments relevant to the identification of phonemes. One's confidence is likely to ebb rapidly when considering the possibilities of a similar analysis for the infinite array of sentences apart from the many problems of correctly stating the internal structure of

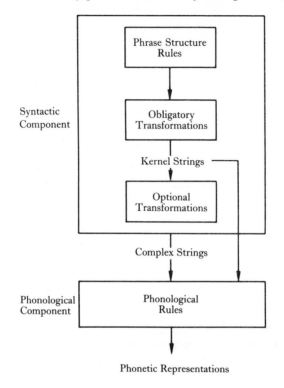

FIGURE 2. *GT-1*

sentences and the relationships among them. The next step is to conclude that the procedures didn't really work very well in the simpler cases either as the problems on the other levels are seen in a new light.

Chomsky's success may be associated with some particular characteristics of linguistics as a discipline as well as with the merits of his position. The fact that human languages had been an object of serious study by many scholars for centuries had created a situation where there was essential agreement on a large body of empirical facts about language and many methods of analysis had been investigated. It was thus possible to measure a new proposal against a rather well developed empirical domain and Chomsky's impact is due in no small part to his ability to offer solutions to a wide range of problems that had been either ignored or handled clumsily by structural methods. One cannot imagine any theorist having had such a rapid and dramatic effect in anthropology or sociology. In addition to the relatively advanced state of linguistic research, the social organization of academic linguistics, especially in the United States, provided (and still provides) an environment where a new idea could enjoy rapid dissemination. There were only a few universities which offered advanced degrees in linguistics and perhaps no more than three major journals (*Language, Word*, and *The International Journal of American Linguistics*) in which purely linguistic research regularly appeared. The Linguistic Society of America was a comparatively small academic group whose members came into frequent personal contact both through winter and summer meetings and the regular summer institutes sponsored by the society at various universities. The traditional policy of presenting only one sequence of papers to the whole society at meetings along with the convention that all papers are open to pointed comment from the floor guaranteed that a new viewpoint would be both presented and publicly debated.[a] Although Chomsky himself was not often an active participant in these affairs, his views were vigorously presented during this early period by his associates, especially Robert B. Lees. One interesting correlative of the rise of transformational grammar to a dominant position has been the striking quantitative expansion of linguistics in American universities. Although a number of general social factors are surely involved in this development, one causal element may well be the increased power and relevance of linguistic theory.

The sure sign of impending revolutionary success is the conversion of the young, and this occurred rapidly as graduate students became transformationalists while their own professors continued to profess structuralism. In addition, many superior students and younger scholars entered linguistics from such fields as philosophy and mathematics. That the originators of paradigms are not immune to such developments will be seen in the discussion of GT-3 where a similar process of revolt seems to be occurring. Just as Chomsky came to disagree with many of the teachings of Zelig Harris, so many of his own students now resist his views. It is, of course, necessary in both cases to emphasize the evolutionary nature of change as well as the more striking revolutionary aspects. At least in linguistics there is a marked continuity of development across the positions described here. I will attempt to show in this section that while Chomsky's innovations in syntax were truly radical, he did retain the central structuralist assumption as to the independence of form and meaning. Further, although his view of meaning differed in a number of important respects from that of Bloomfield and Harris, the full implications of this change were not exploited during this period and, in effect, the status of meaning in a linguistic description remained much the same.

The machinery of a linguistic analysis is now seen as consisting of a set of rules whose goal is to generate, automatically, all and only the grammatical sentences of a language along with a structural description for each which shows how its parts are combined to form the full sentence. Two types of rules define the two major components of a grammar.[b]

(1) *Phrase structure* (or *constituent structure*) rules are of the form XAY ⟶ XBY where A is a single symbol, B is a string of symbols and the environment within which the rule applies consists of X_Y which may be omitted if the rule is not

---

[a] As a result of a greatly expanded membership the Linguistic Society has now begun to present papers in parallel sessions at its winter and summer meetings although specialization has not yet gotten to the point of organizing sessions around some relatively narrow theme.

[b] See Bach (1964), chapters 3 and 4, for a good discussion of these rule types.

subject to environmental constraints (i.e. is 'context free'). Consider the fragment of English defined by the following set of phrase structure rules:

1. S ⟶ NP+VP
2. NP ⟶ NPsing
          NPpl
3. NPsing ⟶ T+N
4. NPpl ⟶ (T)+N
5. VP ⟶ MV+NP
6. MV ⟶ #+V
7. T ⟶ the
8. N ⟶ boy, girl, rabbit
9. V ⟶ chase, like

These rules will permit the derivation of such strings as The+boys+ # like+girls, Girls+ # +like+rabbits, etc. and automically assign a structure to each:

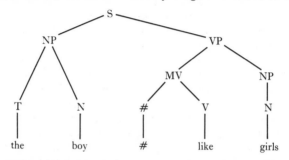

(2) *Transformational* rules are of the form P ⟹ Q where P and Q are strings of symbols and where the rewriting rule may rearrange the symbols of P, delete some of them, or add new symbols, to form Q (simple or singulary transformations). In addition, these rules may combine two strings to form a third: $P_1/P_2 \Longrightarrow Q$ (generalized transformations).

Transformations apply to labeled terminal strings of the phrase structure component in *Syntactic Structures*. In subsequent work (e.g. Chomsky and Miller, 1963) transformations are seen as converting P-markers (P) into a new P-Marker (Q). A P-marker may be regarded as equivalent to a labeled tree as above. For convenience, we will adopt the representation of transformations given in *Syntactic Structures*. Applied to the output of the phrase structure component these rules will produce more complex sentences (or, more properly, the strings underlying them) and state formally the relationship between transformationally derived sentences and the more basic strings from which they are derived.

The output of the phrase structure component will not consist of perfectly well formed strings. In the example above, # is not yet interpreted. Given the constraint that phrase structure rules cannot be of the form A ⟶ ϕ and that they must depend on immediate environments, we require the following two *obligatory* transformations in order to achieve number agreement

$$T_1ob \qquad X\text{--}\#\text{--}Y, \quad \# \begin{cases} S, \text{ where } X = NPsing \\ \phi, \text{ where } X = NPpl \end{cases}$$

$$T_2ob \qquad X\text{-}S\text{-}V \qquad S\text{-}V\text{-}S$$

These will produce *kernel* strings such as:

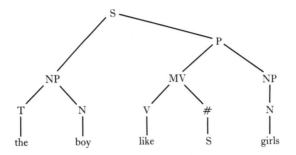

The essential substantive division here is between the kernel of a language consisting of simple active declarative sentences produced by the phrase structure rules and the obligatory transformations and the other more complex sentences formed by applying optional transformations to the set of kernel sentences. Simplified versions of the Relative and Question transformations will illustrate the way in which complex sentences are formed from simple ones:

$$\text{If } NP_1 = \text{boys}, NP_2 = \text{girls}, NP_3 = \text{rabbits}$$
$$V_1 = \text{chase}$$
$$V_2 = \text{like}$$

$$T_Q \quad NP_1 - V_1 - NP_2 \Longrightarrow \begin{array}{l} \text{Who} - V_1 - NP_2 \\ \text{Who} - \text{chase(s)} - \text{girls?} \end{array}$$

(singulary)

$$T_{rel} \quad \left. \begin{array}{l} NP_1 - V_1 - NP_2 \\ NP_1 - V_2 - NP_3 \end{array} \right\} \Longrightarrow \begin{array}{l} NP_1 \text{ who} - V_2 - NP_3 - V_1 - NP_2 \\ \text{boys who like rabbits chase girls} \end{array}$$

(generalized)

The *morphophonemic* component of Chomsky (1957) and Lees (1960) contains rules of the phrase structure type which convert the well formed strings of morphemes produced by the transformational component into a phonemic representation. Thus, the S in the rules above is converted into the proper phonemic form ($/\text{-s}/$, $/\text{-əz}/$, or $/\text{-z}/$) by a straightforward statement of the phonemic environments in which these forms occur. This formulation is of interest here because of its similarities in terminology and content to the phonological component of a structural grammar. It was soon replaced by a full-fledged generative phonological component which applied transformational types of rules to the output of the syntactic component in order to produce phonetic representations (see Chomsky and Miller, 1963; Halle, 1962).

This approach to language both restricts the goals of linguistic analysis and shifts the domain of linguistic theory. The structuralist goal of developing an automatic discovery procedure is rejected in favor of the more modest aim of providing an apparatus for the evaluation of competing grammars of a given language. The domain of linguistic theory becomes the knowledge that native speakers have about the formal properties of their language rather than the set of observable physical utterances which they produce. This has the effect of directing the attention of linguists toward the inner mental equipment of speakers at the expense of a direct concern with their overt behavior. Not very much is made of the distinction between competence and performance in *Syntactic Structures* although references to it are frequent in subsequent work (e.g. Chomsky and Miller, 1963). One of the most striking substantive contrasts with structuralism is the necessity of viewing human language as a phenomenon of immense complexity. This had a reverberating effect in the related areas of language acquisition and use where simplistic models based on behavior theory came under severe attack.[a] It is correct to say that the complexity of language, as viewed by linguists, has been steadily increasing ever since.

Bloomfield's justification for excluding meaning from linguistics rested on his definition of the term which made an account of meaning equivalent to an account of the total social, cultural and individual context of speech. In order for linguistics to have any hope of handling semantics this global definition must be partitioned into a part which is narrowly 'linguistic' as supposed to a part which includes the remainder of human knowledge. Chomsky's rejection of the structuralist definition provides the basis for such a distinction:

It is strange that those who have objected to basing linguistic theory on such formulations as (117i)* should have been accused of disregard for meaning. It appears to be the case, on the contrary, that those who propose some variant of (117i) must be interpreting 'meaning' so broadly that any response to language is called 'meaning'. But to accept this view is to denude the term 'meaning' of any interest or significance. I think that anyone who wishes to save the phrase 'study of meaning' as descriptive of an important aspect of linguistic research must reject this identification of 'meaning' with 'response to language', and along with it, such formulations as (117i). (*Syntactic Structures*, 99–100)

*117i is the assertion that two utterances are phonemically distinct if and only if they differ in meaning.

At the same time, Harris' rejection of meaning as a basis for formal description is retained by Chomsky:

It is, of course, impossible to prove that semantic notions are of no use in grammar, just as it is impossible to prove the irrelevance of any other given set of notions. Investigation of such proposals, however, invariably seems to lead to the conclusion that only a purely formal basis can provide a firm and productive foundation for the construction of grammatical theory. (*Syntactic Structures*, 100)

Like Bloomfield and Harris, Chomsky notes the many correlations between form and meaning but, unlike the structuralists, proposes that semantic factors may function as important criteria for the evaluation of grammars:

The fact that correspondences between formal and semantic features exist, however, cannot be ignored. These correspondences should be studied in some more general theory of language that will include a theory of linguistic form and a theory of the use of language as subparts. In §8 we found that there are, apparently, fairly general types of relations between these two domains that deserve more intensive study. Having determined the syntactic structure of the language, we can study the way in which the syntactic structure is put to use in the actual functioning of language. An investigation of the semantic function of level structure, as suggested briefly in §8, might be a reasonable step towards a theory of the interconnections between syntax and semantics. In

___

[a] See Chomsky (1959) and Miller, Galanter and Pribram (1960). Maclay (1964) describes the general effect of Chomsky's views on psycholinguistic research.

fact, we pointed out in §8 that the correlations between the form and use of language can even provide certain rough criteria of adequacy for a linguistic theory and the grammars to which it leads. We can judge formal theories in terms of their ability to explain and clarify a variety of facts about the way in which sentences are used and understood. In other words, we should like the syntactic framework of the language that is isolated and exhibited by the grammar to be able to support semantic description, and we shall naturally rate more highly a theory of formal structure that leads to grammars that meet this requirement more fully. (*Syntactic Structures*, 102)

It will be noted that, anticipating the competence-performance distinction, meaning is classified as performance ('use'). The subsequent association of semantic interpretation with 'deep structure' is also foreshadowed:

To understand a sentence we must know much more than the analysis of this sentence on each linguistic level. We must also know the reference and meaning of the morphemes or words of which it is composed. These notions form the subject matter for semantics. In describing the meaning of a word it is often expedient, or necessary, to refer to the syntactic framework in which this word is usually embedded; e.g., in describing the meaning of 'hit' we would no doubt describe the agent and object of the action terms of the notions 'subject' and 'object', which are apparently best analyzed as purely formal notions belonging to the theory of grammar.* (*Syntactic Structures*, 104)

* Such a description of the meaning of 'hit' would then account automatically for the use of 'hit' in such transforms as 'Bill was hit by John,' 'hitting Bill was wrong', etc., if we can show in sufficient detail and generality that transforms are 'understood' in terms of the underlying kernel sentences.

In effect, the status of meaning with regard to formal description is much like that found in structuralist accounts. It is outside of linguistics proper and clearly secondary to the description of syntax. However, the attempt to distinguish 'linguistic meaning' and the proposal that grammars are to be more highly evaluated to the extent that they support a valid account of the way in which sentences are 'understood' provides a rationale for the systematic incorporation of meaning into linguistic description. Indeed it requires such a development. If grammars must only distinguish well-formed from ill-formed sentences and provide structural descriptions for the well-formed sentences as stated in Chomsky's discussion of the goals of linguistics, then questions of meaning, however it is defined, may be relegated to some non-linguistic area such as 'use' or 'performance'. But once meaning is accepted as a criterion for the evaluation of grammars, the necessity of systematically describing it becomes clear. Just as an interest in language leads naturally to a concern with syntax, so a commitment to syntax requires that some attention be given to meaning with consequences that are often not obvious at the beginning of such a study.

GT-2

The important paper by Katz and Fodor, 'The Structure of Semantic Theory' (1963) represented the first attempt to make semantics a systematic part of a linguistic description. (Katz' paper in this volume is essentially a restatement of this position.) This move provided the impetus for a number of significant changes in syntax as semantic requirements were imposed on the syntactic component. Initially, Katz and Fodor argued that linguistic knowledge could be separated from the rest of a speaker's knowledge about the world and that the syntactic machinery of

GT-1 could provide a basis for a semantic component which would assign a semantic interpretation consisting of one or more readings to each sentence and in addition, would mark those strings which had no readings as semantically anomalous. The lower bound of a semantic theory is determined by examining those abilities or capacities which all fluent speakers of a language share and which cannot be accounted for by the syntactic component alone. Given a situation where a speaker is exposed to a sentence (S) in isolation we may ask what conclusions he is able to come

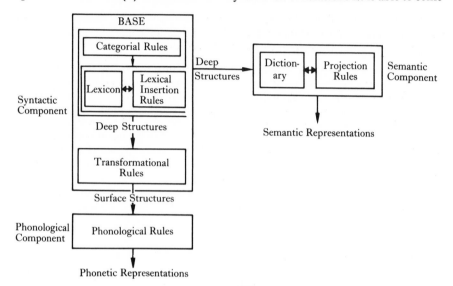

FIGURE 3. *GT-2*

to about S that would not be available to him had he only a syntactic description of S. This procedure is based on the assumption that 'Synchronic linguistic description minus semantics equals grammar'. In general speakers can:

(1) detect non-syntactic ambiguities and characterize the content of each reading of a sentence. If S is the English sentence *The bill is large*, it will be seen that two meanings are possible although only one structural description is available.
(2) determine the number of readings a sentence has by exploiting semantic relations in the sentence to eliminate potential ambiguities. If S is *The bill is large but need not be paid*, one of the readings under (1) is eliminated.
(3) detect anomalous sentences such as *The paint is silent*.
(4) paraphrase sentences by stating for a given S, which other S's have the same meaning.

In this article only (2) and (3) above are given any serious attention.

The upper bound of a semantic theory is determined by the following argument which is reminiscent of many earlier comments on the impossibility of including meaning, as broadly defined, in linguistics:

Since a complete theory of setting selection must represent as part of the setting of an utterance any and every feature of the world which speakers need in order to determine the preferred reading of that utterance, and since, as we have just seen, practically any item of information about the world is essential to some disambiguation, two conclusions follow. First, such a theory cannot in principle distinguish between the speaker's knowledge of his language and his knowledge of the world, because, according to such a theory, part of the characterization of a *linguistic* ability is representation of virtually all knowledge about the world that speakers share. Second, since there is no serious possibility of systematizing all the knowledge of the world that speakers share, and since theory of the kind we have been discussing requires such a systematization, it is ipso facto not a serious model for semantics. However, none of these considerations is intended to rule out the possibility that, by placing relatively strong limitations on the information about the world that a theory can represent in the characterization of a setting, a *limited* theory of selection by sociophysical setting can be constructed. What these considerations do show is that a *complete* theory of this kind is impossible. (Katz and Fodor, 1963, p. 179)

This attempt to present a principled basis for the definition of linguistic meaning is obviously crucial to the incorporation of semantics into grammar. It represents a cautious expansion of the domain of linguistic theories which regards and treats semantics as a completely secondary and subservient component of grammars. In operation, the semantic component takes as input a structural description of some sentence (S). It has available a *dictionary* entry for each word in S which states not only the syntactic classification of the word but also provides by the use of semantic markers, distinguishers and selection restrictions the information necessary for the operation of the *projection rules* which combine the meanings of individual words into the meanings possible for the whole sentence. For example, the dictionary entry for *colorful* will include, according to Katz and Fodor:

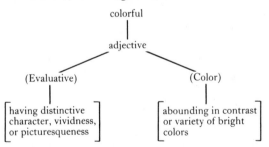

Where markers are enclosed in ( ), distinguishers in [ ] and selection restrictions in ⟨ ⟩. The projection rules, beginning at the bottom of a labeled tree, proceed to amalgamate elements at each higher level until the full sentence has received a semantic interpretation. Should any pair of elements violate the selectional restrictions, their combination will be blocked, e.g. *colorful void* is not possible since *void* will not contain any of the markers listed above for *colorful* whereas *colorful ball* will have four possible readings which combine the two readings for each item. (See Katz and Fodor (1963) and Katz in this volume for a more complete account and Weinreich in this volume for a strongly negative evaluation.)

Two general types of projection rules are discussed. Type 1 rules operate on the kernel sentences produced by phrase structure rules and obligatory transformations. Type 2 rules assign semantic interpretations to more complex sentences formed using optional transformations. Their comments on this issue anticipate the next major change in the syntactic component:

> The basic theoretical question that remains open here is just what proper subsets of the set of sentences are semantically interpreted using type 1 projection rules only. One striking fact about transformations is that a great many of them (perhaps all) produce sentences that are identical in meaning with the sentence(s) out of which the transform was built. In such cases, the semantic interpretation of the transformationally constructed sentence must be identical to the semantic interpretation(s) of the source sentence(s), at least with respect to the readings assigned at the sentence level...It would be theoretically most satisfying if we could take the position that transformations never change meaning. But this generalization is contradicted by the question transformations, the imperative transformation, the negation transformation, and others. Such troublesome cases may be troublesome only because we now formulate these transformations inadequately, or they may represent a real departure from the generalization that meaning is invariant under grammatical transformations. Until we can determine whether any transformations change meaning, and if some do, which do and which do not, we shall not know what sentences should be semantically interpreted with type 2 projection rules and how to formulate such rules. (Katz and Fodor, 1963, p. 206)

Katz and Postal (1964) offer the assumption that singulary transformations do not change meaning. The 'troublesome cases' noted above are handled by proposing that the phrase structure rules be permitted to generate items such as I (imperative), Neg (negative), and Q (question) which provide the conditions for the proper application of the relevant transformations. Rather than converting a string such as *Boys chase girls* into the question *Do boys chase girls?* with a consequent shift of meaning due to the transformation, each of the strings will now have a different underlying structure with Q appearing in the second. The next major step (described in Chomsky, 1966) was to allow phrase structure rules to be of the form A $\longrightarrow$ Sentence which has the effect of eliminating generalized transformations entirely as the phrase structure (or 'base') component of the grammar now generates a generalized phrase marker which states the relationship among all of the simpler sentences which are sources for the final complex sentence.

It will be remembered that the phrase structure component of a GT-1 grammar introduced lexical items as lists following some element such as N or V. Given the hierarchical arrangement of words and phrases required by this component, the statement of selectional restrictions across words and phrases is extremely clumsy. Thus, we can exclude such ungrammatical sentences as *The boy chase the dog* only by setting up nouns which are either singular or plural. But nouns also may be animate or inanimate (i.e. *The rock hated the dog) independently of whether they are singular or plural. Chomsky (1965) suggests that an enormous simplification of the phrase structure component can be achieved by following the model of phonological and semantic analysis and describing each lexical item as a set of features. The syntactic component of a grammar (Figure 3) now consists of two major parts: a *base* component which generates *deep structures* and a *transformational* component which converts deep structures into *surface structures*. Deep structures provide the input to the interpretive semantic component while surface structures are the input to the interpretive phonological component. The base is further analyzed into a *categorial* component and a *lexicon*. The former consists of a set of ordered rewriting rules which provide the recursive power of the grammar and generate trees whose terminal elements consist of grammatical morphemes and empty categories ($\triangle$). The lexicon consists of an unordered set of lexical entries each composed of a set of features. Lexical features are of three kinds: (1) *Category* features merely indicate the general category to which a lexical item must belong if it is to replace some $\triangle$. Thus, an item with the feature [+N] may be inserted only in a $\triangle$ dominated by N as indicated by the presence of a rule of the form $N \longrightarrow \triangle$. (2) *Strict subcategorization* features refer to the categorical environment in which a lexical item may occur, as defined by such terms from the categorial component as NP (noun phrase), VP (verb phrase), PP (prepositional phrase), etc. Thus, an item with the category feature [+V] and the strict subcategorization feature [+ —NP] can replace a $\triangle$ dominated by V and may occur only in environments which include an immediately following NP. (In more familiar terms, the item is a transitive verb.) (3) *Selectional* features refer to the lexical environment in which an item may occur. In particular to those lexical items with which the item in question has a grammatical relation. If an item is marked with the features [+V] and [+ —NP] as above it may also have the selectional feature [+animate] Obj which indicates that it takes only animate objects. The content of selectional features is provided by features such as *animate* or *human* which apply to nouns.

The following drastically oversimplified example will illustrate the operations of the syntactic component of a GT-2 grammar.

I   BASE

A. Categorial component

1. S $\longrightarrow$ NP+VP	4. N $\longrightarrow$ $\triangle$
2. NP $\longrightarrow$ D+N+(S)	5. D $\longrightarrow$ $\triangle$
3. VP $\longrightarrow$ V+NP	6. V $\longrightarrow$ $\triangle$

B. Lexicon

	Category	Strict subcategorization	Selection	Other noun features
boy	[+N]	[+D —]		[+animate] [+human]
mountain	[+N]	[+D —]		[−animate] [−human]
rabbit	[+N]	[+D —]		[+animate] [−human]
praises	[+V]	[+ — NP]	[+human] Subj	
frightens	[+V]	[+ — NP]	[+animate] Obj	
chases	[+V]	[+ — NP]	[+animate] Subj [+animate] Obj	
the	[+O]	[+ — N]		

The categorial component generates structures such as:

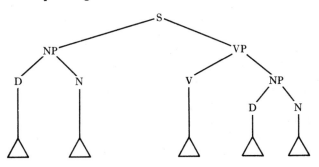

Lexical items are then inserted by substitution transformations which take account of the feature specification of each item. Thus, in the example above the terminal string might be *The mountain frightens the boy* or *The boy chases the rabbit* but not *The rabbit praises the boy* or *The boy chases the mountain*. Because of the recursive power provided by rules such as 2 of the categorial component it is possible to generate infinitely many base or deep structures.

2 TRANSFORMATIONAL COMPONENT

This component takes as input the deep structures generated by the base component and applies singular transformations to them to produce the set of surface structures. Suppose we have a *relative transformation* which has the following effect:

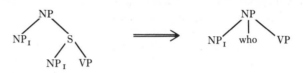

Given the following deep structure:

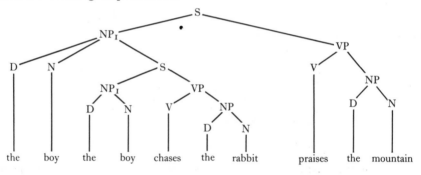

the relative transformation, in effect, substitutes 'who' for the second occurrence of 'the boy' producing *The boy who chases the rabbit praises the mountain*. Thus, complex sentences are formed in a way quite different from the operations of GT-1. A deep structure may be affected by a series of transformations which apply, in order, to the lowest sentence of the tree, then to the next lowest and so on until the highest S has been reached. A more realistic base component would also include other instances of S on the right hand side of rules (i.e. in a rule such as VP ⟶ VB (verbal)+(NP)+(S)) which greatly increases the power and flexibility of the system. The transformational rules act as a 'filter' in that they will block many of the structures generated by the base. For example, had the second noun phrase above been *the girl* the relative transformation could not have been applied. The semantic component must also be able to block deep structures that have no readings. The status of semantics as a proper part of a linguistic description had been anticipated by GT-1. During GT-2 it becomes something more than a poor relation of syntax and something less than a full-fledged partner. It seems obvious in retrospect that the systematic introduction of semantics by Katz and Fodor put various pressures on the syntactic component which were important in inducing the great changes that can be seen when *Aspects of the Theory of Syntax* is compared to *Syntactic Structures*.

Phrase structure rules are redefined, one of the two major types of transformations (generalized) is dropped and a syntactic feature analysis (based on the feature analyses of phonology and semantics) is introduced. It is notable that although the need to view transformations as not affecting meaning arose initially from Katz and Fodor's work in semantics the requirements of the semantic component were not sufficient to justify such a change without independent syntactic motivation. Katz and Postal devote the major portion of their book to a demonstration that the apparent counterexamples to their proposal (that singulary transformations don't change meaning and generalized transformations play only a minor role) may be rejected on purely syntactic grounds. Presumably, if they had failed to show this, their proposal would not have been acceptable regardless of the simplifications it would have introduced to the semantic component. This attitude is clearly shown in Chomsky's comments on their work:

> This principle obviously simplifies very considerably the theory of the semantic component as this was presented in Katz and Fodor ('The structure of a semantic theory'). It is therefore important to observe that there is no question-begging in the Katz-Postal argument. That is, the justification for the principle is not that it simplifies semantic theory, but rather that in each case in which it was apparently violated, syntactic arguments can be produced to show that the analysis was in error on internal, syntactic grounds. In the light of this observation, it is reasonable to formulate the principle tentatively as a general property of grammar. (Chomsky, 1966, p. 37)

One is entitled to suppose that had some dramatic syntactic simplification (such as the introduction of syntactic features in GT-2) been shown to be without semantic justification or even to complicate semantic analysis so much the worse for semantics. This is, of course, not really a case of random prejudice against meaning. Syntax was a very well developed field of study and there seemed no reason to pay much attention to the needs of a badly undernourished semantic component. All of the linguists mentioned here were in essential agreement on this point. Chomsky comments in this volume on the lack of necessary directionality in grammars of the GT-2 type which 'generates quadruples (P, s, d, S) (P a phonetic representation, s a surface structure, d a deep structure, S a semantic representation). It is meaningless to ask whether it does so by "first" generating d, then mapping it into S (on one side) and onto s and then P (on the other).' Nonetheless it cannot be entirely accidental that all known formulations of GT-2 grammars present S (using Chomsky's terminology) as the output of a dependent component whose operation depends on having d as input. The term interpretive applied to the semantic component suggests its secondary status. Thus, whatever the ultimate logical status of the relationships between the components of a grammar may be, it seems evident that for grammars of the GT-2 type, syntax is not only independent of meaning but prior to it both psychologically and operationally. The semantic component is assigned those tasks which cannot be conveniently accomplished syntactically. This is no doubt a carry over from earlier stages where the lack of a semantic component made it necessary to handle many semantic issues syntactically. An instance of this is found in the selection features of the lexicon which will provide a point of attack for the linguists of GT-3B.

 To summarize the status of GT-2 grammars in terms of Table 1: meaning is now an official part of a linguistic description, syntax is independent and divided into deep and surface structures with the former providing the exclusive basis for semantic interpretation. The general goal of a linguistic description is to state the relationships between sound and meaning.

GT-3

All of the papers in this section fall within what we have called the third period of generative-transformational theory and may now be discussed directly.[a] These papers are representative not only of a special interest in semantics on the part of some linguists but of a recent general tendency in theoretical linguistics to emphasize the importance of semantics to linguistic theory. The more or less universal acceptance among transformationalists of the position described in Chomsky's *Aspects of the Theory of Syntax* (GT-2) has given way to a far more heterogeneous situation

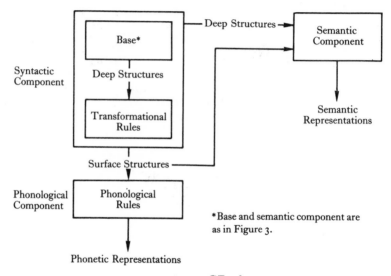

FIGURE 4. *GT-3A*

where an increasing concern with the status and function of meaning in a linguistic description has resulted in various proposed revisions of the theory of *Aspects*, ranging from relatively minor (Chomsky) to quite drastic (Lakoff-1, Fillmore, McCawley). It is clear that in all of these cases semantic considerations are primary in motivating changes in both syntax and semantics thus indicating a shift in the secondary position of the semantic component characteristic of GT-2. This is not to say that all of these papers bear directly on the theoretical issues mentioned in Table 1. Hale, for example, presents a rather straightforward description of a ritual variation of an Australian language. Dixon is especially difficult to classify; while his background is British and Firthian, his intellectual tastes are quite eclectic as his bibliography includes representatives of most of the major schools of linguistic theory in this century: Hjelmslev, Bloomfield, Chomsky, McCawley, Firth, Halliday, etc. What is perfectly clear is his commitment to semantics as a central focus of linguistic theory.

[a] References to papers in this volume will use only the author's name. Lakoff-1 refers to ' On Generative Semantics' and Lakoff-2 to 'Presuppositions and Relative Well-Formedness'.

Chomsky's paper represents the GT-3A position (Figure 4) (with a semantic component as described by Katz). In it, he defines the *standard theory* as being equivalent to the theory of *Aspects.*[a] The main effect of Chomsky's revision of the standary theory (Figure 5) is to modify the requirement that semantic interpretations must be based exclusively on input from the base component in the form of deep structures. On the basis of work by Jackendoff, Kuroda and Emonds he argues that some aspects of meaning can only be accounted for by reference to certain aspects of the surface structure. Crucial to his argument are two semantic notions, *focus* and *presupposition*, which also occur in many of the other papers. The focus of a sentence is that part of it which presents new information and is often marked by stress, while

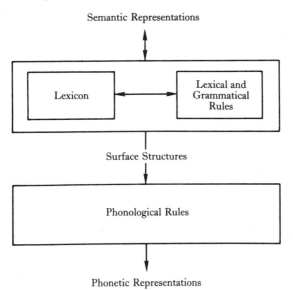

Semantic Representations

Surface Structures

Phonetic Representations

FIGURE 5. *GT-3B*

the presuppositions of a sentence are those propositions not asserted directly, which the sentences presupposes to be true. (See Langedoen, Kiparsky, Lakoff 1, 2, and Fillmore for examples and discussion.) Outside of this change the fundamental assumptions of GT-2 as to the independence of syntax, the interpretive nature of semantics and the independence of syntactic deep structure are retained. Chomsky defines the 'standard theory' and then argues that various possible alternatives as proposed by his critics differ from it only terminologically and not in terms of having different empirical consequences. Thus, the burden of proof rests on those who claim that syntax and semantics cannot be separated or that a separate level of deep structure does not exist.

It is fair to say that Chomsky now seems to occupy a minority position with regard to these issues. The representation of views in this anthology reflects, I think,

[a] Lakoff-1 mistakenly assumes that Chomsky wishes to differentiate these positions. He presents a quotation from Chomsky's paper which seems to indicate that the form of semantic representations in the standard theory is different from such representations in *Aspects*. In fact, this quotation is Chomsky's description of an alternative to the standard theory which he rejects.

the current distribution of opinion among linguists actively working in the area of generative-transformational theory. The ultimate historical source of this divergence seems to lie in the rejection of a broad definition of meaning by Chomsky in *Syntactic Structures* leading to the incorporation of a narrowly defined semantic component by Katz and Fodor. As the change from structuralism to GT-1 was encouraged by a concern with syntax so the changes described here have followed from an interest in meaning.

The linguists of GT-3B do not really constitute a unified school. Their common denominator is a conviction that semantic criteria are at least equal in importance to other factors in justifying solutions to linguistic problems and that semantic problems are an appropriate beginning point for a linguistic investigation. Figure 5 presents a very general picture of what seems to be the position of GT-3B linguists.

Efforts to extend the paradigm of GT-2 to an increasingly wider range of linguistic facts have driven the linguists of GT-3B to raise a variety of objections to a number of the fundamental assumptions on which the standard theory is based. McCawley (1968) provides a striking example of this shift. The major portion of this article accepts (though critically) the standard theory which is then rejected in a postscript where many of the arguments found in McCawley (this volume) are presented. One gets a certain sense of historical inevitability when observing this process. The remarks in Lakoff-1 (note *a*, p. 232) describing the development of generative semantics sound very much like Kuhn's description of the period when an established paradigm is coming under attack for its failure to solve legitimate problems. If this is the beginning of some sort of linguistic revolution, its magnitude should not be overestimated. The battle between Chomsky and his critics is being fought according to rules which Chomsky himself developed[a] and is essentially a sectarian war among scholars who share a common understanding as to the general goals of linguistic analysis. All agree that the aim of a linguistic description is to explain the relationships between two independently specifiable entities: sound and meaning. Further, it is held that descriptions must contain a lexicon and that transformational rules which map phrase-markers on to other phrase-markers are fundamental mechanisms of linguistic theory. Although the existence of a distinct syntactic level of deep structure may be in dispute, no one denies that a distinction has to be made between surface structures and underlying structures. This shared system of values permits confrontations of a very direct and intense sort among linguists who hold different views.

The key issue between Chomsky and the other linguists of GT-3 is the autonomy of syntax. If no principled boundary can be drawn between these two areas, then there can be no distinct level of syntactic deep structure and the question of whether or not semantics is interpretive becomes irrelevant.

Some of the attacks on the independence of syntax are based on arguments having to do with simplicity and generality of a kind previously familiar in syntactic discussions. Weinreich (in this volume) notes the duplication of effort involved in having a dictionary in the semantic component and a separate lexicon in the syntactic component. In addition, there is another duplication, noted here by Langendoen, arising from the fact that the base of the standard theory generates many deep

---

[a] The extent to which the structure of argumentation in Lakoff-1 is modeled on Chomsky is quite striking. For example, he defines a 'basic theory' and then rejects possible alternatives (including Chomsky's standard theory) on the grounds, in part, that they do not imply different empirical consequences.

structures that have no surface realizations and must therefore be blocked by restrictions on the application of transformational rules. This same 'filtering' must also occur in the semantic component. One of the main points of attack has involved the selectional features of the lexicon, which block such strings as the famous (and grammatical in GT-1) sentence *Colorless green ideas sleep furiously*, on the grounds that the use of such features as *Human* or *Animate* in both the syntactic and semantic components is redundant (Weinreich). In addition, McCawley argues that selectional constraints of this kind are, in any case, semantic rather than syntactic. Further evidence for the lack of separation between syntax and semantics follows from the assertion (developed here in McCawley and Lakoff-1) that deep syntactic structures are equivalent, formally, to semantic representations in that both consist of labeled trees (phrase-markers). McCawley assumes that semantic representations are best described using a revised version of standard symbolic logic and that these hierarchical descriptions are equivalent, as formal objects, to the deep structures of the standard theory. While Bierwisch does not reject the general format of the standard theory his discussion of features in similar logical terms tends to support this argument. Bendix, too, turns to symbolic logic for a notation for semantic descriptions. Another sort of semantic intrusion into syntax is presented by the Kiparskys who argue that semantic factors (especially presuppositions) are necessary to account for certain syntactic relationships. They conclude that base structures must contain semantic information. In general, the thrust of many of the arguments in this section is that the well-formedness of sentences cannot be determined solely on formal or syntactic grounds and further, that the syntactic and semantic information necessary for this determination cannot be separated.

Once the autonomy of syntax has been denied, the existence of a distinct level of syntactic deep structure also becomes unacceptable. It would still be possible to propose a distinct level of deep structure with both syntactic and semantic information in it. In the standard theory (and in Chomsky's revision of it) it is mandatory that all lexical insertion occur in a block preceding the application of any transformations. Many of the examples in Lakoff-1 attempt to show that some transformation must be applied *before* some of the lexical material has been inserted and thus that a notion of deep structure defined on this basis is inadequate.

The rejection of interpretive semantics follows immediately from the proposal that syntax and semantics cannot be separated. One style of argument is to show that the relationships between a pair of sentence types cannot be correctly stated if only the purely syntactic deep structures of the standard theory are available.

The semantic motivation of many of the proposed changes in the standard theory has already been noted. The expansion of the domain of linguistic theory, via an expansion of the definition of what is semantic, should also be emphasized. Such semantic notions as *presupposition* and *focus* represent a considerable extension over the narrow semantics of Katz and Fodor which concentrated on the detection of anomaly and the specification of the number of readings which could be assigned to sentences. The open ended definition of a semantic representation in Lakoff-1 as '$SR = (P_1, PR, Top, F, \ldots)$' suggests that many more semantic concepts will begin to function as requirements on a linguistic analysis with consequences that are difficult to foresee now. The radical expansion in the empirical domain of linguistics is seen clearly in Fillmore's paper where it is claimed that a lexicon must include information on the 'the happiness conditions' for the use of a lexical item; that is, the conditions which must be satisfied in order for the item to be used 'aptly'. Bendix explicitly defines his goal as the development of 'a form of semantic description that is oriented to the generation of behavioral data, as well as of sentences'. In Langendoen's paper it is argued that the definition of linguistic competence must be extended to include knowledge about the use of language and, as Weinreich also suggests, the ability of speakers to impose semantic interpretations on sentences that are semantically, and possibly syntactically, deviant. Chomsky's paper contains references to the analysis of 'discourse' and, in general, the papers of this section share the explicit or implicit assumption that the goals of linguistics now must include matters which had previously been entirely ignored or relegated to some non-linguistic area such as 'performance'. One obvious effect of these changes is to erode the distinction between competence and performance although it is argued in Lakoff-2 that this division can be maintained. Perhaps so, but the expansion of competence must necessarily affect the relationship and may result in attempts to impose purely performance constraints (e.g. memory) on competence models and even a denial that any clear line can be drawn between competence and performance. This suggests that a parallel may exist between syntax/semantics and competence/performance in terms of the historical development of their interrelationships. Katz and Fodor expanded the empirical domain of linguistics to include semantics with the ultimate result that the autonomy of syntax is now seriously questioned by linguists. The autonomy of competence may well be the next victim.

Related to the competence/performance dichotomy is another distinction which promises to be an early casualty. The necessity of separating linguistic knowledge from non-linguistic knowledge of the world (the 'dictionary' and the 'encyclopedia') has been emphasized in much contemporary linguistic research. The importance of the independence of linguistic knowledge has rested on the presumed impossibility of handling the total knowledge of speakers in any coherent way. While none of the linguists represented in this anthology are quite willing to give up the possibility of some such distinction the prospects of maintaining a definite boundary do not seem very bright. For example, the content of the presuppositions discussed in Lakoff-2 as to the intelligence and abilities of such objects as goldfish and cats are not strikingly linguistic as this term has been traditionally used.

It is not at all easy to imagine what the final result of this process will be.[a] The effect of *Syntactic Structures* was to narrow the focus of linguistic research to what was felt to be a manageable and relevant part of human language. At the same time, linguistics became a highly empirical science. The conditions for evaluating linguistic theories were explicitly related to the judgments of native speakers as to which strings of words were well-formed as well as to the logical elegance of grammars. It is this empiricism that has been crucial to the changes in the form of linguistic descriptions and to the steady expansion of the proper domain of such descriptions. It is evident that linguistics cannot profitably expand until its domain comes to include everything about human beings that is in some way connected to the acquisition and use of language. The fact that any bit of human knowledge may be involved in the judgments of speakers about the interpretation of sentences is not, in itself, conclusive evidence that such knowledge must be part of a *linguistic* description. Just as individual scholars are motivated to achieve a distinct identity, so disciplines

[a] Yngve (1969) predicts, with some relish, the demise of linguistics as an independent field. He suggests that the more talented survivors may be incorporated into a broader discipline devoted to the study of human linguistic communication and dominated by the 'information sciences'. This is not likely to become a popular view among linguists.

will ultimately seek to define a clear area of study in which they are supreme. Thus, we might speculate that the next major development in linguistic theory will be an attempt to save an overexpanded linguistics by redefining its boundaries relevant to such neighbor areas as psycholinguistics and human communication along lines as yet not clear.

University of Illinois, Champaign

BIBLIOGRAPHY

Bach, Emmon 1964 *An Introduction to Transformational Grammars.* Holt, Rinehart and Winston, New York.
Bloch, B. 1948 A set of postulates for phonemic analysis. *Language* 24: 3–46.
Bloomfield, Leonard 1933 *Language.* Hold, New York.
Chomsky, Noam 1957 *Syntactic Structures.* Mouton and Co., The Hague.
   1959 Review of *Verbal Behavior* by B. F. Skinner. *Language* 35: 26–58. Reprinted in Jakobovits and Miron, 142–71 and Fodor and Katz, 547–78.
   1965 *Aspects of the Theory of Syntax.* MIT Press, Cambridge.
   1966 Topics in the theory of generative grammar. In Sebeok (ed.), *Current Trends in Linguistics,* III, 1–60. Mouton and Co. The Hague.
Chomsky, Noam and George A. Miller 1963 Introduction to the formal analysis of natural languages. In Luce, Bush and Galanter (eds.), *Handbook of Mathematical Psychology* II, 269–321. Wiley, New York.
Fodor, Jerry and Jerrold Katz 1964 *The Structure of Language.* Prentice-Hall, Englewood Cliffs, New Jersey.
Fries, Charles C. 1952 *The Structure of English.* Harcourt-Brace, New York.
Halle, Morris 1962 Phonology in generative grammar. *Word* 18-54-72. Reprinted in Fodor and Katz, 334–54.
Harris, Zelig 1951 *Methods in Structural Linguistics.* University of Chicago Press, Chicago.
   1957 Co-occurrence and transformation in linguistic structure. *Language* 33: 283–340. Reprinted in Fodor and Katz, 155–210.
Jakobovits, Leon A. and Murray S. Miron (eds.) 1967 *Readings in the Psychology of Language.* Prentice-Hall, Englewood Cliffs, New Jersey.
Katz, Jerrold and Jerry A. Fodor 1963 The structure of a semantic theory. *Language* 39: 170–210. Reprinted in Jakobovits and Miron, 398–431 and in Fodor and Katz, II, 479–518.
Katz, Jerrold and Paul M. Postal 1964 *An Integrated Theory of Linguistic Descriptions.* MIT Press, Cambridge.
Kuhn, Thomas 1962 *The Structure of Scientific Revolutions.* University of Chicago Press.
Lees, Robert B. 1957 Review of *Syntactic Structures* by N. Chomsky. *Language* 33: 375–408.
   1960 *The Grammar of English Nominalizations.* Indiana University Press. Reprinted with additional preface 1964, and in 1966 by Mouton, The Hague.
Maclay, Howard 1964 Linguistics and language behavior. *Journal of Communication* 2: 66–73.
McCawley, James D. 1968 The role of semantics in a grammar. In Bach and Harms (eds.), *Universals in Linguistic Theory,* 91–122. Holt, Rinehart and Winston, New York.
Miller, George A. and Noam Chomsky 1963 Finitary models of language users. In Luce, Bush and Galanter (eds.), *Handbook of Mathematical Psychology,* II, 419–91. Wiley, New York.

Miller, George A., Eugene Galanter and Karl Pribram 1960 *Plans and the Structure of Behavior*. Holt, Rinehart and Winston, New York.
Sklar, Robert 1968 Chomsky's revolution in linguistics. *Nation*, September 9, pp. 213–17.
Yngve, Victor H. 1969 On achieving agreement in linguistics. In *Papers from the Fifth Regional Meeting, Chicago Linguistic Society*, 445–62. Department of Linguistics, University of Chicago.

Deep structure, surface structure, and semantic interpretation[a]

NOAM CHOMSKY

In a general way, I will be concerned in this paper with the relation of syntactic structure to semantic representation in generative grammar. I will try to outline a general framework within which much of the discussion of these questions in the past few years can be reformulated, and alternatives compared as to empirical content and justification, and I will discuss some empirical considerations that suggest a choice among these alternatives that is different, in some respects, from either the theory of grammar outlined in Chomsky (1965) or the proposals of a more 'semantically based' grammar that have been developed in the important work of the past few years. Specifically, these modifications have to do with some possible contributions of surface structure to delimiting the meaning of a linguistic expression.

A grammar of a language, in the sense in which I will use this term, can be loosely described as a system of rules that expresses the correspondence between sound and meaning in this language. Let us assume given two universal language-independent systems of representation, a phonetic system for the specification of sound and a semantic system for the specification of meaning. As to the former, there are many concrete proposals: for example, the system described in detail in chapter 7 of Chomsky and Halle (1968). In the domain of semantics there are, needless to say, problems of fact and principle that have barely been approached, and there is no reasonably concrete or well-defined 'theory of semantic representation' to which one can refer. I will, however, assume here that such a system can be developed, and that it makes sense to speak of the ways in which the inherent meaning of a sentence, characterized in some still-to-be-discovered system of representation, is related to various aspects of its form.

Let us assume further that the grammar in some manner specifies an infinite class of surface structures, each of which is mapped onto a phonetic representation by a system of phonological rules. I assume further that the grammar contains a system of grammatical transformations, each a mapping of phrase-markers onto phrase-markers. In ways that need not concern us in detail, the system of grammatical transformations determines an infinite class K of finite sequences of phrase-markers, each such sequence $P_1,\ldots,P_n$ meeting the following conditions:

⟨1⟩ (i) P_n is a surface structure

[a] This paper is reprinted from Professor Shiro Hattori's 60th birthday commemorative volume, *Studies in General and Oriental Linguistics*, TEC Corporation for Language and Educational Research, Tokyo, Japan.

(ii) each P_i is formed by applying a certain transformation to P_{i-1} in a way permitted by the conditions on grammatical rules[a]

(iii) there is no P_0 such that $P_0, P_1, \ldots, P_n$ meets conditions (i) and (ii).

Let us refer to P_1 as a *K-initial* phrase-marker in this case. We refer to the members of K as the *syntactic structures* generated by the grammar. So far, we have described K in terms of the class of surface structures, somehow specified, and the system of grammatical transformations, that is, the grammatical transformations of the language and the conditions on how they apply.

Let us assume further that the grammar contains a lexicon, which we take to be a class of lexical entries each of which specifies the grammatical (i.e., phonological, semantic, and syntactic) properties of some lexical item. The lexicon for English would contain this information for such items as *boy, admire, tall,* and so on. Just how extensive the lexicon must be – equivalently, just to what extent this information is determined by other parts of the grammar – we leave open. We may think of each lexical entry as incorporating a set of transformations that insert the item in question (that is, the complex of features that constitutes it) in phrase-markers. Thus

⟨2⟩ a lexical transformation associated with the lexical item I maps a phrase-marker P containing a substructure Q into a phrase-marker P′ formed by replacing Q by I.

Theories of grammar may differ in the conditions on Q, and more generally, on the nature of these operations.

Suppose, furthermore, that all lexical items are inserted into a phrase-marker before any non-lexical grammatical transformation applies. Thus the grammar meets condition ⟨3⟩:

⟨3⟩ given $(P_1, \ldots, P_n)$ in K, there is an i such that for $j < i$, the transformation used to form P_{j+1} from P_j is lexical, and for $j \geqslant i$, the transformation used to form P_{j+1} from P_j is nonlexical.[b]

In this case, let us define P_i to be the *post-lexical structure* of the sequence $P_1, \ldots P_n$.

Thus a grammar, so conceived, must have rules specifying the class K and relating each member of K to a phonetic and semantic representation. In particular, the grammar will contain a lexicon and grammatical transformations. Within this general framework, we can describe various approaches to the theory of transformational-generative grammar that have been explored during the past few years.

The theory outlined in Chomsky (1965) assumes that in addition to a lexicon, a

---

[a] Some of the conditions may be specific to the grammar (e.g., certain ordering conditions on transformations), and others general (e.g., the principle of the cycle, in the sense of Chomsky, 1965). These conditions will define certain permissible sequences of transformations and determine how a permissible sequence maps a phrase-marker P onto a phrase-marker P′. Hence with each such permissible sequence $T_1, \ldots, T_n$ we can associate the class of all sequences of phrase-markers $P_1, \ldots, P_{n+1}$ such that $T_1, \ldots T_i$, maps P_1 onto P_{i+1} ($1 \leqslant i \leqslant n$) in the manner determined. The class K consists of those sequences of phrase-markers which are so associated with permissible sequences of transformations, which terminate with surface structures and which are maximal in the sense of ⟨liii⟩. Each transformation carries out a certain definite operation on a sub-phrase-marker of the phrase-marker to which it applies; given the principle of the cycle, or others like it, the choice of this sub-phrase-marker may be determined by the position of the transformation in question in the permissible sequence of transformations.

[b] In terms of the previous note, each permissible sequence of transformations can be analyzed as (L, S) where L is a sequence of lexical transformations and S a sequence of nonlexical (i.e., true syntactic) transformations.

system of grammatical transformations, and a system of phonological rules, the grammar contains a system of rules of semantic interpretation and a context-free categorial component with a designated terminal element $\triangle$. The categorial component and the lexicon are referred to as *the base* of the grammar. It is assumed that the grammar meets condition $\langle 3 \rangle$, so that a class of post-lexical structures is defined. A general well-formedness condition is proposed for surface structures. The class K of syntactic structures consists of those sequences $P_1, \ldots, P_i, \ldots P_n$ (P_1 being the K-initial structure, P_i the post-lexical structure, and P_n the surface structure) meeting condition $\langle 1 \rangle$ where, furthermore, P_1 is generated by the categorial component and P_n meets the well-formedness condition for surface structures.[a] Surface structures are mapped into phonetic representations by the phonological rules. Post-lexical structures are mapped into semantic representations by the semantic rules. In this formulation, the post-lexical structures are called *deep structures*. The deep structures contain all lexical items, each with its complement of grammatical features. Furthermore, the configurations of the phrase-marker P_1, which are preserved in the deep structure, can be taken to define grammatical relations and functions in a straightforward manner. It is natural (though, I shall argue, only in part correct) to suppose that the semantic interpretation of a sentence is determined by the intrinsic semantic content of lexical items and the manner in which they are related at the level of deep structure. Supposing this (following, in essence, Katz and Postal, 1964), it would follow that deep structures determine semantic representation under the rules of semantic interpretation.

Thus the deep structures, in this theory, are held to meet several conditions. First, they determine semantic representation. Second, they are mapped into well-formed surface structures by grammatical transformations (without any subsequent insertion of lexical items). Third, they satisfy the set of formal conditions defined by base rules; in particular, the rules of the categorial component define the grammatical functions and order of constituents, and the contextual features of lexical entries determine how lexical items can be entered into such structures.

I will refer to any elaboration of this theory of grammar as a 'standard theory', merely for convenience of discussion and with no intention of implying that it has some unique conceptual or empirical status. Several such elaborations have been proposed and investigated in the past few years.

Observe that a standard theory specifies, for each sentence, a syntactic structure $\Sigma = (P_1, \ldots, P_i, \ldots, P_n)$ (where P_i is the deep, and P_n the surface structure), a semantic representation S, and a phonetic representation P. It asserts, furthermore, that S is determined by P_i and P by P_n under the rules of semantic and phonological interpretation, respectively. More generally, the theory is 'syntactically based' in the sense that it assumes the sound-meaning relation (P, S) to be determined by Σ.

It goes without saying that none of the assumptions in the foregoing exposition is

[a] More specifically, a general principle of lexical insertion is formulated which interprets the features (in particular, the contextual features) of lexical entries as lexical insertion transformations and applies these transformations to P_1 giving, ultimately, P_i. A lexical insertion transformation replaces a particular occurrence of the designated symbol $\triangle$ and the transformation replaces $Q = \triangle$ by I. We may assume, therefore, that the ordering of $P_1, \ldots, P_i$ is immaterial – that is, that we consider as syntactic structures equivalence classes defined by the relation among members of K that differ only by a permutation of $P_1, \ldots, P_i$.

The transformations are said to have a *filtering function* in the sense that the well-formedness condition on surface structures must be met.

Several variants of such a theory are discussed in Chomsky (1965).

self-evident, and that all are open to empirical challenge. Thus, to take perhaps the least controversial example, it might be argued that there is no level of phonetic representation, but that syntactic structures are related directly to the organization of peripheral musculature, sensory organs, and neural structures, by operations that are of an entirely different sort than those of grammar. There is no a priori way to demonstrate that this view is incorrect, or to justify the postulation of the level of phonetic representation, which, in this view, is superfluous. The most that one can hope to show is that an interesting range of phenomena can be accounted for by a theory that incorporates a level of phonetic representation of the sort postulated, that there is no crucial counter-evidence, and that there is no reason to suppose that some alternative form of theory will be more successful. Even stronger doubts can be (and often have been) expressed with respect to the notion of semantic representation. Thus one might argue that nonlinguistic beliefs, intentions of the speaker, and other factors enter into the interpretation of utterances in so intimate – and perhaps so fluctuating and indefinite – a fashion that it is hopeless and misguided to attempt to represent independently the 'purely grammatical' component of meaning, the various 'readings' of expressions in the sense of Katz and Postal (1964) and other versions of the standard theory, and the relation between such readings and a syntactic structure Σ.[a]

If one were to deny the existence of phonetic representation, he might argue that a generative grammar, strictly speaking, is a system of rules relating semantic representation, deep structure, and surface structure, some entirely new sort of theory relating the generated structures to physical signals or perceptual representations. If one were to deny the existence of semantic representation (readings, in the sense of recent discussions), he might argue that a generative grammar is a system of rules relating deep structures, surface structures, and phonetic representation, proposing further that entirely different principles are involved in determining what a person

---

[a] The literature relating to this subject is too extensive for detailed reference. See, for example, Quine (1960) for discussion of the interpenetration of linguistic and nonlinguistic knowledge. Stampe (1968) argues, in part on grammatical grounds, for a 'Gricean view' (see Grice, 1957) that the notion of 'reading' or 'semantic interpretation' must be understood in terms of the more basic notion 'Agent-means-x-by-y', an approach which calls into question the possibility of developing a coherent notion of 'semantic representation' strictly as part of grammar. For conflicting argument, see Katz (1966), Searle (1968).

There are still other sorts of consideration that might lead one to question the notion of 'reading', as construed in recent work. Thus consider such phrases as 'John's picture'. In addition to the readings 'picture of John' and 'picture that John has', the phrase might be interpreted as 'picture that John created', 'picture that John commissioned', and no doubt in other ways. On the other hand, 'John's puppy' is not subject to the latter two interpretations, though it might mean 'puppy to which John (my misnamed pet) gave birth'. On the other hand, it is hardly clear that it is a fact of language that people cannot create or commission the creation of puppies in the way in which they can pictures. Correspondingly, it is unclear whether one can assign to these phrases, by rules of grammar, a set of readings that determine how they figure in, say, correct inference. Or consider such a sentence as 'I am not against *my father*, only against the *labor minister*', spoken recently by a radical Brazilian student. Knowing further that the speaker is the son of the labor minister, we would assign to this utterance a reading in which the italicized phrases are coreferential. On one reading, the sentence is contradictory, but knowing the facts just cited a more natural interpretation would be that the speaker is opposed to what his father does in his capacity as labor minister, and would be accurately paraphrased in this more elaborate way. It is hardly obvious that what we 'read into' sentences in such ways as these – no doubt, in a fairly systematic way – can either be sharply dissociated from grammatically determined readings, on the one hand, or from considerations of fact and belief, on the other.

means by saying so-and-so. Evidently, there is no a priori argument against these views, as there is no a priori necessity for a grammar to define systems of deep and surface structure in the sense of the standard theory. Many of the assumptions in the standard theory are uncontroversial in the sense that they have been adopted, explicitly or implicitly, in those studies that attempt to characterize the notion 'knowledge of a language', and in that there is no known coherent alternative or any reason, empirical or conceptual, to suppose them inadequate. One should not, however, demand the kind of justification that in principle can never be provided.

In summary, I have so far outlined a certain general framework and a 'standard theory' that develops this framework in a specific direction. Furthermore, the literature contains further elaborations of this standard theory, and many realizations of it with respect to particular languages (that is, fragments of grammars of specific languages constructed in terms of the standard theory). At each level, there are reasonable doubts that can be raised, and alternatives can be envisaged. It goes without saying that the investigation of these doubts and the study of alternatives can only be beneficial in the long run, and should be actively pursued. It must also be kept in mind that at each level of discussion, justification can only go so far – in particular, that it can never be conclusive.

Given alternative formulations of a theory of grammar, one must first seek to determine how they differ in empirical consequences, and then try to find ways to compare them in the area of difference. It is easy to be misled into assuming that differently formulated theories actually do differ in empirical consequences, when in fact they are intertranslatable – in a sense, mere notational variants. Suppose, for example, that one were to modify the standard theory, replacing condition ⟨3⟩ by the condition that lexical items are inserted just prior to a transformation affecting the configuration in which they appear. Making this notion precise, we could devise an apparent alternative to the standard theory which, however, does not differ at all in empirical consequences, although the notion 'deep structure' is not defined, at least in anything like the sense above.[a] Given the central character of this notion in the standard theory, the alternative would appear to be significantly different, though in fact it would be only a notational variant. There would be, in other words, no empirical issue as to which formulation is correct or preferable on empirical grounds. Before the standard theory can be compared with this modification, it is necessary to formulate both in such a way that there is an empirical distinction between them.

Similarly, suppose that one were to counterpose to the syntactically based standard theory a 'semantically based' theory of the following sort. Whereas the standard theory supposes that a syntactic structure Σ is mapped onto the pair (P, S) (P a phonetic and S a semantic representation), the new theory supposes that S is mapped onto Σ, which is then mapped onto P as in the standard theory. Clearly, when the matter is formulated in this way,[b] there is no empirical difference between the 'syntactically based' standard theory and the 'semantically based' alternative. The standard theory generates quadruples (P, s, d, S) (P a phonetic representation, s a surface structure, d a deep structure, S a semantic representation). It is meaningless

[a] We might assume that rules of semantic interpretation of the type proposed by Katz in many publications apply cyclically, in parallel with the rules of the cycle of syntactic transformations, assigning readings to successively 'higher' nodes in the process. Thus semantic interpretation would, in effect, match that of the standard theory, though the notion of 'deep structure' is not defined.

[b] As, for example, in Chafe (1967). Chafe also proposes to obliterate the distinction between syntax and semantics, but this, too, is merely a terminological issue, as he formulates it.

to ask whether it does so by 'first' generating d, then mapping it onto S (on one side) and onto s and then P (on the other); or whether it 'first' generates S (selecting however one wishes from the universal set of semantic representations), and then maps it onto d, then s, then P; or, for that matter, whether it 'first' selects the pair (P, d), which is then mapped onto the pair (s, S); etc. At this level of discussion, all of these alternatives are equivalent ways of talking about the same theory. There is no general notion 'direction of a mapping' or 'order of steps of generation' to which one can appeal in attempting to differentiate the 'syntactically based' standard theory from the 'semantically based' alternative, or either from the 'alternative view' which regards the pairing of surface structure and semantic interpretation as determined by the 'independently selected' pairing of phonetic representation and deep structure, etc. Before one can seek to determine whether grammar is 'syntactically based' or 'semantically based' (or whether it is based on 'independent choice' of paired phonetic representation and deep structure, etc.), one must first demonstrate that the alternatives are genuine and not merely variant ways of speaking in a loose and informal manner about the same system of grammar. This is not so easy or obvious a matter as is sometimes supposed in recent discussion.

Perhaps the point can be clarified by reference to a discussion of Katz and Postal (1964, §5.4). Katz and Postal develop a variant of what I have called the standard theory, and then discuss how a model of speech production might be envisioned that incorporates a grammar of this sort. They outline a hypothetical procedure as follows: select a 'message' which is a set of readings, i.e., of semantic representations in the sense discussed above. Select a syntactic structure Σ (in particular, what we have here called the deep structure d in Σ) such that Σ maps onto S by the rules of semantic interpretation of the grammar. However this selection is accomplished, we may regard it as defining a mapping of S onto Σ, and in general, of semantic interpretations onto syntactic structures. Then, map Σ onto a speech signal, making use of the rules of phonological interpretation (giving the phonetic representation P) and rules that relate the latter to a signal. Quite properly, Katz and Postal present this schematic description as an account of a hypothetical *performance* model. In such a model, it makes sense to speak of order of selection of structures, direction of a mapping, and so on. Suppose, however, that we were to interpret this account as an intuitive instruction for using the rules of the grammar to form quadruples (P, s, d, S), i.e., for generating structural descriptions of sentences. Of course, in this case, the notion of 'order of selection of structures' or 'intrinsic direction of a mapping' would have no more than an intuitive, suggestive role; the informal instruction would be one of any number of equivalent instructions for using the rules of the grammar to form structural descriptions. To confuse the two kinds of account would be a category mistake. In short, it is necessary to observe the difference in logical character between performance and competence.

Suppose that we were to develop a modification of the standard theory along the following lines. Using the notation presented earlier the standard theory generates syntactic structures $\Sigma = (P_1, \ldots, P_i, \ldots, P_n)$, where P_1 is a K-initial, P_i a deep, and P_n a surface structure, P_1 being generated by the categorial component, and P_i formed by lexical insertion transformations that replace the substructure Q of P_i by a lexical item, Q always being the designated symbol $\triangle$. P_i is then mapped onto a semantic representation S. Suppose further that we regard S as itself a phrase-marker in some 'semantically primitive' notation. For example, we may think of the lexical entry for 'kill' as specifying somehow a phrase-marker *cause-to-die* that might be related to the phrase-marker that serves as the semantic representation of the phrase 'cause to die'.[a] Suppose now that in forming Σ, we construct P_1 which is, in fact, the semantic representation S of the sentence, and then form $P_2, \ldots, P_i$ by rules of lexical insertion, replacing a substructure Q which is the semantic representation of a lexical item I by I. For example, if P_1 contains $Q = cause\text{-}to\text{-}die$, the lexical entry for 'kill' will permit Q to be replaced by I = 'kill'. Similarly, the lexical entry for 'murder' might indicate that it can be inserted by a lexical transformation for the substructure $Q = cause\text{-}to\text{-}die\text{-}by\text{-}unlawful\text{-}means\text{-}and\text{-}with\text{-}malice\text{-}aforethought$, where the grammatical object is furthermore human; and the entry for *assassinate* might specify further that the object is characterized, elsewhere in the phrase-marker, as a reasonably important person; etc. Similarly, the lexical entry for 'uncle' might specify that it can replace $Q =$ brother of (father-or-mother). And so on, in other cases.

Superficially, this new theory seems significantly different from the standard theory. Thus deep structures are not mapped into semantic representations in the same sense as in the standard theory; rather the converse is true. Furthermore, the rules of lexical insertion operate in a rather different manner, replacing substructures Q, which may be quite complex, by lexical items. We might ask, in such a theory, whether there is any natural break between 'syntax' and 'semantics'. We might, in fact, define certain nonlexical transformations that apply in forming the sequence $(P_1, \ldots, P_i)$, thus violating condition $\langle 3 \rangle$ and eliminating the notion 'post-lexical structure', hence 'deep structure', as defined earlier.[b] Nevertheless, as I have so far formulated the alternatives, it is not at all clear that they are genuine alternatives. It must be determined whether the interpolated 'non-lexical' transformations are other than converses of rules of semantic interpretation, in the standard theory. Furthermore, it is unclear what difference there may be, on empirical grounds, between the two formulations of rules of lexical insertion. Again, before inquiring into the relative merit of alternative systems of grammar, it is necessary to determine in what ways they are empirically distinguishable. To establish that the systems are genuine alternatives, one would have to show, for example, that there is a difference between formulating the lexical insertion operations so that they replace the structure $Q =$ brother of (father-or-mother) (the terms of Q being 'semantically primitive'), on the one hand, and on the other hand, formulating the rules of semantic interpretation so that they assign to 'uncle' a position in the space of concepts (represented in terms of 'semantic primitives') which is the same as that

[a] The relation could not be idientty, however. As has often been remarked, 'causative' verbs such as 'kill', 'raise', 'burn' (as in 'John burned the toast'), etc., differ in meaning from the associated phrases 'cause to die', 'cause to rise', 'cause to burn', etc., in that they imply a directness of connection between the agent and the resulting event that is lacking in the latter case. Thus John's negligence can cause the toast to burn, but it cannot burn the toast. Similarly, I can cause someone to die by arranging for him to drive cross-country with a pathological murderer, but I could not properly be said to have killed him, in this case. The point is discussed in Hall (1965).

[b] Systems of this sort have been developed by McCawley in a number of interesting papers (see bibliography). The specific realizations of such systems proposed by McCawley are genuinely different, on empirical grounds, from the specific realizations of the standard theory that have been proposed for English. However, two questions can be raised: first, are the *systems* genuinely different, or are the genuine differences only in the realizations, which could, therefore, be translated into the other general systems of grammar; second, are the realizations suggested better or worse than the alternatives, on empirical grounds? I will return briefly to the latter question, in a specific case. The former has not been answered, to my satisfaction at least.

assigned, by rules of composition of the sort that Katz has discussed, to the phrase 'brother of (father-or-mother)'. If such a difference can be established, the theories might then be compared, in various ways. For example, one might compare the way in which such related concepts as 'kill', 'murder', 'assassinate' are treated in the two systems, or one might inquire into the nature and generality of the various rules and principles that are presupposed. In general, one might try to show that certain phenomena are explicable in a general way in one system but not in the other. Again, this is not so simple a matter as is sometimes supposed, to judge by recent discussion.

Consider next the following modification of the standard theory. We consider a new set of structures C (for 'case systems') which represent semantically significant relations among phrases such as the relation of agent-action in $\langle 4 \rangle$ and of instrument-action as in $\langle 5 \rangle$:

$\langle 4 \rangle$ John opened the door

$\langle 5 \rangle$ the key opened the door.

Suppose we were to assume, in a realization of the standard theory, that the deep structures of $\langle 4 \rangle$ and $\langle 5 \rangle$ are identical except for lexical entries. Then these deep structures, it might be argued, do not represent the required relations. For example, as grammatical relations are defined in Chomsky (1965), the subject-predicate relation is the relation that holds between 'John' and 'opened the door' in $\langle 4 \rangle$ and between 'the key' and 'opened the door' in $\langle 5 \rangle$; hence the relations of agent-action and instrument-action are not differentiated. Let us therefore construct the structures C_1 and C_2 of C as follows:

$\langle 6 \rangle$ C_1: ([V, open], [Agent, John], [Object, the door])

$\langle 7 \rangle$ C_2: ([V, open], [Instrument, the key], [Object, the door]).

Suppose that the grammar contains a component that generates such structures as C_1 and C_2 and rules that map these onto phrase-markers; for example, the main rule might say that the item specified as Agent takes the position of subject (in the sense of the standard theory), and if there is no agent, this position is occupied by the Instrument, etc. Formalizing these ideas, we might develop a theory in which C is mapped onto a class of phrase-markers which are K-initial in the sense described earlier, further operations being as in the standard theory. However, we drop condition $\langle 3 \rangle$ and relate the lexicon and the rules of semantic interpretation directly to C.[a]

Are case systems, so described, empirically distinguishable from the standard system? It is not at all obvious. Thus consider the example just given. It was argued that if $\langle 4 \rangle$ and $\langle 5 \rangle$ have the same deep structure, apart from lexical entries (let us put aside the question whether this is correct), then the relations indicated in $\langle 6 \rangle$ and $\langle 7 \rangle$ are not represented in these deep structures. However, this argument de-

[a] Case systems of this sort are developed in an important paper by Charles Fillmore (1968). As in the case of the previous note, we may ask (i) whether case systems are genuinely distinct from the standard system, or intertranslatable with it; (ii) whether the specific realizations proposed by Fillmore differ empirically from the specific realizations that have been proposed for the standard system; (iii) if so, how do they compare on empirical grounds? As to the second question, the answer is surely positive. Thus Fillmore's specific proposals do not permit any transformation (e.g., question or relative formation) to apply prior to such transformations as passive, indirect-object-inversion, and others that have been proposed in standard transformational grammars, and there are other specific differences. A serious discussion of question (iii) would take us too far afield. As Fillmore develops these systems, rules of semantic interpretation relate directly both to C and to the K-initial structures into which elements of C are mapped, since this operation is not 'meaning preserving', in the sense that sentences derived from the same element c ϵ C may, as Fillmore observes, differ in meaning.

pends on an assumption, which need not be accepted, regarding rules of semantic interpretation. In fact, the rules mapping C_1 and C_2 onto the deep structures of $\langle 4 \rangle$ and $\langle 5 \rangle$, respectively, can be interpreted as rules of semantic interpretation for these deep structures. Thus one rule (probably universal) will stipulate that for verbs of action, the animate subject is interpreted as the agent; etc. Various qualifications are needed whether we interpret these rules as rules of semantic interpretation or as rules mapping C onto Σ; I see very little difference between them, at this level of discussion, and the same seems to me true in many more complex cases. It might be argued that the case system expresses these facts in a 'direct way' whereas the standard system does so only 'indirectly'.[a] The distinction seems to me meaningless. Without principles of interpretation, a formal system expresses nothing at all. What it expresses, what information it provides, is determined by these principles.

A good part of the critique and elaboration of the standard theory in the past few years has focussed on the notion of deep structure and the relation of semantic representation to syntactic structure. This is quite natural. No area of linguistic theory is more veiled in obscurity and confusion, and it may be that fundamentally new ideas and insights will be needed for substantial progress to be made in bringing some order to this domain. I want to investigate one kind of revision of the standard theory that bears directly on the relation of syntax and semantics, but before doing so, I would like to consider briefly one kind of critique of the standard theory – specifically, concerning the status of deep structure – that seems to me to have been, so far, without consequence, though the general approach is quite legitimate and perhaps hopeful. I have in mind a critique analogous to that developed by Halle and others against the concept of the phoneme, a number of years ago. Halle argued that a generative grammar could provide a level of phonemic representation, in the sense of structural linguistics, only by abandoning otherwise valid generalizations. Analogously, one might ask whether the requirement that deep structures exist in the sense of the standard theory is compatible with otherwise valid generalizations. A negative answer would be highly interesting, and the matter therefore deserves serious investigation. A number of papers have dealt with this matter, but, I think, so far unsuccessfully.

McCawley purports to present such an argument in McCawley (1968b).[b] He considers the following expressions:

$\langle 8 \rangle$ Ax:$x\varepsilon$ {John, Harry} [x loves x's wife]

[a] Similar arguments, equally specious, have been given in support of the view that grammatical relations must be 'directly represented' in underlying structures.

[b] I omit here certain aspects of McCawley's argument that seem to me to impose serious difficulties of interpretation. Not the least of these difficulties is the theory of referential indices that McCawley proposes. To mention just the most serious problem, the idea that every noun phrase must have an intended reference, somehow specified in the underlying structure, seems unreconcilable with the fact that I may perfectly well use noun phrases where I know that there is no reference at all and hence intend no reference (e.g., 'if you are looking for the fountain of youth, you won't find it here', 'he is looking for a man who is taller than himself', etc.). The idea of trying to incorporate 'intended reference' in syntax seems to me misguided. It may clarify matters to point out that in Chomsky (1965), to which McCawley refers in this connection, it is not proposed that reference (actual or intended) be incorporated into syntax, but rather that 'referential expressions' be indexed in a way relevant to the operation of certain syntactic rules, and that the rules that assign semantic interpretation to syntactic structures refer to identity of indices in determining sameness of intended reference. This may or may not be a useful idea, but it is very different from McCawley's proposal that the intended reference of a noun phrase be specified in the grammar by an index, or in his terms, that the index 'be' the intended reference.

⟨9⟩ John loves John's wife and Harry loves Harry's wife
⟨10⟩ John and Harry love John's wife and Harry's wife, respectively
⟨11⟩ John and Harry love their respective wives
⟨12⟩ Ax:x∊M [x loves x's wife]
⟨13⟩ those men love their respective wives
⟨14⟩ that man(x) loves Mary and that man(y) loves Alice
⟨15⟩ that man(x) and that man(y) love Mary and Alice respectively
⟨16⟩ those men love Mary and Alice respectively.

He proposes that ⟨8⟩ and ⟨12⟩ be taken as (approximately) the semantic interpretations of ⟨11⟩ and ⟨13⟩ respectively (where A is the universal quantifier and M is the class of these men). He states further that the transformation which produces ⟨10⟩[a] is 'involved in' the derivation of ⟨11⟩. This transformation, the '*respectively-transformation*', relates ⟨8⟩ to ⟨11⟩, relates ⟨12⟩ to ⟨13⟩, and relates ⟨14⟩ to ⟨16⟩. McCawley furthermore rejects the idea of regarding such sentences as ⟨13⟩ as derived from conjunctions – quite properly: if for no other reason, consider what this proposal would entail for 'the real numbers are smaller than their respective squares'. Furthermore, ⟨16⟩ 'arise[s] from [our ⟨14⟩] by the *respectively*-transformation', which also maps ⟨17⟩ into ⟨18⟩:

⟨17⟩ that man(x) loves Mary and that man(y) loves Alice
⟨18⟩ that man(x) and that man(x) love Mary and Alice, respectively.

The rule of noun phrase collapsing maps ⟨15⟩ into ⟨16⟩ and ⟨18⟩ into ⟨19⟩:

⟨19⟩ that man loves Mary and Alice.

Presumably, then, McCawley intends that the *respectively*-transformation, which is 'involved in' the derivation of ⟨11⟩ from ⟨8⟩, in fact maps ⟨9⟩ into ⟨10⟩ exactly as it maps ⟨14⟩ into ⟨15⟩ and ⟨17⟩ into ⟨18⟩. Combining these various comments, McCawley seems to have in mind the following organization of operations:

$$
\begin{array}{ccc}
\text{I} & \text{R} & \text{R}' \\
⟨20⟩\ ⟨8⟩ \to ⟨9⟩ \to ⟨10⟩ \to ⟨11⟩ \\
\text{I}' \\
⟨12⟩ \to ⟨13⟩ \\
\text{R} & \text{C} \\
⟨14⟩ \to ⟨15⟩ \to ⟨16⟩ \\
\text{R} & \text{C} \\
⟨17⟩ \to ⟨18⟩ \to ⟨19⟩,
\end{array}
$$

where I and I′ are two rules, apparently entirely distinct, relating expressions with quantifiers to phrase-markers of the usual sort; R is a transformation forming sentences with 'respectively'; R′ is a subsequent transformation that forms noun phrases with 'respective'; and C is the rule of noun phrase collapsing.

Having presented this material, McCawley argues as follows. In a standard theory the relation of ⟨8⟩ to ⟨11⟩ and the relation of ⟨12⟩ to ⟨13⟩ must be regarded as semantic (since it involves 'a relationship between a representation involving quantifiers and bound variables and a representation involving ordinary noun phrases'), whereas the relation between ⟨14⟩ and ⟨16⟩ (or ⟨17⟩ and ⟨19⟩) is syntactic, namely, it is expressed by the transformation of conjunction-reduction. McCawley then

---

[a] That is, the 'transformation which produces the sentence ⟨145⟩: "John and Harry love Mary and Alice respectively"', which differs from ⟨10⟩ in deep structure, according to him, only in that where ⟨10⟩ 'has *John's wife* and *Harry's wife*, ⟨145⟩ had *Mary* and *Alice*'.

concludes, without further argument, 'that *respectively* cannot be treated as a unitary phenomenon in a grammar with a level of deep structure and that that conception of grammar must be rejected' in favor of a 'semantically based' theory. This argument is held to be analogous to Halle's argument against the level of phonemic representation.

Even if we accept McCawley's analysis *in toto*, no conclusion follows with respect to the concept of deep structure. His argument is based on an equivocation in the use of the notion '*respectively*-transformation', and collapses when the equivocation is removed. Thus if we use the term '*respectively*-transformation' to refer to the relation of $\langle 8 \rangle$ to $\langle 11 \rangle$, $\langle 12 \rangle$ to $\langle 13 \rangle$, $\langle 14 \rangle$ to $\langle 16 \rangle$, and $\langle 17 \rangle$ to $\langle 19 \rangle$, then this 'transformation' does, as he says, relate semantic to syntactic representations in the first two cases, and syntactic representations to syntactic representations in the latter two. But in the analysis he proposes, namely $\langle 20 \rangle$, the '*respectively*-transformation' carries out four totally different operations; hence it does not express a 'unitary phenomenon'. If, on the other hand, we use the term '*respectively*-transformation' to denote R of $\langle 20 \rangle$, then it does express a 'unitary phenomenon', but it no longer relates semantic to syntactic representation in one case and syntactic to syntactic representation in the other. In fact, $\langle 20 \rangle$ can be formulated in the standard theory, if we take I and I' to be converses of rules of semantic interpretation, and R, R' and C to be syntactic transformations. Therefore McCawley's analysis, right or wrong, is simply a realization of the standard theory, once equivocations of terminology are removed. Consequently, it shows nothing about the level of deep structure. Furthermore, it does not treat the phenomena in question in a 'unitary' manner, since no relation is proposed between I and I'.[a]

I have analyzed McCawley's argument in some detail, both because it is now often referred to as demonstrating the impossibility of incorporating the concept of deep structure in a generative grammar, and because this analysis illustrates clearly some of the difficulties in constructing a genuine alternative to the standard theory.

McCawley observes, quite correctly, that it is necessary to provide some justification for the hypothesis of an 'intermediate' level of deep structure: 'there is no a priori reason why a grammar could not instead[b] consist of, say, a "formation rule" component which specifies the membership of a class of well-formed semantic representations, and a "transformational component" which consists of rules correlating semantic representations with surface syntactic representation...' The

---

[a] A very different interpretation of these phenomena, in a somewhat modified version of the standard theory, is presented in Dougherty (1968). Dougherty's version of the standard theory is close enough to it so that his analysis can be compared on empirical grounds with McCawley's, which is, so far as I can see, entirely within the standard theory (if we drop the matter of indices as intended referents).

[b] The word 'instead', however, begs a number of questions, for reasons already noted. Thus in describing the standard theory one might refer to the deep structures as 'well-formed semantic representations', associating each with the class of semantic representations into which it can be mapped by rules of semantic interpretation. Similarly, one might regard McCawley's 'semantic representations', which, he proposes, be represented as phrase-markers, as nothing other than the deep structures of the standard theory, the 'formation rules' being the rules of the categorial component that form K-initial structures and the lexical rules that form deep structures from them by lexical insertion. McCawley in fact assumes that mutually deducible sentences may have different 'semantic representation' (in his sense), these being related by 'logic', a concept not further specified. To formulate his proposal in the standard theory, we might then take 'logic' to incorporate the rules of semantic interpretation (which express the 'logic of concepts', in one traditional use of this term). In this respect too he fails to differentiate his theory from the standard theory. McCawley deals with these questions in (1968), but inconclusively, I think, in part for reasons mentioned above on p. 189.

same might be said about 'surface structure', 'semantic representation' and 'phonetic representation'. There is only one way to provide some justification for a concept that is defined in terms of some general theory, namely, to show that the theory provides revealing explanations for an interesting range of phenomena and that the concept in question plays a role in these explanations. In this sense, each of the four concepts just mentioned, along with the notion of grammatical transformation and a number of others, receives some justification from the linguistic work that is based on grammars of the standard form. Of course, there is no a priori reason why the standard theory should be correct, so far as it goes in specifying the form of grammar; in fact, I will argue later that it is not. I fail to see what more can be said, at the level of generality at which McCawley develops his critique.

Lakoff has approached the same question – namely, whether deep structures can be defined in the sense of the standard theory without loss of significant generalization – in a more tentative way, in an interesting paper on instrumental adverbs (Lakoff, 1968). He considers such sentences as ⟨21⟩ and ⟨22⟩:

⟨21⟩ Seymour sliced the salami with a knife
⟨22⟩ Seymour used a knife to slice the salami

and gives a number of arguments to show that despite differences of surface structure, the same grammatical and selectional relations appear in these sentences. He argues that the two must have the same, or virtually the same representations in deep structure if selectional features and grammatical relations are to be statable in terms of deep structures, in anything like the sense of the standard theory. He suggests at various points that ⟨22⟩ is much closer to this common deep structure than ⟨21⟩; consequently, instrumental adverbs do not appear in deep structure, and the grammatical relations and selectional features must be associated, for both ⟨21⟩ and ⟨22⟩, with deep structures of roughly the form ⟨23⟩:

⟨22 a⟩   NP   V   NP   [$_S$ NP   V   NP   ]$_S$
          | | ⁀ | | ⁀
       Seymour   used   a knife   Seymour sliced the salami

Alternatively, the concept of deep structure, in the sense of the standard theory, must be abandoned.

Lakoff's argument is indirect; he does not propose underlying structures or grammatical rules, but argues that whatever they are, they must meet a variety of conditions in an adequate grammar, these conditions suggesting either a deep structure such as ⟨23⟩ or the abandonment of the notion deep structure. He points out that if ⟨23⟩ underlies ⟨21⟩, then deep structures must be quite abstract, since ⟨21⟩, which contains only one verb, is based on a structure with an embedded sentence and hence with two verbs. In either case, it would be fair to conclude that a departure from the standard theory is indicated.[a]

However, the argument is weakened – I think, vitiated – by the fact that a number of structures are omitted from consideration that seem highly relevant to the whole matter.[b] Thus alongside of ⟨21⟩ and ⟨22⟩, we have such sentences as ⟨23⟩–⟨26⟩:

[a] In the case of the double verb, what is a departure from more familiar formulations is that in this proposal, the verb *slice* in an embedded underlying sentence becomes the main verb, and the main verb *use* is deleted. On the other hand, it has been suggested many times, in realizations of the standard theory, that items that are in some sense relatively 'empty' of semantic content (such as 'be', have', use', etc.) may be deleted from embedded sentences.

[b] There are also quite a number of relevant factual questions that might be raised. Thus Lakoff assumes that ⟨21⟩ and ⟨22⟩ are synonymous. This is not obvious; compare 'John

⟨23⟩ Seymour used the knife to slice the salami with
⟨24⟩ Seymour used this table to lean the ladder against
⟨25⟩ Seymour used this table to write the letter on
⟨26⟩ Seymour used this car to escape (make his getaway) in.

Such facts as these suggest that underlying ⟨22⟩ is a structure such as ⟨27⟩:

⟨27⟩ Seymour used a knife [ₛSeymour sliced the salami with a knife]ₛ
Seymour used this table [ₛSeymour leaned the ladder against this table]ₛ

The latter might then be related to such sentences as 'Seymour used the knife for a strange purpose', '...in a strange way', etc. To form ⟨23⟩–⟨26⟩ a deletion operation will delete the final NP in the embedded sentence of ⟨27⟩ (an operation analogous, perhaps, to the one used in deriving 'meat is good to eat'). The preposition 'with', furthermore, can optionally be deleted, giving ⟨21⟩ from ⟨23⟩. In ⟨24⟩, 'against' cannot be deleted, but the corresponding prepositions can optionally be deleted (in some styles at least) in ⟨25⟩ and ⟨26⟩, giving ⟨28⟩ and ⟨29⟩ which do not correspond at all to ⟨30⟩ and ⟨31⟩, respectively:

⟨28⟩ Seymour used this table to write the letter, this is the table that Kant used to write the *Critique*, etc.
⟨29⟩ Seymour used this car to escape (make his getaway)
⟨30⟩ Seymour wrote the letter with this table
⟨31⟩ Seymour escaped (made his getaway) with this car (rather, 'in this car').

Very likely, a still more satisfactory analysis can be given, taking other data into account – see note *b*, p. 194. However, the relevant point here is that a wider range of data than Lakoff considered suggests an underlying structure such as ⟨27⟩ for ⟨22⟩; and if this is the case, then the major problems that Lakoff raises dissolve, as can be seen by checking case by case.[a] In particular, deep structures for ⟨21⟩ and carelessly broke the window with a hammer', 'John broke the window carelessly with a hammer', 'John carelessly used a hammer to break the window', 'John used the hammer carelessly to break the window'. The difference of meaning suggests a difference in the meaning of the sentences from which the adverb is omitted. Similarly, consider the many sentences in which 'use' and 'to' have the sense appropriate to this discussion, but which do not correspond to sentences with instrumental adverbs: e.g., 'John used his connections to further his career', 'John used the classroom to propagandize for his favorite doctrines', 'John used the mallet over and over again to reduce the statue to rubble'. Or consider such sentences as (A): 'John used this hammer and that chisel to sculpt the figure.' Believing (A), one would be entitled to give a positive answer to the question 'did John use that chisel to sculpt the figure?' but not to: 'did John sculpt the figure with that chisel?' The matter is even clearer if we consider 'John used this hammer and that chisel in sculpting the figure', which Lakoff considers synonymous with (A) – see p. 12. See Bresnan (1968) for further discussion.

A full analysis would have to bring much other evidence to bear – e.g., such sentences as 'Seymour sliced the salami without (using) a knife', which are not paired with anything like ⟨22⟩, and which suggest that insofar as the deep structures are common, it may be that 'use' is embedded below 'slice' in ⟨21⟩, rather than conversely, as Lakoff suggests.

I do not see how these questions can be resolved without undertaking an analysis of these structures which does propose rules as well as underlying structures, and in this sense, goes well beyond Lakoff's approach to these questions.

[a] In some cases, an explanation can be suggested for facts that would require arbitrary stipulation were the underlying structure to be taken as ⟨23⟩ – e.g., the fact that the complement of 'use' may not contain an instrumental adverb – see p. 21 of Lakoff, ibid. Many of the interesting phenomena that Lakoff notes still demand explanation, of course, but this fact does not help choose among the alternatives, since no explanation or even satisfactory descriptive account is offered in either case.

It is perhaps worth mentioning that the rather similar analysis of manner adverbials presented in Lakoff (1965) is also quite unpersuasive on factual grounds. Lakoff argues that the

$\langle 22 \rangle$, though not identical in this analysis, would nevertheless express the required selectional and grammatical relations in a unified way. And none of Lakoff's general conclusions with regard to deep structure follow if this analysis, or something like it, is correct.

Turning to a somewhat different matter, let us consider once again the problem of constructing a 'semantically based' theory of generative grammar that is a genuine alternative to the standard theory. Reviewing the observations made earlier, the standard theory has the general structure indicated in $\langle 32 \rangle$, where P_1 is the K-initial phrase-marker, P_i the deep structure, and P_n the surface structure of $\Sigma\epsilon K$, and where P is a phonetic and S a semantic representation:

$$\langle 32 \rangle \ \Sigma = (P_1, \ldots, P_i, \ldots, P_n)$$
$$ \downarrow \downarrow$$
$$ S P$$

S is determined from P_i by rules of semantic interpretation, and P from P_n by phonological rules. Only operations of lexical insertion apply prior to P_i, and none apply subsequently; P_1 is generated by the categorial component of the base. Each element of K is formed from the preceding one by transformations, the exact effect of each transformation being determined, by general conditions, by the position of this operation in the sequence of transformational operations that generates K. The grammar, in particular, generates quadruples (S, P_i, P_n, P). As emphasized earlier, there is no precise sense to the question: which of these is selected 'first' and what is the 'direction' of the relations among these formal objects. Consequently, it is senseless to propose as an alternative to $\langle 32 \rangle$ a 'semantically based' conception of grammar in which S is 'selected first' and then mapped onto the surface structure P_n and ultimately P.

manner adverbials too are derived from 'higher predicates', with sentence $\langle i \rangle$, for example, serving as an approximate source for $\langle ii \rangle$:

   $\langle i \rangle$ John is reckless in hanging from trees
   $\langle ii \rangle$ John hangs from trees recklessly.

However, $\langle i \rangle$ is clearly ambiguous, having either the approximate sense of $\langle \alpha \rangle$ of $\langle \beta \rangle$;

   $\langle \alpha \rangle$ John is reckless in that he hangs from trees
   $\langle \beta \rangle$ John is reckless in the way he hangs from trees.

Sentence $\langle ii \rangle$ has only the interpretation $\langle \beta \rangle$. But $\langle \beta \rangle$ itself no doubt derives from something of the form (γ), in which the embedded sentence would be something like $\langle \delta \rangle$, which contains a manner adverbial – in place of 'in that way' one might have 'in a reckless way', 'in a way that is reckless', 'recklessly'.

   $\langle \gamma \rangle$ John is reckless in the way in which he hangs from trees
   $\langle \delta \rangle$ John hangs from trees in that way.

Hence it appears that rather than $\langle i \rangle$ underlying $\langle ii \rangle$, it is more likely that something like $\langle \alpha \rangle$ and $\langle \gamma \rangle$ underlie $\langle i \rangle$ and only $\langle \gamma \rangle$ underlies $\langle ii \rangle$, where $\langle \gamma \rangle$ contains an embedded structure like $\langle \delta \rangle$ with an inherent manner adverbial.

Notice that in $\langle iii \rangle$ and $\langle iv \rangle$ the interpretation is along the lines of $\langle \alpha \rangle$, in $\langle v \rangle$ it is along the lines of $\langle \beta \rangle$, and in $\langle vi \rangle$ it is ambiguous as between $\langle \alpha \rangle$ and $\langle \beta \rangle$:

   $\langle iii \rangle$ clumsily, John trod on the snail
   $\langle iv \rangle$ John trod on the snail, clumsily
   $\langle v \rangle$ John trod on the snail clumsily
   $\langle vi \rangle$ John clumsily trod on the snail.

The examples are discussed in Austin (1961). Such sentences as 'John stupidly stayed in England' are unambiguously interpreted along the lines of $\langle \alpha \rangle$, and, correspondingly, the analogue to $\langle v \rangle$ is ungrammatical. These facts can be accommodated by an approach that takes $\langle \alpha \rangle$ and $\langle \gamma \rangle$ as approximating the underlying sources, but they do not appear consistent with Lakoff's analysis.

Consider once again a theory such as that proposed by McCawley in which P_1 is identified with S and condition $\langle 3 \rangle$ is dropped so that 'deep structure' is undefined. Let us consider again how we might proceed to differentiate this formulation – let us call it 'semantically based grammar' – from the standard theory. Consider such expressions as $\langle 33 \rangle$–$\langle 35 \rangle$:

$\langle 33 \rangle$ John's uncle

$\langle 34 \rangle$ the person who is the brother of John's mother or father or the husband of the sister of John's father or mother

$\langle 35 \rangle$ the person who is the son of one of John's grandparents but is not his father, or the husband of a daughter of one of John's grandparents.

If the concept 'semantic representation' ('reading') is to play any role at all in linguistic theory, then these three expressions must have the same semantic representation. But now consider the context $\langle 36 \rangle$:

$\langle 36 \rangle$ Bill realized that the bank robber was —

and the sentences S_{33}, S_{34}, S_{35} formed by inserting $\langle 33 \rangle$, $\langle 34 \rangle$, $\langle 35 \rangle$, respectively, in $\langle 36 \rangle$. Evidently, the three sentences S_{33}, S_{34}, S_{35} are not paraphrases; it is easy to imagine conditions in which each might be true and the other two false. Hence if the concept 'semantic representation' (or 'reading') is to play any serious role in linguistic theory, the sentences S_{33}, S_{34}, S_{35}, must have different semantic representations (readings). Many such examples can be constructed. The basic point is that what one believes, realizes, etc.,[a] depends not only on the proposition expressed, but also on some aspects of the form in which it is expressed. In particular, then, people can perfectly well have contradictory beliefs, can correctly be said to fail to realize that p even though (in another sense) they know that p, to be aware that p but be unaware that q where p and q are different expressions of the same proposition, etc. Notice that there is nothing in the least paradoxical about these observations. It is the function of such words as 'realize', 'be aware of', etc. to deal with such situations as those just described, which are perfectly common and quite intelligible.

Given these observations, let us return to the standard and semantically based theories. In the standard theory, $\langle 33 \rangle$, $\langle 34 \rangle$, and $\langle 35 \rangle$ would derive from three different deep structures, all mapped onto the same semantic representation. To assign a different meaning to S_{33}, S_{34}, S_{35}, it is necessary to define 'realize' (i.e., assign it intrinsic lexical semantic properties) in such a way that the meaning assigned to 'NP realizes that p' depends not only on the semantic interpretation of p but also on the deep structure of p. In the case in question, at least, there is no contradiction in this requirement, though it remains to meet it in an interesting way.

In the case of the semantically based theory this alternative is of course ruled out. In fact, within the framework of this theory it is impossible to accept all of the following conditions on K-initial structures (semantic representation, in this formulation):

$\langle 37 \rangle$ at the level of K-initial structures:

    (i) $\langle 33 \rangle$, $\langle 34 \rangle$, $\langle 35 \rangle$ have the same representation

    (ii) S_{33}, S_{34}, S_{35} have different representations

---

[a] Similarly, what one can prove, demonstrate, etc. The observation is due to Mates (1950). Scheffler (1955) discusses the matter more generally, and argues that no analysis of synonymy can suffice to explain the possibilities for substitution *salva veritate* in indirect discourse. There has been considerable discussion of these matters, but nothing, so far as I know, to affect the point at issue here.

(iii) the representation of $\langle 36 \rangle$ is independent of which expression appears in the context of $\langle 36 \rangle$ at the level of structure at which these expressions (e.g., $\langle 33 \rangle$–$\langle 35 \rangle$) differ.

In the semantically based theory, these three conditions lead to a contradiction; by $\langle 37 \text{ii} \rangle$, the sentences S_{33}, S_{34}, S_{35} differ in semantic representation (representation at the level of K-initial structures), whereas $\langle 37 \text{i} \rangle$ and $\langle 37 \text{iii} \rangle$ imply that they must be represented identically at this level, the differences of surface form being determined by optional rules that map semantic representations onto linguistic expressions. In the standard theory, the contradiction does not arise. The analogues of $\langle 37 \rangle$ are simultaneously satisfied by (i) rules which assign the same semantic interpretation to $\langle 33 \rangle$–$\langle 35 \rangle$; (ii) rules which make reference to the deep structure of the item appearing in the context of $\langle 36 \rangle$ in determining the meaning of $\langle 36 \rangle$. Condition $\langle 37 \text{iii} \rangle$ then poses no problem.

To reject $\langle 37 \text{i} \rangle$ or $\langle 37 \text{ii} \rangle$ is to abandon the semantically based theory, (or to deny the facts) since K-initial structures will no longer have the properties of semantic representations. Therefore it is necessary to reject $\langle 37 \text{iii} \rangle$, and to assume that the representation of $\langle 36 \rangle$ at the level of K-initial structures (semantic representations) depends on not the meaning but the form of the expression that appears ultimately in the context of $\langle 36 \rangle$. But to do this[a] is in effect to accept the standard theory in a confusing form; differences in deep structure will determine differences of semantic interpretation. In any case, then, the semantically based alternative collapses.

As far as I can see, an argument of this sort can be advanced against any variety of semantically based grammar (what is sometimes called 'generative semantics') that has been discussed, or even vaguely alluded to in the linguistic literature. One has to put this tentatively, because many of the proposals are rather vague. However, at least this much is clear. Any approach to semantically based grammar will have to take account of this problem.

Do considerations of this sort refute the standard theory as well? The example just cited is insufficient to refute the standard theory, since $\langle 33 \rangle$–$\langle 35 \rangle$ differ in deep structure, and it is at least conceivable that 'realize' and similar items can be defined so as to take account of this difference. Interesting questions arise when the matter is pursued further. Thus is it possible for someone to realize that John is believed to be incompetent by everyone without realizing that everyone believes John to be incompetent?, or to realize that Bill saw John but not that John was seen by Bill? Or, suppose that John happens to speak a language just like English in relevant respects, except that it has no word translatable as *uncle*. What, then, is the status of S_{33} as compared to S_{34} and S_{35}? Or, consider such sentences as 'everyone agrees that if John realizes that p, then he realizes that' –, where the space is filled either by p itself or by an expression q distinct from but synonymous with p. No doubt the truth value changes, as q replaces p, indicating that any difference of form of an embedded sentence can, in certain cases at least, play a role in the statement of truth conditions, hence, presumably, the determination of meaning. It remains to be determined whether there is some interesting subclass of such cases in which differences of deep structure suffice to account for the meaning differences, as the standard theory would require. If this could be shown, then the standard theory could still be maintained in a modified form: namely, except for cases in which *any* aspect of form may play a role in determining meaning. Instead of pursuing such questions as this, however,

---

[a] Assuming, that is, that it is possible to give it an intelligible formulation.

I would like to turn to another set of problems that seem to pose serious difficulties for the standard theory. I have in mind cases in which semantic interpretation seems to relate more directly to surface structure than to deep structure.[a]

Consider such sentences as ⟨38⟩:

⟨38⟩ (*a*) is it JOHN who writes poetry?

  (*b*) it isn't JOHN who writes poetry.

Under normal intonation[b] the capitalized word receives main stress and serves as the point of maximal inflection of the pitch contour. A natural response to ⟨38⟩ might be, for example, ⟨39⟩:

⟨39⟩ no, it is BILL who writes poetry.

The sentence ⟨39⟩ is a possible answer to ⟨38*a*⟩ and corroboration of ⟨38*b*⟩. The semantic representation of ⟨38⟩ must indicate, in some manner, that 'John' is the *focus* of the sentence and that the sentence expresses the *presupposition* that someone writes poetry. In the natural response, ⟨39⟩, the presupposition of ⟨38⟩ is again expressed, and only the focus differs. On the other hand, a response such as ⟨40⟩ does not express the presupposition of ⟨38⟩:[c]

⟨40⟩ no, John writes only short STORIES.

In the case of ⟨38⟩, the underlying deep structure might be something like ⟨41⟩:[d]

⟨41⟩ [the one who writes poetry] is John.

If so, then it would be natural to try to determine the focus and presupposition directly from the deep structure, in accordance with the standard theory, the focus being the predicate of the dominant proposition of the deep structure. Alternatively,

---

[a] The material in the remainder of this paper is drawn in large part from lectures given in Tokyo, in the summer of 1966, and prior to that, at MIT and UCLA. I am indebted to many of those who attended for comments and suggestions. Many of these and related topics are discussed by Kraak (1967), where rather similar conclusions are reached independently. I will not consider here some intricate but quite relevant considerations presented in Hall (1968).

[b] The concept 'normal intonation' is far from clear, but I will not try to explicate it here. I am assuming that the phonological component of the grammar contains rules that assign an intonation contour in terms of surface structure, along the lines discussed in Chomsky and Halle (1968). Special grammatical processes of a poorly understood sort may apply in the generation of sentences, marking certain items (perhaps even syllables of lexical items) as bearing specific expressive or contrastive features that will shift the intonation center, as in 'is it John who writes *Poetry*' or 'is it John who *writes* poetry', etc. I am assuming that no such processes apply in ⟨38⟩. Sentences which undergo these processes are distinct in semantic interpretation, and perhaps in syntactic properties as well. Given the obscure nature of these matters, it is difficult to say anything more definite. The matter is further obscured by the fact that these processes, however they are to be described, may assign an extra-heavy stress and extra-dominant pitch to the item that would serve as intonation center under normal intonation – i.e., in the case where these processes do not apply. Quite possibly, these processes are to be described in general as superimposing a new contour on the normal one. Thus in 'It *isn't* John who writes poetry', the word 'John' retains its intonational prominence with respect to the following phrase, exactly as under normal intonation.

[c] A response such as ⟨40⟩ does not deny the presupposition of ⟨38⟩, but rather its relevance. Again, these matters are far from clear, and deserve much fuller study than they have so far received. There is no reason to suppose that a satisfactory characterization of focus and presupposition can be given in purely grammatical terms, but there is little doubt that grammatical structure plays a part in specifying them. For some discussion of these matters in the case of cleft sentences such as ⟨38⟩, see Akmajian (1968).

[d] Following Akmajian, ibid. Alternatively, one might argue that the deep structure is of the form: '[it-one writes poetry] is John', with the rule of extraposition giving 'it is John who writes poetry'. The difference is immaterial, in the context of this discussion.

one might propose that the focus is determined by the surface structure, namely, as the phrase containing the intonation center.

Consider next ⟨42⟩:

⟨42⟩ (a) does John write poetry in his STUDY?
     (b) is it in his STUDY that John writes poetry?
     (c) John doesn't write poetry in his STUDY
     (d) it isn't in his STUDY that John writes poetry.

Again, a natural response might be ⟨43⟩:

⟨43⟩ no, John writes poetry in the GARDEN.

The sentences of ⟨42⟩ have as focus 'study' (or 'in his study') and express the presupposition that John writes poetry somewhere, a presupposition also expressed in the normal response ⟨43⟩. To accommodate these facts within the standard theory, we might take ⟨42b⟩ and ⟨42d⟩ to have a deep structure rather like ⟨41⟩, with the predicate of the dominant sentence being 'in his study', say ⟨44⟩:

⟨44⟩ the place where John writes poetry is in his study.

Again, the predicate expresses the focus and the embedded sentence the presupposition. To extend this analysis to ⟨42a⟩ and ⟨42c⟩, we would have to argue that the underlying structure of 'John writes poetry in his study' is also something like ⟨44⟩, contrary to what is assumed in Chomsky (1965) and many earlier realizations of the standard theory, in which the phrase 'in the study' is taken to be an adverbial modifier in a deep structure containing only one clause.[a]

In the case of ⟨42⟩, once again, an apparent alternative would be to determine focus and presupposition in terms of surface structure: the focus is the phrase containing the intonation center, and the presupposition is determined by replacement of the focus by a variable (we overlook, for the moment, a fundamental equivocation in the latter formulation).

To assist in the choice between these alternatives, it is useful to consider some more complex sentences. Thus consider ⟨45⟩:

⟨45⟩ {was it / it wasn't} {an ex-convict with a red SHIRT (a) / a red-shirted EX-CONVICT (b) / an ex-convict with a shirt that is RED (c)} that he was warned to look out for.

---

[a] This and related proposals are developed, on essentially these grounds, in Lakoff (1965). In more recent publications, other evidence has been cited to support an analysis along the lines of ⟨44⟩ for sentences like ⟨42a⟩, ⟨42c⟩. Thus Lakoff (1967) points out that we can say such things as 'Goldwater won in Arizona, but it couldn't have happened in New York' where 'it' refers to Goldwater's winning, suggesting that 'Goldwater won' is a sentential element in deep structure. However, the force of this argument is weakened by the fact that it would, if followed consistently, also lead us to the conclusion that in simple NVN sentences, the subject and verb constitute a sentence in deep structure (cf. 'John turned the hot dog down flat, but it (that) wouldn't have happened with filet mignon'; 'half the class flunked physics, which would never have happened in English Lit.'). Not only is this an unsatisfactory consequence in itself, but it also leads to an inconsistency, since the same argument yields the conclusion that the verb and object constitute a sentence (cf. 'John turned the hot dog down flat, but it wouldn't have happened with Bill (as recipient)'; 'half the freshman class flunked physics, which would never have happened with the senior class'). Similarly, we would have to conclude that in the sentence '*10 errors were committed* by the Red Sox and the Yankees in the game yesterday, but it (that) would never happen with any two other teams', the italicized expression constitutes a sentence in deep structure. I am aware of no strong argument for the analysis of ⟨42a⟩, ⟨42c⟩ with a deep structure like ⟨44⟩, except for the argument involving presuppositions.

The immediately underlying structure might be ⟨46⟩:

⟨46⟩ the one he was warned to look out for was X,

where X is one of the phrases in the second pair of braces in ⟨45⟩.

In this case, both the predicate phrase of ⟨46⟩ and the embedded clause of the subject must be further analyzed to reach the deep structure.

If it is deep structure that determines focus and presupposition along the lines indicated above, then the focus of the sentences of ⟨45⟩ should be ⟨47⟩, which are close or exact paraphrases of one another, and the presupposition should be ⟨48⟩:

⟨47⟩ (i) an ex-convict with a red shirt
     (ii) a red-shirted ex-convict
     (iii) an ex-convict with a shirt that is red
⟨48⟩ he was warned to look out for someone.

Correspondingly, a natural response to any of ⟨45⟩ would be ⟨49⟩:

⟨49⟩ no, he was warned to look out for an AUTOMOBILE salesman.

This conclusion is quite satisfactory, but there are difficulties when we explore further. Thus consider ⟨50a–c⟩:

⟨50⟩ (a) no, he was warned to look out for an ex-convict with a red TIE
     (b) no, he was warned to look out for a red-shirted AUTOMOBILE salesman
     (c) no, he was warned to look out for an ex-convict with a shirt that is GREEN.

⟨50a⟩, ⟨50b⟩, and ⟨50c⟩ are natural responses to ⟨45a⟩, ⟨45b⟩ and ⟨45c⟩, respectively; however, these are the only natural pairings. Thus ⟨50a⟩ could be a response to ⟨45b⟩ only in the sense in which ⟨40⟩ is a response to ⟨38⟩, that is by a denial of the relevance of the presupposition of ⟨45b⟩. In the case of ⟨42⟩, it was possible to maintain the standard theory by a modification of proposed deep structures. In the case of ⟨45⟩, however, this is quite impossible, without great artificiality. On the other hand, the facts just noted are accounted for directly by the alternative conception of focus and presupposition as determined by the intonation center of surface structure. According to this conception, the focus of ⟨45a⟩ can be taken as any of the phrases ⟨51⟩, and the corresponding presupposition is expressed by replacement of the focus by a variable:

⟨51⟩ (i) an ex-convict with a red shirt
     (ii) with a red shirt
     (iii) a red shirt
     (iv) shirt.

All of the phrases of ⟨51⟩ contain the intonation center in ⟨45a⟩; hence each, in this conception, can be taken as focus. Correspondingly, any of ⟨52⟩ can be a natural response:

⟨52⟩ (i) no, he was warned to look out for an AUTOMOBILE salesman
     (ii) no, he was warned to look out for an ex-convict wearing DUNGAREES
     (iii) no, he was warned to look out for an ex-convict with a CARNATION
     (iv) no, he was warned to look out for an ex-convict with a red TIE

But ⟨50b⟩ and ⟨50c⟩ are not natural responses preserving presupposition in this sense. Similar comments apply to ⟨45b⟩ and ⟨45c⟩.

To shed further light on the matter, consider the sentences ⟨53⟩, which are related to ⟨45a⟩ as ⟨42a, c⟩ are related to ⟨42b, d⟩:

⟨53⟩ $\left(\begin{smallmatrix} \text{was he} \\ \text{he wasn't} \end{smallmatrix} \right\}$ (warned to (look out for (an ex-convict (with (a red (SHIRT)))))).

The phrases enclosed in paired parentheses are the phrases containing the intonation center (certain questions of detail aside). Each of these phrases can be taken as the focus of the sentence, so that natural responses would include, in addition to ⟨52⟩, the following:

⟨54⟩ (i)   no, he was warned to expect a visit from the FBI
       (ii) no, he was simply told to be more cautious
       (iii) no, nothing was said to anyone.[a]

In each case, the presupposition can be determined by replacing what is taken as focus by an appropriate variable. There may be no actual sentence expressing just this presupposition, for grammatical reasons, just as there is no cleft sentence corresponding to the choice of focus, in many cases (hence the qualification on p. 200). For example, ⟨45a⟩ can be interpreted with 'shirt' as focus (so that ⟨50a⟩ is a natural response), but there is no grammatical sentence: 'it was SHIRT that he was warned to look out for an ex-convict with a red'. Similarly, there is no grammatical sentence expressing exactly the presupposition of ⟨45a⟩ with the phrase 'with a red shirt' taken as focus.

Observe, in fact, that the focused phrase need not correspond to a phrase of deep structure at all. This is clear in the case of ⟨53⟩, or, in a simpler case, ⟨55⟩:

$$⟨55⟩ \ (\begin{Bmatrix} \text{is John} \\ \text{John isn't} \end{Bmatrix} \ (\text{certain (to WIN)))}.$$

Natural responses would be any of ⟨56⟩:

⟨56⟩ (a) no, John is certain to LOSE
       (b) no, John is likely not even to be NOMINATED
       (c) no, the election will never take PLACE.

Hence any of the parenthesized phrases of ⟨55⟩ can be taken as focus, but one, 'certain to win', corresponds to no element of deep structure if, as seems correct, the deep structure is something like ⟨57⟩ (with, perhaps, a specification of negation or question):

⟨57⟩ [$_S$ John win]$_S$ is certain.

Similarly, consider the slightly more complex case ⟨58⟩:

$$⟨58⟩ \ \begin{Bmatrix} \text{is John} \\ \text{John isn't} \end{Bmatrix} \ \text{believed to be certain to WIN.}$$

Evidently, 'certain to win' is again a proper choice of focus, in which case what is presupposed is that something is believed of John. If we were to try to construct a cleft sentence corresponding to this interpretation of ⟨58⟩, it would have to be ⟨59⟩, analogous to ⟨60⟩:

⟨59⟩ it is certain to WIN that John is believed to be

$$⟨60⟩ \ \text{it is} \ \begin{Bmatrix} \text{a homicidal MANIAC} \\ \text{INCOMPETENT} \end{Bmatrix} \ \text{that John is believed to be.}$$

---

[a] For naturalness, question and answer (or denial and corroboration) must not only share presuppositions, but also must use as focus items that are somehow related – exactly how, is not clear, but the relation surely involves considerations that extend beyond grammar. Similar considerations arise in the case of natural coordination. For this reason, a pairing of sentences that might be expected on the formal grounds we are discussing may still not be natural, in the intuitive sense we are attempting to explicate. In other words, as in the case of coordination, grammatical (including semantic) considerations can suffice only for partial explication of certain intuitions that clearly involve other cognitive structures as well.

In all such cases, the cleft sentence is very marginal, or even totally unacceptable, from a strictly grammatical point of view, though it is certainly interpretable, presumably by analogy to properly formed sentences. In these deviant sentences as well there is an alternative natural choice of focus, namely 'to win' (in ⟨58⟩) and 'maniac' (in ⟨60⟩).

Continuing to restrict ourselves to normal intonation – that is, the intonation defined by the regular processes described in Chomsky and Halle (1968) – consider the following sentences:

⟨61⟩ did the Red Sox play the YANKEES
⟨62⟩ (i) did the Red Sox beat the YANKEES
      (ii) were the Yankees beaten by the RED SOX.

Sentence ⟨61⟩ can be interpreted as a question about whom the Red Sox played, about what they did, or about what happened. Thus possible answers might be any of ⟨63⟩:

⟨63⟩ (i) no, the TIGERS
      (ii) no, they flew to WASHINGTON
      (iii) no, the game never took PLACE.

Thus ⟨61⟩ is interpreted as presupposing that the Red Sox played someone (but whom?), that they did something (but what?), or that something happened (but what?) – the most natural interpretation perhaps being the first. The phrases containing the intonation center in the surface structure determine focus and presupposition. In the case of ⟨62⟩, there is no reason to suppose that there is any relevant difference in deep structure between (i) and (ii). The expressions of ⟨63⟩ are possible answers to ⟨62i⟩ and ⟨62ii⟩, of course, with different interpretations.[a] It would, for example, be impossible to answer (62ii) by saying: 'No, the Red Sox beat the TIGERS.' Or, to be more precise, this would be an answer only in the sense in which ⟨40⟩ is an answer to ⟨38⟩, that is, by failure to accept the presupposition.

Consider next the sentences ⟨64⟩:

⟨64⟩ (i) did John give the book to BILL
      (ii) did John give Bill the BOOK.

The response 'No, he kept it' is natural in both cases, since in each the phrase 'give ...' is a possible focus; but ⟨65i⟩ is a presupposition-sharing response only for ⟨64i⟩, and ⟨65ii⟩ only for ⟨64ii⟩:

⟨65⟩ (i) no, to someone ELSE
      (ii) no, something ELSE.

Thus although there is no relevant difference in deep structure between ⟨64i⟩ and ⟨64ii⟩, they differ in the range of possible focus and presupposition in the way predicted by the position of intonation center. The same observations hold of pairs such as 'John didn't argue with Bill about MONEY', 'John didn't argue about money with BILL', or 'I didn't buy that car in Italy five YEARS ago', 'I didn't buy that car five years ago in ITALY', etc. Similarly, in the case of such a sentence as 'I didn't buy that car five years ago in a country shaped like a BOOT', there are additional natural responses, conforming to the same principle. The same is true if we consider such sentences as ⟨66⟩:

⟨66⟩ (i) the question is not whether I should argue about money with BILL
      (ii) the question is not whether I should argue with Bill about MONEY.

[a] In this case, ⟨63ii⟩ seems to me the least natural, presumably because of the pairing of the concepts 'win'–'lose'. See note *a*, p. 202.

In the case of either, a natural response is: 'it is whether I should go to England'. But when the focus is taken more narrowly, the sentences are seen to differ in the range of permissible focus and presupposition.

Further support for this general point of view comes from sentences in which, for reasons having to do with particular formatives, the intonation contour shifts. Thus consider ⟨67⟩ and ⟨68⟩:

⟨67⟩ I didn't CATCH him

⟨68⟩ (i) hard work doesn't mature TEEN-agers

(ii) hard work doesn't MATURE people.

In the case of ⟨67⟩, the focus can be 'catch' or 'catch him', as distinct from 'I didn't catch BILL', where it can be 'Bill' or 'catch Bill'. In the case of ⟨68i⟩, the focus can be 'teen-agers' or 'mature teen-agers' ('No, it matures only adults', 'No, it only makes anyone tired'), whereas in the case of ⟨68ii⟩ it can be 'mature' or 'mature people' ('No, it harms them', 'No, it only makes anyone tired'). In fact, even in the simplest sentences similar observations hold. Thus, 'Brutus killed CAESAR' can be used to state what Brutus did or who Brutus killed, whereas 'Brutus KILLED him' can be used to state what Brutus did or what Brutus did to him. And so on, in many other cases.

So far I have restricted attention to cases of 'normal intonation', this being understood tentatively as referring to cases in which the intonation contour is determined by rules of the sort discussed in Chomsky and Halle (1968), with no expressive or contrastive intonation marked in specific expressions by other grammatical processes (see note *b*, p. 199). Turning our attention briefly to cases of the latter sort, it appears that similar conclusions follow. Consider, for example, ⟨69⟩, which differs from ⟨66⟩ in that the intonation center is shifted to the negative element:

⟨69⟩ (i) the question is NOT whether I should argue about money with Bill

(ii) the question is NOT whether I should argue with Bill about money.

Assuming that the intonation is otherwise normal, it still seems to be true, as in the case of ⟨66⟩, that ⟨70i⟩ is a natural response to ⟨69i⟩ but not ⟨69ii⟩, and that ⟨70ii⟩ is a natural response to ⟨69ii⟩ but not ⟨69i⟩:

⟨70⟩ (i) no, (it is whether I should argue about money) with MARY

(ii) no, (it is whether I should argue with Bill) about his trip to EUROPE.

On the other hand, 'No, it is whether I should go to England' is a natural response to either (i) or (ii) of ⟨69⟩. In all cases, the assertion ⟨69⟩ is corroborated. This observation (and the analogous observation in the other instances discussed above) supports the suggestion in note *b*, p. 199 that in some cases, at least, expressive or contrastive stress superimposes a new contour, preserving the arrangement of focus and presupposition defined by the normal intonation.

Consider next such cases as ⟨71⟩:

⟨71⟩ did John give the BOOK to Bill.

In this case, as distinct from the case of normal intonation ⟨64i⟩, the natural response is ⟨65ii⟩, not ⟨65i⟩. On the other hand, the sentence 'No, he kept it' seems much less natural as a response to ⟨71⟩ than to either case of ⟨64⟩. This observation (and its analogue in other cases) suggests that when expressive or contrastive stress shifts intonation center, the same principle applies as in normal cases for determining focus and presupposition, but with the additional proviso that naturalness declines far more sharply as larger and larger phrases containing the intonation center are considered as a possible focus. This would be a very natural interpretation of contrastive or expressive intonation, and it seems consistent with a number of relatively clear cases, at least. Hence it may perhaps be proposed as a first approximation to a general interpretive theory for this phenomenon. The same seems to me to be true when extra-emphasis is given to the item that contains the normal intonation center.

The processes involved in determining contrastive or expressive intonation at the moment do not appear to be germane to this discussion. However, it is worth noting that they cannot be described, at least in any natural way, in terms of deep structure. This becomes most obvious when we consider positions in which there *must* be a contrastive intonation. Thus consider the sentence ⟨72⟩:

⟨72⟩ John is neither EASY to please, nor EAGER to please, nor CERTAIN to please, nor INCLINED to please, nor HAPPY to please,...

In 'parallel constructions', in some sense of this notion that has never been made quite clear, contrastive intonation is necessary. But it is evident, in such examples as ⟨72⟩ at least, that it is a parallelism of surface structure, not deep structure, that is involved. The point is even clearer when we consider such sentences as ⟨73⟩:

⟨73⟩ John is more concerned with AFFirmation than with CONfirmation.

Here, the parallelism requires even a shift in contour within a single word. There are many similar cases.

To summarize these remarks, we seem to have the following situation. Rules of phonological interpretation assign an intonational contour to surface structures. Certain phrases of the surface structure may be marked, by grammatical processes of a poorly understood sort, as receiving expressive or contrastive stress, and these markings also affect the operation of the rules of phonological interpretation. If no such processes have applied, the rules assign the normal intonation. In any event, phrases that contain the intonation center may be interpreted as focus of utterance, the conditions perhaps being somewhat different and more restrictive when the intonation center involves expressive or contrastive stress, as noted. Furthermore, if the latter is superimposed on another, normal contour (in the manner discussed above), then the phrases containing the intonation center of the normal contour continue to determine a possible focus. Choice of focus determines the relation of the utterance to responses, to utterances to which it is a possible response, and to other sentences in the discourse. The notions 'focus', 'presupposition', and 'shared presupposition' (even in cases where the presupposition may not be expressible by a grammatical sentence)[a] must be determinable from the semantic interpretation of sentences, if we are to be able to explain how discourse is constructed and, in general, how language is used.

In many cases, it seems that we can interpret a sentence in these terms, given the intonation center, in the following way. The focus is a phrase containing the intonation center; the presupposition, an expression derived by replacing the focus by a variable. Each sentence, then, is associated with a class of pairs (F, P) where F is a focus and P a presupposition, each such pair corresponding to one possible inter-

---

[a] Note that we are using the term 'presuppostion' to cover a number of notions that should be distinguished. Thus 'it was JOHN who was here' expresses the presupposition that someone was here in the sense that truth of the latter is a prerequisite for the utterance to have a truth value. On the other hand, when we replace one of the foci of 'John gave Bill the BOOK' by a variable, it is not at all clear that the resulting expression determines a presupposition in the same sense, though it does characterize 'what the utterance asserts' and to which utterances it is a proper response, when so understood.

pretation. In terms of these notions we can begin to explicate such notions as proper (presupposition-sharing) response. Thus for a sentence S interpreted as (F, P) to be a proper response to a sentence S' interpreted as (F', P'), it must be the case that P = P'. Furthermore, F and F' must be paired in some 'natural' way, where the relevant concept of 'naturalness' no doubt extends beyond grammar, in the broadest sense of the concept 'grammar'. Further elaborations of these notions are surely in order,[a] but this seems in general a fair first approximation. In the present context, I wish only to emphasize that these notions seem to involve surface structure in an essential way, and thus to provide strong counter-evidence to the standard theory, which stipulates that semantic interpretation must be entirely determined by deep structure.

There is one obvious way to preserve the standard theory in the face of considerations of the sort just discussed, namely, to set the rule ⟨74⟩ as the first rule of the grammar, where F and P are arbitrary structures and S' functions as the initial symbol of the categorial component of the base:

⟨74⟩ S → S' F P

Continuing to generate a full syntactic and phonological structure in accordance with the standard theory, we would then add a new 'filtering rule', namely, that the structure generated is well-formed only if the focus and presupposition, as determined from surface structure, are identical with F and P, respectively. Technically, it would now be the case that deep structure fully determines meaning, even so far as focus and presupposition is concerned.[b] Thus underlying ⟨75 i⟩ we would have structures with the phrase-marker for 'the book', 'give John the book', and 'Bill gives John the book' as focus and corresponding presuppositions; and underlying ⟨75 ii⟩ we would have structures with the phrase-marker for 'John', 'give the book to John' and 'Bill gives the book to John' as focus with corresponding presuppositions; but not conversely, given the well-formedness condition.

⟨75⟩   (i)   did Bill give John the BOOK
           (ii) did Bill give the book to JOHN.

Obviously, this is merely a notational variant of a theory that determines focus and presupposition from the surface structure. In fact, the F and P positions would have to accommodate structures that can only be derived by transformation (as, e.g., in

[a] For example, it should be stipulated that the focus must be composed of full lexical items – more generally, items that make a contribution to the meaning of a sentence that is in some sense independent of anything outside the focus. In particular, then, the syllable containing the intonation center cannot serve as focus when it is part of a larger lexical item (except under the rather different circumstances of contrastive stress, as illustrated by ⟨73⟩). Similarly, in a sentence such as 'Did you call him UP', the item 'up' cannot serve as focus, but only 'call him up' or the full proposition, and in 'Did you take it for GRANTED', neither 'granted' nor 'for granted', but only 'take it for granted' (or the full proposition) can be taken as focus. This is an obvious condition to be placed on the notion of 'focus', given the role it plays in explaining how sentences are used and interpreted. The same can be said of idioms in general. Hence determination of focus must involve reference to the lexicon (and, no doubt, an associated idiom list). This seems to pose no special problem. There are, incidentally, many questions that can be raised about exactly how an idiom list should be related to a grammar, but these, so far as I can see, have no bearing on the topic under discussion; nor is there, for the moment, any interesting suggestion about this matter, to my knowledge. I am grateful to M. Bierwisch for bringing these facts to my attention.

[b] It is worth noting that the proposal discussed earlier to determine the focus as the predicate of the dominant sentence of the deep structure is not very different from this proposal.

cases such as ⟨72⟩ and others where the focus is transformationally derived). The rules ⟨74⟩ and the associated filtering condition are redundant, since they are determined, by a general interpretive principle, from the structure generated in the usual way when these extra formal concepts are eliminated. If we were willing to permit such formal devices, then the claim of the standard theory that deep structure fully determines semantic interpretation would be vacuous; if we do not permit them, it seems to be false.

Observe that these considerations do not touch on one aspect of the standard theory, namely, the hypothesis that the grammatical relations that enter into semantic interpretation are those represented in deep structure. In fact, it seems to me that insofar as the standard theory is plausible in its approach to semantic interpretation, it is with respect to this specific hypothesis. Thus it is natural to suppose that the meaning of a sentence is determined by minimal meaning-bearing elements and the relations into which they enter, these relations being specified in part by the lexicon itself and in part by the deep structure. But this narrower hypothesis remains unchallenged by the consideration of focus and presupposition. On the other hand, the attempt to express the latter concepts in terms of deep structure seems to me to have led to considerable artificiality in the construction of grammars, in recent work.

Turning to related questions, it was suggested a number of years ago by Kuroda (1965) that the position of such elements as 'even' and 'only' is determined by transformational processes, rather than by rules of the base, and that their contribution to the meaning of the sentences in which they appear is determined by their position in surface structure. That their position is determined by transformational processes is suggested by the fact that there are 'global' constraints on their occurrence; for example, 'only' or 'even' can appear in any of the blanks of ⟨76⟩, but it is questionable whether they can appear in more than one of these positions.

⟨76⟩ – John – reads – books on politics.

In particular, neither 'only' nor 'even' can occur in all of these positions. But constraints of this sort are transformational rather than 'phrase-structural' in character. Furthermore, the meaning of the sentence evidently changes as 'even' or 'only' takes one or the other position. Kuroda suggests, then, that there is a certain category of transformations – which he calls 'attachment transformations' – that do affect meaning, in the way indicated.[a]

More recently, Jackendoff has argued in a number of important papers that many semantic phenomena can be explained most readily in terms of features of surface structure. In particular, he suggests (1967) that the scope of logical elements such as negation and quantifiers is determined by surface structure. Thus consider such sentences as ⟨77⟩:

⟨77⟩ (i) not many arrows hit the target
     (ii) many arrows didn't hit the target
     (iii) not many arrows didn't hit the target.

It is plausible to argue that ⟨77 iii⟩ is ungrammatical, though (like many deviant sentences) one can, if required, impose a definite interpretation on it. If so, then placement of negation meets the 'global conditions' that signify that a transforma-

---

[a] His primary examples have to do with the problem of the *wa-ga* distinction in Japanese. Examples such as ⟨76⟩ are somewhat questionable, as Susan Fischer has pointed out, because they also involve critically the placement of contrastive stress. See Fischer (1968), where a different analysis is proposed.

tional process is involved. But, evidently, ⟨77i⟩ and ⟨77ii⟩ are quite different in meaning. Hence if we suppose that the underlying structure is ⟨78⟩ and that ⟨77i⟩ ⟨77ii⟩ are derived by a *not*-placement rule (and ⟨77iii⟩ not directly generated at all), then the deep structure will not determine the meaning.

⟨78⟩ not [many arrows hit the target].

Rather, the scope of negation will be determined by the position of 'not' in surface structure. In ⟨77i⟩, it is the proposition that many arrows hit the target that is denied. In ⟨77ii⟩, many arrows are asserted to have failed to hit the target; i.e., it is the verb phrase that is 'negated'.

In support of this analysis, Jackendoff notes the relation of meaning between active and passive forms involving both quantifiers and negation. Thus he considers the following sentences:

⟨79⟩ the target was not hit by many arrows
⟨80⟩ not many demonstrators were arrested by the police
⟨81⟩ many demonstrators were not arrested by the police
⟨82⟩ John didn't buy many arrows
⟨83⟩ many arrows were not bought by John
⟨84⟩ John bought not many arrows
⟨85⟩ not many arrows were bought by John.

Sentence ⟨79⟩ is a paraphrase of ⟨77i⟩, not ⟨77ii⟩, to which it would be related by the simplest form of the passive operation. Correspondingly, the order of quantifier and negation is the same in the surface structure of the paraphrases ⟨77i⟩ and ⟨79⟩, but different in ⟨77ii⟩. Furthermore, ⟨77ii⟩ has no passive paraphrase. What is suggested by ⟨77⟩–⟨79⟩, then, is that the order of quantifier and negation in the surface structure determines the meaning. Consequently, if the surface subject has a quantifier, then sentence negation (such as ⟨77i⟩) will be different in meaning from verb phrase negation (such as ⟨77ii⟩); but if the quantifier is part of a noun phrase that follows the verb, then the order of negation and quantifier is identical in sentence negation and verb phrase negation, and the meanings will be the same.

This principle is supported by ⟨80⟩, ⟨81⟩. The subject contains a quantifier, and correspondingly the case ⟨80⟩ of sentence negation differs in meaning from the case ⟨81⟩ of verb phrase negation, since the order of quantifier and negation is different. This principle is further supported by examples ⟨82⟩–⟨85⟩. Sentences ⟨82⟩ and ⟨83⟩ are obviously different in meaning, though ⟨84⟩ and ⟨85⟩ are the same in meaning[a] as are ⟨82⟩ and ⟨85⟩. In ⟨82⟩, ⟨84⟩, ⟨85⟩ the order of negation and quantifier is the same; in ⟨83⟩ the order differs. This is as required by the principle just stated.

According to this principle, sentence negation will differ in meaning from verb phrase negation just in case the surface subject contains a quantifier, that is, just in case the order of negation and quantifier differs in the two cases. Since it is the notion 'surface subject' that is involved in determining sameness or difference of meaning, the principle is inconsistent with the standard theory. Furthermore, the

[a] Assuming, that is, that ⟨84⟩ is well-formed. The question is actually irrelevant, having to do with the transformational source of ⟨85⟩ rather than the principle in question. It is sufficient to point out that ⟨82⟩ (under the most natural interpretation) is a paraphrase of ⟨85⟩. Under a less natural, but perhaps possible interpretation, it might be taken as 'there are many arrows that John didn't buy', a possibility that is irrelevant here because it remains consistent with the principle of surface structure determination of scope of negation.

principle of interpretation of surface structures seems clear, and, in addition, the transformations that form passives can be left in a simple form (though they will drastically change meaning, if they change the order of quantifier and negation). These facts, then, provide strong support for the hypothesis that surface structure determines (in part, at least) the scope of logical elements, and serve as strong counter-evidence to the standard theory in its most general form. Conceivably, one might modify the standard theory to accommodate these facts, but this modification would be justified (assuming it possible) just in case it achieved the naturalness and explanatory force of Jackendoff's proposal that negation and quantifiers are associated with phrases of the surface structure, and their interpretation is determined by the phrases in which they appear and their relative order.

Jackendoff's arguments, like those involving focus and presupposition, leave unaffected the hypothesis that the grammatical relations defined in the deep structure are those that determine semantic interpretation. If we modify the standard theory, restricting in this way the contribution of the base to semantic interpretation, we can take account of the fact that many aspects of surface structure appear to play a role in determining semantic interpretation; correspondingly, insofar as some development in syntactic theory is motivated by the demand that these aspects of semantic interpretation be expressed in deep structure, it will have lost its justification. To mention one example, consider the sentences ⟨86⟩:

⟨86⟩   (i)   the sonata is easy to play on this violin
      (ii)   this violin is easy to play the sonata on.

The sentences share a single system of grammatical relations and, in some reasonable sense of paraphrase, may be regarded as paraphrases: they have the same truth conditions, for example. However, they seem different in meaning in that one makes an assertion about the sonata, and the other about the violin. Before this difference is used to motivate a difference in deep structure, however, it must be shown that this aspect of meaning is one expressed in deep rather than surface structure. In the present instance, this conclusion seems at best dubious.[a]

Certain properties of modal auxiliaries also suggest a role for surface structure semantic interpretive rules. Thus J. Emonds has pointed out that 'shall' is interpreted differently in question and corresponding declarative.

⟨87⟩   (i)   I shall go downtown
      (ii)   shall I go downtown
      (iii)   I $\begin{Bmatrix} \text{asked} \\ \text{wonder} \end{Bmatrix}$ whether I shall go downtown.

In ⟨87i⟩ and ⟨87iii⟩, the modal is essentially a tense marker. In ⟨87ii⟩, however, it has a very different meaning, namely the meaning of 'should'. In general, interrogative expressions such as ⟨87ii⟩ have the same meaning as the corresponding embedded expression in sentences of the form ⟨87iii⟩, and, in fact, this observation, appropriately extended, has been used to support the syntactic derivation of interrogatives from embedded interrogative clauses (see, e.g., Katz and Postal, 1964). However, in the case of ⟨87⟩, this expectation is not verified. If we assume that the

---

[a] What is involved, apparently, is a relation of topic-comment which must be distinguished from that of subject-predicate. See Chomsky (1965), for some brief discussion within the framework of the standard theory of a question with a long history. Other arguments for distinguishing ⟨86i⟩ and ⟨86ii⟩ at the deep structure level have been proposed in recent work (e.g., Perlmutter, 1968), but they seem to me unpersuasive.

sentences of ⟨87⟩ are related as are those derived by replacing 'shall' by 'will', or by perfect aspect, etc., then the standard theory in its strongest form is refuted. If, furthermore, we wish to maintain the weaker hypothesis that the semantically functioning grammatical relations are those represented in deep structure, then we must conclude that the relation of 'I' to 'shall' in ⟨87⟩ is not a grammatical relation in this sense – it is not, for example, the subject-predicate relation. This seems a natural enough conclusion.

Other examples involving modals immediately come to mind. Thus it has frequently been noted that ⟨88i⟩ and ⟨88iii⟩ merely predict, whereas ⟨88ii⟩ is ambiguous, in that it may also mean that John refuses to go downtown:

⟨88⟩ (i) John will go downtown
(ii) John won't go downtown
(iii) it is not the case that John will go downtown.

Again, the interplay of negation and modal seems a natural candidate for a principle of surface structure interpretation.[a] Or consider such sentences as ⟨89⟩ (also Emonds):

⟨89⟩ John can't seem to get his homework done on time.

There is no corresponding form without 'not'. Furthermore, the modal is interpreted as associated with an underlying embedded proposition 'John gets his homework done on time'. Hence if 'can' appears in deep structure in association with 'seem', as it appears in association with 'work' in 'John can't work', then a rule of surface structure interpretation is needed to account for its semantic relation to the embedded verbal phrase 'get...' Suppose, on the other hand, that 'can' appears at the deep structure level in association with the embedded sentence 'John gets his homework done on time'.[b] Then a rule is necessary that extracts 'can' from the embedded sentence and assigns it to the matrix sentence – in fact, to exactly the position it occupies in simple sentences. However, note that this extraction is possible only when 'can' is interpreted as indicating ability, not possibility. Thus ⟨89⟩ has approximately the sense of ⟨90⟩, but the sentence ⟨91⟩, if grammatical at all, surely does not have the sense of ⟨92⟩:

⟨90⟩ it seems { that John can't get his homework done on time
{ that John is unable to get his homework done on time

⟨91⟩ the war can't seem to be ended by these means

⟨92⟩ it seems { that the war can't be ended by these means
{ that it is impossible that the war can be ended by these means.

Hence either the extraction operation will have to be sensitive to the difference in sense of two cases of 'can' – an otherwise unmotivated complication – or else the interpretation will have to be 'delayed' until after extraction has taken place. The later choice requires a rule of interpretation that does not apply to deep structure.

Notice that in general rules of semantic interpretation have a 'filtering function' analogous to that of rules of transformation in the standard theory. This is true no matter at what level they apply. Thus a rule of interpretation applying at the deep

[a] Examples such as this have been used to justify the argument that there are two sources for 'will' (and other modals as well). The arguments in general seem to me unconvincing, since an alternative formulation involving rules of interpretation is immediately available. Furthermore, it seems that the phenomena observed are of some generality. Thus the difference in meaning between ⟨88ii⟩ and ⟨88i, iii⟩ is characteristic of the future 'tense' in many languages, and thus has nothing to do, apparently, with the volitional force of the element 'will'.

[b] A conclusion which appears implausible in that in general to-VP constructions, as in ⟨89⟩, exclude modals.

structure level may assign an anomalous interpretation to an otherwise well-formed sentence. A rule of interpretation that applies to other structures of the class K of syntactic structures, say to surface structures, may have the same effect, in principle. Thus a decision that 'can' in ⟨89⟩ appears at the deep structure level in association with 'seem' would not be refuted by the observation that ⟨91⟩ is deviant; rather, the deviance, in this view, would be attributed to the filtering function of a principle of semantic interpretation applying at the surface structure level.

Anaphoric processes constitute another domain where it is reasonable to inquire into the possibility that rules of semantic interpretation operate at the level of surface structure. It was noted by Akmajian and Jackendoff (1968) that stress plays a role in determining how the reference of pronouns can be interpreted. For example, in the sentence ⟨93⟩, 'him' refers to Bill if it is unstressed, but it may refer either to John or to someone other than John or Bill if it is stressed:

⟨93⟩ John hit Bill and then George hit him.

Similarly, in ⟨94⟩, where 'else' is stressed, 'someone else' refers to someone other than John, whereas when 'afraid' is stressed, it refers to John himself:

⟨94⟩ John washed the car; I was afraid someone else would do it.

The same phenomenon can be observed within sentence boundaries. The explanation hinges on the analysis of placement of primary stress, but it is reasonable to suppose, as Akmajian and Jackendoff suggest, that a principle of surface structure interpretation is involved, given what is known about the relation of intonation to surface structure. See also Jackendoff (1967).

Recent observations by Ray Dougherty (1968a, b) lend some support to this proposal. He argues that the interpretive rules of reference must apply after the application of various transformations, using information that is not present at the deep structure level. Thus consider the sentences ⟨95⟩:

⟨95⟩ (i) each of the men hates his brothers
(ii) the men each hate his brothers.

Dougherty gives considerable evidence to support the view that ⟨95ii⟩ is derived from a structure such as ⟨95i⟩, by a rule that moves 'each' to one of several possible positions in a sentence. But clearly (i) and (ii) differ in the range of possible interpretations for the reference of the pronoun 'he'. Thus in (ii), but not (i), it is necessary to interpret 'he' as referring to someone other than the men in question. The deviance of ⟨96ii⟩, then, might be attributed to the filtering effect of rules of surface structure interpretation:

⟨96⟩ (i) each of the men hates his own brothers
(ii) the men each hate his own brothers.

Or, consider the sentences ⟨97⟩:

⟨97⟩ (i) each of Mary's sons hates his brothers
(ii) his brothers are hated by each of Mary's sons
(iii) his brothers hate each of Mary's sons
(iv) each of Mary's sons is hated by his brothers.

The simplest formulation of the passive transformation would derive (ii) from a structure like (i), and (iv) from a structure like (iii). But in (ii) and (iii), 'he' cannot be interpreted as referring to one of Mary's sons, though in (i) and (iv) it can be so interpreted. As Dougherty points out in detail, there are many similar phenomena. The matter is not restricted to pronominalization; thus consider the effect of replacing 'his' by 'the other' in ⟨97⟩. There appears to be, in such cases, a relatively simple rule of interpretation that makes use of surface structure information, and that, with its filtering effect, rules that certain otherwise well-formed sentences are deviant. Such observations as these, then, also lend support to a revision of the standard theory that incorporates such rules.

Turning to still more obscure cases in which semantic interpretation may involve surface properties, consider the curious behavior of perfect aspect in English with respect to the presuppositions it expresses. Quite generally, a sentence such as ⟨98⟩ is taken as presupposing that John is alive:

⟨98⟩ John has lived in Princeton.

Thus knowing that ⟨99⟩ is true, one would not say 'Einstein has lived in Princeton'; rather 'Einstein lived in Princeton':

⟨99⟩ Einstein has died.[a]

But now consider the following sentences:

⟨100⟩ Einstein has visited Princeton
⟨101⟩ Princeton has been visited by Einstein
⟨102⟩ Einstein (among others) has told me that...
⟨103⟩ I have been told by Einstein (among others) that...
⟨104⟩ Einstein has taught me physics
⟨105⟩ I have been taught physics by Einstein.

It seems to me that ⟨100⟩, ⟨102⟩, ⟨104⟩ presuppose the denial of ⟨99⟩, but that ⟨101⟩, ⟨103⟩, and ⟨105⟩ do not. If this is correct, then the semantic interpretation of perfect aspect would appear to depend on certain properties of surface structure.[b]

The problem is still more complex when we consider coordinate constructions. Thus consider the following cases:

⟨106⟩ Hilary has climbed Everest
⟨107⟩ Marco Polo has climbed Everest
⟨108⟩ Marco Polo and Hilary have climbed Everest
⟨109⟩ Marco Polo and many others have climbed Everest
⟨110⟩ Everest has been climbed by Marco Polo (among others)
⟨111⟩ many people have climbed Everest.

[a] As can be seen from ⟨99⟩, it is not invariably true that use of the present perfect aspect as the full auxiliary presupposes that the subject is alive, although ⟨99⟩ would nevertheless only be appropriate under rather special circumstances, e.g., if Einstein's death had just occurred. Where a verb can be used in the historical present, use of the present perfect does not seem to carry the presupposition that the subject is alive. Thus I could not say 'Aristotle has visited Crete' or 'Aristotle visits Crete' (in historical present), but there is no presupposition that Aristotle is alive in 'Aristotle has claimed, investigated, demonstrated,...' (or in 'Aristotle demonstrates in the *Posterior Analytics* that...,' etc.).

[b] Unless it is maintained that the surface subject of the passive is also the deep subject. Although arguments for this view can be advanced (see, e.g., Hasegawa, 1968), it seems to me incorrect, a strong counter-argument being provided by idioms that undergo passivization, moving to the surface subject position noun phrases which cannot otherwise appear as subject – e.g., 'advantage was taken of Bill', 'offense was taken at that remark', 'a great deal of headway was made', etc.

Notice, incidentally, that assumptions about whether the entity referred to by a noun phrase is alive can be related in rather complex ways to the structure of an utterance and the lexical items it contains. Thus if I say that John is a friend of mine or that I get along with John, the presupposition is that he is alive; but if I say that John is a hero of mine or that I admire him, this is no longer presupposed; as, of course, it is not presupposed, in any of these cases, if present tense is replaced by past tense.

Sentences ⟨106⟩ and ⟨107⟩ express the presupposition that Hilary and Marco Polo, respectively, are alive.[a] On the other hand, sentences ⟨108⟩–⟨110⟩ do not express the presupposition that Marco Polo is alive; and ⟨111⟩ expresses no such presupposition with regard to the various climbers of Everest. Intuitions about this matter do not appear too firm, but if the judgments just expressed are accurate, then it seems that surface structure must play a role in determining the presupposition of the utterance in a rather complex manner.

Significant differences in interpretation of sentences as the auxiliary is changed are very easy to demonstrate. Thus sentence ⟨112⟩ presupposes that John is a Watusi, but if we replace 'is' by 'would be', the presupposition is that he is not:

⟨112⟩ John is tall for a Watusi.

Furthermore, ⟨112⟩ presupposes that the Watusi are generally not tall, but if 'even' is inserted after 'tall', the presupposition is that the Watusi are tall, and that John, who is a Watusi, is even taller than expected. If 'even' precedes 'John' in ⟨112⟩, the presupposition is that John, who is a Watusi, is short, as are the Watusi in general. Thus the change in position of 'even' changes the presupposition with regard to the height of John and the standard height of the Watusi.

This by no means exhausts the class of cases where it appears reasonable to postulate rules of interpretation that make use of information not represented in deep structure. These cases suggest that the standard theory is incorrect, and that it should be modified to permit these rules. These considerations may not affect the weaker hypothesis that the grammatical relations represented in deep structure are those that determine semantic interpretation. However, it seems that such matters as focus and presupposition, topic and comment, reference, scope of logical elements, and perhaps other phenomena, are determined in part at least by properties of structures of K other than deep structures in particular, by properties of surface structure. In short, these phenomena suggest that the theory of grammar should be reconstructed along the lines intuitively indicated in ⟨113⟩, using the notation of the earlier discussion:

⟨113⟩ base: $(P_1, \ldots, P_i)$ (P_1 the K-initial, P_i the post-lexical (deep) structure of the syntactic structure which is a member of K)

transformations: $(P_i, \ldots, P_n)$ (P_n the surface structure; $(P_1, \ldots, P_n) \in K$)

phonology: $P_n \to$ phonetic representation

semantics: $(P_i, P_n) \to$ semantic representation (the grammatical relations involved being those of P_i, that is, those represented in P_1).

It is quite possible that other terms in the syntactic structure $(P_1, \ldots, P_n)$ are also relevant for semantic interpretation. Notice, incidentally, that it is, strictly speaking, not P_n that is subject to semantic interpretation but rather the structure determined by phonological interpretation of P_n, with intonation center assigned. We have already noted, in discussing the matter of 'opaque' contexts, that it is impossible to construct a 'semantically based' syntax along the lines that have been proposed in recent discussion. The phenomena that we have now been considering lend further

---

[a] It is even clearer, perhaps, in 'Marco Polo has succeeded in climbing Everest'. However, for some obscure reason, it seems to me that if Hilary had just announced that he had succeeded in climbing Everest, it would have been appropriate, without the presupposition that Marco Polo is alive, to have said: 'But Marco Polo has done it too'.

support to this conclusion (unnecessary support, in that the earlier observations suffice to establish the conclusion). It must be borne in mind, however, that the revision of the standard theory does not imply that grammar is 'syntactically based' in the sense that in generating a sentence one must 'first' form P_1 by the categorial component, 'then' forming P_i, by lexical insertion, 'then' forming the remainder of the syntactic structure $k \in K$ by transformation, 'then' interpreting k by semantic and phonological rules. In fact, this description, whatever its intuitive suggestiveness, has no strict meaning, since the revised theory assigns no 'order' to operations, just as the standard theory assigns no order of application, as already noted. In fact, there is nothing to prevent one from describing the standard theory or the proposed revision as characterizing grammars that map phonetic representation onto triples (deep structure, surface structure, phonetic representation), or as maping pairs (phonetic representation, deep structure) onto pairs (surface structure, semantic representation), etc. In fact, the revision, like the standard theory, characterizes grammars that define a certain relation among these concepts, where the relation has properties determined by the precise nature of base rules, transformations, rules of phonological interpretation, and rules of semantic interpretation.

It may be useful, at this point, to recall the attempts of the past few years to study the relation of syntax and semantics within the framework of transformational-generative grammar. Within this framework, the first attempt to show how the syntactic structure of a sentence contributes to determining its meaning was that of Katz and Fodor (1963), an approach that was modified and extended in Katz and Postal (1964). The basic assumption was that only 'optional' rules of the grammar – 'choice-points', so to speak – could make a contribution to the determination of meaning. In Katz and Fodor (1963) two types of rule of interpretation ('projection rule') were considered, one corresponding to optional phrase structure rules, one to optional transformational rules. In Katz and Postal (1964) it was argued that the only optional rules relevant to determination of meaning were the phrase structure rules, the 'base rules' of the standard theory.

Since surface structure is fully determined by base rules and transformational rules, it might seem natural to suppose that properties of surface structure, not being a matter of 'choice', could not contribute to semantic interpretation. Underlying this assumption, one might discern the remnants of the 'Saussurian' view that a sentence is constructed by a series of successive choices, and that each of these may be related to semantic considerations of some sort. Of course, such talk is only metaphorical when we are concerned with competence rather than performance. It may, however, have occasionally been misleading, suggesting, erroneously, that since surface structure is fully determined by other 'choices', properties of surface structure cannot contribute to semantic interpretation. When we drop the loose and metaphoric use of such notions as 'choice', we see that there is no reason at all why properties of surface structure should not play a role in determining semantic interpretation, and the considerations brought forward earlier suggest that in fact, they do play such a role.

Massachusetts Institute of Technology

REFERENCES

Akmajian, A., 'On the analysis of cleft sentences', mimeographed, M.I.T. (1968).

Akmajian, A. and R. Jackendoff, 'Squib', mimeographed, M.I.T. (1968).

Austin, J. L., 'A plea for excuses', *Proceedings of the Aristotelian Society*, 1956–7. Reprinted in J. D. Urmson and G. J. Warnock (eds.) *John L. Austin's, Philosophical Papers*, pp. 123–52 Oxford, 1961.

Bresnan, J., 'A note on instrumental adverbs and the concept of deep structure', M.I.T. (1968).

Chafe, W. L., 'Language as symbolization', *Language*, XLIII, 57–91 (1967).

Chomsky, N., *Aspects of the Theory of Syntax*, M.I.T. (1965).

and M. Halle, *Sound Pattern of English*, Harper and Row (1968).

Dougherty, R., *A Transformational Grammar of Coordinate Conjoined Structures*, Ph.D. dissertation, M.I.T. (1968a).

'A comparison of two theories of pronominalization', mimeographed, M.I.T. (1968b).

Fillmore, C. J., 'The case for case', in E. Bach and R. Harms (eds.), *Universals in Linguistic Theory*, 1–88, Holt, Rhinehart and Winston (1968).

Fischer, S. D., 'On cleft sentences and contrastive stress', mimeographed, M.I.T. (1968).

Grice, H. P., 'Meaning', *Philosophical Review*, LXVI, 377–88 (1957). Reprinted in this volume.

'Utterer's meaning, sentence-meaning and word-meaning', *Foundations of Language*, IV, 225–42 (1968).

Hall, B. See Partee.

Hasegawa, K., 'The passive construction in English', *Language*, XLIV, 230–43 (1968).

Jackendoff, R., 'An interpretive theory of pronouns and reflexives', mimeographed, M.I.T. (1967).

'An interpretive theory of negation', *Foundations of Language* V (1969), pp. 218–41.

Katz, J. J., *The Philosophy of Language*, Harper and Row (1966).

Katz, J. J. with J. A. Fodor, 'The structure of a semantic theory', *Language*, XXXIX, 170–210 (1963).

Katz, J. J. with P. Postal, *An Integrated Theory of Linguistic Description*, M.I.T. (1964).

Kraak, A., 'Presupposition and the analysis of adverbs', mimeographed, M.I.T. (1967).

Kuroda, S-Y., *Generative Grammatical Studies in the Japanese Language*, Ph.D. dissertation, M.I.T. (1965).

Lakoff, G., *On the Nature of Syntactic Irregularity*, Ph.D. dissertation, Indiana University (1965).

'Pronominalization and the analysis of adverbs', mimeographed, Harvard University (1967).

'Instrumental adverbs and the concept of deep structure', *Foundations of Language*, IV 4–29 (1968).

Mates, B., 'Synonymity', *University of California Publications in Philosophy*, 201–26 (1950).

McCawley, J. D., 'Lexical insertion in a transformational grammar without deep structure', Fourth Regional Meeting, Chicago Linguistic Society, April 1968, Dept. of Linguistics, University of Chicago (1968a).

'The role of semantics in a grammar', in E. Bach and R. T. Harms (eds.), *Universal in Linguistic Theory*, Holt, Rhinehart and Winston, 124–69 (1968b).

'Where do noun phrases come from?', in R. Jacobs and P. Rosenbaum (eds.), *Readings in English Transformational Grammar*, Blaisdell (1970). Reprinted in this volume.

Partee, Barbara Hall, *Subject and Object in Modern English*, Ph.D. dissertation, M.I.T. (1965).
'Negation, conjunction, and quantifiers: syntax vs. semantics', mimeographed, UCLA (1968), presented at the Conference on Mathematical Linguistics, Budapest, September 1968.
Perlmutter, D. M., *Deep and Surface Structure Constraints in Syntax*, Ph.D. dissertation, M.I.T. (1968).
Quine, W. V., *Word and Object*, Wiley and Sons (1960).
Scheffler, I., 'On synonymy and indirect discourse', *Philosophy of Science*, 39–44 (1955).
Searle, J., *Speech Acts: An Essay in the Philosophy of Language*, Cambridge University Press (1968).
Stampe, D. W., 'Toward a grammar of meaning', *Philosophical Review*, LXXVIII, 137–74 (1968).

Where do noun phrases come from?[a]

JAMES D. McCAWLEY

1 Background

The contents of this paper is clearly transformational grammar but not so clearly generative grammar. In discussing English, I will be treating the English language not as a class of sentences but as a code which relates messages (semantic representations) to their encoded forms (surface structures). The possibility or impossibility of a given (surface form of a) sentence depends on two quite separate factors, namely the details of the code and the restrictions on possible messages: a surface structure is possible only if there is a message which the code pairs with it. This paper is transformational grammar in that I maintain that the code has roughly the form of the 'transformational component' of a grammar as discussed in Chomsky (1965), i.e. that the 'encoding' of a message can be regarded as involving a series of intermediate stages, each obtained by applying a transformation to the preceding stage. It is not obviously generative grammar, in that to a large extent I leave open the question of what a 'possible message' is.

There are several kinds of constraints on 'possible message'. One kind of constraint relates to what might be called the 'logical well-formedness' of the message. For example, 'or' is a predicate which may be combined with two or more propositions. However, 'or' by itself is not a 'possible message', nor is 'or' plus a single proposition, nor is 'or' plus things which are not all propositions; the following loosely represented structures are thus not 'possible messages':

or
or (Max drinks vodka martinis)
or (China is industrializing rapidly; the Pope).

These constraints on 'possible message' correspond roughly to what Chomsky (1965) calls 'strict subcategorization'.

In Chomsky (1965), another kind of constraint is discussed, namely 'selectional restrictions', which are supposed to exclude sentences such as

⟨1⟩ *That idea is green with orange stripes.

in which each predicate is combined with the right number of things but the sentence is odd because of 'incompatible' choices of lexical material. Chomsky treats selectional restrictions as idiosyncrasies of lexical items: each lexical item is assigned 'selectional features', which express restrictions on what material it may be combined with in 'deep structures', e.g. the verb *surprise* might have the restriction that it 'requires an

[a] This paper is an extensively revised version of a paper of the same title published in Roderick Jacobs and Peter S. Rosenbaum (eds.), *Readings in English Transformational Grammar* (Blaisdell, 1970); part of its contents first appeared in McCawley (1967). Conclusions very similar to those presented here were arrived at by Postal independently of me and slightly earlier than I arrived at them (Postal, unpublished).

animate object', i.e. that it may only appear in a deep structure in which it is followed by a noun phrase[a] having the property 'animate'. It is not clear that such restrictions, to the extent that they are valid restrictions, have anything to do with deep structures and lexical items rather than with semantic representations and semantic items that appear in them. If it in fact turns out that the 'selectional restrictions' of all lexical items are predictable from their meanings, then they are not restrictions on how lexical items may be combined but rather restrictions on how semantic material may be combined, i.e. restrictions on 'possible message'.[b] It is worthwhile at this point to consider some examples which are occasionally cited as cases where a selectional restriction is not predictable from the meaning of the item in question, i.e. cases in which there are two or more words that allegedly have the same meaning but different selectional restrictions. I maintain that in each case the words actually do not have the same meaning. Consider, for example, the Japanese verbs *kaburu*, *hameru*, *haku*, etc., which one might gloss as 'put on, said of headwear', 'put on, said of gloves', 'put on, said of footwear', etc., thus suggesting that these verbs have the same meaning but different selectional restrictions. Such a description is incorrect, since the verbs in fact refer to the quite different actions involved in putting on the articles of clothes in question, as is demonstrated by the fact that if one puts on an article of clothing in an unnatural manner (e.g. puts a pair of socks on his hands, uses a necktie to hold up his trousers, etc.), the choice of verb is dictated not by the article of clothing but by the manner in which it is put on, e.g. covering one's head with a pair of gloves would be described by *kaburu* rather than *hameru*. Similarly, one might propose defining the English verbs *kick*, *slap*, and *punch* as 'strike with the foot', 'strike with the open hand', and 'strike with the fist', suggesting that they have the meaning 'strike' with different selectional restrictions. However, J. R. Ross (personal communication) has observed that in the bizarre situation in which a person had been subjected to surgery in which his hands and feet were cut off and grafted onto his ankles and wrists respectively, it would be perfectly normal to speak of that person as kicking someone with his fist or slapping someone with his foot. This implies that the verbs refer to the specific motion[c] which the organ in question performs and are thus not simply contextual variants of *strike*.[d] For an excellent

[a] Actually, Chomsky treats the condition as imposed not on the entire noun phrase but only on its head noun. The intenability of that proposal is demonstrated in McCawley (1968a), where I also observe that selectional restrictions cannot be regarded as requiring a noun phrase to possess a property such as 'animate' but only as excluding those noun phrases having semantic representations incompatible with that property (see also McCawley 1968c).

[b] The position that 'category mistakes', which may be the only anomalies that can be correctly described as 'selectional violations', do not correspond to 'possible thoughts' is ably defended in Drange (1966).

[c] Dwain Parrack has called to my attention the fact that the meaning of these verbs includes not merely the motion of the organ but the type of surface contact which results. Parrack points out that *slap* requires contact with a more or less flat surface and that one could thus speak of the soles of someone's feet slapping the surface of the water. Similarly, a certain motion of the arm would be a *punch* if the hand were clenched but a *poke* it it were open and only the fingertips were involved in the contact.

[d] It is occasionally suggested (e.g. Bierwisch 1967:8) that there is a linguistically significant distinction between interpretations which are possible only by imagining some bizarre situation and interpretations which require no such effort of the imagination. However, it is not clear that this criterion really defines a classification of sentences. How easy one finds it to imagine a situation in which a given sentence would be appropriate depends on such extralinguistic factors as his factual knowledge, the strength of his imagination, and the possible presence of LSD in his bloodstream. I suspect that the sentences which one can interpret

compendium of selectional restrictions, some of which are not so obviously predictable from meanings as are those just discussed, I refer the reader to Leisi (1966).

Many so-called selectional restrictions are actually not real restrictions, since
'violations' of them are quite normal in reports of dreams, reports of other people's
beliefs,[a] and science-fiction stories:

⟨2⟩ I dreamed that my toothbrush was pregnant.

⟨3⟩ I dreamed that I poured my mother into an inkwell.

⟨4⟩ I dreamed that I was a proton and fell in love with a shapely green-and-
orange-striped electron.

⟨5⟩ Max thinks that electrons are green with orange stripes.

⟨6⟩ Harry thinks that his toothbrush is trying to kill him.

⟨7⟩ Boris believes that ideas are physical objects and claims to have seen several
that were green with orange stripes.

While some linguists might suggest that a person who says things like

⟨8⟩ My toothbrush is alive and is trying to kill me.

observes different selectional restrictions than normal people do, it is pointless to do
so, since the difference in 'selectional restriction' will correspond exactly to a
difference in beliefs about one's relationship with inanimate objects. A person who
utters ⟨8⟩ should be referred to a psychiatric clinic, not to a remedial English course.

Note, however, that dreams, etc. are not completely free as to how semantic
material may combine. For example, the constraint that the complement of the
progressive *be* must be headed by an 'activity verb' may not be violated even in
sentences such as

⟨9⟩ *I dreamed that Arthur was knowing the answer.

⟨10⟩ *Max believes that Arthur is knowing the answer.

Similarly with the restriction that only a quantity of time can elapse:

⟨11⟩ *I dreamed that my toothbrush elapsed.

These constraints appear to be real constraints on 'possible message'; it is evident
that to enumerate these constraints one will have to tackle some of the classical problems of philosophy, namely the question of categories and the question of the distinction between 'essential' and 'accidental' properties of things.

2 Semantic representation

In referring to the 'logical well-formedness' of a semantic representation, I used the
terms 'proposition' and 'predicate' as they are used in symbolic logic. I will in fact
argue that symbolic logic, subject to certain modifications, provides an appropriate
system for semantic representation within the framework of transformational grammar. I thus hold that the much-criticized title, *The laws of Thought*, which George
Boole gave to the first work on symbolic logic, is actually much more appropriate
than has generally been thought the case.

Since the representations of symbolic logic appear at first glance to be of a quite
different formal nature from the labeled trees which constitute syntactic representation, one might expect that the mechanisms which link semantic representation and

without thinking up some story to embed them in are simply those which it is so easy to
imagine someone's using that it would require no effort to think up such a story.

[a] Jakobson (1941, §§ 26–7) points out that many persons, especially children, associate colors
with sounds and, for example, will not hesitate to say that the vowel [a] is red.

surface syntactic representation would divide into two separate systems, a system of 'semantic rules', which would operate on representations of the one kind, and a system of 'syntactic rules', which would operate on representations of the other kind, and that the two kinds of representation would meet at an intermediate 'level' corresponding to what Chomsky calls 'deep structure'. However, as pointed out by Lakoff (remarks at the Texas Conference on Language Universals, 15 April 1967), the difference in formal nature between syntactic and semantic representation is only apparent. Lakoff observes that some of the traditional categories of symbolic logic are reducible to others (e.g. quantifiers can be considered as two-place predicates, one place corresponding to a propositional function and the other to a set) and that only a small inventory of syntactic categories functions in the 'deeper' stages of the 'derivational history' of sentences (Lakoff 1965; Bach 1968). Many syntactic categories are 'derived' rather than basic; for example, most prepositions originate as parts of verbs, so that prior to a transformation which adjoins the 'prepositional' part of a verb to its object, a verb-plus-PrepP combination has the form verb-plus-NP. Likewise, many category differences which had figured in previous analyses have turned out to hinge merely on whether certain lexical items do or do not 'trigger' certain transformations. For example, there is no need to set up the categories PredP, Aux, and Modal, which appear in Chomsky (1965): one can treat the various auxiliary verbs as simply verbs[a] which (like the verbs *seem, appear*, etc.) trigger a transformation of 'VP-promotion', which detaches the VP from the embedded sentence and puts it after the verb in question:[b]

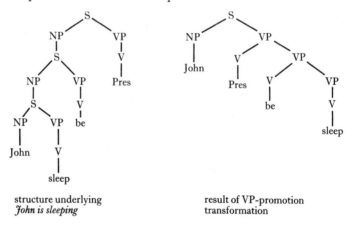

structure underlying   result of VP-promotion
John is sleeping   transformation

and which have the additional peculiarity of being combined with the tense element

[a] This proposal was first made in Ross (1969). A slight change which I have made in Ross's analysis forces me to use the term 'VP-promotion' for the analogue of the transformation which he and Lakoff call '*it*-replacement'. I argue in McCawley (1970) that English has underlying verb-initial order and that that is still the constituent order when this transformation applies. This implies that the transformation is acutally one that raises the subject and not the VP into the next higher clause; indeed, under the newer analysis there is no such constituent as VP.

[b] The analysis of *seem*, etc. as intransitive verbs with a sentence subject is due to Jespersen (1937:57); details of this analysis are given by Rosenbaum (1967: 71–9). The application of 'VP-promotion' to the first of the two trees causes the topmost two NP nodes and the two S nodes in the left branch to vanish by virtue of the 'tree-pruning' principles presented in Ross (1966).

by a fairly early transformation (see Hofmann 1966 for details) and which are thus affected by all subsequent transformations that mention the 'topmost verb' of a clause. Lakoff and Ross concluded (lectures at Harvard and M.I.T., Autumn 1966) that the only 'deep' syntactic categories are Sentence, Noun-Phrase, Verb-Phrase, Conjunction, Noun, and Verb, and that all other traditionally recognized categories are special cases of these categories that correspond to the 'triggering' of transformations by certain lexical items. Bach (1968) then discovered some quite convincing arguments that the Noun–Verb distinction need not be part of this inventory of categories. He argues that all nouns originate in predicate position (e.g. *the anthropologist* arises from a structure paraphrasable as 'the x such that x is an anthropologist') and that the difference between nouns and verbs is that nouns but not verbs trigger a transformation which replaces a relative clause by its predicate element. At the Texas Conference on Language Universals, Lakoff observed that the resulting inventory of categories (Sentence, NP, VP, Conjunction, and 'Contentive' – the term introduced by Bach for the category containing nouns, verbs, and adjectives) matches in almost one-to-one fashion the categories of symbolic logic, the only discrepancy being that the category VP has no corresponding logical category. However, Lakoff argued, there is in fact virtually no evidence for a syntactic category of VP; the various facts that have been cited as evidence for such a category actually have nothing to do with node labels but only with the surface immediate constituent structure, and (as argued in Fillmore 1966 and McCawley (forthcoming)), a surface structure having a constituent consisting of a verb and its objects arises anyway, regardless of whether there is an underlying constituent VP, just as long as tenses and auxiliary verbs are assumed to originate outside of the clauses that they appear in.[a] If one accepts one of the proposals that would do away with VP as an underlying category (and thus do away with the phrase structure rule S → NP VP in favor of a rule S → V NP^n or something such), then not only is the correspondence between 'deep' syntactic categories and the categories of symbolic logic exact, but the 'phrase structure rules' governing the way in which the 'deep' syntactic categories may be combined correspond exactly to 'formation rules' for symbolic logic, e.g. the 'phrase structure rule' that a Sentence consists of a 'Contentive' plus a sequence of Noun Phrases corresponds to the 'formation rule' that a proposition consists of an n-place predicate plus an 'argument' for each of the n places in the predicate.[b]

Since I believe that the correspondence between syntactic and logical categories is slightly different from that proposed by Lakoff (in that I believe that a slightly different kind of symbolic logic is needed for semantic representation), I will not go into the details of the correspondence which Lakoff set up. However, I observe that if such a correspondence is valid, then semantic representations can be considered to be objects of exactly the same formal nature as syntactic representations, namely trees whose non-terminal nodes are labeled by symbols interpretable as syntactic categories. One might object that the trees of syntax are different in formal nature from those which formulas of symbolic logic may be interpreted as, in that the nodes of syntactic trees have a left-to-right ordering relation, whereas it is not clear that there is any left-to-right ordering on the nodes of the trees that I am proposing

[a] Fillmore (1966) does not accept Ross's analysis of auxiliaries as verbs but has an alternative proposal in which (as in Ross's proposal) auxiliaries originate outside of the clause that they appear in.

[b] The rules will, however, probably differ to the extent that in natural languages n will be required to be small, say, at most 4.

as semantic representations. Whether this objection is correct depends on whether one holds that things which mean the same must have the same semantic representation or merely that 'equivalence' of semantic representations can be defined in such a way that things which mean the same have 'equivalent' representations. The former position would, of course, imply that semantic representations cannot have ordered nodes, since

⟨12⟩ John and Harry are similar.
⟨13⟩ Harry and John are similar.

would have to have the same semantic representation, and that representation thus could not have the node corresponding to *John* preceding the node corresponding to *Harry* or vice versa. However, no evidence has as yet been adduced for accepting this position rather than the other one. Until such evidence is found, the question of the ordering of nodes gives no reason for believing semantic representations to be different in formal nature from syntactic representation. I will thus treat the elements of semantic representations as having a linear order and assume that there are rules such as

⟨14⟩ p or q ≡ q or p

which define an equivalence relation on these representations.

These considerations suggest that there is no natural breaking point between a 'syntactic component' and a 'semantic component' of a grammar such as the level of 'deep structure' was envisioned to be in Chomsky (1965)[a] and imply that the burden of proof should be on those who assert that such a breaking point exists. In McCawley (1968a), I argue that setting up a level of 'deep structure' makes it impossible to treat as unitary processes certain phenomena which in fact are unitary processes, in particular, that the use of *respective* and *respectively* in English involves a phenomenon which can be stated as a single rule if there is no level of 'deep structure', but must be divided into special cases, some of which correspond to 'semantic interpretation rules' and others to 'transformations', if a level of 'deep structure' is accepted. Since the argument is rather involved, I will not reproduce it here but refer the reader to McCawley (1968a) for details; the general outline of this argument for rejecting a level of 'deep structure' is, of course, identical to that of Halle's celebrated argument (Halle 1959) for rejecting a 'phonemic' level.

3   Noun phrases and semantic representation

The principal respect in which I find existing versions of symbolic logic insufficient for the representation of meaning has to do with noun phrases. Consider the sentence

⟨15⟩ The man killed the woman.

If one accepts the position (expounded and defended in e.g. McCawley 1968a) that each noun phrase occurrence in a syntactic representation must have attached to it

---

[a] Since the publication of Chomsky (1965), Chomsky has modified his conception of the role of deep structure considerably. He no longer regards 'deep structure' as a 'level' *intermediate between* semantic representation and surface structure but instead holds (Chomsky 1967:407) that 'deep structure completely determines certain highly significant aspects of semantic representation... [but] surface structure also contributes in a restricted but important way to semantic interpretation'.

an 'index', which corresponds to the 'intended reference' of that noun phrase occurrence, then the structure which underlies ⟨15⟩ will have to have some index x_1 attached to *the man* and some index x_2 attached to the woman. The meaning of ⟨15⟩ will then involve the assertion that x_1 participated as agent and x_2 as patient in a certain event y of killing, the assertion that y took place prior to the speech act, and the assertions that x_1 is a man and that x_2 is a woman. One might propose that the semantic representation of ⟨15⟩ is obtained simply by conjoining all these assertions, with perhaps some additional terms to cover the meaning of *the*, which I have ignored:

⟨16⟩ $\text{kill}_y(x_1, x_2) \wedge \text{Past}(y) \wedge \text{Man}(x_1) \wedge \text{Woman}(x_2)$.[a]

However, ⟨16⟩ does not correctly represent the meaning of ⟨15⟩. Note that if one says

⟨17⟩ I deny that the man killed the woman.

he is not denying ⟨16⟩. To deny a conjunction is to assert that at least one of the conjuncts is false. However, in ⟨17⟩ the speaker is not merely asserting that one of the four terms of ⟨16⟩ is false but is asserting specifically that the first term is false and assuming the other three terms to be true: it would not be correct to say ⟨17⟩ when one means that x_1 did in fact kill x_2 but that x_1 is not a man. Similarly, when one asks

⟨18⟩ Did the man kill the woman?,

he is not asking whether the conjunction ⟨16⟩ is true (i.e. whether all four terms are true) but is assuming the truth of the last three terms and asking whether the first term is true. It thus appears that in some sense the meanings of the expressions *the man* and *the woman* play a subordinate role in the meaning of ⟨15⟩.

To represent meaning correctly, symbolic logic will have to be supplemented by a way of representing this type of 'subordination'. The fact that no such device has been used in symbolic logic so far is a result of the fact that symbolic logic has largely been used as a device for representing the content of propositions of mathematics. This 'subordination' relates to an important way in which the sentences of natural languages differ from mathematical propositions. In mathematics one enumerates certain objects which he will talk about, defines other objects in terms of these objects, and confines himself to a discussion of objects which have been either explicitly postulated or explicitly defined and which thus have been assigned explicit names; these names are in effect proper names. However, one does not begin a conversation by giving a list of postulates and definitions. One simply starts talking about whatever he intends to talk about, and the bulk of the things which he talks about will be things for which either there is no proper name (e.g. there is no proper noun *Glarf* meaning 'the nail on the third toe of Lyndon Johnson's left foot') or the speaker does not know any proper name (e.g. an expression such as 'the sexy little redhead that Max was talking to in the coffee shop yesterday' used to refer to someone whose name one does not know). Moreover, people often talk about things which either do not exist or they have identified incorrectly. Indices exist in the mind of the speaker rather than in the real world: they are conceptual entities which the

---

[a] If p and q are propositions, then p ∧ q is the proposition that p and q are both true. One major respect in which ⟨16⟩ fails to be an adequate semantic representation is that it fails to represent the semantic structure underlying *kill*. Details of the correct representation are given in McCawley (1968b).

individual creates in interpreting his experience. Communication between different persons is possible because (1) different individuals often correctly identify things or make similar misidentifications, so that what one speaker says about an item in his mental picture of the universe will jibe with something in his hearer's mental picture of the universe, and (2) the noun phrases which speakers use fulfill a function comparable to that of postulates and definitions in mathematics: they state properties which the speaker assumes to be possessed by the conceptual entities involved in what he is saying and are used chiefly to give the listener sufficient information to identify the things that the speaker is talking about.[a] I conclude that it is necessary for semantic representation to separate a clause into a 'proposition' and a set of noun phrases, which provide the material used in identifying the indices of the 'proposition', e.g.

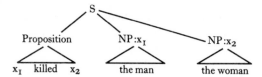

That representations such as the above play a role in grammar is shown by an interesting class of ambiguities which appears to have escaped the notice of linguists until recently, although it has been discussed by philosophers since the middle ages as the distinction between *de dicto* and *de re* interpretation. The sentence

⟨19⟩ Willy said that he has seen the woman who lives at 219 Main St.

is appropriate either to report Willy's having said something such as 'I saw the woman who lives at 219 Main St' (the *de dicto* interpretation) or to report his having said something such as 'I saw Harriet Rabinowitz', where the speaker identifies Harriet Rabinowitz as 'the woman who lives at 219 Main St' (the *de re* interpretation). This ambiguity is brought out by the fact that the sentence can be continued in two ways, each of which allows only one of the two interpretations:

⟨20a⟩ ...but the woman he had in mind really lives in Pine St. [*de dicto*]
⟨20b⟩ ...but he doesn't know that she lives there. [*de re*]

Similarly, while

⟨21⟩ Boris says that he didn't kiss the girl who he kissed.

might conceivably be a *de dicto* report of Boris's having uttered a contradictory sentence such as 'I didn't kiss the girl who I kissed', it is more likely to involve a sentence such as 'I didn't kiss Nancy' reported *de re* by a person who is convinced that Boris really did kiss Nancy. Similarly with

⟨22⟩ Harry admits that he kissed the girl who he kissed.
⟨23⟩ Joe doesn't know that your sister is your sister.

See Castañeda (1967) for further examples of such ambiguities. These facts indicate that in certain kinds of embedded sentences the lexical material relating to noun phrases in the embedded sentence may be semantically either a part of the embedded sentence or part of a higher sentence. The proposal of the last paragraph makes it

[a] See Donnellan (1966) for some highly insightful observations on this use of NPs.

possible for representations to show just such a distinction. For example, the two meanings of ⟨19⟩ can be represented as

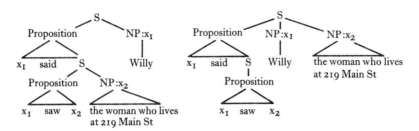

In the first tree *the woman who lives at 219 Main St* is part of what Willy allegedly said; in the second tree it is not. Distinctions relating to what sentence a noun phrase is a constituent of are also involved in ambiguous sentences such as

⟨24⟩ Nancy wants to marry a Norwegian.

which may mean either that there is a Norwegian who Nancy wants to marry or that Nancy wants her future husband to be Norwegian, although she may not yet have found a Norwegian that she would want to marry. In the first case *a Norwegian* is a constituent of the main sentence, and in the second case it is a constituent of the sentence which is the underlying object of *want*. There is a similar ambiguity in

⟨25⟩ John wants to find the man who killed Harry.

In cases of multiple embeddings, it is possible to get multiple ambiguities. For example, Bach (1968) points out that

⟨26⟩ John says that Nancy wants to marry a Norwegian.

is ambiguous between the three senses (i) there is a person who John says Nancy wants to marry and who the speaker identifies as a Norwegian, (ii) John says that Nancy wants to marry a certain person who John identifies as a Norwegian, and (iii) John says that Nancy wants her future husband (whoever he might be) to be Norwegian. It is difficult to see how these three senses could be assigned different 'deep structures' unless those structures allowed noun phrases to occur separate from the propositions that they are involved in and to be constituents of sentences in which those propositions are embedded.[a]

Similarly, the ambiguity of[b]

⟨27⟩ John thinks he is smarter than he is.

between a sense in which John subscribes to the contradiction 'I am smarter than I am' and the more normal sense which asserts that John is not as smart as he thinks he is, is an ambiguity between an underlying structure in which something such as *the extent to which John is smart* is part of the complement of *think* and one in which it is not. In the one case *the extent to which John is smart* will be John's smartness as identified by John, in the other as identified by the speaker. The sentence

⟨28⟩ I wonder why more men don't beat their wives than do.

is ambiguous between (i) I wonder why the number of men who don't beat their wives exceeds the number who do and (ii) I wonder why the number of men who beat their wives is as small as it is. In meaning (ii), the complement of *wonder* is the

---

[a] A valiant attempt at a description of these sentences is given in Bach (1968).
[b] I am grateful to Charles J. Fillmore for calling this sentence to my attention.

question 'Why isn't the number of men who beat their wives greater than n?' and the lexical material corresponding to n (*The number of men who beat their wives* or something such) is an adjunct to the whole sentence rather than part of the complement of *wonder*.

The hypothesis that English sentences are derived from semantic representations of the form proposed above entails the conclusion that English has a transformation which attaches each noun phrase to an occurrence of the corresponding index. In the proposed underlying structure of ⟨19⟩, this transformation would attach *the woman who lives at 219 Main St* to the single occurrence of x_2 and Willy to the first of the two occurrences of x_1. In this case only the first occurrence of x_1 is a possible place to attach Willy: note that in

⟨29⟩ He said that Willy had seen the woman who lives at 219 Main St.

the *he* may not refer to Willy. However, in some cases it is possible to attach a noun phrase to any of several occurrences of the index in question, so that there are alternate surface forms such as

⟨30⟩ After John left his apartment, he went to the pool hall.

⟨31⟩ After he left his apartment, John went to the pool hall.

Since the occurrences of an index to which no full NP is attached are realized as pronouns, no pronominalization transformation as such is needed.[a] The important constraint on pronominalization noted by Ross (1967a) and Langacker (in press) must thus be reformulated as a constraint on the NP-attachment transformation.[b] Ross and Langacker, working from the assumption that pronouns are derived by a transformation which replaces one of two identical noun phrases by a pronoun, concluded that a noun phrase may trigger the pronominalization of another noun phrase either if it precedes it or if it follows it and is in a sentence which the other noun phrase is in a clause subordinate to. Thus, *he* may refer to John in

⟨32⟩ John went to the pool room after he left his apartment.

but not in

⟨33⟩ He went to the pool room after John left his apartment.

However, in ⟨31⟩, where *he* is in a clause subordinate to that containing *John*, it may refer to John. The effect of this constraint can be imposed on the NP-attachment

[a] This is an oversimplification. While ordinary personal pronouns need not be derived by a pronominalization transformation such as that of Ross (1967a) or Langacker (1969), sentence pronominalizations in examples such as

Marvin said that there was a unicorn in the garden, and *so* there was.

Margaret is rumored to have been arrested, but I don't believe *it*.

must in fact be derived through a transformation that replaces one of two identical structures by a pronoun.

[b] One promising alternative is suggested in Lakoff (1968): that the 'antecedent of' relation is marked in surface structures (in addition to the information usually considered to be present in surface structures) and that the Ross–Langacker constraint is an output constraint (in the sense of Ross 1967b) on the antecedent relation.

The treatment of personal pronouns that I proposed above eliminates one important anomaly that pronominalization appeared to have in earlier treatments: it was the only exception to the principle (Ross 1967b:340) that 'feature-changing' rules are 'upper bounded', i.e. that an element may trigger feature changes only in elements of the same clause or of a 'lower' clause. However, this anomaly is eliminated only at the expense of introducing a new anomaly, since NP-attachment would violate Ross's 'coordinate structure constraint', the principle that material may not be moved into or out of a coordinate structure; NP-attachment has to move material into a coordinate structure in cases such as

The girl you like and her brother's roommate have just eloped.

transformation by saying that a noun phrase may be attached to any occurrence of the corresponding index which either precedes or is in a 'higher' sentence than all other occurrences of that index.

The treatment of pronominalization which I have just proposed is supported by sentences of a type first investigated by Emmon Bach and P. S. Peters and also discovered independently by William Woods and Susumu Kuno:

⟨34⟩ A boy who saw her kissed a girl who knew him.

Here *her* is to be interpreted as referring to the girl mentioned in the sentence and *him* referring to the boy. Under the conception of pronominalization which derives a pronoun from a copy of its antecedent, *her* would have to come from a copy of *a girl who knew a boy who saw her*, which would in turn have to come from a copy of *a girl who knew a boy who saw a girl who knew him*, etc., and each of the two noun phrases would have to be derived from an infinitely deep pile of relative clauses. However, under the conception of pronominalization which I propose, this anomaly would vanish. ⟨34⟩ would be derivable from a structure roughly representable as

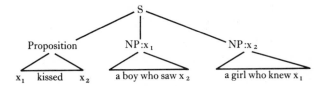

The attachment of noun phrases to index occurrences takes place sequentially. The process may begin with either x_1 or x_2. What results under the Proposition node will be respectively

⟨35⟩ A boy who saw x_2 kissed x_2.
⟨36⟩ x_1 kissed a girl who knew x_1.

In ⟨35⟩, both occurrences of x_2 are possible places for the attachment of the remaining noun phrase; attaching it to the first occurrence of x_2 yields

⟨37⟩ A boy who saw a girl who knew x_1 (=him) kissed x_2 (=her).

and attaching it to the second occurrence of x_2 yields ⟨34⟩. In ⟨36⟩, only the first occurrence of x_1 meets the constraint formulated above, and attaching the noun phrase there yields ⟨34⟩. I call the reader's attention to the fact that there are thus two derivations which convert the tree given above into the surface structure of ⟨34⟩, which may or may not be a defect of this account of pronominalization.

4 Implications

The above treatment of noun phrases necessitates some changes in the 'base component' of a grammar: a distinction between 'Sentence' and 'Proposition' must now be drawn; a Proposition is now a 'Contentive' plus a sequence of indices rather than a 'Contentive' plus a sequence of Noun Phrases, and an overall constraint on semantic representations must be imposed to insure that each representation contains neither too few nor too many noun phrases. An obvious first approximation to this constraint is to say that for each index in a semantic representation there must be at most one corresponding noun phrase, and that that noun phrase must be directly dominated by a S node which dominates all occurrences of that index. Some such constraint is needed anyway to exclude the possibility of saying

⟨38⟩ Napoleon loves Bonaparte's wife.

to mean that Napoleon loves his wife (cf. Gruber 1965). This constraint can be sharpened somewhat. If a personal pronoun occurs in a sentence which does not contain an antecedent for that pronoun, then either the pronoun has an antecedent in some preceding sentence of the discourse (possible a sentence spoken by someone other than the speaker of the sentence in question) or the pronoun is used deictically (i.e. is a direct reference to someone or something physically present as the sentence is uttered) and is stressed and accompanied by a gesture. Since the semantic function of the gesture which accompanies a deictic pronoun is the same as that of the lexical material of an ordinary noun phrase, it is tempting to suggest that in these sentences the gesture *is* a noun phrase, that the attachment transformation attaches that noun phrase to one of the occurrences of the index in question, and that the phonetic reflex of a gesture is stress. In support of this proposal, I note that a deictic pronoun may serve as the antecedent of a pronoun under exactly the same conditions under which an ordinary noun phrase may. For example, in

⟨39⟩ After he left the office, *he* [gesture] went to the pool hall.

the first *he* may have the second as its antecedent. Such is not the case in

⟨40⟩ He went to the pool hall after *he* [gesture] left the office.

This suggests tightening the constraint to make it say that for every index in a semantic representation there is exactly one corresponding noun phrase, except that a noun phrase may be omitted if there is a noun phrase with the same index in an earlier sentence of the discourse.

The proposal that pronouns are derived from index occurrences to which no noun phrase has been attached is further confirmed by the fact that pronouns do not admit ambiguities such as those of sentences ⟨19⟩ and ⟨21⟩–⟨23⟩: in the sentence

⟨41⟩ My friends think that he is a woman.

the choice of *he* rather than *she* can only be on the basis of the speaker's knowledge about the person being talked about and cannot be a report of the friends having said 'He is a woman' of a person who the speaker is convinced really is a woman.

The fact that noun phrases which have a non-restrictive clause are also unambiguously the speaker's contribution:

⟨42⟩ John said that his neighbor, who you met at Arthur's party, has just been sent to Devil's Island for possessing pot.

(note that this could not be followed by '...but the chap that John was talking about really isn't his neighbor') indicates that the formation of non-restrictive clauses takes place before the attachment of noun phrases to their indices.

In German and Latin the ambiguity of sentences like ⟨19⟩ is resolved by the mood of the verb in the relative clause: in the *de dicto* meaning, *lives* is put in the subjunctive mood, and in the *de re* meaning it is put in the indicative mood. This indicates that mood is predictable on the basis of the structure prior to the NP-attachment transformation, specifically that a verb is made subjunctive if it is in a noun phrase which is within the complement of certain verbs at that point of the derivation.[a]

Regarding the distinction between 'Sentence' and 'Proposition', I observe that

[a] On the syntax of Latin subjunctives, see R. Lakoff (1969).

verbs differ as to whether they take a 'Sentence' or a 'Proposition' as complement, i.e. whether the complement may contain noun phrases as well as a proposition. The discussion in section 3 implies that *say*, *deny*, and *want* take Sentences rather than Propositions as their complements. On the other hand, *seem* and *begin* take Propositions as complements: not that the sentences

⟨43⟩ John began beating the man who lives at 219 Main St.

⟨44⟩ Max seems to know the woman who lives at 219 Main St.

do not have the ambiguity which was noted in ⟨19⟩. However, if these sentences are embedded as the complements of verbs such as *say* or *want*, the result has the ambiguity:

⟨45⟩ John wants to begin beating the man who lives at 219 Main St.

⟨46⟩ Harry says that Max seems to know the woman who lives at 219 Main St.

The proposal that sentences are derived from structures in which the adjuncts to 'contentives' are indices rather than full noun phrases renders fairly trivial certain problems which would otherwise present considerable difficulty. For example, it is hard to see how the sentences

⟨47⟩ Everyone loves himself.

⟨48⟩ Everyone loves everyone.

could be assigned different 'deep structures' and how the reflexivization transformation could be formulated without ad-hoc restrictions if syntax were to start with 'deep structures' in which full noun phrases like *everyone* rather than indices were to be the subjects, objects, etc. of verbs and reflexivization were to be contingent on noun phrases being identical. However, if sentences are derived from structures in which the adjuncts of 'Contentives' are indices, as in the semantic representations that I have proposed, ⟨47⟩ and ⟨48⟩ can be derived from the structures

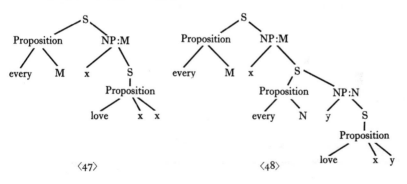

⟨47⟩   ⟨48⟩

Reflexivization is applicable only in the former tree.[a] This proposal also explains why the sequence of words ⟨48⟩, when appropriately stressed, may mean either that everyone has the characteristic of loving everyone or that everyone has the characteristic that everyone loves him.[b] The former meaning corresponds to the tree at the

[a] These trees are a minor modification of the structures proposed in Lakoff and Ross (1967).

[b] These two meanings are 'logically equivalent': if either is true, then so is the other. It is of course reasonable to say that logically equivalent things need not have the same meaning, since any two contradictory propositions are logically equivalent, but one would hardly want to say that the sentences

⟨49⟩ That horse is not a horse.

⟨50⟩ My father had no children.

right above and the latter to

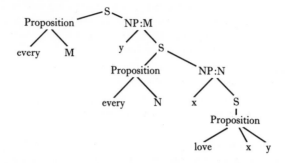

Both structures will yield the same sequence of words since in either case the two indices occurring next to *love* are the only places where the two occurrences of *every* may be attached to their indices by the transformation of quantifier-lowering (Carden 1968).

Finally, I observe that the data which I have discussed here show to be completely untenable the familiar proposal (Chomsky 1957) that each language has a limited repertoire of 'kernel sentences' and that the full range of sentences in the language is obtained by combining and deforming these kernel sentences in various ways. The sentences discussed require analyses in which structures containing less material than would make up a sentence (i.e. structures which have 'slots' for noun phrases but no corresponding lexical material) are embedded in structures which contain more lexical material than there are slots to put it into.

University of Chicago

BIBLIOGRAPHY

Bach, Emmon. 1968. 'Nouns and noun phrases'. Bach and Harms 1968: 90–122.
Bach, Emmon and Robert T. Harms. 1968. *Universals in Linguistic Theory*. New York: Holt, Rinehart and Winston.
Bierwisch, Manfred. 1967. 'Some semantic universals of German adjectivals'. *Foundations of Language*, III, 1–36.

have the same meaning or that they may be translated into German as
⟨49a⟩ Der Kreis ist dreieckig.
⟨50a⟩ Ich küsste ein Mädchen, das ich nie geküsst habe.
respectively. The two meanings of ⟨48⟩ give a much less trivial example of things which are logically equivalent but different in meaning. The following reasons can be cited for regarding the two interpretations of ⟨48⟩ as distinct meanings. (1) To deny the one interpretation is to assert that there is someone who does not love everyone; to deny the other interpretation is to assert that there is someone who not everyone loves. Thus the existence of two quite different things is being *asserted* (although the existence of the one can be *inferred* from the existence of the other). (2) As noted by Peter Geach, the two meanings of
⟨51⟩ Almost everyone loves almost everyone.
are not logically equivalent, since the truth conditions for one meaning may be met at the same time that those for the other meaning are violated. Similarly with the two meanings of
⟨52⟩ All but one of the boys danced with all but one of the girls.
Thus, identical quantifiers cannot in general be permuted *salva veritate*, and formation rules which would provide for the distinction between the two meanings of ⟨51⟩ would provide for the same distinction between meanings involving *every* in place of *almost every* if all quantifiers are treated alike by the formation rules.

Boole, George. 1847. *An investigation of the laws of thought.* London.
Carden, Guy. 1968. 'Quantifiers as higher predicates'. Report NSF-20 of the Harvard University Computation Laboratory.
Castañeda, Hector Neri. 1967. 'Indicators and quasi-indicators'. *American Philosophical Quarterly*, IV, 85–100.
Chomsky, Noam A. 1957. *Syntactic Structures.* The Hague: Mouton.
    1965. *Aspects of the theory of syntax.* Cambridge, Mass.: M.I.T. Press.
    1967. 'The formal nature of language'. An appendix to Eric H. Lenneberg, *Biological foundations of language*, New York: Wiley.
Donnellan, Keith, 1966. 'Reference and definite descriptions'. *Philosophical review*, LXXV, 281–304. Reprinted in this volume, 100–14.
Drange, Theodore. 1966. *Type crossings.* The Hague: Mouton.
Fillmore, Charles J. 1966. 'A proposal concerning English prepositions'. *Georgetown University Monograph Series on Languages and Linguistics*, XIX, 19–34.
    1968. 'The case for case'. Bach and Harms 1968: 1–88.
Gruber, Jeffrey. 1965. *Studies in lexical relations.* M.I.T. dissertation.
Halle, Morris. 1959. *The sound pattern of Russian.* The Hague: Mouton.
Hofmann, T. R. 1966. 'Past tense replacement and the English modal auxiliary system'. In report NSF-17 of the Harvard University Computation Laboratory.
Jakobson, Roman. 1941. *Kindersprache, Aphasie, und allgemeine Lautgesetze.* Reprinted in *Selected writings* I, 328–401.
Jespersen, Otto. 1937. *Analytic syntax.* London: Allen and Unwin.
Lakoff, George. 1965. *On the nature of syntactic irregularity.* Indiana University dissertation.
    1968. 'Pronouns and reference'. Duplicated, Harvard.
Lakoff, George and John Robert Ross. 1967. 'Is deep structure necessary?'. Duplicated, M.I.T.
Lakoff, Robin. 1969. *Abstract syntax and Latin complementation.* Cambridge, Mass.: MIT Press.
Langacker, Ronald. 1969. 'Pronominalization and the chain of command'. In David A. Reibel and Sanford A. Schane (eds.), *Modern Studies in English*, 160–86. Englewood Cliffs, N.J.: Prentice-Hall.
Leisi, Ernst. 1966. *Der Wortinhalt*, 3rd ed. Heidelberg: Quelle u. Meyer.
McCawley, James D. 1967. 'Meaning and the description of languages'. *Kotoba no uchū* (Tokyo), II, nos. 9 (10–18), 10 (38–48), 11 (51–7).
    1968a. 'The role of semantics in a grammar'. Bach and Harms 1968: 124–69.
    1968b. 'Lexical insertion in a transformational grammar without deep structure'. In Darden, Bailey and Davison (editors), *Papers of the 1968 regional meeting, Chicago Linguistic Society*, 71–80. Chicago.
    1968c. Review of Sebeok (ed.), *Current Trends in Linguistics*, III. *Language* XLIV, 556–93.
    1970. 'English as a VSO language'. *Language* XLVI, 286–99.
Postal, Paul M. Unpublished. 'Notes on restrictive relatives'. Yorktown Heights: I.B.M. (1967).
Rosenbaum, Peter S. 1967. *The grammar of English predicate complement constructions.* Cambridge, Mass.: M.I.T. Press.
Ross, John Robert. 1966. 'A proposed principle of tree pruning'. In Report NSF-17 of the Harvard University Computation Laboratory.
    1967a. 'On the cyclic nature of English pronominalization'. *To honor Roman Jakobson* 1669–82. The Hague: Mouton.
    1967b. *Constraints on variables in syntax.* M.I.T. dissertation.
    1969. 'Auxiliaries as main verbs'. *Journal of Philosophical Linguistics* I, no. 1, 77-102.

On generative semantics[a]

GEORGE LAKOFF

1 The basic theory

I would like to discuss some questions having to do with the theory of grammar. I assume that a grammar of a language is a system of rules that relates sounds in the language to their corresponding meanings, and that both phonetic and semantic representations are provided in some language-independent way. I assume that the notion 'possible surface structure' in a possible natural language is defined in terms of 'trees' or 'phrase-markers', with the root S, whose node-labels are taken from a finite set of node-labels: S, NP, V, ... The notion 'tree', or 'phrase-marker', is to be defined in one of the usual ways, in terms of predicates like *precedes*, *dominates*, and *is labeled*. Thus a grammar will define an infinite class of surface structures. In addition I assume that a grammar will contain a system of grammatical transformations mapping phrase-markers onto phrase-markers. Each transformation defines a class of well-formed pairs of successive phrase-markers, P_i and P_{i+1}. These transformations, or well-formedness constraints on successive phrase-markers, P_i and P_{i+1}, define an infinite class K of finite sequences of phrase-markers, each such sequence $P_1, \ldots, P_n$ meeting the conditions:

⟨1⟩ (i) P_n is a surface structure
  (ii) each pair P_i and P_{i+1} meet the well-formedness constraints defined by transformations

[a] Generative semantics is an outgrowth of transformational grammar as developed by Harris, Chomsky, Lees, Klima, Postal, and others. The generative semantics position was arrived at through an attempt on the part of such linguists as Postal, Fillmore, Ross, McCawley, Bach, R. Lakoff, Perlmutter, myself, and others to apply consistently the methodology of transformational grammar to an ever-increasing body of data. We have not all reached the same conclusions, and those presented here are only my own. However, I think it is fair to say that there has developed in recent years a general consensus in this group that semantics plays a central role in syntax. The generative semantics position is, in essence, that syntax and semantics cannot be separated and that the role of transformations, and of derivational constraints in general, is to relate semantic representations and surface structures. As in the case of generative grammar, the term 'generative' should be taken to mean 'complete and precise'.

An earlier version of part of this paper has appeared in Binnick et al. [4]. The present paper is an early draft of some chapters of a book in progress to be called *Generative Semantics* [43]. I would like to thank R. T. Lakoff, J. D. McCawley, D. M. Perlmutter, P. M. Postal, and J. R. Ross for lengthy and informative discussions out of which most of the material discussed here developed. I would also like to thank Drs McCawley, Postal, and Ross for reading an earlier draft of this manuscript and suggesting many improvements. Any mistakes are, of course, my own. I would also like to take this opportunity to express my gratitude to Professor Susumu Kuno of Harvard University, who has done much over the past several years to make my research possible. The work was supported in part by grant GS-1934 from the National Science Foundation to Harvard University.

(iii) there is no P_0 such that $P_0, P_1, \ldots, P_n$ meets conditions (i) and (ii).

The members of K are called the *syntactic structures* generated by the grammar. I will assume that the grammar contains a lexicon, that is, a collection of lexical entries specifying phonological, semantic, and syntactic information. Thus, we assume

⟨2⟩ a lexical transformation associated with a lexical item I maps a phrase-marker P containing a substructure Q which contains no lexical item into a phrase-marker P′ formed by superimposing I over Q.

That is, a lexical transformation is a well-formedness constraint on classes of successive phrase-markers P_i and P_{i+1}, where P_i is identical to P_{i+1} except that where P_i contains a subtree Q, P_{i+1} contains the lexical item in question. Various versions of this framework will differ as to where in the grammar lexical transformations apply, whether they apply in a block, etc.

In this sense, transformations, or well-formedness conditions on successive phrase-markers, may be said to perform a 'filtering function', in that they 'filter out' derivations containing successive phrase-marker pairs (P_i, P_{i+1}) which do not meet some well-formedness condition on such pairs. A system of transformations is essentially a filtering device which defines a class of well-formed sequences of phrase-markers by throwing out all of those sequences which contain pairs (P_i, P_{i+1}) which do not meet some such well-formedness condition, that is, are not related by some transformation. Since transformations define possible derivations only by constraining pairs of successive phrase-markers, I will refer to transformations as 'local derivational constraints'. A local derivational constraint can be defined as follows. Let 'P_i/C_1' mean phrase-marker P_i meets tree-condition C_1. A transformation, or local derivational constraint, is a conjunction of the form P_i/C_1 and P_{i+1}/C_2, as where C_1 and C_2 are tree-conditions defining the class of input trees and class of output trees, respectively. It is assumed that:

$$C_1 = C_1' \text{ and } C_1''$$
$$C_2 = C_2' \text{ and } C_2''$$
$$C_1' = C_2'$$
$$C_1'' \neq C_2''$$
$$C_1' \text{ and } C_2' \text{ are both nonnull}$$
$$C_1'' \text{ and } C_2'' \text{ are not both null}$$

C_1' (which is identical to C_2') will be called the *structural description* (SD) of the transformation. C_1'' and C_2'' will be called the *structural correlates* (SC) of the transformation. The SD of the transformation defines the part of the tree-condition which characterizes both P_i and P_{i+1}. The SC of the transformation defines the minimal difference between P_i and P_{i+1}. Thus, a pair (C_1, C_2) defines a local derivational constraint, or 'transformation'. A derivation will be well-formed only if for all i, $1 \leqslant i < n$, each pair of phrase-markers (P_i, P_{i+1}) is well-formed. Such a pair will, in general, be well-formed if it meets some local derivational constraint. There are two sorts of such constraints: optional and obligatory. To say that a local derivational constraint, or transformation, (C_1, C_2) is *optional* is to say:

(x) $(P_x/C_1 \supset (P_{x+1}/C_2 \supset (P_x P_{x+1}), \text{ is well-formed}))$

To say that (T_1, T_2) is *obligatory* is to say:

(x) $(P_x/C_1 \supset (P_{x+1}/C_2 \equiv (P_x, P_{x+1}) \text{ is well-formed}))$

For a derivation to be well-formed it is necessary (but in general not sufficient) for each pair of successive phrase-markers to be well-formed.

In addition to transformations, or local derivational constraints, a grammar will contain certain 'global derivational contraints'. Rule orderings, for example, are given by global derivational constraints, since they specify where in a given derivation two local derivational constraints can hold relative to one another. Suppose (C_1, C_2) and (C_3, C_4) define local derivational constraints. To say that (C_1, C_2) is ordered before (C_3, C_4) is to state a global derivational constraint of the form:

(i) (j) $((P_i/C_1$ and P_{i+1}/C_2 and P_j/C_3 and $P_{j+1}/C_4) \supset (i < j))$

Another example of a global constraint is Ross' coordinate structure constraint which states that if some coordinate node A^1 dominates node A^2 at some point in the derivatoin P_i, then there can be no P_{i+1} such that A^2 commands A^1 and A^1 does not dominate A^2, where 'command' means 'belong to a higher clause than'. That is, A commands B if and only if the first S-node higher than A dominates B. This is a global derivational constraint of the form:

Let $C_1 = A^i$ dominates *conjunction* $A^{j_1} \ldots A^{j_m}$
$C_2 = A^i$ dominates $X^1 A^k X^2$
$C_3 = A^k$ commands A^i

CSC: (y) $(\sim((P_y/C_1$ and $C_2)$ and $(P_{y+1}/C_3$ and $\sim C_2)))$

What the coordinate structure constraint does is to keep track of the derivational histories of pairs of nodes A^i and A^k. This is just what elementary transformations and their associated rules of derived constituent structure do: they define constraints on successive phrase-markers, keeping all but one or two nodes constant, and then tell what happens to those one or two other nodes in going from the first tree to the second. It seems reasonable on the basis of our present knowledge to limit individual derivational constraints, both local and global, to tracing the histories of at most two nodes – over a derivation in the case of global constraints, and over two successive trees in the case of local constraints. Other examples of global derivational constraints are Ross' other constraints on movement rules (Ross [65]), the theory of exceptions (Lakoff [40] and R. Lakoff [44]), Postal's crossover principle (Postal [57]), output conditions for pronominalization (Lakoff [41]), etc. It should be clear that all theories of transformational grammar have included both local and global derivational constraints. The question arises as to what kinds of local and global derivational constraints exist in natural languages. I will suggest in section 2 below that there is a wider variety than had previously been envisioned. It is assumed that derivational constraints will be restricted to hold either at particular levels in a derivation (semantic representation, surface structure, shallow structure, and deep structure if such exists), or to range over entire derivations or parts of derivations occurring between levels. Constraints holding at particular levels define well-formedness conditions for those levels, and so are analogous to McCawley's node-acceptability conditions [47], which play the role of phrase-structure rules in a theory containing deep structures.

Given a syntactic structure $(P_1, \ldots, P_n)$ we define the semantic representation SR of a sentence as $SR = (P_1, PR, Top, F, \ldots)$, where PR is a conjunction of presuppositions, Top is an indication of the 'topic' of the sentence, and F is the indication of the focus of the sentence.[a] We leave open the question of whether there

---

[a] As I will discuss below, it seems possible to eliminate coordinates for topic and focus in favor of appropriate representations in the presuppositional part of the sentence. Thus, it is conceivable that semantic representations can be limited to ordered pairs (P_1, PR). Such

are other elements of semantic representation that need to be accounted for. Perhaps some examples are in order. Let us start with presuppositions.[a] *Pedro regretted being Norwegian* presupposes thet Pedro is Norwegian. *Sam's murderer reads Reader's Digest* presupposes that Sam was murdered. In general, a sentence may be either true or false only if all its presuppositions are true. Since any proposition (at least nonperformative ones) may be presupposed, it is assumed that the elements of PR are of the same form as those of P_1, and that they are defined by the same well-formedness conditions. The notation given above, with PR as a member of an ordered n-tuple, assumes that presuppositions are structurally independent of P_1. However, Morgan, in an important paper [52], has argued convincingly that there are cases where presuppositions must be linked to certain propositions embedded in P_1, and that such links are identical to, or share properties with, conjunctions. He has also shown that presuppositions may be attributed not merely to the speaker and addressee, but also to the subjects of certain predicates in P_1 (e.g., verbs of saying, thinking, dreaming, etc.). For example, *know* is factive, and presupposes the truth of its complement. *Everyone knows that I am a Martian* presupposes that I am a Martian. *Dream* on the other hand is counterfactual, and presupposes the falsehood of its complement. *I dreamt that I was a Martian* presupposes that I'm not a Martian. Morgan has noticed sentences like: *I dreamt that I was a Martian and that everyone knew that I was a Martian.* If presuppositions were unstructured relative to P_1, this sentence would contain contradictory presuppositions, since *dream* presupposes that I'm not a Martian and *know* presupposes that I am a Martian. Morgan takes this to show that presuppositions cannot be unstructured relative to P_1; rather they must be associated with certain verbs in P_1. Since *know* is embedded in the complement of *dream*, the presupposition of *know* is only assumed to be true of the world of my dream. However, the presupposition of *dream* is assumed to be true of the world of the speaker. There is no contradiction since the presuppositions are true of different possible worlds. The notation given above, which represents the traditional position that presuppositions are unstructured with respect to nonpresupposed elements of meaning, would appear to be false on the basis of Morgan's argument. However, we will keep the above notation throughout the remainder of this work, since the consequences of Morgan's observations are not well understood at

ordered pairs can equivalently be viewed as two-place relationships between the phrase-markers involved. Thus, we might represent (P_1, PR) as $P_1 \longrightarrow PR$, where '$\longrightarrow$' is to be read as 'presupposes', and where PR is drawn from the class of possible P_1's. Under this interpretation, '$\longrightarrow$' can be viewed either as a relation between two parts of a single semantic representation or as a relation between two different semantic representations. Under the latter view, derivational constraints would become transderivational constraints, constraints holding across derivations. It seems to me to be an open question as to whether the theory of grammar needs to be extended to encompass such transderivational constraints. We will assume in the remainder of this paper that it does not. However, it is at least conceivable that transderivational constraints will be necessary and that semantic representations can be limited to P_1's with presuppositional information given transderivationally.

[a] I have used the term presupposition for what some philosophers might prefer to call 'pragmatic implication'. I assume that presuppositions are relative to speakers (and sometimes addressees and other persons mentioned in the sentence). Consider the example given in the text: *Pedro regretted being Norwegian.* A speaker cannot be committed to either the truth or falsity of this sentence without being committed to the truth of *Pedro was Norwegian.* The sentence *Pedro regretted being Norwegian, although I know that he wasn't* is contradictory. When I say that a given sentence presupposes a second sentence, I will mean that a speaker, uttering and committing himself to the truth of the first sentence will be committing himself to the truth of the second sentence.

present and since his examples are not directly relevant to the subsequent discussion.[a]

The notion of 'topic' is an ancient one in the history of grammatical investigation. Grammarians have long recognized that sentences have special devices for indicating what is under discussion. Preposing of topics is common. For example, in *John, Mary hates him, John* is the topic, while in *Mary, she hates John, Mary* is the topic. Clearly, an adequate account of semantic representation must take account of this notion. As with presuppositions, it is usually assumed that topics are structurally independent of the other components of meaning, as our notation indicates. As we will discuss in section 3 below, this may well not be the case. However, throughout what follows, we will assume the traditional position in an attempt to minimize controversy.

'Focus' is another traditional notion in grammar. Halliday [24] describes the information focus as the constituent containing new, rather than assumed information. The information focus often has heavy stress. Thus, in *JOHN washed the car yesterday*, the speaker is assuming that the car was washed yesterday and telling the addressee that the person who did it was John. Again, it is usually assumed that the semantic content of the 'focus' is structurally independent of other components of meaning. Our notation reflects this traditional position, although, as in the case of topic and presupposition, nothing that we have to say depends crucially on the correctness of this position.

I will refer to the above theory of grammar as a 'basic theory', simply for convenience and with no intention of suggesting that there is anything ontologically, psychologically, or conceptually 'basic' about this theory. Most of the work in generative semantics since 1967 has assumed the framework of the basic theory. It should be noted that the basic theory permits a variety of options that were assumed to be unavailable in previous theories. For example, it is not assumed that lexical insertion transformations apply in a block, with no intervening nonlexical transformations. The option that lexical and nonlexical transformations may be interspersed is left open.

As should be obvious from the above discussion, the basic theory does not assume any notion of 'directionality of mapping' from phonetics to semantics or semantics to phonetics.[b] Some writers on transformational grammar have, however, used locu-

---

[a] Assuming '$\longrightarrow$' to indicate the relation 'presupposes', we might represent embedded presuppositions as follows:

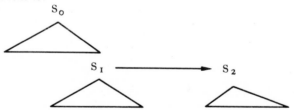

Here the embedded S_1 would be presupposing S_2. In many cases, like those discussed above, it will not be the case that S_0 will also presuppose S_2.

[b] Some readers may have received an incorrect impression of the basic theory with regard to this issue from the confusing discussion of directionality by Chomsky [10]. Chomsky does not claim in that article that any advocates of the basic theory have ever said that directionality matters in any way. However, Chomsky's odd use of quotation marks and technical terms in that paper has led some readers to believe that he had made such a mistaken claim. A close

tions that might mislead readers into believing that they assume some notion of directionality. For example, Chomsky [10] remarks that 'properties of surface structure play a distinctive role in semantic interpretation'. However, as Chomsky points out a number of times in that work, the notion of directionality in a derivation is meaningless, so that Chomsky's locution must be taken as having the same significance as 'Semantic representation plays a distinctive role in determining properties of surface structure' and nothing more. Both statements would have exactly as much significance as the more neutral locution 'Semantic representation and surface structure are related by a system of rules'. The basic theory allows for a notion of transformational cycle in the sense of *Aspects*, so that a sequence of cyclical rules applies 'from the bottom up', first to the lowermost S's, then to the next highest, etc. We assume that the cyclical transformations start applying with P_k and finish applying (to the highest S) at P_1 where k is less than 1. We will say in this case that the cycle applies 'upward toward the surface structure' (though, of course, we could just as well say that it applies 'downward toward the semantic representation', since directionality has no significance).

It should be noted that a transformational cycle defines an 'orientation' on a derivation, and readers should be cautioned from confusing the notion 'cyclical orientation of a derivation' with the notion 'directionality of a derivation'. The former is a real and quite important notion; the latter is meaningless. To say that a cycle is oriented 'upward-toward-the-surface' is the same as to say that it is oriented 'downward-toward-the-semantics', and such terminology makes no claim about where a derivation 'begins'. Most theories of transformational grammar that have been seriously entertained have assumed a cyclical orientation that is upward-toward-the-surface. However, it is possible to envision a theory with an upward-toward-the-semantics cyclic orientation. Moreover, it is possible to imagine theories with more than one cyclic orientation. Consider a sequence of phrase-markers $P_1, \ldots, P_i, \ldots, P_n$. One could imagine a theory such that $P_1, \ldots, P_i$ had an upward-towards-the-semantics cyclic orientation and $P_i, \ldots, P_n$ had an upward-toward-the-surface orientation, or vice versa.

reading of Chomsky's paper should clarify matters. On page 196, Chomsky says: 'let us consider once again the problem of constructing a "semantically based" theory of generative grammar that is a genuine alternative to the standard theory.' He then outlines a theory, in example $\langle 32 \rangle$, containing S (a semantic representation) and P (a phonetic representation), and he correctly notes that it makes no sense to speak of the 'direction' of a derivation from S to P or P to S. He concludes, on the same page: 'Consequently, it is senseless to propose as an alternative to $\langle 32 \rangle$ a "semantically based" conception of grammar in which S is "selected first" and then mapped onto the surface structure P_n and ultimately P.' Here he gives 'directionality' as the defining characteristic of what he calls a '"semantically based" theory of grammar'. His next two sentences are: 'Consider once again a theory such as that proposed by McCawley in which P_1 is identified with S and condition $\langle 3 \rangle$ is dropped so that "deep structure" is undefined. Let us consider again how we might proceed to differentiate this formulation – let us call it "semantically based grammar" – from the standard theory.'

Having first used '"semantically-based" theory of grammar' as a technical term for the 'directionality' position, he then uses '"semantically based grammar"' as a new technical term to describe McCawley's position. Of course, McCawley has never advocated the 'directionality' position and Chomsky has not said that he has. But one can see how such a bewildering use of technical terms might lead readers to such a mistaken conclusion. The only person that Chomsky cites as being a supporter of the 'directionality' position is Chafe [7], who has never been an advocate of the basic theory. However, Chomsky does not cite any page references, and in reading through Chafe's paper, I have been unable to find any claim to the effect that directionality has empirical consequences.

The basic theory does not necessarily include a level of 'deep structure', and the question as to whether such a level exists is an empirical question in the basic theory. We assume that the notion of 'deep structure' is defined in the following way. (i) Assume that all lexical insertion rules apply in a block. (ii) Assume that all upwards-toward-the-surface cyclic rules follow all lexical insertion rules. We define the output of the last lexical insertion rule as 'deep structure'.

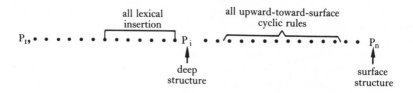

In the following section, I will discuss an example of a global derivational constraint and the evidence it provides for the question of whether 'deep structure' in this sense exists.

2   A derivational constraint involving quantifiers

Let us consider in detail an example of a global derivational constraint,[a]

⟨1⟩ Many men read few books.

⟨2⟩ Few books are read by many men.

In my speech, ⟨1⟩ and ⟨2⟩ are not synonymous.

Sentences like ⟨1⟩ and ⟨2⟩ were brought up in a discussion by Partee [55] with regard to certain inadequacies of a proposal made in Lakoff [40] for the derivation of quantifiers from predicates of higher sentences. That proposal was suggested by the observation that sentences like 'Many men left' are synonymous with those like the archaic 'The men who left were many.' It was proposed that sentences like the former were derived from structures underlying sentences like the latter, with 'many' as an adjective, which is then lowered. Under such a proposal, underlying structures like ⟨3⟩ and ⟨4⟩ would be generated:[b]

---

[a] It should be noted at the outset that all of the sentences discussed in this section are subject to dialect variation. At least one-third of the speakers I have encountered find ⟨1⟩ and ⟨2⟩ both ambiguous. The sentences to be discussed in the remainder of the section are subject to even greater variation, especially when factors like stress and intonation are studied closely. The data I will present below correspond to what I take to be the majority dialect. I take note of cases where my speech differs from that dialect, and a more thorough discussion of dialect differences will be given in the following section, after the discussion of the general constraints. It is especially important to remember throughout this section that the argument to be presented depends on the existence of a single dialect for which the data presented are correct. It is not even important that the dialect described by these data be the majority dialect, though, so far as I can tell, it is.

[b] The structures given in ⟨3⟩ and ⟨4⟩ are not meant to be taken seriously in all details. They are based on a 1965 analysis which mistakenly followed Chomsky in assuming a level of deep structure containing all lexical items and having nodes such as VP Since then such assumptions have been shown to be erroneous. I am using these particular structures only for historical reasons, since Partee [55] uses them in her paper. The main point at issue is whether quantifiers in underlying syntactic representations are in a higher clause than the NPs they quantify (as in predicate calculus) or whether they are part of the NPs they quantify (as they are in surface structure). Expressions such as 'men$_i$' are to be understood as the variable i limited

⟨3⟩

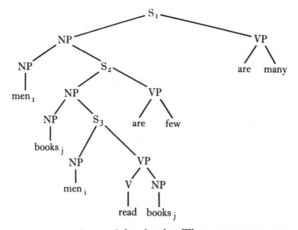

⟨3′⟩ Many are the men who read few books. There are many men who read few books.

⟨4⟩

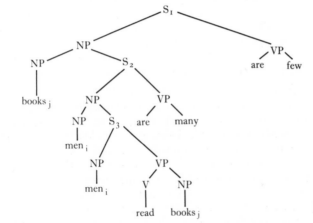

⟨4′⟩ Few are the books that many men read. There are few books that many men read.

In ⟨3⟩, a cyclical rule of quantifier-lowering will apply on the S_2-cycle, yielding *men$_i$ read few books*. The same cyclic rule will apply on the S_1-cycle, lowering *many* to the domain of men. For the purpose of this argument, representations of the following sort (which are somewhat closer to reality) can be used in the place of representations like ⟨3⟩:

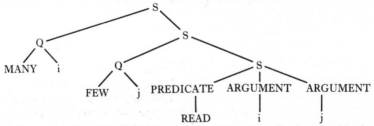

where: *i* is restricted to the domain of men and *j* is restricted to the domain of books.

onto *men*$_i$ and yielding $\langle 1 \rangle$, *many men read few books*. In $\langle 4 \rangle$ let us suppose that the passive, a cyclical rule, applies on S_3, before any quantifier-lowering takes place: this will give us *books*$_j$ *are read by men*$_i$. On the S_2-cycle, *many* will be lowered onto *men*$_i$, yielding *books*$_j$ *are read by many men*. On the S_1-cycle, *few* will be lowered onto *books*$_j$ giving $\langle 2 \rangle$, *few books are read by many men*.

These derivations work as they should and account for the synonymy of $\langle 1 \rangle$ with $\langle 3' \rangle$ and $\langle 2 \rangle$ with $\langle 4' \rangle$. However, if nothing more is said, this proposal will yield incorrect results. For example, if the passive applies to S_3 in $\langle 3 \rangle$, we will get *books*$_j$ *are read by men*$_1$, and then quantifier-lowering on the S_2- and S_1-cycles will yield $\langle 2 \rangle$, *few books are read by many men*. But if $\langle 2 \rangle$ were derived from $\langle 3 \rangle$ in this fashion, it should be synonymous with $\langle 3' \rangle$, *many are the men who read few books*. This is false, and such a derivation must be blocked. Similarly, if the passive were not to apply to S_3 in $\langle 4 \rangle$, the application of quantifier-lowering on the S_2- and S_1-cycles would yield $\langle 1 \rangle$, *many men read few books*, again a mistake, since $\langle 1 \rangle$ does not have the meaning of $\langle 4' \rangle$, *few are the books that many men read*.

Such a proposal would work in the first two cases, but would also predict the occurrence of two derivations that do not occur, at least for the majority of English speakers. If one inspects $\langle 1 \rangle$–$\langle 4 \rangle$, one notices that the correct derivations have the property that the 'higher' quantifiers in $\langle 3 \rangle$ and $\langle 4 \rangle$ are the leftmost quantifiers in $\langle 1 \rangle$ and $\langle 2 \rangle$ respectively. Thus, we might propose a derivational constraint that would say something like: if one quantifier commands another in underlying structure (or rather, P_1), then that quantifier must be leftmost in surface structure. Such a constraint as it stands would be too strong. Consider cases like $\langle 5 \rangle$:

$\langle 5 \rangle$ The books that many men read are few (in number).

$\langle 5 \rangle$ would have an underlying structure like $\langle 4 \rangle$, where *few* is the higher quantifier; however, *few* is to the right of *many* in surface structure. Thus cases like $\langle 5 \rangle$ would have to be ruled out of any such derivational constraint. If one inspects $\langle 2 \rangle$ and $\langle 5 \rangle$, one sees that they differ in the following way. In $\langle 5 \rangle$, *few* commands *many*, but *many*, being in a relative clause, does not command *few*; that is, *few* is higher in the tree than *many* in $\langle 5 \rangle$, just as it is in the underlying structure of $\langle 4 \rangle$. In other words, $\langle 5 \rangle$ preserves the asymmetric command-relationship between the quantifiers that occurs in $\langle 4 \rangle$. In $\langle 2 \rangle$, however, this is not the case. In $\langle 2 \rangle$, neither *few* nor *many* is in a subordinate clause, and so each commands the other and the command-relationship is symmetric. Thus the asymmetry of the command-relationship in the underlying structure, where *few* commands *many* but *many* does not command *few* is lost in $\langle 2 \rangle$. It is exactly these cases where the quantifier that was higher in underlying structure must be leftmost in surface structure. Where the asymmetric command-relationship is lost it must be supplanted by a precede-relationship, which is necessarily asymmetric.

Such a derivational constraint may be stated as follows:

$\langle 6 \rangle$ Let $C_1 = Q^1$ commands Q^2
$\qquad C_2 = Q^2$ commands Q^1
$\qquad C_3 = Q^1$ precedes Q^2 $\qquad$ '/' means 'meets condition'
$\qquad$ Constraint 1: $P_1/C_1 \supset (P_n/C_2 \supset P_n/C_3)$

Constraint 1 states that if two quantifiers Q^1 and Q^2 occur in underlying structure P_1, such that P_1 meets condition C_1, then if the corresponding surface structure P_n meets condition C_2, that surface structure P_n must also meet condition C_3. In short, if an underlying asymmetric command-relationship breaks down in surface structure, a precede-relationship takes over. Constraint 1 is a well-formedness constraint on derivations. Any derivation not meeting it will be blocked. Thus, the derivations ⟨3⟩ ⟶ ⟨1⟩ and ⟨4⟩ ⟶ ⟨2⟩ will be well-formed, but ⟨3⟩ ⟶ ⟨2⟩ and ⟨4⟩ ⟶ ⟨1⟩ will be blocked.

It is important to note that the fact that one of the two quantifiers is in subject position in the sentences we have discussed so far is simply an accident of the data we happened to have looked at. The difference in the interpretation of quantifiers has nothing whatever to do with the fact that in these examples one quantifier is inside the VP while the other is outside the VP. Only the left-to-right order within the clause matters.

⟨7⟩ John talked to few girls about many problems.

⟨8⟩ John talked about many problems to few girls.

These sentences differ in interpretation just as do ⟨1⟩ and ⟨2⟩, that is the leftmost quantifier is understood as the highest in each sentence, though both quantifiers are inside the VP.

Although ⟨1⟩ and ⟨2⟩ are cases where the asymmetry of the underlying command-relationship disappears in surface structure, it happens to be the case in ⟨1⟩ and ⟨2⟩ that condition C_1, which holds in underlying structure, continues to hold in surface structure: that is, Q^1 continues to command Q^2. We might ask if there exist any cases where this does not happen, that is, where Q^1 commands Q^2 in underlying structure, but Q^1 does not command Q^2 in surface structure. A natural place to look for such cases, and perhaps the only one, is in sentences containing complement constructions. Let us begin by considering sentences like ⟨9⟩:

⟨9⟩ Sam claimed that John had dated *few* girls.

⟨9⟩ is open to both of the readings ⟨10⟩ and ⟨11⟩, though ⟨10⟩ is preferable:

⟨10⟩ Sam claimed that the girls who John had dated were few (in number).

⟨11⟩ The girls who Sam claimed that John had dated were few (in number).

⟨10⟩ and ⟨11⟩ would have underlying structures like ⟨10′⟩ and ⟨11′⟩ respectively:

⟨10′⟩

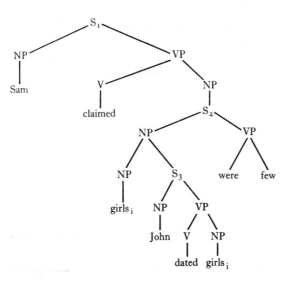

⟨11′⟩

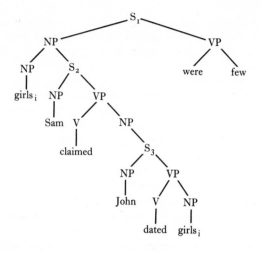

In each case quantifier-lowering will move *few* down to girls$_i$. In ⟨10⟩, *girls$_i$* occurs in the S immediately below *few*; in ⟨11⟩, *girls$_i$* occurs two sentences down from *few*. We are now in a position to test the conjecture that one quantifier may cease to command another in the course of a derivation. Consider ⟨12⟩, where *few* commands *many*:

⟨12⟩

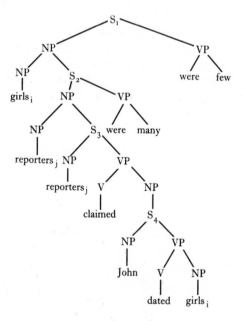

⟨12⟩ would have the meaning of ⟨12′⟩:

⟨12′⟩ Few were the girls who many reporters claimed John dated.

If we allow quantifier-lowering to apply freely to ⟨12⟩, *many* will be lowered to *reporters$_j$* on the S$_2$ cycle, yielding *many reporters claimed that John dated girls$_i$*. The derived structure will now look just like ⟨11⟩, except that it will have the noun phrase *many reporters* instead of *Sam*. As in ⟨11⟩, quantifier-lowering will lower *few* onto *girls*$_1$ yielding ⟨13⟩:

⟨13⟩ *Many* reporters claimed that John dated *few* girls.

In ⟨13⟩, *few* is in a subordinate clause and does not command *many*. Thus we have a case where *few* commands *many* in underlying structure, but not in surface structure. Note, however, that ⟨13⟩ does not have the meaning of ⟨12′⟩. It has the reading of ⟨14⟩:

⟨14⟩ Many were the reporters who claimed that the girls who John dated were few (in number).

where *few* is inside the object complement of *claim* (as in ⟨10⟩) and where *many* would command *few* in underlying structure. Thus, we have a case where a derivation must block if one quantifier commands another in underlying structure, but not in surface structure. To my knowledge this is a typical case, and I know of no counterevidence. Thus, it appears that a derivational constraint of the following sort is needed:

⟨15⟩ Let $C_1 = Q^1$ commands Q^2 '/' means 'meets condition'
    Constraint 2: $(P_1/C_1) \supset (P_n/C_1)$

Constraint 2 says that if Q^1 commands Q^2 in underlying structure P_1, then Q^1 must also command Q^2 in surface structure P_n.

Constraints 1 and 2 are prime candidates for cases where grammatical constraints seem to reflect perceptual strategies. If one considers a perceptual model where surface strings are given as input and semantic representations are produced as output, constraints 1 and 2 guarantee that the relative heights of the quantifiers in the semantic representation of a sentence can be determined by the surface parsing of the sentence. If Q_1 commands Q_2 in surface structure but Q_2 doesn't command Q_1, then Q_1 commands Q_2 in semantic representation. If, on the other hand, Q_1 and Q_2 command each other in surface structure, then the leftmost quantifier commands the rightmost one in semantic representation. If constraints 1 and 2 are reflections in grammar of perceptual strategies, then they would of course be prime candidates for syntactic universals. Unfortunately for such a proposal, there is a lot of idiosyncratic variation with such constraints.

Constraint 2 does not simply hold for quantifiers, but for negatives as well. Consider, for example, ⟨16⟩:

⟨16⟩ Sam didn't claim that Harry had dated many girls.

where *many* does not command *not*. If quantifier-lowering worked freely one would expect that ⟨16⟩ could be derived from all of the following underlying structures:

⟨17⟩ [$_S$ not [$_S$ Sam claimed [$_S$girls$_i$ [Harry dated girls$_i$] were many]]]
⟨18⟩ [$_S$ not [$_S$ girls$_i$ [$_S$Sam claimed [$_S$Harry dated girls$_i$]] were many]]
⟨19⟩ [$_S$girls$_i$ [$_S$not [$_S$Sam claimed [$_S$Harry dated girls$_i$]]] were many]

These have the senses of:

⟨17′⟩ Sam didn't claim that the girls who Harry dated were many.
⟨18′⟩ There weren't many girls who Sam claimed Harry dated.
⟨19′⟩ There were many girls who Sam didn't claim Harry dated.

⟨17⟩ is the normal reading for ⟨16⟩; ⟨18⟩ is possible, but less preferable (like ⟨11⟩); but ⟨19⟩ is impossible. The regularity is just like that of Constraint 2. In ⟨17⟩ and

$\langle 18 \rangle$, *not* commands *many* in underlying structure, just as in surface structure $\langle 16 \rangle$. In $\langle 19 \rangle$, *many* commands *not* in underlying structure, but *many* does not command *not* in surface structure $\langle 16 \rangle$. Thus we can generalize constraint 2 in the following way. Let L stand for a 'logical predicate', either Q or NEG.

$\langle 20 \rangle$ Let $C_1 = L^1$ commands L^2 (L = Q or NEG)
Constraint 2': $(P_1/C_1) \supset (P_n/C_1)$

Conditions of this sort suggest that quantifiers and negatives may form a natural semantic class of predicates. This seems to be confirmed by the fact that Constraint 1 can be generalized in the same fashion, at least for certain dialects of English. Consider, for example, the following sentences discussed by Jackendoff [32], [33]:

$\langle 21 \rangle$ Not many arrows hit the target.
$\langle 22 \rangle$ Many arrows didn't hit the target.
$\langle 23 \rangle$ The target wasn't hit by many arrows.

Jackendoff reports that in his speech $\langle 23 \rangle$ is synonymous with $\langle 21 \rangle$, but not $\langle 22 \rangle$. I and many other speakers find that $\langle 23 \rangle$ has both readings, but that the $\langle 22 \rangle$ reading is 'weaker'; that is, $\langle 23 \rangle$ is less acceptable with the $\langle 22 \rangle$ reading. However, there are a number of speakers whose dialect displays the facts reported on by Jackendoff, and in the remainder of this discussion we will be concerned with the facts of that dialect.

Assuming the framework discussed above, $\langle 21 \rangle$ and $\langle 22 \rangle$ would have underlying structures basically like $\langle 24 \rangle$ and $\langle 25 \rangle$:

$\langle 24 \rangle$ [$_S$ not [$_S$ arrows$_i$ [$_S$ arrows$_i$ hit the target] were many]]
$\langle 24' \rangle$ The arrows that hit the target were not many.
$\langle 25 \rangle$ [$_S$arrows$_i$ [$_S$ not [$_S$arrows$_i$ hit the target]] were many]
$\langle 25' \rangle$ The arrows that didn't hit the target were many.

If constraint 1 is generalized to include 'logical predicates', both negatives and quantifiers, then the facts of $\langle 21 \rangle$–$\langle 23 \rangle$ will automatically be handled by the new constraint 1' given the underlying structures of $\langle 24 \rangle$ and $\langle 25 \rangle$ and the rule of quantifier-lowering. Constraint 1' would be stated as $\langle 26 \rangle$:

$\langle 26 \rangle$ Let $C_1 = L^1$ commands L^2
$C_2 = L^2$ commands L^1
$C_3 = L^1$ precedes L^2 (L = Q or NEG)
Constraint 1': $P_1/C_1 \supset (P_n/C_2 \supset P_n/C_3)$

Any derivation not meeting this condition will be ill-formed.[a]

---

[a] I have ignored the role of stress in this discussion, though it is of course important for many speakers. Many people find that (i)

(i) Many men read few books.

where *few* has extra heavy stress can mean *The books that many men read are few*. Thus, the general principle here seems to be that where the asymmetric command relation is lost in derived structure, then either one or another of what Langacker calls 'primacy relations' must take over. One which I would propose is the relation 'has much heavier stress than'; the other is the relation 'precedes'. These relations seem to form a hierarchy with respect to this phenomenon in such dialects:

1. Commands (but is not commanded by)
2. Has much heavier stress than
3. Precedes.

If one quantifier commands but is not commanded by another in surface structure, then it commands in underlying structure. If neither commands but is not commanded by the other in surface structure, then the one with heavier stress commands in deep structure. And if

⟨21⟩ and ⟨22⟩ work as expected. Take ⟨21⟩: *not* (L^1) commands *many* (L^2) in underlying structure ⟨24⟩, *many* commands *not* in surface structure ⟨21⟩, and *not* precedes *many* in surface structure. So ⟨24⟩ ⟶ ⟨21⟩ meets constraint 1′. Take ⟨22⟩: *many* (L^1) commands *not* (L^2) in underlying structure ⟨25⟩ and in surface structure, *not* commands *many* in surface structure, and *many* precedes *not* in surface structure. So ⟨25⟩ ⟶ ⟨22⟩ meets constraint 1′. Now consider ⟨23⟩, which is the interesting case in this dialect. If one allows the passive transformation to apply to the innermost S of ⟨24⟩ and ⟨25⟩, and then allows quantifier-lowering to apply, both ⟨24⟩ and ⟨25⟩ will yield ⟨23⟩. First consider the derivation ⟨24⟩ ⟶ ⟨23⟩. *Not* (L^1) commands *many* (L^2) in underlying structure ⟨24⟩ and in surface structure, *many* commands *not* in surface structure ⟨21⟩, and *not* precedes *many* in surface structure. Thus, the derivation ⟨24⟩ ⟶ ⟨23⟩ meets constraint 1′. Now consider the derivation ⟨25⟩ ⟶ ⟨23⟩. *Many* (L^1) commands *not* (L^2) in underlying structure and in surface structure, *not* commands *many* in surface structure ⟨23⟩, but *many* (L^1) does not precede *not* (L^2) in surface structure ⟨23⟩. Therefore, the derivation ⟨25⟩ ⟶ ⟨23⟩ does not meet constraint 1′, and so the derivation is blocked. This accounts for the fact that ⟨23⟩ is not synonymous to ⟨21⟩ and that there is no passive corresponding to ⟨25⟩ in this dialect.

It should also be noted that that part of constraint 1′ which says that L^2 must command L^1 in surface structure (P_n/C_2) if the precede-relationship (P_n/C_2) is to come into play, is necessary for the cases discussed. Consider, for example, sentence ⟨25′⟩, *The arrows that didn't hit the target were many*. In ⟨25′⟩, *many* (L^1) commands *not* (L^2) in P_1, and *many* also commands *not* in P_n. Thus, the *if* part of the conditional statement of constraint 1′ is not met, and the fact that *not* (L^2) precedes *many* (L^1) in surface structure (that is, that (P_n/C_3) does not hold) does not matter; since the *if*-condition is not met, the constraint holds and the sentence is grammatical with the reading of ⟨25⟩.

We have assumed so far that constraints 1′ and 2′ mention the *surface* structure. But this is just an illusion which results from considering only simple sentences. Suppose, for example, that we consider complex sentences where deletion has taken place. Consider ⟨27⟩ and ⟨28⟩:

⟨27⟩ Jane isn't liked by many men and Sally isn't liked by many men either.
⟨28⟩ Jane isn't liked by many men and Sally isn't either.

Note that the sentence fragment *Sally isn't either* does not contain *many* in surface structure, but it receives the same interpretation as the full *Sally isn't liked by many men either*, and does not have the reading of *There are many men who Sally isn't liked by*. Constraint 1′ as it is presently stated will not do the job, in that it mentions surface structure P_n rather than some earlier stage of the derivation prior to the deletion of *liked by many men*.

This raises a general problem about constraints like 1′ and 2′. Since they only mention underlying structures P_1 and surface structures P_n, they leave open the

neither has much heavier stress, then the one that precedes in surface structure commands in underlying structure. Letting $C_4 = Q^1$ has much heavier stress than Q^2, the constraint for such a dialect could be stated as follows, though the notation is not an optimum one for stating such a hierarchy:

$$P_1/C_1 \supset (P_n/C_2 \supset (P_n/C_4 \lor P_n/C_3))$$

Since the dialect with this condition is in the minority so far as I have been able to tell from a very small amount of study, I will confine myself to the normal dialect in the remainder of the discussion.

possibility that such constraints might be violated at some intermediate stage of the derivation. My guess is that this will never be the case, and if so, then it should be possible to place much stronger constraints on derivations than 1' and 2' by requiring that all intermediate stages of a derivation P_i meet the constraint, not just the surface structure P_n. Using quantifiers, we can state such a stronger constraint as follows:

⟨29⟩ Let $C_1 = L^1$ commands L^2
$C_2 = L^2$ commands L^1
$C_3 = L_1$ precedes L_2
Constraint 1″: $P_i/C_1 \supset ((i) (P_i/C_2 \supset P_i/C_3))$

⟨29⟩ will now automatically handle cases like ⟨28⟩, since it will hold at all points of the derivation up to the point where the deletion rule applies; after that point, it will hold vacuously. The reason is that *many* will not appear in any phrase-marker after the deletion takes place and so C_2 will not hold in such phrase-markers; and where C_2 does not hold, then C_3 need not hold.

However, this is still insufficient, since Constraint 1″ still requires that if C_2 holds in surface structure, then C_3 must hold in surface structure, as well as in earlier stages of a derivation. But there are late rules which make gross changes in derived structure and produce surface structures in which the constraints do not hold. Compare ⟨30⟩ and ⟨31⟩:

⟨30⟩ Sarah Weinstein isn't fond of many boys.
⟨31⟩ Fond of many boys, Sarah Weinstein isn't.

The rule of Y-movement as discussed in Postal [57] will produce ⟨31⟩ from the structure underlying ⟨30⟩. (Needless to say, ⟨31⟩ is not grammatical in all American dialects. We are considering only those in which ⟨31⟩ is well-formed.) Note that ⟨30⟩ works exactly according to constraint 1″; the reading in which *many* commands *not* in P_1 is blocked since *many* does not precede *not* in surface structure. But ⟨31⟩, where the surface order of *not* and *many* is reversed, shows the same range of blocked and permitted readings as ⟨30⟩. The outputs of Y-movement do not meet constraint 1″, though earlier stages of the derivation do. Thus, it appears that constraint 1″ has some cutoff point prior to the application of Y-movement. That is, there is in each derivation some 'shallow structure' P_a defined in some fixed way such that

Constraint 1‴: $P_i/C_1 \supset ((i) (P_i/C_2 \supset P_i C_3))$, where i ⩽ a ⩽ n

This raises the interesting question of exactly how the 'shallow structure' P_a is to be defined. One possibility is that P_a is the output of the cyclical rules. However, there aren't enough facts known at present to settle the issue for certain. Still, we can draw certain conclusions. Passive, a cyclic rule, must be capable of applying before P_a, since constraint 1‴ must apply to the output of passive. Y-movement must apply after P_a.

Let us now consider the constraints we have been discussing with respect to the process of lexical insertion. Let us constrain the basic theory so that some notion of 'deep structure' can be defined along the lines discussed above. Let it be required that all lexical insertion rules apply in a block and that all upward-toward-the-surface cyclic rules apply *after* lexical insertion. Since passive is a cyclic rule and since passive must be able to apply before P_a is reached, it follows that if there is such a 'deep structure' all lexical insertion must occur before P_a is reached. Thus, it is an empirical question as to whether such a notion of 'deep structure' is correct. If there exist lexical items that must be inserted *after* P_a, then such a notion of 'deep structure' would be untenable since there would exist upward-toward-the-surface cyclic rules (e.g., passive) which could apply *before* some case of lexical insertion.

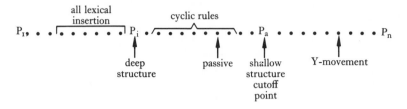

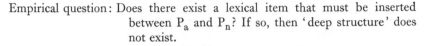

Empirical question: Does there exist a lexical item that must be inserted between P_a and P_n? If so, then 'deep structure' does not exist.

Let us pursue the question somewhat further. Consider the sentences

⟨32⟩ (*a*) I persuaded Bill to date many girls.
     (*b*) I persuaded many girls to date Bill.

⟨33⟩ (*a*) I persuaded Bill not to date many girls.
     (*b*) I persuaded many girls not to date Bill.

⟨34⟩ (*a*) I didn't persuade Bill to date many girls.
     (*b*) I didn't persuade many girls to date Bill.

If we consider the meanings of these sentences, it should be clear that these cases work according to the two derivational constraints that we have stated thus far.

The difference in the occurrence of *not* is crucial in these examples. In ⟨33⟩, *not* in semantic representation would occur inside the complement of *persuade*, while in the semantic representation of ⟨34⟩ *not* will occur in the sentence above *persuade*. That is, in ⟨33⟩ *persuade* commands *not* in SR, while in ⟨34⟩ *not* commands *persuade*. This difference in the occurrence of *not* accounts for the fact that ⟨33 a⟩ is unambiguous, while ⟨32 a⟩ and ⟨34 a⟩ are ambiguous. ⟨32 a⟩ can mean either ⟨35⟩ or ⟨36⟩:

⟨35⟩ There were many girls that I persuaded Bill to date.

⟨36⟩ I persuaded Bill that the number of girls he dates should be large.

⟨34 a⟩ can mean either ⟨37⟩ or ⟨38⟩:

⟨37⟩ There weren't many girls that I persuaded Bill to date.

⟨38⟩ It is not the case that I persuaded Bill that the number of girls he dates should be large.

But ⟨33 a⟩ can mean only ⟨39⟩:

⟨39⟩ I persuaded Bill that the number of girls he dates should not be large.

The reason ⟨33 a⟩ is unambiguous is that since *not* precedes *many* in derived structure, *not* must command *many* in underlying structure (by constraint 1″′). Since *not* originates inside the complement of *persuade*, and *not* must command *many*, *many* must also originate inside the complement of *persuade*. In ⟨32 a⟩ and ⟨34 a⟩, *many* may originate either inside the complement of *persuade* or from a sentence above *persuade*, which accounts for the ambiguity.

Now compare ⟨33 b⟩ and ⟨34 b⟩. In ⟨33 b⟩ *many* both precedes and commands *not* in derived structure; therefore, *many* must command *not* in underlying structure. ⟨33 b⟩ only has the reading of ⟨40⟩:

⟨40⟩ There were many girls that I persuaded not to date Bill.

In $\langle 34b \rangle$, on the other hand, *not* precedes *many* in derived structure and so must command *many* in underlying structure. So $\langle 34b \rangle$ can mean $\langle 41 \rangle$ but not $\langle 40 \rangle$.

$\langle 41 \rangle$ There weren't many girls that I persuaded to date Bill.

Let us now consider the lexical item *dissuade*.[a]

$\langle 42 \rangle$ (a) I dissuaded Bill from dating many girls.

(b) I dissuaded many girls from dating Bill.

In $\langle 42 \rangle$ the word *not* does not appear. The only overt negative element is the prefix *dis-*. Thus, the postlexical structure of $\langle 42 \rangle$ would not have the negative element inside the object complement of *-suade*, but in the same clause, as in $\langle 43 \rangle$:

$\langle 43 \rangle$ (a) I NEG-suaded Bill from dating many girls.

(b) I NEG-suaded many girls from dating Bill.

Moreover, the negative element would precede rather than follow the object of *dissuade*. In terms of precede- and command-relations, the postlexical structure of $\langle 42 \rangle$ would look like $\langle 34 \rangle$ rather than $\langle 33 \rangle$. Suppose P_a were postlexical, that is, suppose that the command-relationship between *not* and *many* in semantic representation were predictable from the precede-relationship at some point in the derivation after the insertion of all lexical items. We would then predict that since *NEG* precedes *many* in $\langle 42 \rangle$, *NEG* must command *many* in the underlying structure of $\langle 42 \rangle$. That is, we would predict that the sentences of $\langle 42 \rangle$ would have the meanings of $\langle 34 \rangle$. But this is false. $\langle 42a \rangle$ and $\langle 42b \rangle$ have the meanings of $\langle 33a \rangle$ and $\langle 33b \rangle$.

Summary of Majority Dialect[b]

Persuade –	$32a$ means 35, 36
Persuade not –	$\begin{cases} 33a \text{ means } 39 \\ 33b \text{ means } 40 \end{cases}$
Not persuade –	$\begin{cases} 34a \text{ means } 37,\ 38 \\ 34b \text{ means } 41 \end{cases}$
Dissuade –	$\begin{cases} 42a \text{ means } 39 \\ 42b \text{ means } 40 \end{cases}$

[a] As noted in footnote a, p. 244, the constraints under discussion require that the quantifiers be unstressed. The facts of $\langle 42 \rangle$, as I describe them for my dialect will, of course, not hold if *many* receives heavy stress. Instead, *many*, as indicated in that footnote, can then be understood as the highest quantifier. This is, of course, equally true in the *dissuade* and *persuade not* cases. The following examples should make this clear.

$\langle \text{I} \rangle$(a) I persuaded Bill not to date many girls, namely, Sue, Sally, Bathsheba, etc.

(b) *I persuaded Bill not to date many girls, namely, Sue, Sally, Bathsheba, etc.

$\langle \text{II} \rangle$(a) I dissuaded Bill from dating many girls, namely, Sue, Sally, Bathsheba, etc.

(b) *I dissuaded Bill from dating many girls, namely, Sue, Sally, Bathsheba, etc.

In the (a) sentences, where *many* is stressed, *many* can be interpreted as commanding *not*. In that case, one is talking about particular girls, and so enumerations with *namely* are possible. When *many* is unstressed, it cannot be interpreted as commanding *not*, and consequently it cannot be taken as picking out particular girls. Hence, enumerations with *namely* are not possible.

[b] Though these facts seem to hold for the majority of speakers I have asked, they by no means hold for all. There is even one speaker I have found for whom the crucial case discussed here does not hold. This speaker finds that *dissuade* does not work like *persuade not* with respect to ambiguities. For him, $\langle 42a \rangle$ can mean not only $\langle 39 \rangle$, but also *There were many girls*

Dissuade means *persuade -NP - not* rather than *not-persuade-NP*, and the constraints on the occurrence of quantifiers in *derived* structure reflect this meaning, and must be stated *prelexically*. The lexical item *dissuade* must be inserted at a point in the derivations of ⟨33⟩ and ⟨34⟩ *after* constraint 1″′ has ceased to operate. Now recall that constraint 1″′ must operate on the output of the passive transformation. Consider ⟨44⟩:

⟨44⟩ (*a*) Many men weren't dissuaded from dating many girls.

(*b*) Not many men were dissuaded from dating many girls.

(*c*) I didn't dissuade many men from dating many girls.

⟨44⟩ shows the characteristics of both ⟨34⟩ and ⟨21⟩–⟨23⟩. In ⟨44⟩, constraint 1″′ must operate both after the passive transformation and before the insertion of *dissuade*. Thus we have cases where an upward-toward-the-surface cyclic rule must apply before the insertion of some lexical item. This shows that any conception of 'deep structure' in which all lexical insertion takes place before any upward-toward-the-surface cyclic rules apply is empirically incorrect. It also shows that the passive transformation may apply to a verb before the overt lexical representation of the verb is inserted, which means that prelexical structures must look pretty much like postlexical structures. In the case of *dissuade*, one might be tempted to try to avoid such a conclusion with the suggestion that *dissuade* is derived by a relatively late transformation from a structure containing the actual lexical item *persuade*. Under such a proposal, *dissuade* would not be introduced by a rule of lexical insertion, but rather by a rule which changes one actual lexical item to another. Such a solution cannot be made general, however, since lexical items like *prohibit*, *prevent*, *keep*, *forbid*, etc., which do not form pairs like *persuade–dissuade*, work just like *dissuade* with respect to the properties we have discussed.

A particularly tempting escape route for those wishing to maintain a level of 'deep structure' might be the claim that the lexical item *dissuade* is inserted precyclically, that *dissuade* requires a *not* in its complement sentence, and that this *not* is deleted *after shallow structure*. Thus, the *not* would be present at the time that the constraints shut off, and all of the above facts would be accounted for. This proposal has some initial plausibility since similar verbs in other languages often have a negative element that appears overtly in its complement sentence. For example, in Latin we have 'Dissuasī Marcō nē iret' (I dissuaded Marcus from going), where *nē*, the morphological alternant of *nōn* in this environment, occurs in the complement sentence.

Let us suppose for the moment that such a solution were possible. This would mean that the complement sentence of *dissuade* would contain a *not* at the level of shallow structure, but not at the level of surface structure. Now consider the following sentences:

⟨45⟩ I dissuaded Mary from marrying no one.

that I dissuaded Bill from dating, even when ⟨42a⟩ does not contain stress on *many*. This speaker seems to have a constraint that holds just at the level of surface structure, and not at the level of shallow structure or above. Interestingly enough, the same speaker has the No-double-negative constraint discussed below in the same form as the majority of other speakers, only for him it holds at surface structure, not shallow structure. For him, ⟨51⟩ below is ungrammatical, but ⟨54⟩ is grammatical. Perlmutter has suggested that such variation in the constraints is due to the fact that children learning their native language are not presented with sufficient data to allow them to determine at which level of the grammar various constraints hold, or exactly which classes of items obey which constraints. Other cases of such variation are discussed in the following section.

⟨46⟩ *I persuaded Mary not to marry no one.

⟨47⟩ *Mary didn't marry no one.

⟨48⟩ I didn't persuade Mary to marry no one.

⟨46⟩ and ⟨47⟩ are ungrammatical in standard English, and in all dialects if the two negatives are both considered as underlying logical negatives (e.g., if ⟨47⟩ has the reading *It is not the case that Mary married no one*). As is well known, this prohibition applies only for negatives in the same clause (cf. example ⟨48⟩, where the negatives are in different clauses). The question arises as to where in the grammar the No-double-negative (NDN) constraint is stated. (*a*) If 'deep structure' exists, it could be stated there; (*b*) it could be stated at shallow structure; (*c*) it could hold at all levels between deep structure and shallow structure; or (*d*) it could hold only at surface structure.

Now consider ⟨45⟩. Under the above proposal for post-shallow-structure deletion of *not* in *dissuade*-sentences, *not* would still be present at the level of shallow structure. (In fact, it would be present at all points between deep structure and shallow structure, that is, all points in the derivation where constraints 1 and 2 hold.) At the level of shallow structure, ⟨45⟩ would have the form:

⟨45'⟩ I dissuaded Mary from not marrying no one.

Thus under the above proposal, the No-double-negative (NDN) constraint could not hold at the level of shallow structure, or at any previous point in the derivation back to deep structure; if it did, ⟨45⟩ would be ruled out. Thus, under the proposal for post-shallow-structure *not*-deletion, the NDN-constraint could only be a constraint on surface structure, not on shallow structure. Let us now take up the question of whether this is possible.

The following sentences are in accord with the NDN-constraint whether it holds at shallow or surface structure, since these sentences have essentially the same representation at both levels.

⟨49⟩ Max said that Sheila Weinstein was spurned by no one.

⟨50⟩ Max didn't say that Sheila Weinstein was spurned by no one.

⟨51⟩ *Max said that Sheila Weinstein wasn't spurned by no one.

Note that in the appropriate dialects, Y-movement can apply to sentences of this form moving the participial phrase of the embedded clause.

⟨52⟩ Max said that Sheila Weinstein wasn't spurned by Harry.

⟨53⟩ Spurned by Harry, Max said that Sheila Weinstein wasn't.

In ⟨53⟩, *Harry* has been moved from a position where it was in the same clause as *n't* to a position where it is in a higher clause. Now consider once more:

⟨51⟩ *Max said that Sheila Weinstein wasn't spurned by no one.

If the NDN-constraint holds for surface, not shallow, structures, then the application of Y-movement to ⟨51⟩ would move *no one* out of the same clause as *n't* and should make the NDN-constraint nonapplicable at surface structure. Thus, ⟨54⟩ should be grammatical:

⟨54⟩ *Spurned by no one, Max said that Sheila Weinstein wasn't.

But ⟨54⟩ is just as bad as ⟨51⟩ – the NDN-constraint applies to both. As we have seen, the NDN-constraint cannot apply to the *surface* structure of ⟨54⟩, since *no one* is not in the same clause as *n't*, having been moved away by Y-movement. Thus, in order to rule out ⟨54⟩, the NDN-constraint must apply before Y-movement; that is, it must apply at the level of shallow structure.

But this contradicts the post-shallow-structure *not*-deletion proposal, since under that proposal ⟨45⟩, which is grammatical, would contain two negatives in the same clause at the level of shallow structure (cf. ⟨45′⟩). Thus, the post-shallow-structure *not*-deletion proposal is incorrect.

A similar proposal might say that prior to shallow structure, *from* replaced *not* with verbs like *dissuade* and *prevent*, while *to* replaced *not* with a verb like *forbid*, and that *from* and *to* 'acted like negatives' (whatever that might mean) with respect to constraints 1 and 2 at shallow structure and above. However, it is clear from sentences like ⟨45⟩ that *from* does not act like a negative with respect to the NDN-constraint at shallow structure. Such a proposal would thus require *from* both to act like a negative and not to act like a negative, a contradiction. Neither of these routes provides an escape from the conclusion that *dissuade* must be inserted after shallow structure.

Another route by which one might attempt to avoid this conclusion would be by claiming that derivational constraints did not operate on the internal structure of lexical items, and that therefore *dissuade* could be inserted before the passive, preserving the notion of 'deep structure'. It would then not interact with the constraints. Such a claim would be false. *Dissuade* does interact with the constraints. As the summary of meaning-correspondences given above shows, *dissuade* acts just like *persuade not*, not like *persuade*. In particular, ⟨42a⟩ is unambiguous, just like ⟨33a⟩. It only has a reading with *many* originating inside the complement, as the constraints predict. If *dissuade* were impervious to the constraints, then we would expect it to act like *persuade*; in particular, we would expect ⟨42a⟩ to be ambiguous, just like ⟨32a⟩, where *many* may be interpreted as originating either inside or outside of the complement. But as we have seen, ⟨42a⟩ does not have the outside reading. *John dissuaded Bill from dating many girls* cannot mean *Many were the girls who John dissuaded Bill from dating*.

It should be noted that this argument does not depend on the details of constraint 1‴ being exactly correct. It would be surprising if further modifications did not have to be made. Nor does this argument depend on any prior assumption that semantic representation must be taken to be phrase-markers, though the discussion was taken up in that context. It only depends on the facts that *persuade-not* and *not-persuade* obey the general constraints on quantifiers and negatives, and that *dissuade* acts like *persuade-not*. Thus, in *any* version of transformational grammar there will have to be stated a general principle relating semantic representations of sentences containing quantifiers and negatives to the left-to-right order of those corresponding quantifiers and negatives in 'derived structure'. If the general principle is to be stated, the notion 'derived structure' will have to be defined as *following* the passive rule, but *preceding* the insertion of *dissuade*. Thus, in no non-ad-hoc transformational grammar which states this general principle will all lexical insertion precede all cyclical rules.

Let us sum up the argument.

(i) Suppose 'deep structure' is defined as a stage in a derivation which follows the application of all lexical insertion rules and precedes the application of any upward-toward-the-surface cyclic rules.

(ii) Evidence was given for derivational constraints 1 and 2, which relate semantic command-relationships to precede- and command-relationships at some level of 'derived structure'.

(iii) But the passive transformation must precede the insertion of lexical items such as *dissuade, prohibit, prevent, keep*, etc.

(iv) Since passive is an upward-toward-the-surface cyclic rule, (iii) shows that the concept of 'deep structure' given in (i) cannot be maintained.

3 Further examples

The constraints discussed in section 2 extend to more cases than just quantifiers and negatives. For example, the same constraints seem to define at least in part the limits of conjunction reduction.

⟨1⟩ John claimed that he robbed the bank and he claimed that he shot Sam.
⟨2⟩ John claimed that he robbed the bank and claimed that he shot Sam.
⟨3⟩ John claimed that he robbed the bank and shot Sam.

In ⟨1⟩ and ⟨2⟩, it is understood that two claims have been made. In ⟨3⟩, it is understood that only one claim has been made. This suggests that ⟨1⟩ and ⟨2⟩ are derived from an underlying structure like ⟨4⟩, while ⟨3⟩ is derived from something like ⟨5⟩.

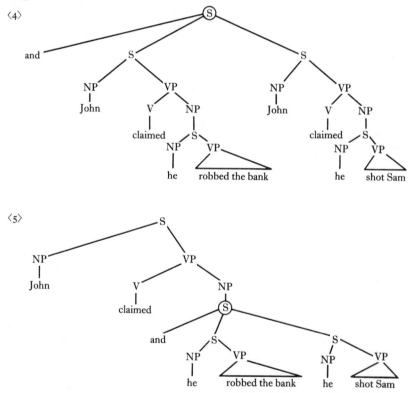

In ⟨4⟩ there are two instances of *claim*, while in ⟨5⟩ there is only one. The following question now arises: If conjunction reduction applies to ⟨4⟩ so that it yields ⟨2⟩, what is to keep it from applying further to yield ⟨5⟩? If it applies freely to yield ⟨5⟩ from ⟨4⟩, we would get the incorrect result that ⟨3⟩ should be synonymous with ⟨1⟩.

Observe that in ⟨4⟩ *and* commands *claim*, but *claim* does not command *and*. In ⟨5⟩ the reverse is true: *claim* commands *and*, but not vice versa. To say that ⟨5⟩

cannot be derived from ⟨4⟩ is to say that the asymmetric command relation between *and* and *claim* cannot be reversed in the course of a derivation. But that is the same as allowing constraint 2 of the previous section to hold for *and* and *claim*.

This constraint will hold for other predicates like *claim* as well, though this is not obvious and to discuss it here would take us far afield.

It should also be noted that, although constraint 2 holds for *and* and verbs like *claim*, constraint 1 does not. For example, consider

⟨6⟩ (*a*) John claimed something outrageous and something quite reasonable.
　　(*b*) John claimed something outrageous and he claimed something quite reasonable.

⟨6⟩ (*a′*)

(*b′*)

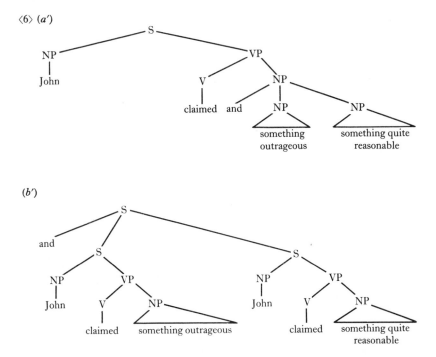

⟨6*a*⟩ involves two claims, and so is synonymous with ⟨6*b*⟩. Let us assume then that ⟨6*a*⟩ is derived from ⟨6*b*⟩ by conjunction reduction. Prior to conjunction reduction, as in ⟨6*b′*⟩, *and* commands the two occurrences of *claim*. After conjunction reduction, *claim* precedes *and*. If *claim* and *and* obeyed constraint 1, this would be impossible. Exceptions to constraints, like this one, are not rare and are discussed below.

The fact that *claim* and *and* obey constraint 2, but not constraint 1, can be made the basis for an explanation of a rather remarkable minimal pair:

⟨7⟩ (*a*) John claimed that he robbed the bank and that Sam shot him.
　　(*b*) John claimed that he robbed the bank and Sam shot him.

In ⟨7*a*⟩ John is making two claims, while in ⟨7*b*⟩ he is making one. How can we account for this fact? Let us assume, as is usual, that the complementizer *that* is

Chomsky-adjoined to the S of an object complement, as in ⟨8a⟩. Other alternatives are given in ⟨8b and c⟩; since they will yield exactly the same result, there is no need to choose among ⟨8a, b, and c⟩ for the sake of this argument. ⟨8a⟩ seems at present to be the least problematic alternative:

⟨8⟩ (a)   (b)   (c)

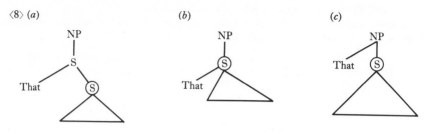

In order to explain the sentences in question, we need to show that the transformation inserting the complementizer *that* inserts only one occurrence of that complementizer per noun phrase complement, even if the complement S is a conjunction. This can be shown readily, if one looks at sentences of the form:

⟨9⟩ NP$_1$ and NP$_2$ are both correct.

'Both' can occur in sentences of this form if there are exactly two NPs conjoined in the subject. Thus, there are no sentences of the form ⟨10⟩ or ⟨11⟩:

⟨10⟩ *NP$_1$ are both correct.
⟨11⟩ *NP$_1$ and NP$_2$ and NP$_3$ and NP$_4$ are both correct.

Let NP$_1$ and NP$_2$ in ⟨9⟩ each contain a pair of conjoined sentences as its complement.

⟨12⟩

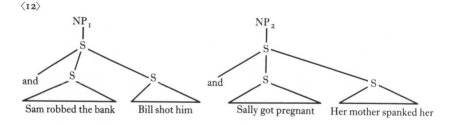

Observe the following facts:

⟨13⟩ *That Sam robbed the bank and Bill shot him are both correct.
⟨14⟩ That Sam robbed the bank and Bill shot him and that Sally got pregnant and her mother spanked her are both correct.
⟨15⟩ *That Sam robbed the bank and that Bill shot him and that Sally got pregnant and that her mother spanked her are both correct.

These sentences indicate that the rule of complementizer placement may introduce at most one occurrence of *that* for each noun phrase complement. ⟨13⟩ provides evidence that *that* is not subject to conjunction reduction. Since the sentence ⟨16⟩

⟨16⟩ That Sam robbed the bank and that Bill shot him are both correct is gramatical, and since ⟨13⟩ would result if conjunction reduction applied to *that*, it appears that conjunction reduction may not apply to *that*. Let us now return to:

⟨7⟩ (*a*) John claimed that he robbed the bank and that Sam shot him.

(*b*) John claimed that he robbed the bank and Sam shot him.

Since ⟨7a⟩ contains two occurrences of *that*, it must contain two noun phrase complements, and since ⟨7b⟩ contains only one occurrence of *that*, it cannot contain two noun phrase complements.

⟨7⟩ (*a'*)

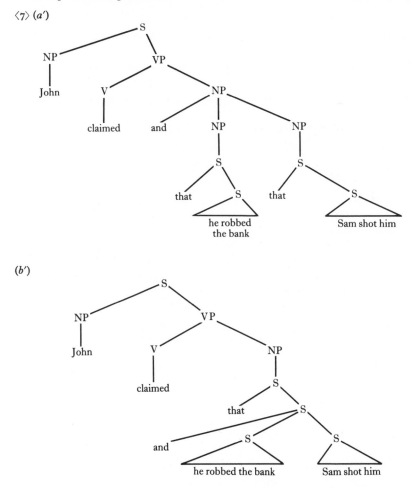

(*b'*)

The essential feature of the analysis in ⟨7a'⟩ is that ⟨7a⟩ is represented as containing a noun phrase conjunction, not a sentence conjunction. Thus, ⟨7a⟩ and ⟨7b⟩ would differ in structure in that the former would contain an NP-conjunction, whereas the latter would contain an S-conjunction. Given this analysis, the difference in meaning between ⟨7a⟩ and ⟨7b⟩ is an automatic consequence of the fact that *claim* and *and* obey constraint 2, but not constraint 1. Since ⟨7a'⟩ does not contain an embedded S-conjunction, the only possible source would be that of ⟨4⟩, where *and* commands *claim* and two claims are indicated. Since constraint 1 does not apply to *claim* and *and*, such a derivation is possible. ⟨7*b*′⟩, on the other hand, contains an embedded conjunction, and so has two conceivable sources: ⟨4⟩ and ⟨5⟩. But since *and* commands *claim* in ⟨4⟩ while *claim* commands *and* (but not vice versa) in ⟨7*b*′⟩, such a derivation is ruled out by the fact that constraint 2 holds for *and* and *claim*. Thus, the only possible derivation for ⟨7*b*′⟩ would be from ⟨5⟩, which indicates only one claim. Thus, the difference in meaning between ⟨7*a*⟩ and ⟨7*b*⟩ is explained by the fact that *and* and *claim* obey constraint 2, but not constraint 1.

Constraint 2 works for *or* as well as for *and*, as the following examples show:

⟨17⟩ (Either) John claimed that he robbed the bank or he claimed that he shot Sam.
⟨18⟩ John either claimed that he robbed the bank or claimed that he shot Sam.
⟨19⟩ John claimed that he either robbed the bank or shot Sam.

These sentences parallel ⟨1⟩–⟨3⟩.[a] Thus, it should be clear that constraint 2 applies to *or*.

Constraint 1 holds for *or*, though not for *and*. Consider the following examples, pointed out by R. Lakoff [46]:

⟨20⟩ Either you may answer the question or not.
⟨21⟩ You may either answer the question or not.

⟨20⟩ and ⟨21⟩ are not synonymous. ⟨20⟩ says that there are two possibilities: either it is the case that you are permitted to answer the question or it is not the case that you are permitted to answer the question. ⟨20⟩ exhausts the range of possibilities. ⟨21⟩, on the other hand, says that you are permitted the choice of answering or not answering. Lakoff points out that the difference in meaning can be accounted for, given Ross' analysis of modals as verbs that take complements. And the difference between ⟨20⟩ and ⟨21⟩ is paralleled by the difference between ⟨22⟩ and ⟨23⟩, where there is an overt verb with a sentential complement.

⟨22⟩ Either you are permitted to answer the question, or not.
⟨23⟩ You are permitted either to answer the question or not.

[a] I assume that *either* arises as follows. The underlying structure of a disjunction is:
(i)

The rule of *or*-copying yields:
(ii)

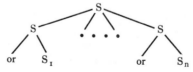

Then the leftmost *or* changes to *either*. In initial position in a sentence, *either* optionally deletes. All *ors* except the last optionally delete. *And* works in a similar fashion.

She proposes that ⟨20⟩ and ⟨21⟩ differ in structure as do ⟨24⟩ and ⟨25⟩:

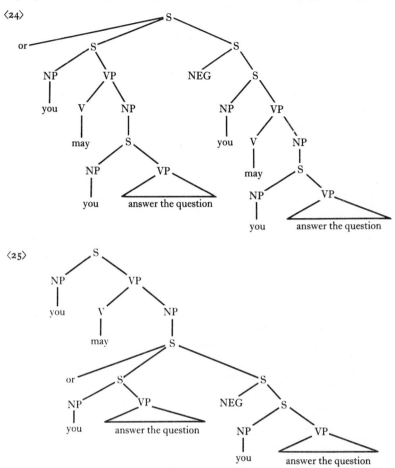

In ⟨24⟩ *or* commands *may* and not vice versa. In ⟨25⟩ *may* commands *or* and not vice versa.

However, in derived structure this asymmetry of command can be neutralized.

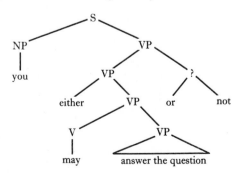

⟨27⟩

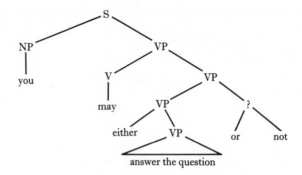

In ⟨26⟩ and ⟨27⟩, the S's have been pruned, and so *may* and *either* command each other in both cases. The assymetric command relation of ⟨24⟩ and ⟨25⟩ is neutralized. However, ⟨26⟩ *you (either) may answer the question or not* has the meaning of ⟨24⟩, while ⟨27⟩ *you may either answer the question or not* has the meaning of ⟨25⟩. The generalization is that if *either* precedes *may* in derived structure, it must command *may* in underlying structure, and conversely. Thus, we have exactly the situation of constraint 1 in the previous section. And under Ross' analysis of modals the fact that the constraints work for modals follows from the fact that the constraints work for the corresponding verbs taking complements (e.g., *permit*).

Adverbs which are understood as predicates that take complements show the same property. Compare ⟨28 and ⟨29⟩:

⟨28⟩ (a) It isn't obvious that John is a communist.
     (b) It is obvious that John isn't a communist.

⟨29⟩ (a) John isn't obviously a communist.
     (b) John obviously isn't a communist.

⟨28 a and b⟩ have underlying structures like:

⟨30⟩ (a)

(b)

If we assume that *obviously* is derived from *obvious* by a rule of adverb-lowering, and if it is assumed that *obvious* is one of those predicates taking a complement for which the constraints of section 2 hold, then it follows that ⟨29*a*⟩ should have the meaning of ⟨28*a*⟩, and that ⟨29*b*⟩ should have the meaning of ⟨28*b*⟩. Since *not* precedes *obvious* in ⟨29*a*⟩, it must command *obvious* in underlying structure ⟨30*a*⟩. Since *obvious* precedes *not* in ⟨29*b*⟩, it must command *not* in underlying structure, as in ⟨30*b*⟩.

The word *only* also obeys the constraints, though this follows automatically from the meaning of *only*. *Only Bill* means *Bill and no one other than Bill*. Since the latter expression contains a quantifier, we would expect the constraints of the previous section to hold. They do.

⟨31⟩ (*a*) John didn't hit Bill and no one else.
     (*b*) Bill and no one else wasn't hit by John.
⟨32⟩ (*a*) John didn't hit only Bill.
     (*b*) Only Bill wasn't hit by John.

The (*a*) sentences contain the reading *It wasn't the case that there was no one other than Bill that John hit*. The (*b*) sentences contain the reading *There wasn't anyone other than Bill that John didn't hit*. This is exactly what the constraints predict.

It should be noted again that the difference between subject and non-subject position in the clause has nothing to do with these constraints. Sentences like

⟨33⟩ I talked to few girls about only those problems.
⟨34⟩ I talked about only those problems to few girls.

show the predicted difference in meaning even though both *few* and *only* are in the VP in both examples.

It should be clear from these examples that at least some global derivational constraints do not serve just to limit the scope of application of a single rule, but rather can limit the applicability of a whole class of rules – in this case, quantifier-lowering, conjunction-reduction, and adverb-lowering, together with rules like passive that interact with them. This result is similar to Ross' findings [65] that certain constraints hold for all movement rules of a certain form, not just for individual rules. Similar results were found by Postal [57] in his investigation of the crossover principle.

It has been known for some time that global derivational constraints have exceptions, as well as being subject to dialectal and idiolectal variation. Consider, for example, Ross' constraints on movement transformations. The coordinate structure constraint, if violated at any point in a derivation, yields ill-formed sentences, e.g., **Someone and John left, but I don't know who and John left*. However, if the coordinate node is later deleted by some transformation, the sentence may be acceptable, e.g., *Someone and John left, but I don't know who*. Thus, the coordinate structure constraint applies throughout derivations, but with the above exception which takes precedence over the constraint. Ross' constraints are also subject to dialectal and idiolectal variation. For the majority of English speakers sentences like *John didn't see the man who had stolen anything* and *John didn't believe the claim that anyone left* are ill-formed as predicted by Ross' complex NP-constraint. However, for a great many speakers the latter sentence is grammatical and for some speakers even the former sentence is grammatical. So, it is clear that the global derivational constraints discovered by Ross are subject to such variation.

It should not be surprising that the global derivational constraints discussed above also have a range of exceptions and are subject to dialectal and idiolectal variation. For example, constraint 2 does not hold for the rule of *not*-transportation, the existence of which has been demonstrated by R. Lakoff [45]. Thus, when L^i is a nonlogical predicate and L^j is a negative, the constraint does not hold. Similarly, constraint 1 does not hold when L^i is a negative and L^j is an auxiliary verb. Thus, *John cannot go* can mean *It is not the case that John can go*. Like Ross' constraints, constraints 1 and 2 admit of a great deal of dialectal and idiolectal variation. There are a great many people (more than one-third of the people I've asked) for whom constraint 1 does not hold for quantifiers and negatives. For such people, *Few books were read by many men* is ambiguous, as is *The target wasn't hit by many arrows*. Other sorts of differences also show up in the constraints. For example, for some people constraints 1 and 2 mention surface structure, not just shallow structure. Individuals with such constraints will find that *I dissuaded Bill from dating many girls* can mean *There are many girls that I dissuaded Bill from dating*. Guy Carden [5] has pointed out that some speakers differ as to whether a constraint can hold at one level as opposed to holding throughout the grammar. McCawley [50] has shown that for some speakers the no-double-negative constraint holds only late in the grammar (at shallow structure). These speakers can get sentences like: *John doesn't like Brahms and Bill doesn't like Brahms, but not Sam – he loves Brahms.* (Before deletion, this would have the structure underlying **Sam doesn't not love Brahms.*) However, many people find such sentences impossible. Carden has pointed out that this could be accounted for if the no-double-negative constraint held throughout the grammar down to shallow structure for such speakers. With other speakers, the same constraint seems to hold over other segments of the derivation (cf. Carden [5]).

Such facts seem to show that constraints 1 and 2 are the norm from which individuals may vary. It is not clear at present how such variations from the norm can best be described, and it would seem that the basic theory will eventually have to be revised to account for such variations on basic constraints.

As the above discussion, as well as those of Postal [57] and Ross [65] have shown, global derivational constraints are pervasive in grammar. For example, interactions between transformational rules and presuppositions are handleable in a natural way using derivational constraints. Consider Kim Burt's observation (cf. Lakoff [43]) that future *will* can optionally delete if it is presupposed that the speaker is sure that the event will happen. Suppose the rule of *will*-deletion is given by the local derivational constraint (C_1, C_2). Suppose tree-condition C_3 describes the presupposition in question. Then Burt's observation can be stated in the form:

$$\langle 35 \rangle \ (P_i/C_1 \ \& \ P_{i+1}/C_2) \supset PR/C_3$$

Presuppositions of coreferentiality can be treated in the same way. For example, consider the shallow-structure constraint that states that a pronoun cannot both precede and command its antecedent. Suppose C_1 states that two NPs are coreferential, C_2 states that the pronoun precedes the antecedent, and C_3 states that the pronoun commands the antecedent. The constraint would then be of the form:

$$\langle 36 \rangle \ PR/C_1 \supset P_a/ \sim (C_2 \ \& \ C_3) \qquad\qquad \text{where } P_a \text{ is a shallow structure.}$$

Another phenomenon that can be handled naturally in the basic theory is Halliday's [24], [25] notion of 'focus'. Halliday [24] describes focus in the following terms:

the information unit, realized as the tone group, represents the speaker's organization of the discourse into message units: the information focus, realized as the location of the tonic, represents his organization of the components of each such unit such that at least one such component, that which is focal, is presented as not being derivable from the preceding discourse. If the information focus is unmarked (focus on the final lexical item), the nonfocal components are unspecified with regard to presupposition, so that the focal is merely cumulative in the message (hence the native speaker's characterization of it as 'emphatic'). If the information focus is marked (focus elsewhere than on the final lexical item), the speaker is treating the non-focal components as presupposed. (Halliday [24], p. 8)

Halliday's account of focus has been adopted by Chomsky [10]. Assume for the moment that Halliday's account of focus as involving the location of stress on surface structure constitutents were correct. Then the content of the sentence would be divided into a presupposed part and a part which is 'new' or focused upon. Recall that derivational constraints enable one to trace the history of nodes throughout a derivation. This is necessary if one is to pick out which parts of P_1 correspond to which surface structure constituents. Given such a notion, the correspondence between PR and FOC, which are part of the semantic representation, and the corresponding surface constituents can be stated by a global derivational constraint. Thus, the Halliday–Chomsky notion of 'focus' can be approached naturally within the basic theory. What is needed to make such representations precise is a precise definition of the notion 'semantic content corresponding to derived structure constituents'.

Of course, the Halliday–Chomsky account of focus is not quite correct. For example, Halliday and Chomsky assume that the constituent bearing main stress in the surface structure is the focus, and therefore that the lexical items in that constituent provide new rather than presupposed information. This is not in general the case. Consider ⟨37⟩:

⟨37⟩ The TALL girl left.

Here the main stress is on TALL, which should be the focus according to Halliday and Chomsky, and should therefore be new, not given, information. However, in ⟨37⟩ TALL is understood as modifying the noun in the same way as the restrictive relative clause *who was tall*. Since restrictive relative clauses are presupposed, it follows that in ⟨37⟩ it is presupposed, not asserted, that the girl being spoken of was tall. Thus, the meaning of the lexical item *TALL* cannot be new information. Another possible candidate for focus might be the whole NP *the tall girl*. But none of the lexical content of this NP is new information, since it is presupposed that the individual under discussion exists, it is presupposed that that individual is a girl and it is presupposed that she is tall. None of this is new. In ⟨37⟩ it is presupposed that some girl left, and it is presupposed that some girl is tall. The new information is that the girl who was presupposed to have left is coreferential with the girl who was presupposed to be tall. The semantic content of the focus is an assertion of coreferentiality. In this very typical example of focus, the lexical-semantic content of the surface structure constituent bearing main stress has nothing whatever to do with the semantic content of the focus.

So far, we have assumed that the Halliday–Chomsky account of focus in terms of surface structure constituents is basically correct. But this too is obviously mistaken. Consider ⟨38⟩:

⟨38⟩ (a) John looked up a girl who he had once met in Chicago.
      (b) John looked a girl up who he had once met in Chicago.

⟨38⟩ (a′)

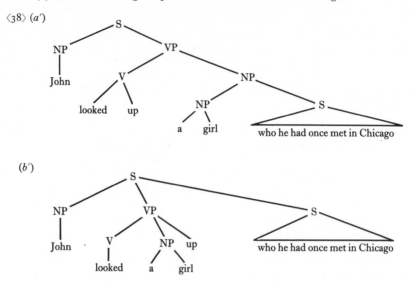

(b′)

(a′) and (b′) have very different surface structure constituents. Assuming that main stress falls on 'Chicago' in both cases, Chomsky and Halliday would predict that these sentences should be different in focus possibilities and in corresponding presuppositions, and that therefore they should answer different questions, and have quite different semantic representations. But it is clear that they do not answer different questions and do not make different presuppositions. Thus, it is clear that focus cannot be defined purely in terms of *surface* structure constitutents. Rather it seems that derived structure at some earlier point in derivations is relevant.

These difficulties notwithstanding, it is clear that the phenomenon of focus does involve global derivational constraints of some sort involving derived structure. Halliday certainly deserves credit for the detailed work he has done in this area, despite the limitations of working only with surface structure. Generative semantics should provide a natural framework for continuing Halliday's line of research.

Another notion which can be handled naturally within the framework of generative semantics is that of 'topic'. Klima has observed that sentences like the following differ as to topic:

⟨39⟩ (a) It is easy to play sonatas on this violin.
      (b) This violin is easy to play sonatas on.
      (c) Sonatas are easy to play on this violin.

(a) is neutral with respect to topic. (b) requires 'this violin' to be the topic, while (c) requires 'sonatas'. There are of course predicates in English which relate topics to the things they are topics of. For example:

⟨40⟩ (a) My story is about this violin.
      (b) That discussion concerned sonatas.

The predicates 'be about' and 'concern' are two-place relations, whose arguments are a description of a proposition or discourse and the item which is the topic of that proposition or discourse. Thus, the (*a*), (*b*), and (*c*) sentences of ⟨41⟩ and ⟨42⟩ are synonymous with respect to topic as well as to the rest of their content:

⟨41⟩ (*a*) Concerning sonatas, it is easy to play them on this violin.
     (*b*) Concerning sonatas, they are easy to play on this violin.
     (*c*) Sonatas are easy to play on this violin.

⟨42⟩ (*a*) About this violin, it is easy to play sonatas on it.
     (*b*) About this violin, it is easy to play sonatas on.
     (*c*) This violin is easy to play sonatas on.

If the topics mentioned in the clause containing 'concern' or 'about' differ from the superficial subjects in these sentences, then there is a conflict of topics and ill-formedness results, unless it is assumed that the sentences can have more than one topic.

⟨43⟩ ?*About sonatas, this violin is easy to play them on.
⟨44⟩ ?*Concerning this violin, sonatas are easy to play on it.

These are well-formed only for those speakers who admit more than one topic in such sentences.

These considerations would indicate that the notion 'topic' of a sentence is to be captured by a two-place relation having the meaning of 'concerns' or 'is about'. If the set of presuppositions contains such a two-place predicate whose arguments are P_1 and some NP, then it will be presupposed that that NP is the topic of P_1. Thus, the notion 'topic' may well turn out to be a special case of a presupposition. Since a semantic specification of 'concerns' and 'is about' is needed on independent grounds, it is possible that the special slot for TOP in semantic representation is unnecessary. Whether all cases of topic will turn out to be handleable in this way remains, of course, to be seen. Whichever turns out to be true, it is clear that the facts of ⟨39⟩ can be handled readily by derivational constraints. Assume that there is a rule which substitutes 'this violin' and 'sonatas' for 'it' in ⟨39⟩. Let (C_1, C_2) describe this operation. Let C_3 describe the topic relation obtaining between P_1 and the NP being substituted. Then the facts of ⟨39⟩ can be described by the following derivational constraint:

⟨45⟩ (i) $(P_i/C_1 \& P_{i+1}/C_2) \supset PR/C_3$

Note that this has exactly the form of the constraint discussed above describing the deletion of future *will*. Global derivational constraints linking transformations and presuppositions have this form. Of course, it may turn out to be the case that a more general characterization of the facts of ⟨39⟩ is possible, namely, that surface subjects in some class of sentences are always topics. In that case, there would be a derivational constraint linking presuppositions and surface structure. In any event the theory of generative semantics seems to provide an adequate framework for further study of the notion 'topic'.

Another sort of phenomenon amenable to treatment in the basic theory is lexical presupposition. As Fillmore (personal communication) has pointed out, 'Leslie is a bachelor' presupposes that Leslie is male, adult and human and asserts that he is unmarried. Similarly, 'Sam assassinated Harry' presupposes that Harry is an important public figure and asserts that Sam killed him. Thus, lexical insertion transformations for 'bachelor' and 'assassinate' must be linked to presuppositional information. This is just the sort of linkage that we discussed above.

Thus far, most of the examples of global derivational constraints we have discussed mention semantic representations in some way. However, this is not true in general. For example, Ross' [65] constraints are purely syntactic. Another example of a purely syntactic global derivational constraint has been discussed by Harold King [38]. King noted that contraction of auxiliaries as in 'John's tall', 'The concert's at 5 o'clock', etc. cannot occur when a constituent immediately following the auxiliary to be contracted has been deleted. For example:

⟨46⟩ (a) Max is happier than Sam is these days.
      (b) *Max is happier than Sam's these days.
⟨47⟩ (a) Rich though John is, I still like him.
      (b) *Rich though John's, I still like him.
⟨48⟩ (a) The concert is this afternoon.
      (b) The concert's this afternoon.
      (c) Tell John that the concert is in the auditorium this afternoon.
      (d) Tell John where the concert is this afternoon.
      (e) *Tell John where the concert's this afternoon.
      (f) Tell John that the concert's this afternoon.

In (e) the locative adverb has been moved from after *is*; in (f) no such movement has taken place.

Since contraction is an automatic consequence of an optional rule of stress-lowering, the general principle is that stress-lowering on an auxiliary cannot take place if at any point earlier in the derivation any rule has deleted a constituent immediately following the auxiliary. Let (C_1, C_2) be the rule of stress-lowering for Aux^j. Let $C_3 = X^i - \text{Aux}^j - A - X^k$ where A is any constituent and $C_4 = X^i - \text{Aux}^j - X^k$. The constraint is:

⟨49⟩ $\sim (\exists\, x)\,(\exists y)\quad (P_x/C_1 \;\&\; P_{x+1}/C_2 \;\&\; P_y/C_3 \;\&\; P_{y+1}/C_4)$

A wide range of examples of global derivational constraints not mentioning semantic representation will be discussed in Lakoff [43]. The redundancy rules discussed by R. Lakoff [44] are further examples of this sort. The exact nature and extent of global derivational constraints is, of course, to be determined through future investigation. It should be clear, however, that a wide variety of such constraints do exist. Thus, the basic theory, in its account of global derivational constraints, goes far beyond the standard theory and the *Aspects* theory, which included only a very limited variety of such constraints.

The basic theory is, of course, not obviously correct, and is open to challenge on empirical issues of all sorts. However, before comparing theories of grammar, one should first check to see that there are empirical differences between the theories. Suppose, for example, one were to counterpose to the basic theory, or generative semantics, an 'interpretive theory' of grammar. Suppose one were to construct such an interpretive theory in the following way. Take the class of sequences of phrase-markers $(P_1, \ldots, P_i, \ldots, P_n)$ where all lexical insertion rules occur in a block between P_1 and P_i and all upward-toward-the-surface cyclic rules apply after P_i. Call P_i 'deep structure'. Assume that $P_i \ldots P_n$ are limited only by local derivational constraints, except for those global constraints that define the cycle and rule ordering. Call $P_i, \ldots, P_n$ the 'syntactic part' of the derivation. Assume in addition that semantic representation $SR = (P_{-m}, PR, TOP, FOC, \ldots)$, where P_{-m} is a phrase-marker in some 'semantically primitive' notation, as suggested by

Chomsky [10] in his account of the 'standard theory'. Then a full derivation will be a sequence of phrase-markers:

$$P_{-m}, \ldots, P_{-j}, \ldots, P_0, P_1, \ldots, P_i, \ldots, P_n$$

Call $P_0 \ldots P_{-m}$ the semantic part of the derivation. Assume that the sequences of phrase-markers $P_0 \ldots P_{-j}$ are defined by local derivational constraints and global derivational constraints that do not mention any stage of the derivation after P_i, the 'deep structure'. Call these constraints 'deep structure interpretation rules'. Assume that the sequences of phrase-markers $P_{-j} \ldots P_{-m}$ are defined by local derivational constraints paired with global derivational constraints that may mention P_i and P_n as well as $P_{-j} \ldots P_{-m}$, PR, TOP, FOC, etc. Call these constraints 'surface structure interpretation rules'. (If such global constraints may mention not only P_i and P_n, but all points in between, then we will call them 'intermediate structure interpretation rules'.) It should be clear that such an 'interpretive theory of grammar' is simply a restricted version of the basic theory. One can look at the deep structure interpretation rules as operations 'going from' P_1 to P_{-j}, which are able to 'look back' only as far as P_i. And we could look at the surface structure interpretation rules as operations going from the 'output' of the deep structure interpretation rules P_{-j} to SR, while being able to 'look back' to P_i and P_n. However, as Chomsky [10] points out, the notion of 'directionality' is meaningless, and so there is no empirical difference between these operations and derivational constraints. Thus, such 'interpretive theories' are no different in empirical consequences than the basic theory, restricted in the above way, provided that such interpretive theories assume that semantic representations are of the same form as phrase-markers or are notational variants thereof. The only empirical differences are the ways in which the basic theory is assumed to be constrained, for example, the question as to whether levels like P_i, P_1, and P_{-j} exist. As we saw above, there is reason to believe that a level P_i does not exist, and no one has ever given any reasons for believing that a level P_{-j} exists, that is, that 'deep structure interpretation rules' are segregated off from 'surface structure interpretation rules'.

So far, no interpretive theory this explicit has been proposed. The only discussion of what might be called an interpretive theory which goes into any detail at all is given by Jackendoff [33], who discusses both surface and intermediate structure interpretation rules. However, Jackendoff explicitly refuses to discuss the nature of semantic representation and what the output of his interpretive rules is supposed to look like, so that it is impossible to determine whether his interpretive theory when completed by the addition of an account of semantic representation will be simply a restricted version of the theory of generative semantics. Jackendoff claims that semantic representations are not identical to syntactic representations ([33] p. 2), but he does not discuss this claim. However, the empirical nature of the issue is clear: Are Jackendoff's interpretation rules simply notational variants of derivational constraints? Do his rules map phrase-markers into phrase-markers? (The only examples he gives do, in fact, do this.) Will the output of his rules be phrase-markers, or notational variants thereof? Of course, such questions are unanswerable in the absence of an account of the form of his rules and the form of their output.

Although Jackendoff does not give any characterization of the output of his envisioned interpretation rules, he does discuss a number of examples in terms of the vague notions 'sentence-scope' and 'VP-scope'. Many of the examples he discusses overlap with those discussed above in connection with global derivational constraints

1 and 2. For example, he discusses sentences like 'Many of the arrows didn't hit the target' and 'The target wasn't hit by many of the arrows', claiming that the difference in interpretation can be accounted for by what he calls a difference in scope, which boils down to the question of whether the element in question is inside the VP or not. If he were to provide some reasonable output for his rules, then his scope-difference proposal might be made to match up with those subcases of constraint 1 where L^i is in subject position and L^j is dominated by VP. The overlap is due to the fact that the subject NP precedes VP. However, there are certain crucial cases which decide between constraint 1 and the extended Jackendoff proposal, namely, cases where the two elements in question are both in the VP. Since they would not differ in VP-scope, the Jackendoff proposal would predict that the relative order of the elements should not affect the meaning. Constraint 1, however, would predict a meaning difference just as in the other cases, where the leftmost element in shallow structure commanded the other element in semantic representation. We have already seen some examples of cases like this:

⟨50⟩ (a) John talked to *few* girls about *many* problems.
     (b) John talked about *many* problems to *few* girls.

⟨51⟩ (a) I talked to *few* girls about *only* those problems.
     (b) I talked about *only* those problems to *few* girls.

These sentences show the meaning difference predicted by constraint 1, but not by Jackendoff's scope-difference proposal. Other examples involve adverbs like *carefully*, *quickly*, and *stupidly*, which he claims occur within the scope of the VP when they have a manner interpretation (as opposed to sentence adverbs like *evidently*, which he says have sentence-scope and are not within the VP). Since Jackendoff permits some adverbs like *stupidly* to have both VP-scope and sentence-scope with differing interpretations, all of the following examples will contain the sentence adverb *evidently* just to force the VP-scope interpretation for the other adverbs, since a sentence may contain only one sentence adverb.

⟨52⟩ (a) John evidently had *carefully* sliced the bagel *quickly*.
     (b) John evidently had *quickly* sliced the bagel *carefully*.

⟨53⟩ (a) John evidently had *carefully* sliced *few* bagels.
     (b) John evidently had sliced *few* bagels *carefully*.

⟨54⟩ (a) John evidently had *stupidly* given *none* of his money away.
     (b) John evidently had given *none* of his money away *stupidly*.

Each of the pairs of underlined words would be within the scope of the VP according to Jackendoff, and so, according to his theory, the (a) and (b) sentences should be synonymous. They obviously are not, and the difference in their meaning is predicted by constraint 1. Thus, constraint 1 handles a range of cases that Jackendoff's scope-difference proposal inherently cannot handle.[a]

But Jackendoff's proposal is inadequate in another respect as well. In order to generate a sentence like 'Not many arrows hit the target' he would need a phrase-structure rule expanding determiner as an optional negative *followed by* a quantifier (Det ⟶ (NEG) Q). The meaning of the NEG relative to the quantifier would be

---

[a] The facts in ⟨52⟩–⟨54⟩ are independent of the phenomenon of attraction to focus discussed by Jackendoff [33]. If heavy stress is placed on 'John' in these sentences, 'John' is made the focus. This does not in any way affect the interpretation of the relative scopes of the adverbs, quantifiers and negatives mentioned in these sentences.

given by his sentence-scope interpretation rule, since 'not many' is part of the subject 'not many arrows' in the above sentence. This interpretation rule makes no use of the fact that *not* happens to precede *many* in this example (and is interpreted as commanding *many*), since Jackendoff's interpretation rule would depend in this case on subject (sentence-scope) position, not the left-to-right order of negative and quantifier. Thus, in Jackendoff's treatment, it is an accident that NEG happens to precede the quantifier with this meaning. Jackendoff's scope rule would give exactly the same result if the NEG had followed the quantifier within the subject, that is, if the impossible **many not arrows* existed. Thus, Jackendoff's phrase-structure rule putting the NEG in front of the quantifier misses the fact that this order is explained by constraint 1.

On the whole I would say that the discussion of surface and intermediate structure interpretation rules found in Chomsky [10], Jackendoff [33] and Partee [55] do not deal with the real issues. As we have seen, such rules are equivalent to transformations plus global derivational constraints, given the assumption that semantic representations can be given in terms of phrase-markers. We know that transformations are needed in any theory of grammar, and we know that global derivational constraints are also needed on independent grounds, as in rule ordering, Ross' constraints [65], R. Lakoff's redundancy rules [44], Harold King's contraction cases [38], and the myriad of other cases discussed in Lakoff [43]. Thus, surface and intermediate structure interpretation rules are simply examples of derivational constraints, local and global, which are needed independently. The real issues raised in such works are (i) Can semantic representation be given in terms of phrase-markers or a notational variant? (ii) Is there a level of 'deep structure' following lexical insertion and preceding all cyclic rules? and (iii) What are the constraints that hold at the levels of shallow structure and surface structure? These are empirical questions. (i) is discussed in Lakoff [43] (forthcoming), where it is shown that, to the limited extent to which we know anything about semantic representations, they can be given in terms of phrase-markers. (ii) was discussed in the previous section, and will be discussed more thoroughly in the following section. (iii) has been discussed in some detail by Perlmutter [56] and Ross [65]. It seems to me that many of the regularities concerning nominalizations that have been noted by Chomsky and other low-level regularities noted by Jackendoff [33] and Emonds [14] are instances of constraints on shallow or surface structure.

4 Autonomous and arbitrary syntax

A field is defined by certain questions. For example,

(i) What are the regularities that govern which linear sequences of words and morphemes of a language are permissible and which sequences are not?

(ii) What are the regularities by which the surface forms of utterances are paired with their meanings?

Early transformational grammar, as initiated by Harris [28], [29] and developed by Chomsky [8], [9], made the assumption that (i) could be answered adequately without also answering (ii), and that the study of syntax was the attempt to answer (i). This assumption defined a field which might well be called 'autonomous syntax', since it assumed that grammatical regularities could be completely characterized without recourse to meaning. Thus, early transformational grammar was a natural outgrowth of American structural linguistics, since it was concerned with discovering the regularities governing the distribution of surface forms.

However, the main reason for the development of interest in transformational grammar was not merely that it led to the discovery of previously unformulated and unformulable distributional regularities, but primarily that, through the study of distributional regularities, transformational grammar provided insights into the semantic organization of language and into the relationship between surface forms and their meanings. If transformational grammar had not led to such insights – if its underlying syntactic structures had turned out to be totally arbitrary or no more revealing of semantic organization than surface structures – then the field would certainly, and justifiably, have been considered dull. It may seem somewhat paradoxical, or perhaps miraculous, that the most important results to come out of a field that assumed that grammar was independent of meaning should be those that provided insights as to how surface grammatical structure was related to meaning.

Intensive investigation into transformational grammar in the years since 1965 has shown why transformational grammar led to such insights. The reason is that the study of the distribution of words and morphemes is inextricably bound up with the study of meanings and how surface forms are related to their meanings. Since 1965, empirical evidence has turned up which seems to show this conclusively. Some of this evidence will be discussed below. Consequently, a thorough-going attempt to answer (i) will inevitably result in providing answers to (ii). The intensive study of transformational grammar has led to the abandonment of the autonomous syntax position, and with it, the establishment of a field defined by the claim that (i) cannot be answered in full without simultaneously answering (ii), at least in part. This field has come to be called 'generative semantics'.

To abandon the autonomous syntax position is to claim that there is a continuum between syntax and semantics. The basic theory has been formulated to enable us to make this notion precise, and to enable us to begin to formulate empirically observed regularities which could not be formulated in a theory of autonomous syntax. Perhaps the empirical issues can be defined more sharply by considering the basic theory vis-a-vis other conceptions of transformational grammar. Suppose one were to restrict the basic theory in the following way. Let PR, TOP, F,...in SR be null. Limit global derivational constraints to those which specify rule order. Limit local derivational constraints to those specifying elementary transformations, as discussed in *Aspects*. Assume that all lexical insertion transformations apply in a block. The resulting restricted version of the basic theory is what Chomsky in [10] describes as a version of the 'standard theory'.

Of course, not all versions of the theory of grammar that have been assumed by researchers in transformational grammar are restricted versions of the basic theory, nor versions of the standard theory. For example, the theory of grammar outlined in *Aspects of the Theory of Syntax* [9] is not a version of either the basic theory or the standard theory. The principal place where the theory of *Aspects* deviates from the standard theory and the basic theory is in its assumption of the inclusion of a non-null Katz–Fodor semantic component, in particular, their conception of semantic readings as being made up of amalgamated paths, which are *strings* of semantic markers and of symbols which are supposed to suggest Boolean operations.[a] They nowhere say that readings are to be defined as phrase-markers made up of the same

---

[a] See Fodor and Katz [16], pp. 503, 531 ff.; also Katz, this volume, pp. 297 ff.

nonterminal nodes as syntactic phrase-markers, nor do they say that projection rules are operations mapping phrase-markers onto phrase-markers, and I am sure that no one could legitimately read such an interpretation into their discussion of amalgamated paths and Boolean operations on markers. Thus a derivation of a sentence, including the derivation of its semantic reading, would be represented in the *Aspects* theory as a sequence $A_m, \ldots, A_0, P_1, \ldots, P_n$, where the P_i's are phrase-markers defined as in the basic theory and the standard theory while the A_j's are amalgamated paths of markers, which are not defined in either the basic theory or the standard theory. (Chomsky [10], p. 188, says: 'Suppose further that we regard S as itself a phrase-marker in some "semantically primitive" notation.... Suppose now that in forming Σ, we construct P_1, which is, in fact, the semantic representation S of the sentence.' In allowing for semantic representations to be phrase-markers, not amalgamated paths, Chomsky is ruling out a nonnull Katz–Fodor semantic component in his new 'standard theory'.) In the *Aspects* model it is assumed that the P_1's are defined by well-formedness constraints (base rules). It is *not* assumed that either the P_n's (surface structures) or the A_m's (semantic readings) are constrained by any additional well-formedness conditions; rather it is assumed that they are completely characterized by the application of transformations and projection rules to the base structures.

Thus, the *Aspects* theory differs in an important respect from the basic theory and the standard theory in the definition of a derivation.[a]

⟨1⟩ (a) *Aspects* theory: $\quad A_m, \ldots\ldots\ldots, A_0, P_1, \ldots\ldots\ldots P_n$
  (b) 'Standard' theory: $P_1, \ldots\ldots\ldots\ldots, P_i, \ldots\ldots\ldots P_n$
           ↑   ↑   ↑
        semantic   deep   surface
      representation   structure   structure

The *Aspects* theory assumes that semantic readings are formal objects of a very different sort than syntactic phrase-markers and that projection rules are formal operations of a very different sort than grammatical transformations. One of the most important innovations of generative semantics, perhaps the most fundamental one since all the others rest on it, has been the claim that semantic representations and syntactic phrase-markers are formal objects of the same kind, and that there exist no projection rules, but only grammatical transformations. In his discussion of his new 'standard theory', Chomsky has therefore adopted without fanfare one of the most fundamental innovations made by the basic theory.

The 'standard' theory is a considerable innovation over the *Aspects* theory in this sense, since it represents an implicit rejection of Katzian semantics and since the difference between having amalgamated paths and phrase-markers as semantic representations is crucial for Chomsky's claim that there exist surface structure interpretation rules. Suppose this were a claim that there are rules that map surface structures onto amalgamated paths containing strings of semantic markers and symbols for Boolean operations. If it were, then such rules would be formal operations which are of an entirely different nature than grammatical transformations. Then such rules could not have those properties of grammatical transformations

[a] In footnote 10 to chapter 1 of *Aspects* (p. 198), Chomsky says: 'Aside from terminology, I follow here the exposition in Katz and Postal (1964). In particular, I shall assume throughout that the semantic component is essentially as they describe it...' In this passage, Chomsky is ruling out the possibility that semantic representations might be phrase-markers.

that depend crucially on the fact that both the input and output of the transformations are phrase-markers. But it has been shown (Lakoff [42]) that, at least in the
case of surface interpretation rules for quantifiers and negatives proposed by
Partee [55] and Jackendoff [33], such interpretation rules must obey Ross' constraints
on movement transformations (Ross [65]). Since Ross' constraints depend crucially
on both the input and output of the rules in question both being phrase-markers
(cf. the account of the coordinate structure constraint in section 1), it can be demonstrated that, if the outputs of surface interpretation rules are not phrase-markers,
then the surface interpretation rule proposals for handling quantifiers and negation
are simply incorrect. Thus, although Chomsky doesn't give any reasons for adopting
this innovation of generative semantics, his doing so is consistent with his views
concerning the existence of surface structure rules of semantic interpretation.[a]

The assumption that there exists a level of deep structure, in the sense of either
the *Aspects* or standard theories defines two possible versions of the autonomous
syntax position; to my knowledge, these are the only two that have been seriously
considered in the context of transformational grammar. We have already seen in
section 2 that there is evidence against ⟨1 b⟩. In what follows, I will discuss just a
few of the wide range of cases that indicate that both the *Aspects* and standard-theory
versions of the autonomous syntax position are open to very serious doubt.

Though Chomsky does not mention the cycle in his discussion of the standard
theory, we saw in section 2 above that it interacts crucially with the claim of the standard theory that all lexical insertion rules occur in a block, since it is shown that
there can be no level of 'deep structure' if such is defined as following all lexical
insertion rules and preceding all upward-toward-the-surface cyclic rules. The
argument of section 2 also showed that there exist some cases of post-transforma-
tional lexical insertion, as was conjectured by McCawley [48] and Gruber [23].
Postal [58] has found a rather remarkable case to confirm McCawley's conjecture.
Postal considers sentences like *John strikes me as being like a gorilla with no teeth* and
John reminds me of a gorilla with no teeth.[b] He notes that both sentences involve a

---

[a] The 'standard theory' is quite different in this respect from the theory assumed by
Jackendoff [33], who insists upon making no assumptions whatever about the nature of semantic representation. Moreover, it is not clear that Chomsky ever seriously maintained the
'standard theory' as described in the passage quoted, since the main innovations of that theory
– allowing semantic representations to be given in terms of phrase-markers (and thus ruling
out Katzian semantics), allowing prelexical transformations, and allowing lexical semantic
readings to be given as substructures of derived phrase-markers – were only made in the context of an argument to the effect that these innovations made by McCawley [48] and others
were not new, but were simply variants of the 'standard theory'. Since Chomsky does not
attempt to justify this innovation, and since he does not mention it outside the context of this
argument, it is not clear that he ever took such an account of the standard theory seriously. In
fact, it is not clear that anyone has ever held the 'standard theory'. Nonetheless, this theory
is useful for pinpointing certain important issues in the theory of grammar, and we will use it
for this purpose in subsequent discussion.

[b] Postal is discussing only the *strike-as-similar* sense of *remind*. There are also two other
senses: the *cause-to-remember* and *make-think-of* senses. Many speakers confuse the *make-
think-of* and *strike-as-similar* senses. They are quite different. For example:

   (i) Talking to you reminds me of my years as a zookeeper.
   (ii) Talking to you makes me think of my years as a zookeeper.
   (iii) *Talking to you strikes me as being similar to my years as a zookeeper.
In addition, the *make-think-of* reading permits passivization, while the *strike-as-similar*
reading does not.

   (iv) John reminded me of a gorilla I once dated.

perception on my part of a similarity between John and a gorilla with no teeth. This is fairly obvious, since a sentence like *John reminds me of a gorilla with no teeth, though I don't perceive any similarity between John and a gorilla with no teeth* is contradictory. Postal suggests that an adequate semantic representation for *remind* in this sense would involve at least two elementary predicates, one of perception and one of similarity. Schematically *SU strikes IO as being like O* and *SU reminds IO of O* would have to contain a representation like:

⟨2⟩ IO [perceive] (SU [similar] O)

where [perceive] is a two-place predicate relating IO and (SU [similar] O) and [similar] is a two-place predicate relating S and O.

⟨2⟩ might be represented as ⟨3⟩:

⟨3⟩

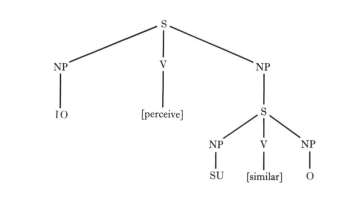

Postal suggests that the semantic representation could be related to the surface structure by the independently needed rules of subject-raising and psych-movement, plus McCawley's rule of predicate-lifting (McCawley [48]). Subject-raising would produce ⟨4⟩:

⟨4⟩

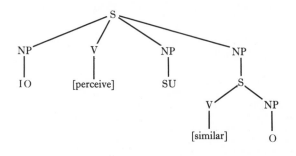

(v) I was reminded by John of a gorilla I once dated.
(iv) is ambiguous between the *strike-as-similar* and *make-think-of* readings, while (v) can only have the *make-think-of* reading.

Psych-movement would yield ⟨5⟩:

⟨5⟩

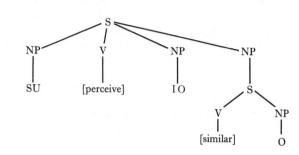

Predicate-lifting would yield ⟨6⟩:

⟨6⟩

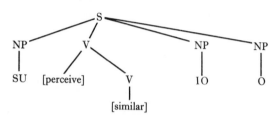

Remind would substitute for [[PERCEIVE] [SIMILAR] ᵥ].[a] The question to be asked is whether there is any transformational evidence for such a derivation. In other words, is there any transformational rule which in general would apply only to sentences with a form like ⟨3⟩, which also apply to *remind* sentences. The existence of such a rule would require that *remind* sentences be given underlying syntactic structures like ⟨3⟩, which reflect the meaning of such sentences. Otherwise, two such rules would be necessary – one for sentences with structures like ⟨3⟩ and one for *remind* sentences.

Postal has discovered just such a rule. It is the rule that deletes subjects in sentences like:

⟨7⟩ To shave oneself is to torture oneself.
⟨8⟩ Shaving oneself is (like) torturing oneself.

---

[a] It should be observed, incidentally, that arguments like Postal's do not depend on complete synonymy (as in the case of *remind* and *perceive-similar*) but only on the inclusion of meaning. As long as the *remind* sentences contain the meaning of ⟨9⟩, the appropriate rules would apply and the argument would go through. If it should turn out to be the case that *remind* contains extra elements of meaning in addition to ⟨9⟩, it would be irrelevant to Postal's argument. A mistake of this sort was made in an otherwise excellent paper by DeRijk [13], who considered examples like *John forgot X* and *John ceased to know X*, with respect to the proposal made by McCawley [48]. DeRijk correctly notes that if *X = his native language*, nonsynonymous sentences result. He concludes that *forget* could not be derived from an underlying structure containing the meaning of *cease to know*. It would be true to say that *forget* cannot be derived from an underlying structure containing only the meaning of *cease to know*, since *forget* means *to cease to know due to a change in the mental state of the subject*. But McCawley's conjecture, like Postal's argument, only requires that the meaning of *cease to know* only be *contained* in the meaning of *forget*, which it is.

Postal observes that the rule applies freely if the subject is the impersonal *one* (or the impersonal *you*). However, there are other, rather restricted circumstances where this rule can apply, namely, when the clause where the deletion takes place is a complement of a verb of saying or thinking and when the NPs to be deleted are coreferential to the subject of that verb of saying or thinking.

⟨9⟩ Bill says that to shave $\begin{Bmatrix} \text{himself} \\ \text{*herself} \end{Bmatrix}$ is to torture $\begin{Bmatrix} \text{himself} \\ \text{*herself} \end{Bmatrix}$.

⟨10⟩ Bill feels that shaving $\begin{Bmatrix} \text{himself} \\ \text{*themselves} \end{Bmatrix}$ is (like) torturing $\begin{Bmatrix} \text{himself} \\ \text{*themselves} \end{Bmatrix}$.

The presence of the reflexive indicates what the deleted NP was. Sentences like ⟨11⟩ show that this rule does not apply in relative clauses in addition to complements, and ⟨12⟩ shows that it does not apply if the deleted NPs are identical only to the subject of a verb of saying or thinking more than one sentence up.

⟨11⟩ *Bill knows a girl who thinks that shaving himself is like torturing himself.

⟨12⟩ *Mary says that Bill thinks that shaving herself is torturing herself.

Postal notes that this rule also applies in the cases of *remind* sentences.

⟨13⟩ Shaving $\begin{Bmatrix} \text{himself} \\ \text{*herself} \end{Bmatrix}$ reminds John of torturing $\begin{Bmatrix} \text{himself} \\ \text{*herself} \end{Bmatrix}$.

If *remind* is derived from a structure like ⟨3⟩, then this fact follows automatically, since ⟨3⟩ contains a complement and a verb of thinking. If ⟨13⟩ is not derived from a structure like ⟨3⟩, then a separate rule would be needed to account for ⟨13⟩. But that would be only half of the difficulty. Recall that the general rule applies when the NPs to be deleted are subjects of the next highest verb of saying or thinking and the clause in question is a complement of that verb, as in ⟨9⟩ and ⟨10⟩. However, this is not true in the case of *remind*.

⟨14⟩ *Mary says that shaving herself reminds Bill of torturing herself.

If *remind* is analyzed as having an underlying structure like ⟨3⟩, this fact follows automatically from the general rule, since then *Mary* would not be the subject of the next-highest verb of saying or thinking, but rather the subject of the verb two sentences up, as in ⟨12⟩. Thus, if *remind* is analyzed as having an underlying syntactic structure like that of ⟨3⟩, one need only state the general rule given above. If *remind*, on the other hand, is analyzed as having a deep structure like its surface structure, with no complement construction as in ⟨3⟩, then one would (i) have to have an extra rule just for *remind* (to account for ⟨13⟩), and (ii) one would have to make *remind* an exception to the general rule (to account for ⟨14⟩). Thus, there is a rather strong transformational argument for deriving *remind* sentences as Postal suggests, which requires lexical insertion to take place following upward-toward-the-surface cyclic rules like subject-raising and psych-movement. This is but one of a considerable number of arguments given for such an analysis by Postal [58].

Postal's claims about *remind*, if correct, provide crucial evidence of one sort against the existence of a level of deep structure in the sense of both the 'standard theory' and the *Aspects* theory, since such a level could be maintained only by giving up linguistically significant generalizations. This would be similar to the argument made by Halle [26] against phonemic representation. Another such argument has been advanced by McCawley [49]. McCawley discusses the phenomenon of *respectively*-sentences, rejecting the claim that *respectively*-sentences are derived from sentence conjunction. Chomsky [10] gives a particularly clear description of the position McCawley rejects. Chomsky discusses the following examples. The angle-bracketed numbers correspond to Chomsky's numbering. The square-bracketed numbers are McCawley's; where McCawley gives no number, the page is listed. Note that not all the examples have square-bracketed numbers or page references, since Chomsky gives more sentences than McCawley does. $\langle 9 \rangle$, $\langle 10 \rangle$, $\langle 15 \rangle$ and $\langle 18 \rangle$ are Chomsky's examples, not McCawley's.

[pp. 164–5] $\langle 8 \rangle$ $Ax:x$ {John, Harry} [x loves x's wife]

$\langle 9 \rangle$ John loves John's wife and Harry loves Harry's wife.

$\langle 10 \rangle$ John and Harry love John's wife and Harry's wife, respectively.

[142] $\langle 11 \rangle$ John and Harry love their respective wives.

[164] $\langle 12 \rangle$ $Ax:x \in M$ [x loves x's wife]

[149] $\langle 13 \rangle$ Those men love their respective wives.

[159] $\langle 14 \rangle$ That man(x) loves Mary and that man(y) loves Alice.

$\langle 15 \rangle$ That man(x) and that man(y) love Mary and Alice respectively.

[158] $\langle 16 \rangle$ Those men love Mary and Alice respectively.

[159] $\langle 17 \rangle$ That man(x) loves Mary and that man(x) loves Alice.

$\langle 18 \rangle$ That man(x) and that man(x) love Mary and Alice, respectively.

[157] $\langle 19 \rangle$ That man loves Mary and Alice.

Chomsky's reconstruction of the position McCawley rejects is given in a diagram which he numbers $\langle 20 \rangle$:

$\langle 20 \rangle$

$$
\begin{array}{ccccc}
 & \text{I} & \text{R} & \text{R}' & \\
\langle 8 \rangle & \longrightarrow \langle 9 \rangle & \longrightarrow \langle 10 \rangle & \longrightarrow \langle 11 \rangle \\
 & \text{I}' & & & \\
\langle 12 \rangle & \longrightarrow \langle 13 \rangle & & & \\
 & \text{R} & \text{C} & & \\
\langle 14 \rangle & \longrightarrow \langle 15 \rangle & \longrightarrow \langle 16 \rangle & & \\
 & \text{R} & \text{C} & & \\
\langle 17 \rangle & \longrightarrow \langle 18 \rangle & \longrightarrow \langle 19 \rangle & &
\end{array}
$$

R′ is the rule that converts *respectively* to *respective* in the appropriate cases, and C is a rule conjunction collapsing. I, R, and R′ are the crucial part of Chomsky's reconstruction of the argument. I and I′ are not mentioned by McCawley at all, and are entirely due to Chomsky. R is what Chomsky refers to as the '*respectively*-transformation' as it is discussed in the transformational literature. Such a rule would map $\langle 9 \rangle$ into $\langle 10 \rangle$, $\langle 14 \rangle$ into $\langle 15 \rangle$, and $\langle 17 \rangle$ into $\langle 18 \rangle$. On pages 163 and 164, McCawley shows that grammars incorporating a rule such as R are inadequate because of their inability to handle sentences containing both plurals and *respectively*. He then remarks (p. 164):

Thus, in order to explain 141–149, it will be necessary to change the formulation of the *respectively* transformation so as to make it applicable to cases where there is no conjunction but there are plural noun phrases, or rather, noun phrases with set indices: *pluralia tantum* do not allow *respectively* unless they have a set index, so that

156. The scissors are respectively sharp and blunt.

can only be interpreted as a reference to two pairs of scissors and not to a single pair of scissors.

McCawley then goes on to outline what he thinks an adequate rule for stating the correct generalization involved in *respectively* sentences might look like. (Incidentally,

Chomsky describes ⟨20⟩ as the position McCawley *accepts* rather than the one he rejects. He then proceeds to point out, as did McCawley, that such a position is untenable because of its inadequate handling of plurals. On the basis of this, he claims to have discredited McCawley's position in particular and generative semantics in general, though in fact he had described neither.)

McCawley gives a rather interesting argument. He begins his discussion of what an adequate account of the *respectively* phenomena might be like as follows (pp. 164–5):

The correct formulation of the *respectively* transformation must thus involve a set index. That, of course, is natural in view of the fact that the effect of the transformation is to 'distribute' a universal quantifier: the sentences involved can all be represented as involving a universal quantifier, and the result of the *respectively* transformation is something in which a reflex of the set over which the quantifier ranges appears in place of occurrences of the variable which was bound by that quantifier.

He then continues:

For example, the semantic representation of 149 is something like

$$\underset{x \,\epsilon\, M}{\forall} [x \text{ loves } x\text{'s wife}],$$

where M is the set of men in question, and 142 can be assigned the semantic representation

$$\underset{x \,\epsilon\, (x_1,\, x_2)}{\forall} [x \text{ loves } x\text{'s wife}],$$

where x_1 corresponds to *John* and x_2 to *Harry*; the resulting sentence has *those men* or *John and Harry* in place of one occurrence of the bound variable, and the corresponding pronominal form *they* in place of the other occurrence. Moreover, *wife* takes a plural form, since after the *respectively* transformation the noun phrase which it heads has for its index the set of all wives corresponding to any *x* in the set in question. The difference between 142 and 145 is that the function which appears in the formula that the quantifier in 142 binds is one which is part of the speaker's linguistic competence ($f(x) = x$'s wife), whereas that in 145 is one created *ad hoc* for the sentence in question ($f(x_1) = $ Mary, $f(x_2) = $ Alice).

Basically, McCawley is saying the following. The sentences *John and Harry love their respective wives* and *Those men love their respective wives* have certain things in common semantically which can be revealed by a common schema for semantic representation, namely,

$$\langle 15 \rangle \quad \underset{x \,\epsilon\, M}{\forall} [x \text{ loves } x\text{'s wife}]$$

The differences between the two sentences come in the specification of the set M. In the former case, M is given by the enumeration of its elements, *John* and *Harry*, whereas in the latter case, M would be specified by a description of the class (the members each have the properties of being a man). Given that these two sentences have a common form, McCawley notes that 'the result of the *respectively* transformation is something in which a reflex of the set over which the quantifier ranges appears in place of the occurrences of the variable which was bound by that quantifier'. Note that he has not proposed a rule; rather he has made the observation that given the open sentence in the above expression

⟨16⟩ *x loves x's wife*

the surface form of the *respectively* sentences is of essentially this form with the nonanaphoric *x*'s filled in in the appropriate fashion – by a description (*those men*) if the set was defined by a description and by a conjunction (*John and Harry*) if the set was given by enumeration. McCawley does not propose a characterization of the necessary operation. He merely points out that there is a generalization to be stated here, and some such unitary operation is needed to state it.

Now McCawley turns to a more interesting case, namely, *John and Harry love Mary and Alice respectively*. He notes that the form of ⟨15⟩ is not sufficiently general to represent this sentence, and observes that there does exist a more general schema in terms of which this sentence and the other two can be represented.

$$⟨17⟩ \quad \bigvee_{x \, \epsilon \, M} \quad [x \text{ loves } f(x)]$$

In cases like ⟨15⟩, $f(x)$ is specified generally in terms of the variable which binds the open sentence, that is, $f(x) = x$'s wife. In cases like the one mentioned above, the function is specified by an enumeration of its values for each of the members of the set M over which it ranges, that is, $f(x_1) =$ Mary and $f(x_2) =$ Alice, where $M = \{x_1, x_2\}$. As before, McCawley notes that all three respectively sentences have the surface form of the open sentence

⟨18⟩ x loves $f(x)$

where x and $f(x)$ have been filled in as specified. Again there is a generalization to be captured, and McCawley suggests that in an adequate grammar there should be a unitary operation that would capture it, though he proposes no such operation.

He then makes the following conclusion:

I conclude from these considerations that the class of representations which functions as input to the *respectively* transformation involves not merely set indices but also quantifiers and thus consists of what one would normally be more inclined to call semantic representations than syntactic representations.

Recall that he is arguing against the *Aspects* theory, not against the standard theory, and in the following paragraph he goes on to propose what the standard theory but not the *Aspects* theory assumes, namely, that semantic representations are given in terms of phrase-markers. In the *Aspects* theory it is assumed that semantic representations and phrase-markers are very different kinds of objects, and McCawley goes on to suggest that if there is a unitary operation relating semantic objects like ⟨17⟩ to the phrase-markers representing *respectively* sentences *and* if, as Postal has suggested, ordinary conjunction reduction (which is assumed to map phrase-markers onto phrase-markers) is just a special case of *respectively* formation (see McCawley [49], pp. 166–7), then *respectively* formation must be a rule that maps phrase-markers onto phrase-markers, and hence semantic representations like ⟨17⟩ must be given in terms of phrase-markers. If McCawley's argument goes through, then it would follow that the concept of deep structure given by the *Aspects* theory (though not necessarily that of the 'standard theory') would be wrong because of its inadequate concept of semantic representation as amalgamated paths. Chomsky's claim in [10] that McCawley has not proposed anything new in this paper is based on an equivocation in his use of the term 'deep structure' and collapses when the equivocation is removed. With respect to the issue of whether or not semantic representations are given by phrase-markers, the notion of 'deep structure' in the *Aspects* theory is drastically different than the notion of 'deep structure' in the 'standard theory'; thus it should be clear that McCawley's argument, if correct, would indeed provide a Halle-type argument against the *Aspects* notion of 'deep structure', as was McCawley's intent.

But McCawley's proposal is interesting from another point of view as well, for he has claimed that the requirement that one must state fully general rules for relating semantic representations to surface structures may have an effect on the choice of adequate semantic representations. In particular, he claims that an adequate semantic representation for *respectively* sentences must have a form essentially equivalent to ⟨17⟩. Such a claim is open to legitimate discussion, and whether it turns out ultimately to be right or wrong, it raises an issue which is important not only for linguistics but for other fields as well. Take, for example, the field of logic. Logic, before Frege, was the study of the forms of valid arguments as they occurred in natural language. In the twentieth century, logic has for the most part become the study of formal deductive systems with only tenuous links to natural language, although there is a recent trend which shows a return to the traditional concerns of logic.[a] In such logical systems, even the latter sort, the only constraints on what the logical form of a given sentence can be are given by the role of that sentence in valid arguments. From the generative semantic point of view, the semantic representation of a sentence is a representation of its inherent logical form, as determined not only by the requirements of logic, but also by purely linguistic considerations, for example the requirement that linguistically significant generalizations be stated. Thus, it seems to me that generative semantics provides an empirical check on various proposals concerning logical form, and can be said in this sense to define a branch of logic which might appropriately be called 'natural logic'.

The imposition of linguistic constraints on the study of logical form has some very interesting consequences. For example, McCawley (in a public lecture at M.I.T., spring 1968) made the following observations: Performative sentences can be conjoined but not disjoined.

⟨19⟩ (a)   I order you to leave and I promise to give you ten dollars.
 (b)   *I order you to leave or I promise to give you ten dollars.

This is also true of performative utterances without overt performative verbs.

⟨20⟩ (a)   To hell with Lyndon Johnson and to hell with Richard Nixon.
 (b)   *To hell with Lyndon Johnson or to hell with Richard Nixon.

The same is true when conjunction reduction has applied.

⟨21⟩ (a)   To hell with Lyndon Johnson and Richard Nixon.
 (b)   *To hell with Lyndon Johnson or Richard Nixon.

McCawley then observes that universal quantifiers pattern in these cases like conjunctions and existential quantifiers like disjunctions.

⟨22⟩ (a)   To hell with everyone.
 (b)   *To hell with someone.

Ross (personal communication) has pointed out that the same is true of vocatives:

⟨23⟩ (a)   John and Bill, the pizza has arrived.
 (b)   *John or Bill, the pizza has arrived.

⟨24⟩ (a)   (Hey) everybody, the pizza has arrived.
 (b)   *(Hey) somebody, the pizza has arrived.

---

[a] For a small (and arbitrarily chosen) sample of such works see Reichenbach [62], Prior [59]–[61], Geach [18]–[22], Montague [51], Parsons [53], [54], Hintikka [30], [31], Davidson [12], Todd [68], Castañeda [6], Føllesdal [17], Rescher [63], [64], Belnap [3], Keenan [37], and Kaplan [35].

McCawley points out that it is no accident that existential quantifiers rather than universal quantifiers pattern like disjunctions,[a] given their meanings. McCawley argues that if general rules governing the *syntactic* phenomena of ⟨19⟩–⟨24⟩ are to be stated, then one must develop, for the sake of stating rules of *grammar* in general form, a system of representation which treats universal quantifiers and conjunctions as a single unified phenomenon, and correspondingly for existential quantifiers and disjunctions.

Further evidence for this has been pointed out by Paul Postal (personal communication). It is well known that repeated coreferential noun phrases are excluded in conjunctions and disjunctions. Thus, ⟨25⟩ and ⟨26⟩ are ill-formed:

⟨25⟩ (a) *Harry$_i$, Sam, and $\begin{Bmatrix} \text{he}_i \\ \text{Harry}_i \end{Bmatrix}$ are tall.

  (b) *Harry, I, Max, and I are tall.

---

[a] The similarities between universal quantifiers and conjunctions on the one hand and existential quantifiers and disjunctions on the other hand have been recognized at least since Pierce, and various notations have been concocted to reflect these similarities. Thus, universal quantification and conjunction might be represented as in ⟨1a⟩ and existential quantification and disjunction as in ⟨1b⟩:

⟨1⟩ (a) (i)
$$\bigwedge_x fx$$

(ii)
$$\bigwedge P_I, \ldots, P_n$$

(b) (i)
$$\bigvee_x fx$$

(ii)
$$\bigvee P_I, \ldots, P_n$$

Such a similarity of notation makes clear part of the obvious relationship between the quantifier equivalence of ⟨11a⟩ and DeMorgan's Law ⟨11b⟩:

⟨11⟩ (a)
$$\sim \bigwedge_x fx \equiv \bigvee_x \sim fx$$

(b)
$$\sim \bigwedge P_I, \ldots, P_n \equiv \bigvee \sim P_I, \ldots, \sim P_n$$

In the case where x ranges over a finite set, ⟨11a⟩ says the same thing as ⟨11b⟩. Yet despite the similarity in notation, the (ii) cases in ⟨1⟩ are not represented as special cases of the (i)'s and ⟨11a⟩ and ⟨11b⟩ are two distinct statements. There is no known notational system in which the (ii)'s are special cases of the (i)'s and in which ⟨11a⟩ and ⟨11b⟩ can be stated as a single equivalence, though it seems that the same thing is going on in ⟨11a⟩ and ⟨11b⟩.

It should also be clear that McCawley's observations are not unrelated to the fact that performative verbs cannot be negated and still remain performative. ('I do not order you to go' is not an order.) Since a conjunction of negatives is equivalent to the negative of a disjunction, by DeMorgan's Laws, it would seem that an adequate account of McCawley's observations should show how the impossibility of disjunctions of performatives follows from the impossibility of negatives of performatives.

⟨26⟩ (a) *Either Harry$_i$, Sam, or $\begin{Bmatrix} \text{he}_i \\ \text{Harry}_j \end{Bmatrix}$ will win.

   (b) *Either Harry, I, Max, or I will win.

Postal notes that conjunctions with *everyone* and disjunctions with *someone* act the same way:

⟨27⟩ *Everyone and Sam left.
⟨28⟩ *Someone or Sam left.[a]

⟨27⟩ and ⟨28⟩ are excluded if *Sam* is assumed to be a member of the set over which *everyone* and *someone* range, though of course not, if other assumptions are made. Postal points out that this is the same phenomenon as occurs in ⟨25⟩ and ⟨26⟩, namely, conjuncts and disjuncts may not be repeated. If there is to be a single general rule covering all of these cases, then the rule must be stated in some notation which treats quantifiers and conjunctions as a single unified phenomenon.

A further argument along these lines has been provided by Robin Lakoff (personal communication). It has long been known that in comparative constructions a conjunction may be expressed by a disjunction. Thus the meaning of ⟨29⟩ may be expressed by ⟨30⟩:

⟨29⟩ Sam likes lox more than herring and whitefish.
⟨30⟩ Sam likes lox more than herring or whitefish.

(Of course, ⟨30⟩ also has a normal disjunctive reading.) Let us assume that there is a transformation changing *and* to *or* in such comparative constructions. Lakoff notes that the same phenomenon occurs with quantifiers:

⟨31⟩ Sam likes canned sardines more than *everything* his wife cooks.
⟨32⟩ Sam likes canned sardines more than *anything* his wife cooks.

The meaning of ⟨31⟩ can be expressed by ⟨32⟩, in which *any* replaces *every*. Again, as she argues, we have the same phenomenon in both cases, and there should be a single general rule to cover both. Thus, the same transformation that maps *and* into *or* must also map *every* into *some/any*. This can only be done if there is a single unified notation for representing quantifiers and conjunctions.

These facts also provide evidence of the sort brought up by Postal in his discussion of *remind* (see McCawley [49]), evidence that indicates that certain transformations must precede the insertion of certain lexical items. Consider *prefer*, which means *like more than*:

⟨33⟩ (a) Sam likes lox more than herring.
   (b) Sam prefers lox to herring.

As we saw in ⟨29⟩–⟨32⟩ above, *and* optionally changes to *or* and *every* to *any* in the *than*-clause of comparative constructions. The same thing happens in the corresponding place in *prefer* constructions, namely, in the *to*-phrase following *prefer*.

⟨34⟩ Sam prefers lox to herring and whitefish.
⟨35⟩ Sam prefers lox to herring or whitefish.
⟨36⟩ Sam prefers canned sardines to everything his wife cooks.
⟨37⟩ Sam prefers canned sardines to anything his wife cooks.

---

[a] Note, however, that sentences like the following are possible:
Someone and Sam left.
Will everyone or just Sam come to the party?

If *prefer* is inserted for *like-more* after the application of the transformation mapping conjunctions into disjunctions, then the fact that this mapping takes place in the *to*-phrase following *prefer* follows as *an automatic consequence of the meaning of 'prefer'*. Otherwise, this phenomenon must be treated in an *ad hoc* fashion, which would be to make the claim that these facts are unrelated to what happens in comparative constructions.

Further evidence for such a derivation of *prefer* comes from facts concerning the 'stranding' of prepositions. The preposition *to* may, in the general case, be either 'stranded' or moved along when the object of the preposition is questioned or relativized.

⟨38⟩ (*a*) Who did John give the book to.
      (*b*) To whom did John give the book

⟨39⟩ (*a*) Who is Max similar to?
      (*b*) To whom is Max similar?

⟨40⟩ (*a*) What city did you travel to?
      (*b*) To what city did you travel?

The preposition *than*, on the other hand, must be stranded, and may move along only in certain archaic-sounding constructions like ⟨43⟩.

⟨41⟩ (*a*) What does Sam like bagels more than?
      (*b*) *Than what does Sam like bagels more?

⟨42⟩ (*a*) Who is Sam taller than?
      (*b*) *Than whom is Sam taller?

⟨43⟩ ? God is that than which nothing is greater.

The preposition *to* following *prefer* does not work like ordinary occurrences of *to*, but instead works just like *than*; it must be stranded where *than* is stranded and may move along in just those archaic-sounding constructions where *than* may.

⟨44⟩ (*a*) What does Sam prefer bagels to?
      (*b*) *To what does Sam prefer bagels?

⟨45⟩ ? God is that to which I prefer only bagels.

Unless *prefer* is derived from *like-more*, these facts cannot be handled in a unified way, and the correlation must be considered accidental. Such facts seem to provide even more evidence against the *Aspects* conception of deep structure.

Let me conclude with one more example. A rather extraordinary case of a syntactic phenomenon whose environment cannot be given in terms of superficial syntactic structure has been reported on by Labov[a] as occurring in the speech of Harlem residents.[b] The rule involves subject-auxiliary inversion, which moves the first word of the auxiliary to the left of the subject in direct questions (Where *has* he gone?), after negatives (Never *have* I seen such a tall boy?), and in certain other environments. However, in the dialect discussed by Labov, this rule operates not simply in direct questions, but in all requests for information, whatever their surface syntactic structure. The following facts obtain in this dialect.

⟨46⟩ (*a*) *Tell me where he went.
      (*b*) Tell me where did he go.

[a] At the UCLA conference on historical linguistics, January 1969.
[b] The same facts obtain for the author's native (Northern New Jersey) dialect. Curme ([11], p. 183) observes that the same phenomenon occurs in popular Irish English.

⟨47⟩ (*a*) *I want to know where he went.

(*b*) I want to know where did he go.

If there is no request for information, inversion does not occur, even with the same verbs.

⟨48⟩ (*a*) Bill told me where he went.

(*b*) *Bill told me where did he go.

⟨49⟩ (*a*) I know where he went.

(*b*) *I know where did he go.

This phenomenon, though it does not occur in standard English, does have its counterpart there in cases like:

⟨50⟩ Where did he go, I want to know.

⟨51⟩ *Where did he go, I know.

⟨52⟩ Where did he go, tell me.

⟨53⟩ *Where did he go, tell Harry.

In this dialect, even such a late syntactic rule as subject-auxiliary-inversion must be stated not in terms of superficial syntactic structure but in terms of the meaning of the sentence; that is, if the generalization is to be captured the subject-auxiliary inversion rule must have in its structural index the information that the sentence in question describes or is a request for information. This is obviously impossible to state in either the *Aspects* or 'standard' theories.

Given the rather considerable array of evidence against the existence of a level of 'deep structure' following all lexical insertion and preceding all upward-toward-the-surface cyclic rules, it is rather remarkable that virtually no arguments have ever been given *for* the existence of such a level. The arguments that one finds in works of the *Aspects* vintage will usually cite pairs of sentences like 'John ordered Harry to leave' and 'John expected Harry to leave', show that they have very different properties, and claim that such properties can be accounted for by assuming some 'higher' level of representation reflecting the different meanings of the sentences ('order' is a three-place predicate; 'expect' is a two-place predicate). Such arguments do seem to show that a 'higher' or 'more abstract' level of representation than surface structure exists, but they do not show that this level is distinct from the level of semantic representation. In particular, such arguments do not show that any intermediate level of 'deep structure' as defined in the precise sense given above exists. It was simply assumed in *Aspects* that this 'higher' level contained all lexical items and preceded all transformations; no arguments were given.[a]

---

[a] There is however some evidence to indicate that well-formedness conditions, perhaps even phrase-structure conditions, apply at shallow structure (some other fairly late place in a derivation). Consider the fact that adjectives take the verb *be* to their left, very often end in -ED, and, if transitive, must have a preposition precede the object NP (as in *fond of*, *surprised at*, etc.). One might guess, given the *Aspects* theory, that such regularities were to be stated at the level of deep structure. Note, however, that the above regularity generalizes to passive participles. Passive participles take the verb *be* to the left, have a -ED ending, and require a following preposition (*by* in English). If there is a single general condition for both adjectives and passive participles, then this condition cannot be stated in an *Aspects*-type deep structure, since the passive transformation would not have applied at this point. Such a condition could only be stated after the application of the passive transformation. For example, it might be statable at the level of shallow structure.

The only attempt to provide such an argument that I have been able to find is a rather recent one. Chomsky [10] cites the context

⟨54⟩ Bill realized that the bank robber was —

and considers the sentences formed by inserting

⟨55⟩ John's uncle

and

⟨56⟩ the person who is the brother of John's mother or father or the husband of the sister of John's mother or father

into the blank in ⟨54⟩. He claims that these sentences would have different semantic representations, and this claim is based on the further claim that 'what one believes, realizes, etc., depends not only on the proposition expressed, but also on some aspects of the form in which it is expressed'. These claims are not obviously true, and have been disputed.[a] But let us assume for the sake of argument that Chomsky is right in this matter. Chomsky does not propose any account of semantic representation to account for such facts. He does, however, suggest (albeit with reservations and without argument) that such examples refute any theory of grammar without a level of 'deep structure', but not a theory with such a level. He continues:

Do considerations of this sort refute the standard theory as well? The example just cited is insufficient to refute the standard theory, since [⟨54⟩–⟨56⟩] differ in deep structure, and it is at least conceivable that 'realize' and similar items can be defined so as to take account of this difference (p. 198).

But this is a *non sequitur*, even given Chomsky's assumptions. ⟨54⟩–⟨56⟩ would show, under Chomsky's assumptions, only that truth values of sentences depend in part on the particular phonological form in which semantic information is expressed. It does not follow that the correlation between phonological forms and the corresponding semantic information must be made *at a single level of grammar*, and certainly not at a level preceding all cyclic rules. It only follows that such correlations must be made at some point or other in the derivations defined by a grammar. So long as such correlations are made somewhere in the grammar, 'it is at least conceivable that "realize" and similar items can be defined so as to take account of' ⟨54⟩–⟨56⟩. Of

---

[a] The original argument was brought up by B. Mates, 'Synonymity', in *Meaning and Interpretation*, Berkeley, 1950, pp. 201–6. Replies were made by A. Church, 'Intensional Isomorphism and Identity of Belief', *Philosophical Studies*, v, 1954, pp. 65–73 and by W. Sellars, 'Putnam on Synonymity and Belief', *Analysis*, xv, 1955, pp. 117–20. I find it hard to see how one could know the meaning of the expression 'man who has never been married' and the meaning of the expression 'bachelor', and how one could believe the proposition expressed by 'John is a bachelor' without also believing the proposition expressed by 'John is a man who has never been married'. I could see how one, upon hearing these *sentences*, might have perceptual or processing difficulties and so might not be able to figure out which propositions correspond to which sentences. But in the normal sense of *believe* and *realize*, one believes or realizes propositions expressed by sentences, not sentences themselves. There is, of course, another sense of *believe* which means roughly *have blind faith in the truth of*. In this sense, a speaker may fail to understand a sentence at all, but may *believe* whatever proposition is supposed to be expressed by that sentence, e.g. 'God is good'. In this sense, one may 'believe' all and only those sentences which have been uttered in a Southern accent by a military officer over the rank of colonel, who has red hair and is smoking a cigar at the time of the utterance. Thus, any aspect of phonetic or contextual difference may matter for this sense of the meaning of 'believe'. If one wants to consider such facts as part of the theory of meaning, one's theory of meaning will have to coincide with one's theory of language use.

course, this is not saying much, since anything that pretends to be a grammar must at the very least show how semantic information correlates with phonological form. All that Chomsky's argument shows is that his examples do not refute any theory of grammar that defines the correlations between semantic information and phonological form, that is, any theory of grammar at all.

This is, as far as I know, the present state of the evidence in favor of the existence of a level of 'deep structure' which contains all lexical items and precedes all cyclic rules. Since the burden of proof must fall on someone who proposes a level of 'deep structure', there is at present no good reason to believe in the existence of such a level and a number of good reasons not to. This of course does not mean that there is no intermediate level at all between semantic representations and surface structures. In fact, as we have seen, it is not unreasonable to believe that there exists a level of 'shallow structure', perhaps following all cyclic rules. I think it is fair to say that at present there is a reasonable amount of evidence disconfirming the autonomous syntax position and none positively confirming it. This is, of course, not strange, since virtually no effort has gone into trying to prove that the autonomous syntax position is correct.

Going hand-in-hand with the position of autonomous syntax is what we might call the position of 'arbitrary syntax'. We might define the arbitrary syntax position as follows: Suppose there is in a language a construction which bears a meaning which is *not* given simply by the meanings of the lexical items in the sentence (e.g., the English question, or the imperative). The question arises as to what the underlying structure of a sentence with such a 'constructional meaning' should be. Since the meaning of the construction must appear in its semantic representation (P_1, in the standard theory) and since this meaning is represented in terms of a phrase-marker (at least in the standard theory), one natural proposal might be that the 'deep structure' phrase-marker of the sentence contains the semantic representation corresponding to the construction directly. Call this the 'natural syntax' position. The arbitrary syntax position is the antithesis of this. It states that the deep structure corresponding to a construction of the sort described *never* contains the phrase-structure configuration of the meaning of the sentence directly. Instead, the deep structure corresponding to the configuration must contain some arbitrary marker. Consider the English imperative as an example. The meaning of the imperative construction in a sentence like *Come here* must be given in terms of a three-place predicate relating the speaker, the addressee, and a sentence describing the action to be performed, as expressed overtly in the sentence *I order you to come here*. Any adequate theory of semantic representation must say at least this much about the meaning of *Come here*. The arbitrary syntax position would maintain in this case that the 'deep structure' of *Come here* would not contain such a three-place predicate, but would instead contain an arbitrary marker. In recent studies such a marker has been given the mnemonic *IMP*, which may tend to hide its arbitrariness. A good name to reveal its true arbitrary nature would be *IRVING*. Under the arbitrary syntax position, the deep structure of *Come here* would contain *IRVING*.[a]

---

[a] The arbitrary syntax position originated in the practice of Katz and Postal [36], and has been adopted by many investigators since then. It is interesting that Katz and Postal considered a position very close to the natural syntax position and saw that it had advantages over the descriptive practice that they decided to adopt. As they say on page 149, note 9: 'On the basis of ⟨41⟩–⟨44⟩ plus the fact that there are no sentences like **I request that you want to go*, **I request that you hope to be famous*, a case can be made for deriving imperatives syntactically

It is possible to show in certain instances that the arbitrary syntax position is incorrect. One of the most telling arguments to this effect has been given by Robin Lakoff in *Abstract Syntax and Latin Complementation* [44].

The Lakoff argument concerns the distribution of the Latin subjunctive and of two morphemes indicating sentence negation. She begins by considering the Latin sentence

⟨57⟩ Venias. (Form: 2nd person singular subjunctive of *venio*, 'to come')

⟨57⟩ is an example of what is called an 'independent subjunctive' in Latin, and it is at least three ways ambiguous, as shown in ⟨58⟩:

⟨58⟩ (i) Come! I order you to come.
(ii) May you come! I want you to come.
(iii) You may come. It is possible that you will come.

There is also in Latin a dependent subjunctive which functions as a complementizer with verbs of certain meaning classes. Some typical examples are:

⟨59⟩ (i) Impero ut venias. 'I order you to come'
(ii) Volo ut venias. 'I want you to come'
(iii) Potest fieri ut venias. 'It is possible that you will come'.

She argues that the sentences of ⟨59⟩ should have underlying structures roughly like ⟨60⟩:

60 (i)

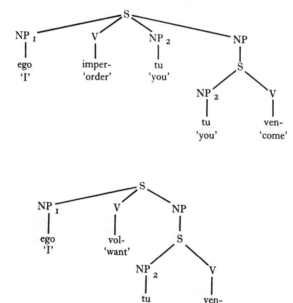

(ii)

from sentences of the form *I Verb*$_{request}$ *that you will Main Verb* by dropping at least the first three elements. This would account not only for ⟨41⟩–⟨44⟩ but also for the facts represented in ⟨35⟩–⟨40⟩. Such a derivation would permit dispensing with *I* and its reading RIM would simplify the semantic component by eliminating one entry. It would also eliminate from the syntax all the necessary heavy selectional restrictions on *I* and the rules that must introduce this element. Although we do not adopt this description here, it certainly deserves further study.'

(iii)

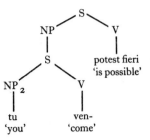

The rule of complementizer-placement will in each case mark the main verb inside a complement structure, that is, one of the form

as subjunctive, given the meaning-class of the next-highest verb.

Lakoff then suggests that it might not be an accident that the sentence of ⟨57⟩, which is odd in that it has a subjunctive in the main clause, has the same range of meanings as the sentences of ⟨59⟩, where the normal rule of subjunctive complementation for verbs of those meanings has applied. She proposes that if it were hypothesized that ⟨57⟩ had three underlying structures just like those in ⟨60⟩, except that in the place of the real predicates *impero*, *volo*, and *potest fieri* there were 'abstract predicates', or nonlexical predicates bearing the corresponding meanings, then the subjunctive in ⟨57⟩ would be derived by the same, independently motivated rule that derives the embedded subjunctives of ⟨59⟩. Since the structures of ⟨60⟩ reflect the meanings of the three senses of ⟨57⟩, such a solution would provide an explanation of why a subjunctive should show up in a main clause with just those meanings. But the arbitrary syntax position would rule out such an explanation. In terms of the arbitrary syntax position, ⟨57⟩ would have three different deep structures, all of them with *venio* as the main verb and with no complement constructions. The difference between the three deep structures could only be given by arbitrary symbols, for example, MARCUS, PUBLIUS, and JULIUS (they might be given mnemonics like IMP, VOL, and POSS, though such would be formally equivalent to three arbitrary names). In such a theory, the three deep structures for ⟨57⟩ would be:

⟨61⟩ (i)

(ii)

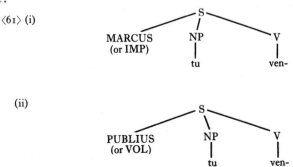

(iii)

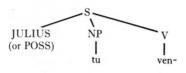

There would then have to be a rule stating that verbs become subjunctive in the environment

⟨62⟩ $\begin{Bmatrix} \text{MARCUS} \\ \text{PUBLIUS} \\ \text{JULIUS} \end{Bmatrix}$ NP —

or equivalently,

$\begin{Bmatrix} \text{IMP} \\ \text{VOL} \\ \text{POSS} \end{Bmatrix}$ NP —

Such a rule would be entirely different than the rule that accounts for the sub-junctives in ⟨59⟩, and having two such different rules is to make the claim that the appearance of the subjunctive in ⟨57⟩ is entirely unrelated to the appearance of the subjunctive in ⟨59⟩, and that the fact that the same endings show up is a fortuitous accident. To claim that it is not a fortuitous accident is to claim that the arbitrary syntax position is wrong in this respect.[a]

Lakoff then goes on to discuss negatives. In Latin, sentence negation may be expressed by one of two morphemes, *nē* and *nōn*. These are, of course, in comple-mentary distribution, and she shows that in sentential complements there is a completely general rule governing their distribution: *nē* occurs in object complements where the verb inside the complement is subjunctive; *nōn* occurs elsewhere, e.g., in nonsubjunctive complements and in subject complements where the main verb is subjunctive. For example, the negatives corresponding to ⟨59⟩ would be ⟨63⟩; the sentences of ⟨64⟩ would be ungrammatical in Latin. *Ut* is optional before *nē*

⟨63⟩ (i)   Imperō (ut) nē veniās   ' I order you not to come '
      (ii) Volō (ut) nē veniās.   ' I want you not to come '
      (iii) Potest fierī ut nōn veniās.   ' It is possible that you won't come '

⟨64⟩ (i)   *Imperō ut nōn veniās.
      (ii) *Volō ut nōn veniās.
      (iii) *Potest fierī (ut) nē veniās.

---

[a] Examples like this also occur in English. Compare:
(i) Ah, to be able to insult my boss!
(ii) Ah, being able to insult my boss!
(i) presupposes that the speaker is not able to insult his boss, while (ii) presupposes that he is able to insult his boss. R. Lakoff has observed that this follows from the fact pointed out by Kiparsky and Kiparsky [39] that factive verbs take poss-ing complementizers, while for-to complementizers occur only with the nonfactive verbs. Since (i) expresses a wish and (ii) a liking, the occurrence of the complementizers could be predicted if abstract verbs bearing those meanings were hypothesized as occurring in the underlying syntactic structure of these sentences. As in the Latin examples, the distribution of the complementizing morphemes depends on the meaning of the understood predicates.

In main clauses without subjunctive main verbs we find *nōn*, not *nē*, just as in the corresponding complement clauses. But in main clauses with subjunctive main verbs, namely, cases like ⟨57⟩, referred to as 'independent subjunctives', we find *both nē and nōn*. That is, both *Nē veniās* and *Nōn veniās* are grammatical in Latin. However, *they do not mean the same thing*. Their meanings are distributed as in ⟨65⟩:

⟨65⟩ (i) Nē veniās. 'Don't come! I order you not to come'
    (ii) Nē veniās. 'May you not come. I want you not to come'
    (iii) Nōn veniās. 'You may not come. It is possible that you won't come'

The distribution of meanings and negatives in ⟨65⟩ corresponds exactly to the distribution in ⟨63⟩. Lakoff argues that this too is no accident. She notes that if the underlying structures for ⟨57⟩ are those of ⟨60⟩, with the appropriate abstract predicates, then the distribution of negatives in ⟨65⟩ follows the ordinary rules specifying the occurrence of *nē* in object complements, not subject complements. This would *explain* the facts of ⟨65⟩. However, if the deep structures of ⟨57⟩ are those of ⟨61⟩, then an entirely different rule would have to be stated, namely: in main clauses with subjunctive main verbs the negative appears as *nē* if either MARCUS (or IMP) or PUBLIUS (or VOL) is present, and *nōn* otherwise. Such a rule would be entirely different from the rule for negatives inside complements, and to have two such different rules is to make the claim that there is no generalization governing the distribution of negatives in ⟨65⟩ and ⟨63⟩, and that the fact that the distribution of forms correlates with the distribution of meanings in these cases is a fortuitous accident. To say that it is not an accident is to say that any theory which rules out abstract predicates and forces such rules to be stated instead in terms of arbitrary markers like MARCUS, IRVING, Q, and IMP is wrong, since linguistically significant generalizations cannot be stated in such a theory.

Now consider what the semantic representations of the three senses of ⟨57⟩ would look like. Sense (i), which is an order, would involve a three-place predicate, specifying the person doing the ordering, the person to whom the order is directed, and the proposition representing the order to be carried out. If one conceives of semantic representation as being given along the lines suggested in Lakoff [43] such a semantic representation would look essentially like ⟨60 i⟩. Similarly, sense (ii) of ⟨57⟩ expresses a desire, and so its semantic representation would have to contain a two-place predicate indicating the person doing the desiring and the proposition expressing the content of the desire. That is, it would essentially have the structure of ⟨60 ii⟩. Sense (iii) of ⟨57⟩ expresses a possibility, and so would have in its semantic representation a one-place predicate containing a proposition, as in ⟨60 iii⟩. In short, if semantic representations are given as in Lakoff [43], then the semantic structures are exactly the structures required for the formulation of general rules introducing the subjunctive complementizer and *nē* in Latin. This seems to me to support the claim that semantic representations are given in terms of syntactic phrase-markers, rather than, say, the amalgamated paths of the *Aspects* theory. It also seems to support the generative semantics position that there is no clear distinction between syntactic phenomena and semantic phenomena. One might, of course, claim that the terms 'syntactic phenomena' and 'semantic phenomena' are sufficiently vague as to render such a statement meaningless. But I think that there are enough clear cases of 'syntactic phenomena' to give the claim substance. It seems to me that if anything falls under the purview of a field called 'syntax' the rules determining the distribution of grammatical morphemes do. To claim that such rules are not 'syntactic phenomena' seems to me to remove all content from term 'syntax'. Thus the general rules determining the distribution of the two negative morphemes *nē* and *nōn* in Latin and the subjunctive morpheme in Latin should be 'syntactic phenomena' if there are any 'syntactic phenomena' at all. Yet as we have seen, the general rules for stating such distributions must be given in terms of structures that reflect the meaning of the sentence rather than the surface grammar of the sentence.

From the fact that the arbitrary syntax position is wrong, it does not necessarily follow that the natural syntax position is right. It is logically possible to hold a 'mixed' position, to the effect that for some such constructions there must be arbitrary markers and for others not. However, since semantic representations for such constructions must be given independently in any adequate theory of grammar, the strongest claim that one could make to limit the class of possible grammars would be to adopt the natural syntax position and to say that there are no arbitrary markers of the sort discussed above. It is conceivable that this is too strong a claim, but it is perhaps the most reasonable position to hold on methodological grounds, for it requires independent justification to be given for choosing each proposed arbitrary marker over the independently motivated semantic representation. To my knowledge, no such justification has ever been given for any arbitrary marker, though of course it remains an open question as to whether any is possible. In the absence of any such justification, we will make the strongest claim, namely, that there exist no such markers.

It should be noted that this is a departure from the methodological assumptions made by researchers in transformational grammar around 1965, when *Aspects* was published. At that time it had been realized that linguistic theory had to make precise claims as to the nature of semantic representations and their relationship to syntax, but existence of semantic representations continued to be largely ignored in syntactic investigations since it was assumed that syntax was autonomous. Moreover, in Katz and Postal [36] and *Aspects*, a precedent had been set for the use of arbitrary markers, though that precedent was never justified. Given such a precedent, it was widely assumed that any deviation from the use of arbitrary markers required justification. However, as soon as one recognizes that (i) semantic representations are required, independent of any assumptions about the nature of grammar, and that they can be represented in terms of phrase-markers and (ii) that the autonomous syntax position is open to serious question, then the methodological question as to what needs to be justified changes. From this point of view, arbitrary markers must be assumed not to exist until they are shown to be necessary, and the autonomous syntax position can no longer be assumed, but rather must be proved.

Lakoff's argument for the existence of abstract predicates was one of the earliest solid arguments not only against arbitrary syntax but also for the claim that the illocutionary force of a sentence is to be represented in underlying syntactic structure by the presence of a performative verb, real or abstract. She has more recently (R. Lakoff [45]) provided strong arguments for the existence of an abstract performative verb of supposing in English. Arguments of essentially the same form have been provided by Ross [66] for the existence of an abstract verb of saying in each declarative sentence of English. Thus, the importance of the argument given above for Latin goes far beyond what it establishes in that particular case, since it provides a form of syntactic argumentation in terms of which further empirical evidence for abstract predicates, performative or otherwise, can be gathered. The basic argument is simple enough:

If the same syntactic phenomena that occur in sentences with certain overt verbs occur in sentences without those verbs, and if those sentences are understood as though those verbs were there, then we conclude (1) a rule has to be stated in the cases where the real verbs occur; (2) since the same phenomenon occurs with the corresponding understood verbs, then there should be a single general rule to cover both cases; (3) since we know what the rule looks like in the case of real verbs, and since the same rule must apply, then the sentences with understood verbs must have a structure sufficiently like that of those with the overt verbs so that the same general rule can apply to both.

If one wishes to avoid the consequences of the Lakoff argument and of other similar arguments, there are two possible ways out. First, one can deny that the form of the argument is valid. Second, one can claim that the generalization is spurious. Let us start with the first way out. Arguments of the above form are central to generative grammar. The empirical foundations of the field rest to a very large extent on arguments of just this form. Take for example the argument that imperative constructions in English are not subjectless in underlying structure, and that they in fact have second person subjects. The sort of evidence on which this claim rests is the following:

⟨66⟩ I shaved $\left\{\begin{array}{l}\text{*me}\\\text{you}\\\text{him}\end{array}\right\}$ $\left\{\begin{array}{l}\text{myself}\\\text{*yourself}\\\text{*himself}\end{array}\right\}$

⟨67⟩ You shaved $\left\{\begin{array}{l}\text{me}\\\text{*you}\\\text{him}\end{array}\right\}$ $\left\{\begin{array}{l}\text{*myself}\\\text{yourself}\\\text{*himself}\end{array}\right\}$

etc.

⟨68⟩ Shave $\left\{\begin{array}{l}\text{me}\\\text{*you}\\\text{him}\end{array}\right\}$ $\left\{\begin{array}{l}\text{*myself}\\\text{yourself}\\\text{*himself}\end{array}\right\}$

. . . .

⟨69⟩ I'll go home, won't $\left\{\begin{array}{l}\text{I}\\\text{*you}\\\text{*he}\end{array}\right\}$

⟨70⟩ You'll go home, won't $\left\{\begin{array}{l}\text{*I}\\\text{you}\\\text{*he}\end{array}\right\}$

etc.

⟨71⟩ Go home, won't $\left\{\begin{array}{l}\text{*I}\\\text{you}\\\text{*he}\end{array}\right\}$

The argument is simple enough. In sentences with overt subjects, we find reflexive pronouns in object position just in case the subjects and objects are coreferential, and nonreflexive pronouns just in case the subjects and objects are noncoreferential. We hypothesize that there is a rule of reflexivization which reflexivizes object pronouns that are coreferential with their subjects. Similarly, in tag questions, we find that as a general principle the pronominal form of the subject of the main clause occurs as the subject of the tag. In imperative sentences we find that a second person subject is understood and that a second person reflexive, but no other, shows up in object position, and that a second person nonreflexive pronoun is excluded in object position. Similarly, we find that the tags for imperative sentences contain second person subjects. We assume that all this is no accident. In order to be able to conclude that imperatives have underlying second person subjects, we need to be able to argue as follows:

If the same syntactic phenomena that occur in sentences with certain overt subjects occur in sentences without those subjects, and if those sentences are understood as though those subjects were there, then we conclude (1) a rule has to be stated in the cases where the overt subjects occur; (2) since the same phenomenon occurs with the corresponding understood subjects, then there should be a single general rule to cover both cases; (3) since we know what the rule looks like in the case of real subjects, and since the same rule must apply, then the sentences with the understood subjects must have a structure sufficiently like that of those with the overt subjects so that the same general rule can apply to both.

If arguments of this form are not valid, then one cannot conclude on the basis of evidence like the above that imperative sentences have underlying second person subjects, or any subjects at all. A considerable number of the results of transformational grammar are based on arguments of just this form. If this form of argument is judged to be invalid, then these results must all be considered invalid. If this form of argument is valid, then the results of R. Lakoff [44], [45] and Ross [66] concerning the existence of abstract performative verbs must be considered valid, since they are based on arguments of the same form. A consistent approach to empirical syntactic evidence requires that abstract performative verbs be accepted or that all results based on arguments of this form be thrown out. Whether much would be left of the field of transformational grammar if this were done is not certain.

Another way out that is open to doubters is to claim that the generalization is spurious. For example, one can claim that the occurrence of a second person reflexive in *Shave yourself* has nothing whatever to do with the occurrence of the second person reflexive pronoun in *You will shave yourself*, and that the fact that the imperative paradigms of ⟨68⟩ and ⟨71⟩ above happen to match up with the second person subject paradigms of ⟨67⟩ and ⟨70⟩ is simply an accident. If the correspondence is accidental, then there is no need to state a single general rule – in fact, to do so would be wrong. And so it would be wrong to conclude that imperative sentences have underlying subjects. If someone takes a position like this in the face of evidence like the above, rational argument ceases.

Let us take another example, the original Halle-type argument. Suppose a diehard structural linguist wanted to maintain that there was a level of taxonomic phonemics in the face of the argument given against such a position by Halle [26] (cf. Chomsky's discussion in Fodor and Katz [16], p. 100). Halle points out that there is in Russian a rule that makes obstruents voiced when followed by a voiced obstruent. He then observes that if there is a level of taxonomic phonemics, this general rule cannot be stated, but must be broken up into two rules, the first relating morphophonemic to phonemic representation (1) obstruents except for c, č and x become voiced before voiced obstruents, and the second relating phonemic to phonetic representation (2) c, č and x become voiced before voiced obstruents. As Chomsky states in his discussion, 'the only effect of assuming that there is a taxonomic phonemic level is to make it impossible to state the generalization'. The diehard structuralist could simply say that the generalization was spurious, that there really were two rules, and that there was no reason for him to give up taxonomic phonemics. In such a case, rational argument is impossible. It is just as rational to believe in taxonomic phonemics on these grounds as it would be to maintain the arbitrary or autonomous syntax positions in the face of the examples discussed above.

It should be noted in conclusion that transformational grammar has in its theoretical apparatus a formal device for expressing the claim that a generalization does not exist. That formal device is expressed by the curly-bracket notation. The curly-bracket notation is used to list a disjunction of environments in which a rule applies. The implicit claim made by the use of this notation is that the items on the list (the elements of the disjunction) do not share any properties relevant to the operation of the rule. From the methodological point of view, curly-brackets are an admission of defeat, since they say that no general rule exists and that we are reduced to simply listing the cases where a rule applies.

Let us take an example. Suppose one wanted to deny that imperative sentences have underlying subjects. One would still have to state a rule accounting for the fact that the only reflexive pronouns that can occur in the same clause as the main verb of an imperative sentence are second-person reflexive pronouns. A natural way to state this is by the use of curly-brackets. Thus, such a person might propose the following reflexivization rule:

⟨72⟩ SD: $\left\{ \begin{matrix} \text{IMP} \\ \text{NP}_1 \end{matrix} \right\}$ — X — NP_2

$\quad\quad$ 1 $\quad\quad$ — 2 — 3
$\quad\quad\quad\quad\quad\quad\quad\quad\quad$ [+REFL]
$\quad\quad$ SC: 1 $\quad\quad$ — 2 — [+PRO]

Conditions: (a) 1 commands 3 and 3 commands 1.
$\quad\quad\quad\quad$ (b) If 1 = IMP, then 3 is second person.
$\quad\quad\quad\quad$ (c) If 1 = NP_1, then NP_1 is coreferential with NP_2.

Such a rule makes a claim about the nonexistence of a generalization. It claims, in effect, that the occurrence of *yourself* in *Shave yourself* has nothing to do with the occurrence of *yourself* in *You will shave yourself*, since the former would arise because of the presence of an IMP marker, while the latter would arise due to the occurrence of a coreferential NP. This, of course, is a claim to the effect that it is an accident that this construction, which allows only second person reflexives, happens to be understood as though it had a second person subject.

This formalism for denying the existence of a generalization could, of course, also be used in the case of the Latin examples cited above. Suppose one wanted to claim that the distribution of *nē* and *nōn* with independent subjunctives had nothing to do with the distribution of *nē* and *nōn* with dependent subjunctives. Then one might attempt (sloppily) to write a rule like the following to account for the occurrence of *nē*:

⟨73⟩ SD: $\left\{ \begin{matrix} \text{NP V (NP)} \\ \left\{ \begin{matrix} \text{IMP} \\ \text{VOL} \end{matrix} \right\} \end{matrix} \right\}$ — [$_\text{S}$ (ut) — *nōn* — NP $\quad$ V $\quad$ X]
$\quad\quad\quad\quad\quad\quad\quad\quad\quad\quad\quad\quad\quad\quad\quad\quad\quad\quad\quad\quad\quad\quad\quad\quad\quad$ [+SUBJ]

$\quad\quad\quad\quad\quad\quad\quad\quad\quad$ 1 $\quad$ — 2 — 3 — $\quad\quad$ 4
$\quad\quad$ SC: $\quad\quad\quad\quad\quad$ 1 $\quad$ — 2 — *nē* — $\quad\quad$ 4

Such a rule would make the claim that the occurrence of *nē* in *Nē venias* has nothing to do with the occurrence of *nē* in *Imperō nē venias*. The former would arise due to the presence of IMP or VOL, while the latter would arise due to the presence of a verb that takes an object complement. The fact that IMP and VOL happen to mean the same thing as verbs that take object complements would have to be considered an accident.

An equivalent formalism for denying the existence of generalizations is the assignment of an arbitrary feature in a disjunctive environment. For example, suppose that instead of deriving *nē* by ⟨73⟩, we broke ⟨73⟩ up into two parts: (i) A rule assigning the arbitrary feature [+IRVING] as follows:

⟨74⟩ SD:
$$\left\{ \begin{matrix} \text{NP} & \text{V} & \text{(NP)} \\ & \left\{ \begin{matrix} \text{IMP} \\ \text{VOL} \end{matrix} \right\} & \end{matrix} \right\} \quad - \quad [_\text{S} \quad (n\bar{o}n) \quad - \quad \underset{[+\text{SUBJ}]}{\text{V}} \quad - \quad \text{X}]$$

$$\begin{array}{ccccccc} & \text{I} & - & 2 & - & 3 & - & 4 \\ \text{SC:} & \text{I} & - & 2 & - & \underset{[+\text{IRVING}]}{3} & - & 4 \end{array}$$

(ii) A rule changing *nōn* to *nē* in sentences with [+IRVING] verbs:

⟨75⟩ SD:
$$\begin{array}{ccccccccc} \text{X} & - & n\bar{o}n & - & \text{NP} & - & \underset{[+\text{IRVING}]}{\text{V}} & - & \text{Y} \\ \text{I} & - & 2 & - & 3 & - & 4 & - & 5 \\ \text{I} & - & n\bar{e} & - & 3 & - & 4 & - & 5 \end{array}$$

One might then claim that one had a completely general rule predicting the occurrence of *nē*: *nē* occurs with [+IRVING] verbs. As should be obvious, assigning arbitrary features in this fashion is just another way of claiming that no general rule can be stated.

An excellent example of the use of such a feature-assignment occurs in Klima's important paper 'Negation in English' (in Fodor and Katz [16]) in the discussion of the feature AFFECT (pp. 313 ff.). Klima notes that two rules occur in certain disjunctive environments (negatives, questions, *only, if, before, than, lest*), which seem vaguely to have something semantically in common. Not being able to provide a precise semantic description of what these environments have in common, he sets up a number of rules which introduce the 'grammatico-semantic feature' AFFECT, whose meaning is not explicated, in just those environments where the rules apply. He then provides a 'general' formulation of the rules in terms of the feature AFFECT. He has, of course, told us no more than that the rules apply in some disjunctive list of environments, those to which the arbitrary feature AFFECT has been assigned. If these environments do have something in common semantically (and I think Klima was right in suggesting that they do), then the general formulation of these rules awaits our understanding of just what, precisely, they do have in common.

Another example is given by Chomsky ([8], p. 39). In his analysis of the auxiliary in English, Chomsky says [the numbering is his]:

⟨29⟩ (ii) Let *Af* stand for any of the affixes *past, S, ϕ, en, ing*.
Let *v* stand for an *M* or *V*, or *have* or *be*..... Then:
$$Af + v \longrightarrow v + Af \ \#$$

Note that '*v*' does not stand for the category 'verb', which is represented by the capital letter '*V*'. '*v*' and '*V*' in this framework are entirely different symbols having nothing whatever in common, as different as '*&*' and '*Z*'.

Af and *v* are arbitrary names and might equally well have been called *SAM* and *PEDRO*, since they have no semantic or universal syntactic significance. In ⟨29ii⟩

Chomsky is stating two rules assigning arbitrary names to disjunctive lists of elements, and one rule which inverts the elements to which those names have been assigned. Since assigning arbitrary features like [AFFECT] is equivalent to assigning arbitrary names, ⟨29 ii⟩ can be stated equivalently in terms of feature-assignment rules. The following rules say *exactly the same thing* as ⟨29 ii⟩:

⟨76⟩

$$\text{SD:} \quad X \; - \; \begin{Bmatrix} M \\ V \\ \textit{have} \\ \textit{be} \end{Bmatrix} \; - \; Y$$

$$\text{SC:} \quad \begin{matrix} 1 & - & 2 & - & 3 \\ 1 & - & 2 & - & 3 \\ & & [+\text{PEDRO}] & & \end{matrix}$$

⟨77⟩

$$\text{SD:} \quad X \; - \; \begin{Bmatrix} \textit{past} \\ S \\ \phi \\ \textit{en} \\ \textit{ing} \end{Bmatrix} \; - \; Y$$

$$\text{SC:} \quad \begin{matrix} 1 & - & 2 & - & 3 \\ 1 & - & 2 & - & 3 \\ & & [+\text{SAM}] & & \end{matrix}$$

⟨78⟩

$$\text{SD:} \quad X \; - \; [+\text{SAM}] \; - \; [+\text{PEDRO}] \; - \; Y$$

$$\text{SC:} \quad \begin{matrix} 1 & - & 2 & - & 3 & - & 4 \\ 1 & - & 3 & - & 2+\# & - & 4 \end{matrix}$$

Such a sequence of rules, in effect, makes the claim that there is no general principle governing the inversion of affixes and verbs. All that can be done is to give two lists, assign arbitrary names to the lists, and state the inversion rule in terms of the arbitrary names.

This, of course, is an absurd claim. There should be a general rule for the inversion of affixes and verbs. What should be said is that there is a universal syntactic category 'verb', of which M, V, *have*, and *be* are instances, and correspondingly, that there is a subcategory 'auxiliary verb' of which M (modals such as *will*, *can*, etc.), *have*, and *be* are instances. One would also need a general characterization of the notion 'affix' rather than just a list. The rule would then state that 'affixes' and 'verbs' invert. But this is not what ⟨29 ii⟩ says. In the *Syntactic Structures* framework, M, V, *have*, and *be* are not all instances of the universal syntactic category 'verb'. They are entirely different entities, having no more in common than *Adverb*, *S*, *windmill*, and *into*. Since the same process applies to all of them, it is impossible to state this process in a nontrivial uniform way unless M, V, *have*, and *be* are instances of the same universal category 'verb', which is what Chomsky's analysis denies.[a] Of course, it is possible to state this process in a trivial uniform way, as in ⟨78⟩.

---

[a] Ross [67] has given an analysis of auxiliaries as instances of the category 'verb', in terms of which general rules mentioning M, V, *have* and *be* can be stated as general rules, rather than as notational variants of lists. Moreover, according to Ross' analysis, the rule of affix-verb inversion can be eliminated in favor of the independently needed rule of complementizer placement.

On the whole, I would say that the major insights of transformational grammar have not come about through embracing rules like $\langle 73 \rangle$, which claim that general statements do not exist, but through eschewing such rules wherever possible and seeking out general principles. Devices like curly-brackets may be useful as heuristics when one is trying to organize data at an early stage of one's work, but they are not something to be proud of. Each time one gives a disjunctive list of the environments where a rule applies, one is making a claim that there are no fully general principles determining the application of that rule. Over the years, curly-brackets have had a tendency to disappear as insights were gained into the nature of the phenomena being described. It may well be the case that they will turn out to be no more than artifacts of the methodological necessity of having to organize data in some preliminary fashion and of the theoretical assumption that syntax is autonomous and arbitrary.

University of Michigan

REFERENCES

[1] Bach, E. and R. Harms, *Universals in Linguistic Theory*, Holt, Rinehart and Winston, 1968.

[2] Bailey, C.-J., B. J. Darden, and A. Davison, *Papers from the Fourth Regional Meeting of the Chicago Linguistic Society*, Linguistics Department, University of Chicago, 1968.

[3] Belnap, Nuel, 'An Analysis of Questions', TM-1287/000/00, Systems Development Corporation, Santa Monica, 1963.

[4] Binnick, R., A. Davison, G. Green, and J. Morgan (eds.), *Papers from the Fifth Regional Meeting of the Chicago Linguistic Society*, Linguistics Department, University of Chicago, 1969.

[5] Carden, Guy, 'Dialects in the Use of Multiple Negations', forthcoming.

[6] Castañeda, Hector-Neri, 'He: A Study in the Logic of Self-consciousness', *Ratio*, December 1966.

[7] Chafe, Wallace, 'Language as Symbolization', *Language*, 1967.

[8] Chomsky, Noam, *Syntactic Structures*, Mouton, 1957.

[9] Chomsky, Noam, *Aspects of the Theory of Syntax*, M.I.T. Press, 1965.

[10] Chomsky, Noam, 'Deep Structure, Surface Structure, and Semantic Interpretation', in this volume.

[11] Curme, George O., *Syntax*, D. C. Heath, 1931.

[12] Davidson, Donald, 'The Logical Form of Action Sentences', in Rescher [63].

[13] DeRijk, Rudolph, 'A Note on Prelexical Predicate Raising', M.I.T. mimeo, 1968.

[14] Emonds, Joseph, 'A Structure-preserving Constraint on NP-movement Transformations', in Binnick et al. [4].

[15] Fillmore, Charles J., 'Types of Lexical Information', in this volume.

[16] Fodor, J. and J. Katz, *The Structure of Language*, Prentice-Hall, 1964.

[17] Føllesdal, Dagfinn, 'Knowledge, Identity, and Existence', *Theoria*, 1967.

[18] Geach, Peter, *Reference and Generality*, Cornell University Press, 1962.

[19] Geach, Peter, 'On Complex Terms', *The Journal of Philosophy*, 7 January 1965.

[20] Geach, Peter, 'Logical Procedures and the Identity of Expressions', *Ratio*, 1965.

[21] Geach, Peter, 'Intentional Identity', *The Journal of Philosophy*, 16 October 1967.

[22] Geach, Peter, 'Quine's Syntactical Insights', *Synthese*, XIX, 1968–9.

[23] Gruber, Jeffrey, 'Functions of the Lexicon in Formal Descriptive Grammars', TM-3770/000/00, Systems Development Corporation, Santa Monica, 1967.

[24] Halle, Morris, *The Sound Pattern of Russian*, Mouton, 1959.
[25] Halliday, M. A. K., 'Some Aspects of the Thematic Organization of the English Clause', RM-5224-PR, Rand corporation, Santa Monica, 1967.
[26] Halliday, M. A. K., 'Notes on Transitivity and Theme in English', *Journal of Linguistics*, 1967.
[27] Harrah, David, 'A Logic of Questions and Answers', *Philosophy of Science*, XXVIII.
[28] Harris, Zellig, 'Discourse Analysis', *Language*, 1952. Also in Fodor and Katz [16].
[29] Harris, Zellig, 'Cooccurrence and Transformation in Linguistic Structure', *Language*, 1957. Also in Fodor and Katz [16].
[30] Hintikka, Jaakko, 'Modality as Referential Multiplicity', *Eripainos Ajatus*, 1957.
[31] Hintikka, Jaakko, *Knowledge and Belief*, Cornell University Press, 1966.
[32] Jackendoff, Ray, 'An Interpretive Theory of Negation', M.I.T. mimeo, 1968.
[33] Jackendoff, Ray, *Some Rules for English Semantic Interpretation*, M.I.T. dissertation, 1969.
[34] Jacobs, R. and P. Rosenbaum, *Readings in English Transformational Grammar*, Blaisdell, 1970.
[35] Kaplan, David, 'Quantifying In', *Synthese*, XIX, 1968–9.
[36] Katz, J. J. and P. M. Postal, *An Integrated Theory of Linguistic Descriptions*, M.I.T. Press, 1964.
[37] Keenan, Edward, *A Logical Base for English*, University of Pennsylvania dissertation, 1969.
[38] King, Harold, 'On Blocking the Rule for Contraction in English', *Linguistic Inquiry*, I, 1970.
[39] Kiparsky, C. A. S. and R. P. V., 'Fact', in this volume.
[40] Lakoff, George, *On the Nature of Syntactic Irregularity*, Indiana University dissertation, 1965; reprinted as *Irregularity in Syntax*, Holt, Rinehart and Winston, 1970.
[41] Lakoff, George, 'Pronouns and Reference', Harvard mimeo, 1968.
[42] Lakoff, George, 'Repartee', *Foundations of Language*, 1970.
[43] Lakoff, George, *Generative Semantics*, Holt, Rinehart and Winston, forthcoming.
[44] Lakoff, Robin, *Abstract Syntax and Latin Complementation*, M.I.T. Press, 1968.
[45] Lakoff, Robin, 'A Syntactic Argument for *Not*-transportation' in Binnick et al., [4].
[46] Lakoff, Robin, 'If's, And's and But's about Conjunction', in Bever and Langendoen, *Studies in Linguistic Semantics*, Holt, Rinehart and Winston, 1971.
[47] McCawley, James D., 'Concerning the Base Component of a Transformational Grammar', *Foundations of Language*, 1968.
[48] McCawley, James D., 'Lexical Insertion in a Transformational Grammar without Deep Structure', in Bailey et al. [2].
[49] McCawley, James D., 'The Role of Semantics in a Grammar', in Bach and Harms [1].
[50] McCawley, James D., 'A Note on Multiple Negations', University of Chicago mimeo, 1968.
[51] Montague, Richard, 'English as a Formal Language', UCLA mimeo, 1968.
[52] Morgan, Jerry L., 'On the Treatment of Presupposition in Transformational Grammar', in Binnick et al. [4].
[53] Parsons, Terence D., 'A Semantics for English', University of Illinois at Chicago Circle mimeo, 1968.
[54] Parsons, Terence D., 'On the Logic of Mass Terms and Quantifiers', University of Illinois at Chicago Circle mimeo, 1968.
[55] Partee, Barbara, 'Negation, Conjunction, and Quantifiers', *Foundations of Language*, to appear.

[56] Perlmutter, David M., *Deep and Surface Structure Constraints in English*, M.I.T. dissertation, 1968.
[57] Postal, Paul, *Crossover Phenomena*, Holt, Rinehart and Winston, 1971.
[58] Postal, Paul, 'On the Surface Verb *Remind*', *Linguistic Inquiry*, I, 1970.
[59] Prior, A. N., *Formal Logic*, Oxford, 1955.
[60] Prior, M. and A. N., 'Erotetic Logic', *Philosophical Review*, LXIV, 1955.
[61] Prior, A. N., *Time and Modality*, Oxford, 1957.
[62] Reichenbach, Hans, *Elements of Symbolic Logic*, Dover, 1947.
[63] Rescher, N. (ed.), *The Logic of Decision and Action*, University of Pittsburgh Press, 1966.
[64] Rescher, N., *The Logic of Commands*, Dover, 1966.
[65] Ross, John R., *Constraints on Variables in Syntax*, M.I.T. dissertation, 1967.
[66] Ross, John R., 'On Declarative Sentences', in Jacobs and Rosenbaum [34].
[67] Ross, John R., 'Auxiliaries as Main Verbs', in Todd [68].
[68] Todd, William, *Philosophical Linguistics*, series I, Great Expectations (Evanston), 1969.

Semantic theory[a]

JERROLD J. KATZ

The semantic component interprets underlying phrase-markers in terms of meaning. It assigns *semantic interpretations* to these phrase-markers which describe messages that can be communicated in the language. That is, whereas the phonological component provides a phonetic shape for a sentence, the semantic component provides a representation of that message which actual utterances having this phonetic shape convey to speakers of the language in normal speech situations.[b] We may thus regard the development of a model of the semantic component as taking up the explanation of a speaker's ability to produce and understand indefinitely many new sentences at the point where the models of the syntactic and phonological components leave off.

If the semantic component is to complete the statement of the principles that provide the speaker with the competence to perform this creative task, it must contain rules that provide a meaning for every sentence generated by the syntactic component. These rules thus explicate the ability a speaker would have were he free of the psychological limitations that restrict him to finite performance. These rules, therefore, explicate an ability to interpret infinitely many sentences. Accordingly, we again face the task of formulating an hypothesis about the nature of a finite mechanism with an infinite output. The hypothesis on which we will base our model of the semantic component is that the process by which a speaker interprets each of the infinitely many sentences is a compositional process in which the meaning of any syntactically compound constituent of a sentence is obtained as a function of the meanings of the parts of the constituent. Hence, for the semantic component to reconstruct the principles underlying the speaker's semantic competence, the rules of the semantic component must simulate the operation of these principles by projecting representations of the meaning of higher level constituents from representations of the meaning of the lower level constituents that comprise them. That is, these rules must first assign semantic representations to the syntactically elementary constituents of a sentence, then, apply to these representations and assign semantic representations to the constituents at the next higher level on the basis of them, and by applications of these rules to representations already derived, produce further derived semantic representations for all higher level constituents, until, at last, they produce ones for the whole sentence.

[a] This paper is reprinted from J. J. Katz, *The Philosophy of Language*, pp. 151–75.
[b] This conception of the semantic component is elaborated in J. J. Katz and J. A. Fodor, 'The Structure of a Semantic Theory', *Language*, XXXIX, 170–210, 1963; reprinted in Fodor and Katz (eds.), *The Structure of Language: Readings in the Philosophy of Language*. It is also elaborated in J. J. Katz, 'Analyticity and Contradiction in Natural Language' (same volume), and in Katz and Postal, *An Integrated Theory of Linguistic Descriptions*

The syntactically elementary constituents in underlying phrase-markers are the terminal symbols in these phrase-markers. Actually, these are the morphemes of the language, but since we are simplifying our discussion, we will consider them to be words. Thus, we may say that the syntactic analysis of constituents into lower level constituents stops at the level of words and that these are therefore the atoms of the syntactic system. Accordingly, the semantic rules will have to start with the meanings of these constituents in order to derive the meanings of other constituents compositionally. This means that the semantic component will have two subcomponents: a *dictionary* that provides a representation of the meaning of each of the words in the language, and a system of *projection rules* that provide the combinatorial machinery for projecting the semantic representation for all supraword constituents in a sentence from the representations that are given in the dictionary for the meanings of the words in the sentence. We will call the result of applying the dictionary and projection rules to a sentence, i.e., the output of the semantic component for that sentence, a *semantic interpretation*. There are, therefore, three concepts to explain in order to formulate the model of the semantic component of a linguistic description: *dictionary, projection rule*, and *semantic interpretation.*[a]

Since the meanings of words are not indivisible entities but, rather, are composed of concepts in certain relations to one another, the job of the dictionary is to represent the conceptual structure in the meanings of words. Accordingly, we may regard the dictionary as a finite list of rules, called 'dictionary entries', each of which pairs a word with a representation of its meaning in some normal form. This normal form must be such that it permits us to represent every piece of information about the meaning of a word required by the projection rules in order for them to operate properly. The information in dictionary entries must be full analyses of word meanings.

The normal form is as follows: first, the phonological (or orthographical) representation of the word, then an arrow, then a set of syntactic markers, and finally, n strings of symbols, which we call *lexical readings*. Each reading will consist of a set of symbols which we call *semantic markers*, and a complex symbol which we call a *selection restriction*. (Here and throughout we enclose semantic markers within parentheses to distinguish them from syntactic markers. Selection restrictions are enclosed with angles.) Thus, a dictionary entry, such as the one below, is a word paired with n readings (for it).

bachelor $\longrightarrow N, N_1, \ldots, N_k$; (i) (Physical Object,) (Living), (Human), (Male), (Adult), (Never Married); (SR).

(ii) (Physical Object), (Living), (Human), (Young), (Knight), (Serving under the standard of another); (SR).

(iii) (Physical Object), (Living), (Human), (Having the academic degree for the completion of the first four years of college); (SR).

(iv) (Physical Object), (Living), (Animal), (Male), (Seal), (Without a mate at breeding time); (SR).

[a] The discussion to follow will explain these three concepts. Roughly, the dictionary stores basic semantic information about the language, the projection rules apply this information in the interpretation of syntactic objects, and semantic interpretations are full representations of the semantic structure of sentences given by the operation of the projection rules.

Each distinct reading in a dictionary entry for a word represents one of the word's senses. Thus, a word with n distinct readings is represented as n-ways semantically ambiguous. For example, the word 'bachelor' is represented as four-ways semantically ambiguous by the above entry.

Just as the meaning of a word is not atomic, a sense of a word is not an undifferentiated whole, but, rather, has a complex conceptual structure. The reading which represents a sense provides an analysis of the structure of that sense which decomposes it into conceptual elements and their interrelations. Semantic markers represent the conceptual elements into which a reading decomposes a sense. They thus provide the theoretical constructs needed to reconstruct the interrelations holding between such conceptual elements in the structure of a sense. It is important to stress that, although the semantic markers are given in the orthography of a natural language, they cannot be identified with the words or expressions of the language used to provide them with suggestive labels. Rather, they are to be regarded as constructs of a linguistic theory, just as terms such as 'force' are regarded as labels for constructs in natural science. There is an analogy between the formula for a chemical compound and a reading (which may be thought of as a formula for a semantic compound). The formula for the chemical compound ethyl alcohol,

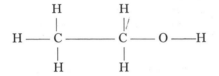

represents the structure of an alcohol molecule in a way analogous to that in which a reading for 'bachelor' represents the conceptual structure of one of its senses. Both representations exhibit the elements out of which the compound is formed and the relations that form it. In the former case, the formula employs the chemical constructs 'Hydrogen molecule', 'Chemical bond', 'Oxygen molecule', etc., while in the latter, the formula employs the linguistic concepts '(Physical Object)', '(Male)', '(Selection Restriction)', etc.

The notion of a reading may be extended so as to designate not only representations of senses of words, but also representations of senses of any constituents up to and including whole sentences. We distinguish between 'lexical readings' and 'derived readings,' but the term 'reading' will be used to cover both. The philosopher's notion of a concept is here reconstructed in terms of the notion of a reading which is either a lexical reading or a derived reading for a constituent less than a whole sentence, while the philosopher's notion of a proposition (or statement) is reconstructed in terms of the notion of a derived reading for a whole declarative sentence.

Just as syntactic markers enable us to formulate empirical generalizations about the syntactic structure of linguistic constructions, so semantic markers enable us to construct empirical generalizations about the meaning of linguistic constructions. For example, the English words 'bachelor', 'man', 'priest', 'bull', 'uncle', 'boy', etc., have a semantic feature in common which is not part of the meaning of any of the words 'child', 'mole', 'mother', 'classmate', 'nuts', 'bolts', 'cow', etc. The first set of words, but not the second, are similar in meaning in that the meaning of each member contains the concept of maleness. If we include the semantic marker (Male) in the lexical readings for each of the words in the first set and exclude it from the lexical entries for each of the words in the second, we thereby express this empirical generalization. Thus, semantic markers make it possible to formulate such generalizations by providing us with the elements in terms of which these generalizations can be stated. Moreover, such semantic generalizations are not restricted to words. Consider the expressions 'happy bachelor', 'my cousin's hired man', 'an orthodox priest I met yesterday', 'the bull who is grazing in the pasture', 'the most unpleasant uncle I have', 'a boy', etc., and contrast them with the expressions 'my favorite child', 'the funny mole on his arm', 'the whole truth', 'your mother', 'his brother's classmate last year', 'those rusty nuts and bolts', 'the cow standing at the corner of the barn', etc. Like the case of the previous sets of words, the members of the first of these sets of expressions are semantically similar in that their meanings share the concept maleness, and the members of the second are not semantically similar in this respect either to each other or to the members of the first set. If the dictionary entries for the words 'bachelor', 'man', 'priest', etc., are formulated so that the semantic marker (Male) appears in each, and if the projection rules assign derived readings to these expressions correctly on the basis of the entries for their words, then the semantic marker (Male) will appear in the derived readings for members of the first set but not in derived readings for members of the second. Again, we will have successfully expressed an empirical generalization about the semantics of a natural language. In general, then, the mode of expressing semantic generalizations is the assignment of readings containing the relevant semantic marker(s) to those linguistic constructions over which the generalizations hold and only those.

Semantic ambiguity, as distinct from syntactic ambiguity and phonological ambiguity, has its source in the homonymy of words. Syntactic ambiguity occurs when a sentence has more than one underlying structure. Phonological ambiguity occurs when surface structures of different sentences are given the same phonological interpretation. Semantic ambiguity, on the other hand, occurs when an underlying structure contains an ambiguous word or words that contribute its (their) multiple senses to the meaning of the whole sentence, thus enabling that sentence to be used to make more than one statement, request, query, etc. Thus, a necessary but not sufficient condition for a syntactically compound constituent or sentence to be semantically ambiguous is that it contain at least one word with two or more senses. For example, the source of the semantic ambiguity in the sentence 'There is no school now' is the lexical ambiguity of 'school' between the sense on which it means sessions of a teaching institution and the sense on which it means the building in which such sessions are held.

But the presence of an ambiguous word is not a sufficient condition for a linguistic construction to be semantically ambiguous. The meaning of other components of the construction can prevent the ambiguous word from contributing more than one of its senses to the meaning of the whole construction. As we have just seen, 'school' is ambiguous in at least two ways. But the sentence 'The school burned up' is not semantically ambiguous because the verb 'burn up' permits its subject noun to bear only senses that contain the concept of a physical object. This selection of senses and exclusion of others is reconstructed in the semantic component by the device referred to above as a selection restriction. Selection restrictions express necessary and sufficient conditions for the readings in which they occur to combine with other readings to form derived readings. Such a condition is a requirement on the content of these other readings. It is to be interpreted as permitting projection rules to combine readings just in case the reading to which the selection restriction applies contains the semantic markers necessary to satisfy it. For example, the reading for 'burn up' will have the selection restriction [(Physical Object)] which permits a reading for a nominal subject of an occurrence of 'burn up' to combine with it just in case that reading has the semantic marker (Physical Object). This, then, explains why 'The school burned up' is unambiguously interpreted to mean that the building burned up. Selection restrictions may be formulated in terms of more complex conditions. We thus allow them to be formulated as Boolean functions of semantic markers. For instance, the selection restriction in the reading for the sense of 'honest' on which it means, roughly, 'characteristically unwilling to appropriate for himself what rightfully belongs to another, avoids lies, deception, etc.' will be the Boolean function [(Human) & $\overline{\text{(Infant)}}$], where the bar over a semantic marker requires that the marker not be present in the reading concerned.

We may note further that semantic anomaly is the limiting case of exclusion by the operation of selection restrictions. Semantically anomalous sentences such as 'It smells itchy' occur when the meanings of the component words of a sentence are such that they cannot combine to form a coherent, directly intelligible sentence. The semantic component of a linguistic description explicates such conceptual incongruence as a case where there are two constituents whose combined meanings are essential to the meaning of the whole sentence but where every possible amalgamation of a reading from one and a reading from the other is excluded by some selection restriction. Constituents below the sentence level can also be semantically anomalous, e.g., 'honest baby' or 'honest worm'. Thus, in general, a constituent $C_1 + C_2$ will be semantically anomalous if, and only if, no reading R_i^1 of C_1 can amalgamate with any reading R_j^2 of C_2, i.e., in each possible case, there is a selection restriction that excludes the derived reading $R_{i,j}$ from being a reading for the constituent $C_1 + C_2$. Consequently, the distinction between semantically anomalous and semantically nonanomalous constituents is made in terms of the existence or non-existence of at least one reading for the constituent. We may observe that the occurrence of a constituent without readings is a necessary but *not* sufficient condition for a sentence to be semantically anomalous. For the sentence 'We would think it queer indeed if someone were to say that he smells itchy', which contains a constituent without readings – which is semantically anomalous – is not itself semantically anomalous.

The projection rules of the semantic component for a language characterize the meaning of all syntactically well-formed constituents of two or more words on the basis of what the dictionary specifies about these words. Thus, these rules provide a reconstruction of the process by which a speaker utilizes his knowledge of the dictionary to obtain the meanings of any syntactically compound constituent, including sentences. But, before such rules can operate, it is necessary to extract the lexical readings for the words of a sentence from the dictionary and make them available to the projection rules. That is, it is first necessary to assign sets of lexical readings to the occurrence of words in the underlying phrase-marker undergoing semantic interpretation so that the projection rules will have the necessary material on which to operate. We will simplify our account at this point in order to avoid technicalities with which we do not need to concern ourselves.[a]

It is at this point that the syntactic information in dictionary entries is utilized. The syntactic markers in the dictionary entry for a word serve to differentiate dif-

---

[a] The assignment of lexical readings is actually accomplished by the same device that introduces the lexical items themselves, but, for the sake of avoiding complications, we shall not go into the formal structure of this device here. Cf. N. Chomsky, *Aspects of the Theory of Syntax*, and J. J. Katz, *Semantic Theory*.

ferent words that have the same phonological (or orthographic) representation. For example, 'store' marked as a verb and 'store' marked as a noun are different words, even though phonologically (or orthographically) these lexical items are identical. Thus, lexical items are distinguished as different words by the fact that the dictionary marks them as belonging to different syntactic categories. Moreover, such pairs of lexical items have different senses so that, in the dictionary, they will be assigned different sets of lexical readings. Thus, it is important to know which of the n different words with the same phonological (or orthographic) representation is the one that occurs at a given point in an underlying phrase-marker because only if we know this can we assign that occurrence of the element in the underlying phrase-marker the right set of lexical readings. Since the underlying phrase-marker will categorize its lowest level elements into their syntactic classes and subclasses, it provides the information needed to decide which of the lexical readings in the dictionary entries for these elements should be assigned to them. All that is required beyond this information is some means of associating each terminal element in an underlying phrase-marker with all and only those lexical readings in its dictionary entry that are compatible with the syntactic categorization it receives in the underlying phrase-marker.

In previous discussions of semantic theory, we suggested a rule of the following sort to fulfill this requirement:

(I) Assign the set of lexical readings R to the terminal symbol σ of an underlying phrase-marker just in case there is an entry for σ in the dictionary that contains R and that syntactically categorizes the symbol σ in the same way as the labeled nodes dominating σ in the underlying phrase-marker.

Iterated application of (I) provides a set of readings for each of the words in an underlying phrase-marker. Thus, for example, given (I) and the fact that there will be two distinct entries for 'store' such that on one it is a noun while on the other it is a verb, 'store' would be assigned a different set of lexical readings in the underlying phrase-marker for 'The store burned up today' and in the underlying phrase-marker for 'The man stores apples in the closet'. When (I) applies no more, the projection rules operate on the underlying phrase-marker.

There is, however, another way of fulfilling this requirement which avoids the postulation of (I). Which way is chosen depends on how the rules in the syntactic component that introduce lexical items into underlying phrase-markers work. If these rules are rewriting rules of the sort illustrated in $(3.1)-(3.10)$,[a] then a rule such as (I) is needed. But if lexical items are inserted into dummy positions in underlying phrase-markers on the basis of a lexical substitution rule in the syntactic component, then we can do away with (I) by allowing the set of readings associated with the lexical item that is substituted to be carried along in the substitution. Since the restrictions on the substitution of a lexical item will be the same as those that are involved in the operation of (I), this procedure will have the same effect as (I). After all lexical substitutions, each lexical item in an underlying phrase-marker will have been assigned its correct set of lexical readings. If it turns out that a lexical substitution rule proves to be better than a set of rewrite rules for introducing lexical items into underlying phrase-markers, the postulation of (I) can be thought of as something that was, but is no longer, necessary to fill a gap that existed in the syntactic component.

---

[a] See J. J. Katz, *The Philosophy of Language*, chapter 3, pp. 126–7. The distinction here is between rules of the sort appearing in *Syntactic Structure* for insertion of lexical items and those appearing in *Aspects of the Theory of Syntax* for the same purpose.

Projection rules operate on underlying phrase-markers that are partially interpreted in the sense of having sets of readings assigned only to the lowest level elements in them. They combine readings already assigned to constituents to form derived readings for constituents which, as yet, have had no readings assigned to them. The decision as to which of the readings assigned to nodes of an underlying phrase-marker can be combined by the projection rules is settled by the bracketing in that phrase-marker. That is to say, a pair of readings is potentially combinable if the two are assigned to elements that are bracketed together. Readings assigned to constituents that are not bracketed together cannot be amalgamated by a projection rule. The projection rules proceed from the bottom to the top of an underlying phrase-marker, combining potentially combinable readings to produce derived readings that are then assigned to the node which immediately dominates the nodes with which the originally combined readings were associated. Derived readings provide a characterization of the meaning of the sequence of words dominated by the node to which they are assigned. Each constituent of an underlying phrase-marker is thus assigned a set of readings, until the highest constituent, the whole sentence, is reached and assigned a set of readings, too.

There is a distinct projection rule for each distinct grammatical relation. Thus, there will be different projection rules in the semantic component of a linguistic description for each of the grammatical relations: subject-predicate, verb-object, modification, etc. The number of projection rules required is, consequently, dictated by the number of grammatical relations defined in the theory of the syntactic component. A given projection rule applies to readings assigned to constituents just in case the grammatical relation holding between these constituents is the one with which this projection rule deals. For example, the projection rule (R 1) is the one designed to deal with modification, viz., the grammatical relation that holds between a modifier and a head, i.e., such pairs as an adjective and a noun, an adverb and a verb, an adverb and an adjective, etc.

(R 1) *Given* two readings,

$R_1: (a_1), (a_2), \ldots, (a_n); (SR_1)$

$R_2: (b_1), (b_2), \ldots, (b_m); (SR_2)$

such that R_1 is assigned to a node X_1 and R_2 is assigned to a node X_2, X_1 dominates a string of words that is a head and X_2 dominates a string that is a modifier, and X_1 and X_2 branch from the same immediately dominating node X,

Then the derived reading,

$R_3: (a_1), (a_2), \ldots, (a_n), (b_1), (b_2), \ldots, (b_m); (SR_1)$

is assigned to the node X just in case the selection restriction (SR_2) is satisfied by R_1. This projection rule expresses the nature of attribution in language, the process whereby new semantically significant constituents are created out of the meanings of modifiers and heads. According to (R 1), in an attributive construction, the semantic properties of the new constituent are those of the head except that the meaning of the new constituent is more determinate than that of the head by itself due to the information contributed by the meaning of the modifier. This rule enforces the selection restriction in the reading of the modifier, thus allowing the embedding of a reading for a modifier into a reading for a head only if the latter has the requisite semantic content.

It should not be supposed that other projection rules are essentially the same as (R 1), in which the derived reading is formed by taking the union of the sets of semantic markers in the two readings on which it operates. If this were the case, there would be no semantically distinct operations to correspond to, and provide an interpretation of, each distinct grammatical relation. Moreover, the semantic interpretations of sentences such as 'Police chase criminals' and 'Criminals chase police' would assign each the same reading for their sentence-constituents, thus falsely marking them as synonymous. That is, since the same sets of semantic markers occur in the lexical readings that are assigned at the level of words in each case, their semantic interpretations would be the same. To indicate the sort of differences that are found among different projection rules, we may consider the projection rules that deal with the grammatical relations of subject-predicate and verb-object in terms of the two examples, 'Police chase criminals' and 'Criminals chase police.'

Neglecting the lexical readings for 'police' and 'criminals' and other senses of 'chase', we may begin by considering the most familiar sense of 'chase':

chase ⟶ Verb, Verb Transitive, ... ; (((Activity) (Nature: (Physical)) of X), ((Movement) (Rate: Fast)) (Character: Following)), (Intention of X: (Trying to catch ((Y) ((Movement) (Rate: (Fast)))); (SR).

But before considering the projection rules that combine senses of 'police' and 'criminal' with the sense reconstructed by this lexical reading, it is necessary to consider the motivation for this lexical reading.

The semantic marker (Activity) distinguishes 'chase' in the intended sense from *state verbs*, such as 'sleep', 'wait', 'suffer', 'believe', etc., and from *process verbs*, such as 'grow', 'freeze', 'dress', 'dry', etc., and classifies 'chase' together with other *activity verbs*, such as 'eat', 'speak', 'walk', 'remember', etc. The semantic marker (Activity) is qualified as to nature by the semantic marker (Physical). This indicates that chasing is a physical activity and distinguishes 'chase' from verbs like 'think' and 'remember' which are appropriately qualified in their lexical readings to indicate that thinking and remembering are mental activities. But (Activity) is not further qualified, so that, *inter alia*, 'chase' can apply to either a group or individual activity. In this respect, 'chase' contrasts with 'mob' which is marked (Type: (Group)) – hence, we can predict the contradictoriness of 'Mary mobbed the movie star (all by herself)', i.e., both one person did something to the star alone and also more than one did that very thing to the star. Also, 'chase' contrasts, in this respect, with 'solo', which is marked (Type: (Individual)) – hence, we can predict that 'They solo in the plane on Friday' has the unique meaning that they each fly the plane by themselves on Friday. Moreover, the semantic marker (Movement) indicates that chasing involves movement of some kind that is left unspecified in the meaning of the verb 'chase', in contrast with a verb such as 'walk', 'motor', or 'swim'. This movement is necessarily fast, as indicated by the semantic marker (Rate: (Fast)) which distinguishes 'chase' from 'creep', 'walk', 'move', etc. Further, this movement has the character of following, as indicated by (Character: (Following)) which distinguishes 'chase' from 'flee', 'wander', etc., and classifies it together with 'pursue', 'trail', etc. Again, for someone to be chasing someone or thing, it is not necessary that the person by moving in any specified direction. This fact is marked by the absence of a qualification on (Movement) of the form (Direction: ( )) which would be needed in the lexical readings for 'descend', 'advance', 'retreat', etc. But it is necessary that the person doing the chasing be trying to catch the person or thing he is chasing, so that 'chase' falls together with the verb 'pursue', on the one hand, and contrast with 'follow', 'trail', etc., on the other. This is indicated by (Intention of X: (Trying to catch $((Y)$ ( )))), where '( )' in the lexical reading under consideration is the semantic marker ((Movement) (Rate: (Fast))) which indicates that the person or thing chased is itself moving fast. It is sometimes held that the person chased must be fleeing from someone or something, but this is mistaken, since a sentence such as 'The police chased the speeding motorist' does not imply that the motorist is fleeing at all. Finally, note that 'chase' is not an achievement verb in the sense of applying to cases where a definite goal is obtained. It is not necessary for the person actually to catch the one he is chasing for him to have actually chased him, as is indicated by the nonanomalousness of sentences such as 'He chased him but did not catch him'. Accordingly, 'chase' contrasts with 'intercept', 'trap', 'deceive', etc.

Now, the slots indicated by the dummy markers 'X' and 'Y' determine, respectively, the positions at which readings of the subject and object of 'chase' can go when sentences containing this verb are semantically interpreted. The projection rule that handles the combination of readings for verbs with readings for their objects embeds the reading for the object of a verb into the Y-slot of the reading for the verb, and the projection rule that handles the combination of readings for predicates and their subjects embeds the reading for the subject into the X-slot of the reading for the predicate. Thus, there are appropriately distinct semantic operations corresponding to the distinct grammatical relations of verb-object and subject-predicate, and the sentence-readings for the two sentences 'Police chase criminals' and 'Criminals chase police' are appropriately different.

Preliminary to defining the concept 'semantic interpretation of the sentence S', we must define the subsidiary notion 'semantically interpreted underlying phrase-marker of S'. We define it as a complete set of pairs, one member of which is a labeled node of the underlying phrase-marker, and the other member of which is a maximal set of readings, each reading in this set being a reading for the sequence of words dominated by the labeled node in question. The set of readings is maximal in the sense that it contains all and only those readings of the sequence of words that belong to it by virtue of the dictionary, the projection rules, and the labeled bracketing in the underlying phrase-marker. The set is complete in the sense that every node of the underlying phrase-marker is paired with a maximal set of readings. In terms of this definition for 'semantically interpreted underlying phrase-markers of S', we can define the notion 'semantic interpretation of an underlying phrase-marker of S' to be (1) a semantically interpreted underlying phrase-marker of S, and (2) the set of statements that follow from (1) by definitions (D1)–(D6) and any further such definitions that are specified in the theory of language,[a] where 'C' is any constituent of any underlying phrase-marker of S,

(D1) C is *semantically anomalous* if and only if the set of readings assigned to C contains no members.

---

[a] These further definitions include all those given in Katz, 'Analyticity and Contradiction in Natural Language', those for concepts such as *presupposition of the question (or imperative) S* and *possible answer to the question S*, which are given in Katz and Postal, *An Integrated Theory of Linguistic Descriptions*, and any other definitions of semantic properties and relations that can be given in terms of configurations of symbols in semantically interpreted underlying phrase-markers.

(D2) C is *semantically unambiguous* if and only if the set of readings assigned to C contains exactly one member.

(D3) C is *semantically ambiguous n-ways* if and only if the set of readings assigned to C contains n members, for n greater than 1.

(D4) C_1 and C_2 are *synonymous on a reading* if and only if the set of readings assigned to C_1 and the set of readings assigned to C_2 have at least one member in common.

(D5) C_1 and C_2 are *fully synonymous* if and only if the set of readings assigned to C_1 and the set of readings assigned to C_2 are identical.

(D6) C_1 and C_2 are *semantically distinct* if and only if each reading in the set assigned to C_1 differs by at least one semantic marker from each reading in the set assigned to C_2.

We can now define the notion 'semantic interpretation of the sentence S' as (1) the set of semantic interpretations of S's underlying phrase-markers, and (2) the set of statements about S that follows from (1) and definitions (D'1)–(D'3) and any further such definitions that are specified in the theory of language, where 'sentence-constituent' refers to the entire string of terminal symbols in an underlying phrase-marker,

(D'1) S is *semantically anomalous* if and only if the sentence-constituent of each semantically interpreted underlying phrase-marker of S is semantically anomalous.

(D'2) S is *semantically unambiguous* if and only if each member of the set of readings which contains all the readings that are assigned to sentence-constituents of semantically interpreted underlying phrase-marker of S is synonymous with each other member.

(D'3) S is *semantically ambiguous n-ways* if and only if the set of all readings assigned to the sentence-constituents of semantically interpreted underlying phrase-markers of S contains n nonequivalent readings, for n greater than 1.

This way of defining semantic concepts, such as 'semantically anomalous sentence in L', 'semantically ambiguous sentence in L', 'synonymous constituents in L', etc., removes any possibility of criticizing the theory of language in which these concepts are introduced on the grounds of circularity in definition of the sort for which Quine had criticized Carnap's attempts to define the same range of concepts. The crucial point about this way of defining these concepts is that their defining condition is stated solely in terms of formal features of semantically interpreted underlying phrase-markers so that none of these concepts themselves need appear in the definition of any of the others. Moreover, these definitions also avoid the charge of empirical vacuity that was correctly made against Carnap's constructions. These definitions enable us to predict semantic properties of syntactically well-formed strings of words in terms of formal features of semantically interpreted underlying phrase-markers. The adequacy of these definitions is thus an empirical question about the structure of language. Their empirical success depends on whether, in conjunction with the set of semantically interpreted underlying phrase-markers for each natural language, they correctly predict the semantic properties and relations of the sentences in each natural language.

The semantic interpretations of sentences produced as the output of a semantic component constitute the component's full account of the semantic structure of the language in whose linguistic description the component appears. The correctness of the account that is given by the set of semantic interpretations of sentences is determined by the correctness of the individual predictions contained in each semantic interpretation. These are tested against the intuitive judgments of fluent speakers of the language. For example, speakers of English judge that the syntactically unique sentence 'I like seals' is semantically ambiguous. An empirically correct semantic component will have to predict this ambiguity on the basis of a semantically interpreted underlying phrase-marker for this sentence in which its sentence-constituent is assigned at least two readings. Also, English speakers judge that 'I saw an honest stone' or 'I smell itchy' are semantically anomalous, as are 'honest stones' and 'itchy smells'. Hence, an adequate semantic component will have to predict these anomalies on the basis of semantically interpreted underlying phrase-markers in which the relevant constituents are assigned no readings. Furthermore, English speakers judge that 'Eye doctors eye blondes', 'Oculists eye blondes', and 'Blondes are eyed by oculists' are synonymous sentences, i.e., paraphrases of each other, while 'Eye doctors eye what gentlemen prefer' is not a paraphrase of any of these sentences. Accordingly, a semantic component will have to predict these facts. In general, a semantic component for a given language is under the empirical constraint to predict the semantic properties of sentences in each case where speakers of the language have strong, clear-cut intuitions about the semantic properties and relations of those sentences. Where there are no strong, clear-cut intuitions, there will be no data one way or the other. But, nevertheless, the semantic component will still make predictions about such cases by interpolating from the clear cases on the basis of the generalizations abstracted from them.

Therefore, the empirical evaluation of the dictionary entries and projection rules of a semantic component is a matter of determining the adequacy of the readings that they assign to constituents, and this, in turn, is a matter of determining the adequacy of the predictions that follow from these readings and the definitions of semantic properties and relations in the theory of language. If the intuitive judgments of speakers are successfully predicted by a semantic component, the component gains proportionately in empirical confirmation. If, on the other hand, the semantic component makes false predictions then, as with any scientific theory, the system has to be revised in such a way that prevents these empirically inadequate predictions from being derivable from the dictionary and projection rules. However, deciding which part(s) of the component have been incorrectly formulated is not something that can always, or even very often, be done mechanically. Rather, it is a matter of further theory construction. We have to make changes which seem necessary to prevent the false predictions and then check out these changes to determine if they avoid the false predictions and do not cause special difficulties of their own. Thus, we are always concerned with the over-all adequacy of the semantic component, and it is on this global basis that we judge the adequacy of its subcomponents. The adequacy of a dictionary entry or a projection rule depends, then, on how well it plays its role within the over-all descriptive system.

Massachusetts Institute of Technology

Explorations in semantic theory[a]

URIEL WEINREICH

1 Introduction

In its current surge of rejuvenation, linguistics faces opportunities long unavailable for reintegrating semantics into the range of its legitimate interests. That sounds associated with meanings are the proper objects of linguistic study has never been denied. But unlike sounds themselves, the meanings with which they are somehow paired are not physically manifest in an utterance or its graphic rendition. And so, when squeamishness about 'mental' data prevailed, particularly in America, the only official role left for the informant was that of an emitter of uninterpreted texts. Semantic material – whether it was imagined to reside in the situational stimulus, or in the speaker's brain, or in another speaker's overt response – was, in any case, inaccessible to observation: it was, in fact, as elusive in the case of living languages as of dead ones. Lexicography carried on in paradisiac innocence without questioning its own theoretical foundations; but for critical linguistics, no theory of meaning was on hand for semantic statements to conform with, and no procedures were in sight for testing semantic claims against finite, surveyable bodies of evidence. As for lay opinions about variant forms – what Bloomfield (1944) dubbed 'tertiary responses' – these were read out of linguistics altogether. 'The linguist's gospel', it was said (Allen 1957: 14), 'comprises every word that proceeds from his informant's mouth – which cannot, by definition, be wrong; but...as a matter of principle, whatever the informant volunteers *about* his language (as opposed to *in* it) must be assumed to be wrong.'

Today many linguists are breaking out of these self-imposed restrictions on the scope of their science. As if fed up with the positivism of the past century, linguists are trying out a bolder stance of much further-ranging accountability. The unedited finite corpus of physical events has lost its paralyzing hold. A concern with informant evaluations of occurring and non-occurring expressions has revolutionized syntax and has opened new perspectives in phonology. 'Tertiary responses' have yielded to systematic sociolinguistic analysis (Labov 1964, 1965), which is liquidating the homogeneity of dialects and the unobservable nature of sound change. The constructs which linguistics is developing in order to deal with language-users' intuitions and attitudes are already more abstract than the constructs of conventional structuralism. When compared with the 'underlying forms' and 'variables' which are the new stock-in-trade of description and dialectology, the conceptual apparatus required for semantics no longer stands out by any glaring degree of non-objectivity.

[a] This paper comprises the first two sections of a paper of the same title in T. A. Sebeok (ed.), *Current Trends in Linguistics* (The Hague, 1966), vol. III. Page, section and chapter numbers given in square brackets after references to later sections of the paper refer to the Sebeok volume.

An aroused curiosity about universals, too (cf. Greenberg 1963), presages a new deal for semantics. For decades every linguistic generalization was hedged with qualifications about the infinite variety of language; the appropriate policy with regard to the definition of 'language' was to reduce it to the bare bones of double articulation and arbitrariness. Today linguists are resuming their search for a far richer characterization of the notion 'human language', and it is apparent that in such a characterization, a detailed statement of the semiotic submechanisms of language will occupy a prominent place.

The fresh opportunities for semantics are, of course, matched by unprecedented requirements regarding the nature of semantic research. Semantics, too, must rise to the Chomskyan challenge of generativeness – the ideal, that is, of fully explicit and literally applicable descriptions. If a semantic theory of a language is to be held accountable for the intuitions of language users as well as for their manifest output, the range of skills for which the theory is responsible must be formulated with great care, and the nature of confirming and disconfirming evidence for theoretical claims in semantics must be determined in advance. Moreover, if semantic theory is to furnish a procedure for evaluating alternative descriptive statements, it must assure the comparability of such statements by specifying the exact form in which they are made.

In several earlier publications (see my Bibliography below) I dealt directly or indirectly with the question of the form of semantic statements as it relates to lexicography. But lexicographic considerations are not the whole story: a full-fledged semantic theory must guarantee that descriptive statements will be compatible with the description of the grammar of a language in all its depth. While the above-mentioned publications were not oblivious to this question, they did not face it in its full complexity. It is the specific purpose of the present paper to explore a semantic theory which might fit into a comprehensive and highly explicit theory of linguistic structure.[a]

A recent attempt to achieve this goal was made by Katz and Fodor (1963). The immediate impact of their work testifies to its importance: it was quickly incorporated into an integrated theory of linguistic descriptions (Katz and Postal 1964) and became a major stimulus for fundamental revisions in transformational syntactic theory (Katz and Postal 1964; Chomsky 1965).[b] In a number of ways, however, the proposals of Katz and Fodor (hereinafter KF) are unsatisfactory. Since an analysis of these inadequacies is a prerequisite to the development of alternative proposals,[c] the first portion of this paper is devoted to a critical discussion of KF [ch. 2]. The next part [ch. 3] develops, in outline, a semantic theory which would contribute to a more satisfying conception of linguistics as a whole. The concluding remarks [ch. 4] compare the two approaches.

[a] In preparing this paper I have profited greatly from discussions with Erica C. Garcia. A number of mistakes were caught by Edward H. Bendix and William Labov; both of them contributed many useful suggestions for improvement – some, unfortunately, too far-reaching to be incorporated in the present version. The research on which the article is based was supported in part by Public Health Service grant MH 05743 from the National Institute of Mental Health.

[b] My indebtedness to Chomsky, which is evident throughout the article, covers far more, of course, than his loan of a pre-publication version of his 1965 monograph. Since his *Aspects of the Theory of Syntax* was not yet in print when the present paper went to press, page references had to be dispensed with; only the most important ones were added in the proofs.

[c] Space limitations prohibit more than passing references to the older literature.

2 The semantic theory KF: a critical analysis

2.1 SCOPE AND COMPONENTS

According to KF, the goal of a semantic theory is to account for certain aspects of human competence with respect to a language. This competence involves the production and understanding of expressions abstracted from the non-verbal setting in which they occur. The domain thus staked out for semantics is relatively narrow; it does not include the human ability to name objects correctly, to distinguish synthetically true statements from synthetically false ones, or to perform other referential tasks. In this respect KF follows the tradition of linguistics and saves the investigation of meaning from the sterile 'reductions' urged upon semantics in recent decades by philosophers of other sciences (cf. Wells 1954; Weinreich, 1968).

But what aspect of linguistic competence is a semantic theory to account for? Programmatically, it aims at nothing less than the ability to interpret sentences.

A semantic theory describes and explains the interpretative ability of speakers [1] by accounting for their performance in determining the number and content of the readings of a sentence; [2] by detecting semantic anomalies; [3] by deciding upon paraphrase relations between sentences; and [4] by marking every other semantic property or relation that plays a role in this ability. [KF, p. 176; bracketed numbers supplied.]

On closer examination, the subject matter of KF turns out to be far less broad. For example, paraphrase relations [3] are touched upon only in passing,[a] and no 'other semantic property or relation', whose explication is promised under [4], is actually considered in the article.[b] Moreover, as will be shown below, the theory cannot deal adequately with the *content* of readings of a sentence. In actuality, KF is concerned with an extremely limited part of semantic competence: the detection of semantic anomalies and the determination of the number of readings of a sentence.

To carry out this goal, KF visualizes a semantic description of a language as consisting of two types of components: a dictionary and a set of 'projection rules'. The dictionary contains statements of meanings of words (or some other suitable entries), each entry being in principle polysemous. The projection rules[c] specify how the meanings of words are combined when the words are grammatically constructed, and, in particular, how the ambiguity of individual words is reduced in context. To express the matter schematically, let us imagine a sentence consisting of words A + B + C. The dictionary gives two meanings for A, three for B, and three for C. By multiplying $2 \times 3 \times 3$, we calculate that the sentence should be 18 ways ambiguous. In fact it turns out, let us say, that the sentence is only three ways ambiguous. The major function of the projection rules is to account for the reduction of the ambiguity

---

[a] Katz and Fodor (1963: 195). The notion 'paraphrase' perhaps remained undeveloped because Fodor (1961) and Katz (1961) disagreed about it. See also the discussion of ⟨86⟩ and ⟨87⟩ below [pp. 448–9].

[b] Katz alone resumed the study of additional relations in a subsequent paper (Katz 1964b). See §3.441 [pp. 446 ff.] for further comments.

[c] This awkward coinage is based on an allusion to the fact that the speaker, to know a language, must *project* 'the finite set of sentences he has fortuitously encountered to the infinite set of sentences of the language' (KF, p. 171). Since every rule of grammar is involved in the projection mechanism, the term fails to identify its specific content. We suggest no terminological replacement, however, since we aim at a more radical revision of the theory [see esp. §3.51].

from 18 to 3. The limiting case is one in which there is *no* interpretation of a sentence, even though its components in isolation do have at least one, and possibly more meanings, each.

In an idealized semiotics signs are regarded as combining expressions and meanings in one-to-one correspondence; the polysemy of words in natural languages is but an awkward deviation from the model. KF conforms to the trend of modern lexicology to eschew this prejudice and to seize on polysemy as a characteristic, and even the most researchable, aspect of natural languages (cf. Weinreich 1963 b). KF is also comfortably traditional with regard to the role of context: the idea of contextual resolution of ambiguities has after all been a commonplace with Neogrammarian as well as descriptivist semanticists.[a] But in assigning this concept so central a place, KF is guilty of two errors. In the first place, it takes no cognizance of the obvious danger that the differentiation of submeanings in a dictionary might continue without limit. (We return to this question in §2.25.) In the second place, one would think, a scientific approach which distinguishes between competence (knowledge of a language) and performance (use of a language) ought to regard the automatic disambiguation of potential ambiguities as a matter of hearer performance.[b] The KF theory only accounts for the construal of unambiguous (or less ambiguous) wholes out of ambiguous parts; it does not undertake to explain, and could not explain, sentences that are *meant* by the speaker *to be* ambiguous. In particular, it cannot represent the ambiguity between a grammatical and a deviant sentence (e.g. *She is well groomed* ' 1. combed and dressed; 2. provided with grooms '), since the theory contains a component (the projection rules) which *automatically* selects the fully grammatical interpretation, provided there *is* one. Thus the theory is too weak to account for figurative usage (except the most hackneyed figures) and for many jokes. Whether there is any point to semantic theories which are accountable only for special cases of speech – namely, humorless, prosaic, banal prose – is highly doubtful (Weinreich 1963 a: 118).

Semantics can take a page out of the book of grammar. The grammar of a language, too, produces ambiguous expression (e.g. *Boiling champagne is interesting, He studied the whole year, Please make her dress fast*). But each such sentence, ambiguous at its surface, corresponds to two distinct, unambiguous deep structures. Its ambiguity arises from the existence of transformational rules which produce identical surface results from different deep sources, and from the simultaneous existence of words which can function in dual syntactic capacities (e.g. *boil* as both a transitive and an intransitive verb). But grammatical theory is *not* required to explain how a hearer of such ambiguous expressions guesses which of two deep structures is represented by a given occurrence of a surface structure, nor is the goal of grammatical theory limited to the calculation of such ambiguities. The preoccupation of KF with disambiguation appears to be an entirely unjustified diversion of effort. Semantic theories can and should be so formulated as to guarantee that deep structures (including their lexical components) are specified as unambiguous in the first place [see §3.1] and proceed from there to account for the interpretation of a complex expression from the known meanings of its components.

---

[a] Although one would not know it from reading KF. See Weinreich (1963 a: n. 48) for references.

[b] This was first pointed out to me by Edward H. Bendix.

2.2 DICTIONARY ENTRIES

If dictionary entries are to be the objects of any formal calculation (by some apparatus such as the 'projection rules'), they must be given in a carefully controlled format.[a] KF proposes the following normal form: every entry contains (i) a syntactic categorization, (ii) a semantic description, and (iii) a statement of restrictions on its occurrences. The syntactic categorization (i) consists of a sequence of one or more 'syntactic markers' such as 'Noun', 'Noun Concrete', 'Verb → Verb Transitive', etc. The semantic description (ii) consists of a sequence of semantic markers and, in some cases, a semantic distinguisher. Semantic markers contain those elements of the meaning of an entry for which the theory is accountable. The semantic markers constitute those elements of a meaning upon which the projection rules act to reduce ambiguity; they are, accordingly, the elements in terms of which the anomalous, self-contradictory, or tautologous nature of an expression is represented. Polysemy of an entry appears in the normal form as a branching in the path of semantic markers (SmM), e.g.:

$$\langle 1 \rangle \qquad\qquad \mathrm{SmM}_1 \to \mathrm{SmM}_2 {\Large\langle} \begin{array}{l} \nearrow \mathrm{SmM}_3 \\ \searrow \mathrm{SmM}_4 \end{array}$$

Correspondingly, reduction of ambiguity is represented as the selection of a particular path (e.g. $\mathrm{SmM}_1 \to \mathrm{SmM}_2 \to \mathrm{SmM}_4$) out of a set of alternatives. The distinguisher contains all the remaining aspects of the meaning of an entry – those, in effect, which do not figure in the calculation of ambiguity reduction. The selection restriction (iii) at the end of an entry (or, in the case of polysemous entries, at the end of each of its alternative paths) specifies the context in which the entry may legitimately appear. The context of an entry W is described in terms of syntactic and semantic markers, either positively (i.e. markers which *must* appear in the paths of entries in the context of W) or negatively (i.e. markers which *may not* appear in the paths of context entries). But the selection restriction does not, of course, refer to distinguishers, since these, by definition, play no role in the distributional potential of the word.

Somewhere in the generative process, the words of a sentence would also have to have their phonological form specified. The omission of such a step in KF is presumably due to reliance on an earlier conception of linguistic theory as a whole which did not anticipate a semantic component and in which the grammar included, as a subcomponent, a lexicon that stated the phonological form and the syntactic category of each word. In an integrated theory, the existence of a lexicon separate from the dictionary is a vestigial absurdity, but one which can be removed without difficulty.[b] We therefore pass over this point and take up the KF conception of normal dictionary entries in detail.

2.2.1 *Form of syntactic markers*

The theoretical status of the syntactic markers in KF is not clear. It is probably fair to understand that the function of the syntactic marker SxM_i is to assure that all

---

[a] On canonical forms of lexicographic definition, see Weinreich (1962: 31 ff.).

[b] Katz and Postal (1964: 161) postulate a 'lexicon' (distinct from the dictionary!) which presumably specifies the phonological form of morphemes. Chomsky (1965) has the underlying phonological shape of morphemes specified by the same component – the lexicon – as the syntactic features.

entries having that marker, and only those, can be introduced[a] into the points of a syntactic frame identified by the category symbol SxM_i. In that case the set of syntactic markers of a dictionary would be just the set of terminal category symbols, or lexical categories (in the sense of Chomsky 1965), of a given grammar.

It is implied in KF that this set of categories is given to the lexicographer by the grammarian. Actually, no complete grammar meeting these requirements has ever been written; on the contrary, since KF was published, a surfeit of arbitrary decisions in grammatical analysis has led syntactic theorists (including Katz) to explore an integrated theory of descriptions in which the semantic component is searched for motivations in setting up syntactic subcategories (Katz and Postal 1964; Chomsky 1965). But before we can deal with the substantive questions of *justifying* particular syntactic features, we ought to consider some issues of presentation – issues relating to the form of these features.

In general, the size and number of lexical categories (traditionally, parts-of-speech) depends on the depth or 'delicacy' of the syntactic subcategorization. (The term 'delicacy' is due to Halliday 1961.) Suppose a category A is subcategorized into B and C. This may be shown superficially by a formula such as $\langle 2i \rangle$; a Latin example would be 'Declinable' subcategorized into 'Noun' and 'Adjective'. However, the specific fact of subcategorization is not itself exhibited here. It would be explicitly shown by either $\langle 2ii \rangle$ or $\langle 2iii \rangle$. In $\langle 2ii \rangle$, $A_1 = B$ and $A_2 = C$; in $\langle 2iii \rangle$ $[+F]$ and $[-F]$ represent values of a variable feature[b] which differentiates the species B and C of the genus A. (An example would be 'Nomen' subcategorized into 'Nomen substantivum' and 'Nomen adjectivum'.) The feature notation has been developed in phonology and has recently been applied to syntax by Chomsky (1965).[c]

$$\langle 2 \rangle \text{ (i)} \quad A \to \begin{Bmatrix} B \\ C \end{Bmatrix} \qquad \text{(ii)} \quad A \to \begin{Bmatrix} A_1 \\ A_2 \end{Bmatrix} \qquad \text{(iii)} \quad A \to A + \begin{Bmatrix} [+F] \\ [-F] \end{Bmatrix}$$

Single, global syntactic markers would correspond to implicit notations, such as $\langle 2i \rangle$; sequences of elementary markers, to a feature notation such as $\langle 2iii \rangle$. The KF approach is eclectic on this point. The sequence of markers 'Verb → Verb transitive' for their sample entry *play* corresponds to principle $\langle 2iii \rangle$; the marker 'Noun concrete' seems to follow the least revealing principle, $\langle 2i \rangle$.[d] To be sure, the examples in KF are intended to be only approximate, but they are surprisingly anecdotal in relation to the state of our knowledge of English syntax; what is more, they are mutually inconsistent.

A revealing notation for syntax clearly has little use for global categories, and we may expect that for the syntactic markers of dictionary entries in normal form, too, only a feature notation would be useful. In our further discussion, we will assume

[a] By a rewrite rule, according to early generative grammar (e.g. Lees 1960), or by a substitution transformation according to Chomsky (1965).

[b] A feature symbol differs from a category symbol in that it does not by itself dominate any segment of a surface string under any derivation; in other words, it is never represented by a distinct phonic segment.

[c] I am informed by Chomsky that the idea of using features was first proposed by G. H. Matthews about 1957, and was independently worked out to some extent by Robert P. Stockwell and his students.

[d] We interpret 'Noun concrete' as a global marker since it is not shown to be analyzed out of a marker 'Noun'.

that on reconsideration KF would have replaced all syntactic markers by sequences expressing subcategorization.

Suppose, then, we conceive of a syntactic marker as of a sequence of symbols (the first being a category symbol and the others, feature symbols). Suppose the dictionary contains entries consisting of partly similar strings, e.g. ⟨3⟩:

⟨3⟩ (i) *land* → Noun → Count → nonAnimate → ... (= 'country')
  land → Noun → nonCount → Concrete → ... (= 'real estate')
  (ii) *cook* → Noun → Count → Animate → ...
  cook → Verb → ...

This partial similarity could be shown explicitly as a branching sequence, as in ⟨4⟩:

⟨4⟩ (i) *land* → Noun ⟨ Count → nonAnimate → ...
                        nonCount → Concrete → ...

  (ii) *cook* ⟨ Noun → Count → Animate → ...
               Verb → ...

But this notation, proposed by KF, does not discriminate between fortuitous homonymy and lexicologically interesting polysemy, for it would also produce entries like ⟨5⟩:

⟨5⟩ *rock* ⟨ Noun → Count → nonAnimate → ...   (= 'stone')
             Verb → ...   (= 'move undulatingly')

KF would therefore have to be extended at least by a requirement that conflated entries with branching of syntactic markers be permitted only if there is a reconvergence of paths at some semantic marker; only, that is, if the dictionary entry shows explicitly that the meanings of the entries are related as in ⟨6⟩.[a] But such

⟨6⟩ *adolescent* ⟨ Noun ⟩ (Human) (nonAdult) (nonChild)
                   Adjective

makeshift remedies, feasible though they are, would still fail to represent class shifting of the type *to explore – an explore, a package – to package* as a (partly) productive process: the KF dictionary would have to store all forms which the speakers of the language can form at will. We return in §3.51 [pp. 455 ff.] to a theory capable of representing this ability adequately.

2.2.2 *Semantic vs. syntactic markers*

The presence of syntactic and semantic markers with identical names (Male, Female, Abstract, etc.) offers strong *prima facie* ground for the suspicion that the distinction between semantic and syntactic markers – a distinction theoretically crucial for KF (pp. 208 ff.; see also §4.1 [pp. 467 ff.]) – is ill-founded. Let us first compare the functions of these putatively separate types of element in the theory.

The function of semantic markers in KF is to express those components of the total meaning of a dictionary entry for which the theory is accountable; more specifically, they express those elements of a meaning of a word upon which the projection rules operate. Hence the semantic markers of words are those elements which, after

---

[a] Even so, there are serious problems in specifying that the point of reconvergence be sufficiently 'low', i.e. that *file* 'record container' and *file* 'abrading instrument', for example, are mere hononyms even though both perhaps share a semantic marker (Physical Object). This important problem is not faced in KF, even though the examples there are all of the significant kind (polysemy). See Weinreich (1963a: 143) for additional comments.

being suitably amalgamated by the projection rules (see §2.3), yield an interpretation of the sentence which is unambiguous, *n* ways ambiguous, or anomalous. A general criterion of economy would presumably require that there be as few markers (primes) as possible; hence, the analyst should aim to add markers only when failure to do so would result in a failure to mark ambiguities or anomalies of sentences. The general principle would seem to be that no semantic marker should appear in the path of any dictionary entry unless it also appears in the selection restrictions of at least one other entry.[a]

Let us take an example. Suppose the difference between the nouns *ball*$_1$ 'gala affair' and *ball*$_2$ 'spherical object' were formulated in terms of distinguishers. Then the theory could not explain the ambiguity of sentence $\langle 7i \rangle$, nor could it mark the anomaly of $\langle 7ii \rangle$ or $\langle 7iii \rangle$. Hence, the dictionary must be revised by the addition of suitable semantic markers, such as (Event) and (Object). And, as we have seen in $\langle 2 \rangle$ above, the addition of a marker (= feature) is equivalent to a step in subcategorization.

$\langle 7 \rangle$ (i)   *I observed the ball.*
      (ii)   *I attended the ball*$_2$.
      (iii)   *I burned the ball*$_1$.

Now what leads a linguist to increase the delicacy of subcategorization in syntax?[b] The reasons turn out to be precisely the same as those for semantics: a subcategorization step is taken if failure to do so would make the grammar generate (*a*) ill-formed expression or (*b*) ambiguous sentences.

(*a*) Suppose a grammar of English were to contain the following rules:

$\langle 8 \rangle$ (i) $S \rightarrow NP + VP$
      (ii) $VP \rightarrow V + (NP)$
      (iii) $NP \rightarrow$ *Tom, Bill*
      (iv) $V \rightarrow$ *liked, waited*

These rules would generate not only *Tom liked Bill* and *Tom waited*, but also *Tom liked and *Tom waited Bill. To prevent the latter, undesirable result, we must reformulate rules $\langle 8ii \rangle$ and $\langle 8iv \rangle$ to show a subcategorization, e.g.:

$\langle 8' \rangle$ (ii) $VP \rightarrow \begin{Bmatrix} V_t + NP \\ V_i \end{Bmatrix}$
      (iv) $V_t \rightarrow$ *liked*
         $V_i \rightarrow$ *waited*

The addition of the syntactic markers *t* and *i* corresponds in form and motivation to the addition of (Event) and (Object) in preventing (= marking as anomalous) such expressions as $\langle 7iii \rangle$ 'I burned the gala affair'.

(*b*) Suppose an English grammar were to allow VPs consisting of Copula + Nomen, and *fat* were a 'Nomen'. This would permit such sentences as *This substance is fat* without exhibiting their ambiguity. One reason[c] for subcategorizing

---

    [a] The marking of ambiguities, without regard to their resolution, is not a sufficient criterion, since an ambiguity can be marked more economically by a branching of the distinguishers after the last semantic marker. Moreover, the only speaker skill for which KF is made accountable is the interpretation of sentences, not the critique of dictionaries. KF alludes to 'maximization of systematic economy' but does not elaborate on this intriguing idea.

    [b] The question of subcategorization, even in pre-transformational syntax, has received little attention. See Xolodovič (1960).

    [c] There are other reasons, too. For instance, we must also show the bifurcation of *fat* in *it looks fat*$_1$ and *it looks like fat*$_2$ in order to prevent the formation of *This one is fat$_2$*ter* and **We have to select the right fat*$_1$*s*.

'Nomen' into Noun and Adjective would be to mark this ambiguity. This is exactly comparable to the introduction of markers for exhibiting the ambiguity of ⟨71⟩.

The typical examples of syntactic ambiguity are of a 'bifocal' kind, e.g. *The statistician studies the whole year* or *He left his car with his girl friend*. That is to say, if an insufficiently delicate subcategorization, as in ⟨9⟩, were to be brought to a degree of

⟨9⟩

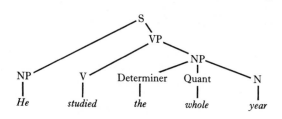

delicacy at which the ambiguity were to be exhibited, *two* interconnected revisions would have to be made: Verbs would have to be divided into transitive and intransitive, and NPs would correspondingly have to be divided into objects, dominated by VP, and adverb-like Temporals. The great rarity of unifocal ambiguities in grammar – even in languages with very poor morphology (cf. Chao 1959) – is itself an interesting comment on the general design of language. However, unifocal syntactic ambiguities do exist, as do bifocal semantic ones.[a]

KF asks if 'the line between grammatical and semantic markers can be drawn in terms of the theoretical function they perform' (p. 209), and comes to the conclusion that a criterion is available: 'grammatical markers mark the formal differences on which the distinction between well-formed and ill-formed strings of morphemes rests, whereas semantic markers have the function of giving each well-formed string the conceptual content that permits it to be represented in terms of the message they communicate to speakers in normal situations'.[b] But this conclusion only begs the question; as we have seen, the distinction between grammatical and semantic anomalies is still unexplained. Instead of being dispelled, the confusion that 'has been generated in the study of language by the search for a line between grammar and semantics' is only increased by the disguised circularity of the KF argument.

The only issue in KF on which the 'metatheoretical' distinction between syntactic and semantic markers has a substantive bearing is the problem of markers of both kinds which 'happen to' have the same names. It is proposed, for example, that *baby* be marked semantically as (Human), but grammatically as nonHuman (hence it is pronominalized by *it*), whereas *ship* is treated in the reverse way. The problem, however, has been solved in a purely grammatical way since Antiquity in terms either of mixed genders or of double gender membership.[c] Besides, it is unlikely that the marking of *baby* as (Human) would solve any semantic problems, since most things predicable of humans who are not babies could no more be predicated of babies than of animals (i.e. nonHumans): is *The baby hates its relatives* any less odd than *The kitten hates its relatives*? Most importantly, however, this solution fails to represent *as a productive process* the reference (especially by men) to lovingly handled objects

---

[a] Taking *throw* as ambiguous ('1. to hurl; 2. to arrange ostentatiously') and given the polysemy of *ball* discussed above, note the bifocal ambiguity of *She threw a ball*. Another example is *He arranged the music* (Weinreich 1963a: 143).

[b] Quoted from p. 518 of the revised version (1964).

[c] E.g. Hockett (1958: 232 ff.). On genders, see also n. 100 below [p. 472].

by means of *she*. The patent fact is that any physical object can in English be referred to by *she* with a special semantic effect. (For a suggestion as to how such a process may be incorporated in a theory, see §3.51 below [pp. 445 ff.].)

To summarize, we have seen that the KF distinction between syntactic and semantic markers is not based, as claimed, on the functions of these entities. The only possibility remaining is that the distinction is based on the content. For example, semantic markers may be claimed to have some denotative content, whereas the syntactic markers would have none. But this would run counter to the spirit of the whole enterprise, which is to explicate intralinguistic semantic phenomena without resort to extraverbal correlations (§2.1). We can only conclude that if formal linguistics is not to be renounced altogether, the distinction between semantic and syntactic markers claimed by KF is non-existent.

2.2.3 *Semantic markers vs. semantic distinguishers*

A desire to analyze a global meaning into components, and to establish a hierarchy among the components, has always been one of the major motivations of semantic research. One criterion for hierarchization has been the isolation of designation or connotation ('lexical meaning', in Hermann Paul's terms; 'distinctive', in Bloomfield's) for study by linguistics, while relegating 'mere' reference or denotation ('occasional' meaning, according to Paul) to some other field.[a] A further criterion – within the elements of designation – has been used in studies of such areas of vocabulary as can be represented as taxonomies: in a classification such as ⟨10⟩, features introduced at the bottom level (*a, b, ... g*) differ from the non-terminal features (1, 2; A) in that each occurs only once.

⟨10⟩

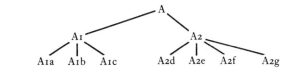

The hierarchization of semantic features into markers and distinguishers in KF does not seem to correspond to either of the conventional criteria, although the discussion is far from clear. Markers are said to 'reflect whatever systematic relations hold between items and the rest of the vocabulary of the language', while distinguishers 'do not enter into theoretical relations'. Now distinguishers cannot correspond to features of denotata, since denotata do not fall within the theory at all. Nor can they correspond to the lowest-level features of a taxonomy, for these – e.g. the features (*a, b, ... g*) in a vocabulary such as ⟨10⟩ – very definitely *do* enter into a theoretical relation; though they are unique, they alone distinguish the coordinate species of the genera A1 and A2.

The whole notion of distinguisher appears to stand on precarious ground when one reflects that there is no motivated way for the describer of a language to decide whether a certain sequence of markers should be followed by a distinguisher or not. Such a decision would presuppose a dictionary definition which is guaranteed to be correct; the critical semanticist would then merely sort listed features into markers and a distinguisher. But this, again, begs the question, especially in view of the

---

[a] See Weinreich (1963a: 152 ff.) for references, and (1964) for a critique of Webster's New International Dictionary, 3rd edition, for neglecting this division.

notoriously anecdotal nature of existing dictionaries (Weinreich 1962, 1964). All suggestions in KF concerning the detailed 'geometry' of distinguishers are similarly unfounded: the theory offers no grounds, for example, for choosing either ⟨11 i⟩ or ⟨11 ii⟩ as the correct statement of a meaning. (SmM$_n$ stands for the last semantic marker in a path; bracketed numbers symbolize distinguishers.)

⟨11⟩ (i)   SmM$_n$

[1]   [2]

(ii)   SmM$_n$

[3]

All KF rules concerning operations on distinguishers (e.g. the 'erasure clause' for identical distinguishers, p. 198) are equally vacuous.

The theory of distinguishers is further weakened when we are told (KF, n. 16) that 'certain semantic relations among lexical items may be expressed in terms of inter-relations between their distinguishers'. Although this contradicts the definition just quoted, one may still suppose that an extension of the system would specify some special relations which may be defined on distinguishers. But the conception topples down completely in Katz's own paper (1964b), where contradictoriness, a relation developed in terms of markers, is found in the sentence *Red is green* as a result of the *distinguishers!*[a] Here the inconsistency has reached fatal proportions. No ad-hoc reclassification of color differences as markers can save the theory, for *any* word in a language could be so used as to produce an anomalous sentence [cf. §3.441].

2.2.4 *Paths vs. selection restrictions*

KF provides that to the terminal element of a path there shall be affixed a string con-sisting of syntactic or semantic markers, or both. The function of this string is to represent conditions on the non-anomalous employment of the word in the meaning represented by that path. For example, the suffix ⟨(Aesthetic Object)⟩ at the end of one of the paths for *colorful* would indicate that the adjective, in the sense corre-sponding to that path, is applicable as a modifier without anomaly only to head nouns which contain the marker (Aesthetic Object) in their paths.

This part of KF, too, is fraught with apparently insuperable difficulties. Consider the case of the adjective *pretty*. It seems to be applicable to inanimates and, among animates, to females. If its selection restriction were stated as ⟨(Inanimate) ∨ (Animate) + (Female)⟩,[b] the normality of *pretty girls* as well as the anomaly of *pretty boys* would be accounted for, since *girls* has the marker (Female) in its path, while *boys* does not. But we can also say *pretty children* without anomaly, even though *child* does not contain (Female) in its path; in fact, English speakers will refer the sex of the neutral *children* from the attribute, and a theory concerned with 'the interpre-tative ability of speakers' must account for this inference. One way of doing so would be to reformulate the selection restriction more carefully, e.g. as ⟨(Inanimate) ∨ ∼

[a] 'There are *n*-tuples of lexical items which are distinguisher-wise antonymous' (Katz 1964b: 532).
[b] If Female necessarily implies Animate, the notation could of course be simplified; see Katz and Postal (1964 16 ff.). But see §4.3 [pp. 471 ff.] on the theoretical implications of such simplifications.

(Male)⟩, to be read: predicable of Inanimates and not predicable of Males. This would explain why *pretty children* is not anomalous, but would not yet show how we infer that the children are girls, since the projection rules only check on whether the conditions of selection restriction are satisfied, but transfer no information from the angle-bracketed position to the amalgamated path. Moreover, the case of Female = ∼ Male is probably quite atypically favorable in that we have two values of a feature dichotomizing a large domain in a relevant way. If, on the other hand, *addled* were to be marked in its selection as restricted to eggs and brains, the restriction would be unlikely to be statable in terms of legitimate markers (without distinguishers); and again we would lack an explanation of how we know that in *It's addled*, the referent of *it* is an egg or a brain. Here a restriction in terms of negatives, e.g. ⟨ ∼ [(∼ Eggs) ∨ (∼ Brains)]⟩, would be intolerably ad-hoc.

Two alternatives may be considered. One would be to regard 'Constructibility with *Z*' as an intrinsic feature of a dictionary entry *W* (i.e. of the 'path' of *W*), and not as a statement external to the path. The other is to adopt a more powerful conception of the semantic interpretation process, in which features of selection restriction of a word *Z* would be transferred into the path of another word, *W*, when it is constructed with *Z*. This is the solution adopted by Katz and Postal (1964: 83) for a special purpose [see §3.51 (*c*)], and it is the general solution which we will elaborate in §3.3 [pp. 429 ff.]. But with respect to KF, it seems safe to conclude that the distinction between paths and selection restrictions is an untenable as its other specifications of the format of dictionary entries.

2.2.5 *The structure of paths*

Before proceeding to a criticism of the notion of projection rules in KF, we have to consider further the algebra of the dictionary entries.

If a dictionary entry, conceived of as a paradigmatic tree, has the form ⟨12i⟩, where A, B, C, D, E, and F are semantic features, there is no reason against reformu-

⟨12⟩ (i)　　　　　　　　　　　(ii)

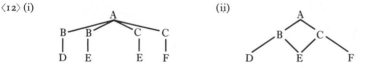

lating it as ⟨12ii⟩ (provided the convention is maintained that all such trees are read 'downwards'). We have already alluded to such cases of reconvergence in the semantic path after branching in the syntactic path, and many examples come to mind of strictly semantic reconvergence, e.g. ⟨13⟩. But it will be noticed that there is no

⟨13⟩ *fox* → (Object) → (Animate) ⟨ (Human) (Animal) → ... ⟩ (Cunning) → ...

a priori order for the markers. Suppose we stipulate that, in a given dictionary, the markers A, B, C, ..., if they occur in the same entry, appear in alphabetical order; then the subpaths A–C–D–G and B–C–H of same entry would be conflated as ⟨14⟩. If, on the other hand, we require (by a metatheoretical convention) that all reconvergence of branches be avoided, the proper form of the complex entry would perhaps as in ⟨15⟩.

⟨14⟩ 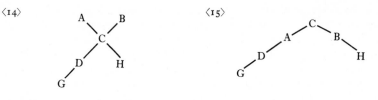 ⟨15⟩

At any rate, it may be useful to realize in advance that the criteria of a fixed order of markers and a fixed form of branching may be mutually irreconcilable.

In contrast to the syntagmatic trees representing the structure of sentences, the purely paradigmatic trees corresponding to polysemous dictionary entries are under no constraints on Analyzability in the sense e.g. of Chomsky and Miller (1963: 301). Hence the descriptive problem reduces itself to finding the most economical pathwork, with a minimum of repetition of features. By exemplifying the problem with trees which are representable as pure taxonomies, KF gives an oversimplified view of the problem. Thus, a classificatory tree such as ⟨16i⟩, in which A, B, C, Q, R, S, and T are features and a path of features connected by lines constitutes a meaning, is equivalent to ⟨16ii⟩, which explicitly represents it as a taxonomy:

⟨16⟩ (i)   (ii)

However, many dictionary entries tend to the form of matrixes of features, as in ⟨17i⟩, and there is no motivated reason to rewrite them as ⟨17ii⟩; the only economy would be achieved by representations such as ⟨17iii–v⟩:

⟨17⟩ (i)   (ii)

(iii)   (iv)   (v)

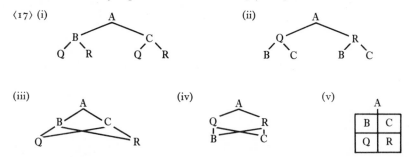

In short, unprejudiced reflection leads to the conclusion that no theoretical motivation is in prospect for specifying the order of features in a path.[a]

A related difficulty arises when we observe the amalgamation of paths which results from the operation of the projection rules. Given the dictionary entries, with their respective paths, $W_a = a_1 + a_2 + \ldots + a_m$ and $W_b = b_1 + b_2 + \ldots + b_n$, the

[a] The prospect that implicational relations among markers, such as those discussed by Katz and Postal (1964) and Chomsky (1965), may automatically yield unique networks of features, is attractive, but it is unlikely to be borne out when non-anecdotal evidence is considered.

compound lexical string $W_a + W_b$ has the compound path $a_1 + a_2 + \ldots + a_m + b_1 + b_2 + \ldots + b_n$; but there is no formal representation of the juncture between that portion of the compound path which came from W_a and that which came from W_b. In KF the distinguishers of W_a, if any, appear at the point where the constituent paths of an amalgamated path have been joined;[a] but as we have seen, this is purely arbitrary with respect to the theory. In fact, the origin of each marker in an amalgamated path is not recoverable; the elements of an amalgamated path in KF, like those of the constituent paths, are strictly unordered sets, and there is no way, valid within the theory, of saying that in the path of $W_a + W_b$, a_m (for example) precedes b_1.

Let us consider some consequences of this property of KF. Given the separate paths for the English words *detective* and *woman*, the constructions *woman detective* and *detective woman* would be represented by identical amalgamated paths, since the order of elements in a path, and hence of subpaths in a path, is theoretically immaterial. KF turns out, in other words, to be unable to represent the interchange of foregrounded and backgrounded information in such pairs.[b] But the dilemma, though awkward, is still relatively benign, since the semantic effect of rearranging the constituents is of very limited significance. A far more pernicious weakness of the theory appears when one considers that the two sentences in ⟨18⟩ would also receive identical semantic interpretations ('readings'). The paths of (1) *cats*, (2) *chase*, and (3) *mice* –

⟨18⟩ (i) *Cats chase mice.*

(ii) *Mice chase cats.*

although amalgamated in the order $1 + (2 + 3)$ in ⟨18i⟩ and as $(2 + 1) + 3$ in ⟨18ii⟩ – would yield the same unordered set of features, $\{1\ 2\ 3\}$, as the amalgamated path; for, as we have seen, there is neither ordering nor bracketing of elements in a KF path. For similar reasons, the theory is unable to mark the distinction between *three cats chased a mouse* and *a cat chased three mice*, between *(bloody + red) + sunset* and *bloody + red + sunset*, and so on for an infinite number of crucial cases.

For KF, the meaning of a complex expression (such as a phrase or a sentence) is an unstructured heap of features – just like the meaning of a single word. The projection rules as formulated in KF *destroy the semantic structure* and reduce the words of a sentence to a heap. Very far from matching a fluent speaker's interpretation of sentences or explicating the way in which the meaning of a sentence is derived from the meaning of its components, KF only hints at the presence, in some unspecified capacity, of some meanings somewhere in the structure of a sentence. It tells us, in effect, that ⟨18i⟩ is an expression which says something about *cats*, *mice* and *chasing*; and that ⟨18ii⟩ does likewise.

It might, of course, be countered that while the set of semantic features in both sentences of ⟨18⟩ is the same, the *grammar* of these expressions is different. But this is precisely the issue: *how* is the difference in grammar concretely related to the difference in total meaning? On this KF is silent. What is particularly ironic is that an enterprise in semantics inspired by the most sophisticated syntactic research ever

---

[a] E.g. on p. 201, under P8, but not consistently (e.g. not on p. 198), and not in accordance with the formal rule stated on p. 198.

[b] Reanalysis of these transformed expressions into their underlying sentences, e.g. *The detective is a woman* vs. *The woman is a detective*, would not resolve matters, since the addition of the same ultimate components can only lead to the same results in the total.

undertaken should end up with a fundamentally asyntactic theory of meaning. In its inability to distinguish ⟨18i⟩ from ⟨18ii⟩, KF is comparable to certain (linguistically useless) psychological accounts of sentences which seek to explain the meaning of a sentence in terms of 'associations' between its component words.[a]

To avoid similar defects in an alternative theory, it may be useful to consider how KF maneuvered itself into a position of bankruptcy on the most essential issue. Apparently, this happened when the authors modeled a theory of linguistic meaning on the concept of the multiplication of classes. As logicians have long known, to express the fact that a *colorful ball* is something which is both colorful and a ball, we may say that *colorful ball* contains the semantic features of both *ball* and *colorful*. The process involved in deriving a compound meaning is expressible as a Boolean class conjunction. One would have thought that with the development of the calculus of many-place predicates, the logic of Boolean (one-place) predicates would be permanently dropped as a model for natural language; yet KF persists in the belief, widespread among nineteenth-century logicians, that Boolean operations are an adequate model for combinatorial semantics. The dire results of such a belief are evident.[b]

2.2.6 *Infinite polysemy*

When one considers the phrases *eat bread* and *eat soup*, one realizes that *eat* has a slightly different meaning in each phrase: in the latter expression, but not in the former, it covers the manipulation of a spoon. Continuing the procedure applied in KF to polysemous items such as *ball* and *colorful*, one would have to represent the dictionary entry for *eat* by a branching path, perhaps as in ⟨20⟩:

$$⟨20⟩ \quad eat \to \ldots \to (\text{Action}) \to \ldots \to (\text{Swallow}) \left\langle \begin{array}{l} (\text{Chew}) \to \ldots ⟨(\text{Solid})⟩ \\ \ldots \to (\text{Spoon}) ⟨(\text{Liquid})⟩ \end{array} \right.$$

The selection restrictions at the end of each subpath would provide the information which makes possible the choice of the correct subpath in the contexts of *bread* and *soup* functioning as object Noun Phrases. But then the activity symbolized by *eat* is also different depending on whether things are eaten with a fork or with one's hands; and even the hand-eating of apples and peanuts, or the fork-eating of peas and spaghetti, are recognizably different. It is apparent, therefore, that a KF-type dictionary is in danger of having to represent an unlimited differentiation of meanings.

Several escapes from this danger can be explored. The most direct one would prohibit branching of paths in a lexical entry except where they represent an experienced ambiguity in some non-ambiguous context. For example, if *file* can be understood as ambiguous (e.g. in the context *I love to — things*: '1. put away for storage; 2. abrade'), the dictionary entry would represent the ambiguity by a branching of paths; on the other hand, if *eat* does not feel ambiguous in a general context such as *I'd like to — something*, the submeanings of *eat* would not be represented in the dictionary. But this will presuppose, as a primitive concept of the theory, an absolute

---

[a] For a relatively recent attempt, cf. Mowrer (1960), ch. 4.

[b] Among linguistic theories growing out of logic, KF is thus a distinct step backward from Reichenbach (1948), who realized the need for the higher functional calculus in a semantic theory applicable to natural languages. But Reichenbach himself was antiquated when compared to Stöhr (1898), who had already appreciated the necessity of supplementing the functional calculus by other models. The logically irreducible character of transitivity was also a continuing interest of Peirce's (e.g. 3.408, and in many other places in his work).

distinction between true ambiguity and mere indefiniteness of reference. The difficulty of validating such a distinction empirically makes its theoretical usefulness rather dubious, although it has been advocated, e.g., by Ziff (1960: 180 ff.).

A more elaborate solution, suggested by Kuryłowicz (1955), could be stated as follows: a dictionary entry W will be shown to have two subpaths (submeanings), W_1 and W_2, if and only if there is in the language a subpath Z_i of some entry Z which is synonymous with W_1 and is not synonymous with W_2. According to Kuryłowicz, the notions of polysemy (path branching) and synonymy are complementary, and neither is theoretically tenable without the other. Thus, the path for *file* would be shown to branch insofar as *file$_1$* is synonymous with *put away*, whereas *file$_2$* is not. However, the condition would have to be strengthened to require the synonyms to be simplex, since it is always possible to have multi-word circumlocutions which are equivalent to indefinitely differentiated submeanings of single words (e.g. *consume as a solid* = *eat$_1$*; *consume as a liquid* = *eat$_2$*). On the notion of lexemic simplicity, see §3.422 below [pp. 450 ff.].

In any case, it is evident that some regard for the experience of previous semantic theorists could have saved KF from an unnecessary trap.

2.3 PROJECTION RULES

The projection rules of KF are a system of rules that operate on full grammatical descriptions of sentences and on dictionary entries to produce semantic interpretations for every sentence of the language. Projection rules are of two types; described informally, projection rules of type 1 (PR1) operate on sentences formed without transformations or with obligatory transformations only; those of type 2 (PR2) operate on sentences formed by optional transformations. It is already anticipated in KF (p. 207) that if the syntactic theory of a language could be formulated without recourse to optional transformations, PR2 could be eliminated. Since the publication of KF the possibilities of a syntax without optional transformations, singulary[a] or generalized,[b] have been shown to be real, so that the need for PR2 no longer exists. Let us then consider the differences among various PR1.

Projection rules in KF differ among each other according to (*a*) the conditions for their application and (*b*) their effect. We take up each factor in turn.

(*a*) The conditions are stated in terms of the grammatical status of constituent strings in a binary (i.e. two-constituent) construction. The specification of the grammatical status of strings in KF is, however, thoroughly eclectic. The terms 'noun' and 'article', to which the rules refer, are lexical categories given by the grammar; similarly, 'verb phrase', 'noun phrase', and 'main verb' are defined as non-lexical (preterminal) categories of the grammar. Such labels, on the other hand, as 'object of the main verb' and 'subject' have a different theoretical status in the syntax which KF takes for granted.[c] Finally, such notions as 'modifier' and 'head', to which PR$_1$ makes reference (p. 198), have no status in the theory at all; they beg a question in disguise [see §3.21] and are probably undefinable without reference to semantics. Although KF gives no indication of the number of PRs in a language (n. 20), it would seem that the procedure would require as many PRs as there are binary

---

[a] Katz and Postal (1964: 31–46).

[b] Chomsky (1965).

[c] A way of defining these syntactic functions derivatively has now been described by Chomsky (1965).

constructions in the grammar. (No treatment for ternary constructions is proposed by KF.)

(*b*) The PRs differ in their effect, such effect being stated in terms of deletions of selection restrictions. Let us represent a construction as ⟨21⟩, where M and N are

$$\langle 21 \rangle \quad A \rightarrow M \langle \mu \rangle + N \langle \nu \rangle$$

lexical strings with their associated sets of syntactic and semantic markers, and μ and ν are their respective selection restrictions. In principle, there are four possible restrictions on the selections of the construction, A, as a whole.

$$
\begin{aligned}
\langle 22 \rangle \quad &\text{(i)} \quad A \langle \mu, \nu \rangle \rightarrow M \langle \mu \rangle + N \langle \nu \rangle \\
&\text{(ii)} \quad A \langle \mu \rangle \rightarrow M \langle \mu \rangle + N \langle \nu \rangle \\
&\text{(iii)} \quad A \langle \nu \rangle \rightarrow M \langle \mu \rangle + N \langle \nu \rangle \\
&\text{(iv)} \quad A \rightarrow M \langle \mu \rangle + N \langle \nu \rangle
\end{aligned}
$$

A may retain the restrictions of *both* constituents (i), or of the left constituent (ii) or of the right constituent (iii); or it may be unrestricted (iv). In KF, projection rule 1 is a rule of type ⟨22 iii⟩; rule 3 is of type ⟨22 ii⟩; rules 2 and 4 are of type ⟨22 iv⟩. No rule of type ⟨22 i⟩ is cited, but there appears no reason to exclude its occurrence in principle.

In sum, the function of the KF projection rules is to classify all binary constructions, terminal as well as preterminal, of a grammar into four types according to the deletion or non-deletion of the selection restrictions of their right and left constituents. Except for the differential effects on selection restrictions, the power of all projection rules is the same: namely, to sum the paths of the constituents. Consequently, the classification of constructions by PRs could easily be shown within the categorial part of the syntax,[a] so that no separate PR 'component' would be necessary.

Before attempting a radically new approach [ch. 3], we must still consider the position of deviant utterances in an explicit linguistic theory. Since KF touches on the problem only tangentially, we must on this point turn to certain other sources which are close to KF in spirit.

2.4 DEVIATIONS FROM GRAMMATICALITY

In the literature on generative grammar, the distinction between grammatical and other kinds of deviance occupies a privileged position, since the very definition of grammar rests on the possibility of differentiating grammatical from ungrammatical expressions. Since ungrammatical formations are a subclass of the class of odd expressions, the difference between ungrammaticality and other kinds of oddity must be represented in the theory of language.

But the problem exemplified by grammatically faultless, yet semantically odd expressions such as *colorless green ideas* has an old history. For two thousand years linguists have striven to limit the accountability of grammar vis-à-vis abnormal constructions of some kinds. Apollonios Dyskolos struggled with the question in second century Alexandria; Bhartṛhari, in ninth century India, argued that *barren woman's son*, despite its semantic abnormality, is a syntactically well-formed expression. His near-contemporary in Iraq, Sîbawaihi, distinguished semantic deviance (e.g. in *I carried the mountain, I came to you tomorrow*) from grammatical deviance, as

---

[a] For example, instead of using '+' in all branching rules (A—→ M + N), we might restrict the plus to rules of types ⟨22 i⟩ and use '+→', '←+', and '←+→', respectively, for rules of type ⟨22 ii–iv⟩.

in *qad Zaidun qâm* for *qad qâm Zaidun* 'Zaid rose' (the particle *qad* must be immediately followed by the verb). The medieval grammarians in Western Europe likewise
conceded that the expression *cappa categorica* 'categorical cloak' is linguistically
faultless (*congrua*), so that its impropriety must be elsewhere than in grammar.[a] The
continuing argument in modern philosophy has a very familiar ring.[b]

The position taken by most writers on generative grammar seems to rest on two
planks: (a) grammatical oddity of expressions is qualitatively different from other
kinds of oddity; and (b) grammatical oddity itself is a matter of degree. Let us consider these two points, in relation to the following examples:

⟨23⟩ *Harry S. Truman was the second king of Oregon.*
⟨24⟩ (i) *Went home he.*
  (ii) *Went home for the holidays.*
  (iii) *He goed home.*
⟨25⟩ (i) *He puts the money.*
  (ii) *He puts into the safe.*
  (iii) *He puts.*
⟨26⟩ (i) *The dog scattered.*
  (ii) *John persuaded the table to move.*
  (iii) *His fear ate him up.*
  (iv) *The cake is slightly delicious.*
  (v) *The water is extremely bluish.*
  (vi) *Five out of three people with me.*
⟨27⟩ (i) *The square is round.*
  (ii) *A square is round.*
  (iii) *The square is loud but careful.*
  (iv) *A square is loud but careful.*

We are not concerned here with any theories of reference which would mark sentence
⟨23⟩ as odd because of its factual falsity; on the contrary, we may take ⟨23⟩ as an
example of a normal sentence, in contrast with which those of ⟨24⟩–⟨27⟩ all contain
something anomalous. The customary approach is to say that ⟨24⟩–⟨26⟩ are deviant
on grammatical grounds, while the examples of ⟨27⟩ are deviant on semantic grounds.
This judgment can be made, however, only in relation to a given grammar $G(L)$ of
the language L; one may then cite the specific rules of $G(L)$ which are violated in
each sentence of ⟨24⟩–⟨27⟩, and indicate what the violation consists of. Whatever
rules are violated by ⟨27⟩, on the other hand, are not in $G(L)$; presumably, they are
in the semantic description of the language, $S(L)$. But as appeared in §2.22, the
demarcation between $G(L)$ and $S(L)$ proposed by KF is spurious, and no viable
criterion has yet been proposed.[c]

---

[a] On Bhartṛhari, see Chakravarti (1933: 117 ff.) and Sastri (1959: 245); unfortunately,
neither source does justice to the subject. The Indian argument goes back at least to Patañjali
(second century B.C.). See also Sîbawaihi (1895: 10 ff.); Thomas of Erfurt (c. 1350: 47).

[b] One is struck, for instance, by the similarity between a recent argument of Ziff's (1964:
206) and a point made by the great Alexandrian, Apollonios, 18 centuries ago (Apollonios
§iv. 3; ed. 1817: 198). Ziff argues that the normality or deviance of *It's raining* cannot depend
on whether it is in fact raining when a token of the sentence is uttered. Apollonios contends
that if a discrepancy between a sentence and its setting were classified as a (type of) solecism,
the occurrence of solecisms would be limited to daylight conditions and to discourse with seeing
persons, since the blind hearer, or any hearer in darkness, could not check a statement for its
conformity with the setting.

[c] The absence of a criterion is all the clearer in a syntax reformulated in feature terms.

In the framework of a theory of syntax in which subcategorization was represented by rewriting rules of the phrase-structure component of a grammar, Chomsky (cf. Miller and Chomsky 1963: 444 f.) proposed to treat degrees of grammaticality roughly in the following way.[a] Suppose we have a grammar, G_0, formulated in terms of categories of words W_1, W_2, ... W_n. We may now formulate a grammar G_1 in which some categories of words – say, W_j and W_k – are treated as interchangeable. A grammar G_2 would be one in which a greater number of word categories are regarded as interchangeable. The limiting case would be a 'grammar' in which all word classes could be freely interchanged. Expressions that conform to the variant grammar G_i would be said to be grammatical at the i level. But it is important to observe that nowhere in this approach are criteria offered for setting up discrete levels; we are not told whether a 'level' should be posited at which, say, W_1 is confused with W_2 or W_2 with W_3; nor can one decide whether the conflation of W_1 and W_2 takes place at the same 'level' as that, say, of W_9 and W_{10}, or at another level. The hope that a quantative approach of this type may lead to a workable reconstruction of the notion of deviancy is therefore, I believe, poorly founded.

A syntax formulated in terms of features (rather than subcategories alone) offers a different approach. We may now distinguish violations of categorial-component rules, as in ⟨24⟩; violations of rules of strict subcategorization, as in ⟨25⟩; and violations of rules of selection, as in ⟨26⟩. The number of rules violated in each expression could be counted, and a numerical coefficient of deviation computed for each sentence. Furthermore, if there should be reason to weight the rules violated (e.g. with reference to the order in which they appear in the grammar), a properly weighted coefficient of deviation would emerge. But although this approach is far more promising than the one described in the preceding paragraph, it does not yet differentiate between the deviations of ⟨24⟩–⟨26⟩ and those of ⟨27⟩. This could be done by postulating a hierarchy of syntactic-semantic features, so that ⟨27⟩ would be said to violate only features low in the hierarchy. It is unknown at this time whether a unique, consistent hierarchization of the semantic features of a language is possible. In §3.51 [pp. 455 ff.], we develop an alternative approach to deviance in which the troublesome question of such a hierarchization loses a good deal of its importance.

Still another conception of deviance has been outlined by Katz (1964a). There it is argued that a semi-sentence, or ungrammatical string, is understood as a result of being associated with a class of grammatical sentences; for example, *Man bit dog* is a semi-sentence which is (partly?) understood by virtue of being associated with the well-formed sentences 'A man bit a dog', 'The man bit some dog', etc.; these form a 'comprehension set'. The comprehension set of a semi-sentence, and the association between the semi-sentence and its comprehension set, are given by a 'transfer rule'. However, the number of possible transfer rules for any grammar is very large: if a grammar makes use of n category symbols and if the average terminal string generated by the grammar (without recursiveness) contains m symbols, there will be $(n-1) \times m$ possible transfer rules based on category substitutions alone; if violations of order by permutation are to be covered, the number of transfer rules soars, and if recursiveness is included in the grammar, their number becomes infinite. The signi-

Chomsky (1965) suggests that syntactic features may be those semantic features which are mentioned in the grammar; but he provides no criteria for deciding when a grammar is complete vis-à-vis the dictionary. See also §4.1 [pp. 467 ff.].

[a] For the sake of perspicuity, we have slightly simplified the original account – hopefully without distorting its intent.

ficant problem is therefore to find some criterion for selecting interesting transfer rules. Katz hopes to isolate those which insure that the semi-sentence is understood. This proposal, it seems to me, is misguided on at least three counts. First, it offers no explication of 'being understood' and implies a reliance on behavioral tests which is illusory. Secondly, it assumes, against all probability, that speakers have the same abilities to understand semi-sentences, regardless of intelligence or other individual differences.[a] Thirdly, it treats deviance only in relation to the hearer who, faced with a noisy channel or a malfunctioning source of messages, has to reconstruct faultless prototypes; Katz's theory is thus completely powerless to deal with intentional deviance as a communicative device. But the overriding weakness of the approach is its treatment of deviance in quantitative terms alone; Katz considers how deviant an expression is, but not what it signifies that cognate non-deviant expressions would *not* signify.

formerly at Columbia University

BIBLIOGRAPHY

W. S. Allen, *On the linguistic study of languages* (Cambridge, 1957).
Apollonios Dyskolos, *Appollonii Alexandrini de constructione orationis libri quattuor*, ed. Immanuel Bekker (Berlin, 1817).
Leonard Bloomfield, 'Secondary and tertiary responses to language', *Language*, XX, 45–55 (1944).
Prabhachandra Chakravarti, *The linguistic speculations of the Hindus* (Calcutta, 1933).
Y. R. Chao, 'Ambiguity in Chinese', in *Studia serica Bernhard Karlgren dedicata* (Copenhagen, 1959).
Noam Chomsky, *Aspects of the theory of syntax* (Cambridge, Mass., 1965).
Noam Chomsky and George A. Miller, 'Introduction to the formal analysis of natural languages', in *Handbook of mathematical psychology*, vol. II, eds. R. D. Luce, R. Bush, and E. Galanter, 269–321 (New York, 1963).
Jerry A. Fodor, 'Projection and paraphrase in semantic analysis', *Analysis*, XXI, 73–7 (1961).
Joseph H. Greenberg, *Universals of language* (Cambridge, Mass., 1963).
M. A. K. Halliday, 'Categories of the theory of grammar', *Word*, XVII, 241–92 (1961).
Charles F. Hockett, *A course in modern linguistics* (New York, 1958).
Jerrold J. Katz, 'A reply to "Projection and paraphrase in semantics" [Fodor (1961)]', *Analysis*, XXII, 36–41 (1961).
'Semi-Sentences', in *Structure of language: readings in the philosophy of language*, eds. J. Fodor and J. Katz, 400–16 (Englewood Cliffs, N.J., 1964a).
'Analyticity and contradiction in natural languages', in Fodor and Katz, 519–43 (1964b).
Jerrold J. Katz and Jerry A. Fodor, 'The structure of a semantic theory', *Language*, XXXIX, 170–210 (1963). Reprinted, with minor revisions, in Fodor and Katz, 479–518 (1964).
Jerrold J. Katz and Paul M. Postal, 'Semantic interpretation of idioms and sentences containing them', Massachusetts Institute of Technology, Research Laboratory of Electronics, *Quarterly Progress Report* no. 70, 275–82 (1963).
An integrated theory of linguistic descriptions (Cambridge, Mass., 1964).

[a] This assumption is apparent from Katz's criticism of another author's approach to ungrammaticality in which differing abilities of individual speakers may be involved (Katz 1964a: 415).

Jerzy Kurylowicz, 'Zametki o značenii slova', *Voprosy jazykoznanija*, 73–81 (1955), no. 3.

William Labov, 'The reflections of social processes in linguistic structures', in *Reader in the sociology of language*, ed. Joshua A. Fishman (The Hague, 1965). *The social stratification of English in New York City* (Washington, D.C.: The Centre for Applied Linguistics, 1966).

Robert B. Lees, *The grammar of English nominalizations* (Bloomington, 1960).

George A. Miller and Noam Chomsky, 'Finitary models of language users', in *Handbook of mathematical psychology*, vol. II, eds. R. D. Luce, R. Bush, and E. Galanter, 419–91 (New York, 1963).

Hobart O. Mowrer, *Learning theory and the symbolic process* (New York, 1960).

Charles Saunders Peirce, *Collected papers*, vol. III (Cambridge, Mass., 1933).

Hans Reichenbach, *Elements of symbolic logic* (New York, 1948).

Gaurinath Sastri, *The philosophy of word and meaning* (Calcutta, 1959).

Sîbawaihi, *Sîbawaihi's Buch über die Grammatik*, tr. G. Jahn (Berlin, 1895).

Adolf Stöhr, *Algebra der Grammatik; ein Beitrag zur Philosophie der Formenlehre und Syntax* (Leipsig and Vienna, 1898).

Thomas of Erfurt (c. 1350), *Grammatica speculativa*, attributed to Duns Scotus and published in his *Opera omnia*, e.g. ed. Paris, bol. I, 1891.

Uriel Weinreich, 'Travels in semantic space', *Word*, XIV, 346–66 (1958).
    'Lexicographic definition in descriptive semantics', in *Problems in Lexicography*, eds. Fred W. Householder and Sol Saporta, 25–43 (Bloomington, 1962).
    'On the semantic structure of language', in *Universals of Language*, ed. Joseph H. Greenberg, 114–71 (Cambridge, Mass, 1963a).
    '[Soviet] Lexicology', in *Current trends in linguistics*, eds. Thomas A. Sebeok, vol. I, 60–93 (The Hague, 1963b).
    Review of Vygotsky (1962), in *American Anthropologist*, LXV, 1401–4 (1963c).
    'Webster's Third: a critique of its semantics', *International Journal of American Linguistics*, XXX, 405–9 (1964); slightly revised Russian version in *Voprosy jazykoznanija*, 128–32 (1965), no. 1.
    'Semantics and Semoitics', in *International Encyclopedia of the Social Sciences*, ed. D. L. Sills, vol. XIV, 164–9 (New York, 1968).

Rulon S. Wells, 'Meaning and Use', *Word*, X, 115–30 (1954).

A. A. Xolodovič, 'Opyt teorii podklassov slov', *Voprosy jazykosnanija*, 32–43 (1960), no. 1.

Paul Ziff, *Semantic analysis* (Ithaca, 1960).

Presupposition and relative well-formedness[a]

GEORGE LAKOFF

I would like to discuss a phenomenon that I think is fairly obvious, but which has been inadequately discussed in the past, and has therefore led to great confusion. It is often assumed that one can speak of the well- or ill-formedness of a sentence in isolation, removed from all presuppositions about the nature of the world. I think it has become clear over the past several years that such a position cannot be maintained. Of course, languages exhibit certain low-level or 'shallow' constraints on the form of sentences. English, for example, requires that, for the most part, verbs must follow their subjects and prepositions must, in general, precede the noun phrases they are associated with. Violation of such constraints does indeed make for ungrammaticality of an absolute sort: '*Hit Sam Irving', '*I went Boston to'. However, there are a great many cases where it makes no sense to speak of the well-formedness or 'grammaticality' of a sentence in isolation. Instead one must speak of relative well-formedness and/or relative grammaticality; that is, in such cases a sentence will be well-formed only with respect to certain presuppositions about the nature of the world. In these cases, the presuppositions are systematically related to the form of the sentence, though they may not appear overtly.

Given a sentence, S, and a set of presuppositions, PR, we will say, in such instances, that S is well-formed only relative to PR. That is, I will claim that the notion of relative well-formedness is needed to replace Chomsky's [1] original notion of strict grammaticality (or degrees thereof), which was applied to a sentence in isolation. It should be pointed out at the outset that such a claim does *not* constitute a position that linguistic knowledge cannot be separated from knowledge of the world. On the contrary, it is a claim that the general principles by which a speaker pairs a sentence with those presuppositions required for it to be well-formed are part of his linguistic knowledge.

Nor should such a claim be considered as blurring the distinction between competence and performance. The study of the relationship between a sentence and those things that it presupposes about the nature of the world by way of systematic rules is part of the study of linguistic competence. Performance is another matter. Suppose that S is well-formed only relative to PR. Then a speaker will make

[a] I would like to thank Al Alvarez, Dwight Bolinger, Kim Burt, R. M. W. Dixon, Georgia Green, Kenneth Hale, and Robin Lakoff for very enlightening discussions of this topic. An earlier version of this paper was presented at the 1968 winter meeting of the Linguistic Society of America under the title 'Selectional Restrictions and Beliefs about the World'. This work was supported in part by grant GS-1934 from the National Science Foundation to Harvard University.

certain *judgments* about the well-formedness or ill-formedness of S which will vary with his extralinguistic knowledge. If the presuppositions of PR do not accord with his factual knowledge, cultural background, or beliefs about the world, then he may *judge* S to be 'odd', 'strange', 'deviant', 'ungrammatical', or simply ill-formed relative to his own presuppositions about the nature of the world. Thus, extra-linguistic factors very often enter in judgments of well-formedness. This is a matter of performance. The linguistic competence underlying this is the ability of a speaker to pair sentences with the presuppositions relative to which they are well-formed. Such facts about performance come in handy for finding cases of relative well-formedness and testing just what presuppositions pair with what sentences. In looking for such cases, it is useful to consider examples where speakers' *judgments* of deviance vary fairly consistently with their factual knowledge and beliefs. The following is a short and rather incomplete survey of cases of this sort.

Chomsky, in *Aspects of the Theory of Syntax* [2], sets up syntactic features such as CONCRETE, ANIMATE, and HUMAN to account for the type of deep structure con-straints he calls 'selectional restrictions'. For example, the difference between sentences like

⟨1⟩ (a) The man is sleeping.
    (b) *The salami is sleeping.

would be accounted for in terms of the feature ANIMATE. The verb *sleep* would require a [+ANIMATE] subject. *Man* would be marked [+ANIMATE] and so would qualify; *salami* would be marked [−ANIMATE] and so would not qualify.

McCawley has argued convincingly that selectional restrictions are semantic and not syntactic in nature. Consider

⟨2⟩ (a) *The corpse is sleeping.
    (b) *The dead man is sleeping.
    (c) *The man who was killed yesterday is sleeping.
⟨3⟩ (a) The man who was killed yesterday but was magically brought back to life is sleeping.
    (b) The man who will be killed tomorrow is sleeping.

The well-formedness of these sentences depends on semantic properties of the entire noun phrase rather than on syntactic properties of the head noun.

One might be tempted to say that such facts are purely semantic and outside the realm of syntax altogether. Syntax, one might say, has to do with such things as the distribution of grammatical morphemes, like *some* and *any*, or *who*, *which*, and *what*, not with the co-occurrence of lexical items. If one takes such a view of the distinction between syntax and semantics, and this is a view taken by most tradi-tional grammarians, then one would still need syntactic features like HUMAN. Traditionally, it has been claimed that *who* is used when speaking of humans and *which* and *what*, when speaking of non-humans.

⟨4⟩ (a) The man *who* I kicked bit me.
    (b) ?*The man *which* I kicked bit me.
⟨5⟩ (a) *The dog *who* I kicked bit me.
    (b) The dog *which* I kicked bit me.

⟨6⟩ (a) *Who* bit you? The man next door.

   (b) ?*What bit you? The man next door.[a]

⟨7⟩ (a) *Who bit you? The dog next door.

   (b) *What* bit you? The dog next door.

However, the use of *who* versus *which* cannot be described in terms of a syntactic feature HUMAN which agrees with the corresponding syntactic property on the head noun of the relative clause. Instead, the choice of *who* and *which* depends on semantic properties of the entire noun phrase. (By the way, the facts given here hold for my own speech and may vary from speaker to speaker.)

⟨8⟩ (a) The human creature *who* I was fighting with was large.

   (b) ?*The human creature *which* I was fighting with was large.

⟨9⟩ (a) *The canine creature *who* I was fighting with was large.

   (b) The canine creature *which* I was fighting with was large.

The occurrence of *who* and *which* is semantically determined, and in fact involves presuppositions. The antecedent noun phrases of *who* must be presupposed to be human.

⟨10⟩ (a) I saw a creature *who* I knew was human.

   (b) *I saw a creature *who* I knew was canine.

⟨11⟩ *I saw a creature *who* I doubted was human.

Know, being a factive verb, presupposes that the creature was human: *doubt* does not.

In addition, the choice of *who* depends on relative chronology.

⟨12⟩ (a) ?*The dead man, *who* I came across in the alley, was covered with blood.[b]

   (b) The dead man, *who* I had once come across at a party in Vienna, now looked a mess.

⟨12b⟩ is all right on the assumption that he was alive at the time I came across him in Vienna.

I do not pretend to understand the conditions under which *who* can be used. Still, it seems clear that the distribution of the grammatical morpheme *who* cannot simply be determined by a syntactic feature like [+HUMAN]; rather, the relative *who* requires, at least, that the person referred to either be presupposed to be alive at the time referred to in the relative clause, or thought of as a human being. Hence, the oddness of *who* in ⟨13a⟩ as opposed to ⟨13b⟩:

⟨13⟩ (a) *We have just found a good name for our child, *who* we hope will be conceived tonight.

   (b) We have just found a good name for our child, *who* we hope will grow up to be a good citizen after he is born.

---

[a] Speakers seem to vary in this case. Some speakers use *what* in such cases with no presuppositions as to human qualities; others seem to require a presupposition of non-humanness.

[b] I find *which* just as bad (or perhaps even worse) in this sentence. *That* is, of course, impossible in nonrestrictive relatives, as is deletion of the relative clause. There seems to be no way to make this sentence completely acceptable. *Who* may well be the best of a number of bad choices. Dwight Bolinger has noted that with interrogative pronouns one could naturally say 'Who is the dead man?' but not 'Who is the corpse?' without being facetious. Similarly, he notes that with 'dead man' one can say 'The dead man whom they brought in was John', but with 'corpse' one is required to say 'The corpse *which* they brought in was *that of* John'. Perhaps *who* is used when the individual is being thought of as a human being.

As in the case of lexical co-occurrence, the occurrence of a grammatical morpheme like *who* is determined by the semantic properties of an entire noun phrase. In this case, one cannot separate the study of the distribution of grammatical morphemes from the study of lexical co-occurrence: semantic information of the same sort is involved in both. Let us now consider what sort of semantic information we are dealing with. If we grant that *who* can only be used to refer to humans, we might suppose that there is a semantic property based on the biological distinction human/non-human.

⟨14⟩ (a) *What   ⎫   ⎧ realizes that I'm a lousy cook?
     (b) Who   ⎬   ⎨ believes that I'm a fool?
     (c) *The desk   ⎪   ⎩ enjoys tormenting me?
     (d) The boy   ⎭

If we assume that *who* and *the boy* must refer to humans, while *what* and *the desk* refer to non-humans, we can account for the facts of ⟨14⟩. But now consider ⟨15⟩:

⟨15⟩ (a) My uncle   ⎫
     (b) My cat   ⎪
     (c) My goldfish ⎪   ⎧ realizes that I'm a lousy cook.
     (d) My pet amoeba⎬   ⎨ believes that I'm a fool.
     (e) My frying pan⎪   ⎩ enjoys tormenting me.
     (f) My sincerity ⎪
     (g) My birth   ⎭

⟨15a⟩ is certainly all right, as it should be. But according to the above hypothesis, ⟨15b⟩ through ⟨15g⟩ should be ungrammatical, since they do not refer to humans. I and many others find ⟨15b⟩ perfectly all right, although some people do not. The reason, I think, is that I and those who agree with my judgment assume that cats have minds, while those who don't find ⟨15b⟩ acceptable don't hold this belief. ⟨15c⟩ and ⟨15d⟩ are stranger, I think, because of the strangeness of the beliefs that goldfish and amoebae have the appropriate mental powers. I suppose that someone who thought his goldfish were capable of such mental activities would find ⟨15c⟩ perfectly acceptable. Similarly, if someone thought his frying pan had a mind, he might find ⟨15e⟩ perfectly all right. If one found such a person, one might send for a psychiatrist, not try to correct his grammar. ⟨15f⟩ and ⟨15g⟩ are another matter. That properties and events have mental powers might seem to be an impossible belief, not just a strange one. If this were true, it would follow that ⟨15f⟩ and ⟨15g⟩ were universally impossible. However, Kenneth Hale informs me that, among the Papagos, events are assumed to have minds (whatever that might mean), and that sentences like ⟨15g⟩ would be perfectly normal. I leave such matters to the anthropologists. Be that as it may, it seems that the subjects of verbs like *realize, believe, enjoy,* etc. are not restricted to humans, but to any beings that the speaker assumes to have the necessary mental abilities. Thus, at least in these cases, one's judgment of the well-formedness of sentences seems to vary with one's beliefs or assumptions. And if one accepts that the distribution of *who* and *which* is a question to be dealt with in a field called 'grammar', then one's judgments of grammaticality seem to vary with one's assumptions and beliefs. Consider

⟨16⟩ My cat, who believes that I'm a fool, enjoys tormenting me.
⟨17⟩ *My cat, which believes that I'm a fool, enjoys tormenting me.

Having had experience with a cunning feline, I find ⟨16⟩ both syntactically and semantically well-formed, while ⟨17⟩ is ungrammatical for me. Thus, *who* seems to refer not simply to humans, but to individuals being thought of as intelligent beings whatever their species. Judgments concerning its proper use will vary with the speaker's beliefs about such matters. A similar argument, concerning the distribution of *some* and *any*, has been given by Robin Lakoff [6].

R. M. W. Dixon and Georgia Green have brought to my attention another class of cases where judgments of well-formedness depend on extralinguistic factors; namely, certain classes of constructions which involve comparisons and contrasts. Consider ⟨18⟩ for example:

⟨18⟩ (a) John insulted Mary and then shé insŭlted hím.[a]

(b) *John insulted Mary and then shé insŭlted him.

When the two verbs are the same, both pronouns must be stressed, unlike normal anaphoric pronouns. Compare ⟨19⟩, where the verbs have opposite meanings; the pronouns cannot both be stressed.

⟨19⟩ (a) *John praised Mary and then shé insŭlted hím.

(b) John praised Mary and then shé insŭlted him.

Judgments about the well-formedness of these sentences involve only a knowledge of one's language: the rules involving stress placement in such constructions and a knowledge of the meanings of *praise* and *insult*. But now consider ⟨20a⟩:

⟨20⟩ (a) John told Mary that she was ugly and then shé insŭlted hím.

In ⟨20a⟩ stress can occur on *him* because telling a woman that she is ugly constitutes an insult. Compare ⟨20b⟩:

⟨20⟩ (b) *John told Mary that she was beautiful and then shé insŭlted hím.

⟨20b⟩ is odd in the same way as ⟨19a⟩, since telling a woman that she is beautiful can only constitute praise in our culture; it cannot constitute an insult. Or consider ⟨21⟩:

⟨21⟩ John called Mary $\begin{cases} \text{a whore} \\ \text{a Republican} \\ \text{a virgin} \\ \text{a lexicalist} \end{cases}$, and then shé insŭlted hím.

I find the sentences of ⟨21⟩ all perfectly well-formed, though those with other beliefs may disagree.

Similar examples have been discussed by Georgia Green [3]. Consider ⟨22⟩:

⟨22⟩ (a) Jane is a sloppy housekeeper and she doesn't take baths either.

(b) ?*Jane is a neat housekeeper and she doesn't take baths either.

The construction, *A and not B either*, carries with it the presupposition that one might expect *A* to entail *not B*. In ⟨22a⟩, such a presupposition is consistent with American cultural values, while in ⟨22b⟩ it would not be. Hence the ill-formedness

---

[a] Dwight Bolinger has observed that this is true only if the individuals involved are being contrasted. Suppose, instead, that the time of the events is being contrasted. Then, one can get:

John insŭlted Mary and THÉN shé insŭlted hím.

of ⟨22b⟩. However, one could easily imagine someone with appropriate cultural values such that he would judge ⟨22a⟩ to be ill-formed and ⟨22b⟩ to be well-formed.

There are other examples of this sort which involve identity constraints. For example, there are certain idiomatic expressions which require two noun phrases in the expression to be coreferential.

⟨23⟩ (a) I have my price.
     (b) *I have your price.[a]

⟨24⟩ (a) I'll take my chances.
     (b) *I'll take your chances.[b]

⟨25⟩ (a) I lost my cool, but I soon regained it.
     (b) *You lost your cool, but I soon regained it.

In cases like ⟨25⟩, where neither noun phrase commands the other, pronominalization is optional and the full noun phrases may be repeated.

⟨26⟩ (a) *Mary* lost her cool, but *she* soon regained it.
     (b) *Mary* lost her cool, but *Mary* soon regained it.

Although the noun phrases may be repeated without pronominalization in my speech, this is possible only in cases where the two noun phrases are presupposed to be coreferential.

⟨26⟩ (c) **Mary* lost her cool, but *Sam* soon regained it.

But since the well-formedness of such sentences is relative to a presupposition of coreferentiality, speakers' judgments will vary with their factual knowledge, beliefs, and information occurring previously in the discourse.

⟨27⟩ *Willie Mays* lost his cool, but the *centerfielder of the Giants* soon regained it.
⟨28⟩ **Willie Mays* lost his cool, but the *quarterback of the Colts* soon regained it.
⟨29⟩ Richard Nixon lost his cool, but $\begin{Bmatrix} the\ new\ president \\ *the\ former\ president \end{Bmatrix}$ soon regained it.[c]

Identity statements made earlier in a discourse also seem to count as presuppositions of coreferentiality. Compare ⟨28⟩ to ⟨30⟩.

⟨30⟩ Upon being informed that he had just been chosen quarterback of the Colts, *Willie Mays* lost his cool, but the *new quarterback of the Colts* soon regained it.

In all of the cases discussed above, the presuppositions involved were attributed to the speaker. However, there are also cases where presuppositions are attributed to some other individual mentioned in the sentence. Consider the difference between verbs like *claim*, *wish*, and *hear* on the one hand, and those like *hope*, *expect*, and *anticipate* on the other. As is well known, stative verbs and adjectives in general do not take the progressive auxiliary in English.

   [a] This may be all right for some speakers, though in a different sense. Suppose I have discovered how much it will take to bribe you, then I can say ⟨23b⟩. ⟨23a⟩ has a very different meaning.
   [b] Some speakers may find this all right in the sense of 'I'll take your chances for you', which is not the sense in which ⟨24a⟩ is understood.
   [c] Some constructions require noncoreferentiality, such as 'X has Y's cooperation'. Thus, we can say 'I have your cooperation', but not '*I have my cooperation', nor '*Sam has his own cooperation'. Similarly, one who assumed that Willie Mays was the centerfielder of the Giants would not find '*Willie Mays has the cooperation of the centerfielder of the Giants' acceptable.

⟨31⟩ (a) I $\left\{\begin{array}{l}\text{*am hearing}\\\text{hear}\end{array}\right\}$ that Sam is a fink.

(b) I $\left\{\begin{array}{l}\text{*am wishing}\\\text{wish}\end{array}\right\}$ that I had a knish.[a]

(c) I $\left\{\begin{array}{l}\text{*am knowing}\\\text{know}\end{array}\right\}$ that Rockefeller really hates blintzes.

(d) I am $\left\{\begin{array}{l}\text{*being amused}\\\text{amused}\end{array}\right\}$ that Sondra has warts.

However, there is a certain subclass of exceptions to this generalization, stative verbs which do take the progressive auxiliary.

⟨32⟩ (a) I $\left\{\begin{array}{l}\text{am expecting}\\\text{expect}\end{array}\right\}$ Schwartz' wife to run off with the butcher.

(b) I $\left\{\begin{array}{l}\text{am hoping}\\\text{hope}\end{array}\right\}$ that my date will turn out not to have warts.

(c) We $\left\{\begin{array}{l}\text{are anticipating}\\\text{anticipate}\end{array}\right\}$ that there will be a great advance in pornolinguistics.

These verbs form a rather interesting semantic class. Consider ⟨33⟩ and ⟨34⟩:

⟨33⟩ (a) Max claimed that his toothbrush was pregnant.
  (b) Max heard that his toothbrush was pregnant.
  (c) Max wished that his toothbrush were pregnant.

⟨34⟩ (a) Max expected his toothbrush to be pregnant.
  (b) Max hoped that his toothbrush was pregnant.
  (c) Max anticipated that his toothbrush would be pregnant.

In the sentences of ⟨33⟩, it is *not* presupposed that Max holds the belief that toothbrushes can reproduce. But in the sentences of ⟨34⟩ such a presupposition is made. That is, it is presupposed in ⟨34⟩ that Max believes that it is possible that his toothbrush could be pregnant. In general, the verbs *expect*, *hope*, and *anticipate* have the property that the sentence in the object complement is not now true, but is possible relative to the beliefs of the subject of the verb. Verbs with this property may optionally take the progressive auxiliary. Thus, there is an overt syntactic correlate of this interesting semantic property.

It is often said that certain aspects of language use are a part of a speaker's linguistic competence. For example, Searle in *Speech Acts* (Cambridge University Press, 1968) adopts the position that a speaker's knowledge of the felicity conditions governing what Austin has called 'illocutionary acts' are part of his linguistic competence, that is, his knowledge of his language. For example, the verb 'christen' as in 'I hereby christen this ship the Jackie Kennedy' has as felicity conditions that the subject of 'christen' is empowered by an appropriate authority to bestow a name on the object of 'christen' at the time of the act of christening, that the ship is present, etc. One might claim that felicity conditions are outside the realm of linguistic competence and are to be studied as part of performance. However, a look at nonperformative uses of potentially performative verbs indicates that is not so, and that Searle is right. Consider ⟨35⟩:

⟨35⟩ Sam smashed a bottle across the bow of the ship, thereby christening it the Jackie Kennedy, although he had no authority to bestow names upon ships.

---

[a] I assume that *wish* is understood here as a stative verb, not the active verb of the same form meaning to *make a wish*. The progressive is, of course, all right with that sense of *wish*.

⟨35⟩ involves a contradiction. It is a contradiction between the assertion that he had no authority to bestow names upon ships and what is presupposed by the verb 'christen', namely, that he had authority to bestow a name on the ship in question. This follows not from any knowledge of the world, but only from a knowledge of the meaning of 'christen'. Any adequate account of semantic representation must show that ⟨35⟩ involves a contradiction between what is asserted in the *although* clause and the presuppositional part of the meaning of 'christen'. Thus, felicity conditions *must* be represented as presuppositions which are part of the meaning of performative verbs if the contradiction involved in ⟨35⟩ is to be represented as part of one's linguistic competence. Thus, a knowledge of the felicity conditions for illocutionary acts turns out to be part of one's knowledge of the regularities by which a grammar pairs presuppositions with sentences, clearly a part of one's linguistic competence. This also indicates that various proposals to extend the notion of 'truth' to illocutionary acts, so that infelicitous acts will be called 'false', has a very sound basis, since that is exactly what must be done in cases like ⟨35⟩ where a potentially performative verb is used nonperformatively and where contradictions (implicitly involving the notion 'truth') can arise from felicity conditions which are presupposed by the verb in question.

Let us review what all this means. It is a fact that a speaker's judgment concerning whether a given sentence is deviant or not will vary with that speaker's factual knowledge, beliefs, etc. In cases like those discussed above, this is a fact about performance. The competence underlying such judgments involves the notion of relative grammaticality. A grammar can be viewed as generating pairs, (PR, S), consisting of a sentence, S, which is grammatical only relative to the presuppositions of PR. This pairing is relatively constant from speaker to speaker and does *not* vary directly with his factual knowledge, cultural background, etc. However, if a speaker is called upon to make a judgment as to whether or not S is 'deviant', then his extralinguistic knowledge enters the picture. Suppose the pair (PR, S) is generated by the grammar of his language. Part of his linguistic knowledge will be that S is well-formed only given PR. If the speaker's factual knowledge contradicts PR, then he may judge S to be 'deviant'.

Let us consider an example. Consider sentences ⟨20a⟩ and ⟨20b⟩:

⟨20⟩ (a) John told Mary that she was ugly and then she insulted him.

(b) John told Mary that she was beautiful and then she insulted him.

In sentences like ⟨20⟩ where reciprocal contrastive stress appears, we find two propositions: f(John, Mary) and g(Mary, John). In ⟨20a⟩, f(John, Mary) = John told Mary that she was ugly. In ⟨20b⟩, f(John, Mary) = John told Mary that she was beautiful. In both, g(Mary, John) = Mary insulted John. Sentences like ⟨20a⟩ and ⟨20b⟩ are well-formed only relative to the following presupposition:

⟨36⟩ f(John, Mary) entails g(John, Mary)

Thus, ⟨20a⟩ and ⟨20b⟩ are well-formed only relative to the presuppositions of ⟨37a⟩ and ⟨37b⟩ respectively.

⟨37⟩ (a) That John told Mary that she was ugly entails that John insulted Mary.

(b) That John told Mary that she was beautiful entails that John insulted Mary.

Thus, the grammar of English will generate the (PR, S) pairs:

$$(\langle 37a\rangle, \langle 20a\rangle) \qquad (\langle 37b\rangle, \langle 20b\rangle)$$

The knowledge that these are well-formed pairs is part of any English speaker's linguistic competence, regardless of his factual knowledge, beliefs, cultural background, etc. However, most English speakers come from a cultural background that makes the following assumptions:

⟨38⟩ (a) Telling a woman that she is ugly constitutes an insult (under normal conditions).

(b) Telling a woman that she is beautiful does not constitute an insult (under normal conditions).

Thus, given no special assumptions about John and Mary in ⟨20⟩, speakers of English whose cultural background assumes ⟨38⟩ will find that ⟨38a⟩ is consistent with ⟨37a⟩, but that ⟨38b⟩ is not consistent with ⟨37b⟩. Although both of the *pairs* (⟨37a⟩, ⟨20a⟩) and (⟨37b⟩, ⟨20b⟩) are well-formed with respect to the grammar of such speakers, those speakers will make the judgment that the *sentence* ⟨20b⟩ is 'deviant' relative to the cultural assumption of ⟨38b⟩, while granting that the *sentence* ⟨20a⟩ is 'not deviant' given the assumption of ⟨38a⟩. Thus, extralinguistic factors do not affect grammatical well-formedness, a notion from the theory of competence which is defined only for (PR, S) pairs; such factors do affect judgments of deviance, which concerns performance, i.e., the use of a sentence in a given context. The failure to observe this distinction has led to considerable confusion in the past decade.

This confusion was, I believe, fostered by Chomsky's use of the notion grammaticalness as relevant to sentences, not to (PR, S) pairs. Now one could, given our notion of well-formedness for (PR, S) pairs, define a different notion of syntactic well-formedness for sentences (not pairs) as follows: for all S, if there exists a PR such that (PR, S) is a well-formed (PR, S) pair, then S is 'syntactically well-formed'. Such a definition would be in the spirit of *Syntactic Structures* and Chomsky's more recent work as well. Since people can make up any definitions they feel like, one ought to ask what would be the point of making up such a definition of 'syntactically well-formed sentence'. Such a definition would define a field of presupposition-free syntax. One might ask then what would be the content of this field, what phenomena would it deal with, would it be interesting? Such a field of presupposition-free syntax would deviate from the traditional study of syntax in that it would no longer involve the study of the distribution of all grammatical morphemes. As we have seen, the distribution of grammatical morphemes like *who* versus *which* cannot be stated in terms of presupposition-free syntax. Since the phenomenon of reflexivization involves the notion of presupposed coreferentiality, the general principle concerning the distribution of reflexive pronouns and of noun phrases that are presupposed to be coreferential could not be stated in terms of presupposition-free syntax. Thus the difference between *I told you to shave yourself* and **I told you to shave you* would not be part of the study of presupposition-free syntax. Similarly, the study of the difference between the sentences *Mary is a girl taller than John* and **Mary is a taller girl than John* would not be part of presupposition-free syntax. And so on. It is not at all clear that very much that is interesting would be part of the study of presupposition-free syntax. It is not even clear that principled grounds could be found for motivating the notion of grammatical transformation within the bounds of such a field. Since selectional restrictions in general

involve presuppositions, any such restrictions could not be used to motivate transformations. If such grounds for motivating transformations were taken away, it is
not clear that very many, if any, of the traditionally assumed transformations could
be motivated within presupposition-free syntax. In fact, it may well turn out that
such a field would be limited to the study of the well-formedness conditions on
possible surface structures of a language. Such a field might well be no more interesting than traditional phrase structure grammar. At present, there is no reason to
believe that it would be.

Let us consider some examples. Al Alvarez and Kenneth Hale have informed me
that in Papago there are two kinds of conjunctions, what might be called the
'proximate' and the 'obviative'. The proximate conjunction appears when the two
conjoined sentences have the same (surface) subject; the obviative appears when the
subjects are different. The obviative conjunction has the phonological form /ku/
and forms a single phonological word with the auxiliary of the following sentence.
The proximate conjunction has the phonological form /k/ in the perfective, and
/c/ in the imperfective; it is phonologically joined to the verb of the *preceding*
sentence when the verb occurs in final position, and otherwise stands between the
two sentences, not joined phonologically to either.

⟨39⟩ Nixon ʔatṣ   wé peg gm hu cúksón wúi him-k
      Nixon aux-quotative first   there Tuscon to go-perfective-proximate

ʔatṣ   ʔamjeḍ gm hu kalihonia wúi hí:
aux-quotative then   there California to go-perfective
Nixon first went to Tuscon and (proximate) then [he] went to California.

⟨40⟩ *Nixon ʔatṣ   wé peg gm hu cúksón wúi hí:
      Nixon aux-quotative first   there Tuscon to go-perfective

kutṣ   ʔamjeḍ gm hu kalihonia wúi hí:
obviative-aux-quotative then there California to go-perfective
*Nixon first went to Tuscon and (obviative) then [he] went to California.

⟨39⟩ is normal in Papago; ⟨40⟩ is ungrammatical – /ku/ simply can't be used that
way in any such sentence. The situation would be reversed if 'Agnew' were inserted
after ' ʔamjed' in the two sentences (call the result ⟨39'⟩ and ⟨40'⟩), since then the
subjects of the sentences would be different. Any adequate account of the grammar
of Papago must take into account the complementary distribution of the proximate
and obviative conjunctions in such cases.

Now consider the following cases:

⟨41⟩ Nixon ʔatṣ   wé peg gm hu cúksón wúi him-k
      Nixon aux-quotative first   there Tuscon to go-perfective-proximate

ʔatṣ   ʔamjeḍ t-gkównal-ga gm hu kalihonia wúi hí:
aux-quotative then   the president there   California to go-perfective
Nixon first went to Tuscon and (proximate) then the president went to
California.

⟨41⟩ is grammatical if it is presupposed that Nixon is the president. The proximate
conjunction must be used in such a case; the obviative would be just as wrong as in
⟨40⟩. ⟨41⟩ would of course be ungrammatical relative to the presupposition that
Nixon is not the president. Prior to January 1969, ⟨41⟩ would have been judged by
a Papago as being just as ungrammatical as ⟨40⟩. Similarly, if /híki' hú/ 'former' is

inserted in ⟨41⟩ before /t-gkównal-ga/ 'president', the result would be judged ungrammatical given the presupposition that Nixon is now president. As soon as Nixon left office, such a sentence would be judged to be grammatical.

The generalization should be clear: the proximate conjunction is used just in case the surface subjects of the conjoined sentences are presupposed to be core-ferential at the time of the speech act, the obviative conjunction is used otherwise. However, this general principle could not be stated in presupposition-free syntax. Clearly ⟨39⟩, ⟨40⟩, ⟨39'⟩ and ⟨40'⟩ fall under the same general principle. Since ⟨40⟩, with a deleted pronoun, and ⟨39'⟩, with a different proper name, are ungrammatical under any set of presuppositions, presupposition-free syntax would have to state a principle ruling those sentences out on the basis of their surface distribution. However, in cases like ⟨41⟩, where the second conjunction has as subject a definite description (e.g., the president), presupposition-free syntax would have to consider all such sentences as grammatical no matter which conjunction was used, since there would always be some presupposition relative to which the sentence would be grammatical. The semantic component corresponding to such a presupposition-free syntax would then have to have an additional principle stating that, with the proxi-mate conjunction, coreferentiality of the two subjects was presupposed. This seman-tic principle would be doing the same work as the syntactic distribution statement. The only effect of assuming that syntax is presupposition-free is to make it impos-sible to state the generalization.

Take another example. Kim Burt has pointed out that the future auxiliary *will* can be deleted in what looks like a very strange set of environments in terms of presupposition-free syntactic structure.

⟨42⟩ (a) The Yankees play the Red Sox tomorrow.

  (b) ?*The Yankees play well tomorrow. The Yankees will play well tomorrow.

⟨43⟩ (a) I get my paycheck tomorrow.

  (b) ?*I get a cold tomorrow. I will get a cold tomorrow.

⟨44⟩ (a) The astronauts return to the earth tomorrow.

  (b) ?*The astronauts return safely tomorrow. The astronauts will return safely tomorrow.

⟨45⟩ (a) Sam gets a day off tomorrow.

  (b) ?*Sam enjoys his day off tomorrow. Sam will enjoy his day off tomorrow.

In terms of presupposition-free syntax, no general principle for the deletion of *will* can be stated. However, Burt has observed that the principle can be stated very simply (or at least approximated) in presuppositional terms: *will* can be deleted just in case it is presupposed that the event is one that the speaker can be sure of. In ⟨42⟩, one can be sure that game will take place, but one can't be sure that Yankees will play well. Since one's knowledge of the world doesn't match up with the presupposition required for the grammaticality of ⟨42b⟩, ⟨42b⟩ is judged odd. Similarly, ⟨43b⟩ is usually judged to be odd since one cannot be sure that one will catch a cold at a particular time. Suppose, however, that I had an appointment to go to some laboratory where I was to be injected with cold germs in a medical experi-ment. With such a presupposition, ⟨43b⟩ would be all right. ⟨44⟩ and ⟨45⟩ also work by the same principle. From such examples it seems clear that any general statement of where *will* can be deleted will have to go beyond the bounds of presupposition-free syntax.

GEORGE LAKOFF

It seems beyond doubt that the principles governing the distribution of morphemes will involve presuppositional information. Where these principles are given
by transformational rules (e.g., *will*-deletion in English), there may be linkages
between presuppositions and the transformational rules. Such linkages are called
'global derivational constraints', and are but special cases of a much more pervasive
phenomenon in grammar (cf. Lakoff [4], [5]).

University of Michigan

REFERENCES

[1] Chomsky, Noam, *Syntactic Structures*, Mouton, 1957.
[2] Chomsky, Noam, *Aspects of the Theory of Syntax*, M.I.T. Press, 1965.
[3] Green, Georgia, 'On *Too* and *Either*, and not just on *Too* and *Either* either',
    Papers from the Fourth Regional Meeting of the Chicago Linguistic Society,
    University of Chicago Linguistics Department, 1968.
[4] Lakoff, George, 'On Generative Semantics', in this volume.
[5] Lakoff, George, *Generative Semantics*, Holt, Rinehart and Winston, forthcoming.
[6] Lakoff, Robin, 'Some Reasons Why There Isn't Any *Some-Any* Rule', *Language*,
    1969.

Presupposition and assertion in the semantic analysis of nouns and verbs in English[a]

D. TERENCE LANGENDOEN

In this paper it will be argued that the presupposition-assertion distinction that is appropriate for the semantic analysis of verbs is inappropriate for the analysis of nouns, and that as a consequence lexical entries for nouns need not take note of this distinction.

The need for the presupposition-assertion distinction in the semantic analysis of verbs may be illustrated by means of examples that have recently been insightfully discussed by Fillmore (this volume). Fillmore points out that if one compares the sentences:

⟨1⟩ Harry criticized Mary for writing the editorial.
⟨2⟩ Harry accused Mary of writing the editorial.

one finds in ⟨1⟩ that Harry presupposed Mary was responsible for writing the editorial and that he asserted that writing the editorial was bad; whereas in ⟨2⟩ that Harry presupposed that writing the editorial was bad and that he asserted Mary was responsible for writing the editorial. In other words, the verbs *criticize* and *accuse* are converses of each other with respect to what is asserted and what is presupposed by the subject when these verbs are used as main verbs in sentences.

The standard test for the claim that such-and-such is presupposed in a sentence is to see whether it is preserved under negation. Thus, if we examine the negative counterparts to ⟨1⟩ and ⟨2⟩, namely:

⟨3⟩ Rocky didn't criticize Max for spending the loot.
⟨4⟩ Rocky didn't accuse Max of spending the loot.

we find that, indeed, the presuppositions of ⟨1⟩ and ⟨2⟩ are preserved. In ⟨3⟩, Rocky still presupposed Max was responsible for spending the loot, and in ⟨4⟩ he still presupposed that spending the loot was bad.

Harris Savin (personal communication) has recently suggested that the negation test can be generalized: presuppositions admit of no adverbial modification whatever, so that the fact that they are unaffected by negation is merely a special case of this more general principle. To see this, consider the examples:

⟨5⟩ Rocky rightfully criticized Max for spending the loot.
⟨6⟩ Rocky rightfully accused Max of spending the loot.

[a] This work was sponsored in part by ARPA, Grant no. DAHC 15 68 G-5 to the Rockefeller University.

Notice that examples ⟨5⟩ and ⟨6⟩ differ in the way in which the adverb *rightfully* modifies the main verb. In ⟨5⟩, Rocky's assertion that spending the loot was bad is further asserted by the speaker to be rightful, whereas in ⟨6⟩ his assertion that Max was responsible for spending the loot is asserted by the speaker to be rightful. In neither case, however, is Rocky's presupposition affected by the adverb. Similarly, in the sentences:

⟨7⟩ Rocky harshly criticized Max for spending the loot.
⟨8⟩ Rocky harshly accused Max of spending the loot.

we note that in ⟨7⟩ it is Rocky's assertion that spending the loot was bad which the speaker considers harsh (implying perhaps that Rocky considered spending the loot to be very bad), whereas in ⟨8⟩ it is Rocky's assertion that Max was responsible for spending the loot which is considered harsh (with no implication that Rocky considered spending the loot to be very bad). Thus again the presuppositions are seen to be unaffected by the adverb.

It has been assumed that the semantic structure of nouns similarly reflects the presupposition-assertion distinction that we have seen to be relevant for verbs (see McCawley, 1968: 267–8; Langendoen, 1970: Chapter 5). Thus, it has been argued, to take a familiar example, that in the sentence:

⟨9⟩ My neighbor is a bachelor.

it is presupposed that my neighbor is an adult, human, male and asserted that my neighbor has never been married. The negation test mentioned above has been used to support this claim, for it has been observed that the sentence:

⟨10⟩ My neighbor isn't a bachelor.

would be an unusual device for communicating information concerning my neighbor's physical maturity, species, or sex, but not concerning his marital status. There is, however, a slight equivocation here; it is said that to use ⟨9⟩ or ⟨10⟩ to assert my neighbor's maturity, species, or sex would be unusual, not ungrammatical, whereas to use *accuse* where one should have *criticize* would result, presumably, in ungrammaticality.

Indeed, we find that our predisposition toward interpreting the predicate noun *bachelor* as presupposing adult, human, male of its subject only holds when that subject refers to an individual, but not when it refers to a group. If we compare example ⟨10⟩ with ⟨11⟩:

⟨10⟩ My neighbor isn't a bachelor.
⟨11⟩ None of my neighbors are bachelors.

we notice that in ⟨11⟩ there is no predisposition on our part to presuppose that the subject refers to a group of males, unlike the situation in ⟨10⟩, in which normally we would presuppose that the subject does refer to a male individual. The reason for this is simply that one is more likely to know an individual's sex rather than his marital status, given that one knows just one of these facts, whereas when dealing with a group, neither characteristic of members of the group is more likely to be known.

Putting matters somewhat differently, we may say that if the presupposition-assertion distinction for nouns is appropriate, there is a considerable degree of freedom that one has in shifting various aspects of their meaning from the assertive side to the presuppositional side, and back again. Thus, we would be obliged to conclude that a sentence such as:

⟨12⟩ My cousin isn't a boy any more.

is ambiguous, depending upon whether the speaker is asserting that his cousin has grown up (presupposing that his cousin is male), or that his cousin has changed sex (presupposing that his cousin is young). Or, that for a word like *alligator*, one can presuppose only that its referent is a physical object, asserting everything else (in the semantic representation of *alligator*), on up to presupposing that its referent has all the properties alligators share with crocodiles, asserting only that it differs from a crocodile just in those respects that they are assumed to differ.

One consequence of our ability to shift the semantic content of nouns with relative freedom is that there can never be two nouns in a language which designate the same objects but which differ in the way in which *criticize* and *accuse* differ. For example, suppose there were a device which poached eggs and which operated by means of an internal combustion engine. We claim that there could be no language with two nouns, say *bnik* and *ftik*, such that when one said:

⟨13⟩ That is a bnik.

one would be presupposing that the device poached eggs and asserting that it is powered by an internal combustion engine, and conversely when one said:

⟨14⟩ That is a ftik.

In order to account for the semantic properties of predicate nouns that we have noted, we find that there is no need to appeal to the presupposition-assertion distinction that is relevant for verbs anyway; these properties are consequences of the syntactic form of the definition of nouns. Typically, a nominal definition takes on the form:

⟨15a⟩ an x that S

or:

⟨15b⟩ a part (piece, etc.) of an x that S

where *x* is a noun of more general meaning and *that S* is a restrictive relative clause. Now, if we examine a negative sentence in which the predicate nominal overtly has the syntactic structure of such a definition, for example:

⟨16⟩ John isn't a headwaiter who begins insulting you the minute you sit down.

we observe that under its most immediate interpretation, it is synonymous with an otherwise identical sentence, in which the negative appears in the relative clause:

⟨17⟩ John is a headwaiter who doesn't begin insulting you the minute you sit down.

Let us suppose that the placement of the negative in ⟨17⟩ corresponds most directly to its deep structure placement. Then, ⟨16⟩ is obtained by the familiar negative-raising transformation which relates also such sentences as:[a]

⟨18⟩ Matthew doesn't seem to appreciate Mark's sense of humor.

⟨19⟩ Matthew seems not to appreciate Mark's sense of humor.

[a] If this analysis is correct, then the complex noun phrase constraint of Ross (1967) will have to be formulated so as to permit the extraction of the negative element from within complex predicate noun phrases.

Returning to simple predicate nominal sentences such as ⟨10⟩:

⟨10⟩ My neighbor isn't a bachelor.

and paraphrasing it by substituting for *bachelor* its definition *man who has never been married*, we obtain:

⟨20⟩ My neighbor isn't a man who has never been married.

If we suppose that ⟨20⟩ has been obtained by negative-raising, just as ⟨16⟩ has, then ⟨20⟩ is synonymous with:

⟨21⟩ My neighbor is a man who has been married at some time.

(the double negation has been removed from the relative clause of ⟨21⟩ to make its surface structure grammatical). But ⟨21⟩ is a paraphrase of the usual sense given to ⟨10⟩ – it affirms that my neighbor is indeed an adult, human, male (i.e., a man) and denies only that he has never been married.

We leave open the question whether negative-raising has actually applied in the derivation of sentences like ⟨10⟩. Within the generative semantics framework, presumably the rule would be said to apply to an abstract structure having the components of ⟨21⟩, yielding ⟨20⟩, which in turn would yield ⟨10⟩ upon lexical insertion. Within interpretive semantics, the rule need not be assumed to have applied; the scope of the negation being a matter for the rules of semantic interpretation to handle.[a]

The flexibility inherent in the interpretation of predicate nominal sentences can be made a consequence of the flexibility of the definition of nouns, a flexibility which is permitted by the syntactic form of those definitions. Rather than saying that a noun is precisely defined by a particular expression of the form ⟨15⟩, let us say that this form is instead a definitional schema, where *x* may be chosen to be any noun in the hierarchical structure of which the noun is part, and where *that S* then specifies the remaining semantic content. It will be observed that this precisely characterizes the flexibility that nouns possess.

Graduate Center and Brooklyn College of the City University of New York

REFERENCES

Fillmore, Charles J. Types of lexical information. This volume.
Langendoen, D. Terence. 1970. *Essentials of English Grammar.* New York: Holt, Rhinehart and Winston.
McCawley, James D. 1968. Concerning the base component of a transformational grammar. *Foundations of Language,* IV, 243–69.
Ross, John R. 1967. *Constraints on Variables in Syntax.* Unpublished M.I.T. Ph.D. dissertation.

[a] Various difficulties have been glossed over in the present account. Notice that although the noun *headwaiter* itself has a definition of the form ⟨15⟩, the negative which appears in ⟨16⟩ can only originate in the overt relative clause, as in ⟨17⟩, not in the covert relative clause in the definition of *headwaiter*. Also, it should be observed that the element corresponding to x in ⟨15⟩ in the definitions of nouns need not be an actually occurring noun in English.

Fact[a]

PAUL KIPARSKY AND CAROL KIPARSKY

The object of this paper is to explore the interrelationship of syntax and semantics in the English complement system. Our thesis is that the choice of complement type is in large measure predictable from a number of basic semantic factors. Among these we single out for special attention *presupposition* by the speaker that the complement of the sentence expresses a true proposition. It will be shown that whether the speaker presupposes the truth of a complement contributes in several important ways to determining the syntactic form in which the complement can appear in the surface structure. A possible explanation for these observations will be suggested.

1 Two syntactic paradigms

The following two lists both contain predicates which take sentences as their subjects. For reasons that will become apparent in a moment, we term them *factive* and *non-factive*.

Factive	*Non-factive*	*Factive*	*Non-factive*
significant	likely	counts	appears
odd	sure	makes sense	happens
tragic	possible	suffices	chances
exciting	true	amuses	turns out
relevant	false	bothers	
matters	seems		

We shall be concerned with the differences in structure between sentences constructed with factive and non-factive predicates, e.g.

Factive: It is significant that he has been found guilty
Non-factive: It is likely that he has been found guilty.

On the surface, the two seem to be identically constructed. But as soon as we replace the *that*-clauses by other kinds of expressions, a series of systematic differences between the factive and non-factive predicates beings to appear.

[a] This work was supported in part by the U.S. Air Force (ESD Contract AF19(628)-2487) and the National Institutes of Health (Grant MH-13390-01). It is reprinted from M. Bierwisch and K. Heidolph (eds.), *Progress in Linguistics*, The Hague: Mouton, 1970.

This paper developed through several revisions out of a paper read in 1967 at Bucharest. These revisions were largely prompted by helpful discussions with many colleagues, among whom we would especially like to thank John Kimball, George Lakoff, Robin Lakoff, Haj Ross, and Timothy Shopen.

(1) Only factive predicates allow the noun *fact* with a sentential complement consisting of a *that*-clause or a gerund to replace the simple *that*-clause. For example,

The fact that the dog barked during the night
The fact of the dog's barking during the night

can be continued by the factive predicates *is significant, bothers me,* but not by the non-factive predicates *is likely, seems to me.*

(2) Only factive predicates allow the full range of gerundial constructions, and adjectival nominalizations in *-ness,* to stand in place of the *that*-clause. For example, the expressions

His being found guilty
John's having died of cancer last week
Their suddenly insisting on very detailed reports
The whiteness of the whale

can be subjects of factive predicates such as *is tragic, makes sense, suffices,* but not of non-factive predicates such as is *sure, seems, turns out.*

(3) On the other hand, there are constructions which are permissible only with non-factive predicates. One such construction is obtained by turning the initial noun phrase of the subordinate clause into the subject of the main clause, and converting the remainder of the subordinate clause into an infinitive phrase. This operation converts structures of the form

It is likely that he will accomplish even more
It seems that there has been a snowstorm

into structures of the form

He is likely to accomplish even more
There seems to have been a snowstorm.

We can do this with many non-factive predicates, although some, like *possible,* are exceptions:

It is possible that he will accomplish even more
*He is possible to accomplish even more.

However, none of the factive predicates can ever be used so:

*He is relevant to accomplish even more
*There is tragic to have been a snowstorm.

(4) For the verbs in the factive group, extraposition[a] is optional, whereas it is obligatory for the verbs in the non-factive group. For example, the following two sentences are optional variants:

That there are porcupines in our basement makes sense to me
It makes sense to me that there are porcupines in our basement.

But in the corresponding non-factive case the sentence with the initial *that*-clause is ungrammatical:

*That there are porcupines in our basement seems to me
It seems to me that there are porcupines in our basement.

---

[a] Extraposition is a term introduced by Jespersen for the placement of a complement at the end of a sentence. For recent transformational discussion of the complexities of this rule, see Ross (1967).

In the much more complex domain of object clauses, these syntactic criteria, and many additional ones, effect a similar division into factive and non-factive predicates. The following lists contain predicates of these two types:

Factive	*Non-factive*	*Factive*	*Non-factive*
regret	suppose	ignore	believe
be aware (of)	assert	make clear	conclude
grasp	allege	mind	conjecture
comprehend	assume	forget (about)	intimate
take into	claim	deplore	deem
consideration	charge	resent	fancy
take into account	maintain	care (about)	figure
bear in mind			

(1) Only factive predicates can have as their objects the noun *fact* with a gerund or *that*-clause:

Factive: I want to make clear the fact that I don't intend to participate
              You have to keep in mind the fact of his having proposed several alternatives
Non-factive: *I assert the fact that I don't intend to participate
              *We may conclude the fact of his having proposed several alternatives.

(2) Gerunds can be objects of factive predicates, but not freely of non-factive predicates:

Factive: Everyone ignored Joan's being completely drunk
              I regret having agreed to the proposal
              I don't mind your saying so
Non-factive: *Everyone supposed Joan's being completely drunk
              *I believe having agreed to the proposal
              *I maintain your saying so.

The gerunds relevant here are what Lees (1960) has termed 'factive nominals'. They occur freely both in the present tense and in the past tense (*having -en*). They take direct accusative objects, and all kinds of adverbs and they occur without any identity restriction on their subject.[a] Other, non-factive, types of gerunds are subject to one or more of these restrictions. One type refers to actions or events:

He avoided getting caught
*He avoided having got caught
*He avoided John's getting caught.

Gerunds also serve as substitutes for infinitives after prepositions:

I plan to enter the primary
I plan on entering the primary
*I plan on having entered the primary last week.

Such gerunds are not at all restricted to factive predicates.

[a] There is, however, one limitation on subjects of factive gerunds:
*It's surprising me that he succeeded dismayed John
*There's being a nut loose disgruntles me.
The restriction is that clauses cannot be subjects of gerunds, and the gerund formation rule precedes extraposition and *there*-insertion.

(3) Only non-factive predicates allow the accusative and infinitive construction:

Non-factive: I believe Mary to have been the one who did it
He fancies himself to be an expert in pottery
I supposed there to have been a mistake somewhere
Factive: *I resent Mary to have been the one who did it
*He comprehends himself to be an expert in pottery
*I took into consideration there to have been a mistake somewhere.

As we earlier found in the case of subject complements, the infinitive construction is excluded, for no apparent reason, even with some non-factive predicates, e.g. *charge*. There is, furthermore, considerable variation from one speaker to another as to which predicates permit the accusative and infinitive construction, a fact which may be connected with its fairly bookish flavor. What is significant, however, is that the accusative and infinitive is not used with factive predicates.

2   Presupposition

These syntactic differences are correlated with a semantic difference. The force of the *that*-clause is not the same in the two sentences

It is odd that it is raining (factive)
It is likely that it is raining (non-factive)

or in the two sentences

I regret that it is raining (factive)
I suppose that it is raining (non-factive).

The first sentence in each pair (the factive sentence) carries with it the presupposition 'it is raining'. The speaker presupposes that the embedded clause expresses a true proposition, and makes some assertion about that proposition. All predicates which behave syntactically as factives have this semantic property, and almost none of those which behave syntactically as non-factives have it.[a] This, we propose, is the basic difference between the two types of predicates. It is important that the following things should be clearly distinguished:

(1) Propositions the speaker asserts, directly or indirectly, to be true
(2) Propositions the speaker presupposes to be true.

Factivity depends on presupposition and not on assertion. For instance, when someone says

It is true that John is ill
John turns out to be ill

---

[a] There are some exceptions ot this second half of our generalization. Verbs like *know*, *realize*, though semantically factive, are syntactically non-factive, so that we cannot say *I know the fact that John is here*, *I know John's being here*, whereas the propositional constructions are acceptable: *I know him to be here*. There are speakers for whom many of the syntactic and semantic distinctions we bring up do not exist at all. Professor Archibald Hill has kindly informed us that for him factive and non-factive predicates behave in most respects alike and that even the word *fact* in his speech has lost its literal meaning and can head clauses for which no presupposition of truth is made. We have chosen to describe a rather restrictive type of speech (that of C.K.) because it yields more insight into the syntactic-semantic problems with which we are concerned.

he is *asserting* that the proposition 'John is ill' is a true proposition, but he is not *presupposing* that it is a true proposition. Hence these sentences do not follow the factive paradigm:

*John's being ill is true
*John's being ill turns out
*The fact of John's being ill is true
*The fact of John's being ill turns out.

The following sentences, on the other hand, are true instances of presupposition:

It is odd that the door is closed
I regret that the door is closed.

The speaker of these sentences presupposes 'the door is closed' and furthermore asserts something else about that presupposed fact. It is this semantically more complex structure involving presupposition that has the syntactic properties we are dealing with here.

When factive predicates have first person subjects it can happen that the top sentence denies what the complement presupposes. Then the expected semantic anomaly results. Except in special situations where two egos are involved, as in the case of an actor describing his part, the following sentences are anomalous:

*I don't realize that he has gone away
*I have no inkling that a surprise is in store for me.[a]

Factivity is only one instance of this very basic and consequential distinction. In formulating the semantic structure of sentences, or, what concerns us more directly here, the lexical entries for predicates, we must posit a special status for pre-suppositions, as opposed to what we are calling assertions. The speaker is said to 'assert' a sentence plus all those propositions which follow from it by virtue of its meaning, not, e.g., through laws of mathematics or physics.[b] Presumably in a seman-tic theory assertions will be represented as the central or 'core' meaning of a sentence – typically a complex proposition involving semantic components like 'S$_1$

---

[a] In some cases what at first sight looks like a strange meaning-shift accompanies negation with first person subjects. The following sentences can be given a non-factive interpretation which prevents the above kind of anomaly in them:

I'm not aware that he has gone away
I don't know that this isn't our car.

It will not do to view these non-factive *that*-clauses as indirect questions:

*I don't know that he has gone away or not.

We advance the hypothesis that they are deliberative clauses, representing the same construc-tion as clauses introduced *but that*:

I don't know but that this is our car.

This accords well with their meaning, and especially with the fact that deliberative *but that*-clauses (in the dialects that permit them at all) are similarly restricted to negative sentences with first person subjects:

*I know but that this is our car
*John doesn't know but that this is our car.

[b] We prefer 'assert' to 'imply' because the latter suggests consequences beyond those based on knowledge of the language. This is not at all to say that linguistic knowledge is disjoint from other knowledge. We are trying to draw a distinction between two statuses a defining proposition can be said to have in the definition of a predicate, or meaning of a sentence, and to describe some consequences of this distinction. This is a question of the semantic structure of words and can be discussed independently of the question of the relationship between the encyclopedia and the dictionary.

cause S₂', 'S become', 'N want S' – plus the propositions that follow from it by redundancy rules involving those components. The formulation of a simple example should help clarify the concepts of assertion and presupposition.

Mary cleaned the room.

The dictionary contains a mapping between the following structures:

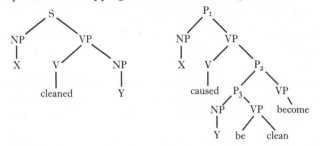

where S refers to the syntactic object 'Sentence' and P to the semantic object 'Proposition'.

A redundancy rule states that the object of 'cause' is itself asserted:

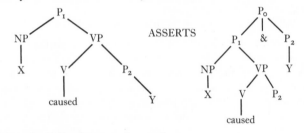

This rule yields the following set of assertions:

X caused[a] Y to become clean

Y became clean.

(Why the conjunction of P_1 and P_2 is subordinated to P_0 will become clear below, especially in (3) and (5).)

Furthermore, there is a presupposition to the effect that the room was dirty before the event described in the sentence. This follows from 'become', which presupposes that its complement has, up to the time of the change referred to by 'become', not been true. This may be expressed as a redundancy rule:

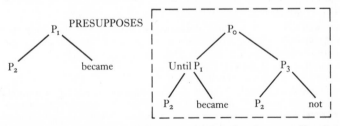

---

[a] Though we cannot go into the question here, it is clear that the tense of a sentence conveys information about the time of its presuppositions as well as of its assertions, direct and indirect. Thus tense (and likewise mood, cf. p. 367 below) is not an 'operator' in the sense that negation and other topics discussed in this section are.

(Presuppositions will be enclosed in dotted lines. Within the context of a tree diagram representing the semantic structure of a sentence presuppositions which follow from a specific semantic component will be connected to it by a dotted line.)

That this, like the factive component in *regret* or *admit*, is a presupposition rather than an assertion can be seen by applying the criteria in the following paragraphs.

(1) Presuppositions are constant under negation. That is, when you negate a sentence you don't negate its presuppositions; rather, what is negated is what the positive sentence asserts. For example,

Mary didn't clean the room

unlike its positive counterpart does not assert either that the room became clean or, if it did, that it was through Mary's agency. On the other hand, negation does not affect the presupposition that it was or has been dirty. Similarly, these sentences with factive predicates (underlined) –

It is not odd that the door is closed

John doesn't regret that the door is closed

presuppose, exactly as do their positive counterparts, that the door is closed.

In fact, if you want to deny a presupposition, you must do it explicity:

Mary didn't clean the room; it wasn't dirty

Legree didn't force them to work; they were willing to

Abe didn't regret that he had forgotten; he had remembered.

The second clause casts the negative of the first into a different level; it's not the straightforward denial of an event or situation, but rather the denial of the appropriateness of the word in question (underlined above). Such negations sound best with the inappropriate word stressed.

(2) Questioning, considered as an operation on a proposition P, indicates 'I do not know whether P'. When I ask

Are you *dismayed* that our money is gone?

I do not convey that I don't know whether it is gone but rather take that for granted and ask about your reaction.

(Note that to see the relation between factivity and questioning only yes–no questions are revealing. A question like

Who is aware that Ram eats meat

already by virtue of questioning an argument of *aware*, rather than the proposition itself; presupposes a corresponding statement:

Someone is aware that Ram eats meat.

Thus, since presupposition is transitive, the *who*-question presupposes all that the *someone*-statement does.)

Other presuppositions are likewise constant under questioning. For instance: a verb might convey someone's evaluation of its complement as a presupposition. To say 'they *deprived* him of a visit to his parents' presupposes that he wanted the visit (vs. 'spare him a visit...'). The presupposition remains in 'Have they deprived him of a...?' What the question indicates is 'I don't know whether they have kept him from...'

(3) It must be emphasized that it is the *set* of assertions that is operated on by question and negation. To see this, compare –

Mary didn't kiss John
Mary didn't clean the house.

They have certain ambiguities which, as has often been noted, are systematic under negation. The first may be equivalent to any of the following more precise sentences:

Someone may have kissed John, but not Mary
Mary may have kissed someone, but not John
Mary may have done something, but not kiss John
Mary may have done something to John, but not kiss him.

And the second:

Someone may have cleaned the house, but not Mary
Mary may have cleaned something, but not the house
Mary may have done something, but not clean the house
Mary may have done something to the house, but not clean it.

All of these readings can be predicted on the basis of the constituent structure:

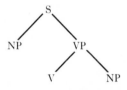

Roughly, each major constituent may be negated.
But the second sentence has still another reading:

Mary may have been cleaning the house,
but it didn't get clean.

That extra reading has no counterpart in the other sentence. *Clean* is semantically more complex than *kiss* in that whereas *kiss* has only one assertion (press the lips against), *clean* has two, as we have seen above. How this affects the meaning of the negative sentence can be seen through a derivation:

(i)

   'Mary didn't clean the house.'

(ii) Application of redundancy rule on 'cause':

'It's not the case that both Mary cleaned the house and the house is clean.'

(iii) DeMorgan's Law yields

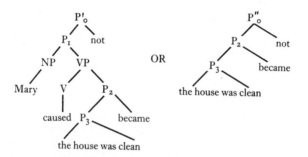

'Either "Mary didn't clean the house" or "the house didn't get clean".'

Thus to say *Mary didn't clean the house* is to make either of the two negative assertions in (iii). The remaining readings arise from distribution of *not* over the constituents of the lexicalized sentence.

Presumably the same factors account for the corresponding ambiguity of *Did Mary clean the house?*

(4) If we take an imperative sentence like

(You) chase that thief!

to indicate something like

I want (you chase that thief)

then what 'I want' doesn't include the presuppositions of S. For example, S pre-supposes that

That thief is evading you

but that situation is hardly part of what 'I want'.

The factive complement in the following example is likewise presupposed independently of the demand:

Point it out to oo6 that the transmitter will function poorly in a cave.

Assume the dictionary contains this mapping:

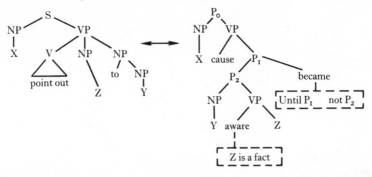

From the causative redundancy rule, which adds the assertion P_1, the definition of *point out* and the fact that *want* distributes over subordinate conjuncts, it follows that the above command indicates

I want oo6 to become aware that the transmitter...

However it doesn't in any way convey

I want the transmitter to function poorly in a cave

nor, of course, that

I want oo6 not to have been aware...

(5) We have been treating negation, questioning and imperative as operations on propositions like implicit 'higher sentences'. Not surprisingly explicit 'higher sentences' also tend to leave presuppositions constant while operating on assertions. Our general claim is that the assertions of a proposition (P_k) are made relative to that proposition within its context of dominating propositions. Presuppositions, on the other hand, are relative to the speaker. This is shown in figures 1 and 2. Fig. 1 shows that the presuppositions of P_k are also presupposed by the whole proposition P_0. In fig. 2 we see that whatever P_0 asserts about P_k it also asserts about the *set* (see (3) above) of propositions that P_k asserts.

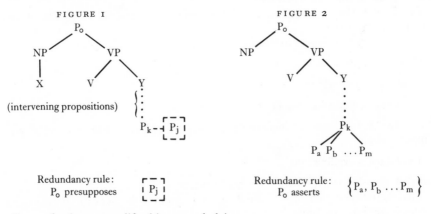

Let us further exemplify this general claim:

John appears to regret evicting his grandmother.

Since *appear* is not factive this sentence neither asserts nor presupposes

John regrets evicting her.

However it does presuppose the complement of the embedded factive verb *regret*, as well as the presupposition of *evict* to the effect that he was her landlord.

It does not matter how deeply the factive complement (italicized) is embedded:

Abe thinks it is possible that Ben is becoming ready to encourage Carl to acknowledge *that he had behaved churlishly.*

This claim holds for presuppositions other than factivity. We are not obliged to conclude from

John refuses to remain a bachelor all his life

that he plans to undergo demasculating surgery, since *bachelor* asserts *unmarried*, but only presupposes *male* and *adult*. Thus (ii) yields:

John refuses to remain unmarried all his life

but not

John refuses to remain male (adult) all his life.

(6) A conjunction of the form S_1 *and* S_2 *too* serves to contrast an item in S_1 with one in S_2 by placing them in contexts which are in some sense not distinct from each other. For instance:

Tigers are ferocious and panthers are (ferocious) too
*Tigers are ferocious and panthers are mild-mannered too.

Abstracting away from the contrasting items, S_1 might be said to semantically include S_2. The important thing for us to notice is that the relevant type of inclusion is *assertion*. Essentially, S_2 corresponds to an assertion of S_1. To see that presupposition is not sufficient consider the following sentences. The second conjunct in each of the starred sentences corresponds to a presupposition of the first conjunct, while in the acceptable sentences there is an assertion relationship.

John deprived the mice of food and the frogs didn't get any either
*John deprived the mice of food and the frogs didn't want any either
John forced the rat to run a maze and the lizard did it too
*John forced the rat to run a maze and the lizard didn't want to either
Mary's refusal flabbergasted Ron, and he was surprised at Betty's refusal too
*Mary's refusal flabbergasted Ron and Betty refused too.

3 A hypothesis

So far, we have presented a set of syntactic-semantic correlations without considering how they might be accounted for. We shall continue by analyzing these facts and others to be pointed out in the course of the discussion, in terms of a tentative explanatory hypothesis, by which the semantic difference between the factive and non-factive complement paradigms can be related to their syntactic differences, and most of the syntactic characteristics of each paradigm can be explained. The hypothesis which we should like to introduce is that presupposition of complements is reflected in their syntactic deep structure. Specifically, we shall explore the possibility that factive and non-factive complements at a deeper level of representation differ as follows:[a]

Factive   Non-factive

If this interpretation is correct, then closest to the factive deep structure are sentences of the type

I regret the fact that John is ill.

The forms in the factive paradigm are derived by two optional transformations: formation of gerunds from *that*-clauses in position after nouns, and deletion of the head noun *fact*. (We do not pause to consider the general rules which take care of the detail involving *that* and *of*.) By gerund-formation alone we get

I regret the fact of John's being ill.

Fact-deletion can apply to this derived structure, giving

I regret John's being ill.

If *fact*-deletion applies directly to the basic form, then the simple *that*-clause is formed:

I regret that John is ill.

Although this last factive sentence has the same superficial form as the non-factive

I believe that John is ill

according to our analysis it differs radically from it in syntactic form, and the two sentences have different deep structures as diagrammed above. Simple *that*-clauses are ambiguous, and constitute the point of overlap (neutralization) of the factive and non-factive paradigms.

If factive clauses have the deep structures proposed by us, these various surface forms in which factive clauses can appear become very easy to derive. That is one piece of support for our hypothesis. The remaining evidence can be grouped under three general headings:

(1) syntactic insulation of factive clauses (section 4)
(2) indifferent and ambiguous predicates (section 5)
(3) pronominalization (section 6).

4   Syntactic insulation of factive clauses

Let us first return in somewhat more detail to infinitive constructions, examining first the derivation of infinitives in general and then of the class of infinitive constructions which we mentioned as being characteristic of non-factive predicates. Basic to our treatment of infinitives is the assumption that non-finite verb forms in all languages are the basic, unmarked forms. Finite verbs, then, are always the result of person and number agreement between subject and verb, and non-finite verbs, in particular, infinitives, come about when agreement does not apply. Infinitives arise regularly when the subject of an embedded sentence is removed by

[a] For further discussion of this point see additional notes on pp. 365–9.

a transformation, or else placed into an oblique case, so that in either case agreement between subject and verb cannot take place. There are several ways in which the subject of an embedded sentence can be removed by a transformation. It can be deleted under identity with a noun phrase in the containing sentence, as in sentences like *I decided to go* and *I forced John to go* (cf. Rosenbaum 1967).

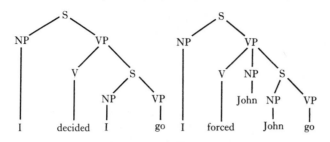

After prepositions, infinitives are automatically converted to gerunds, e.g. *I decided to go* vs. *I decided on going*; or *I forced John to do it* vs. *I forced John into doing it*. These infinitival gerunds should not be confused with the factive gerunds, with which they have in common nothing but their surface form.

A second way in which the subject of an embedded sentence can be removed by a transformation to yield infinitives is through raising of the subject of the embedded sentence into the containing sentence. The remaining verb phrase of the embedded sentence is then automatically left in infinitive form. This subject-raising transformation applies only to non-factive complements, and yields the accusative and infinitive, and nominative and infinitive constructions:

He believes Bacon to be the real author
This seems to be Hoyle's best book.

The operation of the subject-raising rule in object clauses can be diagrammed as follows:

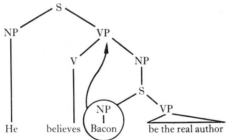

The circled noun phrase is raised into the upper sentence and becomes the surface object of its verb.[a]

We reject, then, as unsuccessful the traditional efforts to derive the uses of the infinitive from its being 'partly a noun, partly a verb', or, perhaps, from some 'basic meaning' supposedly shared by all occurrences of infinitives. We reject, also, the

---

[a] This subject-raising rule has figured in recent work under at least three names: pronoun replacement (Rosenbaum 1967); expletive replacement (Langendoen 1966); and *it*-replacement (Ross 1967). Unfortunately we have had to invent still another, for none of the current names fit the rule as we have reformulated it.

assumption of recent transformational work (cf. Rosenbaum 1967) that all infinitives are *'for-to'* constructions, and that they arise from a 'complementizer placement' rule which inserts *for* and *to* before clauses on the basis of an arbitrary marking on their verbs. Instead, we claim that what infinitives share is only the single, relatively low-level syntactic property of having no surface subject.

Assuming that the subject-raising rule is the source of one particular type of infinitive complements, we return to the fact, mentioned earlier, that factive complements never yield these infinitive complements. We now press for an explanation. Why can one not say

*He regrets Bacon to be the real author
*This makes sense to be Hoyle's best book

although the corresponding *that*-clauses are perfectly acceptable? It is highly unlikely that this could be explained directly by the *semantic* fact that these sentences are constructed with factive predicates. However, the deep structure which we have posited for factive complements makes a syntactic explanation possible.

Ross (1967) has found that transformations are subject to a general constraint, termed by him the Complex Noun Phrase Constraint, which blocks them from taking constituents out of a sentence *S* in the configuration

For example, elements in relative clauses are immune to questioning: *Mary* in *The boy who saw Mary came back* cannot be questioned to give **Who did the boy who saw come back?* The complex noun phrase constraint blocks this type of questioning because relative clauses stand in the illustrated configuration with their head noun.

This complex noun phrase constraint could explain why the subject-raising rule does not apply to factive clauses. This misapplication of the rule is excluded if, as we have assumed, factive clauses are associated with the head noun *fact*. If the optional transformation which drops this head noun applies later than the subject-raising transformation (and nothing seems to contradict that assumption), then the subjects of factive clauses cannot be raised. No special modification of the subject-raising rule is necessary to account for the limitation of infinitive complements to non-factive predicates.

Another movement transformation which is blocked in factive structures in the same way is NEG-raising (Klima 1964), a rule which optionally moves the element NEG(ATIVE) from an embedded sentence into the containing sentence, converting for example the sentences

It's likely that he won't lift a finger until it's too late
I believe that he can't help doing things like that

into the synonymous sentences

It's not likely that he will lift a finger until it's too late
I don't believe that he can help doing things like that.

Since *lift a finger*, punctual *until*, and *can help* occur only in negative sentences, sentences like these prove that a rule of NEG-raising is necessary.

This rule of NEG-raising never applies in the factive cases. We do not get, for example,

*It doesn't bother me that he will lift a finger until it's too late

from

It bothers me that he won't lift a finger until it's too late

or

*I don't regret that he can help doing things like that

from

I regret that he can't help doing things like that.

Given the factive deep structure which we have proposed, the absence of such sentences is explained by the complex noun phrase constraint, which exempts structures having the formal properties of these factive deep structures from undergoing movement transformations.[a]

Factivity also erects a barrier against insertions. It has often been noticed that subordinate clauses in German are not in the subjunctive mood if the truth of the clause is presupposed by the speaker, and that sequence of tenses in English and French also depends partly on this condition. The facts are rather complicated, and to formulate them one must distinguish several functions of the present tense and bring in other conditions which interact with sequence of tenses and subjunctive insertion. But it is sufficient for our purposes to look at minimal pairs which show that one of the elements involved in this phenomenon is factivity. Let us assume that Bill takes it for granted that the earth is round. Then Bill might say:

John claimed that the earth was (*is) flat

with obligatory sequence of tenses, but

John grasped that the earth is (was) round

with optional sequence of tenses. The rule which changes a certain type of present tense into a past tense in an embedded sentence if the containing sentence is past, is obligatory in non-factives but optional in factives. The German subjunctive rule is one notch weaker: it is optional in non-factives and inapplicable in factives:

Er behauptet, dass die Erde flach sei (ist)
Er versteht, dass die Erde rund ist (*sei).

The reason why these changes are in part optional is not clear. The exact way in which they are limited by factivity cannot be determined without a far more detailed investigation of the facts than we have been able to undertake. Nevertheless, it is fairly likely that factivity will play a role in an eventual explanation of these phenomena.[b]

[a] We thought earlier that the oddity of questioning and relativization in some factive clauses was also due to the complex noun phrase constraint:

*How old is it strange that John is?

*I climbed the mountain which it is interesting that Goethe tried to climb.

Leroy Baker (1967) has shown that this idea was wrong, and that the oddity here is not due to the complex noun phrase constraint. Baker has been able to find a semantic formulation of the restriction on questioning which is fairly general and accurate. It appears now that questioning and relativization are rules which follow *fact*-deletion.

[b] This may be related to the fact that (factive) present gerunds can refer to a past state, but (non-factive) present infinitives can not. Thus,

They resented his being away

5   Indifferent and ambiguous predicates

So far, for clarity of exposition, only predicates which are either factive or non-factive have been examined. For this set of cases, the factive and non-factive complement paradigms are in complementary distribution. But there are numerous predicates which take complements of both types. This is analogous to the fact that there are not only verbs which take concrete objects and verbs which take abstract objects but also verbs which take either kind. For example, *hit* requires concrete objects (*boy, table*), *clarify* requires abstract objects (*ideas, fact*), and *like* occurs indifferently with both. Just so we find verbs which occur indifferently with factive and non-factive complements, e.g. *anticipate, acknowledge, suspect, report, remember, emphasize, announce, admit, deduce.* Such verbs have no specification in the lexicon as to whether their complements are factive. On a deeper level, their semantic representations include no specifications as to whether their complement sentences represent presuppositions by the speaker or not. Syntactically, these predicates participate in both complement paradigms.

It is striking evidence for our analysis that they provide minimal pairs for the subtle meaning difference between factive and non-factive complements. Compare, for example, the two sentences

They reported the enemy to have suffered a decisive defeat
They reported the enemy's having suffered a decisive defeat.

The second implies that the report was true in the speaker's opinion, while the first leaves open the possibility that the report was false. This is explained by our derivation of infinitives from non-factives and gerunds from factives. Similarly compare

I remembered him to be bald (so I was surprised to see him with long hair)
I remembered his being bald (so I brought along a wig and disguised him).

Contrast *forget*, which differs from *remember* in that it necessarily presupposes the truth of its object. Although it is logically just as possible to forget a false notion as it is to remember one, language seems to allow for expressing only the latter. We cannot say

*I forgot that he was bald, which was a good thing since it turned out later that he wasn't after all
*I forgot him to be bald.

There is another kind of case. Just as different meanings may accompany subjects or objects differing by a feature like concreteness, as in

The boy struck$_1$ me
The idea struck$_2$ me

is ambiguous as to the time reference of the gerund, and on one prong of the ambiguity is synonymous with
They resented his having been away.
But in
They supposed him to be away
the infinitive can only be understood as contemporaneous with the main verb, and the sentence can never be interpreted as synonymous with
They supposed him to have been away.

so verbs may occur with factive and non-factive complements in different meanings. Compare –

(*a*) I explained Adam's refusing to come to the phone
(*b*) I explained that he was watching his favorite TV show.

In (*a*), the subordinate clause refers to a proposition regarded as a fact. *Explain*, in this case, means 'give reasons for'. When the object is a *that*-clause, as in (*b*), it can be read as non-factive, with *explain that S* understood as meaning 'say that S to explain X'. To account for the differences between (*a*) and (*b*), we might postulate two lexical entries for *explain* (not denying that they are related). In the entry appropriate to (*a*) there would be a presupposition that the subordinated proposition is true. This would require a factive complement (recall that the form of the complement has an associated interpretation) in the same way as the two verbs *strike*₁ and *strike*₂ would receive different kinds of subjects. The entry for (*b*) would have among its presuppositions that the speaker was not committing himself about the truth of the subordinated proposition, so that a factive complement would not fit. Thus, the meaning of the complement form is directly involved in explaining its occurrence with particular verbs.

6 Pronominalization

The pronoun *it* serves as an optional reduction of *the fact*. It can stand directly before *that*-clauses in sentences with factive verbs:

Bill resents it that people are always comparing him to Mozart
They didn't mind it that a crowd was beginning to gather in the street.

Although the difference is a delicate one, and not always clearcut, most speakers find *it* unacceptable in the comparable non-factive cases:

*Bill claims it that people are always comparing him to Mozart
*They supposed it that a crowd was beginning to gather in the street.

This *it*, a reduced form of *the fact*, should be distinguished from the expletive *it*, a semantically empty prop which is automatically introduced in the place of extraposed complements in sentences like

It seems that both queens are trying to wriggle out of their commitments
It is obvious that Muriel has lost her marbles.

Rosenbaum (1967) tried to identify the two and to derive both from an *it* which he postulated in the deep structure of all noun clauses. This was in our opinion a mistake. In the first place, the two *it*'s have different distributions. Expletive *it* comes in regardless of whether a factive or non-factive clause is extraposed, and does not appear to be related to the lexical noun *fact*, as factive *it* is.

The relationship of factive *it* to the lexical noun *fact*, and its distinction from expletive *it*, is brought out rather clearly by a number of transformational processes. For example, the presence of factive *it* blocks the formation of relative clauses just as the lexical noun *fact* does:

*This is the book which you reported it that John plagiarized
*This is the book which you reported the fact that John plagiarized
This is the book which you reported that John plagiarized.

But expletive *it* differs in permitting relativization:

That's the one thing which it is obvious that he hadn't expected
*That's the one thing which the fact is obvious that he hadn't expected.

As Ross (1966) has shown, facts like these create seemingly insoluble problems for a system like Rosenbaum's, in which factive and expletive *it* are derived from the same source. We have not proposed an alternative in anything like sufficient detail, but it is fairly clear that a system of rules constructed along the general lines informally sketched out here, which makes exactly the required syntactic distinction, will not have inherent difficulties in dealing with these facts.

Direct comparison of factive *it* and expletive *it* shows the expected semantic difference. The comparison can be carried out with the verbs which are indifferent as to factivity:

I had expected that there would be a big turnout (but only three people came)
I had expected it that there would be a big turnout (but this is ridiculous—get more chairs).

The second sentence, with *it*, suggests that the expectation was fulfilled, whereas the first is neutral in that respect. On the other hand, expletive *it* adds no factive meaning, and the following sentence is ambiguous as between the factive and non-factive interpretation:

It was expected that there would be a big turnout.

This analysis makes the prediction that cases of *it* which cannot be derived from *fact* will present no obstacle to relativization. This is indeed the case:

Goldbach's conjecture, which I take it that you all know...
The report, which I will personally see to it that you get first thing in the morning...
This secret, which I would hate it if anyone ever revealed...

On the other hand, it is not too clear where these *it*'s do come from. Perhaps their source is the 'vacuous extraposition' postulated by Rosenbaum (1967).[a]

The deep structures which we have posited for the two types of complements also explain the way in which they get pronominalized. In general, both factive and non-factive clauses take the pro-form *it*:

John supposed that Bill had done it, and Mary supposed it, too
John regretted that Bill had done it, and Mary regretted it, too.

But the two differ in that only non-factive clauses are pronominalized by *so:*

John supposed that Bill had done it, and Mary supposed so, too
*John regretted that Bill had done it, and Mary regretted so, too.

These facts can be explained on the basis of the fairly plausible assumptions that *it* is the pro-form of noun phrases, and *so* is the pro-form of sentences. Referring back to the deep structures given in section 3, we see that the only node which exhaustively dominates factive complements is the node NP. For this reason the only pro-form for them is the pro-form for noun phrases, namely, *it*. But non-factive complements are exhaustively dominated by two nodes: NP and S. Accordingly, two pro-forms are available: the pro-form for noun phrases, *it*, and the pro-form for sentences, *so*.

[a] Dean (1967) has presented evidence from German and English that extraposition is the general source of expletive pronouns.

7 Emotives

In the above discussion we rejected Rosenbaum's derivation of infinitive complements like

I believe John to have liked Anselm
I forced John to say 'cheese'

from hypothetical underlying forms with *for–to*:

*I believe for John to have liked Anselm
*I forced John for John to say 'cheese'.

This leaves us with the onus of explaining the *for–to* complements which actually occur on the surface:

It bothers me for John to have hallucinations
I regret for you to be in this fix.

But once the spurious *for–to*'s are stripped away, it becomes clear that the remaining real cases occur with a semantically natural class of predicates. Across the distinction of factivity there cuts orthogonally another semantic distinction, which we term *emotivity*. Emotive complements are those to which the speaker expresses a subjective, emotional, or evaluative reaction. The class of predicates taking emotive complements includes the verbs of emotion of classical grammar, and Klima's affective predicates (Klima 1964), but is larger than either and includes in general all predicates which express the subjective value of a proposition rather than knowledge about it or its truth value. It is this class of predicates to which *for–to* complements are limited. The following list illustrates the wide range of meanings to be found and shows the cross-classification of emotivity and factivity:

	Emotive	*Non-emotive*
FACTIVE EXAMPLES		
Subject clauses	important	well-known
	crazy	clear
	odd	(self-evident)
	relevant	goes without saying
	instructive	
	sad	
	suffice	
	bother	
	alarm	
	fascinate	
	nauseate	
	exhilarate	
	defy comment	
	surpass belief	
	a tragedy	
	no laughing matter	
Object clauses	regret	be aware (of)
	resent	bear in mind
	deplore	make clear
		forget
		take into account

	Emotive	Non-emotive
NON FACTIVE-EXAMPLES		
Subject clauses	improbable	probable
	unlikely	likely
	a pipedream	turn out
	nonsense	seem
future { urgent	imminent
	vital	in the works
Object clauses	intend	predict
	prefer	anticipate
future { reluctant	foresee
	anxious	
	willing	
	eager	
		say
		suppose
		conclude

We have proposed that infinitives are derived in complements whose verbs fail to undergo agreement with a subject. In the infinitives mentioned in section 4, agreement did not take place because the subject was in one or another way eliminated by a transformation. There is a second possible reason for non-agreement. This is that the subject is marked with an oblique case. There seem to be no instances, at least in the Indo-European languages, of verbs agreeing in person and number with anything else than nominative noun phrases. Good illustrations of this point are the German pairs

> Ich werde betrogen 'I am cheated'
> Mir wird geschmeichelt 'I am flattered'
> Ich bin leicht zu betrügen 'I am easy to cheat'
> Mir ist leicht zu schmeicheln 'I am easy to flatter'.

Presumably the same syntactic processes underlie both sentences in each pair. The accusative object of *betrügen* is changed into a nominative, whereas the dative object of *schmeicheln* stays in the dative. But from the viewpoint of agreement, only the nominative counts as a surface subject.

As the source of *for* with the infinitive we assume a transformation which marks the subjects in complements of emotive predicates with *for*, the non-finite verb form being a consequence of the oblique case of the subject.

We can here only list quickly some of the other syntactic properties which emotivity is connected to, giving an unfortunately oversimplified picture of a series of extremely complex and difficult problems. What follows are only suggestive remarks which we plan to pursue at a later time.

First of all, emotives may optionally contain the subjunctive marker *should*:

> It's interesting that you should have said so
> *It's well-known that you should have said so.

(We do not of course mean the *should* of obligation or the *should* of future expectation, which are not limited to emotives).

We assume that a future *should* is optionally deleted by a late rule, leaving a bare infinitive:

I'm anxious that he (should) be found
It's urgent that he (should) be found.

Emotive complements can be identified by their ability to contain a class of exclamatory degree adverbs such as *at all* or (unstressed) *so, such*:

It's interesting that he came at all
*It's well-known that he came at all.

Finally, it seems that one of the conditions which must be placed on relativization by *as* is that the clause be non-emotive although many other factors are certainly involved:

*As is interesting, John is in India
As is well-known, John is in India.

8 Conclusions

Syntactic-semantic interrelationships of this kind form the basis of a system of deep structures and rules which account for the complement system of English, and other languages as well. The importance of a system successfully worked out along the general lines suggested above would lie in its ability to account not only for the syntactic structure of sentential complementation, but also for its semantic structure, and for the relationship between the two. Our analysis of presupposition in the complement system contributes a substantial instance of the relation between syntax and semantics, and enables us to correct an error which has been made in most past work on transformational syntax. The error is that different types of complements (*that*-clauses, gerunds, infinitives) have all been assumed to have the same deep structure, and hence to be semantically equivalent.[a] We have seen that there is good reason to posit a number of different base structures, each mapped by transformations into a syntactic paradigm of semantically equivalent surface structures. The base structures differ semantically along at least two independent dimensions, which express the judgment of the speaker about the content of the complement sentence.

This approach to a theory of complementation is not only more adequate from a semantic point of view. Its purely syntactic advantages are equally significant. It eliminates the need for marking each verb for compatibility with each surface complement type, that is, for treating complementation as basically irregular and unpredicatable. We account for the selection of complement types quite naturally by our proposal that there are several meaningful base structures, whose choice is in large part predictable from the meaning of each predicate. These base structures are subject to various transformations which yield surface structures in which the relation between form and meaning is considerably obscured.

FURTHER NOTES ON FACTIVE AND NON-FACTIVE COMPLEMENTS

We have dealt with the syntactic repercussions of factivity in sentential complementation. This is really an artificially delimited topic (as almost all topics in linguistics

[a] The studies of Lees (1960) and Vendler (1964), however, contain many interesting semantic observations on sentential complementation and nominalization which still await formal description and explanation.

necessarily are). Factivity is relevant to much else in syntax bes.des sentential complementation, and on the other hand, the structure of sentential complementation is naturally governed by different semantic factors which interact with factivity. That is one source of the painful gaps in the above presentation which the reader will surely have noticed. We conclude by listing summarily a couple of possible additional applications of factivity, and some additional semantic factors which determine the form of complements, in order at least to hint at some ways in which the gaps can be filled, and to suggest what seem to us promising extensions of the approach we have taken.

(1) There is a syntactic and semantic correspondence between *truth* and *specific reference*. The verbs which presuppose that their sentential object expresses a true proposition also presuppose that their non-sentential object refers to a specific thing. For example, in the sentences

I ignored an ant on my plate
I imagined an ant on my plate

the factive verb *ignore* presupposes that there was an ant on my plate, but the non-factive verb *imagine* does not. Perhaps this indicates that at some sufficiently abstract level of semantics, truth and specific reference are reducible to the same concept. Frege's speculations that the reference of a sentence is its truth value would thereby receive some confirmation.

Another indication that there is a correspondence between truth of propositions and specific reference of noun phrases is the following. We noted in section 1 that extraposition is obligatory for non-factive subject complements. Compare –

That John has come makes sense (factive)
*That John has come seems (non-factive)

where the second sentence must become

It seems that John has come

unless it undergoes subject-raising. This circumstance appears to reflect a more general tendency for sentence-initial clauses to get understood factively. For example, in saying

The UPI reported that Smith had arrived
It was reported by the UPI that Smith had arrived

the speaker takes no stand on the truth of the report. But

The Smith had arrived was reported by the UPI

normally conveys the meaning that the speaker assumes the report to be true. A non-factive interpretation of this sentence can be teased out in various ways, for example by laying contrastive stress on the agent phrase (*by the UPI, not the AP*). Still, the unforced sense is definitely factive. These examples are interesting because they suggest that the factive vs. non-factive senses of the complement do not really correspond to the application of any particular transformation, but rather to the position of the complement in the surface structure. The interpretation can be non-factive if both passive and extraposition have applied, or if neither of them has applied; if only the passive has applied, we get the factive interpretation. This is very hard to state in terms of a condition on transformations. It is much easier to say that the initial position itself of a clause is in such cases associated with a factive sense.

This is where the parallelism between truth and specific reference comes in. The problem with the well-known pairs like

Everyone in this room speaks two languages
Two languages are spoken by everyone in this room

is exactly that indefinite noun phrases such as *two languages* are understood as referring to specific objects when placed initially ('there are two languages such that...'). Again, it is not on the passive itself that the meaning depends. In the sentence

Two languages are familiar to everyone in this room

the passive has not applied, but *two languages* is again understood as specific because of its initial position.

(2) We also expect that factivity will clarify the structure of other types of subordinate clauses. We have in mind the difference between purpose clauses (non-factive) and result clauses (factive), and different types of conditional and concessive clauses.

(3) There are languages which distinguish factive and non-factive moods in declarative sentences. For example, in Hidatsa (Matthews 1964) there is a factive mood whose use in a sentence implies that the speaker is certain that the sentence is true, and a range of other moods indicating hearsay, doubt, and other judgments of the speaker about the sentence. While this distinction is not overt in English, it seems to us that it may be sensed in an ambiguity of declarative sentences. Consider the statement

He's an idiot.

There is an ambiguity here which may be resolved in several ways. For example, the common question

Is that a fact or is that just your opinion?

(presumably unnecessary in Hidatsa) is directed exactly at disambiguating the statement. The corresponding *why*-question –

Why is he an idiot?

may be answered in two very different ways, e.g.

(*a*) Because his brain lacks oxygen
(*b*) Because he failed this simple test for the third time.

There are thus really two kinds of *why*-questions: requests for *explanation*, which presuppose the truth of the underlying sentence, and requests for *evidence*, which do not. The two may be paraphrased

(*a*) Why is it a fact that he is an idiot?
(*b*) Why do you think that he is an idiot?

(4) Consider the sentences

John's eating them would amaze me
I would like John's doing so.

These sentences do not at all presuppose that the proposition in the complement is true. This indicates a further complexity of the *fact* postulated in the deep structure of factive complements. Like verbs, or predicates in general, it appears to take various tenses or moods. Note that there correspond to the above sentences:

If he were to eat them it would amaze me
I would like it if John were to do so.

These can also be construed as

If it were a fact that he ate them it would amaze me.

A second oversimplification may be our assumption that sentences are embedded in their deep structure form. A case can be made for rejecting this customary approach in favor of one where different verbs take complements at different levels of representation. Consider direct quotation, which appears not to have been treated in generative grammar. The fundamental fact is that what one quotes are surface structures and not deep structures. That is, if John's words were 'Mary saw Bill', then we can correctly report

John said: 'Mary saw Bill'

but we shall have misquoted him if we say

John said: 'Bill was seen by Mary'.

If we set up the deep structure of both sentences simply as

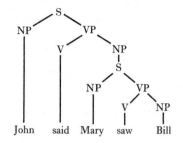

then we have not taken account of this fact. We should be forced to add to this deep structure the specification that the complement either must or cannot undergo the passive, depending on which of the sentences we are quoting. Since sentences of any complexity can be quoted, to whose deep structures the passive and other optional transformations may be applicable an indefinite number of times, it is not enough simply to mark the embedded deep structure of the quoted sentence as a whole for applicability of transformations. What has to be indicated according to this solution is the whole transformational history of the quoted sentence.

A more natural alternative is to let the surface structure itself of the quoted sentence be embedded. This would be the case in general for verbs taking direct quotes. Other classes of verbs would take their complements in different form. We then notice that the initial form of a complement can in general be selected at a linguistically functional level of representation in such a way that the truth value of the whole sentence will not be altered by any rules which are applicable to the complement. Assuming a generative semantics, the complements of verbs of knowing and believing are then semantic representations. From

John thinks that the McCavitys are a quarrelsome bunch of people

it follows that

John thinks that the McCavitys like to pick a fight.

That is, one believes propositions and not sentences. Believing a proposition in fact commits one to believing what it implies: if you believe that Mary cleaned the room you must believe that the room was cleaned. (Verbs like *regret*, although their objects are also propositions, differ in this respect. If you regret that Mary cleaned the room you do not necessarily regret that the room was cleaned.)

At the other extreme would be cases of phonological complementation, illustrated by the context

John went ' — '.

The object here must be some actual noise or a conventional rendering thereof such as *ouch* or *plop*.

A good many verbs can take complements at several levels. A verb like *scream*, which basically takes phonological complements, can be promoted to take direct quotes. *Say* seems to take both of these and propositions as well.

Are there verbs which require their complement sentences to be inserted in deep structure form (in the sense of Chomsky)? Such a verb X would have the property that

John Xed that Bill entered the house,

would imply that

John Xed that the house was entered by Bill

but would not imply that

John Xed that Bill went into the house.

That is, the truth of the sentence would be preserved if the object clause underwent a different set of optional transformations, but not if it was replaced by a paraphrase with another deep structure source. It is an interesting question whether such verbs exist. We have not been able to find any. Unless further search turns up verbs of this kind, we shall have to conclude that, if the general idea proposed here is valid, the levels of semantics, surface structure, and phonology, but not the level of deep structure, can function as the initial representation of complements.

Massachusetts Institute of Technology

BIBLIOGRAPHY

Baker, Leroy (1967). 'A Class of Semantically Deviant Questions'. Unpublished paper.

Dean, Janet (1967). 'Noun Phrase Complementation in English and German'. Unpublished paper.

Klima, Edward S. (1964). 'Negation in English'. In Fodor and Katz (eds.), *The Structure of Language*. Englewood Cliffs, N.J.: Prentice-Hall.

Langendoen, D. T. (1966). 'The Syntax of the English Expletive "IT"'. *Georgetown University Monograph Series on Languages and Linguistics*, no. 19, pp. 207-16.

Lees, Robert B. (1960). *The Grammar of English Nominalizations*. The Hague: Mouton.

Matthews, G. H. (1964). *Hidatsa Syntax*. The Hague: Mouton.

Rosenbaum, P. S. (1967). *The Grammar of English Predicate Complement Constructions*. Cambridge, Mass.: M.I.T. Press.

Ross, John Robert (1966). 'Relativization in Extraposed Clauses (A Problem which Evidence is Presented that Help is Needed to Solve)'. *Mathematical Linguistics and Automatic Translation*, Report no. NSF-17 to the National Science Foundation, Harvard University, Computation Laboratory.

—— (1967). 'Constraints on Variables in Syntax'. Unpublished dissertation, M.I.T.

Vendler, Zeno (1964). *Nominalizations*. Mimeographed, University of Pennsylvania.

Types of lexical information[a]

CHARLES J. FILLMORE

0 The Lexicon

A lexicon viewed as part of the apparatus of a generative grammar must make accessible to its users, for each lexical item,

(i) the nature of the deep-structure syntactic environments into which the item may be inserted;

(ii) the properties of the item to which the rules of grammar are sensitive;

(iii) for an item that can be used as a 'predicate', the number of 'arguments' that it conceptually requires;

(iv) the role(s) which each argument plays in the situation which the item, as predicate, can be used to indicate;

(v) the presuppositions or 'happiness conditions' for the use of the item, the conditions which must be satisfied in order for the item to be used 'aptly';

(vi) the nature of the conceptual or morphological relatedness of the item to other items in the lexicon;

(vii) its meaning; and

(viii) the phonological or orthographic shapes which the item assumes under given grammatical conditions.

In this paper I shall survey, in a very informal manner, the various types of information that needs to be included, in one way or another, in the lexical component of an adequate grammar. I shall, however, have nothing to say about (viii) above, and nothing very reliable to say about (vii).

1 The speech act

I begin by assuming that the semantic description of lexical items capable of functioning as predicates can be expressed as complex statements about properties of, changes in, or relations between entities of the following two sorts: (a) the entities that can serve as arguments in the predicate-argument constructions in which the given lexical item can figure, and (b) various aspects of the speech act itself. In this section I shall deal with the concepts that appear to be necessary for identifying the role of the speech act in semantic theory.

The act of producing a linguistic utterance in a particular situation involves a speaker, an addressee, and a message. It is an act, furthermore, which occurs within a specific time-span, and it is one in which the participants are situated in

[a] Sponsored in part by the National Science Foundation through Grant GN-534.1 from the Office of Science Information Service to the Computer and Information Science Research Center, the Ohio State University.

particular places. Now the time during which a speech act is produced is a span, the participants in the speech act may be moving about during this span, and even the identity of the participants may change during the speech act; but for most purposes the participant-identity and the time-space coordinates of the speech act can be thought of as fixed points. In accepting this fiction, I commit myself to regarding sentences like ⟨1⟩ to ⟨3⟩ as somewhat pathological:

⟨1⟩ I'm not talking to *you*, I'm talking to *you*.

⟨2⟩ I want you to turn the corner...right...*here*!

⟨3⟩ This won't take long, did it?

The producer of a speech act will be called the *locutionary source* (LS), the addressee will be referred to as the *locutionary target* (LT)[a]. The temporal and spatial coordinates of the speech act are the *time of the locutionary act* (TLA), the *place of the locutionary source* (PLS) and the *place of the locutionary target* (PLT).

There are certain verbs in English which refer to instances of speech acts other than the one which is being performed (e.g. *say*), and I shall refer to these as *locutionary verbs*. It is necessary to mention locutionary verbs now because we shall find that linguistic theory requires a distinction between the 'ultimate' speech act and speech acts described or referred to in a sentence. Thus in the linguistic description of some verbs reference is made to either the LS or the agent of a locutionary verb,[b] and these situations must be distinguished from those in which the reference is to the LS alone.

Words in English whose semantic descriptions require reference to some aspect of the locutionary act include *here, this, now, today, come,* and *know*. The word *come*,[c] for example, can refer to movement toward either the PLA or the PLT at either TLA or the time-of-focus identified in the sentence. Thus in ⟨4⟩:

⟨4⟩ He said that she would come to the office Thursday morning

it is understood that the office is the location of the LS or the LT either at TLA or on the said Thursday morning. Uses of the verb *know* allow the LT to infer the factuality of the proposition represented by a following *that*-clause. Thus, in sentence ⟨5⟩:

⟨5⟩ She knows that her brother has resigned

it is understood that the LS at TLA presupposes the factuality of her brother's resignation.

In the semantic description of some verbs reference is made either to the subject of the 'next locutionary verb up', or to the LS just in case the sentence contains no explicit locutionary verb. If we are to believe Ross (see note *b* below), the

---

[a] I have borrowed these terms from philosopher Richard Garner, to whom I am also indebted for a number of suggestions on the content and phrasing of several sections of this paper.

[b] The disjunction in this statement may be unnecessary if we accept John R. Ross's arguments that declarative sentences have phonetically unrealized embedding sentences representable as something like *I declare to you that*... On Ross's view every sentence contains at least one locutionary verb, so that the difference we are after is a difference between references to the 'next higher' locutionary verb and reference to the 'highest' locutionary verbs. See John R. Ross, 'On declarative sentences', to appear in R. A. Jacobs and P. S. Rosenbaum (eds.), *Readings in English Transformational Grammar* (1970), Blaisdell.

[c] See Charles J. Fillmore, 'Deictic categories in the semantics of "come"', *Foundations of Language*, 1966, II, 219–27.

verb *lurk* requires of its subject that it be distinct from the subject of the first commanding locutionary verb. It may look, on just seeing sentences ⟨6⟩–⟨7⟩:

⟨6⟩ He was lurking outside her window
⟨7⟩ *I was lurking outside her window

that what is required is simply non-identity with LS; but this is shown not to be so because of the acceptability of sentence ⟨8⟩:

⟨8⟩ She said I had been lurking outside her window.

From these observations it follows that sentence ⟨9⟩ is ambiguous on whether or not the two pronouns *he* are coreferential, but sentence ⟨10⟩ requires the two *he*'s to be different:

⟨9⟩ He said he had been loitering outside her window.
⟨10⟩ He said he had been lurking outside her window.

There are apparently many speakers of English whose use of *lurk* fails to match the observations I have just reviewed; but for the remainder, this verb provides an example of the distinction we are after.

There is, then, a distinction in semantic descriptions of lexical items between references to properties of the higher clauses that contain them, on the one hand, and to features of or participants in the speech act itself on the other hand. The former situation falls within the area of 'deep structure constraints' (see section 9, below), but the latter requires the availability of concepts related to the speech act.[a]

2 Elementary semantic properties of verbs

I assume that what we might call the basic sense of a word is typically expressible as a set of components, and that while some of these components may be idiosyncratic to particular words, others are common to possibly quite large classes of words. The components themselves may be complex, since they may be required to characterize events or situations that are themselves complex, but the ultimate terms of a semantic description I take to be such presumably biologically given notions as identity, time, space, body, movement, territory, life, fear, etc., as well as undefined terms that directly identify aspects of or objects in the cultural and physical universe in which human beings live. In this section, we shall sample

---

[a] There is an additional use of deictic words, and that is this: in a third person narrative, one can express one's 'identification' with one of the characters in the narrative by letting that character be the focus of words that are primarily appropriate to *hic-nunc-ego*. Thus we may find in an exclusively third-person narrative a passage like (i)

(i) Here was where Francis had always hoped to be, and today was to mark for him the beginning of a new life

in which the words *here* and *today* refer to the place and time focused on in the narrative, not to the place and time associated with the author's act of communication. In what might be referred to as the 'displaced ego' use of deictic words, the author has shown us that he has for the moment assumed Francis's point of view.

I propose that a rather subtle test of psycho-sexual identity can be devised which makes use of a story in which two characters, one male and one female, do a lot of cross-visiting, but in their other activities do nothing that makes one of them clearly more lovable than the other. The subject's task is to listen to the story and then retell it in his own words. The writer of the original story must not use the words *come* and *go*; but if the subject, in retelling the story, states, say, that *Bill came to Mary's house* (using *came* rather than *went*), this fact will reveal that he is experiencing the story from Mary's point of view.

certain elementary semantic properties of verbs, in particular those relating to time, space, movement, and 'will'.

Some verbs refer to activities viewed as necessarily changing in time, others do not; this contrast is frequently referred to with the terms 'momentary' and 'continuative' respectively. *Sleep* is a continuative verb, *wake up* is a momentary verb. A continuing activity, or state, necessarily occupies a span of time, and thus it makes sense to qualify a continuative verb with a complement which identifies one or both of the end-points of such a span, or a distance-measure of the span. Thus while the sentences in ⟨11⟩ make sense, those in ⟨12⟩ do not:

⟨11⟩ She slept $\begin{cases} \text{for three days} \\ \text{until Friday} \end{cases}$

⟨12⟩ *She woke up $\begin{cases} \text{for three days} \\ \text{until Friday} \end{cases}$

On the other hand, the negation of a momentary verb can identify a continuing state; hence, the sentences in ⟨13⟩ make sense:

⟨13⟩ She didn't wake up $\begin{cases} \text{for three days} \\ \text{until Friday} \end{cases}$

Momentary verbs that represent acts that are repeatable may be understood 'iteratively'. *Wake up* is not iterative, as is shown by example ⟨12⟩. *Kick*, however, can be understood iteratively, as we see in ⟨14⟩:

⟨14⟩ He kicked the dog until 5 o'clock.

Momentary verbs that are also 'change-of-state' verbs cannot be used iteratively when a specific object is involved, as we see in example ⟨15⟩:

⟨15⟩ *He broke the vase until 5 o'clock.

But if the same activity can be directed to an unspecified number of objects, then change-of-state verbs can be understood iteratively, too, as we see in ⟨16⟩:

⟨16⟩ He broke vases until 5 o'clock.

It appears, in short, that a lexico-semantic theory will have to deal with those aspects of the meanings of verbs which relate to the occurrence in time of the situations which they identify.

Turning to other types of semantic properties, we note that verbs like *hit* and *touch*, though they differ in that the former is momentary while the latter may be either momentary or continuative, have in common the notion of surface-contact, a property they share with *knock, strike, contact, impinge, smite*, and many others. They differ in that the impact of the described acts is apparently gentler for *touch* than for *hit*.

The verbs *leap* and *jump* agree in implying a momentary change in vertical position (one has to leave the ground in order to perform either of these actions), but they differ in that *leap* seems to imply a change in horizontal position, too. *Slide*, like *leap*, refers to position-changes along a surface, but differs from *leap* in not implying movement away from the surface. *Scuttle*, like *slide*, suggests movement across a surface, but with the assumption that contact with the surface is interrupted and with the further sense that the motion is rapid. *Dart* is like *scuttle* in referring to rapid sudden motion, but fails to share with it any reference to a surface. Verbs of motion, in short, may be described by associating them with properties relating to direction, speed, gravity, surface, etc.[a]

Sometimes a verb has a built-in reference to the outcome of an activity. Conceptually it appears that the actor engages in some activity and though the activity may be directed toward some specific outcome it is the activity itself which (by chance) leads to that outcome. These have been called 'achievement verbs'. One of the tests of an achievement verb is that the modal *may* is usable in construction with such a verb only in its epistemic or predictive sense, not in its pragmatic or permission-granting sense. This is apparently because of the 'by-chance' relationship between the activity and the outcome: one doesn't grant someone permission to have good luck. Hence we find ⟨17⟩ and ⟨19⟩ understandable only in the epistemic sense of *may*, while ⟨18⟩ and ⟨20⟩ can be understood in either the epistemic or the pragmatic sense:

⟨17⟩ He may achieve his goal.
⟨18⟩ He may try to achieve his goal.
⟨19⟩ He may find the eggs.
⟨20⟩ He may look for the eggs.

A final general property of verbs that we may point out in this section has to do with the intentional or non-intentional involvement of one of the participants in the events described by use of the verb. If we compare ⟨21⟩ and ⟨22⟩:

⟨21⟩ John means x by y
⟨22⟩ John understands x by y

we note that in ⟨21⟩, but not in ⟨22⟩, the association between x and y is intentional on John's part. The word *mean* can be used in the sense which *understand* has in ⟨22⟩, but in that case the sentence is differently constructed. ⟨23⟩ is a paraphrase of ⟨22⟩:

⟨23⟩ Y means x to John.

3 Predicate structure

I assume that most of the 'content words' in a language can be characterized in the lexicon in terms of their use as predicates. I take this to be true of nouns, verbs, adjectives,[b] most adverbs, and also a great many conjunctions. Thus a sentence like ⟨24⟩:

⟨24⟩ Harry lives at home because he loves his mother

is evaluated as true or false depending not only on the joint truth-values of the two clauses which flank *because*, but on the truth or falsity of the 'causal' connection between the two situations named by these clauses. The sentence can be interpreted

---

[a] Possibly the richest source of insights into verbs of motion is the recent output of Jeffrey S. Gruber. See his dissertation, *Studies in lexical relations*, M.I.T. (1965, unpublished); 'Look and see', *Language* (1967), XLIII, no. 4, 937–47; and *Functions of the lexicon in formal descriptive grammars* (1967), Technical Memorandum TM-3770/000/00, System Development Corporation, Santa Monica, California.

[b] In other words, I accept the part-of-speech identities argued by George Lakoff in Appendix A of *On the nature of syntactic irregularity* (1965), Report no. NSF-16, Computation Laboratory of Harvard University; as well as the extension of such identities to 'nouns' proposed by Emmon Bach in 'Nouns and noun phrases' (1968), in E. Bach and R. Harms (eds.), *Universals in Linguistic Theory*, Holt, Rinehart and Winston.

as having *because* as its main predicate, a predicate which takes two clauses as its arguments and which is used to assert a 'causal' or 'logical' connection between them.

As predicates, words can be described first of all according to the number of 'arguments' that they take. Thus the verbs *ascend* and *lift* are both motion verbs, they are both used to describe motion upward, but they differ in that while *ascend* is used only of the object that moves upward, *lift* requires conceptually two objects, one the object that is moving upward, the other the object or being that is causing it to move upward. Another way of stating this is: *ascend* is a one-argument predicate, *lift* is a two-argument predicate.[a]

Many verbs are flexible in the number of arguments they take. This is true, for example, of some motion verbs, like *move* and *rotate*, and many change-of-state verbs, like *open* and *break*. *Move*, as can be seen in sentences ⟨25⟩–⟨27⟩, can occur with one, two, or three arguments:

⟨25⟩ The rock moved.
⟨26⟩ The wind moved the rock.
⟨27⟩ I moved the rock (with a stick).

Mention of the object which moves is required of all three uses; the two-argument uses additionally identify either the physical force or object which is directly responsible, or the animate being which is indirectly responsible, for the activity of moving; and the three-argument use identified all three of these (as in ⟨27⟩ with the parenthesized phrase included). The surface-contact verbs *hit, touch, strike*, etc., require conceptually at least two arguments in all of their uses, namely the objects which come into contact, but they accept as a third argument the animate being that is responsible for the coming-into-contact.

The verbs *rob* and *steal* conceptually require three arguments, namely those identifiable as the culprit, the loser, and the loot. The words *buy* and *sell* are each four-argument predicates, the arguments representing the one who receives the goods or services, the one who provides the goods and services, the goods and services themselves, and the sum of money that changes hands.

I have referred in this section to the conceptually required number of arguments. I am distinguishing this from the number of arguments that must be explicitly identified in English sentences. The various ways in which English grammar provides for the omission or suppression of conceptually required arguments is the subject of section 5. To say that conceptually *rob* or *buy* are three- or four-argument predicates respectively is to acknowledge that even when we say merely ⟨28⟩:

⟨28⟩ She robbed the bank

we understand that she took something out of the bank, and when we say ⟨29⟩:

⟨29⟩ She bought it

truthfully, it is necessarily the case that there was somebody who sold it to her and that a sum of money was exchanged.

---

[a] Of course, as motion verbs each of them may take time and space complements as well, as is seen in *The balloons ascended to the rafters just after the speech ended*. Since in general the nature of the time and space complements is predictable from properties of the type discussed in section 2, we may permit ourselves to ignore such matters while discussing the typing of predicates on the basis of the number of arguments they accept.

4 Case structure

In the preceding section I identified the separate arguments associated with the verb *rob* by referring to their roles as 'culprit', 'loser', and 'loot'; in a similar way, I might have identified the three arguments associated with the verb *criticize* as 'critic', 'offender', and 'offense'. It seems to me, however, that this sort of detail is unnecessary, and that what we need are abstractions from these specific role descriptions, abstractions which will allow us to recognize that certain elementary role notions recur in many situations, and which will allow us to acknowledge that differences in detail between partly similar roles are due to differences in the meanings of the associated verbs. Thus we can identify the culprit of *rob* and the critic of *criticize* with the more abstract role of Agent, and interpret the term Agent as referring, wherever it occurs, as the animate instigator of events referred to by the associated verb. Although there are many substantive difficulties in determining the role structure of given expressions, in general it seems to me that for the predicates provided in natural languages, the roles that their arguments play are taken from an inventory of role types fixed by grammatical theory. Since the most readily available terms for these roles are found in the literature of case theory, I have taken to referring to the roles as *case relationships*, or simple *cases*. The combination of cases that might be associated with a given predicate is the *case structure* of that predicate.

In addition to the apparently quite complex collection of complements that identify the limits and extents in space and time that are required by verbs of motion, location, duration, etc., the case notions that are most relevant to the subclassification of verb types include the following:

Agent (A), the instigator of the event
Counter-Agent (C), the force or resistance against which the action is carried out
Object (O), the entity that moves or changes or whose position or existence is in consideration
Result (R), the entity that comes into existence as a result of the action
Instrument (I), the stimulus or immediate physical cause of an event
Source (S), the place from which something moves
Goal (G), the place to which something moves
Experiencer (E), the entity which receives or accepts or experiences or undergoes the effect of an action (earlier called by me 'Dative').

It appears that there is sometimes a one-many relationship between an argument to a predicate and the roles that are associated with it. This can be phrased by saying either that some arguments simultaneously serve in more than one role, or that in some situations the arguments in different roles must (or may) be identical.

Thus verbs like *rise* and *move* can be used intransitively, that is with one noun-phrase complement; the complement may refer just to the thing which is moving upward, or it may simultaneously refer to the being responsible for such motion. Thus in speaking simply of the upward motion of smoke we can say ⟨30⟩:

⟨30⟩ The smoke rose

and in speaking of John's getting up on his own power, we can say ⟨31⟩:

⟨31⟩ John rose.

The case structure of *rise*, then, might be represented diagrammatically as ⟨32⟩:

⟨32⟩

The fact that there are two case lines connecting *rise* with its one argument, and that the line labelled A has its case label in parentheses, reflect the fact that the argument can serve in just one of these roles (O) or simultaneously in both (A and O). *Rise* differs from *arise* in the optionality of A; it differs from *ascend* in having an 'A' line at all, and they all differ from *lift* in that the latter requires two arguments.[a] Thus ⟨33⟩–⟨35⟩:

⟨33⟩

⟨34⟩

⟨35⟩

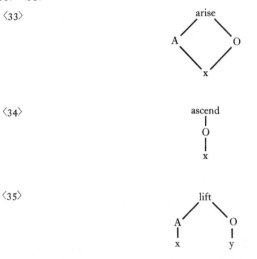

[a] In truth, however, *lift* also requires conceptually the notion of Instrument. That is, lifting requires the use of something (perhaps the Agent's hand) to make something go up. It is conceivable that the basic case structures of *arise* and *lift* are identical, with grammatical requirements on identity and deletion accounting for their differences. Thus the two are shown in (i) and (ii):

For *arise*, however, it is required that x = y = z, and hence only one noun will show up. The meaning expressed by *John arose* is that John willed his getting up, that he used his own body (i.e., its muscles) in getting up, and that it was his body that rose. For *lift*, however, there may be identities between x and z, resulting in a sentence like *John lifted himself*, or there may be identities between y and z, resulting in a sentence like *John lifted his finger*. *Rise*, then, must be described as (iii):

Frequently a linguistically codable event is one which in fact allows more than one individual to be actively or agentively involved. In any given linguistic expression of such an event, however, the Agent role can only be associated with one of these. In such pairs as *buy* and *sell* or *teach* and *learn* we have a Source (of goods or knowledge) and a Goal. When the Source is simultaneously the Agent, one uses *sell* and *teach*; when the Goal is simultaneously the Agent, we use *buy* and *learn*.

It is not true, in other words, that *buy* and *sell*, *teach* and *learn* are simply synonymous verbs that differ from each other in the order in which the arguments are mentioned.[a] There is synonymy in the basic meanings of the verbs (as descriptions of events), but a fact that might be overlooked is that each of these verbs emphasizes the contribution to the event of one of the participants. Since the roles are different, this difference is reflected in the ways in which the actions of different participants in the same event can be qualified. Thus we can say ⟨36⟩:

⟨36⟩ He bought it with silver money

but not ⟨37⟩:

(37) *He sold it with silver money.

Similarly, the adverbs in ⟨38⟩ and ⟨39⟩ do not further describe the activity as a whole, but only one person's end of it:

⟨38⟩ He sells eggs very skilfully.
⟨39⟩ He buys eggs very skilfully.

5 Surface realization of arguments

I suggested in the previous section that the conceptually necessary arguments to a predicate cannot always be matched on a one-to-one basis with the 'cases' that are also associated with the same predicate. It may now be pointed out that there is also no exact correspondence between either of these and the number of obligatorily present syntactic constituents in expressions containing the predicates in question. *Buy*, as we have seen, is a four-argument (but five-case) predicate which can occur in syntactically complete sentences containing two, three or four noun phrases. Thus ⟨40⟩:

⟨40⟩ He bought it (from me) (for four dollars)

in which optionally present segments are marked off by parentheses.

The verb *blame* has associated with it three roles, the accuser (Source), the defendant (Goal), and the offense (Object).

Expressions with this verb can contain reference to all three, as in ⟨41⟩, just two, as in ⟨42⟩ and ⟨43⟩, or only one, as in ⟨44⟩:

⟨41⟩ The boys blamed the girls for the mess.
⟨42⟩ The boys blamed the girls.

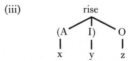

(iii)   rise

where the identity requirements of *arise* obtain just in case x and y are present.

   [a] This has, of course, been suggested in a great many writings on semantic theory; most recently, perhaps, in the exchanges between J. F. Staal and Yehoshua Bar-Hillel in *Foundations of Language*, 1967 and 1968.

⟨43⟩ The girls were blamed for the mess.

⟨44⟩ The girls were blamed.

No sentence with *blame*, however, can mention only the accuser, only the offense, or just the accuser and the offense. See ⟨45⟩–⟨47⟩:

⟨45⟩ *The boys blamed.

⟨46⟩ *The mess was blamed.

⟨47⟩ *The boys blamed (for) the mess.

An examination of ⟨41⟩–⟨47⟩ reveals that the case realized here as *the girls* is obligatory in all expressions containing this verb, and, importantly, that there are two distinct situations in which the speaker may be silent about one of the other arguments. I take sentence ⟨43⟩ as a syntactically complete sentence, in the sense that it can appropriately initiate a discourse (as long as the addressee knows who the girls are and what the mess is). In this case the speaker is merely being indefinite or non-committal about the identity of the accuser. I take sentence ⟨42⟩, however, as one which cannot initiate a conversation and one which is usable only in a context in which the addressee is in a position to know what it is that the girls are being blamed for. Another way of saying this is that ⟨43⟩ is a paraphrase of ⟨43′⟩ and ⟨42⟩ is a paraphrase of ⟨42′⟩:

⟨42′⟩ The boys blamed the girls *for it.*

⟨43′⟩ The girls were blamed for the mess *by someone.*

This distinction can be further illustrated with uses of *hit.* In ⟨48⟩, a paraphrase of ⟨48′⟩, the speaker is merely being indefinite about the implement he used:

⟨48⟩ I hit the dog.

⟨48′⟩ I hit the dog *with something.*

In ⟨49⟩, the paraphrase of ⟨49′⟩, the speaker expects the identity of the 'target' (Goal) to be already known by the addressee:

⟨49⟩ The arrow hit.

⟨49′⟩ The arrow hit *it.*

The two situations correspond, in other words, to definite and indefinite pronominalization.

Sometimes an argument is obligatorily left out of the surface structure because it is subsumed as a part of the meaning of the predicate. This situation has been discussed in great detail by Jeffrey Gruber (see note *a*, p. 374, above) under the label 'incorporation'. An example of a verb with an 'incorporated' Object is *dine*, which is conceptually the same as *eat dinner* but which does not tolerate a direct object.[a]

The verb *doff* has an incorporated Source. If *I doff* something, I remove it from my head, but there is no way of expressing the Source when this verb is used. There is no such sentence as ⟨50⟩:

⟨50⟩ *He doffed his hat from his balding head.

There are other verbs that identify events which typically involve an entity of a fairly specific sort, so that to fail to mention the entity is to be understood as intending the usual situation. It is usually clear that an act of slapping is done with open hands, an act of kicking with legs and/or feet, an act of kissing with both lips;

---

[a] One can, however, indicate the content of the meal in question, as in an expression like *He dined on raisins.*

and the target of an act of spanking seldom needs to be made explicit. For these verbs, however, if the usually omitted item needs to be delimited or qualified in some way, the entity can be mentioned. Hence we find the modified noun-phrase acceptable in such sentences as ⟨51⟩–⟨53⟩:

⟨51⟩ She slapped me with her left hand.[a]
⟨52⟩ She kicked me with her bare foot.
⟨53⟩ She kissed me with chocolate-smeared lips.

Lexical entries for predicate words, as we have seen in this section, should represent information of the following kinds: (1) for certain predicates the nature of one or more of the arguments is taken as part of our understanding of the predicate word: for some of these the argument cannot be given any linguistic expression whatever; for others the argument is linguistically identified only if qualified or quantified in some not fully expected way; (2) for certain predicates, silence ('zero') can replace one of the argument-expressions just in case the speaker wishes to be indefinite or non-committal about the identity of the argument; and (3) for certain predicates, silence can replace one of the argument-expressions just in case the LS believes that the identity of the argument is already known by the LT.

6 Meaning vs. presupposition

Sentences in natural language are used for asking questions, giving commands, making assertions, expressing feelings, etc. In this section I shall deal with a distinction between the presuppositional aspect of the semantic structure of a predicate on the one hand and the 'meaning' proper of the predicate on the other hand. We may identify the presuppositions of a sentence as those conditions which must be satisfied before the sentence can be used in any of the functions just mentioned. Thus the sentence identified as ⟨54⟩:

⟨54⟩ Please open the door

can be used as a command only if the LT is in a position to know what door has been mentioned and only if that door is not, at TLA, open.[b] The test that the existence and specificity of a door and its being in a closed state make up the *presuppositions* of ⟨54⟩ rather than part of its *meaning* is that under negation the sentence is used to give quite different instructions, yet the presuppositional conditions are unaffected.

⟨54'⟩ Please don't open the door.

The presuppositions about the existence and specificity of the door relate to the use of the definite article and have been much discussed in the philosophical literature on referring.[c] The presupposition about the closed state of the door is a property of the verb *open*.

Presuppositions of sentences may be associated with grammatical constructions

---

[a] At least in the case of *slap*, the action can be carried out with objects other than the usual one. Thus it is perfectly acceptable to say *She slapped me with a fish*.

[b] I am only dealing here with those presuppositions which are relatable to the *content* of the utterance. It is also true, of course, that ⟨54⟩ can be used appropriately as a command only if the LT understands English, is believed by the LS to be awake, is not totally paralyzed, etc. These matters have more to do with questions of 'good faith' in speech communication than with information that is to be understood as knowledge about individual lexical items.

[c] P. F. Strawson, 'On referring', *Mind* (1950), LIX, 320–44.

independent of specific predicate words (such as those associated with the word *even* or with the counterfactual-conditional construction),[a] but I shall mention here only those that must be identified with the semantic structure of predicate words. If we limit our considerations to sentences which can be used for making assertions, we can separate the basic meaning of a predicate from its presuppositions by describing the former as being relevant to determining whether as an assertion it is true or false, the latter as being relevant to determining whether the sentence is capable of being an assertion in the first place. If the presuppositional conditions are not satisfied, the sentence is simply not apt; only if these conditions are satisfied can a sentence be appropriately used for asking a question, issuing a command, making an apology, pronouncing a moral or aesthetic judgment, or, in the cases we shall consider, making an assertion.

Let us illustrate the distinction we are after with the verb *prove* in construction with two *that*-clauses. Consider sentence ⟨55⟩:

⟨55⟩ That Harry is still living with his mother proves that he is a bad marriage risk.

It is apparent that if I were to say ⟨55⟩ about somebody who is an orphan, nobody would say that I was speaking falsely, only that I was speaking inappropriately. If *prove* has a *that*-clause subject and a *that*-clause object, we say that the truth of the first *that*-clause is *presupposed*, and that the verb is used to assert a causal or logical connection between the two clauses and thus (when used affirmatively) to imply the truth of the second clause. That this separation is correct may be seen by replacing *prove* in ⟨55⟩ by *doesn't prove* and noting that the presuppositional aspects of ⟨55⟩, concerning the truth of the first *that*-clause, are unaffected by the change.

It is difficult to find pairs of words in which the presuppositional content of one is the meaning content of the other, but a fairly close approximation to this situation is found in the pair of verbs *accuse* and *criticize*. The words differ from each other on other grounds, in that *accuse* is capable of being a 'performative', while *criticize* is not; and *criticize*, unlike *accuse*, is capable of being used in senses where no negative evaluation is intended. In sentences ⟨56⟩ and ⟨57⟩ we are using *accuse* in a non-performative sense and we are using *criticize* as a three-argument predicate in a 'negative-evaluative' sense:

⟨56⟩ Harry accused Mary of writing the editorial.
⟨57⟩ Harry criticized Mary for writing the editorial.

I would say that a speaker who utters ⟨56⟩ presupposes that Harry regarded the editorial-writing activity as 'bad' and asserts that Harry claimed that Mary was the one who did it; while a speaker who utters ⟨57⟩ presupposes that Harry regarded Mary as the writer of the editorial and asserts that Harry claimed the editorial-writing behavior or its result as being 'bad'. The content of the presupposition in each one of these verbs shows up in the assertive meaning of the other.

Certain apparent counter-examples to the claims I have been making about presuppositions can be interpreted as 'semi-quotations', I believe. Some utterances are to be thought of as comments on the appropriate use of words. Uses of the verb *chase* presuppose that the entity identified as the direct object is moving fast. Uses of the verb *escape* presuppose that the entity identified by the subject noun-phrase

---

[a] For some examples see my mistitled paper, 'Entailment rules in a semantic theory', *Project on Linguistic Analysis* Report no. 10, 1965.

was contained somewhere 'by force' previous to the time of focus. These presuppositions, as expected, are unaffected by sentence negation:

⟨58⟩ The dog $\begin{Bmatrix} \text{chased} \\ \text{didn't chase} \end{Bmatrix}$ the cat.

⟨59⟩ He $\begin{Bmatrix} \text{escaped} \\ \text{didn't escape} \end{Bmatrix}$ from the tower.

It seems to me that sentences like ⟨60⟩ and ⟨61⟩ are partly comments on the appropriateness of the words *chase* and *escape* for the situations being described. These are sentences that would most naturally be used in contexts in which the word *chase* or *escape* had just been uttered:

⟨60⟩ I didn't 'chase' the thief; as it happened, he couldn't get his car started.

⟨61⟩ I didn't 'escape' from the prison; they released me.

It is important to realize that the difference between assertion and presupposition is a difference that is not merely to be found in the typical predicate words known as verbs and adjectives. The difference is found in predicatively used nouns as well. In the best-known meaning of *bachelor*, for example, the negation-test reveals that only the property of 'having never been married' is part of the meaning proper. Uses of this word (as predicate) *presuppose* that the entities being described are human, male and adult. We know that this is so because sentence ⟨62⟩:

⟨62⟩ That person is not a bachelor

is only used as a claim that the person is or has been married, never as a claim that the person is a female or a child. That is, it is simply not appropriate to use ⟨62⟩, or its non-negative counterpart, when speaking of anyone who is not a human, male adult.

7 Evaluative and orientative features

Jerrold J. Katz[a] and Manfred Bierwisch[b] have proposed semantic features which treat certain relations between objects and the human beings that deal with these objects. Katz has treated the semantic properties of words that relate to the ways in which objects and events are evaluated, and Bierwisch has proposed ways of associating with words information concerning the ways in which the objects they describe are related to spatial aspects of the language-user's world.

For certain nouns the evaluative information is determinable in a fairly straightforward way from their definitions. This is most clearly true in the case of agentive and instrumental nouns. Many definitions of nouns contain a component which expresses a typical function of the entity the noun can refer to. Thus the lexical entry for *pilot* will contain an expression something like ⟨63⟩:

⟨63⟩ profession: A of [V O A]
          where V is 'navigate', with the presupposition that O is an air vessel

and the lexical entry for *knife* will contain such an expression as ⟨64⟩:

⟨64⟩ use: I of [V O I A]
          where V is 'cut', with the presupposition that O is a physical object.

[a] Jerrold J. Katz, 'Semantic theory and the meaning of "good"' (1964), *Journal of Philosophy*, LXI.

[b] Manfred Bierwisch, 'Some semantic universals of German adjectivals' (1967), *Foundations of Language*, III, 1–36.

For such nouns I assume that the evaluative feature can be automatically specified from the function-identifying part of a definition. A noun which refers to a 'typical' (e.g. 'professional') Agent in an activity is evaluated according to whether the Agent conducts this activity skilfully; a noun which names a typical Instrument in an activity is evaluated according to whether the thing permits the activity to be performed easily. In these ways we can make intelligible our ability to understand expressions like *a good pilot, a good pianist, a good liar, a good knife, a good pencil, a good lock*, etc.

For nouns whose definitions do not identify them as typical Agents or Instruments, the evaluative feature apparently needs to be specified separately. Thus *food* is probably in part defined as ⟨65⟩:

⟨65⟩ function: O of [V O A]
    where V is 'eat'.

Food is evaluated according to properties (namely nutrition and palatability) which are not immediately derivable from the definition of *food*, and this fact apparently needs to be stated separately for this item. To call something *a good photograph* is to evaluate it in terms of its clarity or its ability to elicit positive esthetic responses in the viewer, but neither of these notions can be directly derived from the definition of *photograph*. Here, too, the evaluative feature needs to be stated independently of the definition of the word.

The question a lexicographer must face is whether these matters have to do with what one knows, as a speaker of a language, about the words in that language, or what one knows, as a member of a culture, about the objects, beliefs and practices of that culture. Do we know about books that they are used in our culture to reveal information or elicit certain kinds of esthetic appreciation, or do we know about *the word* 'book' that it contains evaluative features that allow us to interpret the phrase *a good book*? Do we understand the expression *good water* (as water that is safe for drinking) because its semantic description has set aside that one use of water as the use in terms of which water is to be generally evaluated, or because we know that for most purposes (e.g. watering the grass, bathing) any kind of water will do, but for drinking purposes some water is acceptable and some is not? These are serious questions, but we can of course avoid facing them by making, with the typical lexicographer, the decision not to insist on a strict separation between a dictionary and an encyclopedia.[a]

The distinction between lexical information about words and non-lexical information about things must come up in dealing with Bierwischian features too. Let us examine some of the ways in which users of English speak of the horizontal dimensions of pieces of furniture. If we consider a sofa, a table, and a chest of drawers, we note first of all that a sofa or a chest of drawers has one vertical face that can be appropriately called its *front*, but the table does not. For a non-vertically-oriented oblong object that does not have a natural front, its shorter dimension is spoken of as its *width*, the longer dimension as its *length*. For the two items that do

---

[a] As evidence of the linguistic validity of 'evaluative features' I would accept a pair of words which differ *only* in the evaluative features associated with them. If, for example, English *food* and *feed* could always refer to the same objects but served in the expressions *good food* and *good feed* to refer to food that was *palatable* and *nutritious* respectively, such a pair would provide a good argument for the existence of evaluative features as an aspect of linguistic competence.

have a front, the dimension along that front is the *width* (even though it may be the longer of the two dimensions), the dimension perpendicular to the front is its *depth*.

Objects with fronts, furthermore, are typically conceived of as confronted from the outside, as is the case with the chest of drawers, or as viewed from the inside, as with the sofa. The terms *left* and *right* are used according to this inner or outer orientation. Thus the left drawer of a chest of drawers is what would be to our *left* as we faced it, the left arm of a sofa is what would be to our *right* as we face it.

This information is clearly related to facts about the objects themselves and the ways in which they are treated in our culture, and cannot be something that needs to be stated as lexically specific information about the nouns that name them. It seems to me, therefore, that the truly lexical information suggested by these examples is the information that must be assigned to the words *left*, *right*, *wide*, *long* and *deep* (and their derivatives), and that the facts just reviewed about the items of furniture are facts about how these objects are treated by members of our culture and are therefore proper to an encyclopedia rather than a dictionary. It is difficult to imagine a new word being invented which refers to sofas but which fails to recognize one of its facts as its 'front'; and it is likely that if a new item of furniture gets invented, the properties we have been discussing will not be arbitrarily assigned to the noun which provides the name for these objects, but rather the words *wide*, *left*, etc., will be used in accordance with the ways in which people orient themselves to these objects when they use them. That the orientation is a property of the position- and dimension-words is further demonstrated by the fact that the uses I have suggested are not by any means obligatory. If a 3 ft. by 6 ft. table is placed in such a way that one of its 6 ft. sides is inaccessible, with people sitting at and using the opposite side, the table can surely be described as 6 ft. wide and 3 ft. deep. On the other hand, a sofa that is 2 miles 'wide' would probably impress us more as a physical object than as a sofa and would most likely be described as being 2 miles long.

The phenomena I have been mentioning are to be stated as part of the presuppositional components of the lexical entries for the words *left*, *wide*, etc. Uses of the word *wide* presuppose that the object being referred to has at least one (typically) horizontal dimension; and that the dimension which this word is used to quantify or describe is either the main left-to-right extent of the object as human beings conceive their orientation to it, if that is fixed, or it is the shorter of two horizontal dimensions. The adjectives *tall* and *short* (in one sense) presuppose, as *high* and *low* do not, that the object spoken of is vertically oriented and is typically in contact with (or is a projection out of) the ground. Similarly the noun *post*, as opposed to *pole*, presupposes that the object in question is (or is at least intended to be) vertically oriented and in contact with the ground. Many of the features of spatial orientation treated by Bierwisch will take their place, in other words, in the presuppositional components of the semantic descriptions of words usable as predicates.

There are, however, some spatial-orientation features that appear to enter rather basically into the definitions of nouns. Of particular interest are nouns that identify conceptually n-dimensional entities where these are physically realized by m-dimensional objects, where m > n. Thus a *line* is conceptually one-demensional, and a *stripe* is conceptually two-dimensional. If a straight mark on a piece of paper is viewed as a *line*, the dimension perpendicular to its length is its *thickness*, but if it is viewed as a *stripe*, the second dimension is its *width*. If the stripe has a third-dimensional aspect (e.g. if it is drawn with heavy paint), it is that which one speaks of as its *thickness*. These are matters that seem to be related to the 'meaning' of these nouns rather than to presuppositions about the objects they name.

8 Functional shift

Syntactically and semantically different uses of the same word type should be registered in the same lexical entry whenever their differences can be seen as reflecting a general pattern in the lexical structure of the language. I shall call attention to certain situations in which a word that is basically a noun can also be used verbally, and a situation in which a verb of one type is used as a verb of another type.

I have already suggested that the 'sentential' portions of the definitions of agent and instrument nouns serve to provide the evaluative features associated with these nouns. In many cases they also serve in identifying the verbal use of these same nouns. If, for example, the word *pilot* is defined in part as one who flies an airplane, a dictionary entry must show some way of relating this aspect of the meaning of the noun to the meaning of the associated verb. Perhaps, for example, in connection with the activity characterized as ⟨66⟩:

⟨66⟩ V O A
   where V = navigate; presupposition:
                O = air vessel

one might wish to state that the noun *pilot* is the name given to one who professionally serves the Agent role in this activity, the verb *pilot* has the meaning of the verb in this event-description. If the word is further represented as basically a Noun, a general understanding of 'metaphor' will suffice to explain why the verb *pilot* can be used to refer to activities that are 'similar to', not necessarily identical with, the activity of a pilot in flying an airplane.

If the noun *lock* is defined as a device which one uses to make something unopenable, then that is related to the fact that the *verb lock* means to use such a device for such a purpose. If *plastre* is defined as something which one attaches to a surface for a particular range of purposes, then that fact should be shown to be related to uses of the *verb plastre*. I have no proposal on how this is to be done; I merely suggest that when both the verbal and the nominal use of a word refer to events of the same type, the event-description should, other things being equal, appear only once in the lexicon.

Certain fairly interesting instances of verbal polysemy seem to have developed in the following way. Where one kind of activity is a possible way of carrying out another kind of activity, the verb which identifies the former activity has superimposed onto it certain syntactic and semantic properties of the verb which identifies the second or completing activity.

Thus the verb *tie* refers to particular kinds of manipulations of string-like objects, with certain results. In this basic meaning of the verb, it occurs appropriately in sentences ⟨67⟩ and ⟨68⟩:

⟨67⟩ He tied his shoestrings.
⟨68⟩ He tied the knot.

The act of tying things can lead to fastening things, and so an extension of the verb *tie* to uses proper to a verb like *fasten* or *secure* has occurred. The verb can now mean to fasten something by performing tying acts, and it is this which accounts for the acceptability of *tie* in ⟨69⟩:

⟨69⟩ He tied his shoes.

Shoes are simply not in themselves the kinds of objects that one manipulates when tying knots.

In this second use the verb *tie* continues to describe the original activity, but it has been extended to take in the results of such activity. The feature that characterizes this second use, then, will be something like ⟨70⟩:

⟨70⟩ extension: Result (replace *fasten*).

The verb *smear*, to take another example, refers to the activity of applying some near-liquid substance to the surface of some physical object. The activity of smearing something onto a thing can have the result of covering that thing. The word *smear* has, in fact, been extended to take on the syntax and semantics of *cover*. Thus the 'original' and the extended uses of *smear* are exemplified in the following sentences:

⟨71⟩ He smeared mud on the fender.
⟨72⟩ He smeared the fender with mud.

The difference by which ⟨71⟩ and ⟨72⟩ are not quite paraphrases of each other is found in the meaning of sentence ⟨73⟩:

⟨73⟩ He covered the fender with mud.

By claiming that the second use of *smear* is one in which the properties of *cover* have been superimposed we have accounted for the addition to ⟨72⟩ of the meaning of ⟨73⟩, and simultaneously we have accounted for the fact that the extended use of *smear* takes (as does *cover*) the Goal rather than the Object as its direct object, setting the latter aside in a preposition-phrase with *with*.[a]

The verb *load*, let us say, means to transfer objects onto or into a container of some sort. The activity of loading can lead to the filling of that container, and so the verb *load* has taken on the additional syntactic and semantic functions of *fill*. In this way we can account for the use of *load* in sentences ⟨74⟩ and ⟨75⟩ and the similarities between ⟨75⟩ and ⟨76⟩:

⟨74⟩ He loaded bricks onto the truck.
⟨75⟩ He loaded the truck with bricks.
⟨76⟩ He filled the truck with bricks.

The verbs *smear* and *load* have the same co-occurrences in their extended meanings as in their non-extended meanings. The verb *tie* is not like that: in its *fasten*-extension it takes nouns that are not appropriate to its original sense. This means that the description of the feature indicated in the extended use will have to be interpreted (by lexico-semantic rules) in such a way that when the two verbs fail to take the same cases, those of the verb which identifies the resulting action are dominant, the characteristics of the event described by the other verb taking their place among the *presuppositions* of the verb in its extended sense. Thus it is presupposed of the *fasten*-extension of *tie* that the Object is something which can be fastened by an act of tying.

> [a] I believe that ⟨72⟩ can be read as a paraphrase of ⟨71⟩. If this is so, a correct description of the situation must be that in its original meanings, *smear* permits the direct-objects choices seen in either ⟨71⟩ or ⟨72⟩, but in its extended meaning it permits only that exemplified by ⟨72⟩.

If the type of extension that I have been discussing in this section can be shown to be a quite general phenomenon of lexical systems (at present it is little more than a suggestion, the 'evidence' for its correctness being rather hard to come by), then perhaps we can use this concept to eliminate certain problems connected with what Gruber has called 'incorporation'. *Leap* is a verb which takes, in Gruber's system, a phrase in *over*, as seen in ⟨77⟩ and ⟨78⟩:

⟨77⟩ He leaped over the fence.
⟨78⟩ He leaped over the line.

The preposition *over* can be 'incorporated' into *leap*, but only in the understanding that the associated noun is an obstacle;[a] thus, the preposition *over* may be absent from ⟨77⟩, but not from ⟨78⟩. The theoretical issue here has to do with the way in which the process of preposition-incorporation is to be sensitive to the size relationship between the entities identified by the subject and prepositional-object noun-phrases. My interpretation is that there is an *overcome*-extension to the word *leap*, and I claim that this accounts simultaneously for the 'obstacle' presupposition and for the non-occurrence of *over* in the extended-sense sentence.[b]

9 Deep-structure acceptability

Facts about lexical items that relate to the formal properties of sentences can be separated into two sets: requirements on the deep-structure and requirements on the surface structure. The former determine the acceptability of a given word in deep-structures of certain types; the latter specify those grammatical modifications of sentences whose operation is determined by lexical information. The surface conditions are provided in the grammar in the form of the rules which convert deep structures into surface structures (transformational rules), and possibly, in some cases, by the elaboration of special constraints on surface structure.

I shall take the position that content words may all be inserted as predicates, and that their realization as nouns, verbs or adjectives is a matter of the application of rules. Therefore we need not consider part-of-speech classification among the types of information relevant to the lexical insertion into deep structures. What Chomsky has referred to as 'strict subcategorization'[c] corresponds to what I have treated here in terms of the number of arguments a predicate takes and their case structure. What Chomsky has referred to as 'selection'[d] is described here with the concept presupposition and is taken as being more relevant to semantic interpretation than to lexical insertion.

The deep-structure requirements that are of chief interest for this discussion, then, are those of the type Perlmutter has been referring to as 'deep structure constraints' or 'input conditions'.[e]

[a] Gruber (1965), p. 24. The condition for incorporation of *over* into *leap*, in Gruber's words, is this: 'The object of the preposition must be of significant height with respect to the subject.'
[b] This argument is certainly not directed against Gruber's incorporation process as such, only against the proposed need to state separately its applicability to words like *leap*, *jump*, *hop*, etc. The quite literal 'incorporation' of *over* in *overcome* has not escaped my notice.
[c] Noam Chomsky, *Aspects of the Theory of Syntax* (1965), M.I.T. Press, esp. pp. 95–100.
[d] Ibid. pp. 148 ff.
[e] My knowledge of Perlmutter's work on deep-structure and surface-structure constraints comes from Perlmutter's presentations at the January 1968 San Diego Syntax conference and from references in papers and presentations by J. R. Ross.

Examples, due to Perlmutter, are the requirement for *budge* that it occur in a negative sentence, as shown in grammaticality judgments on ⟨79⟩ and ⟨80⟩:

⟨79⟩ *I budged
⟨80⟩ I didn't budge

the requirement for *lurk* (discussed earlier) that its Agent be non-coreferential to the Agent of the 'next higher' locutionary verb; or for *try* that its Agent be coreferential to the (eventually deleted) subject of the 'next lower' sentence, as suggested by ⟨81⟩ and ⟨82⟩:

⟨81⟩ *I tried [for you] to find it.
⟨82⟩ I tried [for me] to find it.

I have included deep-structure constraints in this survey of types of lexical information, but I have nothing new to say about them. I would like to suggest, however, that it may not be necessary to require the extent of detail which Perlmutter envisions or the transformational apparatus which that sometimes entails. Where Perlmutter requires that the Agent of *try* match the *Agent* of the embedded sentence, it may only be necessary to require that the coreferential noun-phrase in the embedded sentence be the one which is destined to be the subject of that sentence. And where Perlmutter requires sentence ⟨83⟩ to be derived transformationally from the structure underlying ⟨84⟩:

⟨83⟩ He tried to be misunderstood
⟨84⟩ He tried to get to be misunderstood

this may not be necessary if *try* is merely described as a verb which expresses, of its Agent subject, the intension and attempt to bring about the situation identified by the embedded sentence. This may be necessary in order to account for the way in which we understand sentence ⟨85⟩:

⟨85⟩ He tried to seem cheerful

a sentence which cannot be straightforwardly paraphrased in such a way as to reveal an underlying agentive notion in the embedded sentence.

10   Government

Once a specific predicate word is inserted into a deep-structure, its presence may call for certain modifications in the rest of the sentence. The typical case of this is what is known as 'government'. For English the operation of the rules establishing 'government' associates prepositions with noun-phrases and 'complementizers' with embedded sentences and their parts.[a] Thus – to consider only the association of prepositions with noun-phrases – we speak of 'giving something *to* somebody', 'accusing somebody *of* something', 'blaming something *on* somebody', 'interesting somebody *in* something', 'acquainting somebody *with* something', and so on. It is certain, of course, that many of the facts about particular choices of prepositions and complementizers are redundantly specified by other independently motivated features of predicates or are determined from the nature of the underlying case relationship, so that a minimally redundant dictionary will not need to indicate

---

[a] The term 'complementizer' is taken from Peter Rosenbaum, *The Grammar of English Predicate Complement Constructions* (1968), M.I.T. Press, and refers to the provision of *that*, *-ing*, etc., in clauses embedded to predicates.

anything about the form of governed constituents directly. Until it is clear just what the needed generalizations are, however, I propose using the brute-force method and specifying the prepositions one at a time for each verb and each case relationship.

11 Transformationally introduced predicators

Certain lexical items can be used as predicates semantically, but cannot themselves occur as surface predicate words. Such words will appear in the syntactic position expected of some other constituent (usually, I think, that of the Result constituent), and there must therefore be lexically associated with them some predicator word, a word that is capable of bearing the tense and aspect properties that can only be attached to verb-like elements. The constructions I have in mind are those of the type '*have* faith in', '*give* credence to', '*be* loyal to', etc., and the predicator words I have in mind include *be, do, give, have, make, take* and a few others.

12 Subject and object

One of the subtlest but at the same time most apparent aspects of the syntax of the predicate is the set of ways in which the subject/verb and verb/object constructions can be set up. These constructions have as their purpose – generally, but not always – a focusing on one terminal of a multi-argument expression, a focusing on one particular role in the situation under discussion.

I have suggested elsewhere[a] that the normal choice of subject is determined by a hierarchy of case types, and only those predicates which require something 'unexpected' as their subjects need to have such information registered in their lexical entry. I believe that, at least in the vast majority of cases, those noun-phrases capable of becoming direct objects of a given verb are at the same time the ones which can appear as the subjects of passive constructions made with the same verb. Thus such information at least does not need to be stated twice, and at best may not need to be stated at all for most verbs, since it is most likely subject to rather general statements about combinations of cases.[b]

There is an extremely interesting set of subject- and object-selection facts that seem to operate in connection with expressions of 'quantity' and 'contents'. The verb *sleep* refers to an 'activity' of an animate being in a particular place, where the one who sleeps is typically mentioned as the subject; but when the focus is on the place and at issue is the *number* of different beings that can sleep in that place, the verb permits the Place noun-phrase to appear as subject. We see this in sentence ⟨86⟩:

⟨86⟩ This houseboat sleeps eight adults or sixteen children.

The verb *feed* has its use as a 'causative' of *eat*, but it can be used to express other kinds of relations with eating, too. To indicate the typical relation between the Agent and a description of the 'contents' of his eating activity, we use sentences like ⟨87⟩:

⟨87⟩ The child feeds on raisins.

[a] 'The case for case' (1968) in E. Bach and R. Harms (eds.), *Universals in Linguistic Theory*, Holt, Rinehart and Winston.
[b] Incidentally, verbs which are obligatorily passive are verbs with one associated noun-phrase designated as 'direct object' but none as 'normal subject'. An automatic consequence of this situation is that the expression will be cast in the passive form.

There is a relation between a recipe (which identifies, among other things, quantities of food) and the number of people who can (with satisfaction) eat the food one prepares by following the recipe, and this is the relation we see expressed in a sentence like ⟨88⟩:

⟨88⟩ This recipe feeds eight adults or four children.

In the case of *feed*, the connection between the Place and a quantity of 'eaters' requires the latter quantity to be a 'rate'. Thus we get sentences like ⟨89⟩:

⟨89⟩ This restaurant feeds four hundred people a day.

It is not clear to me how one can capture facts of the sort I have just suggested in any way short of providing (in effect) separate lexical descriptions for each of these uses. It is clear, anyway, that examples ⟨86⟩–⟨89⟩ cannot be understood as merely exemplifying 'causative' extensions of these verbs.

13 Sample lexical entries

In this last section I exhibit extremely tentative suggestions for the lexical entries for *blame* (in two senses),[a] *accuse*, and *criticize* (in one sense). The feature-pair ± Locutionary indicates whether the verb does or does not refer to a linguistic act; the feature + Performative marks verbs which can be used in first-person simple-present utterances in the performance of an explicitly marked illocutionary act;[b] the feature-pair ± Momentary indicates whether or not the verb can be used, in the affirmative, to refer to an event that takes place at a specific point in time. The variables x, y, and z are argument variables whose case-role and preposition-selection properties are indicated by first, second or third position in the next two items. All are normal transitive verbs, and the normal subject-selection for active sentences is in each case the noun-phrase that fills the Source role. For *blame* the direct object is either the Goal or the Object, since we have both '*blame y for z*' and '*blame z on y*'; for *accuse* the direct object is necessarily the Goal noun-phrase; for *criticize* it is the Goal noun-phrase if that is explicit in the sentence, otherwise the Object noun-phrase. The verbs *accuse* and *criticize* require the Goal noun-phrases to identify human beings; *blame* does not, since one can blame an event on, e.g., an inanimate force. The items identified as 'zero for indefinite' and 'zero for definite' indicate the conditions under which the explicit mention of an argument may be omitted, in the sense mentioned in section 5.

There are many types of lexical information not found in these sample entries, for example those mentioned in sections 7 and 8 above. There are many facts about these particular verbs that are ignored in the descriptions provided here, for example the information that the situation named by the Object noun-phrase for *blame* and *criticize* is necessarily understood as something for which factuality is claimed (notice ⟨90⟩ and ⟨91⟩); that the linguistic event capable of constituting a criticism is necessarily more complicated than the linguistic or gestural event capable of constituting an accusation; and that the presupposed seriousness of the offense for an act of criticism is not as great as that appropriate for acts of accusation

---

[a] There is a third use of *blame* in which it refers to a linguistic or otherwise symbolic act. Thus, if John wrote an offensive letter himself and did something which gave others to believe that I had done it, I could report that John blamed the letter on me.

[b] In the sense of John L. Austin, as presented in his posthumous work *How to do things with words*, Oxford University Press, 1965.

(compare ⟨92⟩ and ⟨93⟩). It is hoped, nevertheless, that from these examples some suggestions can be gleaned for the design of canonical representations in a lexicon.

⟨90⟩ I accused John, who has never left Ohio, of taking part in the Berkeley riots.

⟨91⟩ *I criticized John, who has never left, Ohio, for taking part in the Berkeley riots.

⟨92⟩ I accused John of murdering his mother-in-law.

⟨93⟩ I criticized John for murdering his mother-in-law.

Blame (i): $-$Locutionary, $-$Momentary

arguments:	x, y, z
cases:	Source & Experiencer, Goal, Object
prepositions:	by, on, for
normal subject:	x
direct object:	y or z
presuppositions:	x is human
	z is an activity z' or the result of z'
	x judges [z' is 'bad']
meaning:	x judges [y caused z']
zero for indefinite:	x
zero for definite:	z

Blame (ii): $-$Locutionary, $-$Momentary

arguments:	x, y, z
cases:	Source & Experiencer, Goal, Object
prepositions:	by, on, for
normal subject:	x
direct object:	y
presuppositions:	x is human
	z is an activity z' or the result of z'
	x judges [y caused z']
meaning:	x judges [z' is 'bad']
zero for indefinite:	x
zero for definite:	z

Accuse: $+$Performative, $+$Locutionary, $+$Momentary

arguments:	x, y, z
cases:	Source & Agent, Goal, Object
prepositions:	by, ϕ, of
normal subject:	x
direct object:	y
presuppositions:	x and y are human
	z is an activity
	x judges [z is 'bad']
meaning:	x indicates [y caused z]
zero for indefinite:	x
zero for definite:	z

Criticize: $+$Locutionary, $+$Momentary

arguments:	x, y, z
cases:	Source & Agent, Goal, Object

prepositions: by, ϕ, for
normal subject: x
direct object: z if y = ϕ, otherwise y
presuppositions: x and y are human
                        z is an activity z′ or the result of z′
                        x judges [y caused z′]
meaning: x indicates [z′ is bad]
zero for indefinite: x and z; and y if z is a 'result' and z $\neq$ ϕ
zero for definite: none

Ohio State University

a part. Like the investigator, a native speaker, during enculturation, constructs his model or set of rules on the basis of observed behavior, a model which may or may not accord with that of another speaker. The investigator uses certain methods in his analysis. For example, in accounting for various uses of a form, he may, from a principle of economy, avoid unnecessary polysemy (or homonymy) by trying to generalize to core meanings or lowest common semantic denominators commensurate with paradigmatic distinctiveness. To the extent that his analytic methods represent those used by native speakers, his process of analysis may be said to be a model of a more abstract level of speaker processes, but his resulting description may or may not be congruent with the set of rules actually arrived at by any one native speaker.

5 Notational conventions

The following notations will be used:

(*a*) *Aux(iliary)*, the complex of elements which represents the obligatory choices of tense, number, person, modal auxiliary, assertion, etc. of the verb, is for simplicity symbolized by the simple third-person singular present tense of the verb.

(*b*) X, Y, Z are syntactic variables standing for any string of symbols in the syntactic representation of a sentence.

(*c*) Single quotes enclose meanings such as components, definitions, glosses, translations, or interpretations.

(*d*) For simplicity, A, B, C are used to represent variables or the syntactic position of noun phrase, rather than the more usual indexed NP_1, NP_2, NP_3. A-noun or A-noun phrase refers to the actual morpheme (string) occurring in the position A. When enclosed in single quotes, e.g. 'A' or 'A has B', the letter represents the referent of a token of the A-noun type.[a]

(*e*) Similarly, P, Q, R, or 'P', 'Q', 'R', rather than S_1, S_2, S_3, stand for sentences, schematic sentences, or semantic components in sentence form.

(*f*) P_i, P_j are tokens of P.

(*g*) AF symbolizes a function with variable A. This order, rather than the more usual $f(x)$ or Fx, is used to simplify reading in the subject-predicate order of English. ARB, with variables A and B, stands for a relation between 'A' and 'B'.

(*h*) The existential quantifier 'there is a...' is indicated in a definition by prefixing the indefinite article *a-* (*an-*) to the first occurrence of the quantified variable in the definition.

(*i*) *Not-* is prefixed to a whole sentence to indicate negation with wide scope and directly to a specific item in the sentence for negation with narrow scope. Thus, with narrow scope: 'he not-has every one', corresponding to 'he lacks every one'; with wide scope: 'not- (he has every one)', which can mean 'not-he has every one' or 'he not-has every one' or 'he has not-every one'.

6 Underlying functions

We draw upon some of the methods of symbolic logic for analyzing sentences and reverse them as well to synthesize or generate sentences. Thus *John has a dog* may

---

[a] Symbol tokens are separate observable occurrences in speech. An abstract symbol type is the class of its tokens. A referent is the given thing, event, etc., real or hypothetical, that a given token is made to refer to or stand for by the symbol-using organism as utterer or hearer of the token.

be analyzed into the existential quantifier and functions as 'there is a B' and 'A has B' and 'A = John' and 'B is a dog'. *Dog* is in fact a one-place function, in contrast with functions of two or more places. To show this its representation as a lexical item in a theory (description) of English would not be as one word, but as *A is a dog*. A relational noun such as *son* is a two-place function *A is B's son*. *A gives B to C* has three places. Apparently homonymous items that actually differ in the number of places would show this fact explicitly, such as *A is a child* and *A is B's child*. (Note that 'A is B's child' does not necessarily imply 'A is a child'.) Since difference in number of places correlates with differences in syntactic behavior, such a representation of lexical items as schematic sentences shows the syntactic differences and facilitates the application of appropriate rules to generate utterances. It also does so, for example, for mass nouns *vs.* count nouns *vs.* adjectives, e.g. *A is sugar*, *A is a substance*, *A is sweet*. Thus, to amend what was said above, the unit to be defined is a lexeme as a function.

In the definitions of items, their semantic components are also in the form of schematic sentences or functions. In a theory of a language, then, the definition or meaning of an item is seen as a set of sentences which together translate, or paraphrase, the sentence to be defined (Peirce 1933: pars. 427, 569). A definition is thus a statement of equivalence between the defined sentence and the defining sentences. It corresponds roughly to a similar statement, or schema of statements, in the object language of whose truth native speakers are competent to judge (Weinreich 1962: 42 ff.). It is also the covert major premise in various logical arguments phrased in the object language (Peirce 1933: pars. 176, 179). We can therefore involve informants in testing putative definitions as shown in the discussion of tests below.

7 Some components defined

In Bendix (1966: 61–79) a subset of verbs was isolated from the semantic system of English and analyzed. These were *A gets B*, *A finds B*, *C gives AB*, *C gets AB*, *C lends AB*, *A borrows B from C*, *A takes B from C*, *A gets rid of B*, *A loses B*, and *A keeps B*. Definitions using semantic components were attempted which were sufficient to distinguish the verbs from one another within the subset. Because of their tentative nature, the definitions will not be repeated here. Rather, components will be described, and some definitions will be given in illustrating, eliciting and testing techniques.

In formulating components, certain primitives are used. Thus *at* in *at time T* points to an undivided (but not necessarily indivisible) time, and *time* or *T* and *before* are also primitives, used in defining *immediately before*:

⟨1⟩ T_1 is immediately before T_2 = (i) 'T_1 is before T_2'
(ii) 'not-([T_1 is before a-T_3] and [T_3 is before T_2])'.

Immediately before is required in defining a component of change of state or activity:

⟨2⟩ P comes to be, at T = (i) 'not-P at a-T_1 and at T'
(ii) 'P at a-T_2'
(iii) 'T_1 is immediately before T'
(iv) 'T is immediately before T_2'.

Other types of components, each followed by a rough reading in the object language, are: *A an-F* 'A does or is something'; *B is an-R_h A* 'there is a "have"-state relation between B and A' or simply 'A has B'; *B is A's* 'B belongs to A'. The last two are discussed more fully in 9.2–3 below. In such components, *is* vs. *does* represent state *vs.* activity.

Another primitive is *P causes Q*, as in the definition *C gives B to A* = '(C an-F) causes (B is an-R_h A)'. This label is used when tests indicate the presence of the component that 'C' ('A', etc.) does or is something and, further, that this action or state of 'C' leads to, results in, or may otherwise be said by a speaker to be connected with another action or state in a way that he calls causal.

Ultimately we would want to define such terms of the metalanguage as *cause*, *referent*, or the existential quantifier *there is a...* more exactly within a general linguistic theory as well as, perhaps, in any particular theory specific to a given language. This is not to be misconstrued as an attempt to define causation or existence nor as a commitment to the prescriptions of any particular philosophical school of thought about the nature or definition of what can or cannot be a referent or to the full implications of the quantifier as used in logic. Rather, the interest is in describing how speakers use language and in making clear, for example, what we claim a speaker is asserting when he uses a form whose meaning we define as containing the component 'cause'.

8 Eliciting and testing techniques

There are a number of types of eliciting techniques and tests that can be used in generating the data on which to base a semantic account.[a] These tests assume that the investigator has already obtained rough glosses, translations, and the like for forms in a language and is now interested in locating semantically similar forms more exactly within a paradigm of oppositions. It is also expected that possibilities of collocation with other forms are used as evidence, as well as possible referent situations. The basic goal in all the testing is to isolate and verify the semantic components that mark the oppositions in the paradigm.

8.1 The simplest type of technique, an open-ended interpretation test replicates the paradigm of oppositions by asking the informant to contrast *P* and *Q*, e.g. 'What is the difference in meaning between *He gave it to me* and *He lent it to me?*' Applied to *A has B* and *A gets B*, such a test might lead to a first approximation of a definition *A gets B* = 'A comes to have B' or, using the component labels given above, '(B is an-R_h A) comes to be'. Such test sentences are constructed as minimal pairs varying only in the supposedly single lexemes tested. Differences in meaning are then assumed to be attributable only to the lexemes. In theory, however, the informant is free to imagine different background situations for the two sentences, in which case they would cease to be completely minimal pairs. To control for this, *not-P* and *Q* can be placed in the same sentence, thus replicating paradigmatic opposition with syntagmatic contrast. He could then be asked, 'What does this mean?' or 'If you heard a person say this, what would you think he meant?' Such an interpretation test would have sentences as *He didn't give it to me, he lent it to me* or *He didn't take it, he got it.*

[a] Ariel's (1967) critique of the tests of ch. 2 in Bendix 1966 gives valuable warnings and suggestions for anyone who would use semantic tests uncritically or as operational procedures. His otherwise useful discussion is marred by an a priori approach which would prejudice an empirical descriptive semantics.

Assuming, for this last sentence, the informant interpretation 'He didn't do anything to get it, he just got it', we could account for this response by formulating the component 'cause' in the following model of the informant's process of interpretation:

$\langle 3 \rangle$ (i) $P = Q + R$ [A takes B = A gets B + A does something which causes this; or: (A an-F) causes (A gets B)]

    (ii) not-P, Q   [He didn't take it, he got it; or: not-(A takes B), A gets B]

$\quad\quad = $ not-$(Q+R)$, Q   [not-(A gets B + [A an-F] causes...), A gets B]

$\quad\quad = $ (not-Q + not-R) or (not-Q + R) or (Q + not-R), Q

$\quad\quad = $ Q + not-R, Q

$\quad\quad = $ not-R, Q   [He didn't do anything to get it, he just got it].

The principle is to establish consistency between the first and second clauses. *He didn't take it* ('not-P') is ambiguous for negation of the components 'Q', 'R', or both, but only 'Q + not-R' is consistent with *he got it* ('Q'). In this way the component 'cause' ('R') is isolated in the meaning of *take*. For economy and other reasons the definition could be reduced as *A takes B* = '(A an-F) causes (B is an-R_h A)' but would have to be further revised if reason is found to consider *A takes B* a truncated form of the lexeme *A takes B from C*.

The test frame *not-P, Q* seems to be interpretable when of 'P' and 'Q', one is included in the other or when the unshared components are values along the same dimension. Otherwise (e.g., *He didn't keep it, he got it*), *not-P, Q* seems quite difficult to interpret as anything but a denial of a previous statement P, which is always a possible interpretation in this test. On still other possibilities of informant interpretation than the strict account given above, see the discussion in 8.4.2–3 below.

Such interpretation tests are fairly open-ended and more exploratory and offer the investigator a range of responses from which to induce the possible identity of components. He may then go on to test hypothesized components with more rigorous multiple-choice tests of the types described next.

8.2 In a type of matching test, two minimal-pair sentences, P and Q, are contrasted for greater similarity in meaning to a third sentence R which expresses a presumed component of the meaning of P. In one form, this type involves a syllogistic argument. For example, for 'B is not A's' ('R') as a component of the meaning of *C lends B to A* (P):

$\langle 4 \rangle$ (P) He's only lent it to me $\big\}$ and so (R) it isn't really mine.
     (Q) He's only given it to me

This might also take the form *Since P/Since Q: R*. In the syllogism, R is the proper conclusion if P is the minor premise. The major premise is covert and is the statement of equivalence which defines P = '... + R'.[a]

---

[a] For *lend*, this and other tests contrasting it with other verbs, its possible collocations and referents, etc. might yield the definition *C lends B to A* = 'B is an-R_h C' + 'B not-is A's' + '(C an-F) causes ([B is an-R_h A] + [(B is A's) not-comes to be])', or roughly 'C has B' and 'B is not A's' and 'C causes A to have B without B becoming A's'. Without repetition of the existential quantifier in the second occurrence of 'B is an-R_h A', namely as 'B is R_h A', the false implication would be that 'B' must be in the same relation to 'A' as it is to 'C'. Further, with our formulation, any assertion *C lends B to A* implies that 'B' is such that it can be had in two

This type may also take the form of a ranking test in which an informant is asked which of two sentences makes more sense or is easier to interpret. For example:

⟨5⟩ (i) Since (P) he's only lent it to me, (R) it isn't really mine
Since (Q) he's only given it to me, (R) it isn't really mine

(ii) Since (P) he's only lent it to me, (R) it isn't really mine
Since (P) he's only lent it to me, (not-R) it's really mine

or for 'A has B, (immediately) before T' as a component of *A keeps B, at T*:

⟨6⟩ If he has it, he'll keep it
If he doesn't have it, he'll keep it.

8.3 Informants, thus, are not asked to make absolute judgments that characterize certain sentences as meaningless, senseless, inconsistent, impossible to interpret, etc., but rather to rank sentences as more or less meaningful, easy to interpret, etc. Enough has already been said about the interpretability of 'meaningless' sentences. *If he doesn't have it, he'll keep it* makes sense, for example, if one supplies the background context that 'he' is a stamp collector about to be given a copy of a stamp of which he may or may not already have another copy: 'If he has it, he'll trade it for a stamp he doesn't have; if he doesn't have it, he'll keep it.' The greater number of operations necessary to make a sentence interpretable is what makes it rank lower.

8.4 An important type of ranking test aims at separating criterial semantic components of an item (Weinreich 1962: 33), i.e. those that are necessarily implied by it, from elements of meaning not criterial to it. Again, the only semantic elements tested would be such as had already become likely candidates for being components as the result of interpretation tests and similar sources. Let us assume that the test sentence *She didn't give it to him, she lent it to him* has given rise to the candidates 'she said that he couldn't keep it as long as he wished' and 'she said it wouldn't be his' as components of *she lent it to him*. Compare the following pairs of sentences:

⟨7⟩ (i) (a) She lent it to him but said he could keep it as long as he wished
(b) She lent it to him but said he couldn't keep it as long as he wished

(ii) (a) She lent it to him but said it would be his
(b) She lent it to him but said it wouldn't be his.

In ⟨7i⟩, (a) seems clearly to rank higher than (b); in ⟨7ii⟩ a clear ranking does not seem so easy. Another format does not use *but*:

⟨8⟩ (a) They lent him a book that wasn't his
(b) They lent him a book that was his.

However, we will restrict ourselves to an account of the *but*-test (Weinreich 1963: 154, n. 4).

8.4.1 In the ordinary usage of *P but R*, the *but*-clause negates an element not criterial to *P*. This may be a semantic feature more closely associated as a connotation of *P* or, more broadly, a previous expectation, assertion, etc.:

⟨9⟩ (i) Given: (a) P but R [e.g., He's lost his watch but knows where it is]
(b) P but not-R [e.g., He's lost his watch but doesn't know where it is]

different ways, i.e. that it is possible for 'A' to have 'B' either without 'B' being 'A's or with 'B' being 'A's, thus accounting for the initial anomaly of *C lends A a cold, a black eye*, etc.

(ii) If R negates an element not criterial to P, informants should rank (a) higher; if not-R does, they should choose (b)

(iii) If, of R and not-R, one is a criterial component of P and, thus, the other negates this component, we expect informants to consider both (a) and (b) odd, difficult to interpret, needing rephrasing, etc.

In the case of $\langle 9 \,\text{iii}\rangle$, other means, such as interpretation tests, would make clear whether R or not-R is a component of P. In constructing but-test sentences, a degree of simplicity is maintained by not having not-P as the first clause, for the wide scope of this negation would leave indeterminate which component of P is being negated by the informant. This fact should become clearer from the following exposition of the but-test.

$\langle 10 \rangle$ (i) Let: $P = \text{Type}_p$, $R = \text{Type}_r$, etc.

$\qquad\qquad$ $P_i = \text{Token}_i$ of P, $P_j = \text{Token}_j$ of P, etc.

$\qquad$ where: Type_p, Type_r, etc. each have sentence structure

(ii) $P\ but\ R = \text{'P'} + \text{'(P+not-R) is expected'} + \text{'R'}$.

In the ordinary usage of $P\ but\ R$, the noncriterial element is not-R. For example, *He's lost his watch but knows where it is*, where not-R would be *He doesn't know where it is*. The criterial components may be represented as Q. Where not-R is a connotation of P, this may be represented as $P = Q(not$-$R)$, with the following definition:

$\langle 11 \rangle$ $[P = Q(\text{not-R})] = [(P_1 = Q+\text{not-R})\ \text{and}\ (P_j = Q)]$.

$P = Q(not$-$R)$ must therefore be read "If someone asserts 'P', he is either asserting 'Q+not-R', or he is asserting 'Q'", and not "If someone asserts 'P', he is asserting 'Q+not-R or Q'."[a] In this case, then,

$\langle 12 \rangle$ P but R = Q(not-R) + [(Q+not-R) is expected] + R

so that a speaker may use $P\ but\ R$ to express $Q+R$.

8.4.2 Next there is one test case where Q and not-R are both criterial: $P = Q + not$-R. This would give

$\langle 13 \rangle$ $[P\ but\ R] = [P + \ldots + R] = [Q+\text{not-R}+\ldots+R]$

which is a contradiction, as in *He's lost his watch but did so intentionally*, where *not intentionally* (not-R) is criterial for *lose* (P).[b] Hearing such an utterance in isolation, a person might follow up his initial reaction that it is self-contradictory in one or more of the following ways, among others:

(a) He may reject it on the grounds that the speaker is violating the code or

---

[a] Similarly, for a definition 'P = Q or R', we read "If someone asserts 'P', he is asserting 'Q+R' or 'Q' or 'R'" and not "If someone asserts 'P', he is asserting 'Q or R'." The symbols in a definition stand for types, and not for tokens, and it is tokens which actually convey assertions. Thus, for 'P = Q or R': 'not-P = not-(Q or R) = not-Q or not-R'. In such a disjunctive case, the negation is treated distributively for types and is not read as 'not-(Q or R) = not-Q + not-R'. The latter reading deals with tokens and represents an equivalence of assertions. That is, for 'P = Q or R': '(not-P)$_i$ = not-Q + not-R' and '(not-P)$_j$ = not-Q' and '(not-P)$_k$ = not-R'.

[b] 'Not intentionally' rather than 'unintentionally' since 'A' in *A loses B* can be inanimate: *Her hair has lost its silken sheen*. Wide *vs.* narrow scope of negation indicates no necessary applicability *vs.* applicability respectively of the dimension 'intention' to 'A', as in 'not-(A intends...)' *vs.* 'A not-intends...'. Cf. Zimmer 1964: 21–6.

talking nonsense: e.g. 'It makes no sense' or 'It makes no sense to say that someone loses something intentionally.'

(b) He may interpret it as denying a previous assertion 'not-R' or 'P+not-R': 'Someone must have said it wasn't intentional.'

(c) He may suspect a variant version of the code, namely that for the speaker 'P = Q(not-R)': '...must think that losing something can refer to an intentional act as well.'

(d) He may interpret the contradiction as expressing irony and resolve it by replacing 'P' with 'Q', which gives 'Q+...+R [and thus not-P]'.[a] He may then interpret the irony as 'Someone falsely thinks or says P' and paraphrase *P but R*: 'He's lost his watch, so he says (or so you think), but actually he got rid of it', 'He's lost his watch, so he thinks, but unconsciously he actually got rid of it', etc.

(e) He may suspect a lapse on the part of the speaker and reject 'not-R' and thereby 'P': 'Since he did so intentionally, he didn't really lose it or it wasn't really a case of losing.'

8.4.3 Finally there is another test case where, still, $P = Q + not\text{-}R$, but this time the utterance is *P but not-R*, as in *He's lost his watch but didn't do so intentionally*:

⟨14⟩ P but not-R = P+([P+R] is expected)+not-R = (Q+not-R)+([Q+ not-R+R] is expected)+not-R.

Hearing this in isolation, an informant's first reaction might be 'If P, who would expect R?' since *P* already implies *not-R*. Again, the hearer might resolve the conflict similarly to 8.4.2 (a)–(e) above: (a) He may reject the utterance: 'You must think that (it's possible to say that) a person sometimes loses something intentionally, but that's not so.' (b) Denial of previous assertion: 'Someone must have said that it was intentional.' (c) Variant version of the code. (d) Interpretation of irony: replace 'not-R' with 'R' as in 'He's lost his watch, so he says, but it was intentional.' (e) Interpretation of a lapse: rephrase *P but not-R* as *P and not-R*, thus eliminating '(P+R) is expected', as in 'Well yes, of course: he lost his watch *and* didn't do so intentionally.'

8.4.4 Thus, both sentences of the test pair

⟨15⟩ (a) He lost it but didn't do so intentionally
   (b) He lost it but did so intentionally

are initially confusing since a criterial component is involved, but when pressed for a possible interpretation, an informant might consider ⟨15 b⟩ as describing deceptive behavior. Interpreting deception is easier where such an unobservable as intention is involved. It is harder where intentions are not questioned, as in the initially confusing test pair

⟨16⟩ (a) He got himself one but didn't do anything to get it
   (b) He got himself one but did something to get it

where 'A does something to get B' ('[A an-F] causes [B is anR$_h$ A]') is criterial to *A gets self B*. When presented with such a test pair as ⟨16⟩, it has occurred in work on various languages that an informant, in the confusion of choosing between two unclear sentences, has selected the one in which the *but*-clause does at least negate

---

[a] We describe this replacement as a change from 'The referent of P$_i$ is a member of the class of events defined by P' to 'The referent of P$_i$, which is a member of the class of events defined by Q+R, is (meant to appear) like a member of the class of events defined by P.'

a previously occurring element, even though the latter is a criterial component of the first clause. However, when asked for a possible interpretation or situation of occurrence of the selected sentence, an informant has often concluded that it is actually contradictory, thus supporting the criteriality of the hypothesized component.

8.4.5 We take up another use of the *but*-test, as in:

⟨17⟩ (*a*) He lost it, but it was his
       (*b*) He lost it, but it wasn't his.

'A has B before (time) T' does seem to be a component of *A loses B at T*, but this pair of sentences tests for the more restricted element 'B is A's before T', which is more doubtful, but still a candidate. The effect of the two sentences is not what would be expected if a criterial component were involved, as described in 8.4.2–3 above, yet they do not offer a clearcut choice as to which makes more sense. Each requires a hearer to imagine a background context in order to be interpretable. For ⟨17a⟩, for example, this could be that 'it' might have been someone else's and that it is undesirable to lose others' possessions. For ⟨17b⟩ the background information could be the converse. Whichever background is supplied would make the corresponding sentence rank higher. If 'it was his' were already a connotation of 'he lost it', no background would need to be added, and ⟨17b⟩ should most often be ranked higher by informants. Thus the *but*-test may also be useful for identifying nonconnotations, as well as, perhaps, disjunctive connotations that are contradictories of each other.[a]

8.4.6 A sentence such as *The pickpocket lent the judge a gold watch, but it was the judge's (watch)* illustrates, perhaps in the extreme, how the value of informant judgments can suffer due to interference if other sentence constituents than the tested item (here, *lend*) are not fairly neutral, especially when testing for criteriality. Generality is desirable particularly when using *P but R* and is achieved in the testing of verbs, for example, by employing pronouns or their equivalents rather than nouns.

9 'Have'

The elements *B is R*$_h$ *A* and *B is A's* will now be explored more fully.[b]

---

[a] Ariel (1967: 544) has issued a warning that should be taken seriously when applying the *but*-test: this test could identify disjunctive criterial components as connotations. To use his example, (*b*) would rank higher in: *John is a postman*, (*a*) *but he delivers packages*/(*b*) *but he does not deliver packages*, where *postman* may be defined as 'a man whose job is to deliver letters and/or packages'. Thus, 'deliver packages' could be mistaken as a connotation when, it is contended, it is an essential part of a criterial component that is an inclusive disjunction. The example inadvertently illustrates another point: namely, disjunction can be a function of the analyst's ingenuity. If 'letters and/or packages' is reduced to 'mail', with a unitary definition, *postman* becomes 'a man whose job is to deliver mail', without any disjunction. Then connotational elements of 'mail' must be added to the criterial components of 'mail' to arrive at 'packages'. The problem of descriptive semantics in the context of a prescriptive philosophical discourse is that a descriptive theory may find itself measured against a view of reality when in fact we are dealing with contending theories about a body of data, measured against an evaluating metatheory of description.

[b] For a fuller discussion of *have*, see Bendix 1966: 37–59, 123–34, and Bach 1967. Cf. also Lyons 1967.

9.1 Compare the following sentences with the main verb *have* and their corresponding sentences without *have:*

⟨18⟩ (i) This line has *had* three of its ships run aground this year
Three of the ships of this line have run aground this year
(ii) With the new precautions, we will *have* no more thefts occur in our building
With the new precautions, no more thefts will occur in our building
(iii) He has *had* it said about him that...
It has been said about him that...
(iv) The trees on our street are *having* their lower branches removed
The lower branches of the trees on our street are being removed.

The topic of a passive sentence corresponds to an object of the verb in the equivalent active sentence. (The topic is ordinarily the subject in English.) The above examples show that the topic of a *have*-construction, however, can correspond to a much larger variety of noun-phrase types in sentences without *have*. These can include noun phrases in passive sentences other than that of the agentive *by*+*NP*. It appears then that topicalizing any one of a variety of noun phrases in certain underlying sentence structures requires it to be raised to subject position in the surface structure with introduction of *have*, whereas such noun phrases appear in a variety of other surface-structure positions when not topics.[a]

9.2 There is a special subclass of *have*-sentences whose corresponding sentences without *have* contain the main verb *be*, such as

⟨19⟩ (i) The safe *had* no money in it
There *was* no money in the safe
(ii) John *has* the boy with him
The boy *is* with John
(iii) The neighbors' son will *have* a sitter to watch him⎫
The neighbors will *have* a sitter to watch their son⎭
There will *be* a sitter to watch the neighbors' son
(iv) The top drawer of John's bureau *has* the socks in it⎫
John's bureau *has* the socks in its top drawer⎬
John *has* the socks in the top drawer of his bureau⎭
The socks *are* in the top drawer of John's bureau.

(There are other subclasses of *have-be* correspondences than this one.) For ⟨19 i⟩, for example, the underlying constituents could be roughly represented linearly as follows (omitting *Aux* and the choice of topic markers):

⟨20⟩ (not-[there is a B])+(B is in A)+(B is money)+(A = the safe).[b]

---

[a] This approach has now been made central to the analysis, whereas it was only mentioned as an alternative in the original formulation (Bendix 1966: 102, n. 11) on the basis of a prepublication version of Chomsky 1965.
[b] For the slightly different sentence *The safe didn't have money in it*, negation is applied with wide scope to the totality of constituents: not-([there is a B]+[B is in A]+[B is money]+[A = the safe]). Bach (1967) has suggested that *is*, i.e. *be*, should be excluded from the underlying constituents and introduced by transformation, like *have*, upon topicalization of certain noun phrases. This would leave 'B in A', 'B money' and bring us still closer to the formulations of symbolic logic, which have long omitted such connectives as *be* and *have*. *Is* is included here, however, as a metalinguistic device to indicate 'state', vs. *does* for 'activity'.

'B is in A' is an example of the structure which defines this subclass of *have*-sentences, namely

⟨21⟩ B is XAY, *where*: X = to+Verb+Z

                  or Locative Preposition+Z

                  or with

                  or for

        Y may be null.

        Z may be null.

The subclass, then, has surface structures of the general form *A has BXAY*.

Furthermore, it is related to a class of surface structures, of more immediate interest here, of the general form *A has B* which is realized by suppression of 'XAY'. This will be called the 'accidental *A has B*' for reasons given in 9.4 below. 'B is R_h A', introduced further above in section 7, is a component of 'B is XAY', so that the truncated function 'is X...Y' is a type of 'is R_h'. We can draw examples of the accidental *A has B* from (19 i–iv) above:

⟨22⟩   (i)   The safe has no money

      (ii)   John has the boy

    (iii)   The neighbors' son will have a sitter

    (iv)   The neighbors will have a sitter

     (v)   The top drawer of John's bureau has the socks

    (vi)   John's bureau has the socks

   (vii)   John has the socks.

It can be seen, further, that any one such *A has B* sentence may express an indefinite number of state relations between 'A' and 'B', each of which is made explicit by a particular suppressed 'XAY'.

9.3   There is a class of *A has B* sentences, called 'inherent', which only express one or a restricted number of relations between 'A' and 'B', depending on the particular *B*-noun or *A*- and *B*-nouns occurring. These *B*-nouns are often relational, such as *son, neighbor, assistant, queen, counterpart, end, leg, texture, size*, and will be considered to have the basic form *C is A's B*, e.g. *C is A's son*.[a] The inherent *A has B* is thus related to *C is A's B*, as shown next. Where the *B*-noun is not relational (e.g., *This is John's book*), *C is A's B* corresponds to other sentences which express the nature of the inherent relation between 'A' and 'B/C', e.g. *John owns/wrote this book*.

The *B*-noun phrase of the accidental relation can occur freely with both definite and indefinite articles or determiners: *This box has the diamond, This box has a diamond, The diamond is in this box, There is a diamond (A diamond is) in this box*. The inherent relation does not exhibit the same freedom. Thus: *John has a son, This is John's son*; but awkward: *There is a son of John/of John's/that is John's son* (or *A son is of John/John's [son]*), *John has this son* (except where *B* represents a kind, rather than an individual, e.g. *John has the son we would all like to have*). Given the constituents 'there is a C'+'C is A's B', we would only have a rule that topicalizes *A* in the form *A has a B*. Because of the homonymy, some *A has B* sentences may ambiguously be inherent or accidental, e.g. *John has one arm: He lost the other in the war/And he needs another one for the armchair that he's making*.

   [a] The genitive construction *A's B* will be used throughout to represent, as well, that construction *B of A* which can be pronominalized to *its B, his B, their B*, etc., thus excluding such cases as *book of poetry, coat of paint*, or *crowd of people*.

9.4 *C is A's B* or *B is A's* is marked for a special kind of relation called inherent, whereas *A's B* (or *A's*) alone can correspond to a wider range of expressions for both accidental and inherent relations, e.g. *the shovel that John has*: *John's* (*shovel*); *the weeds in John's garden*: *John's* (*weeds*); as well as *the man that is John's son*: *John's* (*son*). The relations called accidental and inherent are so distinguished because they are logically independent in a way that may be illustrated by the following: (One book is his [book], the other is not.) *He has the book that is his, He has the book that is not his, He doesn't have the book that is his.*

In other words, there is no particular difficulty in trying to interpret sentences which say that what someone does or does not happen to have is or is not his. *C is A's B* or *B is A's* expresses that there is some special relation between 'A' and 'B/C' as opposed to such accidental relations as may be expressed by the *A has B* related to *B is XAY*. In other contexts, the accidental and inherent relations have also been called 'alienable' and 'inalienable'. No claim is made, of course, that the accidental-inherent distinction exists apart from the human view of things, but rather that when a speaker makes an assertion using a form, such as *C is A's B*, which we have defined as expressing inherence, he is making the assertion that there is a relation between 'A' and 'B/C' and that this relation is characterized as being subjectively identified as exclusive and special in some sense. A relation is inherent only when someone says it is or when it cannot be objectively observed as exclusive unless one is told by what objective criteria it is defined as exclusive by a speaker in the given context or by the speakers in the broader context of the culture.

Take again the accidental *A has B*, which is unmarked for inherence, i.e. for whether there is such a special relation between 'A' and 'B'. Imagine a situation in which two people, John and Mary, see a copy of a book lying on a table near Mary and John asks, 'Do you have that copy?' By the definition of *A has B*, we might paraphrase this as 'Is there a relation between you and that copy?' and then expect Mary to be confused on first hearing the question since she can see that John can see that there *is* a relation. The question would have made better sense as 'Is that copy yours?' since this asks 'Is there an inherent relation, a special relation that I cannot see, between you and that copy?', which is usually a relation of ownership for people and books. Of course, *that book* instead of *that copy* would have been easily interpretable since with certain nouns such expressions can also mean 'any other member of the class of that item'.

9.5 There is a third large *A has B* class, in addition to still other lesser ones, which is more derivational and semiproductive than truly generative. We can call this the 'characterizing' *A has B* but may not wish to define it as expressing a relation, depending on the definition of reference adopted. Examples of this class and their corresponding *be* sentences are *A has strength*: *A is strong*; *A has (much) happiness*: *A is (very) happy*; *A has courage*: *A is courageous*.

9.6 There are languages which make a formal distinction between their parallels to these three *A has B* classes and, further, do not introduce a separate verb *have* when *A* is the topic. In Hindi, for example, we have: accidental *A-ke-paas B hai* 'A-near B is, B is near A', inherent *A-kaa B hai* 'A-of B is, B is of A', characterizing *A-ko B hai* 'A-to B is, B is to A'. Japanese makes a similar distinction. In Hindi, topicalization in such constructions is achieved by word order shifts which also function as equivalents of the definite article in English, lacking in Hindi. For example, *A-ke-paas pustak hai* 'A has a book, there is a book near A', *pustak A-ke-paas hai* 'the book is near A, A has the book'. Whether 'have' or 'is near' is the translation depends in part on whether or not 'A' is a person or animate. Compare the Russian *u menja kniga* 'by me book, I have a book', *kniga u menja* 'book by me, I have the book'. As discussed above, *A has B* does not render the inherent relation when *B* is definite: *A-kaa B hai* 'A has a B' (inherently), *B A-kaa hai* 'B is A's' (*C A-kaa B hai* 'C is A's B').

There are many languages which, like Hindi, express the English *have*-relation by means of constructions with a verb *be*, or, where the language has more than one, that *be*-verb expressing 'exist'. Furthermore, this *have/be* is factored out as a component in the inchoative verb *A gets B* and the causative *A gets self B*. However, in Newari, a Tibeto-Burman language, all three are expressed by one verb *dot-*, with inchoative and causative conveyed by separate morphemes: imperfective *du* 'be (exist)/have'; perfective *dot-o* 'came to be/got'; causative (in the perfective) *doe-kol-o* 'caused to be/caused to get'. Still another cross-linguistic parallel supports the idea of *have* and *be* as manifestations of the same thing, namely expressions in different languages for the existential *there is, there exists*. Here languages vary as to whether they express the existential with their equivalent of *have* or *be* or some other verb that we would define as containing a component *have/be*, for example:

⟨23⟩ *have*: Chinese *yǔ*, French *il y a*, Russian *imeetsja*, Spanish *hay* (English *one has*)
be: German *es ist*, Hindi *hai*, Japanese *aru*, Russian *est'*, Italian *c'è*, English *there is*
give: German *es gibt*
find: Russian *naxoditsja*, Swedish *det finns* (English *one finds*).

In Polish *have* and *be* are in complementary distribution with respect to negation: *there is* is translated by *jest* 'is', but *there is no* by an impersonal *niema* 'not-has'.

10   Interpretation

Semantic tests are one source of behavioral data for which a theory can be made to account. More narrowly then, it should predict certain native-speaker responses for given tests. Beyond that, it is assumed that responses reflect the relevant portions of speakers' codes and that the theory will also account for everyday language behavior.

It is important to stress the premise that a hearer will try to interpret or at least to account for a speaker's utterances. He will often accept and make sense even out of sentences that initially appear odd or contradictory when presented to him on semantic tests or in actual discourse, or he may reject them and then proceed to interpret them anyway. On actual tests, informants' responses have shown greater agreement on the interpretations, matchings, and rankings of sentences than on judgments in terms of grosser discriminations of unacceptability, meaninglessness, anomaly, etc. Such discriminations often depend on context and what criteria are used to make them may be inaccessible and possibly irrelevant. Judgments in terms of such discriminations are heuristically useful in analysis, but the central part of a semantic description is constituted by the definitions of lexemes and other syntactic elements. From these, the meanings of compound expressions of unlimited complexity can be correctly inferred. Whether particular combinations, however, are 'impermissible' or 'anomalous' is left to individual speakers – including logicians – to decide. It is not for the description to make such judgments but at most to account for how they may be made.

It appears, then, that more reliable indices of semantic content are provided by tests which require informants to perform tasks derived from the more usual, everyday activity of trying to understand sentences, such as giving interpretations, ranking for interpretability, matching for similarity in meaning, but a theory which uses responses to such tests with discrimination must also account for them by replicating the hearer's manipulation of the code in the process of interpretation. Besides the fact that we make sense of utterances in spite of constantly occurring lapses, the code can be manipulated creatively to produce sentences that are more than everyday prose. Such an advertising line as 'If you were a banker, would you lend you money?' (*N.Y. Times* 1967) successfully communicates what '...would you lend yourself money?' does not. Weinreich (1966) constructed a semantic theory which explicitly includes deviant utterances. Replicas of code manipulation in (re)interpretation of sentences were given in 8.1 and 8.4.1–3 above. Two more examples follow.

Further above, *He has the book that is his* illustrated the independence of the accidental and inherent relations. However, *He has the book that he has* presents a different case. Let us say that sentence structure, as in *He has the book*, conveys the meaning 'this is new information'. On the other hand, the construction definite article plus relative clause of *the book that he has* seems to identify the referent as the one about which the previous new information 'he has a book' has already been given. So it appears 'A has B' is and is not new information. In the *but*-test, irony was one alternative of reinterpretation given to resolve a contradiction; so here also: 'one of these contradictory statements is a lie, namely that this is new information, in order to let you know that I do not want to give you new information in the matter'. The first interpretation of two such occurrences of *A has B* as found here would be that they are tokens of the same type and not of two different, homonymous types. Thus another reinterpretation might be that the second occurrence is a token of the inherent *A has B*.

Consider this causative-*have* construction with embedded passive sentence: *John had the walls whitewashed by the painter*, i.e. *A has B* $V_{tr}+en$ *by C*. For embedded passives, the causative-*have* construction includes the restriction $A \neq C$. To read the semantic import of such a restriction, the theory will include the following rules:

⟨24⟩ (i) Let: (*a*) a Phrase be a member of the class of structural symbols, such as S or NP [here A, B, C represent NPs]
   (*b*) a Type be a member of the class of formal symbol types, such as any particular morpheme symbol or string thereof
   (*c*) a Token be a member of a class of symbol tokens which class is symbolized by a Type.

   Then: for any particular utterance, the referent of Phrase$_i$ = the referent of the Token whose Type is labeled Phrase$_i$ in the structural description of the utterance.

   (ii) A rule 'Phrase$_i$ = Phrase$_j$' should be read as 'referent of Phrase$_i$ = referent of Phrase$_j$'.

   (iii) Referent$_i$ of Phrase$_i$ = referent$_j$ of Phrase$_i$.

Rule ⟨24 iii⟩ states that each Token has only one referent (and is not an allusion to the singular-plural distinction). Thus the restriction $A \neq C$ should be read as 'referent of $A \neq$ referent of C', i.e. 'John is not the painter'. Now assume that someone utters a sentence to us which appears to have the structure A has B $V_{tr}+en$ by C in its first portion: *The painter had the walls whitewashed by himself and not by his assistant*. This utterance appears odd, and we might interpret it in at least two possible ways.

On the one hand we might take it as a lapse and understand that the speaker meant 'The painter himself, and not his assistant, whitewashed the walls'. The account of this interpretation would run as follows: (*a*) The occurrence of the reflexive seems to indicate that the noun phrases *the painter* and *himself* are derived from underlying constructions in which one equals the other, i.e. $A = C$, giving them the same referent. (*b*) The utterance seems to represent the causative-*have* construction with the restriction $A \neq C$, indicating, on the contrary, that the two noun phrases do not have the same referent. (*c*) We resolve the contradiction by ignoring the causative-*have* construction, thus eliminating the rule $A \neq C$.

On the other hand, we might take the speaker more seriously and keep the causative *have*, crediting him with meaning something thereby. We could then suspend rule ⟨24 iii⟩ and assign two referents each to the two noun phrases: 'A_1', 'A_2' and 'C_1', 'C_2'. We equate '$A_1 = C_1$', satisfying the implication of the reflexive, $A = C$, and apply the causative-*have* restriction as '$A_2 \neq C_2$'. This leaves the painter as one individual but two personae, and, using the occurrence of *assistant* as a cue, we might credit the speaker with having meant to convey some such message as 'The painter in his capacity as boss had the walls whitewashed by himself in his capacity as worker and not by his assistant'.

11　Summary

We have been concerned with developing a form of semantic description that is oriented to the generation of behavioral data, as well as of sentences. Thus, eliciting techniques and test sentences have been described which aim at isolating the semantic components that comprise the paradigmatic meanings of items in a language. A semantic component is seen as a schematic sentence or function, a form borrowed from symbolic logic. The items so defined by components are also in the form of schematic sentences or functions and include not only lexical items, but syntactic constructions and such elements of obligatory choice as verb tenses. Definitions are statements of equivalence between the defined and the defining sentences. The results of semantic eliciting techniques and tests can thus be integrated into a transformational semantic grammar which generates surface structures of a language from these underlying sentences or functions. A metatheory based on the intertranslatability of languages would include a stock of semantic components which could in turn be combined to define and generate many of the underlying functions in the description of a particular language. One such universal component discussed at greater length was *be/have*, expressing a state or state relation.

It was also argued that a semantic theory should contain 'rules for breaking rules' to account for code manipulation in such creative uses of language as irony and referent splitting and in the hearer process of interpreting, reinterpreting, and establishing or not establishing logical consistency in sentences.

City University of New York

REFERENCES

Ariel, S. 1967. Semantic tests. *Man*, II, 535–50.

Bach, Emmon. 1967. *Have* and *be* in English syntax. *Language*, XLIII, 462–85.

Bendix, Edward H. 1966. *Componential analysis of general vocabulary*. (Indiana University Research Center in Anthropology, Folklore, and Linguistics, Publication 41.) Bloomington, Ind.

Chomsky, Noam. 1965. *Aspects of the theory of syntax*. Cambridge, Mass., MIT Press.

Fillmore, Charles J. 1969. Review article [of Bendix, 1966]. *General Linguistics*, IX, 41–65.

Lyons, John. 1967. A note on possessive, existential and locative sentences. *Foundations of Language*, III, 390–6.

McCawley, James D. 1968. The role of semantics in a grammar. *Universals in linguistic theory*, ed. Emmon Bach and Robert T. Harms, pp. 124–69. New York, Holt, Rinehart and Winston.

N.Y. Times, 16 March 1967, p. 37.

Peirce, Charles S. 1933. *Collected papers*, V, ed. Charles Hartshorne and Paul Weiss. Cambridge, Mass., Harvard University Press.

Weinreich, Uriel. 1962. Lexicographic definition in descriptive semantics. *Problems in lexicography*, ed. Fred W. Householder and Sol Saporta, pp. 25–43. (Indiana University Research Center in Anthropology, Folklore, and Linguistics, Publication 21.) Bloomington, Ind.

—— 1963. On the semantic structure of language. *Universals of language*, ed. Joseph H. Greenberg, pp. 114–71. Cambridge, Mass., MIT Press.

—— 1966. Explorations in semantic theory. *Current trends in linguistics*, III: *theoretical foundations*, ed. Thomas A. Sebeok, pp. 395–477. The Hague, Mouton. Pp. 395–416 reprinted in this volume.

Zimmer, Karl E. 1964. *Affixal negation in English and other languages: an investigation of restricted productivity*. (Linguistic Circle of New York, Monograph 5.) New York. [= *Word* XX(2) Supplement.]

On classifying semantic features

MANFRED BIERWISCH

1 Introduction

The semantic analysis of natural languages rests crucially on at least the following two assumptions: (i) the meaning of a given sentence can be accounted for on the basis of the words or, more precisely, the dictionary entries of which it consists, and the syntactic relations connecting these items; (ii) the meaning of dictionary entries are not unanalyzable wholes, but can be decomposed into elementary semantic components. These two assumptions are, of course, closely related to each other. The internal organization of the meaning of dictionary entries must be of a form which determines how they enter the composite meaning of more complex constituents according to the syntactic relations within these constituents. The syntactic relations in turn must be specified in such a way that the correct combination of the meanings of related constituents can be determined. A first attempt in this direction has been made by Katz and Fodor (1963) and Katz and Postal (1964). It is based on the assumptions that the syntactic relations in question can be defined in terms of underlying or deep phrase markers as specified in Chomsky (1965), and that the meaning of dictionary entries as well as of more complex constituents is given by strings of basic semantic components. The latter assumption was only a first approximation which turned out to be far too simple. It has been changed rather radically in the meanwhile.[a] In the present paper I shall concentrate on certain problems deriving from assumption (ii) above, more precisely on some aspects of the nature of basic semantic elements. It follows from the introductory remarks that even a discussion of such problems must keep in mind the important interdependence of assumption (i) and (ii).

The set of basic semantic elements has been divided by Katz and Fodor (1963) into two types of elements, called semantic markers and distinguishers. Although Katz has defended this distinction recently (Katz, 1967), it seems to me that it is an outcome of a rather early stage in the development of a semantic theory and must be rejected as theoretically unmotivated. I have presented my arguments for this rejection in Bierwisch (1969, section 5) and will not repeat them here. We are left then with only one type of basic semantic elements which might be called semantic features. These primitive terms, from which semantic descriptions of natural

---

[a] See Katz (1964a, 1964b, 1966, 1967). For a reinterpretation of several of Katz' proposals see Bierwisch (1969). We shall discuss some of the changes involved in more detail below. Weinreich (1966) has sharply criticized the original assumption of mere concatenation – or logical conjunction, for that matter – of semantic components. His counterproposal of distinguishing clusters and configurations of components does not work, however, for reasons briefly discussed in Bierwisch (1968).

languages are constructed, do not, however, form an unstructured, amorphous set. Rather they are classified into several subtypes according to certain aspects, thus constituting ultimately a highly structured system of underlying elements. Some of these aspects of classification will be discussed in the present paper.

2 The formal status of semantic features

Let us first introduce certain basic assumptions concerning the theoretical status of semantic features. It seems to me most natural to consider semantic features essentially as predicate constants in the sense of the predicate calculus as developed in modern logic. This assumption has been stipulated by other authors as well. (See e.g. McCawley, 1968.) It can be motivated by several considerations. I will briefly discuss two of them here.

Firstly it turned out, as already mentioned, that mere concatenation of features as introduced in Katz and Fodor (1963) is not sufficient to account for the semantic interpretation of words and sentences. Thus additional machinery for the combination of semantic features became necessary. One of the means that has been used in this connection, e.g. in Katz (1966, 1967) and Bierwisch (1967), was that of grouping features by simple brackets. Such grouping was meant to express that certain features of a reading are more closely connected than others. It does not specify, however, how differences in grouping are to be interpreted conceptually. In fact, rather different conceptual relations between semantic features have been expressed by identical means in the references quoted.[a] This means that the burden of interpreting the formal relations between semantic elements remains with the conceptual content of particular features. The same formal relation between certain features in a given reading would be interpreted differently depending on the features involved. This is of course quite unsatisfactory. On the one hand, the formal means of semantic representations become rather empty. On the other hand, the problem is postponed, but not solved: the intrinsic interpretation of semantic features would have to provide the formal means giving substance to the empty grouping relations. This in turn would conflict with the assumption that the semantic features are linguistically primitive terms. Katz has tried to avoid difficulties of this type by using certain informal means such as *of*, *at*, etc. in combinations such as ((Condition (Possession *of* Y) *of* X *at* t_1) where X and Y are slots to be filled in by feature complexes and it is a variable over a time interval. These and similar means, introduced only casually, have no clear theoretical status, however. The most natural way to remedy shortcomings of this kind seems to be to construe semantic features as predicates and to reconstruct the more complicated connections among them by using the formal means of the fully fledged predicate calculus, admitting of course adaptations and controlled modifications if necessary. This appears to be at least a promising program. Some steps in this direction have been made in Bierwisch (1969).

The second point to be noted here concerns the fact that semantic features, if

---

[a] Thus in Bierwisch (1967) I used groupings such as (Physical Object (3 Space (Vertical) (Maximal) (Secondary))) to indicate that a given reading refers to a three-dimensional physical object with a vertical, a maximal, and a secondary extension. Obviously, the connection between the element 'Physical Object' and the rest of the configuration represents a conceptual relation completely different from that expressed by the connection between '3 Space' and 'Vertical', 'Maximal', and 'Secondary' respectively. We will return to the problem of a more adequate solution below. Similar problems with respect to several examples given in Katz (1967) are discussed in Bierwisch (1969).

taken as predicates, must be assigned to suitable arguments and that, therefore, these arguments must be part of the semantic representations. They are considered most naturally as variables which are substituted by representations of the referents which are talked about by means of particular occurrences of the expressions to whose readings the variables belong. From this it follows that dictionary entries also must contain appropriate variables which are to be substituted by more specific variables if they occur in particular sentences. In this vein, Weinreich (1962) has pointed out that lexical entries even for single words should have the form of propositional functions, i.e. of (complex) predicates with unbound variables as arguments. The projection rules of Katz and Fodor's theory, which combine the readings of composite constituents according to the relevant syntactic relations, reduce them essentially to the substitution of suitable arguments for the variables occurring in the dictionary entries. This in turn requires that these variables be indexed with respect to the syntactic relations in question. Thus for example transitive verbs such as 'love', 'meet', 'hit', etc. must contain variables indexed for the relations 'Subject of' and 'Object of', for which the pertinent variables occurring in the readings of the Subject-NP and the Object-NP respectively are substituted.[a] Notice that all variables appearing in the reading of a given sentence (more precisely of a sentence type) indicate identity or distinctness of reference. They do not represent particular objects referred to. Otherwise a sentence such as 'I saw a car' would have indefinitely many readings, differing only in the respective arguments.

It is important that essentially the same indication of identical or distinct reference is required for purely syntactic reasons. The relevant phenomena are primarily pronominalization, reflexivization, deletion of identical NP. Using subscripts to represent identical or distinct reference, we have examples of the following type:

$\langle 1 \rangle$ (a) James$_1$ phoned James$_2$. vs. *James$_1$ phoned himself$_2$.

(b) *James $_1$ phoned James$_1$. vs. James$_1$ phoned himself$_1$.

$\langle 2 \rangle$ (a) Lotte Lenya$_1$ sang a song$_2$ and then Satchmo$_3$ sang one$_4$.

(b) Lotte Lenya$_1$ sang a song$_2$ and then Satchmo$_3$ sang it$_2$.

(c) *Lotte Lenya$_1$ sang a song$_2$ and then Satchmo$_3$ sang one$_2$.

(d) *Lotte Lenya$_1$ sang a song$_2$ and then Satchmo$_3$ sang it$_4$.

$\langle 3 \rangle$ (a) Since James$_1$ is married, he$_1$ doesn't stay.

(b) Since he$_1$ is married, James$_1$ doesn't stay.

(c) James$_1$ doesn't stay, since he$_1$ is married.

(d) *He$_1$ doesn't stay, since James$_1$ is married.

(e) He$_2$ doesn't stay, since James$_1$ is married.

$\langle 4 \rangle$ (a) John$_1$ expected him$_2$ to pass the examination.

(b) *John$_1$ expected $\begin{Bmatrix} \text{him}_1 \\ \text{himself}_1 \end{Bmatrix}$ to pass the examination.

(c) John$_1$ expected to pass the examination.

I cannot comment on the details of these phenomena.[b] For the aims of the present paper it is sufficient to realize that certain aspects of reference are crucial for any

---

[a] This process is sketched in somewhat more detail in Bierwisch (1969, section 3). – Note incidentally that the slot variables X, Y, etc. which Katz has introduced, as mentioned above, in his recent writings serve a rather similar purpose. In fact, the revised projection rules based on these variables amount roughly to a substitution of argument expressions, if the semantic markers are reformulated as (complex) predicates.

[b] They are in fact rather complicated and involve several transformational processes and principles governing them. A careful investigation of some of them will be found in Postal (1968).

attempt to account for the facts displayed by $\langle 1 \rangle$–$\langle 4 \rangle$. Chomsky (1965, p. 145) first proposed in this connection to mark certain lexical items as 'referential' and to assign to them integers representing identical or different reference. (For a fairly elaborate application of this idea see Isenberg (1968).) But since reference is actually not a property of lexical categories, it seems more appropriate to assign reference indexes to NPs. Though even that is an oversimplification which ignores certain complications in the analysis of NPs, it will do for the present purpose.

We may assume now that for every NP_i there is a variable X_i functioning as an argument in the semantic representation of the expression of which the NP_i is a part. This argument does not only occur in the reading of the NP_i in question, but is also substituted, by the procedure mentioned above, for variables in the readings of other constituents to which the NP_i bears the relevant syntactic relations.

This provisional sketch of the status of semantic features immediately leads to a first aspect of classification. Semantic features fall into different classes according to the number of arguments required. They may be one-place predicates representing properties, two-place predicates representing two-place relations, etc. We will see below that they must be classified not only with respect to the number, but also with respect to different types of arguments.

3 Some problems of primary arguments

Before pursuing further the classification of semantic features, it may be expedient to consider in somewhat greater detail the interpretation of the variables X_i just introduced, since a perhaps fundamental divergence from the status of variables used in the predicate calculus is necessary. The problems involved are rather complicated: I cannot even list them completely, to say nothing of an adequate solution.

First of all syntactic reference indices indicate reference instances which by no means always consist exactly of one individual as in the examples $\langle 1 \rangle$–$\langle 4 \rangle$. They may comprise finite and even infinite sets of individuals as well:

$\langle 5 \rangle$ (a) [The boys]$_{NP_1}$ hurt [the boys]$_{NP_2}$
(b) [The boys]$_{NP_1}$ hurt [themselves]$_{NP_1}$

$\langle 6 \rangle$ (a) [The even numbers]$_{NP_3}$ can uniquely be mapped on [the odd numbers]$_{NP_4}$
(b) [The even numbers]$_{NP_3}$ can uniquely be mapped on [themselves]$_{NP_3}$.

Thus reference indices in general correspond to sets of individuals, including sets containing one element.

Secondly, two or more sets reflected by respective reference indices can be joined to form a new reference instance:

$\langle 7 \rangle$ [John$_1$, Paul$_2$, Bill$_3$, and [their$_4$ wives]$_{NP_5}$]$_{NP_6}$ squeezed [themselves]$_{NP_6}$ into [one little car]$_{NP_7}$.

Notice that sentences of this type cannot be derived from a conjunction of underlying sentences without conjoined NPs. The index 4 of the possessive NP 'their' obviously joins together the indices 1, 2, and 3, just as these and the index 5 are covered by the index 6.

On the other hand, reference instances may be split up into several particular instances represented by particular indices:

$\langle 8 \rangle$ [The teachers]$_{NP_1}$ disagreed about the proposal$_2$ since [the younger ones]$_{NP_3}$ were optimistic whereas [the others]$_{NP_4}$ were not.

In this example the instance 1 is divided into the separate instances 3 and 4. More complicated cases can easily be found.

Observations of this kind suggest the introduction of certain set theoretical operations over reference indices, and hence also over the corresponding variables occurring in the semantic interpretations.[a] It seems to me that set theoretical sum and difference are necessary and probably sufficient. They may be defined in the following way:

⟨9⟩ Let i, j, k be reference indices, X_i, X_j, K_k the corresponding variables, and $a_1, a_2, \ldots, a_n, b_1, b_2, \ldots, b_m$ ($1 \leqslant n$, $m \leqslant \infty$) variables for single objects to be referred to. Then

(a) $k = i+j$ iff $X_i = \{a_1, \ldots, a_n\}$, $X_j = \{b_1, \ldots, b_m\}$, and $X_k = X_i \cup X_j$;

(b) $k = i-j$ iff $X_i = \{a_1, \ldots, a_n, b_1, \ldots b_m\}$, $X_j = \{b_1, \ldots, b_m\}$, and $X_k = \{a_1, \ldots, a_n\}$.

Notice that even the a_1 through b_m are not representations of particular (real or fictitious) objects, but variables for such objects, since we are still concerned with identity and difference of reference, not with reference itself. With this definition in mind, we can replace ⟨7⟩ and ⟨8⟩ by ⟨10⟩ and ⟨11⟩ respectively:

⟨10⟩ [John₁, Paul₂, Bill₃, and [their$_{1+2+3}$ wives]$_{NP_5}$]$_{NP_{1+2+3+5}}$ squeezed [themselves]$_{NP_{1+2+3+5}}$ into [one little car]$_{NP_7}$.

⟨11⟩ [The teachers]$_{NP_1}$ disagreed about the proposal₂ since [the younger ones] $_{NP_{(1-r)}}$ were optimistic whereas [the others]$_{NP_{1-(1-r)}}$ were not.

The index 1–r in ⟨11⟩ contains a variable r which does not specify a reference instance in itself, but serves only to divide the set represented by 1. This represents the fact that the instance $3 = 1-r$ is constructed from instance 1, and $4 = 1-(1-r)$ from 1 and 3 in the obvious way.[b]

Notice that the variable $X_{1+2+3+5}$ corresponding to the index of the node dominating the whole subject-NP in ⟨10⟩ need not appear within the reading of this NP itself. This reading contains rather the variables X_1, X_2, X_3, X_5, and X_{1+2+3} corresponding to the respective constituent NPs. But the variable $X_{1+2+3+5}$ will be substituted for the subject variable of the verb.

[a] Notice that the availability of such operations is necessary for purely syntactic reasons. To give only a simple example, how could we account in a non-ad-hoc manner for the following facts without knowing that index 3 is made up from 1 and 2?

    (i) John₁ squeezed Bill₂ and himself₁ into a car₄.

    (ii) *John₁ and Bill₂ squeezed himself₁ and himself₂ into a car₄.

    (iii) [John₁ and Bill₂]$_{NP_3}$ squeezed themselves₃ into a car₄.

Similar operations on reference indices are involved in the following examples:

    (iv) [They₁ and John₂]$_{NP_3}$ expected to see $\left\{ \begin{array}{l} \text{[themselves]}_{NP_3} \\ \text{*[themselves]}_{NP_1} \end{array} \right\}$ in that mirror₄.

Such and still more complicated phenomena of pronominalization and deletion have been little explored until now. They obviously require a certain algebra of reference indices.

[b] Similar assumptions with respect to set theoretical operations on reference indices have been made by McCawley (1968). There is only a slight difference. McCawley considers sets of one element as individuals, not as sets. He allows them, however, to be subject to set theoretical operations. I do not see any reason for this peculiarity. McCawley claims that one of the conditions for the choice of number affixes in NPs is the decision as to whether their reference index is a set or an individual. But a rule that can distinguish between a set index and an individual index might be sensible as well for the distinction between sets containing one element and sets containing more then one element. – There are some doubts, by the way, with respect to McCawley's assumption that sets of two or more elements are converted into grammatical plural morphemes. See pp. 430-1, note c for a counterexample.

Turning now to the semantic aspect of the reference indices and the variables corresponding to them, the following interpretation might seem plausible. If P is a semantic feature (or a complex of features) and X_i its argument, then P is predicated of every element contained in X_i. More formally:[a]

$$\langle 12 \rangle \quad [P] \, X_i = {}_{\text{def}} \bigvee_{x \in Xi} P(x).$$

Although this interpretation covers a certain class of cases, it cannot be accepted in general. We will briefly discuss some classes of counterexamples.

Difficulties arise already in such simple cases as $\langle 13 \rangle$:

$\langle 13 \rangle$ (*a*) [The boys]$_{\text{NP}_1}$ saw [the girls]$_{\text{NP}_2}$
      (*b*) [The boys]$_{\text{NP}_1}$ hit [the girls]$_{\text{NP}_2}$.

Ignoring all details not relevant for the present discussion, and using 'B', 'G', 'S', and 'H' as abbreviations for the complexes of semantic features of the readings of 'boy', 'girl', 'saw', and 'hit', respectively, the readings of $\langle 13a \rangle$ and $\langle 13b \rangle$ can be given roughly as $\langle 14a \rangle$ and $\langle 14b \rangle$:

$\langle 14 \rangle$ (*a*) [B] X_1.[G] X_2.[S] $X_1 \, X_2$
      (*b*) [B] X_1.[G] X_2.[H] $X_1 \, X_2$.

Applying definition $\langle 12 \rangle$ we arrive at the representations $\langle 15a \rangle$ and $\langle 15b \rangle$:

$$\langle 15 \rangle \quad (a) \bigvee_{x \in X_1} \bigvee_{y \in X_2} [B(x).G(y).S(x, y)]$$

$$(b) \bigvee_{x \in X_1} \bigvee_{y \in X_2} [B(x).G(y).H(x, y)].$$

Whereas $\langle 15a \rangle$ gives a fairly correct account of the way in which $\langle 13a \rangle$ is understood, this does not hold with respect to $\langle 15b \rangle$ and $\langle 13b \rangle$, since $\langle 13b \rangle$ does not assert that every single boy thrashed every single girl. One might try to remedy this failure within the framework of the predicate calculus, e.g. by introducing a second convention for the interpretation of variables corresponding to reference indices to the effect that the universal quantifier is replaced under certain conditions by the existential quantifier, thereby rendering $\langle 15b \rangle$ into $\langle 16 \rangle$:

$$\langle 16 \rangle \quad \bigvee_{x \in X_1} \exists_{y \in X_2} [B(x).G(y).H(x, y)].$$

Although this doubtless corresponds more closely to the way in which $\langle 13b \rangle$ is understood, it does not work. $\langle 16 \rangle$ does not claim, to be sure, that every boy hit every girl, but it fails to express that, at least in principle, every girl from the group X_2 is thrashed by at least one boy, as implied by $\langle 13b \rangle$. Moreover, any attempt to get along with universal and existential quantifiers will be dependent much too strongly on idiosyncratic properties of the dictionary entries involved. And in many cases it forces completely unmotivated decisions. Consider sentences such as 'The boys loved the girls', 'The boys chased the girls', etc.

A still more serious obstacle is hidden in sentences like $\langle 7 \rangle$, where the reading of 'squeeze' cannot be distributed separately on the individuals contained in the reference instance $X_6 = X_{1+2+3+5}$. The squeezing is understood obviously as an action brought about by the group as a whole. This is still more obvious in sentences like:

$\langle 17 \rangle$ [The policemen]$_{\text{NP}_1}$ rounded up [the demonstration]$_{\text{NP}_2}$.

---

[a] I use the notation '[P] $X_1 \, X_2 \ldots X_k$' for a k-place predicate assigned to its k arguments. For some motivation of this notation see Bierwisch (1969).

The problem involved here is connected with what McCawley (1968) has tried to explain by the feature 'Joint'. In cases of jointness the predication does not apply separately to the individuals referred to.[a]

A related, but still different problem arises in sentences like:

⟨18⟩ [The whites]$_{NP_1}$ oppress [the negroes]$_{NP_2}$

⟨19⟩ [The Romans]$_{NP_1}$ destroyed [Carthage]$_{NP_3}$.

Like ⟨7⟩ and ⟨17⟩ these examples do not apply to the elements of the relevant reference instances individually. But beyond this it is understood that not all individuals comprised under NP_1 and NP_2 of ⟨18⟩ and ⟨19⟩ are in fact involved in the relevant actions or relations. Rather the groups referred to are understood as a plurality whose individuals are not singled out with respect to participation or nonparticipation in the states and processes in question. It is claimed only that the group as a whole is concerned. Notice that such considerations apply also to apparently trivial cases of non-joint verbs like 'know':

⟨20⟩ [The Chinese of the seventh century]$_{NP_1}$ knew porcelain.

By uttering ⟨20⟩ one does not claim that everybody in the China of the seventh century had the relevant knowledge. ⟨20⟩ is true also under the weaker condition that the Chinese people as a whole knew porcelain.

From these and similar considerations I conclude that an interpretation of reference indices and corresponding arguments along the lines indicated by ⟨12⟩ must be given up. Instead of this I propose to consider an argument X_i as a variable to be substituted by the representation of a fraction of the (real or fictitious) universe talked about. This fraction is made up from one or more equivalent objects or individuals which are singled out for separate predication only under specific conditions. These conditions are either part of the predicates to be applied or expressed by particular specifiers and quantifiers such as 'every', 'all', 'two', 'many', etc. (see section 4). Further factors may possibly be involved. This proposal would account in a natural way for the 'indeterminacy' of the distribution of the relevant predicates on the particular individuals in cases like ⟨13⟩ and ⟨17⟩–⟨20⟩, where no individualizing features are involved. I do not try to formalize this proposal concerning the interpretation of reference instances, since any serious formal account of these problems would have to deal with many problems which are too poorly understood at present. I suspect however that the proposed concept of 'global reference' cannot be explained in terms of more basic notions of a semantic theory, but must be taken as a primitive notion itself. I presume, in other words, that in this respect the quantification theory of modern logic and the linguistic semantic theory are radically different in that they take the inverse direction: whereas quantification theory takes individuals as the starting point from which exhaustive and partial sets are constructed by means of universal and existential quantification, linguistic semantics probably has to start with sets as primitive terms which may further be specified with respect to the participation of their elements in particular state of affairs. This means that what is a particular case in linguistic semantics has been taken as the basic one within the quantification theory according to the special purpose of its construction. But these are vague speculations which must be motivated much more carefully.

---

[a] The shortcomings of McCawley's analysis of this problem and the different aspects of jointness have been discussed carefully by Lang (1967). I cannot go into the details here.

these details here. We simply assume for the present discussion that the Q in the operators (QX_i) is a variable not only over Def, Indef, and Spec, but also over combinations of these with numerals, the precise nature of this combination being left open.

Indefinite quantifications like 'many', 'few', 'some' are similar to numerals in many respects, the main difference being that they provide no absolute quantification, but one which is relative to a certain norm.

Notice that features representing absolute or relative quantification can to a certain extent be combined with true predicative features in a way characteristic for such predicative elements. Thus we have 'very many books' alongside with 'very good books'. Other cases are 'more than fifty books', 'at most fifty books', etc. This complicates the internal semantic structure of the (QX_i) still further.

A particular combination of quantifying features and those specifying reference instances appear in the readings of 'all', 'every', 'each', 'any', and the determiner constituents of the different types of generic NPs. What they have in common seems to be the fact that they may or must relate the set $X_i x$ to the complete class characterized by the pertinent predicative features, a relation which is not involved in absolute and relative quantification. This is accomplished however in completely different ways resulting in similarly different consequences. Consider e.g. the following sentences:

⟨29⟩ (a) [Every logician]$_{NP_1}$ would infer that.
    (b) [All logicians]$_{NP_1}$ would infer that.
    (c) [Logicians]$_{NP_1}$ would infer that.
    (d) [A logician]$_{NP_1}$ would infer that.
    (e) [The logician]$_{NP_1}$ would infer that.

One might perhaps argue that these sentences are synonymous in one reading in the sense that identical truth conditions are associated with this reading. I have some doubts in this respect, however. In any case, the truth conditions are arrived at by means of completely different operations. The X_1 of ⟨29a⟩, e.g., is a set consisting of one element, and the delimiting features of 'every' represent an instruction which might be paraphrased as 'substitute for this element the representation of each single object such that the conditions expressed by "logician" are met'. The X_1 of ⟨29b⟩ on the other hand consists of a representation of the class of logicians and the delimiting features of 'all' represent an instruction to the effect that predicates assigned to X_1 apply to all elements of this class. That these loosely indicated operations may have different results is brought out quite clearly by the following examples:

⟨30⟩ (a) [Every student of our college]$_{NP_1}$ likes baseball.
    (b) [All students of our college]$_{NP_1}$ like baseball.
⟨31⟩ (a) [Every student of our college]$_{NP_1}$ performs a play.
    (b) [All students of our college]$_{NP_1}$ perform a play.

Whereas ⟨30a⟩ and ⟨30b⟩ are synonymous, ⟨31a⟩ and ⟨31b⟩ are not, ⟨31a⟩ requiring as many one-man-performances as there are students. I will not pursue the problems of universal and generic sentences any further, as they constitute a large topic on their own. Let me add only that the operators involved behave differently again with respect to negation: the sentences ⟨29b⟩ and ⟨29c⟩ are at least similar in meaning in a way in which $\langle 32a \rangle$ and $\langle 32b \rangle$ are not, these being synonymous with $\langle 33a \rangle$ and $\langle 33b \rangle$ respectively:

$\langle 32 \rangle$ (a) It is not true that all logicians would infer that.

       (b) It is not true that logicians would infer that.

$\langle 33 \rangle$ (a) Not all logicians would infer that.

       (b) Logicians would not infer that.

Let me summarize briefly the lengthy and inconclusive discussion of delimitations. We first assumed that the reference indices i correspond to variables X_i functioning as arguments of the predicative features. These variables range over sets of possible objects which must meet the conditions represented by the predicative features. We then faced the necessity of elements delimiting the sets to be substituted for the X_i. These delimiting elements have a status completely different from that of the predicative elements. The delimiting features form operators of the form (QX_i) where the Q is a variable over (complexes of) those features. For each NP_i of the deep structure there is precisely one (QX_i) prefixed to the reading either of that NP_i or of a constituent containing that NP_i.[a] The delimiting features in turn subdivide into quantifying and specifying elements. The former represent absolute, relative, and universal characteristics of the extension or scope of the set to be substituted for the X_i. The latter represent instructions to pick up the set referred to in one or the other way. Specifying and quantifying features are combined in a still unexplored way to form the operators (QX_i). The quantifying features may also be modified by certain predicative features, this modification being governed by the principles pertinent to the combination of predicative features in general.

It is worth noting that our interpretation of the arguments X_i and the pertinent delimitations, which is based so far on observations concerning countable objects, accounts at the same time in a very plausible way for mass objects:

$\langle 34 \rangle$ (a) [The wine you gave me]$_{NP_1}$ was excellent.

       (b) [Water]$_{NP_1}$ is still missing.

       (c) Would you like [some coffee]$_{NP_1}$?

       (d) [All iron]$_{NP_1}$ rusts.

The X_i of these examples is to be substituted simply by the representation of improper sets, as it were, which cannot be individualized at all, since they are not conceived as consisting of individuals. From this it follows immediately that certain quantifications do not apply. (This restriction must be derived from the inherent semantic and/or syntactic features of the respective nouns in a way not to be discussed here.) All other decisions however remain unchanged: the predicative features apply to the elements of the set X_i referred to, i.e. to a single improper object in this case. Quantifying features concern the set as a whole. And the selection of the appropriate referent represented by the specifying features is as usual. Thus the features Def, Spec, and Indef account for $\langle 34a \rangle$, $\langle 34b \rangle$, and $\langle 34c \rangle$, respectively. And the operation represented by 'all' works for $\langle 34d \rangle$ in the same way as for $\langle 29b \rangle$, $\langle 30b \rangle$, and $\langle 31b \rangle$.

---

[a] An exception to this statement are predicative NPs in sentences like 'John is a doctor' which do not identify but describe the referent of the subject NP. Such predicate nouns do not provide a new reference instance. They differ from all other NPs in other respects as well and must be treated separately in any case.

5 Problems of relational features

Having settled provisionally the problem of arguments and their delimitation, we return to the predicative features. We already mentioned that these are to be classified in part according to the number of their arguments. This then raises questions of the following type: is there an upper limit for the number k of k-place predicates to be included in the inventory of basic semantic elements? Is there a fixed number of arguments for each predicate or are there those with a varying number of arguments? We will see furthermore that predicative features may differ not only with respect to the number, but also to the type of arguments. Let us consider these problems by means of some examples.

It is obvious from the foregoing discussion that the number of arguments of a predicative feature must be somehow connected to the number of NPs required by the lexical entry whose reading contains that feature. Thus the verbs 'hit' and 'give', for example, are often considered as expressing two- and three-place relations, respectively, 'buy' and 'sell' on the other hand as expressing relations with two, or three, or even four arguments:

$\langle 35 \rangle$ [John]$_{NP_1}$ hit [Bill]$_{NP_2}$

$\langle 36 \rangle$ [John]$_{NP_1}$ gave [Bill]$_{NP_2}$ [a book]$_{NP_3}$

$\langle 37 \rangle$ [John]$_{NP_1}$ sold ([Bill]$_{NP_2}$) [a book]$_{NP_3}$ (for [three florins]$_{NP_4}$)

The fact that 'give' requires three NPs to form a non-elliptic sentence, however, does not imply that its reading contains a three-place predicate. In fact, it has been shown by Bendix (1966) that the meaning of 'give' should be analyzed into two two-place relations in the following way:

$\langle 38 \rangle$ C causes (A has B)

where C represents the referent of the subject-NP, B that of the direct object, and A that of the indirect object. If we assume that X_S, X_D, and X_I represent variables of a dictionary reading to be substituted by the X_i of the subject, direct object, and indirect object respectively, then $\langle 38 \rangle$ can be replaced by $\langle 39 \rangle$:

$\langle 39 \rangle$ [Cause] X_S ([Have] $X_I X_D$)

Such a combination of features, where the second argument of the predicate Causation is not a variable over a set of individuals but a proposition, raises several problems that will be discussed below.

With respect to 'buy' and 'sell', Katz (1967) has proposed an analysis that reduces the apparent four-place relation essentially to a combination of several two-place features of the form 'X possesses Y' with varying arguments and different time specifications, thus characterizing the process of exchange of goods and money between the subject and the indirect object referent of 'buy' and 'sell'. There are several possibilities to represent the necessary time specifications. One of them is the introduction of an abstract relation '[R]t(P)' where t is a variable over time intervals and P a proposition representing a process or a state occurring at time t. With this convention we can represent Katz' analysis of 'sell' in the following way, using X_F as the variable to be substituted by the X_i of the prepositional phrase with 'for':

$\langle 40 \rangle$ [R] t ([Poss] $X_S X_D$.[Poss] $X_I X_F$).[R] t' ([Poss] $X_S X_F$.[Poss] $X_I X_D$)

The time interval t' immediately follows t[a].

---

[a] This analysis is oversimplified and in many respects not relevant for the present discussion. Thus I ignore here and throughout this paper the problem of selection restrictions or

What I have tried to show by means of these examples, which could easily be extended, is that many – if not all – lexical entries requiring or admitting three or more NP_1 are to be analyzed into appropriate combinations of two-place features. Thus the number k of possible arguments for elementary predicative features can certainly be strongly restricted. In fact, one might hypothesize that in general only one- and two-place relations are required.[a]

It seems reasonable furthermore to assume that each particular feature always has the same number of arguments. The varying number of NPs permitted in sentences such as $\langle 37 \rangle$ does not imply that there are features with a varying number of arguments, but merely that some of these arguments are not specified by additional lexical items and thus do not appear in the surface structure in the form of separate constituents. (Whether a lexically unspecified argument must leave at least a pronoun as its trace or whether it can be dropped altogether or left unspecified even in the deep structure representation is part of the syntactic characterization of particular lexical items.)

We must now consider somewhat more detailed features like 'Caus', which take propositions as one of their arguments. First of all, the existence of such features leads to a further classification of features, which, then, must be specified not only with respect to the number, but also the type of their arguments. Though 'Caus' and 'Have' are both two-place relations, they belong obviously to different types of features. This classification, however, must be specified still further. Consider the following sentences:

$\langle 41 \rangle$ [John]$_{NP_1}$ regrets [that [Paul was waiting]$_S$]$_{NP_2}$

$\langle 42 \rangle$ [John]$_{NP_1}$ believes [that [Paul was waiting]$_S$]$_{NP_2}$

The reading of 'regret' and 'believe' must contain, among other things, relational

presuppositions. In fact, the lexical reading of 'sell' must contain at least the following selectional features: '[Human] X_S.[Human] X_I.[Money] X_F.[Physical Object v Artifact] X_D'. I doubt, furthermore, that Katz' analysis is correct, insofar as it is based on the assumption that 'buy' and 'sell' are in a sense inverse relations, i.e. that (i) and (ii) are synonymous sentences:

(i) John sold Bill a book. (ii) Bill bought a book from John.

This claim would imply that (iii) and (iv) are also synonymous:

(iii) Bill was sold a book. (iv) Bill bought a book.

But that this is at least controversial can be seen from the following examples:

(v) Bill was sold a book in order to give it to Mary.
(vi) Bill bought a book in order to give it to Mary.

Therefore $\langle 40 \rangle$ should be replaced, presumably, by something like:

(vii) [R] t ([Poss] $X_S X_D$.[Poss] $X_I X_F$).[R] t ([Caus] X_S ([Poss] $X_S X_F$.[Poss] $X_I X_D$))

See Katz (1967) and Bierwisch (1969) for further discussion. Notice incidentally that also 'give' must contain a time specification. Thus $\langle 39 \rangle$ should be rendered into (viii):

(viii) [R] t ([Caus] X_S ([Have] $X_I X_D$)).

[a] An apparent counterexample to this speculation is the preposition 'between' and its equivalents in other languages:

(i) John$_1$ was seated between Bill$_2$, Paul$_3$, Sam$_4$,..., and Peter$_n$.

Sentences like

(ii) John$_1$ was seated between them$_7$

suggest, however, that NP_2 through NP_n in (i) must be analyzed as dominated by a single NP, whose reference index is $2+3+4+\ldots+n$. The corresponding variable $X_{2+3+\ldots+n}$ would then be the second argument of the feature characterizing 'between', just as X_7 is its second argument in (ii). If this assumption is right, 'between' would contain a two-place predicate which requires its second argument variable to be a set of more than one object.

features representing a certain attitude of their first argument – the X_1 provided by the NP_1 'John' in $\langle 41 \rangle$ and $\langle 42 \rangle$ – towards their second argument, provided by the complement 'that Paul was waiting'. These features, which we may abbreviate by 'Believe' and 'Regret' for the sake of simplicity, are similar to 'Caus' in that they take a proposition as their second argument. But whereas 'Believe' represents indeed an attitude towards a proposition, viz. that the individuals represented by the first argument are inclined to take it as a true proposition, 'Regret' represents an attitude not towards a proposition, but towards a fact represented by the proposition in question. In this respect, 'Caus' must be classed with 'Regret', as it is the fact described by its argument proposition, not that proposition itself, which is brought about by the first argument of 'Caus'. Thus 'John gave Mary some wine' does not mean that John causes the proposition expressed by 'Mary has some wine', but the fact represented by that proposition. Let us call features of the 'Caus' and 'Regret' type 'fact-features' – or for short 'F-features', and those of the 'Believe' type 'proposition-features' or 'P-features'. It seems to me that the difference between F-features and P-features is closely connected to the problem which Reichenbach (1966, pp. 268–74) tries to explicate by means of so-called fact variables. For every proposition, Reichenbach claims, there may be a fact such that the proposition describes that fact.[a] This suggests that P-features take a proposition as their second argument, whereas F-features take the name of a fact.[b] The difference between these two types of arguments must be represented explicitly for both syntactic and semantic reasons. For syntactic consequences I refer to Kiparsky and Kiparsky (1970). The revealing distinction between Factive and Non-factive predicates proposed there has at least one of its most important sources in the distinction of F-features and P-features. (It would be premature to claim that both distinctions are identical, though one might at least look for arguments in this direction.) Among the semantic reasons for the necessity of an explicit distinction between factive and propositional arguments are at least the problems of negation and of opaque contexts. As to the first problem, there are several of the P-features/where the negation of that feature and of its argument are synonymous, whereas no such possibility exists for F-features. A rough illustration is given by the negations of $\langle 41 \rangle$ and $\langle 42 \rangle$:

$\langle 43 \rangle$ (*a*) John does not believe that Paul was waiting.

        (*b*) John believes that Paul was not waiting.

[a] Reichenbach introduces the following notation: $p \equiv (\exists v)\, p^*(v)$. In other words: p is true if and only if there is a fact v such that v is described by p. The asterisk on p transforms the proposition p in a predicate assigned to a fact variable. Since this p uniquely specifies a particular fact, the name of that fact can always be given by means of a definite description: $(\imath v)\, p^*(v)$, meaning 'the fact that p'.

[b] Actually the distinction between fact arguments and proposition arguments, discussed here with respect to relational features, need not be restricted to these. There may be one-place predicates taking propositions and fact names respectively as their only arguments. Adjectives like 'important', 'odd', 'tragic', etc. in one of their readings contain one-place F-features, 'true', 'possible', etc. corresponding P-features. The same holds for verbs like 'count' and 'seem', respectively. I am somewhat cautious with respect to one-place F- and P-features, because it is not a priori clear whether most of the relevant items do not contain an implicit second argument representing the speaker of the utterance. This means that it might turn out that 'That he comes, is true' must be analyzed as something like 'The proposition expressed by "he comes" is true for me'. Notice, incidentally, that F-features and P-features need not necessarily be disjoint sets. There may be features taking alternatively propositions and fact names as arguments.

⟨44⟩ (a) John does not regret that Paul was waiting.

(b) John regrets that Paul was not waiting.

Whereas the sentences under ⟨43⟩ are synonymous, those under ⟨44⟩ are obviously not. I cannot go into the details here of the problems involved, a characterization of those P-features which have the property in question and those which have not, etc.

The problem of opaque contexts can be stated – with many oversimplifications – as follows. Given two propositions P_1 and P_2 which are logically equivalent, but not synonymous. (We might say that P_1 and P_2 describe in a sense the same fact but in different ways.) Then a context is opaque, if P_1 cannot be substituted for P_2 *salva veritate*.[a]

Assume, for example, that, given a certain setting, 'Paul was waiting' is true if and only if 'Paul was wasting his time' is true. Then both sentences describe the same fact. In this case, ⟨46⟩ is true if and only if ⟨42⟩ is true, whereas ⟨45⟩ might be false, even though ⟨41⟩ is true, and vice versa:

⟨45⟩ John believes that Paul was wasting his time.

⟨46⟩ John regrets that Paul was wasting his time.

Hence ⟨41⟩ is an opaque context, while ⟨42⟩ is not. The relevant fact is that P-features can constitute opaque contexts, while F-features cannot.

It follows from these remarks that a formal explication of the distinction in question and a suitable notation to represent it depends on several problems of semantic representations. Most of these problems are scarcely understood, to say nothing about their solution. If I nevertheless propose a tentative notation, it is only in order to facilitate some additional remarks. What any account of the problem in question must provide seems to be the following: given a proposition P, i.e. a well defined combination of delimiting and predicative features with suitable arguments, the name of the fact associated with P must be derivable. Using the notation proposed so far, we may write this name as '(Def X_i) ([P]X_i)'.[b] The way in which such a fact name provides the argument for an F-feature depends on whether alternative (i) or (ii) discussed above with respect to the operators (QX_i) is accepted. Let P be the reading of 'Paul was waiting' and '(Def X_1) ([J]X_1)' the reading of 'John$_1$'. Assume furthermore that the reading of 'regrets' consists only of the F-feature 'Regret', then we get ⟨47⟩ and ⟨48⟩, corresponding to ⟨23⟩ and ⟨24⟩ respectively, as the reading of ⟨41⟩:

⟨47⟩ (Def X_1) ([J]X_1.(Def X_2) ([P]X_2.[Regret] X_1 X_2))

⟨48⟩ [Regret] (Def X_1) ([J]X_1) (Def X_2) ([P]X_2)

The propositional argument of a P-feature, on the other hand, can be represented by the proposition itself. Assuming that the reading of 'believe' consists only of the feature 'Believe' (and its argument variables), we get either ⟨49⟩ or ⟨50⟩ as the reading of ⟨42⟩:

⟨49⟩ (Def X_1) ([J]X_1.[Believe]X_1(P))

⟨50⟩ [Believe] (Def X_1) ([J]X_1) (P)

---

[a] The problem of opacity has been discussed mainly in connection with believe-sentences. See e.g. Carnap (1964) and references quoted therein.

[b] This notation is, of course, analogous to that of Reichenbach mentioned in note a, p. 425, above as I have argued earlier that the operator (Def X) corresponds, mutatis mutandis, to (ʔ x). I omit the asterisk of Reichenbach's notation, since in fact a confusion cannot arise.

Consider now the hypothesis of Paul and Carol Kiparsky (1969) that the complements of Factive and Non-factive predicates have the underlying structure $\langle 51a \rangle$ and $\langle 51b \rangle$, respectively:

$\langle 51 \rangle$ (a)   (b)

This assumption, motivated by several syntactic phenomena, suggests that the noun 'fact' of $\langle 51a \rangle$ has semantically the effect of deriving the fact name corresponding to the proposition expressed by S. In other words, the reading of 'the fact' in one of its meanings has the form '(Def X_i) ([P]X_i)' where P is a variable over propositions to be substituted by the reading of the complement sentence and i is the reference index of the dominating NP-node. This seems to me a fairly reasonable account of the semantic connection between 'the fact' and its attributive clause in $\langle 52a \rangle$, which is unlike that of the restrictive clause to 'the fact' in $\langle 52b \rangle$:

$\langle 52 \rangle$ (a) John$_1$ regrets [the fact that [Paul talked about it$_3$]$_S$]$_{NP_2}$
       (b) John$_1$ regrets [the fact [that Paul talked about]$_S$]$_{NP_2}$

It must now be noted that fact arguments appear not only in the syntactic function of complements to Factive verbs or adjectives, but also within the readings of single lexical items. The reading of 'give' is an obvious example. $\langle 39 \rangle$ must therefore be reformulated as:

$\langle 53 \rangle$ [Caus]X_S (Def X_1) ([[Have] X_I X_D]X_i).

The fact in question is represented here by X_i. Similarly, propositional arguments occur both as explicit complements and as internal elements of lexical readings. An example of the latter would be 'esteem' in one of its readings. This might be given as:

$\langle 54 \rangle$ [Believe]X_S ([High Value]X_D)

where 'High Value' is short for features representing positive evaluation.

So far we have discussed problems of relational features occurring in the readings of verbs. These considerations apply directly to adjectives governing objects, e.g. 'free', 'due', etc., or complements, such as 'eager', 'able', etc. I will now touch briefly on relational features occurring in readings of relational nouns such as 'father', 'friend', 'side', 'roof', etc. Nouns of this type express social roles, the part-whole-relation, etc. Semantically the representation of such relations is straightforward, by means of such two-place predicates as 'Parent', 'Part', etc. The problem here is the syntactic index required for the variables in the lexical reading. Notice first of all that every noun must contain in its lexical reading a variable which is to be replaced by X_i, if the noun in question functions as the head noun of an NP$_i$. This variable will be indexed for the relation 'Subject-of', since it is precisely this variable which is to be substituted by the argument expression provided by the subject NP, if a noun is used in the predicate NP of a copula sentence. The question then is: what is the syntactic index of the second argument occurring in relational nouns? This question is directly related to the problem of the structure underlying such NPs as 'John's father', 'the left side of a chair', etc. It has been assumed, at least implicitly, by many writers that possessive genitives should be derived from underlying relative clauses with 'have'. This seems to me quite unnatural, however, in the case of inherent relations pertinent to relational nouns of the type discussed here. It seems more reasonable to assume that relational nouns 'govern' an object NP in much the same way as verbs and adjectives do. Thus the structure underlying 'John's father' should be something like

⟨55⟩

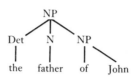

A similar proposal has recently been made by Fillmore (1968).[a] If we abbreviate the syntactic function of an attributive NP of the type in question by the index G, the reading of 'father' can be given as:

⟨56⟩ $[Male]X_S . [Parent] X_S X_G$

From this lexical reading, ⟨58⟩ can be derived as the semantic interpretation of ⟨57⟩:

⟨57⟩ $[[John's]_{NP_1} father]_{NP_2}$
⟨58⟩ $(Def X_2) ([Male]X_2 . [Parent]X_2 (Def X_1) ([J]X_1))$

An example of unalienable possession would be 'hand', whose lexical reading is given in ⟨59⟩ with 'Hand' as an abbreviation for a rather complex set of semantic features:

⟨59⟩ $[Hand]X_S . [Human]X_G . [Part]X_S X_G$

6 Parasitic arguments

We have discussed so far predicative features whose arguments are provided essentially by NPs bearing the relevant syntactic relation to the constituent containing the features in question. I will now consider some phenomena which seem to require a different sort of argument, inducing thereby a further classification of semantic features. The need for such subsidiary arguments can be illustrated by means of the so-called relative adjectives such as 'heavy', 'old', 'loud', 'big', 'long', 'high', and their antonyms.

Relative adjectives specify a certain parameter and indicate that the object(s) referred to exceed (or fall short of) a certain point within that parameter. This point can be indicated explicitly in comparative constructions such as ⟨60⟩, it is understood implicitly in cases like ⟨61⟩ and must hence be part of their readings too:

⟨60⟩ John has a bigger house than Bill (has).
⟨61⟩ John has a big house.

Analyzing a set of German adjectives describing spatial dimensions, which are a proper subset of relative adjectives, I tried to account for the facts just mentioned

---

[a] Fillmore discusses several details of the particular syntactic behavior of NPs in the case of unalienable possession, which is an instance of inherent relation of the type discussed here. Even if one does not agree with all the details of the 'Case-Grammar' proposed by Fillmore, his observations give clear evidence for the assumption that relational nouns admit (or require) a governed NP.

by means of simple markers such as (Vertical), (Maximal), (Secondary), etc. specifying the relevant parameters, and the pair of markers ($+$Pol) and ($-$Pol) to indicate the extension with respect to the specified parameter. (For details see Bierwisch, 1967.) This analysis is unsatisfactory for at least two reasons. Firstly, it does not represent correctly the connections between the different features occurring in a reading (cf. note *a*, p. 411, above). Secondly, it cannot capture certain important empirical facts. Consider the following examples:

⟨62⟩ (*a*) Towers are high.
     (*b*) Towers are high for buildings.
     (*c*) Towers are higher than average buildings.

⟨63⟩ (*a*) These towers are high.
     (*b*) These towers are high for towers.
     (*c*) These towers are higher than average towers.

The sentences in ⟨62⟩ and again in ⟨63⟩ are certainly close paraphrases. This shows that even the positive form of relational adjectives must have a reading similar to that of the comparative, the term for comparison being provided by the average elements of a certain class. This class is that of the subject NP if it is not generic, it is the next larger class, i.e. the genus proximum, if the NP is generic.[a] What is important here is that we need a relational feature representing the comparison involved in relative adjectives. Thus we will replace the elements ($+$Pol) and ($-$Pol) by the relation 'Greater than' and its converse, represented as '[Gr] YZ' and '[Gr^{-1}] YZ', respectively. But what is represented by the arguments Y and Z? Obviously, Y refers to the extent to which an object X_1 occupies a certain parameter which is specified by the rest of the adjective's reading. In other words, Y represents, in the present example, a certain spatial extension of the objects referred to by X_1. The second argument of 'Gr', then, indicates the corresponding extension of the compared objects. Thus the arguments of 'Gr' are of a rather different type than the X_1 considered so far. In fact, they represent not objects, but properties of objects, where these properties are treated as a kind of parasitic or subsidiary argument. They must be established by a further type of relational features extracting

[a] This differentiation between generic and nongeneric sentences with respect to the class used for comparison is obviously necessary, and raises a problem that cannot be dealt with by means of the relative semantic marker proposed by Katz (1967) in this connection. According to Katz' analysis the comparison is made always with respect to the lowest category in the subject's reading, i.e. its genus proximum. But the differentiation introduced above is still an oversimplification in several respects. Thus the sentence

(i) The high towers of the town will be reconstructed

is ambiguous, according to the restrictive or non-restrictive interpretation of the adjective 'high':

(ii) The towers of the town which are high will be reconstructed.
(iii) The towers of the town, which are high, will be reconstructed.

And these imply different norms for comparison, as can be seen from these paraphrases:

(iv) The towers of the town which are higher than the average towers of the town will be reconstructed.
(v) The towers of the town, which are higher than average towers, will be reconstructed.

In other words: restrictive modifiers behave like non-generic sentences, non-restrictive modifiers like generic ones with respect to the choice of the class for comparison. These and several other problems connected with the correct determination of the class to which the subject is compared deserve a careful analysis of both syntactic and semantic conditions which go far beyond the present aims. In the following discussion, I will simply presume that the class is somehow determined.

extensions (or other parameters) from the primary arguments. '[Vert] $X_i Y$' with the interpretation 'Y is the vertical extension of X_i' and '[Max] $X_i Y$' with the interpretation 'Y is the maximal extension of X_i', for example, may specify the relevant parameters of 'high' and 'long', respectively.[a]

The argument Y in these examples represents stretches. In other cases, e.g. in one reading of 'big', we need global extensions, i.e. something like the product of all relevant extensions. These might be introduced by a feature '[Vol] $X_i Y$'.

Since the extensions introduced so far are constituted by the corresponding dimensionality of the primary objects, they must occur not only in the adjectives which assign to them a certain value (greater than a certain average, e.g.), but also in the reading of nouns which specify the dimensionality of the objects referred to. Thus the fact that 'line', 'square', and 'house' designate conceptually one, two, and three dimensional objects, respectively, might be expressed by the features '[1 Ext] $X_i Y_1$', '[2 Ext] $X_i Y_1 Y_2$', and '[3 Ext] $X_i Y_1 Y_2 Y_3$'.[b] Features of this type are part of the explication of the hitherto used global marker (Physical Object). They provide the arguments which must be substituted for the corresponding variables within the readings of relative adjectives.

Given the conventions discussed so far, the reading of 'high' can be given provisionally as $\langle 64 \rangle$:

$\langle 64 \rangle$ [Vert] $X_S Y_i . [Gr] Y_i Z$

Z is still used as an unexplicated abbreviation for the vertical extension of the compared average object. (In comparative constructions it must be substituted by the representation of the concerned extension of the compared NP.) Y_i is a variable that is to be substituted by an extension of the reading of the subject-NP that does not contradict the condition 'Vert'. We thus get $\langle 65 \rangle$ as the reading of 'The table$_1$ is high', if we abbreviate all the features of 'table' not relevant here by 'Table':

$\langle 65 \rangle$ (Def X_1) ([Table]$X_1 . [3$ Ext] $X_1 Y_1 Y_2 Y_3 . [$Vert] $X_1 Y_1 . [Gr] Y_1 Z$)

Notice that the parasitic arguments that we have introduced represent sets of extensions (or other parameters), just as the primary arguments represent sets of objects: a sentence as 'These tables are high' says something about the set of vertical extensions of the set of objects involved.[c]

---

[a] This is, of course, an oversimplification. Certain additional qualifications are required. The situation is still more complicated in other cases, such as 'wide', 'broad', 'deep', etc. My present purpose, however, is not a detailed analysis of the adjectives in question, but a discussion of certain formal aspects of their readings. For the details ignored here see Bierwisch (1967) and Teller (1969).

[b] We thus have a set of features of the general form '[n Ext] $X_i Y_1 \ldots Y_n$' which would replace the element (n Space) used in Bierwisch (1967). Notice that 'n Ext' would be an exception to the above speculation that at most two-place features are necessary to account for the semantic structure of natural languages. This would force us to restrict the tentative claim to primary arguments. Another possibility would be to split up the $n+1$-place predicate 'n Ext' into n two-place features '[1 Ext] $X_i Y_1$',..., '[n Ext] $X_i Y_n$' where the direction of Y_j is orthogonal to Y_k for $1 \leqslant k \leqslant j-1$. '1 Ext' in turn might be replaced by 'Vert', in cases where the noun in question has a designated vertical axis. Similarly for 'Max', etc. In this case 'j Ext' would be the indication of an extension without any further specification. All these are details of an adequate axiomatization of the conceptual structure according to which spatial objects are organized by human perception. I am not concerned here with substantive questions of this kind.

[c] This can be seen more clearly in nominalizations such as 'the height of these tables' where the parasitic arguments are turned into primary ones in a way not to be discussed here.

We must still go one step further in specifying the subsidiary arguments. According to our previous assumptions, the 'normal height' Z in the reading of 'high' must be replaced by the vertical extension of NP_2 in a sentence like $\langle 66 \rangle$:

$\langle 66 \rangle$ [This table]$_{NP_1}$ is higher than [that one]$_{NP_2}$

Hence the reading of $\langle 66 \rangle$ would contain a component '[Gr] Y_1 Y_1''. But this is obviously wrong. What we need is an indexing of the subsidiary arguments with respect to the primary arguments whose parameters they represent. Instead of the Y_1 we must have something like $Y^i_{X_S}$ where the subscripts must be substituted by the same X_1 which replaces the primary argument variable X_S. This then would yield the correct component '[Gr] $Y^1_{X_1} Y^1_{X_2}$' instead of the rejected '[Gr] Y_1 Y_1'' in the reading of $\langle 66 \rangle$.

A moment's reflection shows that this double indexing is in a sense redundant, since it simply repeats information already expressed by the relational features 'Vert', 'Max', etc., which specify extensions of particular objects, not arbitrary ones. Our notation thus obscures somehow the relevant facts. We may remedy this deficiency by adapting another notion developed in modern logic, viz. that of relational descriptions.

Given a two-place predicate 'R(x, y)' whose first argument is always uniquely specified if the second argument is given, an individual can be identified as the one which bears R to a given y. A relational description of this type is usually written as 'R'y'. The element R' introduced here can be understood as a function mapping every y on a particular x, roughly speaking. Hence the following equivalence holds:

$\langle 67 \rangle$ R(x, y) $\equiv$ R'y = x

R'y is an individual of the same type as x, more precisely, it is a name of x. Notice now, that the features 'Vert', 'Max', and several others have precisely the property discussed here, except that it is the second argument which uniquely depends on the first, not vice versa. But this is only a notational arbitrariness. Based on the feature 'Vert' we may thus introduce a corresponding function by the following definition:

$\langle 68 \rangle$ [Vert] XY = $_{def}$ Y = [Vert]'X

Similar functions can be derived from 'Max', 'Vol', etc. From $\langle 68 \rangle$ it follows that an expression of the form '[Vert]'X_1' is the name of the same set of extensions formerly represented as a variable with double index. Hence '[Vert]'X_1' has the status of an argument and must enter readings in precisely this role. If we assume that X_N represents the above mentioned average object with respect to which the modified noun is compared, the reading of 'high' can now be given as follows:

$\langle 69 \rangle$ [Gr] [Vert]'X_S [Vert]'X_N

Functions of the type discussed can also be derived from combinations of relational features. Thus '[Vert.Max] X_1 Y' would say that Y is an extension of X_1 which is both vertical and maximal. Hence '[Vert.Max]'X_1' designates this extension. Combinations of this type are required for several adjectives. I cannot go into these details here.

What has been said with respect to relative adjectives, in particular to those referring to spatial extensions, applies to other parameters as well. A case of particular

---

'The height' in this case clearly refers to a set of extensions, not to a single one, as can be seen from examples like 'The height of the tables is different'. In spite of the fact that a set containing more than one object is referred to, it is treated morphologically as singular.

interest is time. The relation 'R' that was introduced provisionally to express the time-placement of a fact described by a proposition relates a fact to a time interval just as 'Vert' relates an object to its vertical extension. The variables representing time intervals are thus subsidiary arguments of the same kind as the variables over space intervals. Time designations such as 'morning', 'hour', 'year', etc. are then nouns where these subsidiary variables are turned into primary ones – just as in nominalized adjectives like 'length', 'height', 'width', etc. Problems of this type, however, are topics of separate investigations.

7 Conclusions

Presupposing the general framework of transformational generative grammar, we have discussed some formal aspects of semantic representations. These are considered as complex structures that are built up from basic primitive terms and assigned to syntactic surface representations by the syntactic rules of the grammar of a given language. We assumed that an uninterpreted grouping of the basic semantic elements, proposed previously e.g. in Katz (1966, 1967), or Bierwisch (1967), is not sufficient. We claimed that an adaptation of certain means developed in modern logic might be at least a promising program. Pursuing this proposal along certain fairly different lines, we arrived at the following provisional conclusions:

Semantic representations presuppose at least the following basic requirements (or their equivalents in certain other notation):

(1) A set of variables representing sets of objects to which linguistic expressions can refer. Variables occurring in readings of sentences are indexed with respect to sameness or difference of reference.

(2) A set of semantic features which are classified according to several aspects into at least the following subsets:

(2.1) Predicative features representing properties and relations ascribed to the elements of the sets represented by the variables in (1). Predicative features are further subclassified according to the number of their arguments into one-place predicates, two-place predicates, etc. The possibility was suggested that the number of arguments might be restricted to two. Another subclassification of predicative features might be induced by the type of arguments they take: those representing true objects, facts, propositions, parasitic objects or parameters of objects.[a] Predicates may also take predicates as arguments, thus leading to higher level predication. This possibility obviously bears on the classification of semantic features. It has not been dealt with in the present paper. For some discussion see Bierwisch (1969).

(2.2) Delimiting features characterizing the sets of objects referred to with respect

---

[a] To what extent the classification according to types of arguments is a purely formal aspect of semantic features is a dubious question, since it presupposes a subclassification of arguments. But whether e.g. primary arguments and parasitic arguments can be distinguished on purely formal grounds, i.e. without dragging ontological questions into the semantic theory, is not at all obvious. Since I know, for the time being, of no empirical consequences, the question will be left open here. Notice, however, that one has to be careful to avoid arbitrary subdivisions of the argument variables. Assume, for example, we were to introduce a set of primary variables and a set of variables ranging only over parasitic objects, i.e. properties or parameters. We might then subdivide the former into those ranging over abstract entities and those ranging over concrete, physical objects, the latter being subdivided in variables over animate and unanimate objects, etc. This would either duplicate the information expressed by the semantic features or even transfer it from the semantic features to the argument variables. But it is the set of semantic features that should formally reconstruct the substantive content of meaning.

to quantity and to the operations that constitute these sets. Delimiting features fall into two classes: features specifying reference instances and features indicating absolute and relative quantity.

The predicative semantic features (2.1) correspond to predicate constants, the delimiting features (2.2) to quantifiers and certain other operators of modern logic. Whereas the variables in (1) have no substantial content and serve only the purpose of representing identical and different reference, the semantic features in (2) must have a constant interpretation in terms of cognitive and perceptual conditions inherent in human organisms and governing their interaction with the physical and social environment.

(3) The elements listed under (1) and (2) are combined according to a set of rules or conventions to form semantic representations. These rules are clearly connected to the different classes of semantic features discussed above. We have touched on them only marginally in the present paper.

The requirements (1), (2), and (3) specify an infinite set of semantic representations, where the rules postulated in (3) may be considered as the syntax of such representations. These requirements then indicate, at least partially, what semantic representations look like. They are, of course, by no means a full semantic theory of natural languages. An empirically motivated formal classification of the basic elements of a semantic theory is, however, a necessary part of such a theory.

The formal classification of semantic features discussed here is not the only type of structure organizing the set of primitive semantic terms. A fairly complex system of mutual inclusion, exclusion, hierarchical subordination, etc. must be assumed to govern this basic inventory. The nature of this structure, which reflects the substantive content of semantic features, is as yet poorly understood. Some of its aspects may be expressed by redundancy rules, which are similar in certain respects to the meaning postulates of modern logic.

According to our basic assumption, semantic representations must be connected to syntactic surface representations. Let us assume that for this purpose it is sufficient to specify the connection between syntactic deep structures and semantic representations, the former being mapped by transformational rules on the appropriate surface structures. The connection between syntactic deep structures, which specify a set of basic syntactic relations, and semantic representations is provided by the lexicon, which associates semantic representations with basic syntactic formatives, and general conventions specifying the semantic effect of the basic syntactic relations.

Semantic representations of lexical entries are of the same character as those of syntactically complex expressions, with one exception: lexical entries do not contain referentially indexed variables, since lexical entries do not refer to particular sets of individuals.[a] Instead they contain variables that are categorized with respect to the syntactic relations defined by the syntactic deep structure. The general conventions specifying the semantic effect of syntactic relations are based on precisely these categorized variables. We have touched upon two alternative principles according to which these conventions might operate.

---

[a] They do not even refer to the whole class of individuals specified by the concept in question. A word like 'father' can be used in generic sentences to refer to the class of all fathers, but it does not refer to it as a dictionary entry. The fact that unique terms like 'the sun', 'New York', etc. always refer to a particular object, is in this respect a fact of encyclopedic knowledge, not of the lexicon of a given language.

I have ignored throughout this paper the fact that lexical readings are generally subdivided into two parts: the proper semantic content and the selection restriction – or more generally: the presuppositions for its appropriate use. This topic involves a number of different problems. It has no bearing, however, on the classification of features. Every semantic primitive that can be part of the proper conceptual content of a lexical reading can also be part of a selection restriction, and vice versa.[a]

I have made only very vague claims about the precise content of the syntactic deep structure and the relations defined by it. A more precise formulation of the classification of semantic features than that given here must obviously be related to certain specific assumptions with respect to the syntactic representations. This concerns e.g. the syntactic status of quantifiers, determiners, negation, and even of noun phrases. Thus the specification of the form of semantic representations depends to a fairly large extent on the presupposed syntactic analysis. But it is equally obvious, on the other hand, that the syntactic behavior of particular lexical entries can be predicted to a large extent on the basis of its semantic structure, if this is made explicit in an appropriate way.

Deutsche Akademie der Wissenschaften zu Berlin

REFERENCES

Bach, Emmon and Harms, Robert T. (1968), *Universals in Linguistic Theory*, Holt, Rinehart and Winston, New York.
Bellert, Irena (1969), 'On the Problem of a Semantic Interpretation of Subject and Predicate in Sentences of Particular Reference', in: Bierwisch and Heidolph (1969).
Bendix, Edward Herman (1966), *Componential Analysis of General Vocabulary*, Indiana University, Bloomington.
Bierwisch, Manfred (1967), 'Some Semantic Universals of German Adjectivals', in: *Foundations of Language*, III, 1–36.
  (1969), 'On Certain Problems of Semantic Representations', in: *Foundations of Language*, V, 153–84.
  (1968), Review of Weinreich, 'Explorations in Semantic Theory', in: *Current Anthropology*, IX, 160–1.
Bierwisch, Manfred and Heidolph, Karl Erich, eds. (1969), *Progress in Linguistics*, Mouton, The Hague.
Carnap, Rudolf (1964), 'On Belief-Sentences', reprinted in: Rudolf Carnap, *Meaning and Necessity*, University of Chicago Press.
Chomsky, Noam (1965), *Aspects of the Theory of Syntax*, MIT Press, Cambridge, Mass.
Fillmore, Charles J. (1968), 'The Case for Case', in: Bach and Harms (1968), 1–88.
Isenberg, Horst (1968), *Das direkte Objekt im Spanischen*, Studia Grammatica IX, Berlin.
Karttunen, Lauri (1968), 'What do Referential Indices Refer to?', mimeographed by the Rand Corporation.
Katz, Jerrold J. (1964a), 'Analyticity and Contradiction in Natural Language', in: Fodor and Katz, eds., *The Structure of Language*, Prentice-Hall, New York, 519–43.

---

[a] Katz (1967) has made a different claim, viz. that only semantic markers can appear in selectional restrictions, whereas distinguishers cannot. I have argued in Bierwisch (1967) that this claim, together with the whole notion of distinguisher, should be given up.

(1964b), 'Semantic Theory and the Meaning of "Good"', in: *The Journal of Philosophy*, LXI, 739–66.

(1966), *The Philosophy of Language*, Harper and Row, New York.

(1967), 'Recent Issues in Semantic Theory', in: *Foundations of Language*, III, 124–94.

Katz, Jerrold J. and Fodor, Jerry A. (1963), 'The Structure of a Semantic Theory', in: *Language*, XXXIX, 170–210.

Katz, Jerrold J. and Postal, Paul M. (1964), *An Integrated Theory of Linguistic Descriptions*, MIT Press, Cambridge, Mass.

Kiparsky, Paul and Kiparsky, Carol (1970), 'Fact', in: Bierwisch and Heidolph (1970). Reprinted in this volume.

Lang, Ewald (1967), '"Jointness" und Koordination', mimeographed by Deutsche Akademie der Wissenschaften zu Berlin.

McCawley, James D. (1968), 'The Role of Semantics in a Grammar', in: Bach and Harms (1968), 125–69.

Perlmutter, David M. (1970), 'On the Article in English', in: Bierwisch and Heidolph (1970).

Postal, Paul M. (1968), 'Cross-Over Phenomena, A Study in the Grammar of Conference', mineographed by The Thomas J. Watson Research Center, IBM Corporation, New York.

Reichenbach, Hans (1966), *Elements of Symbolic Logic*, reprinted by the Free Press, New York.

Teller, Paul (1969), 'Some Discussion and Extension of Manfred Bierwisch's Work on German Adjectivals', in: *Foundations of Language*, V, 185–217.

Weinreich, Uriel (1966), 'Explorations in Semantic Theory', in: Sebeok, ed., *Current Trends in Linguistics*, III, 395–477, Mouton, The Hague. Pp. 395–416 reprinted in this volume.

1962, 'Lexicographic Definition in Descriptive Semantics', in: Householder and Saporta (eds.), *Problems in Lexicography*, Indiana University, Bloomington.

A method of semantic description

R. M. W. DIXON

The North Queensland language Dyirbal includes a special 'mother-in-law language' which provides a unique set of data suggesting the form that the semantic description of the verbs of a natural language should take. §1 below describes the 'mother-in-law language'. §2 explains and justifies the method of semantic description that is suggested by the complex correspondences that exist between mother-in-law vocabulary and the vocabulary of the everyday Dyirbal language. §3 applies the method to a semantic description of Dyirbal verbs. The reader without sufficient time to study this paper in full could extract the jist by reading §1.1 (omitting the last paragraph), §2.1, §2.2 and §2.4.[a]

Two well-known approaches to semantic description – the componential method and the definitional method – each provide certain insights, but each also has rather serious drawbacks (discussed in §2.1). The method of semantic description described in this paper combines the insights of these two approaches without at the same time taking over any of their drawbacks. Briefly, the lexical verbs of a language are held to fall naturally into two mutually exclusive sets: *nuclear* verbs and *non-nuclear* verbs. *Componential* semantic descriptions can be provided for the nuclear verbs, in terms of a small set of rather general and well-motivated semantic features (some of which are also likely to underlie categories in the grammar of the language). The semantic content of non-nuclear verbs can be *defined* in terms of semantic descriptions of nuclear verbs (or of previously defined non-nuclear words), and the syntactic apparatus of the language.

1 The Dyirbal 'mother-in-law language'

1.1 WORD CORRESPONDENCES BETWEEN EVERYDAY AND MOTHER-IN-LAW LANGUAGES

Every user of Dyirbal had at his disposal two distinct languages:

(1) Guwal, the unmarked or 'everyday' language; and

(2) Dyalŋuy, a special (so called 'mother-in-law' – see Capell 1962, Thomson 1935) language used in the presence of certain 'taboo' relatives.

[a] Field work in 1963, 1964 and 1967 was supported by the Australian Institute of Aboriginal Studies and the Central Research Fund of the University of London. This paper is an up-dated condensation of the semantics section of my Ph.D. thesis (Dixon 1968a) which was written whilst I was at University College London; the final version of the paper was written whilst I was visiting lecturer at Harvard University, 1968/9 and was supported in part by NSF grant GS-1934. Whatever value it has is due largely to the intelligence, patience and willingness of Chloe Grant and George Watson, the two main informants. I am also grateful to Michael Silverstein for his critical comments on the paper.

When a man was talking within hearing distance of his mother-in-law, for instance, he had exclusively to use Dyalŋuy, for talking on any topic; when man and wife were conversing alone they could use only Guwal. The use of one language or the other was entirely determined by whether or not someone in proscribed relation to the speaker was present or near-by; there was never any choice involved.

Dyalŋuy has identical phonology and almost exactly the same grammar as Guwal.[a] However, it has an entirely different vocabulary, there being not a single lexical word common to Dyalŋuy and Guwal.

Confronted with a Guwal word a speaker will give a unique Dyalŋuy 'equivalent'. And for any Dyalŋuy word he will give one or more corresponding Guwal words. It thus appears that the two vocabularies are in a one-to-many correspondence: each Dyalŋuy word corresponds to one or more Guwal words (and the words so related are in almost all cases not cognate with each other). For instance, Guwal contains a number of names for types of grubs, including d^yambun 'long wood grub', *bugulum* 'small round bark grub', $mandid^ya$ 'milky pine grub', gid^ya 'candlenut tree grub', *gaban* 'acacia tree grub'; there is no Guwal generic term covering all five types of grub. In Dyalŋuy there is a single noun d^yamuy 'grub' corresponding to the five Guwal nouns; greater specificity can only be achieved in Dyalŋuy by adding a qualifying phrase or clause – describing say the colour, habitat or behaviour of a particular type of grub – to d^yamuy.

The ease with which Guwal–Dyalŋuy correspondences reveal semantic groupings can be seen from the following example.

A bilingual informant who offered single word glosses for Guwal verbs identified *nudin*, *gunban* and ban^yin all as 'cut'. But Dyalŋuy correspondents are:

Guwal	Dyalŋuy
nudin	$d^yalŋgan$
gunban	$d^yalŋgan$
ban^yin	*bubaman*

bubaman is also the correspondent of Guwal *baygun* and d^yindan. The meanings of these verbs can be explained as follows:

nudin 'cut deeply, sever'
gunban 'cut less deeply, cut a piece out' } Dyalŋuy $d^yalŋgan$ 'cut'

baygun 'vigorously shake or wave, bash something on something else'
d^yindan 'gently wave or bash, e.g. blaze bark, rain falling gently'
ban^yin 'split a soft or rotten log by embedding a tomahawk in the log and then bashing the log against a tree' } Dyalŋuy *bubaman* 'shake, wave or bash'

ban^yin, which at first appeared to be related to *nudin* and *gunban*, is thus seen, on the basis of Dyalŋuy data, to be related to *baygun* and d^yindan, as further specifications of the Dyirbal concept 'shake, wave or bash'; whereas *nudin* and *gunban* are further specifications of the quite different concept 'cut'.

Dyalŋuy contains far fewer words than Guwal – something of the order of a quarter as many. Whereas Guwal has considerable hypertrophy, Dyalŋuy is

---

[a] Grammatical differences are only differences of degree: Dyalŋuy makes much more use of verbalisation and nominalisation processes than does Guwal; and so on.

characterised by an extreme parsimony. Every possible syntactic and semantic device is exploited in Dyalŋuy in order to keep its vocabulary to a minimum, it still being possible to say in Dyalŋuy everything that can be said in Guwal. The resulting often rather complex correspondences between Guwal and Dyalŋuy vocabularies are suggestive of the underlying semantic relations and dependencies for the language.

In some cases a Dyalŋuy word corresponds to Guwal items belonging to different word classes. The major word classes in Dyirbal are, on syntactic grounds, noun, whose members refer to objects, spirits, language and noise; adjective, referring to quantities, qualities, states and values of objects; verb, referring to action and perception; adverbal, referring to qualities and values of action and perception; and time qualifiers. Adjectives modify nouns and agree with them in case; in exactly the same way adverbals modify verbs and agree with them in transitivity, and in tense or other final inflection. Now some adjectives refer to qualities of objects that are exactly analogous to qualities of actions, dealt with through adverbals. Dyalŋuy exploits this similarity by having a single word where Guwal has both an adjective and a non-cognate adverbal. Thus Guwal adjective *wunay* 'slow' has Dyalŋuy equivalent *wurgal*; Guwal intransitive adverbal *wundin*^{y}*u* 'do slowly' is rendered in Dyalŋuy by the intransitively verbalised form of the adjective *wurgal*, that is *wurgalbin*. In other cases a state, described by a Guwal adjective, may be the result of an action, described in Guwal by a verb. Dyalŋuy will exploit this relation of 'consequence' (cf. Lyons 1963.73), having either just a verb or just an adjective where Guwal has both verb and adjective. Thus Guwal adjective *yagi* 'broken, split, torn' has Dyalŋuy correspondent *yilgil*; Guwal verb *ṛulban* 'to split' is in Dyalŋuy *yilgilman*, the transitively verbalised form of the adjective *yilgil*. And Guwal transitive verb *n*^{y}*ad*^{y}*un* 'to cook' is in Dyalŋuy *durman*; Guwal adjective *n*^{y}*amu* 'cooked' has Dyalŋuy correspondent *durmanmi*, the perfective participle of the verb *durman*. Guwal–Dyalŋuy word correspondences thus reveal the semantic connection between certain quality adjectives and certain quality adverbals; and between certain state adjectives and certain 'affect' verbs.

1.2 SOCIOLINGUISTIC BACKGROUND

A full account of which relatives were 'taboo' (and in whose presence the use of Dyalŋuy was obligatory) would require a more complete understanding of the Dyirbal kinship system than the writer has. For a woman, father-in-law was taboo; similarly for a man, mother-in-law, and also father's sister's daughter and mother's brother's daughter but not mother's sister's daughter. The relation was symmetrical: if X was taboo to Y then Y was taboo to X.[a]

The name 'Dyirbal language', as used here, covers the 'languages' of three tribes: Mamu, Dyirbal and Giramay.[b] These are grammatically almost identical and have over seventy per cent common vocabulary so that it is convenient to refer to them as three dialects of a single language, and to give a single overall description of their grammar (Dixon 1968a). Each of these tribes had its own Guwal and Dyalŋuy

---

[a] Children were promised in marriage at an early age, thus acquiring a full set of taboo relatives; they seem to have learnt Dyalŋuy in the same way that children normally learn a language.

[b] Dyiru and Gulŋay – both now pretty well extinct – could probably have been considered other dialects of this language. Gulŋay was spoken by the Malanpara tribe, referred to a great deal by Roth (1901–5, 1907–10).

(and similarly for at least half-a-dozen surrounding tribes). Although the Guwals are still actively spoken, the Dyalŋuys are not. Mamu and Dyirbal Dyalŋuys, for instance, ceased being used about 1930; since then just the Guwals have been used, even in the presence of taboo relatives. However, older speakers remember a great deal of their Dyalŋuy, and speak it readily when asked. The writer gathered a large amount of Dyalŋuy data independently from a number of different Dyirbal and Mamu informants; the consistency of this data indicated that the language was being accurately remembered.

Each particular dialect had no lexical words common to its Guwal and its Dyalŋuy. However in a number of cases a word in the Dyalŋuy of one tribe occurs, with a similar meaning, in the Guwal of a neighbouring tribe. For instance, the word for 'back' is *mambu* in Dyirbal Guwal, *dʸudʸa* in Dyirbal Dyalŋuy, *dʸudʸa* again in Mamu Guwal and *ŋabil* in Mamu Dyalŋuy.

English loan words were taken directly into Guwal, suitably cast into the mould of Dyirbal phonology: thus, *bigi* 'pig', *bulugi* 'cattle', *dʸuga* 'sugar', *dʸarudʸa* 'trousers', *mani* 'money' and so on. We have seen that everything must have a different name in Dyalŋuy from that which it has in Guwal; and we have also noted that Dyalŋuy is extremely parsimonious, keeping its vocabulary as small as possible. Consistent with this, Dyalŋuy most often dealt with loans by extending the range of meaning of an existing Dyalŋuy term; only in one or two isolated cases did Dyalŋuy take on a new word to correspond to loans in Guwal. Thus *bigi* 'pig' is translated into Dyalŋuy by *ginga*, which was already the Dyalŋuy correspondent of Guwal *gumbiyan* 'porcupine (echidna)': pig is assigned to the same taxonomic group as porcupine. Cattle are named in Dyalŋuy according to their main functional characteristic, *ŋunŋun* 'breast'. Sugar is, by its texture, identified in Dyalŋuy with 'sand', *warunʸ*; whilst trousers are given the same name as the part of the body they cover, Dyalŋuy *dabara*, 'thigh'.[a] However, no existing Dyalŋuy item could be extended to cover 'money' and in this case a new word was introduced into Dyalŋuy: *walba*.

2 The method of semantic description

2.1 COMPONENTIAL AND DEFINITIONAL APPROACHES

In the componential approach each word is semantically described in terms of a number of basic components or features; the features involved in the semantic description of the words of a language may be arranged in a number of systems.[b] For instance *man* might be describable through features 'human' – from the

---

[a] Loans were taken into Dyirbal as nouns, adjectives or time words, never as verbs. Thus 'work' becomes an adjective, *wagi*; however this normally occurs in verbalised form *wagibin* (cf. §2.3).

[b] The features of a system are referentially complementary; also, each semantic description can choose one and only one feature from a given system. Systems typically enter into dependency trees, involving conditions of the type: if a semantic description chooses feature x_1 from system $\{x_1, x_2\}$ then it must go on to make a choice of one feature from system $\{y_1, y_2, y_3\}$; if x_2 is chosen then a choice must be made from system $\{z_1, z_2\}$. For an example of such a dependency tree – but involving syntactic features – see Chomsky 1965.83. The idea of linguistic system, in the sense used here, has been stressed by Firth (1957) and Halliday (1961); it has been made much of in recent work by Halliday (1967–8) – where the term 'system network' is applied to dependency trees. Little or no use is made of the idea of system in the work of Lamb (1964), Katz and Fodor (1963), for instance, and these writers appear to set up rather ad hoc features in dealing with the examples they consider. Compare Weinreich's

system {human, non-human} – 'male' – from system {male, female} – and 'old' – from system {young, old} – see Hjelmslev and Uldall (1957.45–6). This method of semantic description has been applied most successfully to kinship terms (for instance Goodenough 1956, Lounsbury 1956, 1964a, 1964b). In the definitional approach the meaning of a word is defined in terms of the meanings of other words, related together through any of the grammatical constructions of the language (this method does not deal in terms of elements of meaning smaller than the meanings of words). This approach has been employed – though not always very consistently or well – in monolingual dictionaries; for a discussion and critique see Weinreich (1962).

The componential approach has the advantage of breaking down word meanings into more basic units. Words of similar meaning are now directly related through their semantic descriptions having one or more features in common. This approach is unsatisfactory in that it does not refer much to the grammatical resources of the language, and the grammatical patterning is after all the rich heart of a language, that we would expect a semantic description to refer to and to exploit in a variety of ways. A further drawback of the componential approach is that very many semantic features must be posited – a number of the same order as the number of words to be semantically described.[a] Many features occur just in the semantic description of a single word. A componential description thus appears to gain little in economy over a simple list of the words of the language, each considered as a semantic primitive in its own right.

The definitional approach has the advantage that it does exploit grammatical relationships. It does not however in all cases clearly demonstrate the relationship between synonyms, near-synonyms and antonyms, as does the componential approach. The most severe drawback to the definitional approach lies in vicious circles – in following up the definitions of the words occurring in the definition of a given word, and so on, one will sooner or later be confronted with a word that has been encountered earlier in the chain; an advantage of the componential approach is that vicious circles are avoided.

2.2   THE HYPOTHESIS

For optimum semantic description of the verbs of a natural language we suggest using a mixture of the componential and definitional methods – minimising the drawbacks associated with each method. We divide the lexical verbs of a language into two groups: *nuclear* and *non-nuclear*. Componential definitions of nuclear

comment (1962.30): 'the consistency of lexicography could be improved if dictionary makers were held to the assumption that the terms of a language are, on the whole, complementary. This assumption implies that the most important case to deal with in semantic description is one in which, where the signification of one term ends, that of another begins. On the whole, a semantic description should not aim at "absolute" definitions, but at definitions which delimit the meaning of a term from that of terms with similar meanings (synonyms).' It should be noted that what is here called a 'feature' corresponds to a 'value of a feature' in Jakobson–Halle–Chomsky terminology, our 'system' corresponding to their 'feature'. Thus where we would refer to a binary system with features 'to' and 'from', they would instead talk of a feature 'to', with possible values ' +to' (corresponding to our feature 'to') and ' –to' (corresponding to our 'from'). This difference is entirely terminological. A non-trivial difference is that we do not restrict ourselves to binary systems. Empirically, most systems do have only two features but there are some with three or more features: thus {sit, stand, lie}, which cannot be reduced in any non-arbitrary way to a pair of binary systems.

[a] Cf. Lamb 1966.566: 'it seems obvious that the number of such minimal [semantic] components for any real language will number in the thousands'.

verbs are generated from systems of primitive semantic features; non-nuclear verbs are defined in terms of nuclear verbs (or of already defined non-nuclear verbs) utilising the full grammatical possibilities of the language in the formulation of these definitions.[a]

For componential definition of nuclear verbs only a fairly small number of rather general systems of semantic features are required. A primitive feature will normally recur in the semantic descriptions of several nuclear verbs; a system of such features may also occur in the grammatical description of the language.

For a quick, rough and ready illustration of nuclear and non-nuclear verbs consider *look* and *stare* in English. *Look* is a nuclear verb, whereas *stare* is non-nuclear; *stare* could semantically be defined as, say, *look hard* – that is, the meaning of *stare* could be defined in terms of the meanings of *look* and *hard*, related together through a typical verb-plus-adverb construction. *Look* cannot in the same way be defined in terms of some more general verb; however, *look* does show some semantic similarities of various types to a number of other nuclear verbs – for instance, *look* is to *listen* as *see* is to *hear* – and these similarities can be captured through appropriate componential descriptions of the verbs.

This method of semantic description eliminates the disadvantages associated with the pure componential and pure definitional approaches. Great use is made of grammatical relations in formulating the definitions of non-nuclear words. Vicious circles are eliminated. Pairs of words that are synonyms or near synonyms are explicitly related through being either (1) non-nuclear words defined in terms of the same nuclear word, or (2) a nuclear word, and a non-nuclear word defined in terms of that nuclear word, or (3) two nuclear words whose componential descriptions involve a common feature or features. Antonyms are similarly explicitly related through being, or being defined in terms of, nuclear words whose semantic descriptions have a common core of features and differ only in choosing different features from a certain semantic system. Further, only a small number of rather general semantic features are required; there is no need for ad hoc features that each occur only in the semantic description of a single word.

The division of the verbs of a language into nuclear and non-nuclear sets is not in any sense an arbitrary division, but an entirely natural one; this naturalness will be justified in the remainder of this section, in §2.3, and especially in §2.4. The nuclear verbs can be componentially described in terms of a small number of rather general features. To also provide completely componential descriptions of non-nuclear verbs we would have to set up additional, ad hoc systems, whose features would each occur only in a single semantic description (cf. the discussion of *bilin*[y]*u* and *bumiran*[y]*u* in §2.3). And nuclear words *must* be componentially described since it appears that they are not susceptible to definition in terms of other nuclear words, in the way that non-nuclear verbs are (see §2.3).

Suppose that there were a language which had the requirement that its lexicon

[a] Nuclear words tend to have greater frequency than non-nuclear items. This is not to say that the least frequent nuclear verb is more frequent than the most common non-nuclear one; rather that in almost every case a certain nuclear word will have greater frequency than non-nuclear words that are related to it (for instance, by being defined in terms of it). Compare with Weinreich's suggestion (1962.37) that 'one way of progressively reducing the defining meta-language in richness would be to require that the definition of a term X be formulated only in words of frequency greater than that of X'. However, relative frequency is not a criterion in the present method, merely an empirical accompaniment of the nuclear/non-nuclear distinction.

contain an absolute minimum number of verbs. Such a language need not contain any non-nuclear verbs. In place of a putative non-nuclear verb it could simply use a 'definition': thus instead of *stare* it could have *look hard*. The language would, however, have to contain a full set of nuclear verbs, since nuclear items cannot be replaced by definitions as can non-nuclear verbs. Dyalŋuy behaves almost exactly like this.

We have noted that Dyalŋuy is extremely parsimonious, that it has as small a vocabulary as possible consistent with it being possible to express in Dyalŋuy everything that can be expressed in Guwal. Guwal has several hundred nuclear verbs, and in addition several thousand non-nuclear verbs. Dyalŋuy has a full set of nuclear verbs (in a one-to-one correspondence with the set of Guwal nuclear verbs) but it has almost no non-nuclear items.

Since nuclear verbs cannot be defined in terms of other lexical words, we can see that it is essential for Dyalŋuy to have a direct correspondent for each Guwal nuclear verb, if it is to function as a full natural language. However, it need not have any non-nuclear words: Guwal non-nuclear verbs can in Dyalŋuy be 'defined' in terms of one or more nuclear verbs, and the grammatical relations of the language. When a single-word Dyalŋuy correspondent of a Guwal non-nuclear verb is elicited, the nuclear verb that is the 'head' of the full definition is given.

Dyalŋuy occasionally allows itself the luxury of a direct correspondent for a Guwal non-nuclear verb. But most often in such cases one dialect has a direct correspondent for a non-nuclear word and the other does not, making do with the usual 'definition in terms of nuclear words'. For instance, in the Mamu dialect nuclear verb *bawalbin* was given as the Dyalŋuy correspondent of two Guwal verbs: nuclear *yanu* 'go' and non-nuclear *manman^{y}u* 'shift camp'; a fuller specification of the meaning of *manman^{y}u* was given by the Dyalŋuy 'definition' *naŋguɲunu bawalbin* 'go from the camp'. The Dyirbal dialect also has Dyalŋuy *bawalbin* corresponding to *yanu* but in this instance has a non-nuclear Dyalŋuy verb, *bargan^{y}u*, as direct correspondent of *manman^{y}u*. In the case of the non-nuclear verb *gunin^{y}u* 'search', Mamu has a direct Dyalŋuy correspondent *banman* 'search' whereas Dyirbal Dyalŋuy just defines it as *n^{y}uɾimarind^{y}an^{y}u* 'look repeatedly' (*n^{y}uɾiman* is in both dialects the Dyalŋuy correspondent of nuclear *buɾan* 'look' and also of eleven other non-nuclear Guwal verbs).

The next section, §2.3, discusses the Dyalŋuy correspondents of some of the more frequent verbs concerned with motion, presenting data which is directly suggestive of the mixed componential-plus-definitional approach. §2.4 deals with the procedure followed in obtaining Guwal–Dyalŋuy correspondences, the results of which underline the naturalness of the nuclear/non-nuclear division and indicate unequivocally which items from the vocabulary of Guwal fall into the nuclear set. §2.3 goes into a fair amount of syntactic detail; the reader anxious for the shortest and strongest justification of the nuclear/non-nuclear distinction should go directly to §2.4.

Before discussing Guwal–Dyalŋuy correspondences in detail, we first make the following assumptions:

(1) if a Guwal word and a Dyalŋuy word are in one-to-one correspondence then a relation of synonymy exists between them: they have the same semantic description;

(2) if several Guwal words are in a many-to-one relation to a Dyalŋuy word then a relation of hyponymy exists between them: the semantic descriptions of the Guwal words each include the semantic description of the Dyalŋuy word (that is, the

Dyalŋuy semantic description is the highest common factor of the Guwal semantic descriptions).

Here 'synonymy' and 'hyponymy' relations are recognised between words in two different languages (rather than between words all in one language, as is usual) – or rather, between words in two different 'realisations' of the same underlying semantic system.

It should be noted that assumption (2) does not apply to certain instances of metaphoric-type Dyalŋuy naming (see Dixon 1968a.246 ff.); these however all concern nouns. The class of verbs is far more semantically complex and highly structured than other classes; it is thus the ideal domain in which to test the effectiveness of any method of semantic description. The method outlined is, in the present paper, applied only to description of Dyirbal verbs.[a]

2.3 DYALŊUY CORRESPONDENTS OF SOME VERBS OF MOTION

We must first explain some fragments of the grammar of Dyirbal.

Every head noun in an NP is normally accompanied by a 'noun marker'; this agrees with the noun in case, indicates which of the four noun classes the noun belongs to, and specifies the visibility/location of the referent of the noun. The first element in a noun marker is one of the three forms:

bala-	'there (and visible)'
yala	'here (and visible)'
ŋala-	'not visible (but audible)'

bala- is also the unmarked form, used when no implication of location/visibility is intended. The second element is a case morpheme: zero for nominative, *-ŋgu-* for ergative, *-gu-* for dative, *-ŋu-* for genitive. Finally there is indication of noun class: *-l* for class I, *-n* class II, *-m* class III and zero for class IV (for a full discussion of the noun classes, and their semantic content, see Dixon 1968b).

A noun marker can optionally be followed by one of a set of twelve bound forms, further specifying the location of the referent of the noun:[b]

bayd^yi	short distance downhill	*dayi*	short distance uphill
bayd^ya	medium distance downhill	*daya*	medium distance uphill
bayd^yu	long distance downhill	*dayu*	long distance uphill
balbala	medium distance downriver	*dawala*	medium distance upriver
balbulu	long distance downriver	*dawulu*	long distance upriver
	guya	across the river	
	bawal	long way (in any direction)	

[a] Nouns, with the exception of subsets of kinship and agegroup (boy, man, etc.) terms, do not lend themselves to componential analysis and definition in the way that verbs do – see Dixon 1968a for detailed discussion; the nuclear/non-nuclear distinction is not, generally, very useful in the semantic description of adjectives (see Dixon, forthcoming, a, for discussion).

[b] There is a thirteenth form, *ŋaru* 'behind', in the Mamu dialect only. There is also a

(Dyirbal is spoken in the Rain Forest just south of Cairns, in the hilly, well-watered area between the east coast and the dividing range; hence the rich set of terms for referring to 'up', 'down', with respect to 'river', 'hill', etc.)

Thus:

⟨1⟩ *balanbayd*ʸ*i*   *d*ʸ*ugumbil*   *miyandan*ʸ*u*

there-nom-class II-short distance down hill woman-nom laugh-pres/past
The woman a short distance downhill there is laughing.

A verb can be accompanied by one of a set of eight verb markers. The first element in a verb marker is one of the forms:

bala-   'there'
yala-   'here'

and the second element is a locational case morpheme: *-ru* 'to (with respect to a place)', *-ri* 'to (with respect to a direction)', *-ŋum* 'from', or *-y* 'at'. Allative (*-ru*, *-ri*) and ablative (*-ŋum*) forms are usually shortened by the deletion of the *-la-* of *bala-* or *yala-*, the *-r-* of the allative inflections being simultaneously changed to *-l-*: thus *balu*, *bali*, *baŋum* but *balay*. A verb marker may optionally be followed by one of the twelve bound forms in the table above. Thus:

⟨2⟩ *baladawala*   *diban*   *balbalin*ʸ*u*   *yali*

there-nom-class IV-medium distance upriver stone-nom roll-pres/past here-to (direction)
The stone a medium distance there upriver is rolling in this direction.

⟨3⟩ *balanguya*   *d*ʸ*ugumbil*   *banin*ʸ*u*   *baŋumbawal*

there-nom-class II-across river woman-nom come-pres/past there-from-long way
The woman on the other side of the river there is coming from a long way out there.

⟨4⟩ *balan*   *guda*   *n*ʸ*inan*ʸ*u*   *balaybayd*ʸ*i*

there-nom-class II dog-nom sit-pres/past there-at-short distance downhill
The dog there is sitting down a short distance there downhill.

Simple verbs concerned with position fall into two classes: 'motion' verbs that occur only with 'to' and 'from' verb markers, and 'rest' verbs that occur only with 'at' markers. Typical members of these classes are:

	Motion	*Rest*
intransitive	*yanu* 'go'	*n*ʸ*inan*ʸ*u* 'stay, sit'
	*banin*ʸ*u* 'come'	*d*ʸ*anan*ʸ*u* 'stand'
	*wayn*ʸ*d*ʸ*in* 'motion uphill'	*bungin* 'lie'
	wandin 'motion upriver'	
	buŋan 'motion downhill'	
	*dadan*ʸ*u* 'motion downriver'	
	mabin 'cross the river'	
	mayin 'come out'	
	marin 'follow'	

further set of bound forms: *gali* 'vertically down', *gala* 'vertically up', *galu* 'directly in front'. A noun or verb marker can optionally be accompanied by one form from the *bayd*ʸ*i* set and/or one form from the *gali* set (in that order).

transitive	*mundan* 'lead'	*galgan* 'leave'
	bundin 'take out'	*d^{y}aran* 'stand something up'
	band^{y}an 'follow'	*yuban* 'put (lying) down'

Any noun or adjective stem may be verbalised in three distinct ways: it may be transformed into an intransitive verbal stem by the addition of *-bin*, or into a transitive verbal stem by the addition of *-man* (causative) or *-(m)ban* (non-causative). For instance, *bulgan* 'big' can modify a noun in an NP within an intransitive sentence:

⟨5⟩ *balan d^{y}ugumbil bulgan banin^{y}u.* The big woman there is coming.

Or we can have a perfectly proper sentence involving just the noun, marker and adjective (there is no verb 'to be' in Dyirbal):

⟨6⟩ *balan d^{y}ugumbil bulgan.*   The woman there is big.

We can also have:

⟨7⟩ *balan d^{y}ugumbil bulganbin.*   The woman there has become big
       (i.e. has grown big recently).

The difference between ⟨7⟩ and ⟨6⟩ lies entirely in the active/stative contrast (in the traditional sense) between verbal form *bulganbin* in ⟨7⟩ and nominal form *bulgan* in ⟨6⟩. Addition of verbaliser *-bin* does not affect the semantic content of *bulgan*, but merely marks it as an intransitive verbal.

Allative ('to') and locative ('at') verb markers can be verbalised to *-bin* or *-(m)ban* forms (but not to causative *-man* forms); ablative ('from') markers cannot be verbalised. Verb markers can be verbalised whether or not they are augmented by bound forms. Thus we can have *yalubin, yaluguyabin, balubawalbin* and so on; but not **baŋumbin*. Thus:

⟨8⟩ *baladawala diban balbalin^{y}u yalibin*
       The stone a medium distance there upriver is rolling in this direction.

⟨9⟩ *balan guda n^{y}inan^{y}u balaybayd^{y}ibin*
       The dog there is sitting down a short distance there downhill.

In ⟨2, 4⟩ the verb markers are non-inflecting verbal qualifiers. In ⟨8, 9⟩ the verbalised markers must agree with the head verb in transitivity and tense, and can take the full range of suffixes open to verbs. Speakers of Dyirbal recognise a semantic difference between ⟨2, 4⟩ and ⟨8, 9⟩, but this cannot easily be brought out through English translation – for discussion see Appendix 2.

Now Guwal–Dyalŋuy correspondences include the following (the Guwal verbs are all nuclear):

	Guwal	Dyalŋuy
'go'	*yanu*	*bawalbin*
'motion uphill'	*wayn^{y}d^{y}in*	*dayubin*
'motion downhill'	*buŋan*	*bayd^{y}ubin*
'motion upriver'	*wandin*	*dawulubin*
'motion downriver'	*dadan^{y}u*	*balbulubin*
'motion across river'	*mabin*	*guyabin*

The Dyalŋuy correspondents each consist of a bound form and the intransitive verbalising affix *-bin*; these bound forms can not by themselves be verbalised in Guwal, although an allative or locative verb marker to which they are bound can.

From our assumption that a one-to-one Guwal–Dyalŋuy correspondence implies a relation of synonymy between the words concerned, and the fact that verbaliser *-bin* does not change the semantic content of a word, but just marks it as an intransitive verbal, it follows that the semantic content of *wayn^{v}d^{v}in* must be the same as that of *dayu*, and so on. (That is, leaving aside the feature 'long way' in the semantic description of *dayu*; verbs of motion do not specify distance and in fact Dyalŋuy correspondents involve the unmarked feature from the system {long, medium, short distance}.)

Verbs markers can be described in terms of semantic features as follows:

(1) a single system specifies 'here' or 'there' (there is no third 'heard but not seen' feature for verb markers, as there is for noun markers):

(2) a tree of systems specifies case inflections:

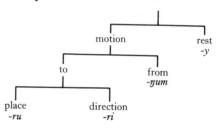

-ru and *-ri* are grouped together as 'to' since they (unlike 'from' and 'at') can cooccur with nouns and adjectives in allative inflection; 'to' and 'from' are grouped together, as 'motion', since one class of verbs normally occur only with motion markers, and another class normally occur only with rest markers.

(3) bound forms *-bayd^{v}i* etc. can be described by the tree:[a]

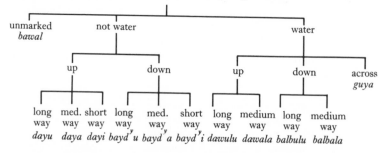

[a] We can give phonological realisations for the features in the tree, and thus generate ten of these forms (the exceptions are *bawal* and *guya*). 'Up' is associated with phonological specification *daS*, 'down' with *baLP*, where S, P and L stand for 'semivowel', 'plosive' and 'linking consonant (after a vowel and before an intersyllabic nasal and/or plosive)' respectively. 'Not water' is associated with a prosody of palatalisation and 'water' with one of labialisation; the prosodies extend over all segments not already fully specified, i.e. over S, P and L in the

The fact that simple verbs concerned with position[a] fall into two classes, 'motion' and 'rest', suggests that the semantic descriptions of these verbs should be taken to include the appropriate feature: 'motion' or 'rest'. From this and the Dyalŋuy data we can generate semantic descriptions:

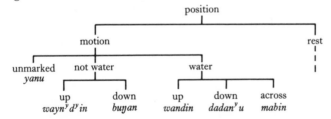

Verbs of 'rest' can be further distinguished through the widely-occurring – if not universal – system {stand, sit, lie}.[b]

We see that in the semantic descriptions of verb markers, the choice of a feature from the system {water, not water, unmarked} is independent of the choice made from system {motion, rest}. In the case of verbs, however, a feature from the system {water, not water, unmarked} can only be included in a semantic description together with the feature 'motion', and not with the feature 'rest'. That is, there are verbs specifying 'whereness' of motion but not 'whereness' of rest; whereas verb markers can specify 'whereness' of both motion and rest.

Looking now at the Dyalŋuy correspondents of *yanu* and *mundan*:

	Guwal	Dyalŋuy
'go'	*yanu*	*bawalbin*
'lead'	*mundan*	*bawalmban*

we see that we have the (non-causative) transitive verbaliser -(m)ban, plus *bawal*, for *mundan*, as against intransitive verbaliser -*bin* plus *bawal* for *yanu*. This would suggest that *yanu* and *mundan* have the same semantic content, and differ only in that *mundan* belongs to the class of transitive verbs and *yanu* to that of intransitive verbs.

We have thus been able to provide semantic descriptions of a number of verbs of motion entirely in terms of (1) features from systems which underlie grammatical categories in Dyirbal, and (2) specification of transitivity.

Dyirbal verbs are very strictly categorised as transitive or intransitive; there are

forms above. *S* plus palatalisation is the segment *y*; *S* plus labialisation is *w*. *P* plus palatalisation is *d^y*; plus labialisation is *b*. The linking consonant can be *y*, *l*, *r* or *ṛ*; *L* plus palatalisation is *y*. There is no specifically bilabial linking consonant, so that *L* plus labialisation specifies quite naturally the 'unmarked' linker *l*. Thus forms *day-*, *daw-*, *bayd^y-*, *balb-* are generated. 'Short', 'medium' and 'long distance are associated with -*i*, -*a* and -*u* respectively. (For a discussion of the extra affix -*lu* ~ -*la* on 'water' forms, of phonological marking, and of the full phonological structure of Dyirbal, see Dixon 1968a.)

[a] There is no obvious term to refer to both verbs of motion and verbs of rest; 'position' is used here as something that seems no worse than any other possibility.

[b] The semantic description of *yanu* – but not those of *wayn^yd^yin, buŋan, wandin, dadan^yu* and *mabin* – also involves the unmarked features 'to' and 'there', contrasting with the semantic description of *banin^yu* 'come' which involves 'to' and 'here'. The Dyalŋuy correspondent of *banin^yu* is *yalibin*, i.e. the verbalised form of the 'to (direction) here' marker (and to keep Dyalŋuy different from Guwal, *yali* – unlike other verb markers *bali, yalu, balu, yalay, balay* – does not occur in intransitively verbalised form in Guwal).

a number of pairs of verbs that appear to have the same or almost the same semantic content and to differ only in transitivity, but (with one exception: *ŋaban^yu* 'bathe'; *ŋaban* 'immerse in water') the members of such pairs are non-cognate. In almost all these cases there is a single transitive verb in Dyalŋuy; the reflexive form of this verb is given as Dyalŋuy correspondent of the intransitive member of the Guwal pair.[a] For instance:

		Guwal	Dyalŋuy
{	*tr* 'take out'	*bundin*	*yilwun*
	intr 'come out'	*mayin*	*yilwuyirin^yu*
{	*tr* 'stand [it] up'	*d^yaran*	*dindan*
	intr 'stand up'	*d^yanan^yu*	*dindayirin^yu*
{	*tr* 'follow [someone]'	*band^yan*	*d^yumbamban*
	intr 'follow'	*marin*	*d^yumbambayirin^yu*

(see also the discussion of these examples in §3.3). The Dyalŋuy data provides clear evidence that *d^yanan^yu* and *d^yaran* have the same semantic content, and differ only in transitivity; and similarly in the other cases.

Initially *dayubin* was given as correspondent not only of *wayn^yd^yin* 'motion up, usually uphill' but also of *bilin^yu* 'climb a tree (without any aid)' and of *bumiran^yu* 'climb a tree with the help of a length of loya vine (*gamin*)'.[b] At a later stage of elicitation the informant volunteered fuller correspondents:

Guwal	Dyalŋuy
wayn^yd^yin	*dayubin*
bilin^yu	*dayubin danduŋga*
bumiran^yu	*dayubin d^yuyibila*

where *dandu* corresponds to Guwal *yugu* 'tree, stick' and *d^yuyi* to Guwal *gamin* 'loya vine'. In the case of *bilin^yu*, *dayubin* is qualified by *dandu* in locative inflection (i.e. 'climb [at] a tree'); for *bumiran^yu*, *d^yuyibila* 'with a *gamin*' is specified as qualification of the subject of the verb.

Now we could extend the tree on page 447 by including an extra system $\{x_1, x_2, x_3\}$:

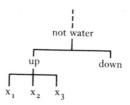

where x_1 refers to climbing a tree with a *gamin*, x_2 to climbing a tree (without a *gamin*) and x_3 is the unmarked feature; in fact, *wayn^yd^yin* can either have a generic meaning, including the particular meanings of *bilin^yu* and *bumiran^yu*, or else a more specific meaning, complementary to those of the other two words ('climb something other than a tree, e.g. a hill, steep bank or cliff'). However, x_1, x_2 and x_3 would be entirely ad hoc: these features would not recur in the semantic descriptions of any

---

[a] The addition of affix *-yiriy* ~ *-mariy* to a transitive stem produces a reflexive form, that functions like an intransitive stem.

[b] For a full account (and picture) of the method of tree-climbing with a *gamin* implement, see Lumholtz 1889, 89–90.

other words. We would not be providing any semantic insight by setting up a three-feature system merely semantically to distinguish three words. And in fact *yugu* and *gamin* are 'nuclear' nouns, dealt with by the semantic description of the class of nouns; *dʸuyi* and *dandu* are in normal syntactic relations with *dayubin* in the Dyalŋuy correspondents. We can thus take *waynʸdʸin* as a nuclear verb, with semantic description (position, motion, not water, up; intransitive) and *bilinʸu* and *bumiranʸu* as non-nuclear verbs, semantically defined in terms of the semantic descriptions of *waynʸdʸin* and of two nouns, through established grammatical relations. There is no need to set up the ad hoc system $\{x_1, x_2, x_3\}$. (However, we cannot dispense with any of the features in the tree on page 447; *waynʸdʸin*, *mabin*, etc. can *not* be defined in terms of any other words; and in any case these features do occur both in the semantic descriptions of a number of lexical words, and within the grammar.)

The examples above have shown how Guwal–Dyalŋuy correspondences provide an immediate rationale for the division into nuclear and non-nuclear verbs; and for the techniques of componential description of nuclear words and of definition of non-nuclear words.

2.4 PROCEDURE FOLLOWED

An account of the exact procedure followed in eliciting data from informants will provide important evidence for the recognition of which Guwal verbs fall into the nuclear set. The procedure involved two stages: Guwal–Dyalŋuy, then Dyalŋuy–Guwal:

(1) First, each Guwal word in turn was put to the informant and he was asked for its Dyalŋuy equivalent. A spontaneous reaction was looked for here – the informants were encouraged to say the first thing they thought of, and not to think hard checking if it was correct, or complete enough. Most often a single word correspondent was given; sometimes this was qualified by a number of other words (in grammatical relation to it).

(2) A card was then made out for each Dyalŋuy word, and each Guwal word that the Dyalŋuy word had been given as correspondent for was listed on the card. Cards contained from one to about twenty Guwal words. The informants were then asked each Dyalŋuy word in turn, and asked to give its Guwal equivalent. In each case a single Guwal word was given as the main (or 'central' – indicated by C) equivalent. The informant was asked for any other Guwal equivalents, and the words on the card checked to see that this was their correct Dyalŋuy correspondent (errors made at stage 1 were corrected here). Taking each Guwal word in turn, the informant was then asked to add something to the Dyalŋuy word in order to distinguish in Dyalŋuy between the Guwal words; he was at this stage encouraged to take his time and think to get a full and correct correspondent.

For example, in stage 1 *wuyuban* had been given as Dyalŋuy correspondent of, amongst others:

buwanʸu 'tell'
dʸinganʸu 'tell a particular piece of news'
gindimban 'warn'
ŋaran 'tell someone one hasn't a certain thing (e.g. food) when one has'

In stage 2 the informant was asked to verify that *wuyuban* was a bona fide Dyalŋuy item; he was then asked how he would render *wuyuban* in Guwal. There were at this stage a number of possibilities open to the informant. He could have said '*buwan^{y}u, d^{y}ingan^{y}u, gindimban, ŋaran...*', giving all the items for which he had said *wuyuban* in stage 1; or he could have just mentioned *one* of the Guwal verbs. In fact his response was of the second type: he simply gave *buwan^{y}u* as the equivalent of *wuyuban*; and he had to be prompted – in the second stage – to say that *d^{y}ingan^{y}u, gindimban, ŋaran*, and so on, were also Guwal equivalents of Dyalŋuy *wuyuban*.

The writer then went into the second stage of elicitation with the other main informant. Exactly the same results were obtained – again just *buwan^{y}u* was given as Guwal correspondent of *wuyuban*. And the same thing happened for every one of the Dyalŋuy verbs. Although from two to twenty verbs were listed on each card – those Guwal items for which the Dyalŋuy verb had been given as correspondent in stage 1 – each informant gave just one of these as Guwal equivalent in stage 2; and in each case the two informants picked *the same item*.

It is fairly obvious what is happening here. Each card contained one nuclear Guwal verb and a number of non-nuclear verbs. When the nuclear Dyalŋuy verb was put to an informant, he *always* chose the nuclear Guwal verb as its correspondent, *never* one of the non-nuclear verbs. These results provide not only justification for the nuclear/non-nuclear distinction, but also a procedure for checking which of the everyday language verbs are nuclear.

The writer next checked with the informants that Guwal verbs *buwan^{y}u* and *d^{y}ingan^{y}u* did differ in meaning, despite the fact that they had both been rendered by the same Dyalŋuy item, *wuyuban*, in stage 1. The informant was asked how this difference in meaning could be expressed in Dyalŋuy, if it were necessary to do so. He replied that *buwan^{y}u* would just be rendered by *wuyuban* but that *d^{y}ingan^{y}u* would be expressed by *wuyuwuyuban*, with the verb reduplicated. Verbal reduplication in Dyirbal means 'do it to excess'; thus *wuyuwuyuban* perfectly conveys the meaning of *d^{y}ingan^{y}u* – calling everyone to gather round as one *rather deliberately tells* some story or news item. Similarly, when confronted by *buwan^{y}u* and *gindimban*, the informant said that for *buwan^{y}u* the Dyalŋuy translation would be just *wuyuban*, but that for *gindimban* it would be *n^{y}ungulmban wuyuban*, involving a transitively verbalised form of the number adjective *n^{y}ungul* 'one'. *n^{y}ungulmban wuyuban* is literally 'tell once', an adequate 'definition' of *gindimban* within the context of Dyirbal culture. In the case of the pair *buwan^{y}u* and *ŋaran* the informant again gave just *wuyuban* for *buwan^{y}u* but volunteered *wuyuban d^{y}ilbuŋga* for *ŋaran*. *d^{y}ilbu* means 'nothing' and *-ŋga* is the locative inflection; *wuyuban d^{y}ilbuŋga* is literally 'tell concerning (i.e. that there is) nothing'. At this stage the field notebook read:

> *wuyuban*
>
> | C *buwan^{y}u* | *wuyuban* |
> |---|---|
> | *d^{y}ingan^{y}u* | *wuyuwuyuban* |
> | *gindimban* | *n^{y}ungulmban wuyuban* |
> | *ŋaran* | *wuyuban d^{y}ilbuŋga* |

Effectively, the non-nuclear verbs *d^{y}ingan^{y}u, gindimban* and *ŋaran* had been defined in terms of nuclear *buwan^{y}u/wuyuban*; nothing had been added to the nuclear verb when distinguishing between Guwal pairs in Dyalŋuy – each time *buwan^{y}u* had been left simply with correspondent *wuyuban*. The same results were obtained for all the cards, for both informants. The everyday nuclear verb was always left with just the mother-in-law verb as correspondent, and 'definitions' were given for the non-nuclear verbs. The 'definitions' can be seen to be of different syntactic types – for a full account see §3.2. Informants sometimes gave identical definitions for a non-nuclear verb, but often rather different ones. However, they agreed in *always* giving *some* definition for a non-nuclear verb, and *never* attempting one for a nuclear word.

Since Dyalŋuy has not been spoken much since 1930 it sometimes required an effort for informants to remember the less frequent Dyalŋuy words. For instance, *dʸubumban* was given for *nʸuganʸu* 'grind', before *yurwinʸu* was remembered. *yurwinʸu* and *nʸuganʸu* are in one-to-one correspondence; *dʸubumban* is the correspondent of *bidʸin* 'punch (generally: hit with a rounded object)', *dudan* 'mash', *dʸilwan* 'kick, or shove with knee' and *dalinʸu* 'fall on top of' (see §3.3). The initially suggested correspondence *dʸubumban* for *nʸuganʸu* did serve to show the affinity, for speakers of Dyirbal, between the concepts of 'grind' (as in *yurwinʸu*) and of 'deliver a blow with a rounded object' (as in *dʸubumban*). In many other instances initial spontaneous 'incorrect' correspondents – that were later withdrawn or amended as a more specific Dyalŋuy word was remembered – served to provide valuable insights into the workings of informants' minds (their internalised semantics); clues to semantic relationships from this source were almost always reinforced from other data, and could then be used in assigning semantic descriptions.

An argument between the main Mamu informant and his wife, a speaker of the Dyirbal dialect, over the correct Dyalŋuy correspondent of *darbin* 'shake [something] off a blanket' provided additional evidence for the very general Dyirbal concept 'set in motion in a trajectory'. The concept is further specified as either 'set in motion in a trajectory, leaving go of (= throw)', Dyalŋuy *nayɲun*, or else 'set in motion in a trajectory, holding on to (= shake, wave or bash)', Dyalŋuy *bubaman*. The Mamu informant suggested *bubaman* for *darbin*, on the grounds that the blanket is held onto; his wife preferred *nayɲun* since the crumbs (or whatever is being shaken off) do leave the blanket. *darbin* was clearly identified by both speakers with the general concept 'set in motion in a trajectory' but they interpreted the action involved from different points of view, in applying the 'hold on to/let go of' criterion, in order to decide on the Dyalŋuy correspondent.[a]

2.5 SEMANTICS AND SITUATION

There are two sides to linguistic meaning: sense and reference. Taking Lyons's definitions, the *sense* of a word is 'its place in a system of relationships which it contracts with other words in the vocabulary'. By semantic description, in this paper, we mean sense description; a word is semantically related to other words through their sense descriptions involving common features or else different features from the same system, or through their being defined one in terms of another or all in terms of some other, nuclear, word. *Reference* is 'the relationship which holds between words and the things, events, actions and qualities they "stand for"' (Lyons 1968.427, 424).

We have suggested sense descriptions in terms of systems of semantic features. Now each feature has reference, but this reference is *relative to* the reference of the other features in the system. Using 'situation' to refer to non-language

---

[a] The object of *madan/nayɲun* is normally 'that which is thrown' and of *baygun/bubaman* 'that which is shaken'. Since the object of *darbin* is normally 'the blanket' (or whatever is shaken) and not 'the crumbs' (or whatever is shaken off), *bubaman* is probably the correct Dyalŋuy correspondent.

phenomena, we can talk of the relative 'situational realisations' (rather than reference) of the features in a semantic system; this has some similarities to the statement of phonetic realisations for phonological features.[a] For example (oversimplifying a little), *balgan* 'hit with a stick or other long rigid object, held in the hand' and *bund^{y}un* 'hit with the hand (unclenched) or with a long flexible object such as a bramble or strap [held in the hand]' differ only in the features 'long rigid implement' and 'long flexible implement', having in common the features 'affect by striking' and 'held onto'.[b] The situational realisations are relative: *balgan* implies a *more* rigid object and *bund^{y}un* a *less* rigid one. The realisation of *balgan* cannot just be given, in isolation, in terms of 'rigid object'; but since 'rigid object' *is* specified in the semantic description of *balgan* it is implied that a *different* word is used in the case of relatively '*non*-rigid objects', i.e. *bund^{y}un*. That is, *balgan* and its situational realisation are, on one dimension, complementary to *bund^{y}un* and its situational realisation; the situational realisation of *balgan* can only significantly be given *with respect to* the situational realisation of *bund^{y}un*.

Situational realisations are stated for elements in the semantic description of Dyirbal. Given the semantic description of a sentence (together with its interrelating grammatical description) we can provide a situational interpretation, that specifies the types of events that the sentence could refer to. We cannot work the other way round: given a situational description, we cannot from this produce a semantic description. That is, we can have 'situational interpretations' of semantic descriptions but *not* 'situational feed-in' to semantics, generating a suitable semantic description. The reason for this lies partly in the fact that a particular situational phenomenon may be, in different instances, the realisation of different semantic categories (that is, there is overlapping of situational realisations) depending on (1) the situational environment, and/or (2) the semantic oppositions in a particular text. Taking these in turn:

(1) a certain intensity of gaze – measured perhaps by size of eye and pupil, rate of blinking, time focussed on a single object – could realise either *buɾan* 'look' or *ŋaɾn^{y}d^{y}an^{y}u* 'stare' depending on the situational environment. An intensity of gaze that in a casual, informal setting might be considered a little piercing could well be the realisation of *ŋaɾn^{y}d^{y}an^{y}u* relative to a less intense look as realisation of *buɾan*. But in a more formal, ceremonial say, setting, this might be a quite usual intensity and be the realisation of *buɾan*, relative to a *really* intense stare being the realisation of *ŋaɾn^{y}d^{y}an^{y}u*.

(2) *n^{y}inan^{y}u* has a generic meaning 'stay (in camp)' (whatever the various postures adopted – sitting, standing, lying) as opposed to motion verbs such as *yanu* 'go'; and also a particular meaning 'sit' as opposed to *d^{y}anan^{y}u* 'stand' and *bungin* 'lie'.

---

[a] For instance, in Dyirbal [ɛ] can be the phonetic realisation of either /a/ or /i/ – of /a/ in a palatal environment and of /i/ in a non-palatal environment. Only statements of relative realisation can be given for the members of a phonological system. Thus we can only say that the realisation of /i/ is relatively less open than the realisation of /a/ – that is, in a particular environment /i/ will be realised by a less open vowel than /a/. But the most open realisation (over all possible environments) of /i/ is in fact more open than the least open realisation (over all possible environments) of /a/.

[b] More strictly, *balgan* implies 'hit with a long rigid object, held on to', *minban* 'hit with a long rigid object, let go of', and *bun^{y}d^{y}un* 'hit with a long flexible object'; *bun^{y}d^{y}un* does not involve any specification 'held on to' as opposed to 'let go of' but in fact the long flexible objects used by the Dyirbal are always held onto since they would be ineffective thrown.

Thus:

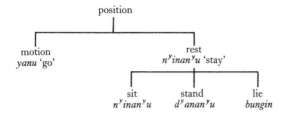

The semantic description of $n^y inan^y u$ is (position, rest; intr) for its generic sense; (position, rest, sit; intr) for the more specific sense. Now suppose that everyone, bar one man and one woman, is leaving a camp; the man is standing watching the others leave. An observer could appropriately say *balamangan yanu/bayi yaṛa n^y inan^y u* 'they are all going but the man is staying', comparing the events of the people leaving, and the man remaining in the camp. Or else the observer could equally appropriately say *balan d^y ugumbil n^y inan^y u/bayi yaṛa d^y anan^y u* 'the woman is sitting down and the man standing up', comparing the events of the woman sitting in the camp and the man standing there. Here a single event – the man standing – can be the realisation of $n^y inan^y u$ (relative to the realisation of *yanu*) or else of $d^y anan^y u$ (relatives to the realisation of $n^y inan^y u$), depending entirely on the semantic oppositions in the text.[a]

It will be seen from these examples that it is not possible to take an isolated situational event, and demand that the linguistic description of a language should be able to indicate a language description of it. (Thus the view that language essentially works by being given 'real events' and providing word pictures of them is, on the most literal level, erroneous.) A particular correspondence between a situational event, and an occurrence of a piece of language, is relative to (and determined by) the grammatical and semantic systems of the language, the semantic oppositions in the text to which the piece of language belongs, and the '(anthropological/sociological) situational system' of the culture (such as that dealing with 'formal' and 'informal' settings, and so on). What can be stated is a range of (relative) situational realisations for a semantic category, and thus for a sentence. This range can be reduced by considering the particular semantic oppositions in a text that includes the sentence. It may be still further reduced by considering the description of the situation in which the text occurs, or to which it refers (if it is a narrative, say) and so on, in terms of some suitable theory of situational organisation (if such a theory should ever be put forward).

2.6 RANGE OF MEANING, AND NON-UNIQUENESS OF DEFINITIONS

The situational realisation of a nuclear verb will be approximately the sum of the situational realisations of its component features. But only approximately – each

---

[a] The semantic oppositions here (*yanu* contrasting with $n^y inan^y u$, or $n^y inan^y u$ contrasting with $d^y anan^y u$) are *contrastive correlations* of the sort discussed in detail in Dixon 1965.111–4 ff. A text in any language can be exhaustively analysed into *contrastive* and *replacement correlations* that specify the semantic form and plot of the text. Such textual correlations involve limited selection from the full semantic possibilities of the language. A considerable quantity of Dyirbal text material was analysed into contrastive and replacement correlations and the results used – together with Dyalŋuy and other types of data – in deciding upon correct semantic descriptions. For more details see Dixon 1968a.251–4; and see also n. *a*, p. 456 and n. *a*, p. 468, below.

word also has its own character, its own special personality traits as it were, that mould from the meanings of the constituent features a distinctive concept. (It will thus be seen that semantic features are, to some extent, slightly idealised generalisations.)

Each word has a *range* of situational realisations.[a] It may have some 'central' realisation, and be then extended to cover situational phenomena that are situationally similar to the 'central' phenomenon. In this way the semantic system of a language can cover all relevant situational patterns, first focussing on 'central' patterns and then extending outwards from these to cover the intermediate territory. 'Ambiguous' types of situational phenomena, that can be covered by extension from two different foci, say, may be the subject of taboo (see Douglas 1966, Leach 1964).[b] Central phenomena seem, empirically, most likely to be concerned with aspects of humans and human behaviour; phenomena relating to geographical features and happenings, and to artifacts, are often dealt with as extensions from central phenomena.

For instance, neither Dyirbal nor its neighbour language Mbabaram have any verbs whose central situational realisation involves rain falling. Certain verbs are used to refer to rainfall (as a non-central part of their meaning) according as the action of rain falling seems situationally similar to the central situational realisation of the verb. Dyirbal is spoken in an area of devastatingly heavy rainfall (150 in. per year) whereas the area in which Mbabaram is spoken, although only 50 miles distant, has a rainfall of only about 30 in. per year. Dyirbal has two verbs, both transitive, used to refer to rainfall: *bid^{v}in* 'hit with a rounded object, e.g. punch, throw a stone at' for the extremely heavy wet season storms, and *d^{v}indan* 'gently wave or bash, .e.g. blaze bark, sharpen pencil' for the only moderately heavy storms throughout the rest of the year. Mbabaram, however, refers to rainfall exclusively through the intransitive verb *aɲanuŋ* 'fall down'. The different types of rainfall in the two regions lead to situational analogies being established to quite different 'human actions', that are the central realisations of dissimilar verbs in the two languages.

The referential relation between a nuclear word and corresponding non-nuclear words appears to be of the same type as that between the central meaning of a word and its extensional meanings. First we have nuclear words, whose central situational realisations are as it were the 'major foci'. Then non-nuclear words, whose central situational realisations are minor foci, each related to (that is, a defined extension from) some major focus. Now all situational phenomena can be described in terms of these major and minor foci; a point in situational space which is not itself a focus is treated as an extensional or peripheral realisation of that word whose focus it is situationally most similar to.

[a] For further discussion of 'range of meaning' see McIntosh 1961, Nida 1964. Bloomfield (1933.149) uses the terms *normal* (or *central*) and *marginal* (*metaphoric* or *transferred*) meaning.

[b] For instance, Leach (1964.40–2) mentions that speakers of English classify animals in certain ways, and on the basis of this linguistic classification decide what to eat and what not to eat. Leach considers the basic discrimination to rest in three words: (1) *fish*, creatures that live in water – this category is fairly elastic, including shell-fish; (2) *birds*, two-legged animals with wings that lay eggs (they need not necessarily fly); (3) *beasts*, four-legged mammals living on land. All the creatures considered edible by us are fish, birds or beasts. There is a large residue (reptiles and insects) rated as not food by us, although rated as food by many other peoples. The hostile taboo is applied most strongly to creatures that are most anomalous in respect of the major categories, for example, snakes – land animals with no legs that lay eggs.

For instance, two of the nuclear verbs in Dyirbal are *bid^yin* 'hit with a rounded object' and *baygun* 'shake, wave or bash'. The componential semantic descriptions of these two verbs are (affect, strike, rounded object; tr) and (position, affect, motion, direction, hold on to; tr) respectively. There is a non-nuclear verb *d^yindan* 'gently wave or bash' that is defined semantically in terms of the semantic description of *baygun* and the specification 'with agent exercising full control over the action' or something similar. The situational phenomenon of rain falling is not a major or minor focus – that is, it is not the central realisation of any verb, nuclear or non-nuclear. The two different kinds of rainfall are, to the Dyirbal, most similar to the central realisations of nuclear *bid^yin* and non-nuclear *d^yindan* respectively, and reference to rainfall can thus be looked upon as secondary or extensional meanings of these two verbs.

We referred above to Leach's and Douglas's discussion of phenomena which have some degree of similarity to two different foci; such a phenomenon cannot be classified as clearly an extension from a single focus and may as a result be considered taboo. Something of the same sort applies for non-nuclear words, considered as 'extensions' of nuclear words (although in this case, of course, there is no taboo attached). A word may be clearly non-nuclear, but its reference may be as it were 'equidistant' from a number of major foci: that is, it can be adequately defined in several different ways, each in terms of a different nuclear word. The next two paragraphs illustrate this with Dyalŋuy data, involving first a non-nuclear Guwal verb, and then a noun.

The situational realisation of the non-nuclear verb *wirban* 'sweep ground with bramble stick or weed to clear away leaves' appears to be equally similar to several major foci. Three different mother-in-law correspondents were given at different times, involving three different nuclear verbs – Dyalŋuy *wirygan^yu* corresponding to Guwal nuclear verb *giban^yu* 'scrape, trim' with semantic description (affect, strike, sharp-bladed implement, low degree; tr); Dyalŋuy *burganman*, Guwal nuclear *digun* 'clear away, rake away, dig – using hands – on the surface', semantic description (position, affect, motion, place, from, here; tr); and Dyalŋuy *nayŋun*, Guwal nuclear *madan* 'throw', semantic description (position, affect, motion, direction, let go of; tr). Thus the assessment of similarity between the realisation of *wirban* and some major focus picks on either the action of the bramble stick on the ground (in the 'scrape' case), or what the bramble stick does to the leaves (in the 'clear away' – i.e. move position away from here – case), or the way in which the stick does this to the leaves (in the 'throw' case). The action has in Guwal a single appropriate name, *wirban*, but in Dyalŋuy it can be described equally well by either *wirygan^yu*, *burganman* or *nayŋun*.

There are only around a quarter as many nouns describing types of trees in Dyalŋuy as in Guwal. One Dyalŋuy noun corresponds to three or four Guwal nouns, whose referents all have the same sort of grain, or the same type of sap, or the hardest wood, or can be used for making shields, or water-bottles, and so on. The seven Guwal nouns for trees bearing fruit that requires preparation (roasting, grinding, soaking, etc.) all have one-to-one Dyalŋuy correspondents. But there is a single Dyalŋuy noun, *gumulam*, for the two dozen or so Guwal nouns referring to trees whose fruit can be eaten raw. Now the moreton bay fig tree, *magura* in Guwal, has fruit that can be eaten raw, and so can be described as *gumulam* in Dyalŋuy. But shields are made from the wood of this tree and so it can also be called *gigiba* in Dyalŋuy, as can the slippery blue fig, Guwal *milbir*. Thus this type of tree has

in Guwal a single appropriate name, but in Dyalŋuy it can be described equally
well by either *gumulam* or *gigiba*.[a]

These examples suggest that a non-nuclear word *need not have a unique definition*.
That is, it may be possible to provide several definitions, in terms of different
nuclear words, each of which gives a sufficient specification of the semantic content
of the non-nuclear word. Such cases will probably not be too common – most non-
nuclear words have just one obvious definition. Where several definitions *do* seem
equally appropriate it would probably be best to include them all in the semantic
description of the word, even though any one would theoretically be sufficient.

There may also be cases where several definitions are possible for a non-nuclear
word, and where no one definition is a sufficient semantic description;[b] the word
will only be fully specifiable semantically by the logical conjunction of two or more
definitions. No examples of this have yet been encountered but it seems rather likely
that they will be, as this method of semantic description is applied more widely.

We conclude then that there may not be just one unique definition that fully
specifies the semantic content of a non-nuclear word. The semanticist's task is to
investigate possible definitions and to give that one that is clearly the most appropriate, or if several seem equally appropriate to state them all. Detailed mother-in-
law correspondences for non-nuclear verbs were often the optimum (or one of the
optimum) definition(s); examples of types of mother-in-law definition are given
in §3.2 (It should be noted that, unlike non-nuclear verbs, nuclear verbs do appear
each to have a single possible semantic description.)

3 Application to Dyirbal verbs

§3.1 discusses some of the systems involved in the semantic descriptions of nuclear
verbs; §3.2 gives examples of some of the types of definition of non-nuclear verbs;
finally, §3.3 discusses criteria for grouping verbs, and gives some examples of semantic description.

3.1 SOME SEMANTIC SYSTEMS

There follows a list of some of the systems underlying the semantic descriptions of
nuclear verbs. The first six systems are needed for the grammatical description
of Dyirbal, and also underlie one or more lexical verbs. If one feature can be regarded as unmarked then it is given in SMALL CAPITALS; for instance 'THERE' is
unmarked with respect to 'here', since *yala*- verb and noun markers always imply
'here' whilst *bala*- markers imply either 'there', or else that no information is given

---

[a] In fact, the name used in Dyalŋuy to describe this tree is likely to depend on the context of
reference: *gumulam* would be used if the reference to the tree concerned its fruit, but *gigiba*
if its wood were being talked about. Similarly, the choice of one of the three possible Dyalŋuy
descriptions for the action of 'sweeping the ground with a bramble stick or weed to clear away
leaves' (see preceding paragraph) is likely to be determined by the contextual correlations
within a particular text (see Dixon 1965, and n. *a*, p. 453, above, and n. *a*, p. 468, below).
Dyalŋuy data does not indicate any nuclear/non-nuclear distinction amongst nouns, as it does
with verbs; instead it simply reveals the underlying semantic taxonomy of nouns. But the
data reported in this paragraph, concerning *magura*, seems to involve the same type of non-
uniqueness – definition of a specific Guwal term in terms of either of two generic Dyalŋuy
words – as does that in the previous paragraph concerning *wirban*.

[b] A semantic description of a word is sufficient if it applies to that word and no other; a
non-sufficient description is one that is equally applicable to two or more words.

as to the location/visibility of the referent of the noun (and *bala-* forms are approximately ten times as frequent as *yala-* ones); and so on. For some systems there is no evidence indicating that one feature should be regarded as unmarked with respect to the others.

(1) {here, THERE}. This underlies the semantic descriptions of noun and verb markers (interrelating with system {seen, unseen} in noun markers) and of verbs *yanu* 'go' (i.e. 'to' plus 'there') and *banin^{y}u* 'come' (i.e. 'to' plus 'here').

(2) {from, TO}. This underlies the ablative and allative case inflections on verb markers, nouns and adjectives; and the semantic descriptions of *yanu* and *banin^{y}u*, which involve specification of 'to' (plus 'THERE'/'here'), as against *wayn^{y}d^{y}in* 'motion uphill', *waymban^{y}u* 'go walkabout (i.e. unspecified motion)' etc. that do not. It is also related to 'go in' and 'come out' – which could be described as 'TO' plus 'interior' and 'from' plus 'interior' – in nuclear *gundan* 'put in' and *buŋgan* 'empty out'. 'Go in' also serves to distinguish non-nuclear *ṛuygin^{y}u* 'look in at' from nuclear *buṛan* 'look at'.

(3) {motion, REST}. This underlies the semantic descriptions of verb markers and of all verbs of motion and of rest (see §2.3). It also serves to distinguish non-nuclear *bilan* 'give by taking or sending' from nuclear *wugan* 'give (without moving)'

(4) {up, down}. This system underlies the semantic description of some of the bound forms *bayd^{y}i* etc., and of certain corresponding verbs of motion (see §2.3). 'Up' also serves to distinguish non-nuclear *waban* 'look up at' from nuclear *buṛan* 'look at'.[a]

(5) {seen, unseen}. This system underlies the semantic descriptions of noun markers, and of nuclear verbs *buṛan* 'see, look at' and *ŋamban* 'hear, listen to'; and of non-nuclear verbs *d^{y}urgan^{y}u* 'spear, with the actor holding onto the spear and being able to see what he is spearing' and *wagan^{y}u* 'spear, with the actor holding onto the spear and being unable to see what he is spearing'.

(6) {water, NOT WATER}. This system is contained in the system {unmarked, not water, water} set up for *bayd^{y}i* etc. forms (§2.3): the larger system could be written {unspecified, {unmarked, water}}. That is, *bawal* is unmarked with respect to *dayu* and *dawulu*; of the latter two *dayu* is unmarked with respect to *dawulu*. Thus *dawulu* only refers to 'upriver' but *dayu*, although its central meaning is 'uphill', can also refer to 'up a cliff', 'up a tree' or 'up in the air'. The same system underlies the semantic descriptions of certain verbs of motion – *wayn^{y}d^{y}in* etc. (§2.3); it also distinguishes *yu ŋaran^{y}u* 'swim' from *waymban^{y}u* 'go walkabout' and *walŋgan* 'float' from *n^{y}inan^{y}u* 'stay' (§3.3). The same system distinguishes non-nuclear verb *d^{y}inban* 'spear, throwing the spear, at something in the water' from *bagan* 'spear, throwing the spear, at something on land'.

(7) {leave go of, hold on to}. This system is not required for the grammatical description of Dyirbal but distinguishes nuclear *madan* 'throw' from *baygun* 'shake, wave or bash'; nuclear *minban* 'hit with thrown long rigid object' from *balgan* 'hit with long rigid object held in the hand'; and – together with systems {seen, unseen} and {water, NOT WATER} it distinguishes non-nuclear *d^{y}urgan^{y}u*, *wagan^{y}u* 'spear, holding on to the spear' from *d^{y}inban*, *bagan* 'spear, throwing the spear'.

---

[a] Generally, there is no reason for considering either 'up' or 'down' as the unmarked feature in this system. However, in the particular context of 'look', 'down' appears to be unmarked; *buṛan* has the meanings 'look, look down' whereas there is a separate verb *waban* 'look up'.

3.2 TYPES OF DEFINITION

Definitions of non-nuclear verbs can either be (1) in terms of a single nuclear verb, or (2) in terms of two or more nuclear verbs, or (3) in terms of a verbalised noun or adjective. In some cases informants gave mother-in-law correspondences which only defined a part of the meaning of a non-nuclear verb; occasionally they had to resort to mime to complete the specification of a meaning.

(1) *Definition in terms of a single nuclear verb*, that receives some special modification or qualification, or some reassignment of syntactic functions.

(i) addition of a grammatical particle

Nuclear *gayban* 'call': Dyalŋuy correspondent *mayban*[a]

Non-nuclear *ŋaɽin* 'answer': *ŋurinᵞdᵞi mayban* 'call in turn'

(ii) reduplication (usually implying that the action is performed to excess)

Non-nuclear *yadᵞan* 'speaker asks someone to accompany him': *bunman*

Non-nuclear *gurŋalman* 'speaker keeps on asking a person to accompany him when the person has already refused (the person being someone who is a friend of the speaker, and normally accompanies him)': *bunmabunman*

(iii) addition of an aspectual affix[b], for instance *-ganiy*, implying that the event is repeated, at longish intervals

Nuclear *midᵞun* 'take no notice of': *nᵞanᵞdᵞun*

Non-nuclear *budᵞilmban* 'completely ignore (someone who is trying to attract one's attention)': *nᵞanᵞdᵞulganinᵞu*

(iv) addition of a derivational affix:

(*a*) reflexive – see examples in §2.3, §3.3.

(*b*) reciprocal affix *-(n)bariy* (plus obligatory reduplication)

Nuclear *wugan* 'give': *dᵞayman*

Non-nuclear *munᵞdᵞan* 'divide': *dᵞaymaldᵞaymalbarinᵞu* (lit 'give to each other')

(*c*) causative affix *-m(b)al*

Nuclear *yilmbun* 'pull, pull up': *yilwun*

Non-nuclear *yiwulumban* 'pull branch down so that fruit can be pulled off it': *yilwulman*

(v) addition of an adverbal

Nuclear verb *bidᵞin* 'hit with rounded implement': *dᵞubumban*

Nuclear adverbal *ŋunin* 'start doing, try, test': *ŋuɽbin*

Non-nuclear verb *ɽuldᵞun* 'test by banging, e.g. bang the reverse of a tomahawk head on a log to test for hollowness': *ŋuɽbin dᵞubumban*

(vi) addition of locational specification

Nuclear *mabin* 'cross river': *guyabin*

Non-nuclear *balŋgan* 'cross by walking on log thrown across river': *dalmbiɽaru guyabin*

(*dalmbir* is the Dyalŋuy correspondent of Guwal *yugu* 'tree, stick, log'.

---

[a] Throughout these definitions, a Guwal verb is given on the left, and its Dyalŋuy definition on the right of the colon.

[b] In the wide sense of 'aspectual' (here, covering all non-inflectional affixes that do not change the syntactic status of a stem); there are four such affixes in Dyirbal: *-ganiy* 'event repeated at longish intervals', *-dᵞay* 'event repeated many times within short time span', *-yaray* 'start to do/do a bit more' and *-galiy* ~ *-(n)bal* 'do quickly'. Informants gave definitions involving the first three of these affixes.

The locative inflection is *dalmbiɽa*, and *-ru* is an additional suffix indicating motion, thus *dalmbiɽaru* 'along a log'.)

(vii) addition of some specification of an NP in (subject or object) grammatical relation to the verb

(*a*) specification of NP head (Mamu dialect example)

Nuclear *buŋgan* 'empty out, capsize': *yiyaban*

Non-nuclear *buybun* 'spit': *ŋudʸuŋgu yiyaban*

(Guwal *nʸumba* 'spittle' has the Dyalŋuy correspondent *ŋudʸu*; *-ŋgu* is the ergative/instrumental inflection.)

(*b*) specification of NP modifier – see *bumiranʸu* in §2.3

(*c*) specification of relative clause qualifying NP head

Nuclear *dʸuran* 'rub': *dʸurmbayban*

Non-nuclear *yidʸin* 'an aboriginal doctor rubbing a sick person (often using sweat from his armpits) to cure him': *maŋgaybiŋu dʸurmbayban* (*maŋgay* is the Dyalŋuy correspondent of Guwal *wuygi* 'no good'; thus intransitively verbalised *maŋgaybin* 'feel ill' from which is formed the relative clause *maŋgaybiŋu* 'who feels ill'.)

(viii) reassignment of syntactic function.

For nuclear *buwanʸu* 'tell' the subject is the teller and the object the person to whom the news is told; for non-nuclear *dʸinganʸu* the subject is again the teller but the object is what is being talked about (the person to whom it is told can, optionally, be included as indirect object). Both were given Dyalŋuy correspondent *wuyuban* in stage 1 of elicitation, and in stage 2 *dʸinganʸu* was further specified as *wuyuwuyuban* 'tell to excess' (§2.4). In this case Dyalŋuy exposes the difference in semantic content ('tell *some particular news or story*') but not the difference in syntactic function (what is object and what indirect object). See also the discussion of 'follow' and 'lead' in §3.3.

(ix) addition of a semantic feature

In addition to the types of definition shown by Dyalŋuy correspondences, a non-nuclear Guwal verb can also be defined in terms of a nuclear verb plus a single semantic feature. In such cases Dyalŋuy correspondents cannot reflect the situation accurately: for instance, the semantic description of *waban* 'look up' is that of *buɽan* 'look' (C-equivalent of Dyalŋuy *nʸuɽiman*) plus that of *gala* 'up'; however *nʸuɽiman gala* is not a permissible VP (since *gala* is a bound form that can only occur with a verb marker – see n. *b*, p. 443): informants gave *nʸuɽiman yalugalamban* as Dyalŋuy correspondent of *waban*.

(2) *Definition in terms of more than one nuclear verb.* Some non-nuclear verbs can be regarded as expressing the logical intersection of two (or sometimes more) basic concepts, each of which corresponds to a single nuclear verb. For example, there are concepts 'clear away', Dyalŋuy *burganban* with Guwal C-equivalent *digun*; and 'look at, look for, see', Dyalŋuy *nʸuɽiman* with Guwal C-equivalent *buɽan*. Guwal non-nuclear *nuwan* 'sort through things searching for something – i.e. rummage' is the intersection of these two concepts. A simple one-word Dyalŋuy correspondent cannot be given for *nuwan* since it is equally related to *nʸuɽiman* and to *burganban*; in such cases normal Dyalŋuy elicitation procedure breaks down: an informant will give one Dyalŋuy word one time and the other another time, or else combine them, as *burganban nʸuɽimali* 'sort through in order to look' (in fact

a mark of such double-concept non-nuclear words is the unusual inconsistency of Dyalŋuy correspondents given for them). A further example is non-nuclear $d^yaymban$ 'find', which is the intersection of nuclear concepts $buran$ 'look at, look for, see' and $maŋgan$ 'pick up' (Dyalŋuy $mulwan$); and so on.[a]

(3) *Definition in terms of a verbalised adjective or noun.* Non-nuclear verb *maban* 'set fire to; light fire or lamp' was given Dyalŋuy correspondent *ŋarganamban*, the transitively verbalised form of Dyalŋuy noun *ŋargana*, itself the correspondent of Guwal n^yara 'flame' (thus *maban* is defined as 'make (become) a flame'). An example involving a verbalised adjective was given in §1.1.

Turning now to the few cases where Dyalŋuy does not provide a full (or fairly full) definition: it seems that some Guwal verbs cannot adequately be distinguished within Dyalŋuy as a meta-language. There are three possibilities here:

(1) The defining parameter distinguishing two verbs may not be describable (or not easily describable). Nuclear d^yuran 'rub', non-nuclear *baŋgan* 'paint or draw with finger (and nowadays: write)' and non-nuclear n^yamban 'paint with the flat of the hand' were all given correspondent $d^yurmbayban$ in stage 1 of elicitation. But neither Guwal nor Dyalŋuy has a term for 'finger' as opposed to 'hand': Guwal *mala*, Dyalŋuy *manburu* cover both meanings. Thus stage 2 elicitation was not able to specify 'finger as agent' for *baŋgan*, for instance; instead $d^yirin^yd^yirin^yd^yu$ $d^yurmbayban$ 'rub with white clay' was given (in fact white clay is almost always applied with the finger). Here, the defining characteristic could not be specified and so a secondary characteristic of the type of action (in this case, the material used) was specified. If a conversation in Dyalŋuy did try to describe the (unusual) application of white clay with the flat of the hand, mime would probably be resorted to.

(2) In some cases the full range of meaning of a Guwal verb is not specifiable in Dyalŋuy: different detailed Dyalŋuy correspondents have to be given according to the different specific meanings in instances of use. For example:

Nuclear *ŋanban* 'ask': Dyalŋuy *baŋarmban*
Non-nuclear *nungan* 'speaker asks or encourages someone to assist him in doing something bad (e.g. fighting or stealing)': this can only be distinguished from *ŋanban* within Dyalŋuy by specifying a particular bad action, e.g. Dyalŋuy *baŋarmban bagul duyilaygu* 'ask to fight [a third man]'; and so on.

(3) In some cases, words are only distinguishable in Dyalŋuy through an accompanying mime, or through some paralinguistic information. For example, in stage 1 elicitation *gundumman* 'bring together' was given for both nuclear *bugaman* 'chase, run down' and non-nuclear d^yulman 'squeeze (and mix flour with ingredients in baking, etc.)'. In stage 2 elicitation d^yulman could only be distinguished in terms of *gundumman yalaban* together with a squeezing mime, made by grasping the hands together. (Dyalŋuy *yalaban* is the correspondent of Guwal *yalaman* 'do it like this'.) As another example, the Dyirbal informant gave $burmbun^yman$ as the Dyalŋuy correspondent of both n^yund^yan 'kiss' and *buybun* 'spit at'; the informant was unable further to define either of these verbs within Dyalŋuy and when asked how one would distinguish in Dyalŋuy between 'the man kissed the girl' and 'the man

---

[a] There is an important difference between these examples and the *wirban* example of §2.6. Here the semantic description of the non-nuclear verb is exactly the logical intersection of two nuclear descriptions. Several different definitional descriptions could be given for *wirban*, each itself a full and adequate definition (and these definitions involve different nuclear verbs).

spat at the girl' demonstrated different voice timbres that clearly set off the two sentences: a warm, inviting timbre for the 'kiss' sentence and a higher, harsher timbre for 'spit'.

3.3 EXAMPLES OF SEMANTIC DESCRIPTION

An outline of the semantic structure of the class of Dyirbal verbs, with full semantic descriptions of about 250 members of the class, is in Dixon 1968a. In this section we discuss criteria for semantic grouping within the verb class, and give semantic descriptions of a small sample of verbs.

Dyirbal verbs fall, on syntactic and semantic grounds, into seven sets (the sixth of which is a residue set):

(1) Verbs of position, including 'go', 'sit', 'lead', 'take', 'throw', 'pick up', 'hold', 'empty out', and so on.

(2) Verbs of affect, including 'pierce', 'hit', 'rub', 'burn', and so on.

(3) Verbs of giving.

(4) Verbs of attention: 'look', 'listen', 'take no notice'.

(5) Verbs of speaking and gesturing: 'tell', 'ask', 'call', 'sing', and so on.

(6) Verbs dealing with other bodily activities; a residue set including 'cry', 'laugh', 'blow', 'copulate', 'cough', and so on.

(7) Verbs of breaking, that are 'meta' with respect to verbs in other sets, and can also have specific meanings: 'break', 'fall', 'peel'.

Each set (except the sixth) is semantically homogeneous; that is, a collection of semantic systems, arranged in a single dependency tree, serves to generate the componential descriptions of nuclear members of the set.

A syntactic criterion, setting off set 1 from the other sets, concerns the identification of syntactic function between transitive and intransitive sentences. We will use A, O and S as abbreviations for the functions 'subject of transitive verb', 'object of transitive verb', 'subject of intransitive verb' respectively. Now verbs in some sets (2, 3, 4) are all[a] transitive. In other sets both transitive and intransitive verbs occur; a transitive verb in a mixed transitivity subset is often related to an intransitive verb (in the same subset) by having the same semantic content, and differing only in transitivity. Such transitive verbs are of two distinct types according as, when an intransitive verb and a transitive verb with the same semantic content describe the same event, the realisation of the subject of the intransitive verb coincides with:

(1) the realisation of the object of the transitive verb (the case $S \equiv O$)

or

(2) the realisation of the subject of the transitive verb (the case $S \equiv A$).

Further, the subject of the reflexive form of a transitive verb of type (1) is equivalent to the object of the original transitive form; and for a verb of type (2), to the transitive subject.

Dyalŋuy data demonstrates the types $S \equiv O$ and $S \equiv A$. In almost every case Dyalŋuy has just one transitive verb where Guwal has a transitive-and-intransitive

---

[a] The few exceptions are intransitive roots that are obviously historically derived from transitive roots – a fossilised reflexive in set 4 and one fossilised reciprocal in each of sets 2 and 3.

pair; the reflexive form of the Dyalŋuy verb is given as correspondent of the Guwal intransitive verb. Examples are (cf. §2.3):

(1) S ≡ O type

Guwal *dʸaran*, tr; *dʸananʸu*, intr 'stand'

balan   baŋgul   dʸaran
she(class II)-nom he(I)-erg stand
'He stands her up'

balan   dʸananʸu
she(II)-nom stand
'She stands'

Dyalŋuy *dindan*, tr 'stand'

balan   baŋgul   dindan
she(II)-nom he(I)-erg stand
'He stands her up'

balan   dindayirinʸu
she(II)-nom stand-reflexive
'She stands'

(2) S ≡ A type

Guwal *dʸaŋganʸu*, tr; *manʸdʸanʸu*, intr 'eat'

balam   baŋgun   dʸaŋganʸu
food(III)-nom she (II)-erg eat
'she is eating the food'

balan   manʸdʸanʸu
she(II)-nom eat
'She is eating'

Dyalŋuy *yulminʸu*, tr 'eat'

balam   baŋgun   yulminʸu
food(III)-nom she(II)-erg eat
'She is eating the food'

balan   yulmimarinʸu
she(II)-nom eat-reflexive
'She is eating'

Other S ≡ O pairs include take out/come out (he takes her out/she comes out), waken (he wakens her/she wakens); S ≡ A pairs include follow (he follows her/he follows), tell/talk (he tells her/he talks).

Another syntactic correlation concerns 'accompanitive forms'; any intransitive verb can be made into an 'accompanitive form' by adding affix -*m(b)al*; it then has the syntactic possibilities of a transitive verb. The accompanitive form of an intransitive verb that has the same semantic content as a transitive verb of type S ≡ A is syntactically and semantically equivalent to the transitive verb; this is not so for type S ≡ O verbs. Thus corresponding to *balan manʸdʸanʸu* 'she is eating' we have accompanitive *balam baŋgun manʸdʸayman* 'she is eating food' which could be realised by the same event as *balam baŋgun dʸaŋganʸu; manʸdʸayman* and *dʸaŋganʸu* appear to be exact synonyms.[a] However, the accompanitive form of *balan dʸananʸu* 'she is standing' would be a construction with the 'woman' NP in the ergative case, say *bala baŋgun dʸanayman* 'she is standing with it (e.g. a stick in her hand)', whereas for the transitive equivalent of *dʸananʸu* the 'woman' NP must be in nominative case: *balan baŋgul dʸaran* 'he is standing her up (if she is a child that cannot stand well, say, or a sick woman)'.

All transitive verbs, except verbs of giving, occur in reflexive form and thus can be typed as S ≡ A or S ≡ O from consideration of the syntax of the reflexive, even if they do not belong to a transitive/intransitive pair.

Verbs in set 1 are all of type S ≡ O. Verbs in sets 2, 4, 5, 6 and 7 are of type S ≡ A. Verbs in set 3 cannot be categorised as either type. There is an apparent exception, a transitive/intransitive pair 'follow' that is concerned with motion (and thus should be in set 1) but is of type S ≡ A (see §2.3). This pair is best regarded

---

[a] We are deliberately oversimplifying here, since there is in fact a slight difference in semantic content for the 'eat' pair. *manʸdʸanʸu* is 'eat fruit or vegetables to appease hunger' while *dʸaŋganʸu* is 'eat fruit or vegetables (whether or not particularly hungry)'; there is a different verb, intransitive *ɽubinʸu*, for the eating of fish or meat. However, for all the other pairs mentioned here transitive and intransitive members have exactly the same semantic content.

as the *converse* (Lyons 1963.72) of the S $\equiv$ O pair 'lead/go' (see §2.3) – thus we have *she goes/he leads her/she follows him/she follows*. The relation 'converse' involves interchange of subject and object functions. The 'follow' pair are thus basically (i.e. in deep representation) of type S $\equiv$ O, like other verbs in set 1.

Verbs of 'position' clearly fall into two selectional sets: verbs of motion that can only occur with 'to' or 'from' qualifiers, and verbs of rest that are restricted to 'at' qualification. The selection is so strong that it justifies including an appropriate semantic feature – 'motion' or 'rest' – in the description of each nuclear verb of position. Verbs in other sets can, in plausible situational circumstances, receive unmarked 'rest' qualification (e.g. hit/laugh/give *at* the camp), but cannot occur with 'to' or 'from' verb markers. (Appendix 2 discusses some exceptions to this: verbs of seeing.) All this suggests that a verb marker accompanying a position verb should be within the VP whereas one accompanying a verb not concerned with position should be outside the VP.

Set 1 is thus syntactically set off from the other sets by (1) including all and only verbs of type S $\equiv$ O, (2) its members strongly selecting either 'motion' or 'rest' verb markers.

The first subset of set 1 can be called 'simple motion'.[a] The componential descriptions all include the feature 'motion'; they continue with 'here' or 'there' (with unmarked 'to') for *banin^{y}u* 'come', *yanu* 'go'; or with 'not water' plus 'down' or 'up', or 'water' plus 'down' or 'up' or 'across' for the specialised verbs of motion (see §2.3). 'Transitive' or 'intransitive' must be added to complete a semantic description. Most verbs of simple motion are intransitive but there is, for instance, *dangan* 'river or flood washes something away', the transitive correspondent of intransitive *dadan^{y}u* 'motion downriver' (and their Dyalŋuy correspondents are *balbulumban*, *balbulubin* respectively – see §2.3). The very general verb *waymban^{y}u* 'go walkabout' can be given semantic description just (position, motion; intr); *d^{y}ad^{y}an* 'hop' – which has the same Dyalŋuy correspondent, *d^{y}urbarin^{y}u*, as *waymban^{y}u* – is defined in terms of *waymban^{y}u* and the selection of a subject that is a member of the kangaroo genus (definition type 1-vii-a in §3.2).

The next subset, 'simple rest', includes a number of transitive and intransitive verbs with the feature 'rest', and then 'sit', 'stand' or 'lie'. We then have verbs dealing with position where immersion in water is involved. Since we have already had cause to set up the system {water, NOT WATER} this can be exploited here: we can assign semantic description (position, water; intr) to the very general verb *ŋaban^{y}u* 'bathe' and descriptions (position, water, motion; intr) and (position, water, rest; intr) to *yuŋaran^{y}u* 'swim' and *walŋgan* 'float', the 'water' equivalents of *waymban^{y}u* 'go walkabout' and *n^{y}inan^{y}u* 'stay' respectively. The inclusion of the feature 'water' in the componential descriptions of these three verbs implies the inclusion of 'NOT WATER' for verbs of simple motion and simple rest. However in this and other cases we can shorten statements of semantic description by omitting UNMARKED features (such as 'NOT WATER');[b] the fact that there *is* a verb with semantic description (position, water, motion; intr) indicates that *waymban^{y}u*, stated as (position, motion; intr), implicitly involves the feature 'NOT WATER'.

A componential semantic description is generated from a dependency tree of

---

[a] The notes that follow mention only *some* of the nuclear verbs in each set.

[b] The notion of semantic markedness could be developed further (beyond the tentative comments in the present paper) along lines similar to Chomsky and Halle's discussion of phonological markedness, in chapter 9 of *The sound pattern of English* (1968).

systems, such as that on page 447; one feature is chosen from the topmost system, one feature from the system depending on that choice, and so on, until some node at the bottom of the tree is reached. The features in a componential description are *ordered* according to their positions in the dependency tree. Thus *wandin* is (position, motion, water, up; intr). The importance of the ordering can be seen if we compare *wandin* with *yuŋaran^yu*, description (position, water, motion; intr).[a] The placc of 'water' in the ordered string of features is vital: by the conventions which arc used here 'water' following 'motion' implies motion with respect to water (hence *wandin*, 'motion upriver'), whereas 'water' following 'position' implies position in water (hence *yuŋaran^yu*, 'swim').[b]

Set 1 also includes subsets of 'induced motion' and 'induced rest'. For instance, 'carry' and 'pull' relate to induced motion, whereas 'lead' does not involve any inducing. Induced motion can be regarded as the intersection of features 'affect' and 'motion' (for 'affect' see also the discussion below of set 2). It can be of two types 'induced change of position' and 'induced state of motion': the difference is dealt with through the system {place, direction} already needed for the grammatical description (see the two kinds of allative verb marker – §2.3). Similarly, induced rest involves 'affect' and 'rest'.

Of the 'induced change of position' verbs, *gundan* 'put in' is (position, affect, motion, place, interior, to; tr); its complement *buŋgan* 'empty out' has 'from' in place of 'to', *budin* 'take or bring by carrying in hand' and *dimban^yu* 'carry, not with the hand – e.g. carry a dilly-bag with the strap across one's forehead, wear a hat, wear a feather loincloth' have features 'person', 'hand' and 'person', 'not hand' respectively, in addition to 'position', 'affect', 'motion', 'place'. Complements *yilmbun* 'pull, drag' and *digun* 'clear away, rake away, dig – using hands – on the surface' require 'here', 'to' and 'here', 'from' respectively in addition to 'position', 'affect', 'motion' and 'place'.

There are four main 'induced state of motion' verbs. The general verbs, transitive *duman* 'shake' and intransitive *gawurin^yu* 'move generally, e.g. toss and turn, fidget, "just going along"',[c] whose descriptions involve just 'position', 'affect', 'motion', 'direction'. And the more specific *madan* 'throw' and *baygun* 'shake, wave or bash' that also involve 'let go of' and 'hold on to' respectively.

The main 'induced state of rest' verb is *n^yiman* 'catch hold of' with description (position, affect, rest, person; tr). There are also verbs that appear to be nuclear and which combine the notions of induced rest and induced motion: *maŋgan* 'pick up' can be described as 'induced motion *in order to be* in an induced state of rest (relative to a person)' where the *in order to be* is the purposive construction of Dyirbal grammar (see Dixon 1968a). Cases like this suggest that the hypothesis of this paper -- nuclear equals componential, non-nuclear equals definitional

---

[a] Here a single system – {water, NOT WATER} – occurs twice, at different positions, in the dependency tree.

[b] These two semantic descriptions also differ in that *wandin* involves 'up'. The description of Dyirbal does not yield a true minimal pair – two semantic descriptions involving the same features, differing only in feature ordering; however, such a minimal pair is likely to be encountered as this method of semantic description is applied more widely.

[c] *gawurin^yu* can refer to any sort of induced state of motion, of anything. Consider the sentence *bayind^yana gingin, mala gulud^yilu baygul, gawurimban bagu maŋagu guralmali* (he-nom-emphatic dirty-nom, hand-nom not-emphatic he-ergative, move-accompanitive-pres/past it-dat ear-dat clean-causative-purposive) 'he's certainly got a dirty [ear], and can't even [be bothered to] lift (*gawurin^yu*) [his] hand to clean [his] ear'.

description–is an oversimplification. Another complexity–for which no provision was made in §2 – is the undoubtedly nuclear verb *banagan^{y}u* 'return' whose semantic description involves (position, motion; intr) and also reference to some place that the subject was at rest at before beginning a series of interconnected actions that he has just completed; *banagan^{y}u* has a wide range of meaning and can imply return to the place left a few seconds ago, or to the camp left three days ago.

Set 2, of affect verbs, are all of type S ≡ A, on the evidence of reflexives, and are neutral with respect to locational qualification. Unlike set 1, which includes a good number of intransitive verbs, set 2 verbs are all transitive (the single exception – *bundalan^{y}u* 'fight' – is probably a fossilised reciprocal form and requires plural subject). The actions realising verbs from set 2 induce a 'state' in the object, that can be described by a (normally non-cognate) adjective – see §1.1; this is the main property distinguishing these verbs from those of other sets.

Members of the first subset include feature 'affect' and also 'strike'; there is then specification of 'rounded object', 'long rigid object' (this choice requires an obligatory further choice from the system {let go of, hold on to}), 'long flexible object', 'sharp pointed implement' (for verbs 'pierce', 'spear', etc.) or 'sharp bladed implement' (for 'cut', 'scrape', etc.). Other verbs in set 2 involve features 'rub', 'wrap/tie', 'fill in', 'cover', 'burn'.

There are several stages to making a hut, and a special verb describes each in Guwal:

(1) *gadan* 'build up the frame by sticking posts into the ground' – this is a non-nuclear verb defined in terms of nuclear *bagan* 'pierce', description (affect, strike, sharp pointed implement; tr).

(2) *bard^{y}an* 'tie the posts together with loya vine, making the frame firm', defined in terms of non-nuclear *d^{y}unin* 'tie together, make tight', which is itself defined in terms of nuclear *d^{y}aguman* 'tie up, wrap up', description (affect, wrap; tr).

(3) *wamban* 'fill in between the frame with leaves, bark or grass, taking especial care to make the bottom snakeproof', defined in terms of nuclear *guban*, (affect, cover; tr).

Now in Dyalŋuy there is one holophrastic verb that covers all three stages, i.e. that refers to the complete complex action of building a hut; Dyalŋuy *wurwan^{y}u* is the correspondent of Guwal *gadan*, *bard^{y}an* and *wamban*. Normally a Dyalŋuy verb corresponds to a Guwal nuclear verb and a number of related Guwal non-nuclear verbs; here *wurwan^{y}u* refers to three non-nuclear verbs each of which is related to a different nuclear verb (and each of the three nuclear verbs has its own Dyalŋuy correspondent). This is the only instance in the writer's data of a holophrastic verb of this type occurring in Dyalŋuy.

Set 3 includes nuclear *wugan* 'give' and seven non-nuclear items. Semantically these are derived from the syntactic dummy *poss* that in the grammar underlies possessive constructions (although *poss* itself is not realised as a 'surface verb' in Dyirbal; instead it triggers off an 'affix transfer' transformation and is then deleted – for details see Dixon, 1969).

Set 4, verbs of attention, involves first a choice from the system {take notice, take no notice}. Nuclear verb *mid^{y}un* 'take no notice of, ignore' has description (attention, take no notice; tr); the reflexive form of *mid^{y}un* carries the meaning 'wait'. Feature 'take notice' involves a further choice, from system {seen, unseen (but heard)}: nuclear verbs *buṛan* 'see, look at' and *ŋamban* 'hear, listen to'. The reflexive of *ŋamban* can be translated 'think' (lit. 'hear oneself').

Verbs in set 5 refer to language activity. There are six main nuclear verbs, *buwan^{y}u* 'tell', *ŋanban* 'ask', *gayban* 'call, by voice or gesture', *bayan* 'sing', and the complements *gigan* 'tell to do, let do' and *d^{y}abin* 'stop someone doing something'; the reflexive of *d^{y}abin* bears the meaning 'refuse'. *ŋaɽin* 'answer' is a non-nuclear verb defined in terms of *gayban* 'call' – literally 'call in turn' (see definition type 1-i in §3.6).

Verbs which do not fall into any of the other sets are grouped together in set 6. These verbs have no common semantic feature, but they do appear to have a vague overall resemblance – they are all concerned with 'bodily activity' (other than seeing, hearing and speaking): 'eating', 'kissing', 'laughing', 'coughing' and so on. These verbs deal with particularly human physiological activities, so that we would be unlikely to find any exactly corresponding verbs in non-human languages; the semantic features required for the other verb sets, however, might well recur (albeit in different dependencies, and so on) in the semantic description of a 'language' spoken by non-human aliens. Set 6 includes the only Dyirbal intransitive verbs that are not semantically related (as synonyms, near-synonyms or antonyms, leaving aside differences of transitivity) to some transitive verbs.

Verbs in set 7 refer to the 'breaking off' of some state or action; they are thus in a sense 'meta' with respect to verbs in other sets. If *gayn^{y}d^{y}an* occurs in a VP with verb X it means 'the action referred to by X was broken off, i.e. stopped'; in a VP by itself the unmarked realisation of *gayn^{y}d^{y}an* is 'something – e.g. a branch – breaks'. A feature 'break off' can be set up, that can occur in a semantic description together with either 'motion' or 'rest'; *gayn^{y}d^{y}an* is thus (break off, rest; intr). 'Break off' plus 'motion', without any further specification, implies 'fall down' (nuclear *bad^{y}in^{y}u*); with specification 'down on the ground' it implies 'be bogged (in mud)' (non-nuclear *buman^{y}u*). There is a single Dyalŋuy verb, *dagaran^{y}u*, corresponding to Guwal *gayn^{y}d^{y}an*, *bad^{y}in^{y}u* and *buman^{y}u*, supporting the description of these words as all involving feature 'break off'. *dagaran^{y}u* is also the correspondent for Guwal verbs *gun^{y}d^{y}in* 'take all the peel off, take clothes off, take a necklace off, etc.', *ginbin* 'peel off just the top layer of skin (from a fruit)' and *d^{y}inin* 'break a shell (of an egg or nut)'.

The semantic descriptions of some words may appear surprising to speakers of English. Thus 'come out' might be thought to involve simple motion, rather than induced motion, and to have as complement 'go in'. In Dyirbal, nuclear *gundan* 'put in' (and *gundabin* 'go in') has as complement *buŋgan* 'empty out'; *bundin* 'take out' and *mayin* 'come out' are non-nuclear verbs defined in terms of nuclear *yilmbun* 'pull, drag' (that has as complement *digun* 'clear away, rake away'); all these verbs involve induced motion. Dyalŋuy gives evidence for these groupings, which are examples of the mildly different 'world view' of the Dyirbal.

Each of the nuclear verbs mentioned in the brief survey above (but for a very few items, such as *miyandan^{y}u* 'laugh') has associated non-nuclear verbs that are semantically defined in terms of the nuclear verb. As examples of the types of further specification that non-nuclear items may provide for a single nuclear verb, we give below three nuclear verbs and their associated non-nuclear verbs. These examples have been chosen to involve only straightforward definitions, avoiding complexities and indeterminacies. Parentheses round a word refer to the semantic description of that word.

(1) Nuclear *maŋgan* 'pick up'; semantic description (position, affect, motion, place; tr) joined by the purposive grammatical construction 'in order to' with

(position, affect, rest, person; tr). Non-nuclear *ŋanginⁱu* 'scrap about – e.g. on floor, in dish or in dilly-bag – picking up crumbs'; defined as (*maŋgan*) plus affix *-dⁱay* 'repeat event many times within short time-span'. Non-nuclear *dⁱalŋinⁱu* 'mock, i.e. imitate a person's voice and words'; defined as (*ŋanginⁱu*) plus specification of *guwal* 'language' in object function to the verb. There is also non-nuclear *dⁱaymban* 'find', the intersection of (*maŋgan*) and (*buɽan*) 'look' – see definition type 2 in §3.2.

(2) Nuclear *bidⁱin* 'hit with a rounded object, e.g. punch with fist, hit with a thrown stone, a woman banging a skin drum stretched across her knees, a car crashing into a bank, heavy rain falling'; semantic description (affect, strike, rounded object; tr). Non-nuclear *dⁱilwan* 'kick, or shove with knee'; defined as (*bidⁱin*) with specification of *dⁱina* 'foot' or *buŋgu* 'knee' in transitive subject/instrument function to the verb. Non-nuclear *dudan* 'mash [food] with a stone'; defined as (*bidⁱin*) plus specification of either *diban* 'stone' as subject and/or *wudⁱu* 'food' as object. Non-nuclear *dalinⁱu* 'deliver a blow to someone lying down, e.g. fall on someone' involves the intersection of (*bidⁱin*) and (*badⁱinⁱu*), i.e. (break off, motion; intr).

(3) Nuclear *buɽan* 'see, look at'; semantic description (attention, take notice, seen; tr). Non-nuclear *waban* 'look up at'; defined as (*buɽan*) plus 'up', as in bound form *-gala* (see n. *b*, p. 443). *barmin* 'look back at'; (*buɽan*) plus 'behind', as in bound form *-ŋaru* (see n. *b*, p. 443). *walginⁱu* 'look over or round [something] at [something else], e.g. look at someone hiding somewhere, look at a baby inside a cradle, look round a tree at, look out through a window at'; (*buɽan*) plus X with locative inflection and the further affix *-ru* 'motion', where X is a noun referring to any object – i.e. 'look round an object'. *ɽuyginⁱu* 'look in at, e.g. look in through a window at, open a suitcase and look inside'; (*buɽan*) plus 'interior', 'to'. *ɽugan* 'look on at, i.e. watch someone going out in front'; (*buɽan*) plus specification of the object as moving (by a relative clause involving the unmarked verb *yanu* 'go') and/or plus 'out in front', bound form *galu* (see n. *b*, p. 443). *wamin* 'look at someone, without the person being aware that he is being watched'; (*buɽan*) with subject X and object Y, together with (*gulu buɽan*) – *gulu* is the operator 'not' – with subject Y and object X. *ŋaɽnⁱdⁱanⁱu* 'stare at'; (*buɽan*) plus qualifier *dⁱunguru* 'strongly'. *gindan* 'look with the aid of artificial light [at night]'; (*buɽan*) with specification of *nⁱaɽa* 'light' as part of the subject/instrument NP. *yaŋan* 'look over land for something, e.g. look for good water hole, or for forked stick for house frame, or look at water to see if there are any fish in it' – note that the object of *yaŋan* will be the place that is looked at, whereas for *buɽan* it is most frequently the thing that is looked for; (*buɽan*) with object some 'place', joined by the purposive construction 'in order to' with (*buɽan*) with object 'some needed object'. *ŋuŋgan* 'look at something from the past, e.g. go back and search for something that was found before but left for later, or look at a tree, say, and recognise it'; (*buɽan*) plus specification of the object 'from past time'.

Detailed Dyalŋuy correspondences support these and other definitions – details are in Dixon 1968a.

Nothing in this paper should be taken as definitive. §2 is an initial, exploratory account of a method of semantic description; further work is likely to reveal it to be grossly oversimplified. Similarly, the descriptions of Dyirbal verbs in §3 are in most cases only tentative, and require further checking in the field, as well as more detailed assessment for theoretical consistency and sufficiency. There are points of similarity with recent work on generative grammar – particularly Fillmore (1968) and unpublished work of Lakoff, McCawley, Postal and Ross. It would be instructive to develop further some of the ideas in this paper on the basis of this work; this has not been attempted here.

I have tried in this paper to stick fairly closely to semantic descriptions suggested by mother-in-law language data, and not to extrapolate or generalise too far beyond this. Further work would have to range more widely and more deeply – for instance, nuclearity may be a rather 'deep' notion, and each nuclear 'concept' need not necessarily have a simple surface representation, as has been implied here.

A next step would be to attempt to apply the method of semantic description outlined here to other languages. Explanatory investigations of English verbs from this point of view have proved encouraging (see Dixon, forthcoming, b).

APPENDIX 1 – NEGATION

The negative particle, *gulu*, can be looked upon as a semantic operator, that effectively replaces a feature in a semantic description by another feature from the same system. A simple example will illustrate this.

In text xxxi*b*, a girl goes out after dark in search of a mythical spirit, despite repeated warnings not to. There is in the text a pattern of semantic replacement[a] involving essentially *gulu ŋamban* (*bala baŋumanganu guwal*) 'didn't listen (to what they were all saying)' and *mid^{y}uganin^{y}u* 'repeatedly took no notice'. In this instance the semantic replacement marks a relation of effective synonymy (in a particular context): *gulu ŋamban* has the same meaning as would *mid^{y}un; mid^{y}uganin^{y}u* merely repeats, with the addition of *-ganiy* 'kept on doing it'.

The dependency tree for set 4 is:

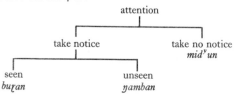

Now we are regarding *gulu* as an operator that replaces a feature in a semantic description by another feature from the same system (*the* other feature if it is a binary system). So that here *gulu* plus (attention, take notice, unseen; tr) is effectively (attention, take no notice; tr): *gulu* operates on feature 'take notice', replacing it by 'take no notice' (no choice from system {seen, unseen} is possible after a 'take no notice' choice). However *gulu ŋamban* 'didn't hear' and *buṛan* 'saw' have also been observed in semantic replacement relation in Dyirbal texts. In this case *gulu* plus (attention, take notice, unseen; tr) is effectively (attention, take notice, seen; tr): *gulu* leaves alone 'take notice' but operates on 'unseen', replacing it by 'seen'.

Generally, *gulu* 'not' can operate on *any* feature in a semantic description. Its precise operation in a particular instance can be inferred from semantic oppositions and replacements in the text (contextual correlations in the sense of Dixon 1965.111–7), or from the situational or other information available to a listener.

APPENDIX 2 – BRACKETING

We have so far talked only of the semantic descriptions of words. It is worthwhile asking whether the semantic description of a phrase is to be regarded merely as the

---

[a] That is, there is a replacement correlation – see Dixon 1965.115–7 ff.; and compare n. *a*, p. 453 and n. *a*, p. 456, above.

sum of the semantic descriptions of its constituents,[a] or whether some of the syntactic bracketing of the phrase should be retained in the semantic description. For example, bracketing could tell us whether a certain feature in the semantic description of a VP was supplied by a verb, or by a locational adjunct; in the absence of bracketing we would not be able to tell the syntactic origin of a particular feature.

It appears that *some* syntactic bracketing must be retained in the semantic descriptions of phrases. The following example demonstrates this need.

We mentioned that each verb of position includes in its semantic description one of the features 'motion' or 'rest', and can only select appropriate locational qualifiers; locational qualifiers can be verb markers and/or nouns and adjectives in allative/ ablative or locative inflection. Verbs in other sets do not include 'motion' or 'rest' in their semantic descriptions, and can only select unmarked locative ('at') qualifiers. Exceptions are verbs of 'seeing' which can occur freely with either allative or locative qualification: 'look towards', 'look at', etc.; in fact these are the only verbs that can take either motion or rest qualification.

Verb markers and nouns and adjectives in allative or locative (but not in ablative) inflection can be verbalised – see §2.3. Such a verbalised form has exactly the same possibilities of aspectual modification, inflection, etc. as the other verbs in the VP (in fact a VP in Dyirbal can consist of any number of verbs – that have simultaneous reference – provided they all agree in surface transitivity and in final inflection).

Thus we have VP:

⟨1⟩ *buṛan   yaludayi*
    look-pres/past here-to(place)-up-not water-short distance
    'looks up here'

or, with transitive verbaliser -(m)ban:

⟨2⟩ *buṛan yaludayimban*
    'looks up here'

We can also have:

⟨3⟩ *buṛan   balay   gayuŋga*
    look-pres/past there-at cradle-at
    'looks at the cradle there'

or else a VP like ⟨3⟩ but without *balay* or without *gayunga*.

Now in text xv.16 there occurs:

⟨4⟩ *buṛan yaludayimban gayuŋga*
    'looks up at the cradle'

However, we cannot have:

⟨5⟩ **buṛan yaludayi gayuŋga*

That is, although *buṛan* can occur with 'to' or 'at' qualification, it cannot take both together; however, *buṛan* and a verbalised 'to' qualifier can take 'at' qualification.

It seems desirable to show that there is a *semantic* difference between ⟨1⟩ and ⟨2⟩, and between ⟨4⟩ and ⟨5⟩ – a difference that is intuitively felt by speakers of Dyirbal but which is difficult to bring out in English translation – and at the same time to account semantically for the non-acceptability of ⟨5⟩. Now the features in any semantic system are mutually exclusive; the semantic description of a word can contain only one feature from any system. The non-acceptability of ⟨5⟩ can be accounted for in terms of this constraint, if we say that within a VP the semantic information coming from verbs and verbalised forms is bracketed off from that coming from qualifiers. The constraint

[a] For the discussion here we are effectively restricting attention to nuclear words, which have componential semantic descriptions. The semantic description of a non-nuclear word is likely itself to involve a phrase marker (as in recent unpublished work of McCawley).

is now that more than one feature from a given system cannot occur *within the same brackets* in a semantic description. Thus for ⟨4⟩ the semantic information coming from *buṛan* and *yaludayimban* is bracketed off from that coming from *gayuŋga*; features 'motion' – from *yalu* – and 'rest' – from the locative inflection on *gayu* 'cradle' – do not appear within the same brackets; ⟨4⟩ is thus an acceptable VP. In the case of ⟨5⟩, however, the semantic description of *buṛan* is bracketed off from that of *yaludayi gayuŋga*; in this case the incompatible features 'motion' and 'rest' appear within the same brackets, and the VP is thus unacceptable.

The example has shown that *some* syntactic bracketing must be carried over into semantic descriptions,[a] if we are to be able to make correct predictions concerning acceptability. Probably not all syntactic bracketing has to be retained; each type of bracketing would have to be considered individually, and its semantic relevance assessed.

Australian National University

REFERENCES

Bloomfield, Leonard, *Language*, New York (1933).
Capell, A. 'Language and social distinction in aboriginal Australia', *Mankind*, v, 514–22 (1962).
Chomsky, Noam, *Aspects of the theory of syntax*, Cambridge, Mass. (1965).
Chomsky, Noam and Halle, Morris, *The sound pattern of English*, New York (1968).
Dixon, R. M. W., *What is language?*, London (1965).
    The Dyirbal language of North Queensland, University of London Ph.D. thesis (1968a) [a revision of the sections on grammar and phonology is to be published by Cambridge University Press].
    'Noun classes', *Lingua*, xxi, 104–25 (1968b).
    'Relative clauses and possessive phrases in two Australian languages', *Language*, xlv, 35–44 (1969).
    'Where have all the adjectives gone?' (forthcoming, a).
    'The semantics of giving', to appear in the Proceedings of the International Colloquium on Formalisation and Models of Linguistics, held at Paris, 27/29 April 1970 (forthcoming, b).
Douglas, Mary, *Purity and Danger*, London (1966).
Fillmore, Charles J., 'The case for case', pp. 1–89 of *Universals in linguistic theory* (ed. Emmon Bach and Robert T. Harms), New York (1968).
Firth, J. R. *Papers in linguistics, 1934–1951*, London (1957).
Goodenough, Ward H., 'Componential analysis and the study of meaning', *Language*, xxxii, 195–216 (1956).
Halliday, M. A. K., 'Categories of the theory of grammar', *Word*, xvii, 241–92 (1961).
    'Notes on transitivity and theme in English', *Journal of Linguistics*, iii, 37–81 and 199–244, iv, 179–215 (1967–8).

[a] In some cases the semantic description of a non-nuclear verb may involve the same features as the semantic description of a VP containing a nuclear verb together with some locational qualification. For example, *waban* 'look up' and *buṛan* [*yalu*] *gala* (see definition type 1–ix, §3.b). The difference here is that the semantic description of *waban* involves no bracketing, whereas for the VP the semantic features coming from *buṛan* are bracketed off from the one coming from *gala*. (Note that if it were thought desirable semantically to distinguish *waban* from *buṛan* [*yalu*] *galamban* – where the locational qualifier is verbalised and there is thus no bracketing in the semantic description of the VP – then this would indicate the need for further bracketing, separating off the features coming from a full verb, and those coming from a verbalised locational adjunct.)

Hjelmslev, L. and Uldall, H. J., *Outline of glossematics*, I, Copenhagen (1957).

Katz, Jerrold J. and Fodor, Jerry A., 'The structure of a semantic theory', *Language*, XXXIX, 170–210 (1963).

Lamb, Sidney M., 'The sememic approach to structural semantics', *American Anthropologist*, LXVI, 3, part 2, 57–78 (1964).

'Epilegomena to a theory of language', *Romance Philology*, XIX, 531–73 (1966).

Leach, Edmund, 'Anthropological aspects of language: animals categories and verbal abuse' in *New directions in the study of language* (ed. Eric H. Lenneberg), Cambridge, Mass. (1964).

Lounsbury, Floyd G., 'A semantic analysis of Pawnee kinship usage', *Language*, XXXII, 158–94 (1956).

'A formal account of the Crow- and Omaha-type kinship terminologies', in *Explorations in cultural anthropology* (ed. W. H. Goodenough), New York (1964a).

'The structural analysis of kinship semantics', in *Proceedings of the ninth international congress of linguists* (ed. H. G. Lunt), The Hague (1964b).

Lumholtz, Carl, *Among cannibals*, London (1889).

Lyons, John, *Structural semantics*, Oxford (1963).

Introduction to theoretical linguistics, Cambridge (1968),

McIntosh, Angus, 'Patterns and ranges', *Language*, XXXVII, 325–37 (1961).

Nida, E. A., *Towards a science of translating*, Leiden (1964).

Roth, W. E., *North Queensland ethnography – Bulletins 1–8*, Brisbane (1901–5); Bulletins 9–18 in *Records of the Australian Museum* (1907–10).

Thomson, D. F., 'The joking relationship and organised obscenity in North Queensland', *American Anthropologist*, XXXVII, 460–538 (1935).

Weinreich, Uriel, 'Lexicographic definition in descriptive semantics', in *Problems in Lexicography* (ed. Fred W. Householder and Sol Saporta), Bloomington (1962).

A note on a Walbiri tradition of antonymy

KENNETH HALE

This paper will be concerned with a particular linguistic tradition learned by Walbiri men at a certain point in their integration into ritual life.[a] The Walbiri are an aboriginal people of Central Australia now living on several settlements and cattle stations distributed over a rather vast area in the western part of the Northern Territory. They have received no little attention from anthropologists, and there exists an excellent account of their society by M. J. Meggitt (1962). My own work with the Walbiri has been primarily linguistic. But, in the Walbiri community, it becomes immediately evident that much of what is observed in linguistic usage is intimately related to such cultural spheres as kinship and ritual.

The linguistic tradition with which I will be concerned here is associated with advanced initiation rituals of the class which the Walbiri loosely refer to as *kankalu* ('high, above'), a general term embracing the *katjiri* described by Meggitt in a recent monograph (Meggitt, 1967) and a set of about seven ritual complexes termed *waduwadu* (or *kumpaltja*) by the Walbiri at Yuendumu. The linguistic remarks which follow should not, perhaps, be generalized beyond the Yuendumu Walbiri or beyond the *waduwadu* subclass of *kankalu* rituals; however, Walbiri men have asserted that they are valid for all Walbiri and for all *kankalu*.

Before continuing, I must ask the reader to cooperate with a specific condition which Walbiri men place on material of this sort. All knowledge relating centrally to *kankalu* ritual is exclusively the property of men who have gone through the rituals. While many Walbiri are eager to have the material recorded and published as a matter of scientific record, they request strongly that it be handled with care. Specifically, they request that none of the knowledge be discussed with uninitiated Walbiri men or with Walbiri women and children. If the reader should have an opportunity to discuss the material in this note with a Walbiri man, it is important that he first determine that the man concerned is a *maliyara* (see below and Meggitt, 1962, p. 234) and that he is willing to discuss matters relating to *kankalu* ritual.

Walbiri youths normally become junior *kankalu* novices shortly after their first initiation, i.e., shortly after circumcision (at an average age of 13). And it is in the

[a] This work was supported in part by the National Science Foundation (NSF Grant No. GS-1127) and in part by the National Institutes of Health (Grant MH-13390-01).

My debt to the Walbiri people is immeasurable, and I am particularly grateful to those Walbiri men who brought the *tjiliwiri* activity to my attention. I hope that some day they will be able to derive some benefit from my exploitation of their linguistic knowledge.

context of their first *kankaḻu* novitiate that they acquire the linguistic skills to be described in the following paragraphs.

Men who are advanced in *kankaḻu* ritual are referred to as *maḻiyara*, a term widespread in Central and Northwestern Australia. But they are also referred to as *tjiliwiri*, a term which in its profane usage means 'funny' or 'clown'. The term *tjiliwiri* is also extended to refer specifically to formalized 'clowning' during *kankaḻu* ritual and to the men who serve as guardians of junior *kankaḻu* novices.

It is another extension of this term that is relevant here. In apparent close association with the concept of ritual clowning is a type of semantic pig-Latin spoken by guardians in the presence of junior novices. This is also referred to as *tjiliwiri* and will be so designated in what follows. It is a 'secret language' in the sense that it is spoken, theoretically, and probably in fact, only in the context of *kankaḻu* ritual and beyond the hearing of uninitiated Walbiri.

Walbiri men sometimes refer to *tjiliwiri* as 'up-side-down Walbiri'. They say that, to speak *tjiliwiri*, one turns ordinary Walbiri 'up-side-down' (*ŋaḍalj-kitji-ni*), a skill which junior novices are expected to acquire by observation while they are in seclusion in connection with their first *kankaḻu* involvement.

In very general terms, the rule for speaking *tjiliwiri* is as follows: replace each noun, verb, and pronoun of ordinary Walbiri by an 'antonym'. Thus, for example, if a *tjiliwiri* speaker intends to convey the meaning 'I am sitting on the ground', he replaces 'I' with '(an)other', 'sit' with 'stand', and 'ground' with 'sky'. Similarly, if he intends to convey the meaning 'You are tall', he replaces 'you' with 'I', and 'tall' with 'short'. To illustrate, I cite simple sentences in Walbiri (the (a)-form) together with their *tjiliwiri* equivalents (the (b)-forms) and their English translations.[a]

⟨1⟩ (a) ŋatju ka-ṇa walja-ŋka njina-mi.
    (I present-I ground-locative sit-nonpast)
   (b) kaṛi ka-Z ŋuṛu-ŋka kari-mi.
    (other present-he sky-loc stand-nonpast)
    'I am sitting on the ground'.

⟨2⟩ (a) njuntu A-npa kiriḍi.
    (you stative-you tall)
   (b) ŋatju A-ṇa ḍaŋkaḻpa.
    (I stative-I short)
    'You are tall'.

⟨3⟩ (a) njampu A-Z wiṛi.
    (this stative-it big)
   (b) pinka A-Z wita.
    (distant stative-it small)
    'This (one) is big'.

[a] The alphabetic notation used in the examples has the following values: bilabial, apico-alveolar, apico-domal, lamino-alveolar, and dorso-velar stops /p t ṭ tj k/; nasals matching the stops in position of articulation /m n ṇ nj ŋ/; laterals matching the non-peripheral stops and nasals /l ḷ lj/; plain and retroflexed flaps /r ḍ/; bilabial, apico-domal, and lamino-alveolar glides /w ṛ y/; and vowels: high front /i/, high back (rounded) /u/, low central /a/. Nominal and verbal words are stressed on the first syllable; the clitic auxiliaries (e.g., /ka-ṇa/ 'present-I') are unstressed, though written as separate words in the examples. Some auxiliary stems are phonologically vacuous – these are represented by A in parenthetic glosses. The auxiliaries are inflected for person; the third person singular person markers, and the second person singular in imperatives, are phonologically null – these are represented by Z(ero) in the glosses.

⟨4⟩ (a) yatitjara A-Z ya-nta.
       (north definite-you go-imperative)
   (b) kuḷira A-ṇa njina-mi-ra.
       (south definite-I stay-nonpast-hence)
       'Go north!'

⟨5⟩ (a) waljpali ka-Z ya-ni-ṇi njampu-kura.
       (European present-he come-nonpast-hither this-to)
   (b) puntu-njayiṇi ka-Z njina-mi-ṇi pinka-kura.
       (kinsman-real present-he stay-nonpast-hither distant-to)
       'A European is coming here'.

⟨6⟩ (a) tjuṇpuṇpu njampu A-Z kulupaṇṭa.
       (hill-tobacco this stative-it strong)
   (b) yaṟunpa pinka A-Z ṇaljtjimpayi.
       (plains-tobacco distant stative-it weak)
       'This hill-tobacco is strong'.

Ordinary Walbiri is spoken rapidly, and apparently, *tjiliwiri* speech is equally
fast. Novices are not taught the semantic principle directly; rather, they must learn
it by observation. According to my consultants, the novices are exposed to rapid
dialogues between guardians. In these dialogues, one guardian speaks *tjiliwiri*
while the other answers, or rather interprets the message, in Walbiri. Each exchange
in the dialogue is punctuated by the expletive *yupa*, an exclusively *tjiliwiri* word.
In the following illustrative dialogue, the first individual, *A*, speaks *tjiliwiri*, and
the second, *B*, responds in Walbiri. The free translation of the *tjiliwiri* speech
represents the intended message; the parenthetic gloss is literal.

⟨7⟩ *A:* kaṟi ka-Z kakaraṟa njina-mi-ra.
        (other present-he east stay-nonpast-hence)
        'I am going west'.

    B: njuntu ka-npa kaḷara ya-ni.
        (you present-you west go-nonpast)
        'You are going west'.

    A: yupa.
        (yes)
        'Yes'.
        waḷu A-ṇa-ḷa puṟumaḍa-ni kaṟi-ki.
        (fire definite-I-him withhold-nonpast
        other-dative)
        'Give me water'.

    B: ṇapa A-ṇa-ṇku yi-nji njuntu-ku.
        (water definite-I-you give-nonpast you-dative)
        'I should give you water'.

    A: yupa.
        (yes)
        'Yes'.
        kaṟi A-Z yaljaki.
        (other stative-he quenched)
        'I am thirsty'.

> *B:* njuntu A-npa puraku.
> (you stative-you thirsty)
> 'You are thirsty'.
>
> *A:* yupa.
> (yes)
> 'Yes'.
> waḷu ka-Z njina-mi-ṇi ḍiwaṛatji-kiḷi.
> (fire present-it stay-nonpast-hither
> calm-having)
> 'Rain and wind are approaching'.
>
> *B:* ŋapa ka-Z ya-ni-ṇi waḷpa-kuḷu.
> (rain present-it come-nonpast-hither
> wind-having)
> 'Rain and wind are approaching'.
>
> *A:* yupa.
> (yes)
> 'Yes'.

Novices are said to learn to speak and understand *tjiliwiri* in two to four weeks. Walbiri men stated that when they first heard *tjiliwiri*, its principle eluded them for some time; eventually, they explained, it came as a sudden flash of insight. I would like now to consider briefly the nature of this principle and, in particular, the problem of determining a possible *tjiliwiri* antonym of a given Walbiri lexical item.

An obvious antonymy principle is that of polarity, i.e., the principle involved in oppositions of the type represented by English good/bad, big/little, strong/weak. This principle is used wherever possible to determine *tjiliwiri* equivalents, as in:

⟨8⟩ ŋurtju/matju (good/bad);
　　wiṛi/wita (large/small);
　　palka/ḷawa (positive/negative, present/absent);
　　kiriḍi/ḍaŋkaḷpa (long/short);
　　wakuṭuḍu/ṛampaku (strong/weak);
　　pirtjiḍi/ṛampaku (heavy/light);
　　pina/ŋurpa (knowing/ignorant);
　　piḍaku/yaṇunjtjuku (sated/hungry);
　　yaljaki/puraku (quenched/thirsty).

It is evident, however, that the rather obvious and very suggestive polarity principle of antonymy is unsuited to many lexical domains. Within the *tjiliwiri* tradition, polarity is simply a special case of a vastly more general principle. Consider, for example, lexical items whose semantic interrelationships are typically taxonomic:

⟨9⟩ wawiri/kanjala (kangaroo/euro);
　　wakuljari/yulkaminji (rock-wallaby/wallaby sp.);
　　waḷawuru/puḷapuḷa (eaglehawk/kite);
　　puḷapuḷa/wi·njwi·njpa (kite/grey falcon);
　　kirkaḷanjtji/waṛukupalupalu (whistling eagle/hawk sp.);
　　kiḷilkiḷilpa/ŋaŋkamaḍa (galah/cockatoo sp.);
　　manjtja/ŋalkiḍi (mulga/witchetty-bush);

ŋapiṛi/wapuṇuŋku (red-gum, *E. camaldulensis*/ghost-gum, *E. papuanis*);
yakatjiri/warakaḷukaḷu (desert raisin, *S. ellipticum*/*S. petrophilum*).

It is apparent that the principle employed here is that of opposing entities which, according to some taxonomic arrangement or other, are most similar, i.e., in formal terms, entities which are immediately dominated by the same node in a taxonomic tree. Thus, members of a given class of objects are opposed to other members of the same immediate class – a large macropod is opposed to another large macropod (kangaroo/euro), a eucalypt is opposed to another eucalypt (red-gum/ghost-gum), and so on. A great deal of variability can be observed in lexical domains whose structure is taxonomic, since considerable latitude is permitted in the hierarchical arrangement of oppositions within well defined classes. Accordingly, on a given occasion, a *tjiliwiri* speaker may oppose macropods according to habitat, leaving size and other attributes constant insofar as it is possible to do so; on another occasion, he may oppose them on the basis of size, leaving constant the other attributes. In general, however, an effort is made to oppose entities which are minimally distinct. The accuracy with which a *tjiliwiri* speaker is able to do this determines the ease with which his dialogue partner will be able to interpret the message.

Consider now the principle employed in opposing the Walbiri subsection terms:

⟨10⟩ tjapananka/tjuŋarayi (A_1/A_2);
    tjapaŋaḍi/tjapaltjari (D_2/D_1);
    tjupurula/tjaŋala (B_1/B_2);
    tjampitjinpa/tjakamara (C_2/C_1).

The parenthetic glosses here are the traditional ones employed in the anthropological literature (e.g., Meggitt, 1962, pp. 164–87). These subsections are arranged in more inclusive groupings: there are two patrimoieties (AD, BC), two matrimoieties (AC, BD), two merged alternate generation moieties (AB, CD), and four sections ($A_1 D_2$, $D_1 A_2$, $B_1 C_1$, $B_2 C_2$). There exists a system of terminology corresponding to each of these groupings. It can be seen that the *tjiliwiri* opposite of a given subsection X is that subsection Y whose members belong simultaneously to the same patri-, matri-, and merged alternate generation level moieties as the members of X. Equivalently, the *tjiliwiri* opposite of X is that subsection Y whose members (collectively) are distinct from those of X *solely* on the basis of section membership (i.e., in biologically based terms, on the basis of agnation). Furthermore, it is obvious that section membership (or agnation) is the only dimension which can *minimally* oppose subsection terms, a fact that requires some contemplation to appreciate. By any pairing other than the actual *tjiliwiri* one, the paired terms differ in at least two dimensions. This and the earlier examples indicate that *minimal opposition* is basic to the *tjiliwiri* principle of antonymy.

Kinship terms, as in ⟨11⟩, are opposed on the basis of seniority and generation (ascending as opposed to descending). The two principles are complementary, since the first is relevant only *within* Ego's generation, the second only *relative* to Ego's generation:

⟨11⟩ papaḍi/kukuṇu (OBr/YBr);
    kapiḍi/ŋawuru (OSi/YSi);
    ŋumpaṇa/kaṇṭiya (senior WiBr/junior WiBr);
    kiḍana/ŋalapi (Fa/So);

pimiḍi/ŋalapi (FaSi/Da);
ŋati/kuḍuna (Mo/SiDa);
ŋamiṇi/kuḍuna (MoBr/SiSo);
tjamiḍi/wankili (MoFa/MoBrSo).

Notice that if an equally operative principle such as cross/parallel (cf. Kay, 1965) were employed, it would not be possible to oppose all kinship terms minimally – i.e., it is not possible to pair Walbiri kinship terms in such a way that the members of each pair are distinct only on the cross/parallel dimension. To be sure, there is at least one other distinction which could serve to oppose Walbiri kinship categories minimally, namely, sex antonymy. However, if sex of kinsman were used here, it would violate an apparent requirement that *tjiliwiri* opposites be phonologically distinct. While sex antonymy plays an important role in the Walbiri terminology, it is defective, and many categories which are distinct only on the sex dimension are represented by a single, phonologically unified kinship term (e.g., /ŋalapi/ (So, Da), /tjatja/ (MoMo, MoMoBr), etc.).

The observation just made leads to another one which reveals rather well the abstract nature of the *tjiliwiri* activity. It could conceivably be argued that phonologically unified elements of the type represented by /ŋalapi/ (So, Da) are not semantically ambiguous in the way I have implied; i.e., one might argue that the term /ŋalapi/ is simply the name for the class of Ego's agnates in the first descending generation and that the principle of sex antonymy plays no role at that point in the Walbiri kinship system. We know that this is not the case, since the kinship system as a whole requires recognition of a class of male /ŋalapi/ as opposed to a class of female /ŋalapi/ – this is shown, e.g., by the fact that the offspring of Ego's male /ŋalapi/ are his parallel (in fact, agnate) kinsmen, while the offspring of his female /ŋalapi/ are not; furthermore, there is a technical usage according to which the female, but not the male /ŋalapi/ is referred to by the special term /yuṇṭalpa/ (cf. Meggitt, 1962, p. 115). The male and female /ŋalapi/ are clearly distinct in *tjiliwiri* usage – the male is opposed to /kiḍana/ (Fa), while the female is opposed to /pimiḍi/ (FaSi). And, in general, polysemy is reflected in *tjiliwiri* practice:

⟨12⟩ (*a*) kuḍuna (SiSo, SiDa)/ŋamiṇi (MoBr), ŋati (Mo).
    (*b*) pirtjiḍi (heavy, hard)/ṛampaku (light), manja (soft).
    (*c*) yu- (to feed, to give)/tjatja- (to eat off of someone), puṛumaḍa- (to withhold).
    (*d*) kitji- (to cause to topple, to throw)/pirtjiḍi-ma- (to make fast), mampu-ma- (a coinage of uncertain derivation).

As expected, synonymy is also reflected:

⟨13⟩ (*a*) ŋanaŋana, tjuwaṛi (natural water hole)/multju (dug well).
    (*b*) maŋulpa, wurumpuṛu (lance)/kuḷaḍa (spear).
    (*c*) njinji, punjunju (junior *kankaḷu* novice)/tjaḷu(-paḍu) (elder).
    (*d*) wawiri, maḷu (kangaroo)/kanjala (euro).

It is clear from these examples that the *tjiliwiri* principle of antonymy is semantically based, i.e., that the process of turning Walbiri 'up-side-down' is fundamentally a process of opposing abstract semantic objects rather than a process of opposing lexical items in the grossest and most superficial sense. This abstract nature of the *tjiliwiri* activity is particularly evident in cases of the type represented by ⟨14 *a–e*⟩

in which an opposition reflects the partial synonymy of lexical items – that is, cases in which one semantic object is opposed to another which enters into the make-up of two distinct though semantically related lexical items:

⟨14⟩ (a) ḍaku (hole, trench), karu (creek bed)/ḍartja (even ground).

(b) ŋaka (after, a short time hence), njuru (before, a short time ago)/tjalaŋu (now).

(c) ŋapa (water, rain), ŋawara (running water)/waḷu (fire).

(d) minjminj-ma- (to wet, moisten), mapa- (to paint, anoint)/paḍuna-ma- (to dry).

(e) waḷpa (wind in circular motion), payi (wind moving in one direction, breath, air)/ḍiwaṛatji (calm).

It is, of course, not surprising that *tjiliwiri* is semantically based in the particular sense that it makes use of abstract semantic structures – it would be difficult, otherwise, to conceive of how an individual could learn the system in a relatively short time, through exposure to a severely limited and highly unsystematic sample of actual speech. It is abundantly clear that a general principle is learned and that, once acquired, the principle enables the learner to create novel *tjiliwiri* sentences and to understand any well-formed novel sentence spoken by another. This general principle of antonymy is determined by a semantic theory which we know, on other grounds, must be shared by the speakers of Walbiri. The *tjiliwiri* activity provides us with a surprisingly uncluttered view of certain aspects of this semantic theory and is certainly not irrelevant to the much discussed, though occasionally incoherent, question of whether the semantic structures we, as students of language or culture, imagine to exist do, in fact, have any 'reality' for the speakers who use the system. In this connection, it is interesting to note that the semantic relationships which *tjiliwiri* reveals are not inconsistent with semantic structures which have been recognized throughout the history of semantic studies – these include not only such obvious concepts as polarity, synonymy, and antonymy, but also the more general conception that semantic objects are componential in nature, that lexical items often share semantic components and belong to well defined domains which exhibit particular types of internal structure (e.g., the tanoxomic structure of the biological and botanical domains; the paradigmatic structure of the kinship domain; and so on). None of these observations is surprising, to be sure, given the semantic basis of *tjiliwiri*, but it is rather supportive to learn that even very specific principles, postulated on independent grounds, are actively manipulated in *tjiliwiri* – e.g., the principle of agnatic kinship, generation antonymy, and the like.

Abstract systems of relationship among lexical items are reflected in a particularly striking manner in certain cases. Consider, for example, the conception (whose validity is apparent on a variety of grounds – semantic, syntactic, and morphological) that the Walbiri verbs of perception belong to a special lexical subset:

⟨15⟩ nja- (to see);
puḍa-nja (to hear, feel);
paṇṭi-nja- (to smell).

What is of interest here is the fact that an important aspect of the internal structure of this lexical subset can be revealed, in standard Walbiri usage, only by recourse to a syntactic device, i.e., negation. The abstract semantic structure of the subset must provide that each of the three perception verbs be paired by an antonym – it is relatively apparent that the three verbs cannot themselves be contrasted with one another in any way which is obviously consistent with the principle of minimal opposition. In a rather clear sense, then, the Walbiri domain of perception contains six terms, rather than three – in other words, it contains three pairs of opposed terms. Since Walbiri has no standard antonyms for the perception verbs, the domain is lexically defective and is completed only by negating the existing forms in one way or another:

⟨16⟩ nja-njtja-waŋu (not seeing), kula-...nja- (not to see);
puḍa-nja-njtja-waŋu (not hearing, not feeling), kula-...puḍa-nja- (not to hear, feel);
paṇṭi-nja-njtja-waŋu (not smelling), kula-...paṇṭi-nja- (not to smell).

This recourse is not available in *tjiliwiri*, however, due to the general convention that the negative may not be used to create *tjiliwiri* opposites. In cases where Walbiri lacks a suitable antonym for a particular lexical item, an opposite must be coined for *tjiliwiri*. In *tjiliwiri*, the class of perception verbs is complete in the sense that it corresponds more or less exactly to what we have reason to assume is the actual abstract structure of the lexical subset:

⟨17⟩ yuḍu^2-tjari- (to see)/nja- (not to see);
tjutu^2-tjari- (to hear)/puḍa-nja- (not to hear);
ḍulpu^2-tjari- (to smell)/paṇṭi-nja- (not to smell).

Notice further that the internal cohesion of the domain is preserved in the form of the *tjiliwiri* coinages – i.e., all share the morphological peculiarity that they are composed of a reduplicated root preposed to the verbal formative /-tjari-/ (inchoative).[a]

This example is one of many that might have been given to illustrate the fact that the objects which are manipulated in *tjiliwiri* practice are abstract in the special sense that they are rarely represented overtly in any systematic way in the ordinary Walbiri lexicon. In effect, the *tjiliwiri* activity involves an analysis of lexical items and lexical domains in terms of their atomic semantic components – the situations in which resort must be had to special coinages are especially interesting in this regard in as much as the semantic components are then represented directly in the *tjiliwiri* forms. In the set of first person pronouns, for example, the constant feature, i.e., 'first person' or 'speaker', is represented directly by the *tjiliwiri* term /kaṛi/ 'other'. That is to say, the *tjiliwiri* term /kaṛi/ represents the meaning which is shared by the Walbiri forms /ŋatju/ 'I', /ŋali(tjara)/ 'we dual inclusive', /ŋatjara/ 'we dual exclusive', /ŋalipa/ 'we plural inclusive', and /ŋanimpa/ 'we plural exclusive'. The additional semantic dimensions associated with the Walbiri forms (i.e., number, exclusion of the addressee, and the like) can be represented optionally by suffixes, as in /kaṛi-tjara/ (other-dual) 'we dual', /kaṛi-mipa/ (other-only) 'we exclusive'. In any event, /kaṛi/ is constant and identifies the class of first person pronouns. In addition to the rather interesting antonymy principle implied by the use of /kaṛi/ – i.e., the opposition of ego and alter – the source of the form itself

---

[a] The etymologies of the *tjiliwiri* perception verbs, except for /tjutu^2-tjari-/, are not altogether clear. The form /tjutu/ refers to stoppage, closure, and to deafness. The form /waṛuŋka^2-ṭari-/ is sometimes used in *tjiliwiri* in place of /tjutu^2-tjari-/. The form /waṛuŋka/ refers to deafness, heedlessness, and to one form of insanity.

attests to the abstractness of *tjiliwiri* antonymy.[a] In ordinary Walbiri usage, /kaṛi/
is not a free form but a suffix, and was apparently abstracted for *tjiliwiri* from its
function in such standard Walbiri expressions as /tjinta-kaṛi/ (one-other) 'another',
/yapa-kaṛi/ (person-other) 'another person', /ŋura-kaṛi/ (place-other) 'another
place'. Another example of this same general type is found in the *tjiliwiri* treatment
of the demonstrative determiners. The standard Walbiri forms are: /njampu/ 'this';
/yalumpu/ 'that near addressee, that near'; /yali/ 'that away from speaker and
addressee, that distant'; /yinja/ 'that beyond field of reference, that ultra-distal'.
From this set, *tjiliwiri* abstracts the proximate-nonproximate dimension and represents it by means of the opposition entailed in the pair /pinka/-/kutu/ (far-near).
Standard Walbiri /njampu/ and /yalumpu/, both 'proximate', are represented in
tjiliwiri by the single form /pinka/ (far). What is clearly involved here is a semantic
analysis of the Walbiri paradigm; the *tjiliwiri* opposition is based on the abstract
dimensions inherent in that analysis – i.e., /pinka/ represents the abstract proximity
feature shared by /njampu/ and /yalumpu/. The abstractness of this representation,
as in the case of /kaṛi/ above, is evidenced further by the fact that a grammatical
readjustment is required – the forms /pinka/ and /kuta/, which of all Walbiri forms
permit perhaps the most direct and unencumbered reference to the semantic
opposition proximate-nonproximate, are adverbials, not determiners in standard
Walbiri usage.

The existence of auxiliary speech-forms in Australia is well known (see, for example,
Robert Dixon's detailed account of the so-called mother-in-law language elsewhere
in this volume), but the particular form represented by *tjiliwiri* has, to my knowledge,
not been reported. Unfortunately, I am able to give only a brief account of this form
due to the circumstance that I did not become aware of its existence until I was
invited to attend the *kankaḷu* initiation at Yuendumu in 1967, very near the end of
my stay there. Since I was at that time involved in an intensive study of Walbiri
syntax, which required most of my attention, I was able to investigate *tjiliwiri* to
an extremely limited extent only. Nonetheless, I believe the material to be of sufficient interest to warrant this preliminary report, particularly in view of the fact
that an extensive study is still very much possible in the Walbiri community.

The theoretical potential of *tjiliwiri* is rather obvious. I have indicated a number
of ways in which it reveals aspects of the semantic structure shared by speakers of
Walbiri – typically, it confirms observations which can be made in the course of
conventional grammatical and semantic analysis, but occasionally it reveals categories
which are by no means obvious. For example, it would appear, superficially, that
the use of the verbalizing suffixes /-tjari-/ (inchoative) and /-ma-/ (causative,
transitive) in *tjiliwiri* coinages, is determined by the purely syntactic transitivity of
the Walbiri equivalent:

⟨18⟩ wanti- (to fall)/pirtjiḍi-tjari-;
　　　waŋka- (to speak)/wuḍuŋu[2]-tjari-;
　　　.............................
　　　kitji- (to cause to fall)/pirtjiḍi-ma-;
　　　patji- (to cut)/paḷpuṛu-ma-;
　　　.........................

---

[a] The *tjiliwiri* equivalent of Walbiri /njuntu/ 'you' is /ŋatju/ 'I'. That is, second person is
opposed to first person. This principle is sometimes extended to Walbiri /ŋatju/ as well – i.e.,
in place of the ego-alter principle normally used.

with /-tjari-/ forming intransitives and /-ma/ forming transitives. It appears, however, that the use of these suffixes is determined by a factor which is much less apparent than this. Thus, /-ma-/ forms transitive verbs of affecting – i.e., verbs denoting actions which produce effects upon the entities denoted by their objects. While it is true to say that /-tjari-/ forms intransitives, the suffix is also used in forming *tjiliwiri* equivalents of verbs which are syntactically transitive in Walbiri but which are not verbs of affecting in the above sense (e.g., the perception verbs of ⟨17⟩ – the object of perception is not 'affected' in any sense akin to that in which the object of, say, *cutting* is 'affected'). This is not the only way in which this elusive and covert category is reflected in *tjiliwiri* usage. The entire class of verbs of affecting, and only those, can be opposed to the single verb /yampi-/ 'to refrain from affecting, to leave alone'. That is, where the context is sufficiently clear, it is possible to abstract the feature shared by all verbs of affecting and represent it by its opposite /yampi-/, as in expressions like:

⟨19⟩ pinti-ŋki yampi-
(skin-instrumental leave)
'to cut by means of a knife';

yintjiṛi-ḷi yampi-
(speargrass-instrumental leave)
'to spear by means of a spear';

waḷu yampi-
(fire leave)
'to drink water';

waḷu-ŋku yampi[2]-
(fire-ergative leave (reduplicated))
'to be rained on, wetted by rain'.

The examples which I have used in this brief account illustrate the potential relevance of the *tjiliwiri* activity to the study of purely linguistic matters. There is, however, another aspect of the tradition which I have not touched on but which promises to be of considerable anthropological interest. The majority of Walbiri-*tjiliwiri* equivalences conform to a general principle of antonymy which is relatively consistent throughout and which can be characterized in terms of highly familiar, universal semantic notions – this is the principle which Walbiri speakers refer to by the term /ŋaḍalj-kitji-/ 'to invert, turn over, reverse'. There remain, however, many oppositions which can be fully understood only in reference to other aspects of Walbiri culture. The opposition of fire and water, for example, an opposition which is not uncommon elsewhere in the world, is an important theme in Walbiri ritual and epic. Many of the most important epics have an episode relating to fire, or burning, and another relating to rain, or flooding. In taboo-laden and culturally loaded areas, equivalences are, from the strictly linguistic point of view, idiosyn-cratic – e.g., the oppositions: kuṛa/ŋuḷu (semen, coitus/mulga seed); kuna/wamulu (excrement/animal or vegetable down used in ritual decoration); kuḍitji/ḷuŋkaḍa (shield, circumcision ground/blue-tongue lizard); ṭaṇŋa/maṛaljpa (always, eternally/totem center); kuyu/tjutju (animal, meat/sacred object, dangerous entity, useless object or material). Although each individual equivalence in such cases may not be explainable in terms of a single general principle, it is unlikely that the segment of the vocabulary which exhibits this seemingly idiosyncratic behavior is itself an arbitrary assemblage of lexical items. Rather universal considerations of taboo (cf. Leach, 1964) may well make it possible to predict which Walbiri terms will have idiosyncratic equivalents in *tjiliwiri* – e.g., /maliki/ 'dog', an animal living in close association with humans, is paired idiosyncratically with /tjatjina/ 'crest-tailed marsupial mouse', while wild animals are opposed according to general taxonomic principles. Furthermore, it is most likely that individual associations of this type can be explained in terms of highly general, nonlinguistic, principles of opposition of the sort studied by Lévi-Strauss (Lévi-Strauss, 1964).

Despite the serious ritual associations of *tjiliwiri*, Walbiri men consider it proper to enjoy it thoroughly. They derive much pleasure from it. When I began to understand the principle, I remarked to one Walbiri man: 'You certainly have something here!' He replied: 'Indeed we have!'

Massachusetts Institute of Technology

REFERENCES

Kay, Paul. 1965. 'A generalization of the cross/parallel distinction'. *American Anthropologist*, LXVII, 30–43.
Leach, E. 1964. 'Anthropological aspects of language: animal categories and verbal abuse'. In Lenneberg, E. H. (ed.), *New Directions in the Study of Language*, Cambridge.
Lévi-Strauss, C. 1964. *Le Cru et le Cuit. Mythologiques*, Paris.
Meggitt, M. J. 1962. *Desert People*, Sydney.
    1967. 'Gadjari among the Walbiri Aborigines of Central Australia'. *The Oceania Monographs*, no. 14.

PART III

PSYCHOLOGY

Overview[a]

DANNY STEINBERG

Until quite recently psychology has had relatively little to offer toward an understanding of semantic problems. It has generally been the case that while notions of meaning have been entertained by psychologists they were not done so for the purpose of explicating anything about language but for the purpose of explicating something about learning principles. Ever since the work of Ebbinghaus in the late nineteenth century, words or nonsense words have served as convenient stimuli and responses for psychologists exploring aspects of the learning process. Ebbinghaus' own preference for using the nonsense word became predominant in research and it was not until the 1930s that natural words began to be employed to any significant degree. Attention then became focused on the use of meaning as an independent variable with the thought that variations in the meaning of words might have a differential learning effect.

Almost all research utilizing meaning as a variable since that time has centered about the question, Does similarity of meaning facilitate (or inhibit) the learning of responses to stimuli? Such a question appeared basic since a sentence was conceived of as a sequence of words. The problem was to explain under what conditions the words became associated to one another. Experiments were (and still are) designed to determine the basis of word association and a variety of paradigms were used such as those of paired associates, transfer, and retroactive inhibition.

Actually it has only been since the work of Osgood, Miller, Skinner, Brown and Carroll in the 1950s that psychologists have become a part of the twentieth century intellectual boom in the study of language. The work of the anthropologist, the philosopher, and of the linguist, in particular, has confronted the psychologist with an array of language phenomena which he had heretofore not considered. Since such phenomena require explication within some theoretical framework, psychological theorists have been active in attempting to account for them. All of the papers in the present section strongly reflect in some way or other interdisciplinary influences. Those by Osgood, Fodor, Lenneberg, McNeill, and Bever and Rosenbaum attempt to provide a theoretical framework with which to deal with semantic problems, while Miller's offers experimental procedures which attempt to determine semantically relevant units.

Theorists have typically attempted to meet the challenge of new problems in three different and opposing ways; (1) by accounting for the phenomena with existing behavioristic theory, e.g., Skinner; (2) by modifying existing behavioristic theory, e.g., Osgood and Staats; or (3) by proposing new theory, e.g., Chomsky and his

[a] I would like to thank C. E. Osgood of the University of Illinois and Masanori Higa of the University of Hawaii for their many helpful suggestions.

colleagues Miller, Lenneberg, Fodor, McNeill, Bever and Garrett. The introduction of new problems and alternative theories (especially Chomsky's) has thus resulted in a serious questioning of the credibility of a behavioristic explanation of language phenomena.

The essence of the Chomskyan position is perhaps best summarized in a recent paper by Chomsky (1967a) in which he considers the problem of designing a model of language-acquisition, i.e., an abstract language acquisition device that duplicates certain aspects of the achievement of the person who succeeds in acquiring linguistic competence. Chomsky views himself as holding a rationalist conception of acquisition of knowledge where 'we take the essence of this view to be that the general character of knowledge, the categories in which it is expressed or internally represented, and the basic principles that underlie it, are determined by the nature of the mind. In our case, the schematism assigned as an innate property to the language acquisition device determines the form of knowledge (in one of the many traditional senses of "form"). The role of experience is only to cause the innate schematism to be activated and then to be differentiated and specified in a particular manner' (p. 9).

The innate schematism which Chomsky refers to incorporates 'a phonetic theory that defines the class of possible phonetic interpretations; a semantic theory that defines the class of possible semantic representations; a schema that defines the class of possible grammars; a general method for interpreting grammar that assigns a semantic and phonetic interpretation to each sentence, given a grammar; a method of evaluation that assigns some measure of "complexity" to grammars. ...[Thus] the given schema for grammar specifies the class of possible hypotheses; the method of interpretation permits each hypothesis to be tested against the input data; the evaluation measure selects the highest valued grammar compatible with the data' (pp. 7–8).

Chomsky thus posits that certain ideas (knowledge) are inherent in man's nature and that when such innate ideas are activated through experience they become functional for him. When Descartes proposed this doctrine in the seventeenth century, he was vigorously criticized by the British empiricists, particularly Locke. Locke held that all knowledge is derived from experience, and that sensation (from objects in the external world) and reflection (an awareness of operations in our own mind) were the only two sources of experience. He denied the notion that the mind incorporates innate ideas which are independent of experience. Nothing could be in the intellect which was not first in the senses. Simple ideas, the result of sensation and reflection, were considered the basis of thought and the units with which all complex ideas were composed. Locke, however, did not deny the existence of abstract ideas. The denial of abstract ideas was espoused by Berkeley and other empiricist philosophers like James Mill. He postulated instead a strict elementarism based on sensation in which complex ideas were the simple sum of simple ideas. For Mill, simple ideas became complex ones through principles of association. Certain notions of Bentham's utilitarianism, that human activity could be accounted for in terms of seeking pleasure and avoiding pain, were also incorporated into Mill's philosophy. Some of the essential doctrines of behaviorism are thus represented in Mill: elementarism with respect to ideas (knowledge), the principles of association, and the Law of Effect (or the principle of reinforcement) to account for learning when sheer frequency and contiguity of association were considered insufficient.

The focus upon behavior in psychology was largely the result of Watson's writing in the early part of this century. For Watson, mind was considered something extranatural and illusory. Consciousness and mental processes were rejected as subjects of study and their very existence doubted because they could not be directly observed by the senses. What was real was matter. Mind was a superstition and any investigation concerning it was believed to be unscientific. Human psychology was to be accounted for in terms which were wholly behavioral. The description of behavior was to be made in terms of physiological processes or, preferably, in terms of the muscular or glandular activities which comprised such processes. Only the language of the physical world could provide objectivity.

Behavior was assumed to be composed of simple units like reflexes and all larger behavior units were assumed to be integrations of a number of stimulus-response connections. Such connections are formed through the process of conditioning which involves the operation of associationist principles. Since Watson held that conditioning is the simplest form of learning and the elementary process to which all learning is reducible, everything that a person may learn in a lifetime must therefore be derived from the simple muscular and glandular responses which the child produces in infancy. Thus, in place of the classical doctrine of the association of ideas, behaviorism substitutes the association of motor responses. An idea, if the term is to have any meaning at all for the behaviorist, is a unit of behavior. What a child inherits are physical bodily structures and their modes of functioning. It has neither general intelligence nor any mental traits. Emotions are not considered as matters of feeling but as bodily reactions.

Without mind, what then was 'thought' for Watson? Thoughts were laryngeal habits and such habits, he held, were developed from random unlearned vocalizations by conditioning. At first language is overt and then it becomes sub-vocal, at which point it becomes thinking. Thinking then is the making of implicit verbal reactions which, while not spoken speech, is still behavior. It is behavior but in the form of changes in tension among the various speech mechanisms. Such changes in tension were regarded as duplicating the movement involved in overt speech. No distinction is therefore made between language and thought. They are one and the same in accord with a materialist conception where only matter is considered real.

With respect to language, words were considered by Watson as substitutes for objects and concrete situations since it was believed that they evoked the same responses that the object and situations themselves elicited. (Both the Osgood and Fodor papers discuss the problems involved in this conception.) The meaning of a word is simply the conditioned response to that word. For example, suppose the word 'bottle' is uttered whenever a child is presented a bottle. Suppose, too, that the child reaches for the bottle when it is presented. After a number of repetitions of 'bottle' uttered, and the bottle presented and reached for, the child will produce the reaching response when only the word 'bottle' is uttered. The child has then, in Watson's view, learned the meaning of the word, its meaning being the behavior which is elicited by it. Sentences were regarded as an association of words, a series of behavioral responses.

Skinner is one psychologist who has deviated very little from the original Watsonian behaviorism. Most other behaviorists have deviated, and some in relatively significant respects. One of the most significant changes (resulting in the prefix 'neo' to behaviorism) has involved the positing of unobserved events, termed intervening variables or hypothetical constructs by such psychologists as Hull and his student, Osgood. Still true to the materialist tenets of Watsonian behaviorism, however, such concepts are required to be specified in operational terms. That is, all concepts must

be derived from empirical ultimates. For neo-behaviorists like Osgood, such ultimates are the data of observation, physical response units.

According to Osgood the cognitive content or meaning of perceived objects, events
and words are derivable from behavior. If a stimulus object such as a bottle does not
elicit a distinctive pattern of behavior it has no meaning, but will derive meaning
through conditioning when it occurs with a stimulus which does produce a distinctive
pattern of behavior. Such a potent stimulus might be a nipple in the mouth, as
Osgood suggests, which elicits such behavior as sucking movements, milk ingestion,
and autonomic and musculature changes. Figure 1 (from Osgood, 1964) illustrates
the basis of the system.

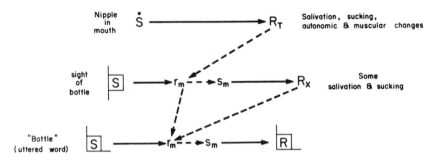

FIGURE I. *The development of meaning according to Osgood*

The repeated presentation of the conditioned stimulus (Osgood's $\boxed{S}$ in Figure,
i.e., the sight of a bottle) which originally does not elicit a distinctive pattern of
behavior in contiguity with the unconditioned stimulus (S in Figure, i.e., the nipple
of bottle touching the mouth) which does elicit a distinctive pattern of behavior
(R_T, i.e., salivation, sucking, etc.) results in the conditioned stimulus producing
unobserved (r_m) and observed (R_X) behaviors. Both r_m and R_X are considered to be
portions of the total behavior, R_T. It is r_m which Osgood designates as the meaning
of the perceived object ('perceptual sign'), bottle. Behaviors like glandular changes
and minimal postural adjustments are those which Osgood has suggested (in earlier
papers) are the most easily conditioned to become components of r_m.

The meaning of words is derived in a similar fashion as the meaning of perceptual
signs, that is, through conditioning. The meanings of words ($\boxed{S}$) can be derived
directly from observed behavior (R_T), or from the previously conditioned meaning
(r_m) and observed behavior (R_X) of perceptual signs ($\boxed{S}$). Words which have taken on
meaning can also serve to provide meaning to other words ('assign' learning). In all
cases, meaning is derived as a part of R_T, R_X, or r_m. Such a system of meaning,
Osgood points out, does not involve any circularity since all meaning is derived from
observable behavior. Thus the meaning of all words including such so-called abstract
words as *imagination, thought, god, theory, generative, necessity*, and *proposition* is
accounted for in the ultimate terms of physical response elements, as is the meaning
of sentences. Whether or not this mediation view (the unobserved response, r_m,
'mediates' between the observable stimulus and response) is to be considered as a
reference theory of meaning is a question which Fodor discusses in his paper.

Fodor's argument is essentially as follows. Single-stage conditioning accounts of language[a] are inadequate. This is so, because one absurd implication of the single-stage position is that a person would respond to words as he would to the referents of those words, e.g., a person would respond to the word *fire* as he does to actual fire. Although mediation theorists attempt to avoid this pitfall by the positing of an unobserved response (r_m), nevertheless, Fodor claims mediation theory is reducible to single-stage theory. He contends that such a reduction must be made because mediationists hold to the postulate that there is a one-to-one correspondence between mediating (r_m) and gross responses (R_T) for the purpose of providing a sufficient condition for making reference. Thus, either the mediationist's view of language is completely referential, or it is not referential at all. Either outcome, Fodor holds, is fatal to mediation theory.

For Fodor, acceptance of the one-to-one postulate implies that (observability aside) r_m must equal R_T. Thus, he is able to conclude that single-stage and mediation theories differ only in terms of observability and not in explanatory value. While Osgood agrees with Fodor that single-stage theory is inadequate and agrees to the holding of the one-to-one postulate, he does not agree that mediation theory can be reduced to single-stage theory. He holds while there is a one-to-one correspondence between r_m and R_T, nevertheless, r_m does not equal R_T (observability aside) and need not equal R_T for a referring function to be made. To determine just what implications follow from acceptance of the one-to-one postulate is then the problem which must be resolved.[b]

The foregoing brief description of some of the essentials of Osgood's theory is, I believe, basically representative of neo-behaviorist theorizing. Other neo-behaviorists, such as Staats (1967) and Berlyne, differ in many respects but not in any significant way especially when their views are compared with those of the Chomskyans. All neo-behavioristic theories derive 'ideas' from behavior with the utilization of principles of association. This is in sharp contrast to the Chomskyans who base the acquisition of knowledge on the notion of innate ideas and then rely essentially on some sort of matching principles which interact with experience in order to activate those ideas. The views of both of these camps contrast with those of such empiricists as Piaget and de-Zwart. These psychologists derive ideas from experience with the use of a *modus operandi* which a person inherits. Cognitive structures are generated as a result of this intellectual functioning.

For the behaviorist, ideas are ultimately composed of elements of behavior and in this sense are not abstract. On the other hand, neither the Chomskyans nor the Genevans (Piaget, de-Zwart, etc.) limit themselves to ideas based on the elements of the physical world, for they posit ideas which are purely abstract or mental. Ideas like transformations, deep structure, phrase marker, etc. cannot, they maintain, be reduced to physical terms. Given that behaviorists exclude such abstract ideas from their theory, it is not surprising that they sharply disagree with Chomsky's view of language as a set of rules since most of Chomsky's rules are abstract and involve a

---

[a] A single-stage conditioning model may be diagrammed as follows:

$$\text{UCS} \longrightarrow \text{UCR}$$
$$\text{CS} \longrightarrow \text{CR}$$

where UCS is the unconditioned stimulus, an object; UCR is the unconditioned response to the unconditioned stimulus; CS is the conditioned stimulus, a word form; and CR is the unconditioned response. Two different single-stage models are possible here: (1) where UCR = CR, i.e., both responses are identical; and (2) where UCR ≠ CR, i.e., both responses are not identical. The model both Fodor and Osgood criticize is (1), which is a substitution notion.

[b] For further discussion concerning the dispute see Osgood, Fodor and Berlyne, all (1966).

number of abstract entities. In the behaviorist view, sentences are constructed not by the application of rules but by the association of meaning constructs. The adequacy of the behaviorist's system which relies solely upon such a construct as r_m and upon associationist principles has been challenged by Chomsky (1957) and by others such as Bever, Fodor and Garrett (1968).[a] In his paper, in this section, Osgood attempts to justify the behaviorist system. This attempt cannot be regarded as successful, however, since Osgood does not specify precisely how any of the interesting language phenomena which he discusses (presuppositions, center-embedding, etc.) may be accounted for within his behaviorist system.

A summary of the essential characteristics of the various theoretical positions which have been discussed are indicated in Table 1. A question mark (?) has been placed next to those theorists whose position on all matters is uncertain.

TABLE 1. *Characteristics of some psychological theories*

| | Theorists | | |
Characteristics	Chomsky, McNeill, Lenneberg, Miller?, Fodor?	Osgood?, Skinner, Staats	Bever, Piaget?, De-Zwart?, Brown?
PHILOSOPHICAL ORIENTATION			
Rationalist	+		
Empiricist		+	+
BASIS OF KNOWLEDGE			
Innate Ideas	+		
Experience			+
Physical Response		+	
OPERATIONS TO YIELD IDEAS			
Matching	+		
Strategies			+
Association		+	
NATURE OF IDEAS			
Abstract	+		+
Response Elementarism		+	
NATURE OF LANGUAGE			
Generative	+		+
Markovian		+	
LANGUAGE AND COGNITION			
Language Independent	+		
Language Equals Cognition		+	
Language Dependent			+
LANGUAGE KNOWLEDGE AND USE			
Distinction Utilized	+		+
Distinction Not Utilized		+	

Even though Piaget is an empiricist (see Piaget, 1968), it does seem that he could incorporate a generative grammar into his psychology. Some theorists, however, are of the opinion that one must adopt a neo-rationalist view of the acquisition of knowledge (innate ideas, etc.) if one wishes to adopt a generative grammar. De-Zwart

---

[a] For other challenges to the sufficiency of the principle of association, see Lashley (1951), Bever, Fodor and Garrett (1968), McNeill (1968) and Asch (1969).

(1969), for example, has stated that 'Because of this epistemological position [Chomsky's innateness proposals] a psycholinguist who accepts Chomsky's theories of generative grammars for particular languages, cannot at the same time reject his, be it tentative, model of language acquisition' (p. 328). On the other hand, Goodman (1969) has stated that 'I can never follow the argument that starts with interesting differences in behavior between parallel phrases such as "eager to please" and "easy to please"; that characterizes these differences as matters of "deep" rather than "surface" structure; and that moves on to innate ideas' (p. 138).

An *empiricist generative* position would be a viable one from Goodman's point of view. In accord with such a position one could maintain that knowledge is derived from experience, and hold with Putnam (1967) and others (including perhaps Roger Brown) that any concept utilized in a grammar has a counterpart in general cognition and that such a counterpart is realized prior to language. Such notions as underlying structure, surface structure, transformation, grammatical categories and sentence could be viewed developing, therefore, through induction in non-linguistic contexts. The development of language is thus dependent on cognition.

In his paper, in this section, McNeill discusses the problem of whether language is entirely a function of general cognition (the condition for what McNeill terms a 'weak' linguistic universal) or whether it is a function of general cognition *and* a special linguistic ability (the condition for a 'strong' linguistic universal). While McNeill thinks it possible that both 'weak' and 'strong' universals may obtain, he propounds the thesis (as do the other Chomskyans) that humans possess and develop a special linguistic ability independent of general cognition.

Given the state of knowledge in psychology with respect to such topics as 'thought' and 'cognitive processes' and given the amount of uncertainty in linguistics with respect to what is called 'language universals', certainly no psychologist is in a position to gainsay McNeill's formulation regarding the possibility of a special linguistic ability. However, even if our knowledge was further advanced, one is presented with unusual difficulties in attempting a satisfactory empirical resolution of the question as posed by McNeill. The fundamental obstacle to such a goal is what may be termed the translation problem. For in order to test the hypothesis that 'sentence' is or is not solely restricted to language, we must attempt a translation from linguistic constructs to non-linguistic constructs. We must know what to look for when we try to observe the possible occurrence of the concept in areas other than language. The question of what would constitute an adequate translation, however, is not an easy matter to determine. Considering the magnitude of the difficulties involved with regard to gathering relevant evidence, it is likely that McNeill's notion of a special linguistic ability will be a viable hypothesis for a long while to come.

The neo-behaviorist, however, does not separate language from cognition. Osgood, for instance, regards the semantic component of language also as the cognitive system of man. The conception of the neo-behaviorists concerning the relations between language and cognition with respect to the understanding and production of speech is depicted in Figure 2. The conception of the Chomskyans is also shown. In sharp distinction to the behaviorists, the Chomskyans regard language (LANGUAGE KNOWLEDGE in Figure includes semantics) as a system which functions as a vehicle for thought (COGNITION in Figure), i.e., language is used to express and understand thought. The cognitive processes which result in or constitute thought are, however, unspecified.

A further distinction made by the Chomskyans, but not by the neo-behaviorists, is that of a separation between language knowledge and language use. Such a distinction is the basis of Chomsky's competence-performance distinction where a theory of language is incorporated into a theory of language use.[a] It is a theory of language use which involves the processes of understanding and production. Language knowledge (competence) interacts with rules pertaining to the use of such knowledge (Chomsky, this volume) as well as with conditions regarding memory, interest and the like to produce sentences or the understanding of them. With respect to production, a cognitive process yields a thought which in turn is expressed in language and uttered in speech. The reverse process would be that of understanding.

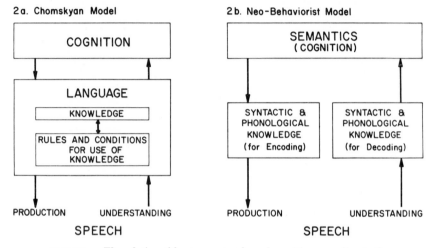

FIGURE 2. *The relation of language, speech, and cognition according to the Chomskyans and the neo-behaviorists*

Chomsky has maintained that it would be 'counter-intuitive' and absurd to view a competence model such as the one he describes, as one of performance. He states (Chomsky, 1967b), 'In fact, in implying that the speaker selects the general properties of sentence structure before selecting lexical items (before deciding what he is going to talk about), such a proposal seems not only without justification but entirely counter to whatever vague intuitions one may have about the processes that underlie production' (p. 436). Chomsky thus regards language knowledge (competence) as a *resource for use* rather than as a *process of use*. It is information about our language which we tap in order to produce and understand sentences, and such information in and of itself is inadequate for such performance tasks.

Language knowledge (a system of rules) is regarded by Chomsky as actual knowledge that a person has. As such the content or the structure of this knowledge is a psychological characteristic of a person. It is the discovery of this underlying system of rules, this 'mental reality underlying actual behavior' (Chomsky, 1965) with which linguistic theory is concerned. The linguistic universals which Chomsky posits, deep structure, grammatical categories, etc., are held to be psychologically real since they specify characteristics of mind (Chomsky, 1968). Thus, while the

---

[a] For further discussion on this topic see the papers by Harman (1967, 1969), Chomsky (1969), Schwartz (1969), Fodor (1969) and those in Lyons and Wales (1966).

linguist's characterization of competence may be neutral in a sense with respect to use, it is evidently not neutral with respect to precisely how that knowledge is stored or organized in the person.

As an example of how a competence model may be utilized in performance, Chomsky (this volume) cites Katz and Postal. In their procedure for production, a semantic representation from the competence model is first selected, then a syntactic structure, then a phonetic representation. Such a set of selection rules would provide an account of a hypothetical performance model with respect to production. Another set of rules would be required to deal with the understanding or interpretation of sentences.

A problem arises, however, with regard to distinguishing the rules of 'performance' with those of 'competence'. The basis for such a distinction is not immediately evident since both alleged types appear to qualify as language knowledge. For example, along with Katz (this volume), Chomsky regards those rules of the semantic component which are concerned with the disambiguation of sentences as belonging to competence. On the other hand, Weinreich (also this volume) views such rules as involved in the use of language. He says that we 'ought to regard the automatic disambiguation of potential ambiguities as a matter of hearer performance'. If the distinction between competence and performance is to be a viable one, more definitive criteria than those provided at present appear necessary.

In contrast to the empiricist notions of Osgood where meaning is conceived to be derived from behavior, Lenneberg holds a view of meaning that is essentially rationalistic. Words are viewed as tags for concept categories or categorization processes which are considered to be innate, i.e., the words of the lexicon are regarded as labels of innate categories. Such categories vary from species to species and the criteria or the nature of categorization, Lenneberg maintains, have to be determined empirically for each species.

In accord with his theme that language is species-specific, Lenneberg contends that there is a biologically based capacity for naming that is specific to man. He claims in the paper in this section that it is not possible for an animal to do the 'name-specific stimulus generalization' that every child does automatically, i.e., to regard some words as class names, and not specific object names. Lenneberg says: 'The hound who has learned to point to the tree, the gate, the house in the trainer's yard will perform quite erratically when given the same command with respect to similar but physically different objects, in an unfamiliar environment.' With respect to this claim the reader may wonder about the following case. Many dogs learn to respond to such verbal commands as 'Forward', 'Left', 'Right', 'Sit'. Given the command 'Forward', a seeing-eye dog, for example, will go forward under a variety of environmental conditions, e.g., in a house, in a yard, in a street, or on a bus. The question is whether such a case would qualify as an instance of 'name-specific stimulus generalization'. Unfortunately, Lenneberg offers but a brief sketch of his ideas[a] before passing on to present a series of interesting and valuable empirical studies involving various aspects of relating words and objects with respect to physical dimensions.

The paper by Miller provides a very useful survey of various empirical methods which have been used in the study of meaning. While reading this review it should

---

[a] For an interesting discussion related to a consideration of Lenneberg's view that names with unique referents may be incorporated into discourse but are not considered part of the lexicon, see the papers (particularly that of Linsky) in the Reference section of Philosophy.

be borne in mind that the psychologist's early venture into meaning was dominated, in the main, by the hope that meaning would provide a basis of differentiation among words, and thus account for how words became associated to form sentences. Measurement scales were constructed on the basis of such alleged semantic dimensions as: the number of associations given to a word within a unit of time, amount of salivation, intensity of galvanic skin response, or degree of favorability. The obvious inadequacy of any approach where but a single dimension of meaning is assumed to differentiate all words was recognized by many theorists, particularly Osgood and Deese. Osgood tried to remedy the situation with his multi-dimensional technique, the Semantic Differential; Deese with his structural description of association. Unfortunately no theorist who has been involved in the measurement of meaning has been able to isolate experimentally all of those *varied* aspects of meaning necessary for a relatively complete solution to semantic problems like those concerned with the formation of phrases and sentences and others considered in this volume.[a]

Of particular importance in the Miller paper is the presentation of a new method which Miller and his colleagues have developed. This method uses word classification as a discovery procedure through which elements of meaning ('semantic markers') may be identified. The words to be classified are typed on cards and subjects are asked to sort them into piles on the basis of 'similarity and meaning'. The data then are tabulated in a matrix form on the basis of co-occurrence pairs so that an analysis which yields semantic clusters may be made. The method produces rather credible results as a glance at the Miller figure in the paper shows.

Since the technique yields a hierarchical arrangement of semantic markers, Miller recognizes that it is applicable only to those lexical domains where semantic markers structure to form a hierarchy.[b] Miller has noted that there are lexical domains such as kinship terms where the semantic features which characterize it are not structured in a hierarchical fashion. One assumption underlying the method is that lexical items which have been judged as 'similar' (by being classified together) have been done so on the *same* basis of meaning. Such an assumption is necessary since there would be little semantic value to the finding that a group of subjects judged certain items as 'similar' but that it was based on judgments involving different semantic dimensions.[c]

[a] I do not wish to imply that units which have been experimentally isolated by various theorists are necessarily semantically irrelevant. The *Evaluation* factor isolated by Osgood's Semantic Differential, for instance, is undoubtedly of great semantic significance.

[b] The method is also inapplicable in those cases where the semantic markers composing a hierarchy must be cross-referenced. This restriction is necessitated by the fact that no lexical item can appear in a final cluster analysis more than once. For example, with reference to the Miller figure it does not appear possible to deal with lexical items that have both 'social' *and* 'personal' characteristics, or both 'personal' *and* 'quantitative' characteristics, etc.

[c] To illustrate this point, suppose that subjects have been given a set of kinship terms to classify, and among them are the items 'daughter' and 'son'. Further, suppose that some of the subjects think of 'generation' as a meaningful feature and classify daughter and son together because they are of the same generation, while the other subjects think of 'lineality' (relatedness by blood) as a meaningful feature and classify daughter and son together because they are of the same lineage. Since these data are collapsed together in the co-occurrence matrix, a distinction between such features of meaning as 'generation' and 'lineality' would not be maintained. The matrix of co-occurrence would merely indicate that daughter and son are 'similar' to one another. Furthermore, that daughter and son will not be regarded as 'similar' if a subject makes a classification on the basis of 'sex' also raises a problem since judgments of 'similar' and 'dissimilar' are treated by Miller as mutually exclusive. That is to say, the method cannot show that daughter and son are similar in some respects (generation lineality) while at the same time dissimilar in others (sex).

The paper of Bever and Rosenbaum is also concerned with the determination of semantic structures. However, their focus is not so much upon the methodological problem of discovering semantic structure as it is upon the explanatory value of such structures. They posit certain featural and lexical hierarchical structures so that they may provide an explanation for why it is that certain sentences are regarded as semantically appropriate while others are regarded as semantically deviant. With reference to the first type of hierarchy, the authors postulate that elements of meaning (semantic features) underlie lexical items, e.g., a lexical item such as 'person' is characterized by such features as +*human* and +*animate*. They hold that features are related to one another in a hierarchical fashion, e.g., +*human* always implies +*animate*. The relations obtaining between features is conceived to be one of class inclusion. The assignment of semantic features to lexical items in terms of contextual privilege of occurrence enables the authors to provide a theoretical basis according to which acceptable and deviant sentences may be distinguished.[a]

The postulation of a second type of hierarchy, that of lexical items (to be distinguished from hierarchies of features),[b] also allows the authors to account for judgmental differences concerning the acceptability of sentences. The *be* and *have* (inalienable inclusion) lexical hierarchies provide especially valuable insights for semantic theory construction and thus deserve the close attention of the reader.[c]

In closing, it is worth noting that all of the papers discussed above, besides contributing to our understanding of certain semantic problems, also illustrate that the impact of Chomsky's ideas has been profound upon psychologists working in the area of language. The number of notable psychologists who have abandoned behaviorist theory has risen now to the extent that one of the most viable counter-behaviorist forces since the advent of behaviorism has been created. While thus far it is mainly in the area of language that behaviorism has been put on the defensive, the distinct possibility remains that all areas of psychology will be revolutionized by Chomskyan ideas (especially those of a transformational generative nature). It is still too early to tell however. In this regard one is reminded of the impact which Gestalists have had upon psychology. Their anti-behavioristic notions failed to penetrate much beyond their own area of specialization, that of perception.

University of Hawaii

---

[a] For a detailed consideration of some of the methodological problems involved in such an approach see the Alston paper in the Philosophy section.

[b] Dixon (in the Linguistics section) hypothesizes a distinction somewhat similar to that of Bever and Rosenbaum with his concepts of 'nuclear' and 'non-nuclear'.

[c] A question may be raised, however, with respect to the consistency with which postulated deviancy criteria have been applied to the various sets of sentences. For example, it seems odd that sentence $\langle 4\,ei\rangle$ 'the ant has an arm' is viewed as not deviant while one like $\langle 22\,c\rangle$ 'a shotgun shoots bullets' is viewed as deviant. According to the authors, $\langle 4\,ei\rangle$, 'the ant has an arm', is not deviant because it 'doesn't happen to be true' but $\langle 4\,eii\rangle$, 'the arm has an ant', is deviant because it 'couldn't possibly be true'. While this deviancy criterion is applied to these sentences and some others, it is not clear if this same criterion is also utilized when other sets of sentences are considered, such as those which include items as $\langle 7\rangle$, 'The sword shoots bullets', and $\langle 22\,c\rangle$ 'A shotgun shoots bullets'. Both of these sentences are regarded as deviant, i.e., they 'couldn't possibly be true'. Yet it does seem possible that a sword or a shotgun could be constructed to fire bullets, i.e., that the sentences in question actually 'do not happen to be true'. Further empirical validation may be in order here.

REFERENCES

Asch, S. E. A reformulation of the problem of associations. *American psychologist*, 1969, XXIV, no. 2.

Berlyne, D. Mediating responses: a note on Fodor's criticisms. *Journal of verbal learning and verbal behavior*, 1966, V, 408–11.

Bever, T. The cognitive basis for linguistic structure. In *Cognition and the development of language*. J. R. Hayes (ed.). New York: Wiley, 1970.

Bever, T., Fodor, J. and Garrett, M. A formal limitation of associationism. In *Verbal behavior and general behavior theory*, T. R. Dixon and D. L. Horton (eds.). Englewood Cliffs, N.J.: Prentice-Hall, 1968, 582–5.

Chomsky, N. *Syntactic structures*. The Hague: Mouton, 1957.
Aspects of the theory of syntax. Cambridge, Mass.: MIT Press, 1965.
Recent contributions to the theory of innate ideas. *Synthese*, 1967a, XVII, 2–11.
The formal nature of language. In *Biological foundations of language*, E. Lenneberg. New York: Wiley, 1967b, 397–442.
Language and mind. New York: Harcourt, Brace & World, 1968.
Linguistics and philosophy. In *Language and philosophy*, S. Hook (ed.). New York: New York University Press, 1969, 51–94.

De-Zwart, H. S. Developmental psycholinguistics. In *Studies in cognitive development*, D. Elkind and J. Flavell (eds.). New York: Oxford University Press, 1969, 315–36.

Fodor, J. More about mediators: a reply to Berlyne and Osgood. *Journal of verbal learning and verbal behavior*, 1966, V, 412–15.
Current approaches to syntax recognition. Cambridge Mass.: MIT, Department of Linguistics, 1969, Mimeo.

Goodman, N. The emperor's new ideas. In *Language and philosophy*, S. Hook (ed.). New York: New York University Press, 1969, 138–42.

Harman, G. Psychological aspects of the theory of syntax. *Journal of philosophy*, 1967, LXIV, 75–87.
Linguistic competence and empiricism. In *Language and philosophy*, S. Hook (ed.). New York: New York University Press, 1969, 143–51.

Lashley, K. S. The problem of serial order in behavior. In *Psycholinguistics*, S. Saporta (ed.). New York: Holt, Rhinehart and Winston, 1951, 180–98.

Lyons, J. and Wales, R. J. *Psycholinguistics papers*. Edinburgh: Edinburgh University Press, 1966.

McNeill, D. On theories of language acquisition. In *Verbal behavior and general behavior theory*, T. R. Dixon and D. L. Horton (eds.). Englewood Cliffs, N.J.: Prentice-Hall, 1968, 406–20.

Osgood, C. E. A behavioristic analysis of perception and language as cognitive phenomena. In *The cognitive processes*, R. Harper, C. Anderson, C. Christensen and S. Hunka (eds.). Englewood Cliffs, N.J.: Prentice-Hall, 1964, 184–210.
Meaning cannot be an r_m? *Journal of verbal learning and verbal behavior*, 1966, V, 402–7.

Piaget, J. Quantification, conservation, and nativism. *Science*, 1968, CVXII, 976–79.

Putnam, H. The 'innateness hypothesis' and explanatory models in linguistics. *Synthese*, 1967, XVII, 12–22.

Schwartz, R. On knowing a grammar. In *Language and philosophy*, S. Hook (ed.). New York: New York University Press, 1969, 183–90.

Staats, A. Linguistic mentalistic theory versus an explanatory S-R learning theory of language development. Honolulu: University of Hawaii, 1967, Mimeo.

Where do sentences come from?

CHARLES E. OSGOOD

The other day, intent on demonstrating something obvious (but not trivial) about the language of Simply Describing, I went to my graduate seminar in psycho-linguistics armed with a plastic ring (orange), one ball (small and black) and two plastic cups (one red and one green), all in a shopping bag. I told the some thirty students that I would ask them to close their eyes when I gave the number of each of five little demonstrations, then to open them when I said 'open' and to close them again when I said 'close' – and then they were to simply describe, in a single sentence that a hypothetical six-year-old boy 'just outside the door' would understand, what they observed during the eyes-open period. The only props were a bare table (desk) and myself, whom I instructed them to refer to as 'the man'. The following sequence of events transpired:

1. 'Close your eyes for #1.' I place *the orange ring in the middle of the table.* I say 'open' and then in a few seconds 'close', and I ask them to 'simply describe what you saw in a sentence'.
2. 'Close for #2.' I remove the plastic ring, take the black ball, and *I stand holding the ball* in front of me about shoulder high. I say 'open', then 'close' and 'simply describe what you saw'.
3. 'Close for #3.' I place *the black ball in the middle of the table.* And then again, 'open' (pause) 'close' and 'simply describe'.
4. 'Close for #4.' I remove the ball, take the *red* cup, and *I stand holding the red plastic cup in my hand.* 'Open' (pause) 'close' and 'describe'.
5. 'Close for #5.' I place the red cup back in the shopping bag, take the *green* one, and I place the *green plastic cup in the middle of the table.* Then finally, 'open' (pause) 'close' and 'simply describe'.

Note that each of the three critical demonstrations, #1, #3 and #5, is identical except for the particular object which is in the middle of the otherwise bare table, an orange ring, a black ball, or a green cup. Yet the types of sentences they produced varied markedly.

The sentences produced by #1 were typically either

⟨1⟩ *An* orange ring is on the table   or

⟨2⟩ There is *an* orange ring on the table

with at least one and often more adjectives included. Sentences with definite articles.

⟨3⟩ *The* ring on the table is orange

or sentences making explicit the adjectival transformation

⟨4⟩ A ring is on the table *and it is orange*

almost never occurred. On the other hand, after seeing #2 (usually described as *the man is holding a small black ball*), demonstration #3 did regularly yield sentences with the definite article along with adjectival pronominalization

⟨5⟩ *The* black ball is on the table

but (except for a few philosophers in the class) types ⟨1⟩ and ⟨2⟩ rarely occurred and

⟨6⟩ The ball on the table *is black*

never occurred. Yet demonstration #5, following *the man is holding a red cup*, did typically produce

⟨7⟩ The cup on the table *is green*

but *never* sentences analogous to ⟨5⟩, which in this case would have been

⟨8⟩ The green cup is on the table.

It is obvious that non-linguistic, perceptual antecedents can create *cognitive presuppositions* in the same way that previously heard or uttered sentences do and that these presuppositions influence the form that descriptive sentences take. If a speaker has already seen a particular BLACK BALL,[a] and assumes that his listener is familiar with it also, then it is absurd for him to say *the ball on the table is black*, since it is its new location which is now informative, not its color or size. Perhaps not so obvious is the fact that such demonstrations call into question the semantic equivalence of even such common transformations as N *is* A into AN. In demonstration #1 above, not a single sentence included an N *is* A assertion (*ring is orange*), yet this was the first exposure to this or any other object in the situation. Furthermore, it is apparent that *the green cup is on the table* is not semantically equivalent to *the cup on the table is green*; demonstration #3 produces the former type but not the latter, whereas #5 produced the latter type but not the former. The answer *yes* to the question *is the cup on the table green?* refers to the location of the object (its color being presupposed).

Where do sentences come from? In a generative grammar (linguistic competence model) the symbol S represents the set of all possible grammatical sentences in language L, and S can be re-written or expanded in all of these possible ways – infinitely given recursive devices.[b] But linguists ordinarily start with some actually uttered or potentially utterable sentence or non-sentence and procede to analyse it in *a posteriori* fashion. The term 'generative' can thus be misleading. In any case, it is clear that any theory of language behavior (psycholinguistic performance model) must inquire into the antecedents of S and relate these antecedents to the forms and contents of particular sentences. Neither the syntactic bone nor the lexical flesh of sentences created by real speakers is independent of the non-linguistic contexts in which they occur. The body of this paper will be concerned with demonstrating, in some detail, how the form as well as the content of sentences can be influenced by manipulating the perceptual context in which they are produced.

Philosophers and linguists have given precious little attention to pre-linguistic cognitive behavior, in any of its three senses – prior to language in the evolution of human species, prior to language in the development of the individual or prior to language in the generation of particular sentences by ordinary speakers. A philo-

---

[a] Throughout this paper I will use all small capitals when referring to perceptual signs and their interactions in perceived events.

[b] Given the proliferation of subordinate embedded Ss characteristic of contemporary linguistic analysis, and the mutual constraints placed by one S upon others, I wonder how this symbol actually could have a constant reference.

sopher-linguist may claim that there can be no thinking without language,[a] but it is difficult to imagine how pre-linguistic humans types survived without 'thinking', to say nothing of dogs and untrained deaf-mutes. Lenneberg, in his *Biological Foundations of Language* (1968), devotes exactly three pages (276–8) to pre-linguistic development, and this exclusively to articulation; philosophers of language typically talk about 'objects' (e.g., the traditional APPLE) as physical givens when they are really talking about perceptual signs which themselves have to be learned. And only quite recently have philosophers of ordinary language – and only very recently linguists – begun to take an interest in the cognitive pre-conditions for uttering sentences of certain types. Psychologists like Piaget, on the other hand, have perhaps paid too much attention to cognitive development and too little to language development (cf. Mehler and Bever, 1967, vs. Piaget, 1968, in *Science*).

My arguments in this paper will be as follows: (1) That pre-linguistic perceptuo-motor behavior displays many (even if not all) of the characteristics of linguistic behavior, including semantics (meanings of perceptual signs like APPLE or BALL), syntactics (appropriate interpretations of DADDY HITS BALL vs. BALL HITS DADDY) and pragmatics (appropriate behaviors with respect to gestured TAKE BALL vs. GIVE BALL). (2) That the ability to paraphrase perceptuo-motor events in language (Simply Describing), and vice versa (Simply Acting Out), implies both (*a*) that perceptual signs and events must have meaning and structure (one cannot paraphrase a meaningless, structureless string like *colorless furiously ideas sleep green*) and (*b*) that perceptual and linguistic signs and sequences must, at some level, share a common representational (semantic) system and a common set of organizational (syntactic) rules.[b] (3) That that which is shared by both sign systems is not linguistic but cognitive in nature, and therefore that the constructs and rules of generative grammars are, in principle, incapable of dealing with it.[c] And, finally, (4) that this non-linguistic cognitive system is 'where sentences come from' in sentence creating by speakers and 'where sentences (finally) go to' in sentence understanding by listeners.

The new look at presuppositions

When did linguists begin to look at the presuppositions that lie behind the production of sentences? The first comments I heard about it were at the Linguistic Institute at Illinois in the summer of 1968, but by the second Institute at Illinois in 1969 the classrooms and lecture halls were full of presuppositions, and, as the reader will have noted, more than half of the contributions to the Linguistics section of this volume deal in one way or another with the same topic. I do not know just who introduced the term 'presupposition' to the linguistic literature, or exactly when, but Fillmore, Lakoff, McCawley or Ross would certainly be good candidates for the honor. Where did this new psycholinguistic notion come from? Fillmore (1969) traces his indebtedness to certain of the philosophers of ordinary language, particularly Austin (1962) and Alston (1964) with their concerns about 'illocutionary acts' and the 'happiness

[a] As Zeno Vendler did at a conference on *Language and Thought* at Tucson in 1968 (see J. Cowan (ed.), *Studies of Language and Thought*, 1970).

[b] This is not, of course, a claim that perceptual and linguistic semantics and syntactics are co-extensive with each other, merely that the one must at least overlap the other: presumably there are some things which one can 'think' with language which he cannot 'think' without it.

[c] I shall also argue that this cognitive system is to be identified with the 'semantic component' of grammars and the latter in turn with the 'deep structure' level of syntax – not an outrageous argument any longer, apparently.

or felicity conditions' for the use of certain sentences, but also to Strawson's paper 'On Referring' (1950) and most remotely to Frege (1892).

In recent linguistic papers the role of presuppositions in a wide variety of sentence phenomena has been described; as a small sample, in the choice among verbs of judging (*accuse, criticize, blame*) by Fillmore (1969), in the choice between *some* and *any* by R. Lakoff (1969), in the use of pronouns and reflexives (McCawley, 1968; Sampson, 1969). Also, the kinds of problems in grammatical description that led Chomsky (1965) to introduce the notion of *indexing* (e.g., choice among the *teacher*$_1$ *hates himself*$_1$ and *the teacher*$_1$ *hates the teacher*$_2$) come down to special instances of presupposing (cf. Sampson, 1969). Finally, we have those situations, like the demonstration with which I began this paper, where presuppositions are established by non-linguistic contexts. Olson (1969), moving along very similar lines, has described how varying comparison objects (BLACK CIRCLE vs. WHITE SQUARE) can influence the way a speaker will refer to a given object (WHITE CIRCLE), saying...*under the white one* for the first contrast and...*under the round one* for the second, for example.

This preoccupation with presupposition has led to debate, within as well as between linguists, as to the nature of such phenomena. Are presuppositions linguistic in nature (prior, even if implicit, sentences) or are they psychological in nature (cognitive states or sets)? Morgan (1969) says that 'many kinds of presuppositions "act like" previously uttered sentences...but to describe both with the same set of rules, it is necessary to represent presuppositions as trees to the left of the sentences they are associated with' (pp. 66–7). Lakoff (1969), in explaining why there can't be any *some-any* rule,[a] states that 'something that is present only at the most abstract levels of grammar must play a part in determining the form which surface structures must take...when *some* is used, the presupposition is necessarily positive; when *any* appears, it may be negative or neutral'. But she also says that 'the beliefs and expectations of the speaker are reflected in his choice of *some* or *any*, and the meaning of the sentence is correspondingly changed'. And Lakoff concludes that 'whether [representation of presuppositions] would be done by assuming separate sentences (e.g., *I hope this S is true, or I believe that the subject of this sentence is true*), or in another way, is not clear...' (pp. 612–13).

Other linguists seem to have accepted the non-linguistic, cognitive nature of presuppositions. McCawley (this volume), for example, in discussing indexing, states that 'indices exist in the mind of the speaker rather than in the real world...and...the noun phrases which speakers use fulfill a function comparable to that of postulates and definitions in mathematics: they state properties which the speaker assumes to be possessed by the conceptual entities involved in what he is saying'. Similarly Sampson (1969): 'I suggest that the mind of a person entering into a speech situation should be regarded as containing a number of entities which I shall call "referents" [cf. McCawley's term, "intended referent"] (p. 4)...so we must consider the hearer's stock of referents to be arranged in some kind of mental space (p. 7) [and finally]...I believe this is a genuine problem which would sooner or later have to be faced by any theory of language' (p. 9). Schlesinger (forthcoming) builds a general theory of language production in which a non-linguistic 'intentional' system is assumed to antedate strictly linguistic operations.

---

[a] Among Lakoff's compelling examples are *Who wants some beans?* (speaker expects someone does want beans) vs. *Who wants any beans?* (speaker expects no one wants beans) and *If you eat any candy, I'll whip you* (speaker is threatening you) vs. *If you eat some spinach, I'll give you ten dollars* (speaker is promising you).

Psychologists, of course, have generally assumed that non-linguistic cognitions were the sources and destinations of sentences. James Deese (1969), for example, speaks of 'understanding' as occurring when a listener can make some external information, linguistic or perceptual, conform to some conceptual category, this category providing the abstract structure of an interpretation. And on the encoding side, according to Deese 'the primordial thought that gives rise to a linguistic interpretation is not itself linguistic' (p. 519). He believes that these cognitive categories are derived from underlying perceptual relations, including such gestalt-like properties as *grouping, contrast* and *similarity* as well as *classification* in terms of superordination and subordination. Donald Campbell (1966), after making the astute observation that 'you can't teach a language by telephone', asserts that 'it is the child's already available entifications of the physical environment which provide his trial meanings', and by 'entification' he is also referring to the gestalt-like perceptual tendencies toward grouping in terms of qualitative similarities and contrasts and the *common fate* (trans-positions as wholes) which characterize the things which ordinarily get named (*cup* and *saucer*, but probably not *cupandsaucer* and certainly not *cupandlittlefingerandsoutheastcornerofroom*). As Campbell points out, languages tend to follow Aristotle's advice and 'cut nature at her joints'.

To indicate what I consider to be the absurdity of treating presuppositions as prior, implicit sentences, let me borrow an example from Fillmore (1969) and expand on it. For the imperative sentence ⟨9⟩

⟨9⟩ Please shut the door

the 'happiness conditions' (presuppositions) include at least the following: (*a*) speaker and addressee have a role relation which permits such requests; (*b*) the addressee is capable of shutting the door; (*c*) the speaker has in mind some particular door and assumes the addressee can identify this door (saying *go shut a door* would be like saying *go fly a kite!*); (*d*) the door in question is presently open; and (*e*) the speaker wants this door to become closed. Now let us phrase appropriate prior implicit sentences for this speaker.

⟨10⟩ I am your father   (*a*)
⟨11⟩ You are able to shut doors   (*b*)
⟨12⟩ There is a door here as we both know   (*c*)
⟨13⟩ That door is open   (*d*)
⟨14⟩ I desire the door to become closed.   (*e*)

I submit that it is ludicrous to postulate that a speaker and a listener (to interpret fully) must riffle through (generate) such a series of sentences in order to create or understand ⟨9⟩ – and this is a comparatively *simple* sentence. Not only is it intuitively much more satisfying to assume that situational (perceptual) signs are shared by speaker and hearer and create the presuppositions, but processing of the above sort would be incredibly inefficient. Fillmore points out that we can *infer* such presuppositions from certain structural and semantic characteristics of the sentence, but then adds: 'more importantly, however, it needs to be pointed out that some of these conditions are really *preconditions* [italics mine] for the use of the sentence' (p. 96). In other words, such non-linguistic cognitions are *predictive* of the form and content of the sentence.

In an attempt to bring such presuppositions and indexings into the realm of familiar linguistic analysis, it seems to me that many linguists are now carrying their methods to ridiculous lengths. For example, Sadock (1969) proposes that there are *super-hypersentences* that are required to handle sentences like ⟨15⟩

⟨15⟩ I promise that this is the end

with the highest node being a super-hypersentence of the form *Speaker – V – Addressee – S* (where *V* is + performative complement), with this *S* dominating an ordinary hypersentence of the form *Speaker – V – Addressee – S* (where *V* is + promise) and with this *S* dominating in turn the sentence *Speaker promise Address that this is the end*. When will we have *Extra-superduper-super-hypersentences?* In a similar, if not so extreme, vein, we find McCawley (as reported in Morgan, 1969) analysing the semantically rather simple verb *kill* as a derived constituent CAUSE (BECOME (NOT (ALIVE))) as part of his attempt to give semantic representation in the form of tree-like structures. By a rule of 'predicate-raising' a series of embedded sentences becomes *x causes become not alive y* which is considered to be, if not synonymous with, at least an adequate paraphrase of *x kill y*. The underlying motivation here, as I see it, is to apply familiar linguistic constructs and rules to the semantic component, but if it can be shown that purely perceptual signs and their interrelations utilize the same semantic system (as when John perceives PAUL KILLS TOM) then I think this attempt is doomed to failure.

Nevertheless, presuppositions do influence the structure of sentences and, as Fillmore puts it (1969, p, 98), 'any complete account of the grammatical description of a language will need to bring in presuppositional facts at many points'. Yet, if presuppositions cannot be handled with strictly linguistic constructs and rules, several fundamental issues arise: (1) *Is the study of presuppositions a linguistic or a psycholinguistic business?* Or, put another way, can presuppositions be considered as part of the grammar of a language, as Fillmore claims? (2) *Can syntax be formally separated from semantics?* As Maclay presents the situation in his overview of the linguistic contributions to this volume, 'the thrust of many of the arguments in this section is that the well-formedness of sentences cannot be determined solely on formal or syntactic grounds...[and] once the autonomy of syntax has been denied, the existence of a distinct level of syntactic deep structure also becomes unacceptable'. (3) *Can any clear distinction between competence and performance be maintained under these conditions?* According to Maclay, 'Katz and Fodor expanded the empirical domain of linguistics to include semantics with the ultimate result that the autonomy of syntax is now seriously questioned by linguists. The autonomy of competence may well be the next victim'. Such autonomy will clearly be questioned by the data on paraphrasing 'things' which I will present in the next section of this paper. It would appear that Chomsky was building better than he knew when, in more or less of an aside on the first page of his *Language and Mind* (1968), he transformed linguists into psychologists: 'I think there is more of a healthy ferment in cognitive psychology – and in the particular branch of psychology known as linguistics – than there has been for many years.' Welcome to the fold!

Paraphrasing things

In the fall of 1968 I arranged this demonstration-experiment for 26 native speakers of English.[a] A sequence of 32 'scenes' were presented, preceded by the following

[a] Subjects 1–9 were staff and graduate assistants in the Center for Comparative Psycholinguistics; subjects 10–26 were graduate students in my seminar on psycholinguistics. Several people who are native speakers of a language other than English participated, but they did their describing in their own native language and their data were not included for present purposes.

instructions: 'Start each numbered item (1–32) with your eyes closed; when I say 'open' observe what is transpiring and then close your eyes when I say 'close'; describe in a single sentence just what you see while your eyes are open; imagine that you are describing things for a hypothetical six-year-old child standing behind a hypothetical screen – that is, use simple ordinary language, not fancy scientific or philosophical jargon; refer to me as *the man* and to other objects and events as you see fit. Any questions? All right, close your eyes and get ready for #1.' Ideally, of course, these perceptual 'scenes' should have been prepared on color film, and this will be done for cross-cultural studies with children of various ages that we are planning.

For convenience in discussion, I will refer to the various objects used in the demonstrations (including myself as THE MAN) as *entities* (perceptual signs analogous to 'words') and to the interactions and relations produced among these perceived entities as *events* (perceptual structures analogous to 'sentences'). Table 1 presents the sequence of perceptual events[a] along with the psycholinguistic intentions presumably involved – both my intentions as the experimenter and, if successful, the encoding intentions of the subjects created by the perceptual inputs. These intentions are the predictions being made in this experiment, and the sentences compared in testing them will be detailed in course. Needless to say, I ordered the sequence of perceptual events with presuppositional 'malice aforethought'; for example, #3 (THE BALL IS ROLLING ON THE TABLE) followed by #4 (THE BALL IS ROLLING SLOWLY ON THE TABLE) or #15 (A BALL, A SPOON AND A POKER CHIP ARE ON THE PLATE) followed by #16 (THE PLATE IS EMPTY).

Note that events #12, #13 and #14 (THE BLACK BALL HITS THE BLUE BALL AND THEN THE BLUE BALL HITS THE ORANGE BALL) were actually only performed once, with the subjects asked 'to describe it in a different sentence' (#13) and then 'in yet a different sentence' (#14); the same applies to events #18 through #20 (THE BALL IS ON THE TUBE AND THE TUBE IS ON THE PLATE). These two demonstrations were designed to get at surface transformations of the same perceptually-anchored deep or semantic structure within the same speaker as well as across speakers (which occurs for every demonstration). While the subjects were writing their sentences I prepared for the next event, but nothing was done or placed on the table until they had closed their eyes.

Although the major purpose in this little experiment was to demonstrate the influence of perceptual presuppositions upon the *form* of sentences, there was also very obvious (but not trivial) influence of perceptual input upon the lexical *content* of sentences. Table 2 presents the 26 sentences produced by Event #10 (A BLACK BALL ROLLS AND HITS THE BLUE BALL).[b] We note that every speaker with a single exception encodes both a *black ball* and *blue ball* in his sentence (no poker chips, tubes or plates), the former indicated as Actor (Instrumental case) and the latter as Recipient (Objective case). The one exception (marked X, subject #13) apparently opened her eyes too late! Similarly for event #15 (A BALL, A SPOON AND A POKER CHIP ARE ON THE PLATE), although the lexical items selected varied somewhat (e.g., *chip* vs. *disc*) as did the presence of qualifying adjectives, all subjects but two refer to all four entities and the proper relation of the plate (*lid, dish, ashtray, etc.*) to the other three.

---

[a] Although the wordings of these events in Table 1 were largely descriptive, to serve as my own guide in executing the demonstrations, they also reflect to some extent my own intentions, e.g., #7 ONE BALL ROLLS AND HITS THE OTHER (reflexive), #8 THE BALL THE MAN BOUNCES HITS THE PLATE (center-embedding).

[b] Space does not permit inclusion of the sentences produced in all 32 demonstrations. These raw data can be obtained from the author on request.

TABLE I

Perceptual events	Psycholinguistic intentions						
	(1) V/T	(2) CASES	(3) DET	(4) PN	(5) ADJ	(6) PASS	(7) MISC
1. THE MAN IS HOLDING A BLACK BALL	I	A, O	I	−	+	−	
2. THE BALL IS ON THE TABLE	III	I, L	D	−		−	
3. THE BALL IS ROLLING ON THE TABLE	I	I, L	D	−		−	
4. THE BALL IS ROLLING SLOWLY ON THE TABLE	I	I, L	D	−		−	ADV
5. THE MAN PUTS THE BALL ON A PLATE	II	A, O, L	D, I	−	−, +	+	
6. THE MAN IS HOLDING TWO BLACK BALLS	I	A, O	O	+	+	−	
7. ONE BALL ROLLS AND HITS THE OTHER	I, IV	I, O	I, D	+		−	REFLEXIVE CONJUNCT
8. THE BALL THE MAN BOUNCES HITS THE PLATE	II, IV	A, I, O	D, D, D	+	−	−	EMBEDDING CONJUNCT
9. THE MAN IS HOLDING A BLUE BALL	I	A, O	I	−	+	+	COMPAR
10. A BLACK BALL ROLLS AND HITS THE BLUE BALL	I, IV	I, O	I, D	+	+, +	+	CONJUNCT
11. A BIG ORANGE BALL IS HIT BY A BLACK BALL	IV	I, O	I, I	−	+, +	+	COMPAR
12. THE BLACK BALL HITS THE BLUE BALL AND THEN THE BLUE BALL HITS THE ORANGE BALL							EMBEDDING
13. (SAME AS #12)	IV (2)	I, O (2)	D, D, D	+	+, +, +	+	CONJUNCT WH-
14. (SAME AS #12)							
15. A BALL, A SPOON AND A POKER CHIP ARE ON THE PLATE	III	I (3), L	I (3), D	−	−	−	CONJUNCT
16. THE PLATE IS EMPTY	III	I	D	−	−	−	NEG
17. A CARDBOARD TUBE IS ON THE PLATE	III	I, L	I, D	−	+, −	−	
18. A BALL IS ON THE TUBE AND THE TUBE IS ON THE PLATE			I, D, D				EMBEDDING

TABLE I (cont.)

	III (2)	I, L (2)	D, D, D	+	−	−	CONJUNCT PPh
19. (SAME AS #18)						−	
20. (SAME AS #18)			D, D, D			−	
21. THE ORANGE BALL HITS THE TUBE	IV	I, O	D, D	−	+, −	−	PRESENT T.
22. THE BALL HIT THE TUBE	IV	I, O	D, D	−	−, −	−	PAST T.
23. THE BALL WILL HIT THE TUBE	IV	I, O	D, D	−	−, −	−	FUTURE T.
24. THE BALL MISSES THE TUBE	IV	I, O	D, D	−	−, −	−	NEG
25. THE BALL WILL MISS THE TUBE	IV	I, O	D, D	−	−, −	−	FUTURE T. NEG
26. SOME POKER CHIPS ARE ON THE TABLE	III	I, L	I, D	−	+	−	
27. THE MAN TAKES SOME CHIPS	II	A, O	D, I	−	−	−	
28. THE MAN DOESN'T TAKE ANY CHIPS	II	A, O	D, I	+	−	−	NEG
29. THE MAN TAKES THE BLUE CHIP	II	A, O	D, D	−	+	−	
30. THE MAN TAKES A WHITE CHIP	II	A, O	D, I	−	+	−	
31. THE MAN TAKES THE MIDDLE CHIP	II	A, O, L	D, D	−	+	−	
32. THERE IS NOTHING ON THE TABLE	III	I?, L	D	−	+	−	NEG

The two exceptions were apparently fixated on only the ball in the plate. Also obvious is the influence of perceiving two or more entities of the same nominal class upon producing plurals, specifically in #6 where all but two subjects refer to *two balls* and in #26, 27 and 28 where all speakers without exceptions use the plural (*chips, discs, buttons*). One could quite correctly say 'what else could they say?' when constrained by these perceptual events.

TABLE 2

Demonstration 10
A BLACK BALL ROLLS AND HITS THE BLUE BALL
Subject #

1. The small dark ball hit the large blue ball.
2. The man rolled one black rubber ball along the table, hitting the blue plastic ball.
3. A small black ball was rolled against the larger blue ball.
4. A blue rubber ball, which had stood in the path of a small black rubber ball, was knocked away by the small ball.
5. The small black ball hit the stationary large blue ball and caused it to move also.
6. The man rolled a small black ball into a large blue ball that was sitting on the table.
7. The black ball collides with the blue ball, moving it from resting position.
8. The man shot the small black ball at the stationary blue ball.
9. A small black ball was rolled by the man along the table to a stationary blue ball and the two balls collided.
10. One small black ball rolled into a large blue ball on the table.
11. The small black ball rolls and hits the medium blue ball.
12. The black ball rolled into the blue ball.
X13. The blue ball was rolling slowly across the table.
14. A blue ball was struck by a smaller black ball.
15. A little black ball rolling across a table strikes and moves a bigger blue ball.
16. Man rolls small black ball on table and it hits larger blue ball.
17. Black ball hitting larger blue ball.
18. A squash ball is hitting a blue ball and then going in opposite direction.
19. The black ball rolls and hits the blue one lying on the table.
20. A rolling black ball hits a stationary blue ball and sets it in motion.
21. The black ball rolled down the table and hit the big shiny blue one.
22. A rolling black ball hits a blue ball.
23. The small black ball rolls into the large blue ball and moves it.
24. The black ball rolled against the larger blue one.
25. The small black ball rolled and hit the blue ball.
26. A small black ball rolls into a larger blue one.

The general procedure for testing specific predictions about the effects of perceptual events upon the forms of sentences describing them will be to compare the sets of sentences generated by those events in which a given presupposition is assumed to be operating with sets of sentences where it is not. The latter 'control' events will be matched as closely as possible in other respects. For example, it is predicted that an action involving two categorically identical objects (#7, where one black ball rolls and hits *another*) will have a higher probability of generating a reflexive form than a control event involving two categorically different objects (#10, where a black ball rolls and hits the blue ball); thus event #10 serves as a control for event #7 with respect to this particular prediction (it may have other functions with respect to other predictions). The results of the tests will be expressed in simple percentages, finer statistical manipulations being unnecessary for present purposes.

VERBS AND TIMES

In a chapter of his *Linguistics in Philosophy* (1967) bearing this title, Zeno Vendler derives a two-feature system for English verbs with respect to the time dimension. The supraordinate feature is Action vs. State: one can say significantly *I am PUSHING it* (Action verb), but it is strange to say *I am KNOWING it* (State verb). The subordinate feature is Interminal vs. Terminal (my terms, not Vendler's): one can reasonably ask *for how long did you PUSH?* (Interminal Action Verb) and *how long did it take you to DRESS?* (Terminal Action verb), but hardly the converse questions, *for how long did you DRESS?* or *how long did it take you to PUSH?*; and similarly for State verbs, *for how long did you KNOW the girl?* (Interminal State verb) and *at what time did you MEET the girl?* (Terminal State verb), but hardly the converse, *for how long did you MEET the girl?* In column (1) of Table 1 all of the perceptual events are assigned to these four Vendler categories, Interminal Action events, I (e.g., THE MAN IS HOLDING A BALL), Terminal Action events, II (e.g., THE MAN PUTS THE BALL ON A PLATE), Interminal State events, III (e.g., THE BALL IS ON THE TABLE), and Terminal State events, IV (e.g., THE BALL HITS THE TUBE).

Two types of sentential phemonema (here, lexical) were predicted: (1) that usage of forms of the verb *to be*, either as auxiliary (*is holding*) or alone (*is on the table*),[a] would be relatively more frequent (*a*) for Stative than for Action perceptual events and (*b*) for Interminal than for Terminal events; (2) that usage of the present tense relative to the past tense would be greater for Interminal than for Terminal perceptual events. In demonstrations where two events were involved (e.g., ONE BALL ROLLS AND HITS ANOTHER), the verb phrases for each were considered separately. Table 3 summarizes the results over all 32 demonstrations. As can be seen the effect of Interminal as compared with Terminal events upon producing usage of both the verb *to be* and the present tense rather than the past is marked, particularly the former (Terminal events very rarely are encoded with *to be* forms, e.g., *the ball is hitting the pole*). Stative events produce slightly more *to be* and present tense verbs than Action, but the differences would not be significant. In any case, we can conclude that events which are perceived as on-going (Interminal) tend to be encoded as existing in the subjective present as compared with events perceived as completed (Terminal), which tend to be encoded as being in the past and no longer existing.

TABLE 3. *Verbs 'to be' and present (vs. past) tense as percentages of all verbs as a function of Vendler's verb types*

	% to be	% present
Interminal Action	0·46	0·54
Interminal State	0·53	0·55
Terminal Action	0·04	0·34
Terminal State	0·11	0·36

A CASE FOR PERCEPTUAL CASE

It seems perfectly obvious that both pre-linguistic children and household pets behave with respect to perceived events in ways indicative that they 'know' that

[a] Occurrences of forms of *to be* in passive constructions were *not* counted, since the *is* in the passive form *is hit by* is part of an obligatory transformation and is certainly not semantically equivalent to the *is* in the active *is hitting*.

DADDY (agent) THREW THE BALL (object), the reverse, that THE BALL (instrument) BROKE THE WINDOW (object) and not the reverse, that THE NEW BALL IS IN THE PAPER BAG (locative), and so on *ad infinitum* through the busy days of child and puppyhood. But a demonstration of this obvious may be in order. In column (2) of Table 1 are given my assignments of Fillmore's case relations of all entities to the actions or states in all 32 perceptual events. According to Fillmore (1968), the Agentive case (A) is the animate, responsible source of the action (here, THE MAN); the Instrumental case (I) is the inanimate force or object contributing to the action or state (here the various BALLS); the Dative case (D) is the animate being affected by the action or state (not applicable here); the Locative case (L) is that which identifies the location or spatial orientation of the action (here THE TABLE, THE PLATE); the Objective case (O) is any entity whose rule in the action or state is identified by the action or state (I am not sure I understand this one, but I have coded all 'acted upon' entities this way); and the Resultive case (R) is the entity resulting from the action (not applicable here).

For the three cases with which I felt I could deal confidently (Agentive, Instrumental and Locative), the entities involved were compared with all entities *not* assigned this case relation to see if, in the sentences produced, the corresponding noun phrases did display that case relation. For example, in events where THE BLUE BALL is the initiator of action (...AND THE BLUE BALL HITS THE ORANGE BALL) it is described in the Instrumental case as compared with events where it is the recipient of action (THE BLACK BALL HITS THE BLUE BALL). Table 4 presents the results. We note that in 84% of the sentences describing events where THE MAN is the actor, we get *the man verbed*, whereas in only 8% of the sentences describing events where THE MAN is not the actor does this happen – and these are all instances where the presupposed Osgood is indicated as the unseen agent! Similar results obtain for the Instrumental and Locative cases. The reason for the relatively high frequency of locatives in sentences for events assumed not to include this relation (64%) is that speakers often included such information *gratis* (e.g., for #9, the man is holding a blue ball *in his hand*). The obvious effect of perceptual 'case' upon sentential case seems to be confirmed.

TABLE 4. *Percent sentential cases of given types as function of perceptual entity–action relation*

	Sentential case (%)	Sentential non-case (%)
Agentive (THE MAN...)	84	8
Instrumental (THE BALL...)	88	19
Locative (...ON THE TABLE)	95	64

DETERMINATION OF DETERMINERS

Sampson (1969) suggests that there are two major types of noun phrases, 'establishing' and 're-identifying', these being obviously related to the grammarian's distinction between Indefinite and Definite. When one says *I bought a red car yesterday*, this, according to Sampson, is an instruction to the listener to add a new referent to his stock; but when one says *I bought the red car yesterday*, this represents an attempt by the speaker to single out an already existing referent (presupposed) in the listener's mind. Sampson then proceeds to criticize those linguists who have treated this distinction as if it were a semantic feature, particularly Carlota Smith (1964) who maintained that definiteness is actually a semantic gradient. My data on perceptual presuppositions clearly supports Smith's view. However, the determination of *a* vs. *the* is complicated by another type of presupposition – namely, whether or not the entity represented by the noun phrase is a member of a set (*a white chip*) or unique (*the blue chip*). A familiar entity which is perceived as a member of a set will still be accorded the indefinite status with *a*, as in #30 (THE MAN TAKES A WHITE CHIP) after the WHITE CHIPS have already been 'established' as a referent. I shall use the more general (psychologically) terms 'identifying' (for the indefinite condition) and 'recognizing' (for the definite condition), thereby eliminating the listener as a necessary condition.

To determine the effect of entity repetition upon choice of determiner, I selected four entities which had repeated appearances in perceptual events, BLACK BALL, PLATE, BLUE BALL and TUBE. Table 5 gives the percentage of indefinite *a* to total determiners (i.e., the percentage of definite *the* is the value given subtracted from 1·00). First, we may note the sharp drop in % indefinite from first appearance to all subsequent appearances; second, we note that any interval filled with perceived events not involving the entity in question tends to produce an increase in the probability of the indefinite *a* (e.g., demonstrations #8 to #15 for second and third appearances of PLATE). The BLACK BALL, having appearances in demonstrations #1, #2 and #3 provides the purest test of the gradient hypothesis. This gradient in determination of determiners suggests that this is the general condition for semantic feature coding, with discrete, binary coding being a special case.

TABLE 5. *Determiners 'a' relative to 'the' as a function of first, second and third appearances of the entity*

	1st appearance		2nd appearance		3rd appearance	
	#	%	#	%	#	%
BLACK BALL	1	73*a*	2	54*a*	3	35*a*
PLATE	5	80*a*	8	40*a*	15	56*a*
BLUE BALL	9	96*a*	10	42*a*	12	50*a*
TUBE	17	88*a*	18	42*a*	21	42*a*

The influence of perceived uniqueness vs. membership in a set of entities is clearly indicated by comparing #29 (THE MAN TAKES THE BLUE CHIP) and #31 (THE MAN TAKES THE MIDDLE CHIP) with #30 (THE MAN TAKES A WHITE CHIP).[a] The uniquely perceived entities produce only 48 % *a* (#29) and 29 % *a* (#31), whereas the member of a set (ONE OF THE WHITE CHIPS IN THE RING) produces 100 % indefinite *a* (if we exclude *one of the* and *one*) – there were 57 % *a* and 0 % *the* with transforms involving *one* excluded.

PERCEIVED REDUNDANCY AND THE USE OF PRONOUNS

Use of definite *the* involves an implicit kind of redundancy – the entity in question is assumed by the speaker to be 'familiar' to the hearer. Such redundancy is more

---

[a] Events #29 through #31 involved a circle of chips, mostly white but including one blue and one red, with a single white chip in its middle.

explicit in the encoding of pronouns. When I say *she also dyes her hair* I am expressing the presupposition that a particular female entity is the agent (either by virtue of a previous utterance, *Mary goes in for mod clothes* or of a previous perceptual experience, PARTICULAR FEMALE ENTITY NAMED MARY IS WEARING A MINI-MINI-SKIRT). We might, therefore, expect series of events involving the same entities to produce an increasing probability of pronouns as well as Definite determiners. But since the instructions for these demonstrations served to isolate the separate, numbered events from each other – for example, no one described #2 (THE BALL IS ON THE TABLE) as *it is on the table* immediately after having seen #1 (THE MAN IS HOLDING A BLACK BALL) – we must satisfy another general condition: namely, that the same entity be perceived as participating in two (or more) actions or stative relations if, on its subsequent occurrence, the NP referring to the entity is to be replaced by a pronoun.[a] Olson (1969) gives an extended justification of why semantic decisions about pronouns are based upon cognitions – presupposed knowledge of the intended referent – and not upon rules internal to language.

Table 6 provides several different tests of these hypotheses. 6(A) illustrates the general case, comparing #8 (THE BALL THE MAN BOUNCES HITS THE PLATE) in which the same entity (THE BALL) participates in two actions (MAN BOUNCES BALL and BALL HITS PLATE) with #5 (THE MAN PUTS THE BALL ON A PLATE) in which this same entity participates in only one action. The former produces 19 % *it* replacements for *the ball* (5 speakers) and the latter only 4 % (one speaker – who wrote *there is a can top with a small black ball which is later placed in it by the man*!). 6(B) presents three similar events which observers could perceive either as two actions or a single action (e.g., #22 either as THE BALL ROLLED AND HIT THE TUBE or as THE BALL HIT THE TUBE) and compares the frequencies of pronouns produced in two-verb (two actions perceived) vs. one-verb (single action perceived) descriptions. Whereas about half of the observers who perceived two actions of the same entity used *it* for the second reference to THE BALL, none of the single-action perceivers produced a pronoun.

Table 6 provides additional evidence for participation of the same entity in more than one action or state relation as a condition for pronoun production, as well as an indication of transformational equivalence of WH- and PN under certain conditions. In demonstrations #12–14 only THE BLUE BALL participates in two action relations (HIT BY BLACK and HITTING ORANGE) and in demonstrations #18–20 only THE TUBE participates in two stative relations (SUPPORTING BALL and BEING SUPPORTED BY PLATE). As can be seen, the summed percentages of pronoun *it* and relative *which* are much higher for BLUE BALL and for TUBE than for the two other single-event entities in each set. What is also notable is that the entity which is initially the 'focus of attention' and usually referred to first (BLACK BALL in both #12 and #18) almost never is replaced by either *it* or *which*. The common usage trends of *it* and *which* are apparent in these data, as is the fact that they are 'allosemes' (where one speaker says *the black ball hits the blue ball AND IT hits the orange ball*, another speaker will say *the black ball hits the blue ball WHICH hits the orange ball*); whether *it* and *which* are in free variation or are in some manner conditioned I cannot determine from the data.

Sampson (1969) deals with the kinds of pre-conditions determining both definite

---

[a] It can be noted in passing that pronouns retain only those features (gender, number, person) usually sufficient for disambiguation of the intended referent. *The man rolled the ball and IT hit the tray* is entirely disambiguated by the gender feature; interestingly enough, *one ball hit another and IT hit a third* is also unambiguous by virtue of the second ball being 'the nearest one to the focus of attention' as Sampson (1969) puts it.

TABLE 6. *Tests of pronoun usage as a function of entity redundancy*

(A): *Illustration of general case*	% PN
#8 (THE BALL THE MAN BOUNCES HITS THE PLATE)	0·19
vs.	
#5 (THE MAN PUTS THE BALL ON A PLATE)	0·04

(B): *As a function of manner of perception (double vs. single action)*

	% PN,	TWO VERBS	% PN,	ONE VERB
#21 (THE ORANGE BALL HITS THE TUBE)	0·53	(8/15)	0·00	(0/11)
#22 (THE BALL HIT THE TUBE)	0·33	(4/12)	0·00	(0/14)
#23 (THE BALL WILL HIT THE TUBE)	0·58	(7/12)	0·00	(0/14)

(C): *As a function of entity participation in two actions or states: also demonstration of transformational equivalence of PN and WH-*

	BLUE BALL (two			BLACK BALL (one)			ORANGE BALL (one)		
	it	which	total %	it	which	total %	it	which	total %
#12	2	9	0·42	0	0	0·00	3	0	0·11
#13	1	8	0·34	0	0	0·00	4	1	0·19
#14	1	10	0·42	1	0	0·04	1	0	0·04

	TUBE (two)			BALL (one)			PLATE (one)		
#18	3	9	0·46	0	0	0·00	1	1	0·08
#19	7	11	0·68	0	0	0·00	1	2	0·11
#20	7	6	0·50	0	1	0·04	2	4	0·23

(D): *Reflexive with respect to membership in same semantic category*

	% NP$_1$ = 'one'	% NP$_2$ = reflexive
#7 (ONE BALL ROLLS AND HITS THE OTHER)	0·64	0·57
#10 (A BLACK BALL ROLLS AND HITS THE BLUE BALL)	0·08	0·15

articles and pronouns as involving a general 'reflexive' (redundancy) principle, with which I would agree. 'If the referent has already occurred in the kernel sentence, the NPP [noun-phrase proper] takes the feature "reflexive". This will account for the usage of the reflexive in "the teacher hates himself" and also for the emphatic usage in, for instance, "the landlord himself constructed the building"' (p. 7). Although my demonstrations did not include any complete reflexives (e.g., THE MAN SLAPPED HIMSELF), where the intended referent of the pronoun is identical with the animate noun being replaced, they do include at least one case where Actor and Object of action *are in the same semantic category, even though not identified as 'the same'* – and this serves to extend further the functioning of a generalized Reflexive feature. As shown in Table 6(D), perceptual event #7 (ONE BALL ROLLS AND HITS THE OTHER), both balls being squash, produces 64% substitutions of *one* for the indefinite *a* for subject NP$_1$ and 57% substitutions of some reflexive form (*the other, another, a second*) for the indefinite *a* for the object NP$_2$; these high values can be compared with 8% *one* for NP$_1$ and 15% reflexive for NP$_2$ in #10 (A BLACK BALL ROLLS AND HITS THE BLUE BALL) – and all reflexive NP$_2$ in this case were *one* (*a blue one* substituted for *a blue ball*). Perhaps *one, another, the other* and the like should be called 'semi-reflexives' in that they take account of (reflect) only the shared, non-distinguishing semantic features of the set of entities – here, the common small-black-ballness in event #7 and the ballness in #10.

REDUCING UNCERTAINTY WITH MODIFIERS

Usage of adjectives and adverbs clearly reflects presuppositions on the part of the speaker as to the amount of information possessed by the listener about the entity being referred to. (1) If the entity is presumed to be novel to the listener, and hence must be identified ('established' as a referent according to Sampson, 1969), the speaker is under pressure to elaborate the noun phrase; if it is presumed to be familiar already, and hence need merely be recognized ('re-identified' according to Sampson) by the listener, economy dictates stripping the noun phrase to the bare nominal bone.[a] (2) If the entity is a member of a set, to all of which a selected noun applies, and if either (a) the intended entity must be distinguished from others previously identified (implicit contrast over time) or (b) the intended entity must be distinguished from others presently being identified (explicit contemporary contrast), then the speaker is under pressure to use distinguishing modifiers to reduce the listener's uncertainty.[b] These contrastive conditions for modifying noun phrases are well described by Olson (1969) in what he calls the 'paradigm case' – a gold star being placed under the same ROUND WHITE WOODEN BLOCK, but the observer describing this simple event as under *the white one* (if a ROUND BLACK BLOCK is also present), *under the round white one* (if both ROUND BLACK and SQUARE WHITE are present), and so forth.[c]

These two functions of modifiers – identifying vs. distinguishing – cross-cut the functions of Indefinite and Definite determiners and pronouns. The identifying function parallels the use of indefinite *a*. Table 7 presents both the number of adjective *tokens* (totals) and the number of adjective *types* (different adjectives) for the same reappearing entities as reported in Table 5 for determiners (BLACK BALL, PLATE, BLUE BALL and TUBE), in terms of both order of appearance (columns) and complexity of noun phrases (rows) – the latter being (0), only noun, and (1), (2) or even (3) modifying levels are in bold-face type. First we note, again, that the big drop in frequency and diversity of qualifying is between first and subsequent appearances, which is the clearest separation between 'identifying' and 'recognizing'. Also, again, when there is a big time gap between successive appearances (e.g., PLATE for second, event #8, and third, event #15) the tendency to 'identify' reasserts itself, suggesting that this is a continuous semantic feature. Secondly, it can be seen

---

[a] Sampson relates 'establishing' and 're-identifying' to both selection of indefinite vs. definite articles and to usage of pronouns, but he does not extend these same cognitive preconditions to the use of adjectives. That all three aspects of noun phrases are influenced commonly indicates that postulation of such cognitive pre-conditions for their form and content, e.g., where noun phrases come from (cf. McCawley, this volume), is well-motivated, not *ad hoc*.

[b] Of course, the same underlying cognitive conditions can also lead to the selection of a more finely differentiating noun as head of an NP (e.g., *husband* rather than *married man*) or verb as head of a VP (*meander* rather than *walk slowly and aimlessly*).

[c] Apparently because in this paradigmatic case speakers substitute *one* for *wooden block* (where *all* entities are WOODEN BLOCKS), Olson claims that 'words do not name things' (p. 16). Not one of my observers said *the man is holding a small black one* for demonstration #1! Surely, by the same general principle of uncertainty reduction which Olson stresses, the use of *ball* to name (distinguish) this sub-set of palpable objects from many others (blocks, marbles, earrings, etc. and etc.) is equally informative – indeed, rather grandly so, since whole bundles of semantic features are being selected rather than a single feature being discriminated.

TABLE 7. *Frequency (tokens), diversity (types) and complexity (number a's in NP) of adjectives as a function of successive appearances of entity*

	a's in NP	1st appearance TOKENS	TYPES	2nd appearance TOKENS	TYPES	3rd appearance TOKENS	TYPES
BLACK BALL		#1		#2		#3	
	(0)	(4)		(9)		(11)	
	(1)	9		11		9	
	(2)	11 31	5	6 23	5	6 21	5
	(3)						
PLATE		#5		#8		#15	
	(0)	(10)		(18)		(13)	
	(1)	9		4		7	
	(2)	5 22	16	2 11	7	5 20	11
	(3)	1		1		1	
BLUE BALL		#9		#10		#12	
	(0)	(0)		(0)		(2)	
	(1)	7		11		13	
	(2)	14 53	11	14 42	9	8 32	7
	(3)	6		1		1	
TUBE		#17		#18		#19	
	(0)	(3)		(5)		(8)	
	(1)	16		14		11	
	(2)	9 40	16	7 28	12	6 26	13
	(3)	2		0		1	
Totals		146	48	104	33	99	36
Means		37	12	27	8	25	9

that, very consistently for successive appearances, the complexity of noun phrases referring to the entities decreases; for example, for BLUE BALL there are 6 triple adjectives, 14 double and only 7 single for the first appearance, whereas by the third there is only one triple, 8 doubles and 13 single adjectives. (In all cases of single adjective, it was *blue*.) Quite evidently, as the speaker presupposes more familiarity with the referent, the less the frequency, diversity and complexity of his adumbrations on the head noun.

Conversely – and independent of whether the intended referent is being identified (novel) or being recognized (familiar) – if the perceptual entity must be contrasted with others of its class, then we find an increased likelihood of adjectival modification. Table 8 (A) provides a demonstration of the effects of *implicit* contrast, comparing #9 (THE MAN IS HOLDING A BLUE BALL), with #6 (THE MAN IS HOLDING TWO BALLS); the sudden contrast of a larger blue ball with only small black balls which had preceded this event creates a very high frequency of both color and size adjectives, as compared with #6, and no sentences without adjectives. I have added to this table the effects of TWO BLACK BALLS upon producing sentences with *two* #6, just to hammer in the obvious. Table 8(B) provides a parallel demonstration of the effects of *explicit* contrast, comparing #10 (A BLACK BALL ROLLS AND HITS THE BLUE BALL) and #11 (A BIG ORANGE BALL IS HIT BY A BLACK BALL) with #7 (ONE BALL ROLLS AND HITS THE OTHER); the simultaneous perceptual presence of two balls constrastive in both size and color produces the predicted adjectival effects, and there are no sentences without adjectives as compared with 46% zero ADJ in #7. Here I add the frequency of *quantifiers*: the sudden appearance of the orange ball (over double the size of the blue ball) for the first time in #11 produced 53% sentences with *much* and *very*; the one observer of #10 who quantified wrote *medium blue ball*, and no one used a quantifier for #7.

TABLE 8. *Effects of implicit (A) and explicit (B) contrast upon modification*

		Size ADJ (%)	Color ADJ (%)	Zero ADJ (%)	'two' (%)
(A): *Implicit contrast (successive events in time)*					
#9	vs	64	100	0	0
#6		15	68	23	84
(B): *Explicit contrast (simultaneous events)*		Size ADJ (%)	Color ADJ (%)	Zero ADJ (%)	Quant. (%)
#10		57	100	0	4
#11	vs.	100	95	0	53
#7		11	46	46	0

Manner *adverbs* serve the same identifying and contrasting functions for perceived actions as adjectives do for perceived entities. When an action deviates from an implicit norm (contrastive over past experience) and thereby contradicts a presupposition about that kind of action, the probability of adverbial modification increases. No deliberate test of this proposition was made, but it could easily be demonstrated, e.g., by THE MAN REACHING WITH EXCRUCIATING SLOWNESS FOR A BALL. However, when an action deviates from a standard established in immediately prior perception, adverbial modification is also predicted. Demonstration #3 (THE BALL IS ROLLING ON THE TABLE) produced only 8% adverbials (both sentences containing *quickly*) and no other 'manner' expressions, but the immediately following #4 (THE BALL IS ROLLING SLOWLY ON THE TABLE) produced 68% *slowly* and 11% other 'manner' expressions.

NEGATION AND UNFULFILLED EXPECTATIONS

The general condition for use of some form of *Neg* in describing things appears to be another type of contrast – contrast between what was expected or predicted by the speaker and what he actually observed. The unfulfilled expectation may concern the absence or altered condition of an entity, in which case *Neg* would be expressed in a noun phrase; or it may concern the absence or altered character of an expected action or state, in which case *Neg* would appear in the verb phrase. The forms that *Neg* can take are many (cf. the 'classic' paper on negation by Klima, 1964) and the rules by which *Neg* forms combine with each other are very subtle (cf. a paper by Baker, 1969, on double negatives). Linguists have noted that 'positive polarity' and 'negative polarity' expressions tend to occur in pairs: for example, *always/never, some/any, either/neither, always/never* and certain verbs like *expect/doubt* (note that *I expect that you will find some gold* and *I doubt that you will find any gold* are entirely acceptable but *I expect that you will find any gold* and *I doubt that you will find some gold* are questionable at best). With regard to predicates like *I am surprised that, I am relieved that* and *It is lucky that*, Baker (1969) says that 'speaking intuitively, we can say that each of these predicates expressed a relation of "contrariness" between a certain fact and some mental or emotional state' (p. 33). Thus, *I am surprised that anyone could sleep* but hardly **I am surprised that someone could sleep*.

As a test of the unfulfillment prediction with respect to presence or absence of expected entities, we may compare events #16 (THE PLATE IS EMPTY) and #32 (THERE IS NOTHING ON THE TABLE) with events #15 (A BALL, A SPOON AND A POKER CHIP ARE ON THE PLATE) and #26 (SOME POKER CHIPS ARE ON THE TABLE). For the control events (#15 and #26) none of the sentences produced contained *Neg* in any form, but for the test events (#16 and #32) there were 62% and 85% *Neg* respectively. Most *Neg* in #16 were the adjectival form, *empty*; in #32 *Neg* divided about equally between adjectives (*bare, naked*) and the negative nominal form, *nothing*. To test the unfulfillment prediction with respect to absence of an expected action, we may compare events #24 (THE BALL MISSES THE TUBE) and #25 (THE BALL WILL MISS THE TUBE) with control events #22 (THE BALL HIT THE TUBE) and #23 (THE BALL WILL HIT THE TUBE). Again, there were no *Neg* of any type for the control events, whereas the test events yielded 26% and 30% *Neg* respectively—usually either *not* with a positive verb (*not hit*) or a negative verb (*miss*), although one *without hitting* was offered.

SUBJECTIVE TIME-ZERO AND VERB TENSE

It seems a reasonable proposal that events comprehended as occurring during, prior to, and after the period of perceptual observation will tend to be encoded in the present, past, and future tenses respectively. Events #21 (THE ORANGE BALL HITS THE TUBE), #22 (THE BALL HIT THE TUBE) and #23 (THE BALL WILL HIT THE TUBE) were intended to test it. Although the resulting sentences revealed some rather interesting phenomena, the main effect predicted was not as strong as anticipated. This was in part due to difficulties in engineering these time relations – eyes 'open' during actual time of contact between ball and tube, 'open' just after time of contact, and 'open' prior to, but closed just before, time of contact – and also due to the fact that, from the observers' point of view, two verbs were involved in these demonstrations, inter-minal *rolling* and terminal *hitting*.

It also became evident in scanning all 32 demonstrations that there was a division between perceiving events as contemporaneous (present tense) and completed (past tense). What was not expected was that individual observer-speakers would display rather constant and characteristic 'time-zero' reference points. In describing terminal actions (Vendler's 'accomplishments') and terminal states (Vendler's 'achievements'), some individual speakers display a characteristic 'set' for the present (subjective time-zero equals termination of the perceived event) and other speakers a 'set' for the past tense (subjective time-zero equals their own time of reporting the event). To test the degree of within-individual consistency in time orientation two widely separated 'accomplishment' events (#5, THE MAN PUTS THE BALL ON A PLATE; #27, THE MAN TAKES SOME CHIPS) and 'achievement' events (#11, A BIG ORANGE BALL IS HIT BY A BLACK BALL; #21, THE ORANGE BALL HITS THE TUBE) were examined. Table 9 gives the percentage of speakers who are consistent (present-present or past-past) or inconsistent (present-past or past-present) for the two 'accomplishment' events (A) and two 'achievement' events (B). It can be seen that the vast majority of speakers are internally consistent (92% for both types of events). Not shown here is the fact that, with very few exceptions, the same speakers who use a given tense for 'accomplishments' also use the same tense for 'achievements' i.e., time 'sets' are not conditioned differentially by types of events.

Using the ratio of present-to-past 'sets' for our speakers (approximately 40% present to 60% past), we can now ask to what extent the attempted manipulations of time orientation in events #21–23 were successful in producing deviations from this base-line. As shown in Table 10, the terminal achievement verb *hit* does show the predicted trends: #21 (present) has a present/past split at about the base-line; #22 (past) is biased more toward past tense (although the over-all frequency of *hit* goes down); and #23 (future) shows much reduced past tense and 8% future. The interminal action verb *roll* shows much higher frequencies of occurrences for #22 (past) and #23 (future) – events for which contact between BALL and TUBE was *not* observed but much 'rolling' was – yet the ratio of present to past stays quite constant near the base-line level. Also shown in Table 10 are relative frequencies of certain prepositional phrases accompanying the verb *rolled* (in either present or past tense); rolling *at* or *into* occur when the event is seen, rolling *from* or *away* when it has already happened, and rolling *toward* when the event is anticipated. That the preposition *toward* carries a future orientation is evident from the 53% occurrences when the event is anticipated but not seen, and of these most were accompaniments of past-tense *rolled*. There appears to be semantic equivalence between *will hit* and *rolled toward* in expressing a future-orientation feature.

TABLE 9. *Consistency of subjective time-zero across events*

		A: Accomplishments #27				B: Achievements #21	
		pres %	past %			pres %	past %
#5	pres	34	4	#11	pres	38	8
	past	4	58		past	0	54

TABLE 10. *Some effects of varying time relations of events to perceptions*

	#21 (present) %	#22 (past) %	#23 (future) %
hits	34	8	11
hit	64	30	11
will hit	0	0	8
rolls	19	38	34
rolled	34	50	53
will roll	0	0	0
roll(ed) at, into	15	0	4
roll(ed) from, away	0	15	0
roll(ed) toward	0	0	53

TESTS OF SOME WHOLE SENTENCE EFFECTS

Conjunction and disjunction.

When two or more contrastive entities are perceived as participating in a common action or state, the corresponding NPs are likely to be conjoined by *and*. Comparison of #15 (A BALL, A SPOON AND A POKER CHIP ARE ON THE PLATE) with #17 (A CARDBOARD TUBE IS ON THE PLATE) tests the multiplicity-of-entities requirement and comparison

Transformations.

The still-dominant view among generative linguists seems to be that surface forms of sentences are transforms of deeper forms which are themselves sentential in nature. Originally these deeper structures were thought to be 'kernel' sentences of active, declarative form; during the past decade they have gradually become more abstract, but still forms of sentences. The implication of the very recent work on presuppositions, as well as of my little demonstrations, would seem to be that what is 'transformed' into a surface sentence is not another 'sentence' (hyper or otherwise) but rather a momentary cognitive state which is not linguistic at all yet has its own complex semantic structure.

Entering the human information-processing system via purely perceptual signs and events eliminates any obvious dependence, at least, upon prior sentences. Furthermore, *ideally*, such perceptual inputs serve to 'anchor' the states of the cognitive systems of different speakers so that transformational equivalents can be observed directly when they Simply Describe. I have already noted in passing several instances of surface transforms – for example, the combination of a present or even past-tense verb with *toward* to express the same Future feature as *will* (*rolled toward* substituting for *will hit*) or the interchangeability of *and it* (*hits*, etc.) and *which* (*hits*, etc.).

The reason I said above that 'ideally' speaker cognitive states are anchored by common perceptual input is that in these demonstrations – as compared with color film strips, for example – the observers may actually be perceiving somewhat different events and they certainly may be attending to different aspects of events selectively. Careful inspection of the 26 sentences produced by demonstration #10 (Table 2) clearly indicates this: for example, subject #1 says *the small dark ball hit the large blue ball* (reporting only the contact event), whereas #2 says *the man rolled one black ball along the table, hitting the blue plastic ball* (reporting the presupposed Agent, ROLLING as well as HITTING events, and even the TABLE locale). But we can also note genuine transformations of the 'same' semantic information across different describers: #3 implies with a passive construction (*a small black ball was rolled against the larger blue ball*) the same presupposed Agent that #2 made explicit; #14's sentence is the passive version of #1's simple active (*a blue ball was struck by a smaller black ball*) #9 expresses MAN ROLLING BALL and BALLS COLLIDING in two sentences conjoined by *and*: the compound noun *squash-ball* expresses the same information for speaker #18 as does *small black rubber ball* for other speakers. Any distinction between syntactic and lexical transformations would be hard to draw – what, for example, about the obvious equivalence of #7's *collides with*, #10's *rolls into*, and #11's *rolls and hits*?

The two demonstrations requiring repeated sentencings were designed deliberately to get at transformations, both across repetitions and within single speakers. As an illustration of transformations across repetitions, I checked the relative frequency of passive constructions across THE BLACK BALL HITS THE BLUE BALL AND THEN THE BLUE BALL HITS THE ORANGE BALL (#12–14); there were only 19 % passives for #12, which rose to 50 % for #13, and then fell again to 34 % for #14 – clearly suggesting that this construction is a non-preferred but readily available transformation.

A few examples of within-speaker (across-repetitions) transformation must suffice: for #12–14, speaker #1 says *the small dark ball hit the blue ball which hit the orange ball*, then provides us with one of the rare center-embeddings *the blue ball the dark ball hit hit the orange ball*, and finally gives us a reverse entity ordering *the orange ball was hit by the blue ball which had been hit by the dark ball,* all necessarily cast into the passive form; for speaker #6 sentence 1 is the forward active and sentence 3 is the backward passive, but sentence 2 is an interesting combination of directions, *the blue ball was hit by the black ball and hit the orange ball.* For #18–20, speaker #15 tells us first that *a little black ball is perched on the end of a long tube which in turn is standing inside a can cover,* then that *three objects are piled atop each other – from bottom up, a can cover, a tube and a ball,* and finally, with extraordinary efficiency, that *a ball on a tube is in a can*; speaker #22 does a kind of phrase-shuffling act, moving from *a ball sits on top of the cylinder which stands in the lid,* into *a cylinder holding at its top a ball stands on the lid,* and finally to *a lid has standing in it a cylinder which is topped by a ball.*

If such sentences are all transformations of the same cognitions established by the perceptual input – and they certainly do seem to be informationally equivalent – then they would seem to pose some interesting problems for the generative grammarian. For example, if they are semantically equivalent transformations, then they should, at some depth level, trace back to the 'same' hyper-sentence. But what would this be? Or another question: what, in a generative grammar, 'triggers' one surface structure as opposed to some other? To say, for example, that some *Manner T-passive* is attached to the entire, deep-structure tree of speaker #1's third effort above is entirely *post hoc* and begs the real question, at least from the point of view of a performance model.

One thing suggested by these examples and the total data is that particular transformations are 'triggered' by the entity which is the focus of the speaker's momentary interest or attention and is therefore likely to be represented in his first encoded noun phrase. Once an observer of perceptual event #12 has begun encoding *the orange ball...*, he is, so to speak, stuck syntactically by this decision in conjunction with the cognitive 'facts of the world' – that THE ORANGE BALL WAS HIT BY THE BLUE BALL (not the reverse), and that THE BLUE BALL WAS PREVIOUSLY HIT BY THE BLACK BALL (not the reverse), and so forth. This would suggest that the job of syntax is not central but rather peripheral in ordinary language – merely accommodating lexical decisions made on the basis of the fleeting interests and motivation entertained by speakers – a notion that will surely raise the hackles of many linguists.

Some implications for psycholinguistic theory

Most linguists and many psycholinguists, convinced by Chomsky's review (1959) of Skinner's *Verbal Behavior* (1957) and Fodor's critique of neo-behaviorism's mediation models (1965, 1966) that contemporary learning theories are incapable 'in principle' of handling language behavior, have been looking elsewhere for a 'more powerful' theory of performance. Not finding anything else of interest in psychology, it was natural to turn to generative grammar itself not only as a theory of language competence but also as a model for a performance theory as well. This approach has proven quite sterile, as Chomsky himself (1961) predicted it would: 'The attempt to develop a reasonable account of the speaker has, I believe, been hampered by the prevalent and utterly mistaken view that a generative grammar itself provides or is related in some obvious way to a model for the speaker' (1961, footnote #16). Not only can it be easily shown that any contemporary generative grammar is itself incapable 'in principle' of handling language performance, but I have tried to demonstrate elsewhere (Osgood, 1966, 1970) that Fodor's critique of representational mediation theory is untenable.

INSUFFICIENCY OF GENERATIVE GRAMMAR AS A THEORY OF PERFORMANCE

If any extant generative grammar *were* seriously proposed as a model of speaker-hearer performance, it would immediately face several obvious insufficiencies in principle: (1) Grammars, by virtue of their recursive properties, can produce sentences of indefinite length and complexity; speakers and hearers are finite devices, and their limitations must be incorporated in any performance theory. (2) Grammars are not time-bound in generating sentences; speakers and hearers operate within time constraints, and sentences must be created and understood on a 'left-to-right' basis. (3) Grammars deal with idealized speaker-hearers who have perfect knowledge of the rules of their language (la langue); performance theories must deal with real speaker-hearers who are both fallible and variable, i.e., perform probablistically (la parole). (4) Grammars contain no learning principles; given the obvious arbitrariness of both lexical and grammatical rules for particular languages, which must be acquired via experience by their speakers, any performance theory must account for such learning. (5) Grammars do not (and need not) provide any account of selection among alternative expansions of the left-hand terms in re-write rules (e.g., of the fact that NP may be rewritten as $T+N$, $T+A+N$, $T+N+WH+S$ and so on *ad libidum*); any performance model must account for the antecedents of such 'decisions'. (6) Grammars do not (and need not) account for the non-linguistic antecedents in speaking or subsequents in understanding of the superordinate symbol S in the generation of sentences; performance theories of creating and interpreting sentences must provide an account of such non-linguistic antecedents and subsequents of sentences.

This last requirement for performance theories is, of course, the question of 'where sentences come from and go to'. The process of Simply Describing, as I have illustrated it here, is a very ordinary and familiar psycholinguistic performance of real speakers. It involves the ability to paraphrase, in language, entities and events (perceptual signs and their interrelations) which are essentially non-linguistic in nature. That speakers of a language do have this ability is evident in the highly similar information content across the sentences produced by the 26 speakers in each of the preceding demonstrations. Possession of this paraphrasing ability implies both (*a*) that perceptual signs and events must themselves have meaning and (*b*) that perceptual and linguistic signs must share a common representational (semantic) and organizational (deep syntactic) system – otherwise, signals in the perceptual and linguistic systems would 'pass each other like ships in the night'. Furthermore, both the recent linguistic evidence on 'presuppositions' (reviewed earlier in this paper) and the detailed ways in which perceptual antecedents in my demonstrations literally 'drive' the form and content of sentential describings strongly suggest that this shared representational and organizational system is not linguistic but cognitive in nature.

If this is the case, then the constructs and rules of generative grammars are, in principle, incapable of accounting for Simply Describing Things. The only conceivable approach via generative linguistics would be an extreme form of Whorfian psycholinguistic relativity – the assumption that people literally impose their 'sentences' upon the real world, perceiving entities and events in it in terms of what their grammars permit them to say about them. Not only is it counter-intuitive (to me, at least) that I must generate an implicit sentence, *the man is holding a small black ball*, before I can perceive and understand that this is in fact the state of affairs in the real world, but even casual observations of the behaviors of pre-linguistic children and non-linguistic higher animals contradict any such linguistic restriction upon ability to comprehend. More than this, such extreme psycholinguistic relativism would imply that the basic cognitive abilities of perceptual grouping, contrasting and categorizing are determined by language rather than being determinants of language. As I indicated earlier, the absurdity of this position becomes apparent when one conjures with the number of antecedent sentences which would be required to fulfill the 'happiness conditions' for a simple sentence like *Please shut the door.*

REPRESENTATIONAL MEDIATION THEORY

It must be understood that *learning theory* is not synonymous with, but rather is a necessary part of, general *behavior theory.*[a] General behavior theory would include many innately determined properties and functions of the nervous system, including wired-in sensory and motor projection systems, the existence of unlearned reflexes and instincts, and the possession of certain innately determined cognitive abilities by any given species. In other words, behavior theory does *not* necessarily assume that the 'mind' is a *tabula rasa* at birth. Among the innately given of human cognitive abilities would undoubtedly be: (1) *the gestalt-like tendencies to establish perceptual entities* in terms of such principles as common fate, qualitative sensory similarity, sensory proximity, and contour completeness of continuity; (2) *the tendency to organize behavior*, both perceptuo-motor and linguistic, *hierarchically* in terms of levels of units-within-units-within-units; (3) *the tendency to organize or differentiate the units within each level componentially*, so that a large number of alternatives can be differentiated in terms of a relatively small number of elements or features. Although it is true that these tendencies find expression in, i.e., are universal to, all human languages, this is not because they are linguistic in nature but rather because linguistic systems develop within the constraints (or potentialities) of human cognitive abilities more generally.

Representational mediation theory is the only learning approach that has seriously attempted to incorporate the *symbolic processes* in general and *meaning* in particular within an S-R associationistic model. It is traceable directly to the germinal work of Clark L. Hull (1930, 1943), although he utilized it primarily in drive and reinforcement connections (the fractional anticipatory goal reaction, r_g). It was extended as a general rather than occasional explanatory device in the writings of people like Miller and Dollard (1941), Mowrer (1960), Goss (1961), the Kendlers (1962) and many others, including the writer (Osgood, 1953 and subsequently). In its present form it has come to be known as two-stage learning theory or *neo-behaviorism* (and r_g has become r_m, or representational mediation process).[b]

My own use of representational mediation theory as the basis for a behavioristic theory of meaning has made explicit the origins of mediation (symbolic) processes in non-linguistic, perceptuo-motor behaviors (cf. my 1957 paper on perception and

---

[a] In a way, the learning-theory component of general behavior theory is analogous to the syntactic component of a generative grammar; it is a set of 'rules' for generating increasingly complex behaviors from more primitive, innately given, elements and processes.

[b] Actually, my own version of neo-behaviorism is a three stage (or level) theory, including a level of sensory (s-s) and motor (r-r) integration designed to account for gestalt-like phenomena of perceptual organization (e.g., closure) and motor skill or programming phenomena. See Osgood, 'A Behavioristic Analysis of Perception and Language as Cognitive Phenomena', 1957.

language as cognitive processes). Significates (referents or things signified) are simply those patterns of stimulation, including previously learned signs, which regularly and reliably produce distinctive patterns of behavior. Thus, early in infant life, presence of a nipple in the mouth reliably produces a complex pattern of sucking movements, milk ingestion, reduction of hunger pangs, autonomic changes characteristic of gratifying states of affairs, relaxation of voluntary musculature, and so forth. Sight of the BOTTLE (a perceptual sign) regularly antedates presence of nipple-in-mouth, and this is a necessary and sufficient condition for *sign learning*. The primary postulate of this theory of meaning is that *a stimulus pattern which is not the same physical event as the thing signified will become a sign of that significate when it becomes conditioned to a representational mediation process, this process (a) being some distinctive part of the total behavior produced by the significate and (b) serving the mediate overt behaviors which are appropriate to* ('take account of') *the significate.* Thus the perceived entity, BOTTLE, becomes a perceptual sign of MILK FOOD-OBJECT and, appropriately, the infant will now salivate and make anticipatory sucking movements to the sight of BOTTLE.

Obviously, this is only the crudest beginning of a semantic system, but it is a beginning nevertheless, and by the time LAD is launching himself into linguistic sign learning most of the entities in his familiar environment (utensils, food objects, pets, toys, faces that smile and faces that frown) have been meaningfully differentiated in terms of distinctive mediators (representations, cognitions). In this pre-linguistic process, not only are particular representations established for perceptual signs, which can then serve as 'prefabricated' mediators in learning correlated linguistic signs, but even more importantly many of the distinctive features of meaning (components of r_m in behavior theory terms) are being differentiated.

Although this is only the roughest sketch of the neo-behaviorist theory of meaning, it should at least be evident that, in principle, it both provides for the meaningfulness of non-linguistic, perceptual signs and indicates that perceptual and linguistic signs share the same semantic system. Debate over the adequacy of representational mediation theory as the foundation for a theory of meaning (Fodor, this volume, 1966; Osgood, 1966, 1970) has focused on two major issues – the nature of mediating processes and their source.

With regard to the *nature* of representational mediation processes in behavior, I have stated repeatedly that r_m and its automatic consequence, s_m, are defined *functionally* in S-R terms in order to incorporate mediation processes within the larger body of learning theory and facilitate transfer of such single-stage principles as habit-strength, generalization and inhibition to this two-stage model. Representing the mediation process with the symbol M, we can say that M has response-like functions as a dependent event (decoding, understanding) and stimulus-like functions as an antecedent event (encoding, creating). In theory r_m's have the status of *hypothetical constructs* (with potential existential properties) rather than of intervening variables (convenient summarizing fictions). I assume that, certainly in the adult human language user, these mediating, symbolic events have become purely cortical processes – processes whose neurological nature and locus will not be known for a very long time.

With regard to the *source* of representational mediation processes, it is true that the theory postulates – for the historical *origins* of meanings – derivation of r_m's (mediating reactions to signs) from R_T's (overt reactions to the things signified), and Fodor has posed (this volume) what he believes to be a dilemma: either r_m's must stand in unique one-to-one relations with their R_T's (in which case mediation theory is reducable to single-stage learning theory and subject to Chomsky's (1959) strictures) or r_m's must stand in one-to-many relations to R_T's (in which case the theory fails in principle to account for unambiguous reference). I have accepted the first horn of this dilemma, but argue that it is not really a 'horn' at all. Let me summarize my arguments briefly.

(1) Since r_m's are hypothetical constructs, they bear a part-to-whole relation to R_T's only in the sense of being 'derived from' and being 'distinctively representational of' R_T's. They are not 'parts of' in the literal sense of being a sub-set of the overt R's making up R_T.

(2) Since r_m's evoked by previously learned signs can also serve as R_T's in the establishment of subsequent signs, there is no requirement in theory that every sign be based upon significate-produced behaviors. Thus listeners and readers can construct increasingly elaborate systems of meanings for abstractions (*gravity*, *justice*), for things not experienced directly (*zebras*, *Viet Cong*) and even things non-existent (*unicorns* and *elves*), from the context of other words which already have meanings. I have referred to this process as *assign learning*. Needless to say, a very large proportion of the lexicons of adults are assigns. The distinctive semantic features (components of r_m) established for primary perceptual and linguistic signs are capable of complex, novel and, indeed, 'emergent' interactions – a crucial point to which I shall return momentarily.

(3) Representational mediators are assumed to be *componential* in character. A relatively small number of independent, bipolar reaction components, by virtue of their combination in diverse simultaneous patterns, can serve to differentiate a very large number of distinct r_m's, each related to its R_T uniquely – but uniquely as a whole, not in terms of unique components. This is what I have referred to as the 'emic' principle of neo-behaviorism (cf. Osgood, 1970). Like the *phoneme* (or, more properly, the *sememe*) in linguistic theory, the r_m in neo-behaviorism (*a*) renders functionally equivalent classes of different behavioral events, either signs having the same significance for the receiver of behaviors expressing the same intention for the source, (*b*) is an abstract entity, unobservable itself but necessary for interpretation of what is observed, and (*c*) is resolvable into a 'simultaneous bundle' of distinctive features or components which serve to differentiate among classes of meanings (significances or intentions). In exactly the same sense that one cannot substitute some particular phone (which one?) for a phoneme in linguistic theory, one cannot substitute any particular overt R (which one?) for an r_m in behavior theory. Thus representational mediation theory does *not* reduce to single-stage S-R theory, as Fodor claims.

It is important to point out that r_m's are not mere replicas of their R_T's; rather, they are representations of those features of R_T's which have made a difference in appropriateness of behaving toward the things signified by signs and have therefore been differentially reinforced – again in close analogy to the distinctive features of phonemes. Thus the locomotor movements by which an organism moves from place A to place B (walking, running, swimming, etc.) may be non-distinctive with respect to avoiding signs of danger at A or approaching signs of safety at B, but the *affective* features of the diverse ways of avoiding vs. approaching will be common to the different R_T's and hence be differentially reinforced as components of the meanings of the signs of A vs. B.

In what sense can such a representational system be said to have 'emergent' properties? The study of cognitive dynamic has a considerable, if often controversial, literature in psychology (cf. Osgood, 1960, for a review). This research has been concerned with laws determining changes in meaning of signs when they interact contextually with other signs (attitude change toward sources and topics, interaction between perceptual and linguistic signs, prediction of the meanings of phrases from the meanings of their component words, and so forth). According to representational mediation theory, each feature of meaning corresponds to a single bipolar reaction system (component of r_m) which can only be in one state at any one time. If a component is simultaneously driven in opposed directions (e.g., as in *sincerity admires* on an Abstract/Concrete feature or as in *manipulate generously* on an Ego/Alter orientation feature), then – exactly as is the case for reciprocally antagonistic overt response systems – that component 'freezes' and what we refer to as awareness of *semantic anomaly* is experienced. Similarly, driving a component simultaneously in the same direction produces *semantic intensification* (as in *manipulate slyly* or in *sudden surprise*), and if one sign has coding on a feature for which another sign does not, then the meaning of the combination is *semantically modified* or enriched (as in *plead with sincerely* on a Moral/Immoral feature).

By virtue of having their own rules of interaction, these components of r_m's to signs do display certain emergent properties: (1) They enable the listener to react appropriately to (i.e., interpret) novel utterances; although one probably has never encountered the sentence *She will make someone a nice husband* before, the masculinization of *she* via interaction with *husband* serves to appropriately characterize the woman in question. (2) Meanings for new lexical items can be built out of the interactions of the features of contextual words which already have meanings (assign learning); in concept formation, superordinate terms like *fruit* will retain those features shared by their subordinate instances (*lemon, banana, apple*) but become non-differential on conflictual features (e.g., Shape or Sweetness). (3) Interaction among semantic components allows for selection among alternative senses of words (cf. the 'projection rules' of Katz and Fodor, 1963); for example, selection of two different senses of the word *play* in the sentences *he wrote a good play* vs. *he made a good play* is based upon congruence with the contrastive features of *write* and *make*. (4) Presence or absence of anomaly in this behavioral sense parallels the linguistic notion of 'selection rules' (cf. Chomsky, 1965); overriding such rules, when sufficient numbers of shared features are present, creates metaphors like *the thunder shouted down the mountainside* – where the −Human feature of *thunder* is countermanded by the other shared features between *thunder* and +Human *shouted* (i.e., *thunder* is personified).

The most deep-set resistance to this neo-behaviorist account of meaning, I think, is to its anchoring of r_m's to R_T's – that is, to having semantic distinctions originate in differences in the *behaviors* to Things rather than simply in differences in the *perceptions* of Things. Not only does this seem too restrictive as a source of features, but it reduces the gap between rational (talking) man and mechanistic (non-talking) animal quantitatively and certainly eliminates it qualitatively.[a] In my experience, this 'metaphysical repugnance' persists even when it is understood that this anchoring is *historical*, applying to the acquisition of primary perceptual and linguistic signs, and not contemporary for the adult speaker.

[a] I am sure that both Chomsky and I, despite our metaphysical differences, wish that people would behave more rationally with respect to other people – but this would be *more* like most animals!

From a behavioristic point of view, this anchoring of meanings to subsequent as well as antecedent observables serves at least two functions: (1) It eliminates, in principle, the sheer *circularity* of defining semantic features in terms of perceptible distinctions – which *would* erase all but terminological differences between mentalistic and behavioristic positions. (2) It places rather severe constraints upon *what* differences that are perceptibly discriminable will actually come to make a difference in meaning. The human sensory systems are capable of numbers of perceptible discriminations which are literally incredible; any psycholinguistic theory must be able to account for the relatively small number (and nature) of those discriminations which do in fact make a difference in meaning. Without such constraints, representational mediation theory would indeed become a universal Turing Machine – like God's Will, capable of explaining anything but patently incapable of disproof. As is currently being charged against contemporary generative grammar, it would be *too powerful*.

LANGUAGE UNIVERSALS

All this has obvious implications for the Nativist–Empiricist controversy over language universals. Chomsky (1968, ch. 3, and earlier) and developmental psycholinguists who have adopted his views take a strong nativist position. According to Chomsky, 'Thus...[language] appears to be a species-specific capacity that is essentially independent of intelligence...We must postulate an innate structure that is rich enough to account for the disparity between experience and knowledge, one that can account for the construction of the empirically justified generative grammars within the given limitations of time and access to data...The factual situation is obscure enough to leave room for much difference of opinion over the true nature of this innate mental structure that makes acquisition of language possible' (1968, pp. 68–9).

A number of philosophers have taken issue with Chomsky's nativist claims, and among them Nelson Goodman's arguments (1967), couched in a dialogue between Jason and Anticus (who thinks that what Jason has brought back from Outer Cantabridgia is more fleece than golden), are most relevant to our present concern.

'Anticus: ...Don't you think, Jason, that before anyone acquires a language, he has had an abundance of practice in developing and using rudimentary prelinguistic symbolic systems in which gestures and sensory and perceptual occurrences of all sorts function as signs...

Jason: True, but surely you do not call those rudimentary systems languages.

Anticus: No; but I submit that our facility in going from one symbolic system to another is not much affected by whether each or either or neither is called a language ...if the process of acquiring the first language is thought of as beginning with the first use of symbols, then it must begin virtually at birth and takes a long time' (pp. 25–6).

Chomsky's reply to this (1968) was as follows: 'This argument might have some force if it were possible to show that the specific properties of grammar...were present in some form in these already acquired prelinguistic "symbolic system". But since there is not the slightest reason to believe that this is so, the argument collapses' (pp. 70–1).

The evidence offered in this paper would seem to give quite ample 'reason to believe' that many properties of grammar are present in some form in pre-linguistic perceptuo-motor behavior. Not only does the sheer ability of Paraphrase Things in language imply that perceptual and linguistic signs must share the same underlying cognitive or semantic system, but the detailed ways in which non-linguistic, perceptual 'presuppositions' determine the form and content of descriptive sentences imply a very intimate interaction between these channels.[a] *If*, as seems perfectly obvious, the learning of the significance of perceptual signs and their relations in perceived events is prior to the learning of language *per se*, and *if* the development of perceptuo-motor communications systems in the human species was prior to the development of human languages *per se* – as seems obvious also, unless we assume that humanness and language were simultaneous mutations as humanoids dropped from the trees and stood erect! – then it seems perfectly reasonable to conclude that much, if not all, that is universal in human languages is attributable to underlying cognitive structures and processes and not either species-specific or language-specific.

Of course, one could argue that all I have done is to push the Nativist–Empiricist issue down a notch or two: This is true, but it is about the nativism of human linguistic competence that the strongest claims have been made. Chomsky's second line of defence of Nativism moves precisely in this direction: 'If it were possible to show that these prelinguistic symbolic systems share certain significant properties with natural language...we would then face the problem of explaining how the prelinguistic systems developed these properties' (1968, p. 71). Even assuming that language behavior is learned on the foundation of pre-linguistic cognitive structures and learning, we still will have to determine to what degree and in what ways this foundation is itself innate or learned.

I have already suggested that behavior theory (as distinguished from learning theory) must accept certain cognitive abilities as innate, although not necessarily species-specific for humans: the ability to form perceptual entities on the basis of certain gestalt-like tendencies; the ability to organize behavior hierarchically; the ability to differentiate units within each hierarchical level componentially. My perceptual demonstrations have suggested certain other tendencies which could be innate. The distinction between Entities and Actions and their integration as Events certainly seem fundamental (cf. Fillmore's case-relations), and, as I understand it, these are equally fundamental in 'deep structure'. The perception of Entities as novel or as unique or as requiring differentiation from others (vs. redundant) was found to have predictable effects upon a range of linguistic phenomena – determiner, pronoun, reflexive, adjective and adverb usage. Unfulfilled cognitive expectations were found to underlie diverse manifestations of negation (as has also been recently stressed by linguists, of course). The rather delicate dependence of verb tense upon the relation of event-time to subjective time-zero was also demonstrated. Various whole-sentence effects, including transformations, were shown to be influenced by the structures of perceptual situations, e.g., as sequentially ordered actions vs. simultaneously ordered relations.

[a] That linguists are becoming increasingly aware of this interaction is evidenced by, for example, Lakoff's (1969) discussion of the extra-linguistic (situational) determinants of choice among *some* and *any* and McCawley's (this volume) discussion of the deictic use of pronouns, where their antecedents are perceptually present entities and not prior noun phrases or sentences.

I think that the investigation of perceptuo-linguistic paraphrasing – *particularly cross-culturally and down the age scale to as young speakers of diverse languages as possible* – may be able to shed further light on human universals.

University of Illinois

REFERENCES

1. Alston, W. P. *The philosophy of language.* Englewood Cliffs, N.J.: Prentice-Hall, 1964.
2. Austin, J. L. *How to do things with words.* London: Oxford U. Press, 1962.
3. Baker, C. L. Double negatives. *Papers in Linguistics*, 1969, I, 16–40.
4. Campbell, D. T. Ostensive instances and entitativity in language learning and linguistic relativism. Paper presented at *The Center for Advanced Study in the Behavioral Sciences* (Palo Alto, Cal.), 1966.
5. Chomsky, N. Verbal behavior (a review). *Language*, 1959, XXXV, 26–58.
6. Chomsky, N. On the notion 'rule of grammar'. In *Proceedings of symposia in applied mathematics, American Mathematical Society*, 1961, XII, 6–24.
7. Chomsky, N. *Aspects of the theory of syntax.* Cambridge, Mass.: M.I.T. Press, 1965.
8. Chomsky, N. *Language and mind.* New York: Harcourt, Brace & World, Inc., 1968.
9. Cowan, J. (ed.) *Studies of language and thought.* Tucson, Arizona: U. Arizona Press, 1970.
10. Deese, J. Behavior and fact. *Amer. Psychologist* 1969, XXIV, 515–22.
11. Fillmore, C. J. The case for case. In E. Bach and R. Harms (eds.), *Universals in linguistic theory.* New York: Holt, Rinehart and Winston, 1968.
12. Fillmore, C. J. Verbs of judging: An exercise in semantic description. *Papers in Linguistics*, 1969, I, 91–117.
13. Fodor, J. A. Could meaning be an r_m? *J. verb. Learn. verb. Behav.*, 1965, IV, 73–81. Reprinted in this volume.
14. Fodor, J. A. More about mediators: a reply to Berlyne and Osgood. *J. verb. Learn. verb. Behav.*, 1966, V, 412–15.
15. Frege, G. Ueber Sinn und Bedeutung. *Zeitschrift für Philosophie und Philosophische Kritik*, 1892, C, 25–50.
16. Goodman, N. The epistemological argument. *Synthese*, 1967, XVII, 23–8.
17. Goss, A. E. Verbal mediation response and concept formation. *Psychol. Rev.*, 1961, LXI, 248–274.
18. Hull, C. L. Knowledge and purpose as habit mechanisms. *Psychol. Rev.*, 1930, XXXVII, 511–525.
19. Hull, C. L. *Principles of behavior.* New York: Appleton-Century-Crofts, 1943.
20. Huttenlocher, Janellen, Eisenberg, Karen and Strauss, Susan. Comprehension: Relation between perceived actor and logical subject. *J. verb. Learn. verb. Behav.*, 1968, VII, 527–30.
21. Katz, J. J. and Fodor, J. A. The structure of a semantic theory. *Language*, 1963, XXXIX, 170–210.
22. Kendler, H. H. and Kendler, T. S. Vertical and horizontal processes in problem solving. *Psychol. Rev.*, 1962, LXIX, 1–16.
23. Klima, E. S. Negation in English. In J. A. Fodor and J. J. Katz (eds.), *The structure of language: Readings in the philosophy of language.* Englewood Cliffs, N.J.: Prentice-Hall, 1964.
24. Lakoff, R. Some reasons why there can't be any *some-any* rule. *Language*, 1969, XLV, 608–15.

25. Lenneberg, E. H. *Biological foundations of language*. New York: Wiley & Sons, 1968.
26. Maclay, H. S. Overview of linguistic contributions. This volume.
27. McCawley, J. D. The role of semantics in a grammar. In E. Bach and R. T. Harms (eds.), *Universals in linguistic theory*. New York: Holt, Rinehart and Winston, 1968.
28. McCawley, J. D. Where do noun phrases come from? This volume.
29. Mehler, J. and Bever, T. On the cognitive capacities of young children. *Science*, 1967, CLVIII, 141; 1968, CLXII, 979–81.
30. Miller, H. E. and Dollard, J. *Social learning and imitation*. New Haven, Conn.: Yale U. Press, 1941.
31. Morgan, J. L. On arguing about semantics. *Papers in Linguistics*, 1969, I, 49–70.
32. Mowrer, O. H. *Learning theory and the symbolic processes*. New York: John Wiley & Sons, 1960.
33. Olson, D. R. Language and thought: Aspects of a cognitive theory of semantics. Paper presented at the *Conference on Human Learning*, Prague, Czechoslovakia, July 1969.
34. Osgood, C. E. *Method and theory in experimental psychology*. New York: Oxford U. Press, 1953.
35. Osgood, C. E. A behavioristic analysis of perception and language as cognitive phenomena. In *Contemporary approaches to cognition*. Cambridge, Mass.: Harvard U. Press, 1957.
36. Osgood, C. E. Cognitive dynamics in the conduct of human affairs. *Pub. Opin. Quart.*, 1960, XXIV, 341–65.
37. Osgood, C. E. Meaning cannot be an r_m? *J. verb. Learn. verb. Behav.*, 1966, V, 402–7.
38. Osgood, C. E. Is neo-behaviorism up a blind alley? (submitted for publication in 1970, *Psychol. Review*).
39. Piaget, J. Quantification, conservation, and nativism. *Science*, 1968, CLXII, 976–9.
40. Sadock, J. M. Super-hypersentences. *Papers in Linguistics*, 1969, I, 1–15.
41. Sampson, G. Noun-phrase indexing, pronouns, and the 'definite article'. Mimeo, *Linguistic Automation Project*, Yale University, February 1969.
42. Schlesinger, I. M. Production of utterances and language acquisition. In D. I. Slobin, ed., *The ontogenesis of grammar: Some facts and several theories*, forthcoming.
43. Skinner, B. F. *Verbal behavior*. New York: Appleton-Century-Crofts, 1957.
44. Smith, Carlota S. Determiners and relative clauses in generative grammar of English. *Language*, 1964, XL, 37–52.
45. Strawson, P. F. On referring. *Mind*, 1950, LIX, 320–44.
46. Vendler, Z. *Linguistics in philosophy*. Ithaca, N.Y.: Cornell U. Press, 1967.

Are there specifically linguistic universals?[a]

DAVID McNEILL

This paper presents the argument that it is quite impossible to say, at the moment, if the structure of thought influences the structure of language. One might reasonably ask why such an argument needs to be presented. For obviously thought affects speech, at least under favorable conditions, and the structure of English (for instance) obviously is not the same as the structure of thought. There would thus seem to be no room for argument. That thought influences language is either obviously right or it is obviously wrong.

There is, however, a sense in which the influence of thought on the structure of language is not obvious. In fact, in this further sense, the question remains entirely open and no one can yet say what connection, if any, there is between thought and language. But that is the argument of the paper.

To explain the argument it is necessary first to describe how language appears to be acquired. The account will necessarily be brief; a more complete description can be found elsewhere (McNeill, 1970).

The structure of language is largely abstract. Most of what one knows about sentence structure is not indicated in any acoustic signal. It is not indicated even in the arrangement of words as uttered in sentences. Everyone knows, for example, that *Mary* and *see* go together differently in *Mary is eager to see* and *Mary is easy to see*. The grammatical relation between the two words differs in these sentences but the difference is not present to the senses. It is abstract. Since this is the case, and since in general each of us has access to linguistic information that cannot be displayed, a question arises as to how we when children ever acquired such undisplayable information. Let us call this the problem of abstraction. It is through an attempt to solve the problem of abstraction that the effect of thought on language becomes an important theoretical possibility.

The problem of abstraction is not, of course, unique to language development. It appears also in perceptual development and in the development of manual skill, to mention just two situations. It appears whenever the result of experience differs systematically from experience itself. A grammar is systematically different from the sentences a child experiences. The visual perception of three dimensions and the ability to organize a series of movements into a single act of reaching (Bruner, 1968) are systematically different from the visual and tactual stimulation that support them. The way abstraction is overcome in development, however, may not be the same in

---

[a] This paper is based on a paper entitled 'Explaining linguistic universals' in J. Morton (ed.), *Biological and social factors in psycholinguistics*. London: Logos Press, 1971.

these situations. The point of this paper is that such questions cannot be prejudged. What follows, therefore, might be unique in all respects to language; we return to this equivocation at the end, for it has an origin more profound than simple caution in the face of surprising Nature.

A hypothesis about the acquisition of language, focusing directly on the problem of abstraction, holds that the 'concept of a sentence' appears at the beginning of language acquisition. It is not something arrived at after a long period of learning. It is, rather, something that organizes and directs even the most primitive child speech. The facts of language acquisition cannot be understood in any other way. The concept of a sentence is a method of organizing linguistic information, including semantic information, into unified structures. Words fall into grammatical categories and the categories are related to one another by specific grammatical functions, such as subject, predicate, object of verb, etc. The concept of a sentence can appear so early, before grammatical learning has taken place, because it reflects specific linguistic predispositions, some of which might be innate.

The successive stages of language acquisition in this view result from a progression of linguistic hypotheses, which children invent, regarding the way the concept of a sentence is expressed in their language. The term 'hypothesis' is used here in the sense of a tentative explanation subject to empirical confirmation; it does not necessarily imply that the processes of invention and test take place consciously (but cf. Weir, 1962, for examples of apparently deliberate hypothesis testing).

Children everywhere begin at 12 to 14 months with the same initial hypothesis: the concept of a sentence is conveyed by single words. By 18 or 24 months, typically, this hypothesis is abandoned for another: the concept of a sentence is expressed through combinations of two and three words. This hypothesis is in turn abandoned and new ones are adopted, presumably under the pressure of adult speech, which at the outset consists entirely of negative examples. There is in this way a constant elaboration of the forms through which the more or less fixed concept of a sentence is expressed. The consequence is that the concept of a sentence becomes steadily more abstract. Thus, the concept of a sentence does not *have* to be displayed in the form of specimens, for children themselves make it abstract. We should note that the process of language acquisition gives to adult grammar a certain necessary arrangement. The part of the underlying structure that defines the concept of a sentence will exist in every language, but it will be related via transformations to surface structures that contain many idiosyncracies. Such is the universal form of grammar, according to the linguistic theory developed by transformational grammarians.

Three examples will be cited in connection with this view of language acquisition. Two come from different places on the developmental line that children follow in expressing the concept of a sentence. The third comes from a developmental line that appears to run in a completely different direction. Let us first consider the holophrastic period of development. Many linguists and psychologists have observed that young children express something like the content of full sentences in single-word utterances. But what has not been realized is that the concept of a sentence undergoes a continuing evolution throughout the holophrastic period. Children do not express 'something like' sentences with single words, therefore, they express precisely the relations that define the concept of a sentence, these relations appearing one by one during the second year of life. My examples are drawn from a diary kept by Dr P. Greenfield, who was to my knowledge the first to notice this phenomenon. The initial step seems to be the assertion of properties. At 12 months and 20 days

(12; 20) the child said *hi* when something hot was presented to her, and at 13; 20 she said *ha* to an *empty* coffee cup – clearly a statement of a property since the appropriate thermal stimulus was absent. At 14; 28 the child first used words to refer to the location of things as well as to their properties. She pointed to the top of a refrigerator, the accustomed place for finding bananas, and said *nana*. The utterance was locative, not a label or an assertion of a property, since there were no bananas on the refrigerator at the time. A short while later a number of other grammatical relations appeared in the child's speech. By 15; 21 she used *door* as the object of a 'verb' (meaning 'close the door'), *eye* as the object of a 'preposition' (after some water had been squirted in her eye), and *baby* as the subject of a 'sentence' (after the baby had fallen down).

By the time words are actually combined several grammatical relations have become available and have been available for several months. Greenfield's daughter first combined words at 17 months but at 15 months had expressed such relational concepts as attribution, location, subject, and object. Combining words is a new method for expressing these same relations. Thus the expression of sentences changes while the concept of a sentence itself does not. Far from being the product of grammatical learning, therefore, the concept of a sentence is the original organizing principle for grammatical learning. There is a constant attempt by a child to enlarge the linguistic space through which this concept passes. It is impossible to say on current understanding why children begin to combine words. Doing so reflects a fundamentally different view of the nature of language: whereas previously any word could appear in almost any relation, now many words appear in only one relation.

The result contrasts very sharply with what had existed before. There is a definite restriction on the patterns that appear as word combinations. One child, Adam, whose speech was first studied at 26 months, will serve as a convenient example. A distributional analysis of Adam's speech resulted in three grammatical classes: modifiers, nouns, and verbs. With three grammatical classes there are $(3)^2 = 9$ possible two-word combinations and $(3)^3 = 27$ possible three-word combinations. Of this set of possibilities only 4 two-word and 8 three-word combinations actually occurred. Every one of the occurring combinations corresponded to one or more of the formal definitions of the basic grammatical relations, as contained in linguistic theory (for further discussion, see McNeill, 1966). There were utterances such as *Mommy come* and *eat dinner* but there were none such as *come eat dinner*. The first corresponds to the subject-predicate relation, the second to the verb-object relation, and the third to no relation – it is the result of conjoining two independent sentences, each of which separately embodies a basic grammatical relation. It is worth pointing out that surface structures like *come and eat dinner* are abundantly exemplified in the speech of adults, but the abstract relations underlying such surface sentences can be understood only if the transformational operation of conjunction is understood. Adam at 26 months had not discovered this operation.

Between Greenfield's diary and Adam's speech at 26 months a complete change has taken place in the expression of the concept of a sentence. There is however continuity in the concept expressed. It is fairly clear, moreover, that children possess a specific *ability* to express the concept of a sentence through combinations of words; it does not result merely from uttering words in combination. Gardner and Gardner (1969) have begun a fascinating attempt to teach a chimpanzee to 'speak' English by means of deaf-mute signs. By the use of signs Gardner and Gardner hoped to overcome the entirely peripheral limitations that they supposed had obstructed previous attempts to teach chimpanzees to speak (e.g., Kellogg and Kellogg, 1933), and there can be little doubt that this supposition is correct. Washoe, as the chimpanzee is called, has a well established vocabulary of some 30 words at 25 months. The words apparently are used as genuine symbols and their number compares favorably to the number of words children use at the same chronological (though not the same developmental) age. Gardner and Gardner and their assistants address Washoe in sentences, much as parents address children in sentences, and Washoe has begun to combine signs into what Gardner and Gardner accept also as sentences. It is the process of combination, however, that shows a basic difference between children and Washoe. Unlike Adam or any other child Washoe combines words *without* restriction. The grammatical puritanism that hails the beginning of true syntax in children is thus utterly lacking in Washoe. The following are examples of Washoe's combinations (each English word corresponds to a deaf-mute sign):

go in	(at some distance from a door)
go out	
go sweet	(while carried to a raspberry bush)
open flower	(to be let into a flower garden)
open key	(for a locked door)
listen eat	(at the sound of the dinner bell)
listen dog	(at the sound of an unseen dog barking)

In none of these utterances is the relation the same, with the exception of 'go in' and 'go out' – which unlike the others could be imitations. In 'open flower', for instance, the relation is movement toward something, whereas in 'open key' the relation is one of instrumentality. It is possible that Washoe actually draws such distinctions, in which event we should conclude that she is far in advance of children at a comparable length of utterance, but it is more probable that Washoe does not organize words into grammatical relations at all. She produces instead whatever signs are appropriate to her situation when she happens to know them, but successive signs are independent of each other. It is consistent with this interpretation that she does not produce signs in fixed order. She can, as Gardner and Gardner supposed, acquire words and she is not limited to uttering a single word at a time. But as far as these few examples reveal she cannot organize meanings in the particular ways routinely and universally available to children.

I have summarized, most briefly, a few reasons for supposing that the concept of a sentence is not a product of learning. It can be detected during the holophrastic period, before words have ever been put into combination, and even can be seen undergoing development at that time. It covers completely all instances of word combination, when these appear. And it fails to cover word combinations when the concept is not available for reasons of species specificity. The concept of a sentence apparently reflects specific abilities, presumably innate to man, which children possess as a necessary condition for the acquisition of language.

Faced with the prospect of crediting children with this balance of a priori structure a number of psycholinguists have felt an urge to explain it. Explanation is of course commendable. It is the ultimate goal of all of us. My argument merely is that certain basic distinctions cannot be overlooked if such explanations are to be found. It is worthwhile examining the situation closely. According to Schlesinger (in press) the underlying structure of sentences 'are determined by the innate *cognitive* capacity of the child. There is nothing specifically linguistic about this capacity.' And Sinclair-deZwart (1970) has written that 'Linguistic universals exist precisely because thought structures are universal'. Both Schlesinger and Sinclair-deZwart are discussing the abstract linguistic structures referred to here as the concept of a sentence. The influence of cognition, in their view, is visible in the universal underlying structure of language.

Contrary to these claims, however, the abstract structure of sentences may equally reflect specific linguistic abilities. A distinction must be drawn between at least two kinds of linguistic universal. It is possible that further divisions will prove useful, which would move us even further from the monolithic situation implied by Schlesinger and Sinclair in the statements quoted above, but the following distinction will serve as a beginning:

Weak linguistic universals have a necessary and sufficient cause in one or more universals of cognition or perception.

Strong linguistic universals may have a necessary cause in cognition or perception, but because another purely linguistic ability also is necessary, cognition is not a sufficient cause.

This distinction is purely psychological. Linguistics has nothing to do with it. Linguistic theory, where it is postulated what is universal in language, gives no hint of the causes of linguistic universals. Consider, for example, the grammatical categories of nouns and verbs. They appear universally in the underlying structure of language because linguistic descriptions are universally impossible without them. According to the theory of language acquisition described earlier in this paper, they appear universally because children spontaneously organize sentences in terms of such categories. Now, in addition, we wish to explain this activity of children. If nouns and verbs are weak linguistic universals they have a necessary and sufficient cause in cognitive development. Most theories of cognition would provide something like the notions of actor, action, and recipient of action, as categories of intellectual functioning that are available to young children. One might argue that nouns and verbs are the reflection in language of these cognitive categories. If, however, nouns and verbs are strong universals, some further ability is necessary in addition to the cognitive categories of actor, action, and recipient of action. The latter may be necessary but they are not sufficient to cause the appearance of nouns and verbs in child grammar.

The essence of the strong-weak distinction is the psychological status of the linguistic terms 'noun' and 'verb'. Do they refer to processes of cognition, which are given special names in the case of language, or do 'noun' and 'verb' refer also to specific linguistic processes, which have an independent ontogenesis? There is no way to prejudge this question; contrary to Sinclair and Schlesinger, there is no a priori answer. The only way to proceed is to raise the question for each linguistic universal in turn. A proper investigation, of course, would not always and necessarily lead to the same answer with every universal. It is conceivable that language is a mixture of weak and strong universals.

As an example of how the distinction between weak and strong universals can be approached, consider an observation reported by Braine (1970). He taught his two-and-a-half year old daughter two made-up words: the name of a kitchen utensil (*niss*) and the name of walking with the fingers (*seb*). The child had no word for either the object or the action before Braine taught her *niss* and *seb*. Neither word was used in a grammatical context by an adult, but the child used both in the appropriate places.

There were sentences using *niss* as a noun, such as *that niss* and *my niss*, and sentences using *seb* as a verb, such as *more seb* and *seb Teddy*. More important in the present context there were sentences also with *seb* as a noun, *that seb* and *my seb*, but none with *niss* as a verb. The asymmetry suggests that verbs are weak universals and nouns strong. Association with an action is necessary and sufficient for a word to be a verb but some additional linguistic property can make a word into a noun. Association with an action does not block the classification. The situation is not as perfect as we might like it, however, for association with an object may also be sufficient for a word to become a noun. If this is true, we should further subdivide linguistic universals into weak, strong, and 'erratic' types. An erractic universal is one in which a linguistic universal has two sufficient causes and therefore no necessary ones. Both the cognitive category of object and a linguistic ability can lead a word to become a noun.

I would argue that no progress can be made in the explanation of linguistic universals unless these basic distinctions are honored. Sweeping remarks, that all universals of language are in reality universals of cognition, cross the line separating science from dogma.

University of Chicago

REFERENCES

Braine, M. D. S. The acquistion of language in infant and child. In C. Reed (ed.), *The learning of language: Essays in honour of David H. Russell*. New York: Appleton-Century-Crofts, 1970.
Bruner, J. S. Eye, hand, and mind. Heinz Werner Memorial Lecures, Clark University, 1968 (also in the 1968 Annual Report of the Center for Cognitive Studies, Harvard University).
Gardner, R. A. and Beatrice T. Gardner. Teaching sign language to a chimpanzee. *Science*, 1969, 165, 664–72.
Kellogg, W. N. and L. H. Kellogg. *The ape and the child*. New York: McGraw-Hill, 1933.
McNeill, D. Developmental psycholinguistics. In F. Smith and G. A. Miller (eds.). *The genesis of language*. Cambridge, Mass.: MIT Press, 1966.
The Acquisition of language. New York: Harper and Row, 1970.
Schlesinger, I. M. Production of utterances and language acquisition. In D. I. Slobin (ed.), *The ontogenesis of grammar*. New York: Academic Press, in press.
Sinclair-deZwart, Hermine. Sensorimotor action schemes as a condition of the acquisition of syntax. In R. Huxley and E. Ingram (eds.), *Mechanisms of language development*. New York and London: Academic Press, 1970.
Weir, Ruth. *Language in the crib*. The Hague: Mouton, 1962.

Language and cognition[a]

ERIC H. LENNEBERG

1 The problem

That the capacity for naming has a biological dimension may be seen from the difficulties that animals experience in this respect. For instance, it is possible to train a hunting dog to 'point', and it may be quite possible to teach him to point to a specific set of objects in a specific environment upon appropriate command in a natural language. But it does not appear to be possible to teach a dog to do the 'name-specific stimulus generalization' that every child does automatically. The hound who has learned to 'point to the tree, the gate, the house' in the trainer's yard will perform quite erratically when given the same command with respect to similar but physically different objects, in an unfamiliar environment. The correctness of the animal's responses may even vary with such extra-linguistic cues as the geographical position, posture, and bodily movements of his master, the time of day, or the clothes that people are wearing while he is being exercised. There is no convincing evidence that any animal below man has ever learned to relate any given word to the same range of stimuli that is covered by that word in common language-usage. So-called proof to the contrary always lacks proper controls on interpretation. For instance, there is a report on a parrot who could say good-by (in German) and who supposedly knew what this word meant or when it was properly used. Once the bird was also heard to say good-by upon the arrival of some friends of the family; the proud owner judged this to be a sign that his pet did not merely know the meaning of the word but was even using it to produce a desired effect: to send the just-arrived friends away, presumably because he had taken a dislike to them.

It may be well to stress once more that our concern is with the capacity for (natural, human) language which, ordinarily, leads to the understanding of a de-finably structured type of utterance; or, in other words, with *knowing* a language. The infant who has a repertoire of three tricks (wave by-by, show me your tongue, show me how tall you are) which he can perform upon the appropriate commands but who can understand no other sentence of the same grammatical, structural type has not yet begun to acquire language. The essence of language is its productivity; in the realm of perception and understanding of sentences, it is the capacity to recognize structural similarities between familiar and entirely novel word patterns. Thus our criterion for knowing language is not dependent upon demonstrations that an individual can talk or that he goes through some stereotyped performance upon hearing certain words, but upon evidence that he can analyze novel utterances through the application of structural principles. It is the purpose of this chapter to show that

[a] This paper is reprinted from E. H. Lenneberg, *Biological Foundations of Language*, Wiley, New York, 1967.

the understanding of the word-object relationship, the learning and acquisition of reference, is also dependent upon certain cognitive, analytic skills, much the way understanding sentences is. The problem of reference cannot be discussed without simultaneous considerations of the relationship between language and cognition.

Evidence for understanding language may be supplied by different kinds of response. It is not necessary that the subject has the anatomical and physiological prerequisites for actual speech production. In the case of man, we may cite children who have learned to understand language but who cannot speak; compare this to children who have the anatomical equipment for speech production but whose cognitive apparatus is so poorly developed that only the primordia for language are detectable but not fullfledged comprehension. In the case of animals, we have birds who can talk but who give no evidence of language understanding and we have the famous case of Clever Hans, the horse, who had a nonacoustic response repertoire (stamping of hoofs) that, unfortunately, gave the erroneous impression of a coding system for the German language. Had the horse actually had the cognitive capacity for acquiring a natural language, his motor response limitations would have been no obstacle to his giving evidence for language comprehension. A similar argument could be made for the physical nature of the input data. Language acquisition is not dependent in man upon processing of acoustic patterns. There are many instances today of deaf-and-blind people who have built up language capacities on tactually perceived stimulus configurations.

2 Toward a biological conception of semantics

The activity of *naming* or, in general, of using words may be seen as the human peculiarity of making explicit a process that is quite universal among higher animals, namely, the organization of sensory data. All vertebrates are equipped to superimpose categories of functional equivalence upon stimulus configurations, to classify objects in such a way that a single type of response is given to any one member of a particular stimulus category. The criteria or nature of categorization has to be determined empirically for each species. Frogs may jump to a great variety of flies and also to a specific range of dummy-stimuli, provided the stimuli preserve specifiable characteristics of the 'real thing'.

Furthermore, most higher animals have a certain capacity for discrimination. They may learn or spontaneously begin to differentiate certain aspects within the first global category, perhaps by having their attention directed to certain details or by sharpening their power of observation. In this differentiation process initial categories may become subdivided and become mutually exclusive, or a number of coexisting general and specific categories or partially overlapping categories may result. Again, the extent of a species' *differentiation capacity* is biologically given and must be ascertained empirically for each species. Rats cannot make the same range of distinctions that dogs can make, and the latter are different in this respect from monkeys. The interspecific differences cannot merely be explained by differences in peripheral sensory thresholds. Apparently, a function of higher, central processes is involved that has to do with cognitive organization.

Most primates and probably many species in other mammalian orders have the capacity to relate various categories to one another and thus to respond to *relations* between things rather than to things themselves; an example is 'to respond to the largest of any collection of things'. Once more, it is a matter of empirical research to discover the limits of relations that a species is capable of responding to.

In summary, most animals organize the sensory world by a process of *categorization*, and from this basic mode of organization two further processes derive: *differentiation* or discrimination, and interrelating of categories or the perception of and tolerance for *transformations*. In man these organizational activities are usually called concept-formation; but it is clear that there is no formal difference between man's concept-formation and an animal's propensity for responding to categories of stimuli. There is, however, a substantive difference. The total possibilities for categorization are clearly not identical across species.

2.1 WORDS AS LABELS FOR CATEGORIZATION PROCESSES

The words that constitute the dictionary of a natural language are a sample of labels of categories natural to our species; they are not tags of specific objects. When names have unique referents, such as Michelangelo, Matterhorn, Waterloo, they may be incorporated into discourse but are not considered part of the lexicon.[a] Thus most words may be said to label realms of concepts rather than physical things. This must be true, for otherwise we should have great difficulty in explaining why words refer to *open* classes. We cannot define the category labeled *house* by enumerating all objects that are given that name. Any new object that satisfies certain criteria (and there is an infinity of such objects) may be assigned that label. It is easier to say what such criteria are *not* than to say what they are. They are not a finite set of objectively measurable variables such as physical dimensions, texture, color, acidity, etc. (except for a few words, which constitute a special case; these will be discussed under the heading 'The Language of Experience'). We cannot predict accurately which object might be named *house* and which not by looking only at the physical measurements of those objects. Therefore, categorization and the possibility of word-assignment must usually be founded on something more abstract.

The abstractness underlying meanings in general, which has been the focus of so much philosophizing since antiquity, may best be understood by considering concept-formation the primary cognitive process, and naming (as well as acquiring a name) the secondary cognitive process. Concepts are superimpositions upon the physically given; they are modes of ordering or dealing with sensory data. They are not so much the *product* of man's cognition; conceptualization is the *cognitive process itself*. Although this process is not peculiar to man (because it essentially results from the mode of operation of a mechanism that can only respond in limited ways to a wide variety of inputs), man has developed the behavioral peculiarity of attaching words to certain types of concept formation. The words (which persist through time because they may be repeated) make the underlying conceptualization process look much more static than it actually is, as we shall demonstrate presently. Cognition must be the psychological manifestation of a physiological process. It does not appear to be a mosaic of static concepts, or a storehouse of thoughts, or an archive of memorized sense-impressions. The task of cognitive organization never comes to an end and is never completed 'in order to be used later'. Words are not the labels of concepts completed earlier and stored away; they are the labels of a *categorization process or family of such processes*. Because of the dynamic nature of the underlying process, the referents of words can so easily change, meanings can be extended, and categories

[a] For instance the proper name *Benjamin Pfohl* will ordinarily not be considered to be part of the dictionary of the English language.

are always open. *Words tag the processes by which the species deals cognitively with its environment.*

This theoretical position also elucidates the problem of translation or the equation of meanings across natural languages. If words label modes of cognizing, we would expect that all semantic systems would have certain formal commonalities. For instance, if we hear a given word used in connection with a given object or phenomenon, we are able to intuit the general usage of that word – it does not have to be paired with two hundred similar objects or phenomena before we can make predictions whether the name applies to a new object. Man's cognition functions within biologically given limits. On the other hand, there is also freedom within these limits. Thus every individual may have highly idiosyncratic thoughts or conceptualize in a peculiar way or, in fact, may choose somewhat different modes of cognitive organization at different times faced with identical sensory stimuli. His vocabularly, which is much more limited and unchangeable than his capacity for conceptualizing, can be made to cover the novel conceptual processes, and other men, by virtue of having essentially the same cognitive capacities, can understand the semantics of his utterances, even though the words cover new or slightly different conceptualizations. Given this degree of freedom, it becomes reasonable to assume that natural languages always have universally understandable types of semantics, but may easily have different extensions of meanings, and that, therefore, specific semantic categories are not coterminous across languages.

It does not follow from this that differences in semantics are signs of obligatory differences in thought processes, as assumed by Whorf (1956) and many others. The modes of conceptualization that happen to be tagged by a given language need not, and apparently do not, exert restrictions upon an individual's freedom of conceptualizing. This will be discussed subsequently.

3 The empirical study of naming: the language of experience

In section 2, I have proposed that words tag conceptualization processes, not things directly. But since the input to at least some conceptualization processes is the sensory perception of things, there is, of course, also a relationship between words and objects. This relationship may be studied empirically, under laboratory conditions. For this purpose, it is desirable to choose words that have simple referents, in the sense that they can be exhaustively described by objective measurement of *all* of their physical properties. This condition is fulfilled in only a small class of words, namely those that describe the sensation of certain physical properties themselves. Let us refer to these words collectively as the *language of experience*; examples are words for temperature, taste, hearing, or vision.

3.1 DESCRIPTION OF THE REFERENTS

The language of experience is particularly well suited for research because its referents have four advantages over the referents of most other types of words: first, they may be ordered by objective, logical criteria (for example, the centigrade scale for temperatures; the frequency scale of pure tones of equal intensity, etc.), whereas furniture, relatives, or most other concrete referents have no such logical and unique orders. Second, the referents have continuity in nature (for example, within certain limits any degree of temperature may be encountered or any sound frequency heard, whereas the domain of chairs does not grade into the domain of, say, benches, nor do uncles grade into aunts). Third, words in the language of experience refer to *closed classes*; our sensory thresholds set limits to perception and thus there is a bound to the range of phenomena that may be called hot, loud, green, etc. Fourth, the referents are simple in the sense that each instant may be completely specified by a fixed and very small number of measurements. Tempera- ture is specified by just one measurement; pure tones by two – intensity and fre- quency; colors by three – for example, the Munsell scales of hue, brightness, and saturation.

3.2   REFERENT SPACES

If it is possible to describe phenomena completely and accurately in terms of certain dimensions, it is also possible to construct coordinate systems in which every point corresponds to one specific phenomenon. Every point on a thermometric scale corresponds to one specific temperature; every point within a pitch-intensity co- ordinate system corresponds to one specific pure tone. Sometimes such coordinate

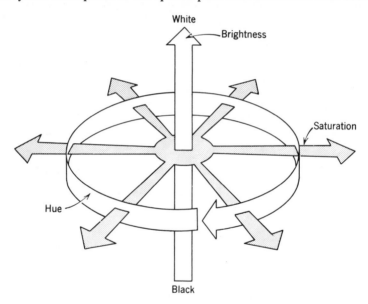

FIGURE I. *Schematic diagram showing relations between hue, brightness, saturation in the Munsell Color System*

systems are called *spaces*. The coordinate systems mentioned so far are unidimen- sional and two-dimensional spaces; for the specification of color we need a three- dimensional space, and since these *spaces* are merely abstract, mathematical concepts, one can also have four-, five-, and *n*-dimensional spaces. The three-dimensional color space is most easily visualized. Figure I shows the arrangement of the coordi- nates and Fig. 2 shows an actual model of a color space. The choice of a polar co- ordinate system is arbitrary. It offers some conveniences, but the three dimensions might have been arranged differently.

With the introduction of the notion of a color space we might review once more why we wish to arrange physical phenomena in this way. Our aim is to study how words relate to objects. But we cannot study the behavior of *words* unless we keep some control over the objects. We must do everything we can so as to bring some order into physical 'reality'. We have chosen phenomena that can be measured and ordered with respect to each other. The color space is simply an ordering device that allows us to assign every possible color a specific position or point. The entire

FIGURE 2. *Psychological color solid. Left: for colors perceived under good visual color matching conditions such as those of a textile color matcher; right: for colors perceived under supraliminal conditions, as when using a good instrument. (From Nickerson and Newhall, 1943)*

world of color is encompassed in the color space. Our next step now is to discover how the color words of a given language, say English or Navaho, fit into this space. Where are the colors in our space that answer to the name *red*? Obtaining replies to this type of question will be called *mapping color terms into the color space*. Naturally, every language is likely to have somewhat different maps; but the color space, which merely describes the psychophysical properties of colors, is constant for all of mankind.

3.3 NAMES MAPPED INTO REFERENT SPACES

Let us now consider how the words of the language of experience apply to the referents as we have ordered them into spaces. Much work has been done on color terminology and its relationship to the color space. Our task is twofold: we might start with a collection of color words in use by English speakers (for instance, by asking a sample of speakers to write down all the words for colors they can think of) and then try to assign each word a region in the color space; or we might sample the color space itself by selecting two hundred colors, evenly distributed throughout the space, and then show each color to a representative sample of speakers of English and ask them to write down the English word or words that best describe that color. Let us call the first procedure *Approach A* and the second *Approach B*.

When *Approach A* is used, we obtain a name-map that reflects the meaning of certain words in the lexicon of a natural language. This type of information must be distinguished from the actual use that speakers make of these words. It does not completely reveal the mechanism of naming. This becomes most obvious from the fact that *Approach A* leaves certain parts of the color space unmapped. There will be certain colors to which none of the words collected earlier (in the absence of color samples) properly refers. The operations that lead to such results are quite simple. Subjects are presented with comprehensive color charts (Brown and Lenneberg, 1954, used the Munsell book of colors) and are asked to point to all the colors that might be called *x* (*brown, lavender, orange*, etc.). There will be some words which some subjects cannot locate at all (say, heliotrope); he may not know what the word means, or there may be wide disagreement between subjects what the proper location of a word is (say, magenta). After having given all the words in our compilation to all our subjects we will discover that there is a residue of physical colors to which no one assigned any of the names – the 'innominate regions'.

Now, when *Approach B* is used the 'innominate regions' of the color space all but disappear. Although one or another subject may say 'I don't know what one calls that color', there will be others who quite readily assign some descriptive phrase to any color shown. *Approach B* reflects much better the versatility or basic productivity inherent in the act of naming, while it provides at the same time some statistic on the actual usage in a given language.

Since subjects will differ among themselves in their precise naming-habit and since there will be colors that are ambiguous with respect to common names in English, colors will differ in their probability of being called *x*. In fact, many colors will be called by more than one name, each name having a different probability of being assigned to a given color.

I have shown experimentally (Lenneberg, 1957) that speakers of English do not only distribute their words – in this case for colors – in a characteristic and language-specific way over the stimulus continuum, but that one can also elicit from subjects correct estimates of probability that a given stimulus will be given a specific name.

If we take the color space and ask how does the word *red* fit into this continuum, it will appear that a certain circumscribed region of points (each point represents a different shade of color) is covered by this word. However, only some points are felt to be 'good *reds*'; other points are more *orangy reds* or *brownish reds*. That means that the map or region of the word *red* has a focus or center (points that are the most typical examples of the word *red*) and that the borders of the map are never sharp; they fade out and are overlapped by at least one, usually several other word maps.

It is difficult to visualize the fourth dimension of probability in the three-dimensional maps of words inside the color space. But the overlapping, encompassing, and subdividing nature of the probabilistic gradients of name maps is clearly seen if one takes out of the color space a linear array of colors (for example, stimuli that vary only in hue, holding brightness and saturation constant) so as to produce a unidimensional subsample of the color space, and then maps the vocabulary on to this continuum as shown in Fig. 3. The procedure for obtaining this map was that of *Approach B*. Notice that there are certain colors which all subjects call by essentially the same name whereas other colors are sometimes called by this and sometimes by that name. Colors nos. 4 and 13 in Fig. 3 were given so many individual names that it was not possible to represent them all by a distinct curve.

Thus, when words are mapped into a referent space, a probabilistic variable emerges that we might call *name-determinacy*. The loci in the referent space vary with regard to the name-determinacy. The determinacy is greatest in the center of the foci and decreases centrifugally from these points; it is lowest at the intercept of name maps. The region which appeared to be 'innominate' under *Approach A* is one of great and widespread name-indeterminacy. Name-determinacy is, of course, a social phenomenon.

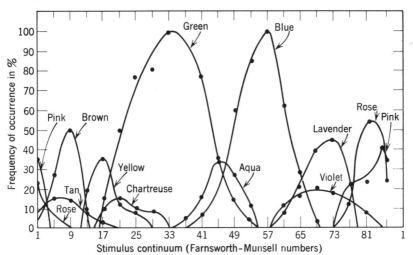

FIGURE 3. *Frequency distributions of selected English color words over the Farnsworth–Munsell colors.* N = 27. *Approach B was used.*

The procedures outlined here are a convenient way of demonstrating some of the semantic differences between languages. For instance, color name maps have been worked out for Zuni (Lenneberg and Roberts, 1956) and partially for Navaho (Landar et al., 1960), and Conklin (1955) adopted the Lenneberg–Roberts technique for the study of Hanunoo color categories.

This type of empirical research on semantic relationships in various languages can only be carried out within a very small part of the total vocabulary of a language, namely only on the language of experience. Therefore, it is not possible to verify empirically the far-reaching claims that have been made about the 'total incomparability of semantics' among unrelated languages. The languages investigated all have names for colors, and apart from finding mergers of two of our English color maps into one, or the substructuring of one of our maps, or shifts in border lines, nothing particularly astonishing has been discovered. We gather that man, everywhere, can make reference to colors and communicate about color fairly efficiently.

3.4 CONTEXTUAL DETERMINANTS IN COMMON NAMING

The insight gained through the application of *Approach B* still gives an inaccurate picture of how the naming of phenomena proceeds. The structure of the name maps might induce us to believe that the English word *red* can only be used for a very few colors (the focus of the region 'red'); and that as we move away from the focus the word would always have to be qualified as for instance *yellowish-red*, *dirty red*, etc. However, when one refers to the color of hair, or the color of a cow, or the skin color of the American Indian the word *red* is used without qualifiers. Conversely, the physical color which is named *brownish-orange* when it appears in the Munsell *Book of Colors*, and *red* when it is the color of a cow, may be called *ochre* when it appears on the walls of a Roman villa.

Quite clearly the choice of a name depends on the context, on the number of distinctions that must be made in a particular situation, and on many other factors that have little to do with the semantic structure of a given language *per se*; for example, the speaker's intent, the type of person he is addressing himself to, or the nature of the social occasion may easily affect the choice of a name.

Only proper names are relatively immune from these extra-linguistic determining factors. But although the attachment of the name Albert Einstein to one particular person remains completely constant, it implies that in instances such as these the most characteristic aspect of language is eliminated, namely the creative versatility – the dynamics – and it may well be due to this reduction that proper names are not 'felt' to be part of a natural language. Once more we see that there are usually no direct associative bonds between words and physical objects.

It is also possible to study empirically the naming behavior that is specific to a given physical context. This was first done by Lantz and Stefflre (1964), who used a procedure which we may call *Approach C*. In this method subjects are instructed to communicate with each other about specific referents. There are several variants for a laboratory set-up. The easiest is as follows: two subjects are each given an identical collection of loose color chips. The subjects may talk to each other but cannot see each other. Subject *A* chooses one color at a time and describes it in such a way that subject *B* can identify the color chip. For instance, subject *A* picks one color and describes it as 'the color of burned pea soup'. Subject *B* inspects his sample and makes a guess which color the other subject might have picked; the experimenter records the identification number of both colors and measures the magnitude of the mistake (if any) in terms of the physical distance between the colors.[a] If this procedure is done on a sufficiently large number of subject-pairs one may statistically reduce individual competence factors and come up with a measure, called *communication accuracy*, that indicates how well each of the colors in the particular context presented may be identified in the course of naming behavior.

Note that *Approach C*, leading to the estimation of communication accuracy, no longer indicates exclusively the nature of the relationship between particular words and objects and is thus somewhat marginal to the problem of reference. For instance, a given color chip, say a gray with a blue-green tinge, may obtain a very low communication accuracy score if it is presented together with fifty other grays, greens, blues, and intermediate shades, but a very high score if it is presented with fifty shades all of them in the lavender, pink, red, orange, and yellow range. Although this approach emphasizes the creative element of naming, it probably does not distinguish between the peculiar semantic properties of one natural language over another. The difference between the three approaches is summarized in Table 1.

[a] This is possible if we choose a linear stimulus array with perceptually equidistant stimuli as, for instance, described by Lenneberg, 1957.

TABLE 1. *Three approaches to the study of naming*

	Yields a variable called	Gives information on	Assuming that the stimulus array is	Has following un-realistic elements	Cannot predict
Approach A	(No name)	The common reference of specific words found in the lexicon of a natural language	A large sample of points homogeneously distributed through the color space	Assumes dictionary meanings to be a set of rigid word-thing associations, ignoring that naming is a process that changes with situation	What name a person actually will use in any instant
Approach B	Codability (originally this word referred to a combination of highly correlated variables); for simplicity we are equating codability here with what is called in the text name-determinacy	The probability of various words that might be used by a language community with respect to a given stimulus	A small sample of points distributed through the color space in a predetermined way	Essentially same assumption as above	What name a person actually will use in any condition *other* than a particular stimulus array and under specific experimental conditions
Approach C	Communication accuracy	How efficiently a community may identify one stimulus in a collection of others, using words	No assumptions about nature of sample; however, results vary with every sample and therefore communication accuracy is not a quality imposed invariantly upon a stimulus by a natural language	Does not reveal semantic peculiarities of any natural language; does not distinguish clearly between commonality among subjects due to culture as against due to language	How efficiently a color may be identified in an unspecified or unknown context

4 Naming and cognitive processes

4.1 GENERAL STRATEGIES

How certain can we be that naming is actually the consequence of categorization, as claimed in the introduction, instead of its cause? If there is freedom (within limits) to categorize and recategorize, could the semantic structure of a natural language restrict the biological freedom? Is our cognitive structure influenced by the reference relationships of certain words? What would cognition be like in the absence of language?

Questions of this sort may be partially answered by following either of two strategies. We may use various features of natural languages as the independent variable and study how these affect certain features of cognitive processes; or we may use the relative presence or absence of primary language as the independent variable and see to what extent the development of cognition is dependent on language acquisition. Congenitally deaf children are the most interesting subjects if the latter approach is used.

The former approach harbors a problem that must be dealt with explicitly. It is necessary that the relationship between individual words and natural phenomena can be studied empirically so that we have an objective measure of how well or how poorly the language actually deals with one or the other phenomenon. In the previous section we have discussed this matter in detail and have given reasons why the best types of words to be used in this kind of study are those that refer to sensation, in short the language of experience.

Colors have been the favorite stimulus material because their physical nature can be described relatively easily, standard stimulus material is readily available, the relative frequency of occurrence in the environment is not too likely to affect subjects' reactions in an experiment, and perceptual qualities may also be controlled relatively easily.

4.2 ACUITY OF DISCRIMINATION

There is a collection of colors, known as the Farnsworth–Munsell Test (Farnsworth, 1949), consisting of 86 chips of equal brightness and saturation but differing in hue. These colors were chosen by the authors of the test so as to constitute an instrument for the testing of hue discrimination. Subjects are presented with two chips, say one green and one blue, and are asked to put a disarranged collection of shades between these two colors into the proper order so that an array results with all chips finely graded from green to blue. Similar tasks are required with shades between blue and purple, purple and red, and red and green. When the test is administered to a noncolor-blind standard population of young American adults, a certain average number of sorting mistakes occur, but the mistakes occur with equal likelihood anywhere in the spectrum (and the mistakes for any one color chip have a frequency distribution that is the same for all colors).

The question now arises whether the construction and standardization of the test might have been biased by the language habits of its authors who were English speaking individuals. Since the color vocabulary superimposes a classification upon the physical color continuum might it not have predisposed us to pay more attention to the borderline cases, thus sharpening our acuity across wordclass boundary, and to pay relatively less attention to the clear-cut cases, thus dulling our acuity within the word classes?

Lenneberg and J. Bastian (unpublished data) administered the Farnsworth–Munsell Hue Discrimination Test to a group of Zuni and Navaho Indians whose color vocabulary had been mapped into the stimulus continuum by Lenneberg and Roberts (1956) and Landar et al. (1960) respectively. The different locations of the Zuni and Navaho color-name boundaries did not predict in either case systematic difference in discriminatory acuity between the Indians and a control group of Anglo-American farmers living in the same region. Thus there is little ground to assume that the peculiarities of English color words have affected the construction of the Farnsworth–Munsell Test in any basic way.

There is, however, another type of evidence that may cause us to reconsider our conclusions. Beare (1963) elicited English color words for a series of monochromatic lights. Here stimulus material had the advantage of being perfectly controlled in its physical properties and each color could be specified in terms of wavelength. She required subjects to name every stimulus presented as red, orange, yellow, green, blue, violet and to make the response as fast as possible. After an appropriate decision had been made the subject was allowed to qualify the name given by further adjectives. If a stimulus fell within a name map the response was made relatively quickly and few or no qualifiers were used. A stimulus that fell between maps had a longer response latency and elicited a greater number of qualifiers. Thus, she used *Approach B* to map words into the particular referent space represented by her stimuli.

The same stimulus material had been used earlier for another experiment (Judd, 1932) designed to discover acuity of hue-discrimination. In that experiment it was found that the differential limen was not constant for the visible spectrum. In some spectral areas our eye was shown to discover smaller hue differences than in other spectral areas. In Fig. 4 Judd's hue-discrimination curve is shown; low points indicate relatively high acuity. The other curve plotted in Fig. 4 is based on Beare's naming data; it is the frequency of occurrence of name-qualifiers. This frequency is high when a stimulus is ambiguous with respect to its word-class assignment. The curve reflects what we have called name-indeterminacy. Figure 4 suggests that there is a relationship between the two variables. Within the large, clear-cut word classes acuity seems to be slightly less sharp than in the areas of relatively large name-indeterminacy. These findings are merely suggestive, but not conclusive.

Beare used a slightly different procedure to compare her naming data with Judd's acuity data, but she was generally skeptical about a causal relationship between the two. In her own words: 'what fit there is between the two functions is tenuous, and it is possible that the configuration of the naming curve will change (if a different stimulus array is used), or with changes in instructions with regard to categories of judgment'. Even if Beare shall eventually be proved to have been too modest in her conclusions, we can still not be certain whether the name boundaries of the English words are entirely arbitrary; only if other languages can be shown to have markedly different boundaries for this particular stimulus array and if the speakers distribute their loci of greatest acuity in a predictable direction may we conclude that non-physiological factors have affected hue-discrimination. For the time being, we must withhold judgment.

4.3 MEMORY AND RECOGNITION

The first experiment on the effects of certain language habits on memory and recognition was carried out by Brown and Lenneberg (1954). Until that time investigators had attempted to manipulate language variables by quite ephemeral conditions such as teaching subjects nonsense names for nonsense objects or influencing their verbal habits by instructions given immediately prior to the cognitive task (Carmichael, Hogan and Walter, 1932; Kurtz and Hovland, 1953; and others cited in Carroll, 1964), but the semantic structure of the subjects' native language was neither known to the experimenter nor utilized as a variable in the experiment. Yet the habits induced by one's native language have acted upon an individual all his life and are, therefore, much more relevant to the basic problem that concerns us here than the verbal habits that can be established during a single experimental session. Thus, experiments that build upon the properties of a natural language are in a sense more crucial than the other type.

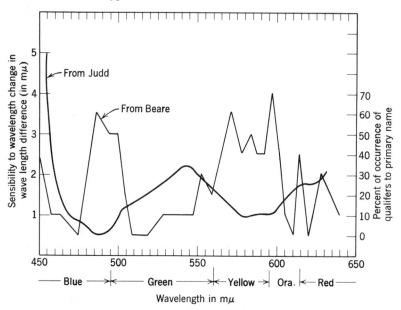

FIGURE 4. *Relationship between Judd's (1932) data on sensibility to wavelength change and Beare's data (1963) on English speakers' likelihood of using qualifiers in their color names. Judd's curve shows that acuity of color discrimination varies with wavelength. It is sharpest (minima in curve) in those areas of the spectrum where our color categories overlap. (Data redrawn to scale)*

The Brown and Lenneberg experiment need not be reported here in any great detail. Subjects were screened for color-blindness and had to be native born American-English speakers. Their task consisted of correctly recognizing certain colors. They were shown one color (or in some experimental conditions four colors) at a time for a short period; these were the *stimulus colors.*

After a timed interval they viewed a large color chart (to which we shall refer as the *color context*) from which they chose the colors they had been shown before. The colors were identified by pointing so that no descriptive words were used either by the experimenter or the subject.

In earlier investigations the foci and borders of English name maps had been determined through *Approach A*; basing ourselves on this information we selected a small sample of colors (the stimulus colors) consisting of all foci and an example of each border. A few additional colors were added (only the region of high saturation was included). This yielded a sample of 24 colors. Next we used *Approach B* in order to obtain lists of words and descriptive phrases actually used by English speakers in order to refer to the colors in our particular sample and context. *Approaches A* and *B* together provided us with background knowledge of the language habits that English-speaking subjects might bring to bear upon the task of color recognition in our experiment. We hypothesized that there would be a relationship between the two.

The measure of name-determinacy, obtained through *Approach B*, was found to be highly correlated with response latency and shortness of response (as confirmed by Beare, 1963) and was called by Brown and Lenneberg *codability*. Codability is essentially a measure of how well people agree in giving a name to a stimulus, in our case a color. Good agreement (that is, high codability) may be due to two independent factors: the language of the speakers may provide, through its vocabulary, a highly characteristic, unique, and unambiguous word for a highly specific stimulus (for example, the physical color of blood); or the stimulus may in fact be quite non-descript in the language, but it is given special salience in a physical context (for instance red hair). Codability does not distinguish these two factors.

We found that codability of a color did not predict its recognizability when the recognition task was easy (one color to be identified after a seven second waiting period); however, as the task was made more difficult codability began to correlate with recognizability and the relationship was most clearly seen in the most difficult of the tasks (four colors had to be recognized after a three minute waiting period during which subjects were given irrelevant tasks to perform).

At first these results looked like experimental proof that under certain conditions a person's native language may facilitate or handicap a memory-function. However, it became apparent later that the role of physical context needed to be studied further (see also Krauss and Weinheimer, 1965). Burnham and Clark (1955) conducted a color recognition experiment employing essentially the same procedure as Brown and Lenneberg but using a different sample of colors (namely the Farnsworth–Munsell Test Colors instead of the high saturation colors from the Munsell *Book of Colors*). They found that colors differed in their recognizability as shown in Fig. 5 with mistakes occurring in specific directions, clearly pointing to some systematic bias in the subjects' performance. Lenneberg (1957) obtained naming data on the same stimulus material and a comparison of these data with those of Burnham and Clark (Fig. 3) seemed to show that the bias may be related to the semantic structure 'built into' the subjects by virtue of their native English. The only difficulty with this interpretation is this. In the Brown and Lenneberg experiment a positive correlation was found between codability and recognizability, whereas the recognizability of the Burnham and Clark colors was negatively correlated with their codability. This divergence is not due to experimental artifact. Lantz (unpublished data) repeated the Burnham and Clark experiment twice with different groups of adults in different parts of the country and Lantz and Stefflre (1964) replicated the Brown and Lenneberg experiment, all with similar results.

There are strong reasons to attribute the difference in outcome of the two experiments to the peculiarities of the stimulus arrays and the physical context of the individual colors.

Burnham and Clark had hoped to eliminate endpoints by using a circular stimulus array. But *S*s' naming habits restructured the material so as to furnish certain anchoring points: namely, the boundaries between name maps. Thus, a few colors acquired a distinctive feature, whereas most of the colors, all that fell within a name region, tended to be assimilated in memory. To remember the word 'green' in this task was of no help since there were so many greens; but to remember that the color was on the border of 'green' and 'not-green' enabled the subject to localize the stimulus color with great accuracy in the color context.

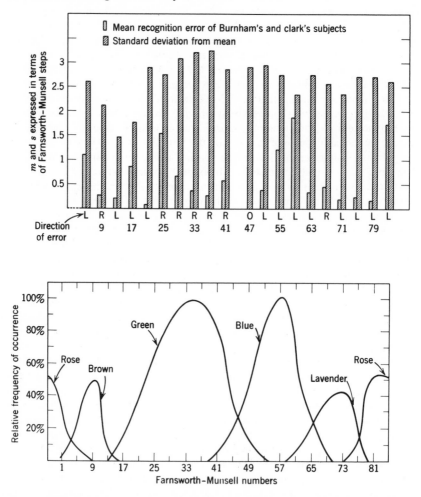

FIGURE 5. *Distribution of errors in color recognition as reported by Burnham and Clark (1955).* Below: *frequency distribution of the most common English color names over the Farnsworth–Munsell color series. Approach B was used*

The Brown and Lenneberg study did not provide subjects with any two or three outstanding landmarks for recognition. Since there was no more than one color within individual name categories, the possibility of assimilating colors (or of grouping colors into classes of stimulus equivalence) was drastically reduced if not eliminated. On the other hand, so many colors marked boundaries, that *betwixtness* as such was no longer a distinctive feature.

These two experiments suggest that semantic structure influences recognition only under certain experimental circumstances, namely when the task is difficult and the stimuli are chosen in a certain way. If it is possible to obtain a positive as well as a negative correlation between codability and recognizability, it will also be possible to select stimuli in such a way that codability has zero correlation with recognizability of a color. In other words, both the Brown and Lenneberg experiment and the Burnham and Clark experiment are special cases.

Lantz and Stefflre (1964) have shown that it is actually not the semantic characteristics of the vocabulary of a natural language that determine the cognitive process *recognition* but the peculiar use subjects will make of language in a particular situation. Instead of using the information from either *Approach A* or *B* (which primarily brings out language peculiarities), they used *Approach C* which measures the

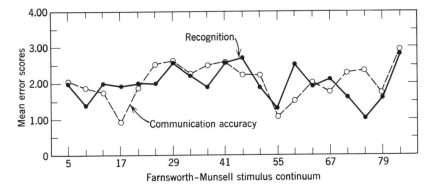

FIGURE 6. *When communication accuracy is determined for every color in a specific stimulus array, it predicts recognizability of that color in that setting well. (Data for this graph based on DeLee Lantz and E. H. Lenneberg, 1966)*

accuracy of communication in a specific setting, without evaluating the words that subjects choose to use; their independent variable was *communication accuracy* which reflects efficiency of the process but does not specify by what means it is done. This is quite proper in that the *how* is largely left to the creativity of the individual; he is, in fact, *not* bound by the semantics of his natural language; there is little evidence of the tyrannical grip of words on cognition.

When communication accuracy is determined for every color in a specific stimulus array it predicts recognition of that color quite well as may be seen from Fig. 6. Codability, on the other hand, predicts recognizability only in special contexts and stimulus arrays. From this we may infer that subjects make use of the ready-made reference facilities offered them through their vocabulary, only under certain circumstances. The rigid or standard use of these words, without creative qualifications, is in many circumstances not conducive to efficient communication. Communication accuracy or efficiency will depend frequently on individual ingenuity rather than on the language spoken by the communicator.

The Lantz and Stefflre experiment points to an interesting circumstance. The variable, communication accuracy, is a distinctly social phenomenon. But recognition of colors is an entirely intrapersonal process. Lantz and Stefflre suggest that there are situations in which the individual communicates with himself over time.

This is a fruitful way of looking at the experimental results and one that also has important implications regarding human communication in more general terms. It stresses, once again, the probability that human communication is made possible by the identity of cognitive processing within each individual. The social aspect of communication seems to reflect an internal cognitive process. We are tempted to ask now, 'What is prior – the social or the internal process?' At present, there is no clear answer to this, but the cognitive functioning of congenitally deaf children, before they have learned to read, write, or lip-read (in short, before they have language) appears to be, by and large, similar enough to their hearing contemporaries to lead me to believe that the internal process is the condition for the social process, although certain influences of the social environment upon intrapersonal cognitive development cannot be denied.

What preliminary conclusions may we draw then from these empirical investigations? Four major facts emerge. First, the semantic structure of a given language only has a mildly biasing effect upon recognition under special circumstances; limitations of vocabulary may be largely overcome by the creative use of descriptive words. Second, a study of the efficiency of communication in a social setting (of healthy individuals) may give clues to intrapersonal processes. Third, efficiency of communication is mostly dependent upon such extra-semantic factors as the number of and perceptual distance between discriminanda. Fourth, the social communication measures become more predictive of the intrapersonal processes as the difficulty of the individual's task increases either by taxing memory or by reduction of cues (cf. also Frijda and Van de Geer, 1961; Van de Geer, 1960; Glanzer and Clark, 1962; Krauss and Weinheimer, 1965).

4.4 CONCEPT FORMATION

Concept formation is considered to be synonymous with categorization. A natural language was already said to tag certain types of conceptual processes by means of words. We may ask now whether the existence of such a word (say in the speaker's language knowledge) is responsible for making a given concept particularly salient and easy to attain.

A survey of reports of such tests (Carroll, 1964) does, indeed, give that impression; in most instances concepts that may be named, or where the principle may be formulated easily in the native language of the subjects, are felt to be easier to attain in experiments than when this is not the case. However, these findings do not necessarily indicate that natural language is a biasing factor in the formation of concepts in general. The concepts tagged by the vocabulary of natural languages are not completely arbitrary as may be seen from the large degree of semantic correspondences between languages. It is true that translation always brings out some absence of correspondence between two languages. However, the experience of the physical environment finds expression in all languages. It is mostly the aspect or mode of reference and the metaphorical extensions that vary. Comparative studies in the language of experience indicate that those phenomena that have perceptual or cognitive salience in the environment (for our species) always are particularly amenable to reference, regardless of the natural language. Therefore, if namability tends to be coupled with cognitive salience, it is not certain whether results in concept formation experiments are due to subjects' naming habits or whether both the naming *and* the concept attainment are due to a more basic factor such as biologically given cognitive organization.

Certain other considerations should make us doubtful about any strong claim upon the 'constraint' of words upon the speaker's cognitive capacities. A wide range of human activities is based upon concept formation that must have taken place in the absence of naturally occurring words. Examples are the development of mathematics (where a language is simply created *ad hoc* as the concepts are developed), or of music, or of the visual arts or of science in general. Peoples in underdeveloped countries who are suddenly introduced to a new technology for which there is no terminology in their language can learn the new concepts by simply introducing foreign words into their vernacular or by making new use of old words. Once more, this may be explained by seeing naming as a creative process, not a rigid convention.

The most dramatic semantic difference between languages may be found in the realm of feelings and attitudes. Here, indeed, translation is often a total impossibility. Are we able to demonstrate that our awareness of personal feelings is selectively enhanced by words handed down to us through semantic traditions? Empirical demonstrations would be difficult and I am, frankly, dubious about their promise. Consider the difficulty to describe accurately the nature of feelings, for instance during a psychiatric interview; or our awareness of how coarse and nondescript some of the words for feelings and attitudes are, for example, honor, love, pride, etc. In many instances, our emotions appear to be more subtle than can be indicated by the use of these threadbare terms.

5 Postscript to so-called language relativity

The term language relativity is due to B. L. Whorf, although similar notions had been expressed before by many others (Basilius, 1952). In his studies of American Indian languages, Whorf was impressed with the general difficulties encountered in translating American Indian languages into European languages. To him there seemed to be little or no isomorphism between his native English and languages such as Hopi or Nootka. He posed the question, as many had done before him, of whether the divergences encountered reflected a comparable divergence of thought on the part of the speakers. He left the final answer open, subject to further research, but it appears from the tenor of his articles that he believed that this was so, that differences in language are expressions of differences in thought. It has been pointed out since (Black, 1959; Feuer, 1953; but see also Fishman, 1960) that there is actually little *a priori* basis for such beliefs.

The empirical research discussed in previous sections (partly stimulated by Whorf's own imaginative ideas) indicate that the cognitive processes studied so far are largely independent from peculiarities of any natural language and, in fact, that cognition can develop to a certain extent even in the absence of knowledge of any language. The reverse does not hold true; the growth and development of language does appear to require a certain minimum state of maturity and specificity of cognition. Could it be that some languages require 'less mature cognition' than others, perhaps because they are still more primitive? In recent years this notion has been thoroughly discredited by virtually all students of language. It is obvious that on the surface every language has its own peculiarities but it is possible, in fact assumed throughout this monograph, that languages are different patterns produced by identical basic principles (much the way sentences are different patterns – infinitely so – produced by identical principles). The crucial question, therefore, is whether we are dealing with a universal and unique process that generates a unique *type* of pattern. This appears to be entirely possible. In syntax we seem to be dealing always with the same formal *type* of rule and in the realm of semantics we have proposed that the *type* of relationship between word and object is quite invariant across all users of words. It is due to this uniformity that any human may learn any natural language within certain age limits.

When language learning is at its biological optimum, namely in childhood, the degree of relatedness between first and second language is quite irrelevant to the ease of learning that second language. Apparently, the differences in surface structure are ignored and the similarity of the generative principles is maximally explored at this age. Until rigorous proof is submitted to the contrary, it is more reasonable to assume that all natural languages are of equal complexity and versatility and the choice of this assumption detracts much from the so-called relativity theory.

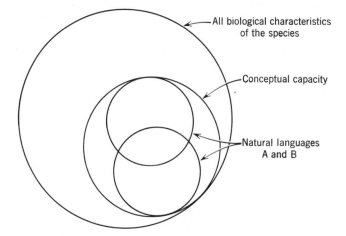

All biological characteristics of the species

Conceptual capacity

Natural languages A and B

FIGURE 7. *The relationship of natural languages to the human capacity for conceptualization*

Since the use of words is a creative process, the static reference relationships, such as are apparent through *Approach A* or as they are recorded in a dictionary, are of no great consequence for the actual use of words. However, the differences between languages that impressed Whorf so much are entirely restricted to these static aspects and have little effect upon the creative process itself. The diagram of Fig. 7 may illustrate the point. Common to all mankind are the general biological characteristics of the species (outer circle) among which is a peculiar mode and capacity for conceptualization or categorization. Languages tag some selective cognitive modes but they differ in the selection. This selectivity does not cripple or bind the speaker because he can make his language, or his vocabulary, or his power of word-creation, or his freedom in idiosyncratic usages of words do any duty that he chooses, and he may do this to a large extent without danger of rendering himself unintelligible because his fellow men have similar capacities and freedoms which also extend to understanding.

6 Summary

Words are not labels attached to objects by conditioning; strictly associative bonds between a visual stimulus and an auditory stimulus are difficult to demonstrate in language. It is usually through the relative inhibition in the formation of such bonds that the labeling of open, abstract categories can take place, which is the hall-mark of the human semantic system. The meaning-bearing elements of language do not, generally, stand for specific objects (proper names are a special case), and strictly speaking, not even for invariant classes of objects. Apparently they stand for a cognitive process, that is, the *act* of categorization or the *formation* of concepts. Operationally, such a process may be characterized as the ability to make a similar response to different stimulus situations within given limits,[a] which rests on the individual's capacity to recognize common denominators or similarities among ranges of physical phenomena. Even though there may be a large overlap among species in their capacity for seeing certain similarities, one also encounters species-specificities. No other creature but man seems to have just that constellation of capacities that makes naming possible such as is found in any natural language.

Natural languages differ in the particular conceptualization processes that are reflected in their vocabulary. However, since speakers use words freely to label *their own conceptualization processes*, the static dictionary meaning of words does not appear to restrict speakers in their cognitive activities; thus it is not appropriate to use the vocabulary meanings as the basis for an estimation of cognitive capacities.

The process of concept formation must be regulated to some extent by biological determinants; therefore, naming in all languages should have fairly similar formal properties. The basis for metaphorizing in all languages is usually transparent to everyone; it never seems totally arbitrary or unnatural. Also, we have excellent intuitions about what might be namable in a foreign language once one knows something about that community's culture, technology, or religion. Even the demarcation lines of semantic classes are more frequently 'obvious' to a foreign speaker than a matter of complete surprise. When we learn new words in a foreign language we do not have to determine laboriously the extent of the naming class for every word we learn; there will be only a few instances where demarcations are not obvious. The incongruences between languages become marked only in certain types of grammatical classifications such as animate-inanimate, male-female, plural-singular, to'ness-from'ness, etc. But, curiously enough, on this level of abstraction, where the semantics of the form-classes often does take on an entirely arbitrary character, cognitive processes seem least affected. There is no evidence that gender in German or the declinational systems of Russian, or the noun-classification systems in the Bantu languages affect thought processes differentially.

Because of methodological obstacles, the relationship between language and cognition cannot be investigated experimentally except for a very restricted realm of the lexicon, namely the words for sensory experience. Three approaches were described that reveal (1) the dictionary meaning of sensory words: (2) the use speakers make of the dictionary meanings in a given situation; (3) the efficiency of communication in a given situation, granted complete freedom of expression. The first approach is the most faithful representation of the reference relationships that are peculiar to a given natural language, and the other two approaches are progressively less so; the third approach is the most faithful representation of the communicative freedom enjoyed by speakers of any language, whereas the other two approaches fail to reflect this aspect.

Semantic aspects of words may aid memory functions under certain special

[a] I recognize that this formulation leaves the problems of synonymity and homonymity unsolved; but I am not aware of a satisfactory solution through any other formulation.

conditions, but the bias thus introduced by properties of a natural language are slight and they are minimized or eliminated by other, more potent factors (that have nothing to do with the semantic structure of a language) in most situations.

Cornell University

REFERENCES

Basilius, H. (1952), Neo-Humboldtian ethnolinguistics, *Word*, VIII, 95–105.

Beare, A. C. (1963), Color-name as a function of wave-length, *Am. J. Psychol.*, LXXVI, 248–56.

Black, M. (1959), Linguistic relativity: the views of Benjamin Lee Whorf, *Philosoph. Rev.*, LXVIII, 228–38.

Brown, R. W. and Lenneberg, E. H. (1954), A study in language and cognition, *J. abnorm. soc. Psychol.*, XLIX, 454–62.

Burnham, R. W. and Clark, J. R. (1955), A test of hue memory, *J. appl. Psychol.*, XXXIX, 164–72.

Carmichael, L., Hogan, H. P. and Walter, A. A. (1932), An experimental study of the effect of language on the reproduction of visually perceived form, *J. exp. Psychol.*, XV, 73–86.

Carroll, J. B. (1964), *Language and Thought*. Prentice-Hall, Englewood Cliffs, New Jersey.

Conklin, H. C. (1955), Hanunoo color categories, *Southwestern J. Anthrop.*, XI (4), 339–44.

Farnsworth, D. (1949), *The Farnsworth-Munsell 100 Hue Test for the Examination of Color Discrimination; Manual*. Munsell Color Co., Baltimore.

Feuer, L. S. (1953), Sociological aspects of the relation between language and psychology, *Philosoph. Sci.*, XX, 85–100.

Fishman, J. A. (1960), A systematization of the Whorfian hypothesis, *Behavioral Sci.*, V, 323–39.

Frijda, N. H. and Van de Geer, J. P. (1961), Codability and recognition: an experiment with facial expressions, *Acta Psychol.*, XVIII, 360–7.

Glanzer, M. and Clark, W. H. (1962), Accuracy of perceptual recall: an analysis of organization, *J. verb. Learn. verb. Behav.*, I, 289–99.

Judd, D. B. (1932), Chromaticity sensibility to stimulus differences, *J. opt. Soc. Am.*, XXII, 72–108.

Krauss, R. M. and Weinheimer, S. (1965), The effect of referent array and communication mode on verbal encoding (unpublished).

Kurtz, K. H. and Hovland, C. I. (1953), The effect of verbalization during observation of stimulus objects upon accuracy of recognition and recall, *J. exp. Psychol.*, XLV, 157–64.

Landar, H. J., Ervin, S. M. and Horowitz, A. E. (1960), Navaho color categories, *Language*, XXXVI, 368–82.

Lantz, DeL. and Lenneberg, E. H. (1966), Verbal communication and color memory in the deaf and hearing, *Child Development*, XXXVII, 765–79.

Lantz, DeL. and Stefflre, V. (1964), Language and cognition revisited, *J. abnorm. soc. Psychol.*, LXIX, 472–81.

Lenneberg, E. H. (1957), A probabilistic approach to language learning, *Behavioral Sci.*, II, 1–13.

Lenneberg, E. H. and Roberts, J. M. (1956), *The Language of Experience: a Study in Methodology*. Memoir 13, Indiana University Publications in Anthropology and Linguistics.

Nickerson, D. and Newhall, S. M. (1943), A psychological color solid, *J. opt. Soc. Am.*, XXXIII, 419–22.

Van de Geer, J. P. (1960), *Studies in Codability: 1. Identification and Recognition of Colors*. Report no. E001–60, 1960, State University of Leyden, Psychol. Institute, The Netherlands.

Whorf, B. L. (1956), *Language, Thought and Reality*, J. B. Carroll (ed.). Technology Press of M.I.T., Cambridge, Massachusetts and John Wiley and Sons, New York.

Could meaning be an r_m?

JERRY A. FODOR

This paper is primarily devoted to an investigation of the cogency of mediational accounts of linguistic reference.[a] As Chomsky (1959) has extensively discussed the inadequacies of so-called 'single-stage' accounts of language, I shall not review at great length the history of attempts to accommodate verbal behavior within learning theory by equating learning the meaning of a word with establishing a simple S-R connection. Since, however, it has been the failure of such attempts that has prompted psychologists to adopt mediational models, a brief review of the nature of their inadequacies is in order. Suffice it to recall that single-stage models have invariably suffered from three defects that are jointly sufficient to establish the inadequacy of simple S-R theories of meaning.

First, though there can be no objection to considering the verbalizations of fluent speakers to be 'linguistic responses', one must not suppose that, in this context, 'response' means what it usually means: 'A stimulus-occasioned act. An (act) correlated with stimuli, whether the correlation is untrained or the result of training' (Kimble, 1961, glossary). On the contrary, a striking feature of linguistic behavior is its freedom from the control of specifiable local stimuli or independently identifiable drive states. In typical situations, what is said may have no obvious correlation with conditions in the immediate locality of the speaker or with his recent history of deprivation or reward. Conversely, the situations in which such correlations do obtain (the man dying of thirst who predictably gasps 'Water') are intuitively highly atypical.

Indeed, the evidence for the claim that linguistic responses are responses *in the strict sense* would appear to be non-existent. There is no more reason to believe that the probability of an utterance of 'book' is a function of the number of books in the immediate locale than there is to believe that the probability of an utterance of the word 'person' or 'thing' is a function of the number of persons or things in view. Lacking such evidence, what prompts one to these beliefs is, first, a confusion of the strict sense of 'response' ('a stimulus-correlated act') with the loose sense in which the term applies to *any* bit of behavior, stimulus-correlated or otherwise; and second, a philosophy of science which erroneously supposes that unless all behavior is held to consist of responses in the strict sense some fundamental canon of scientific method

[a] This work was supported in part by the United States Army, Navy, and Air Force, under Contract D.A. 36-039-AMC-03200 (E), The National Science Foundation (Grant G.P.-2495), The National Institutes of Health (MH-04737-04), The National Aeronautics and Space Administration (N.s.G.-496); and in part by the United States Air Force (E.S.D. Contract A.F. 19(628)-2487). This paper although based on work sponsored by the United States Air Force, has not been approved or disapproved by that Agency. It is reprinted from *Journal of Verbal Learning and Verbal Behavior*, IV, 1965, 73–81.

is violated. But the claim that behavior consists solely of responses cannot be established on methodological grounds alone. On the contrary, such a claim constitutes an extremely geneal empirical hypothesis about the degree to which behavior is under the control of specifiable local stimulation.

The second inadequacy of simple S-R models of language is also a consequence of the identification of verbalizations with responses. In laboratory situations, an organism is said to have mastered a response when it can be shown that it produces any of an indefinite number of functionally equivalent acts under the appropriate stimulus conditions. That some reasonable notion of functional equivalence can be specified is essential, since we cannot, in general, require of two actions that they be identical either in observable properties or in their physiological basis in order to be manifestations of the same response. Thus, a rat has 'got' the bar-press response if and only if it habitually presses the bar upon food deprivation. Whether it presses with its left or right front paw or with three or six grams of pressure is, or may be, irrelevant. Training is to some previously determined criterion of homogeneity of performance, which is another way of saying that what we are primarily concerned with are functional aspects of the organism's behavior. We permit variation among the actions belonging to a response so long as each of the variants is functionally equivalent to each of the others. In short, a response is so characterized as to establish an equivalence relation among the actions which can belong to it. Any response for which such a relation has *not* been established is *ipso facto* inadequately described.

We have just suggested that it is not in general possible to determine stimuli with which verbal responses are correlated. It may now be remarked that it is not in general possible to determine when two utterances are functionally equivalent, i.e., when they are instances of the same verbal response. This point is easily overlooked since it is natural to suppose that functional equivalence of verbal responses can be established on the basis of phonetic or phonemic identity. This is, however, untrue. Just as two physiologically distinct actions may both be instances of a bar-press response, so two phonemically distinct utterances may be functionally equivalent for a given speaker or in a given language. Examples include such synonymous expressions as 'bachelor' and 'unmarried man', 'perhaps' and 'maybe', etc. Conversely, just as an action is not an instance of a bar-press response (however much it may resemble actions that are) unless it bears the correct functional relation to the bar, so two phonemically identical utterances – as, for example, utterances of 'bank' in $\langle 1 \rangle$ and $\langle 2 \rangle$ – may, when syntactic and semantic considerations are taken into account, prove to be instances of quite different verbal responses:

$\langle 1 \rangle$ The bank is around the corner.
$\langle 2 \rangle$ The plane banked at forty-five degrees.

It appears, then, that the claim that verbal behavior is to be accounted for in terms of S-R connections has been made good at neither the stimulus nor the response end. Not only are we generally unable to identify the stimuli which elicit verbal responses, we are also unable to say when two bits of behavior are manifestations of the same response and when they are not. That this is no small difficulty is evident when we notice that the problem of characterizing functional equivalence for verbal responses is closely related to the problem of characterizing such semantic relations as synonymy; a problem for which no solution is at present known (cf. Katz and Fodor, 1963).

Finally, the identification of verbalizations with responses suffers from the difficulty that verbalizations do not admit of such indices of response strength as frequency and intensity. It is obvious, but nevertheless pertinent, that verbal responses which are equally part of the speaker's repertoire may differ vastly in their relative frequency of occurrence ('heliotrope' and 'and' are examples in the case of the idiolect of this writer), and that intensity and frequency do not covary (the morpheme in an utterance receiving emphatic stress is not particularly likely to be a conjunction, article, preposition, etc., yet these 'grammatical' morphemes are easily the most frequently occurring ones). What is perhaps true is that one can vary the frequency or intensity of a verbal response by the usual techniques of selective reinforcement.[a] This shows that verbal behavior can be conditioned but lends little support to the hypothesis that conditioning is essential to verbal behavior.

Faced with these and other difficulties, a number of learning theorists have acknowledged the necessity of abandoning the simpler S-R accounts of language. What has not been abandoned, however, is the belief that *some* version of conditioning theory will prove adequate to explain the characteristic features of verbal behavior, and, in particular, the referential functions of language. I wish to consider the extent to which more complicated versions of conditioning theory may succeed where simple S-R models have failed.

The strength of single-stage S-R theories lies in the fact that they produce an unequivocal account of what it is for a word to refer to something. According to such theories, w refers to x just in case:

⟨3⟩ There is a response R such that the presentation of x increases the probability of R, and such that R is non-verbal;

⟨4⟩ the utterance of w increases the probability of R being produced by a hearer;

⟨5⟩ the presentation of x increases the probability of the utterance of w.

Thus, on the S-R account, 'apple' refers to apples just because

⟨6⟩ there is some non-verbal response (reaching, eating, salivating, etc.) the probability of which is increased by the presentation of apples, and

⟨7⟩ the utterance of 'apple' tends to increase the probability of that response being produced by a hearer;

⟨8⟩ the presentation of apples tends to increase the probability of the utterance of 'apple'.

The weakness of the single-stage S-R theory lies largely in the failure of words and their referents to satisfy conditions ⟨4⟩ and ⟨5⟩. That is, utterances of words do not, in general, serve as stimuli for gross, non-verbal responses, nor can such utterances in general be viewed as responses to specifiable stimuli. What is needed, then, is a version of conditioning theory which provides an account of reference as explicit as that given by single-stage S-R theories; but does not require of verbalizations that they satisfy such postulates as ⟨4⟩ and ⟨5⟩. It has been the goal of much recent theorizing in psycholinguistics to provide such an account in terms of mediating responses.

Consider, for example, the following passage from Mowrer (1960, p. 144):

As Fig. 1 shows, the word 'Tom' acquired its meaning, presumably, by being associated with, and occurring in the context of, Tom as a real person. Tom himself has elicited in John, Charles, and others who have firsthand contact with him a total reaction that we can label R_t, of which r_t is a component. And as a result of the paired

[a] It is possible to raise doubts about the possibility of operant conditioning of verbal behavior. Cf. Krasner (1958).

presentation of occurrence of 'Tom'-the-word and Tom-the-person, the component, or 'detachable', reaction, r_t, is shifted from the latter to the former.

It is clear that this theory differs from S-R theories primarily in the introduction of the class of constructs of which r_t is a member. In particular, it is with r_t the 'detached component', and not with R_t, the gross response, that Mowrer identifies the meaning of the word 'Tom'.

FIGURE 1. *Pattern of conditioning responsible for the understanding of the word 'Tom' (after Mowrer, 1960)*

The differences between r_t and R_t thus determine the difference between Mowrer's approach to verbal behavior and single-stage theories. In particular,

⟨9⟩ while R_t is an overt response, or, at any event, a response that may have overt components, r_t is a theoretical entity; a construct postulated by the psychologist in order to explain the relation between such gross responses as R_t and such stimuli as utterances of 'Tom'. Hence,

⟨10⟩ occurrences of mediating responses, unlike occurrences of overt responses, are not, in general, directly observable. Though it *may* be assumed that progress in physiology should uncover states of organisms the occurrence of which may be identified with occurrences of mediating responses, r_t and other such mediators are, in the first instance, functionally characterized. Hence, whatever evidence we have for the correctness of the explanatory models in which they play a role, is *ipso facto* evidence for their existence.

⟨11⟩ The relation of r_t to R_t is that of proper part to whole. It is important to notice that for each r_i and R_i, $r_i \neq R_i$ is a basic assumption of this theory. In each case where it does not hold, mediation theory is indistinguishable from single-stage theory. To put it the other way around, single-stage theory is the special case of mediation theory where $r_i = R_i$.

It is clear that the introduction of mediating responses such as r_t will render a learning-theoretic approach to meaning impervious to some of the objections that can be brought against single-stage theories. For example, the failure of verbalization to satisfy ⟨4⟩ and ⟨5⟩ is not an objection to theories employing mediating responses, since it may be claimed that verbalizations do satisfy ⟨12⟩ and ⟨13⟩.

⟨12⟩ If r_i is a mediating response related to w_i as r_t is related to 'Tom'; then the utterance of w_i increases the probability of r_i.

⟨13⟩ If S_i is an object related to R_i as Tom is related to R_t, and if r_i is a mediating response related to R_i as r_t is related to R_t; then the presentation of S_i increases the probability of r_i. Since mediating responses are not directly observable, and since the psychologist need make no claim as to their probable physiological basis, such postulates as ⟨12⟩ and ⟨13⟩ may prove very difficult to refute.

The result of postulating such mediating events as r_t in the explanation of verbal behavior is, then, that an account of reference may be given that parallels the explanation provided by single stage S-R theories except that $\langle 4 \rangle$ and $\langle 5 \rangle$ are replaced by $\langle 12 \rangle$ and $\langle 13 \rangle$. Roughly speaking, a word refers to an object just in case first, utterances of the word produce in hearers a mediating response that is part of the gross response that presentation of the object produces and second, presentation of the object produces in speakers a mediating response which has been conditioned to occurrences of the relevant word.

A rather natural extension of Mowrer's theory carries us from an explanation of how words function as symbols to an account of the psychological mechanisms underlying the understanding and production of sentences:

> What, then, is the function of the sentence 'Tom is a thief'? Most simply and most basically, it appears to be this. 'Thief' is a sort of 'unconditioned stimulus'...which can be depended upon to call forth an internal reaction which can be translated into, or defined by, the phrase 'a person who cannot be trusted', one who 'takes things, steals'. When, therefore, we put the word, or sign, 'Tom' in front of the sign 'thief', as shown in Fig. 2, we create a situation from which we can predict a fairly definite result. On the basis of the familiar principle of conditioning, we would expect that some of the reaction evoked by the second sign, 'thief', would be shifted to the first sign, 'Tom', so that Charles, the hearer of the sentence, would thereafter respond to the word 'Tom', *somewhat as he had previously responded to the word 'thief'* (p. 139).

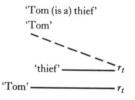

FIGURE 2. *Pattern of conditioning responsible for the understanding of the sentence 'Tom is a thief' (after Mowrer, 1960)*

Or, as Mowrer puts it slightly further on,

> [The word 'thief' is] presumed to have acquired its distinctive meaning by having been used in the presence of, or to have been, as we say, 'associated with' actual thieves. Therefore, when we make the sentence, 'Tom (is a) thief', it is no way surprising or incomprehensible that the r_t reaction gets shifted...from the word, 'thief' to the word, 'Tom' (p. 144).

It thus appears to be Mowrer's view that precisely the same process that explains the shifting of r_t from occasions typified by the presence of Tom to occasions typified by the utterance of 'Tom' may be invoked to account for the association of the semantic content of the predicate with that of the subject in such assertive sentences as 'Tom is a thief'. In either case, we are supposed to be dealing with the conditioning of a mediating response to a new stimulus. The mechanism of predication differs from the mechanism of reference only in that in the former case the mediating response is conditioned from a word to a word and in the latter case from an object to a word.

Thus, a view of meaning according to which it is identified with a mediating response has several persuasive arguments in its favor. First, it yields a theory that avoids a number of the major objections that can be brought against single stage

S-R theories. Second, it appears to provide an account not only of the meaning of individual words, but also of the nature of such essential semantic relations as predication. Third, like single-stage S-R theories, it yields a conception of meaning that is generally consonant with concepts employed in other areas of learning theory. It thus suggests the possibility that the development of an adequate psycholinguistics will only require the employment of principles already invoked in explaining non-verbal behavior. Finally, these benefits are to be purchased at no higher price than a slight increase in the abstractness permitted in psychological explanations. We are allowed to interpose *s-r* chains of any desired length between the S's and R's that form the observation base of our theory. Though this constitutes a proliferation of unobservables, it must be said that the unobservables postulated are not different *in kind* from the S's and R's in terms of which single-stage theories are articulated. It follows that, though we are barred from direct, observational verification of statements about mediating responses, it should prove possible to infer many of their characteristics from those that S's and R's are observed to have.

Nevertheless, the theory of meaning implicit in the quotations from Mowrer cited above is thoroughly unsatisfactory. Nor is it obvious that this inadequacy is specific to the version of the mediation theory that Mowrer espouses. Rather, I shall try to show that there are chronic weaknesses which infect theories that rely upon mediating responses to explicate such notions as reference, denotation, or meaning. If this is correct, then the introduction of such mediators in learning-theoretic accounts of language cannot serve to provide a satisfactory answer to the charges that have been brought against single-stage theories.

In the first place, we must notice that Mowrer's theory of predication at best fails to be fully general. Mowrer is doubtless on to something important when he insists that a theory of predication must show how the meaning of a sentence is built up out of the meanings of its components. But it is clear that the simple conditioning of the response to the predicate to the response to the subject, in terms of which Mowrer wishes to explain the mechanism of composition, is hopelessly inadequate to deal with sentences of any degree of structural complexity. Thus, consider:

⟨14⟩ 'Tom is a perfect idiot.'

Since what is said of Tom here is not that he has the properties of perfection and idiocy, it is clear that the understanding of ⟨14⟩ cannot consist in associating with 'Tom' the responses previously conditioned to 'perfect' and 'idiot'. It is true that the meaning of ⟨14⟩ *is*, in some sense, a function of the meaning of its parts. But to understand ⟨14⟩ one must understand that 'perfect' modifies 'idiot' and not 'Tom', and Mowrer's transfer of conditioning model fails to explain either how the speaker obtains this sort of knowledge, or how he employs it in understanding sentences. Analogous difficulties can be raised about

⟨15⟩ 'Tom is not a thief.'

Lack of generality is, however, scarcely the most serious charge that can be brought against Mowrer's theory. Upon closer inspection, it appears that the theory fails even to account for the examples invoked to illustrate it. To begin with, the very fact that the same mechanism is supposed to account for the meaning of names such as 'Tom' and of predicates such as 'thief' ought to provide grounds for suspicion, since the theory offers no explanation for the obvious fact that names and predicates are quite different kinds of words. Notice, for example, that we ask:

⟨16⟩ 'What does "thief" mean?'

but not

⟨17⟩ 'What does "Tom" mean?'

and

⟨18⟩ 'Who is Tom?'

but not

⟨19⟩ 'Who is thief?'

All this indicates what the traditional nomenclature already marks: there is a distinction to be drawn between names and predicates. A theory of meaning which fails to draw this distinction is surely less than satisfactory.

Again, it is certainly not the case that, for the vast majority of speakers, 'thief' acquired its meaning by 'having been used in the presence of...actual thieves'. Most well brought up children know what a thief is long before they meet one, and are adequately informed about dragons and elves though encounters with such fabulous creatures are, presumably, very rare indeed. Since it is clearly unnecessary to keep bad company fully to master the meaning of the word 'thief', a theory of language assimilation must account for communication being possible between speakers *who have in fact learned items of their language under quite different conditions.* Whatever it is we learn when we learn what 'thief' means, it is something that can be learned by associating with thieves, reading about thefts, asking a well-informed English speaker what 'thief' means, etc. Though nothing prevents Mowrer from appealing to some higher-level conditioning to account for verbal learning of linguistic items, it is most unclear how such a model would explain the fact that commonality of response can be produced by such strikingly varied histories of conditioning.

Nor does the mediation theory of predication appear to be satisfactory. It is generally the case, for example, that conditioning takes practice, and that continued practice increases the availability of the response. But, first, there is no reason to suppose that the repetition of simple assertive sentences is required for understanding, or even that it normally facilitates understanding. On the contrary, it would appear that speakers are capable of understanding sentences they have never heard before without difficulty, that the latency for response to novel sentences is not strikingly higher than the latency for response to previously encountered sentences (assuming, of course, comparability of length, grammatical structure, familiarity of vocabulary, etc.), and that repetition of sentential material produces 'semantic satiation' rather than increase of comprehension. Second, the conditioning model makes it difficult to account for the fact that we do not always believe what we are told. Hearing someone utter 'Tom is a thief' may, of course, lead me to react to Tom as I am accustomed to react to thieves; but I am unlikely to lock up the valuable *unless I have some reason to believe that what I hear is true.* Though this fact is perfectly obvious, it is unclear how it is to be accounted for on a model that holds that my understanding of a sentence consists in having previously distinct responses conditioned to one another. It might, of course, be maintained that the efficacy of conditioning in humans is somehow dependent upon cognitive attitude. This reply would not, however, be of aid to Mowrer, since it would commit him to the patently absurd conclusion that a precondition for understanding a sentence is believing that it is true; and, conversely, that it is impossible to understand a sentence one does not believe. Third, if my understanding of 'Tom is a thief' consists of a transference of a previously established reaction to 'thief' to a previously established reaction to 'Tom', it is difficult to account for the fact that I, a speaker who does not know Tom and therefore has no previously established response to utterances of the word 'Tom', am perfectly capable of grasping that sentence. It is clear what the story ought to be; I know that 'Tom' is a name (a bit of information which, by the way, I did *not* pick up by hearing 'Tom' uttered 'in the presence of...or, as we say "associated with"' names), and I therefore know that the sentence 'Tom is a thief' claims that someone named Tom is a thief. What is not clear, however, is how a story of this sort is to be translated into a conditioning model.

Some more serious difficulties with mediation theories of reference now need to be investigated.

According to our reconstruction, the mediation theorist wishes to hold, in effect, that a sufficient condition of w_i standing for S_i is that it elicits a response r_i such that r_i is part of R_i. It is clear, however, that such a theory will require at least one additional postulate: namely, that each fractional response which mediates the reference of an unambiguous sign belongs to one and only one gross response. For, consider the consequence of rejecting this postulate. Let us suppose that the unambiguous word w_i evokes the mediating response r_i and that r_i is a member of both R_i and R_j, the functionally distinct unconditioned gross responses to S_i and S_j, respectively. The assumptions that R_i and R_j are functionally distinct and that w_i is univocal, jointly entail that S_i and S_j cannot both be referents of w_i. But if S_i and S_j are *not* both referents of w_i, then the principle that membership of a fractional response in a gross response is a sufficient condition for the former to mediate references to stimuli that elicit the latter has been violated. For, since r_i belongs to both R_i and R_j, applying this principle would require us to say that w_i refers both to S_i and S_j. Hence, if we are to retain the central doctrine that a sufficient condition for any w to refer to any S is that it elicit a part of the relevant R, we will have to hold that the fractional responses which mediate unambiguous signs are themselves unambiguous, i.e. that they belong to only one gross response. If, conversely, we do *not* adopt this condition – if we leave open the possibility that r_i might belong to some gross response other than R_i – then clearly the elicitation of r_i by w_i would at best be a *necessary* condition for w_i referring to S_i.

The same point can be made in a slightly different way. All the learning-theoretic accounts of meaning we have been discussing agree in holding that 'The essential characteristic of *sign behavior* is that the organism behaves towards the sign...in a way that is somehow "appropriate to" something other than itself' (Osgood, 1957, p. 354), i.e. that signification depends upon transfer of behavior from significate to sign. The distinctive characteristic of mediation theories is simply that they take it to be sufficient for signification if *part* of the behavior appropriate to the significate transfers to the sign. Partial identity is held to be adequate because the mediating response is simply those 'reactions in the total behavior produced by the significate ...[that] are more readily conditionable, less susceptible to extinctive inhibition, and hence will be called forth more readily in anticipatory fashion by the previously neutral [sign] stimulus' (p. 355). In short, mediators are types of anticipatory responses and they act as representatives for the gross responses they anticipate.

But notice that, just as the single-stage theorist must postulate a distinct sort of appropriate behavior (i.e. a distinct gross response) for each distinct significate, so the mediation theorist must postulate a distinct anticipatory response (i.e. a distinct mediator) for each distinct gross response. If this condition were not satisfied, there would exist mediators which can anticipate many different gross responses. Given an occurrence of such an ambiguous mediator, there would be no way of determining *which* gross response it is anticipating. From this it follows that there would be no way of determining the significance of the sign that elicited the mediator. For the theory only tells us that the significate is that object to which the gross response anticipated by the mediator is appropriate, and this characterization of the significate is informative only if the relevant gross response can be unequivocally specified, given that one knows which mediator has occurred.

Having established the result that the membership of r_i in more than one R is incompatible with the assumption that w_i is unambiguous, we can now generalize the argument to include the case of ambiguous signs. Since no sign is *essentially* ambiguous (i.e. no sign is in principle incapable of being disambiguated) there must always be some set of conditions that, if they obtained, would determine which R is being anticipated by a mediating response elicited by an ambiguous sign. This fact permits us to treat each ambiguous w as a family of univocal homonyms. Each member of such a family is assigned a distinct r such that the specification of that r includes whatever information would be required to choose it as the correct disambiguation of a given occurrence of w. Thus a word like 'bat' would be associated with two mediators the specification of one of which would involve information adequate to certify an occurrence of that word as referring to the winged mammal, while the specification of the other would involve information adequate to certify an occurrence of that word as referring to sticks used in ball games. Each of these mediators would thus correspond to precisely one R.

The treatment of ambiguous signs thus differs in no essential respects from the treatment of univocal ones; an ambiguous sign is simply a set of univocal signs all of which happen to have the same acoustic shape. It thus follows that the same arguments that show that the relation of r's to R's must be one to one if mediation theories are to provide a coherent analysis of the reference of univocal signs also show that the relation must be one to one in the case of the r's that mediate the reference of each of the homonymous members of an ambiguous sign. In either case the assumption of a one to one correspondence avoids the possibility that r_i could be part of a response other than that elicited by S_i, since for each linguistically relevant r it is now assumed that there is one and only one R of which it is a part. We can now securely adopt the cardinal principle of mediation theories of reference, viz. that w_i refers to S_i just in case r_i belongs to R_i.

Though the adoption of the postulate of a one to one correspondence between mediating and gross responses is clearly necessary if mediation theories are to propose sufficient conditions for linguistic reference, that postulate is subject to two extremely serious criticisms.

In the first place, it would appear that the sorts of response components that become 'detached' from gross responses, and which are thus candidates for the position of r_m, are not of a type likely to be associated with R's in a one to one fashion. Rather, they appear to be rather broad, affective reactions that, judging from the results obtained with such testing instruments as the semantic differential, would seem to be common to very many distinct overt responses. Second, even if it should turn out that the detachable components of R's can be placed in one to one correspondence with them, the following difficulty arises on the assumption that such a correspondence obtains.

If we assume a one to one relation between gross and mediating responses, the formal differences between single-stage and mediation theories disappears. So long as each r_i belongs to one and only one R_i, the *only* distinction that can be made between mediation and single-stage views is that, according to the former but not the latter, some of the members of stimulus-response chains invoked in explanations of verbal behavior are supposed to be unobserved. But clearly this property is irrelevant to the explanatory power of the theories concerned. It is thus not possible in principle that mediation theories could represent a significant generalization of single-stage theories so long as the mediating responses in terms of which they are articulated are required to be in one to one correspondence with the observable responses employed in single-stage S-R theories.

To put it slightly differently, two theories that explain the event e_n by claiming that it is causally contingent upon a set of prior events $e_1, e_2, \ldots e_{n-1}$, give substantially identical explanations of e_n if they differ only on the question whether all members of $e_1 \ldots e_{n-1}$ are observable. But it would appear that once a one to one association is postulated between mediating responses and the gross responses of which they are a part, precisely this relation obtains between single-stage and response-mediation explanations of verbal behavior. To put it still differently, once we grant a one to one relation between r's and R's, we insure that the former lack the 'surplus meaning' characteristic or terms designating *bona fide* theoretical entities. Hence, *only* observability is at issue when we argue about whether our conditioning diagrams ought to be drawn as in Fig. 3*a* or as in Fig. 3*b*.

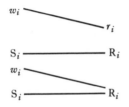

FIGURES 3*a*, 3*b*. *Schematic representation of difference between mediational and single-stage theories of the pattern of conditioning underlying the understanding of the sign* w_i. *See text*

If this argument is correct, it ought to be the case that, granting a one to one correspondence between r's and R's, anything that can be said on the mediation-theoretic view can be simply translated into a single-stage language. This is indeed the case since, in principle, nothing prevents the single-stage theorist from identifying the meaning of a word with the most easily elicited part of the response to the object for which the word stands (viz., instead of with the whole of that response). Should the single-stage theorist choose to make such a move, his theory and mediation theory would be literally indistinguishable.

In short, the mediation theorist appears to be faced with a dilemma. If his theory is to provide a sufficient condition of linguistic reference, he must make the very strong assumption that a one to one correspondence obtains between linguistically relevant mediators and total responses. On the other hand, the assumption that each mediator belongs to one and only one total response appears to destroy any formal distinction between mediation and single-stage theories, since, upon this assumption, the explanations the two sorts of theories provide are distinguished *solely* by the observability of the responses they invoke. It is, of course, possible that some way may be found for the mediation theorist simultaneously to avoid both horns of this dilemma, but it is unclear that this has been achieved by any version of mediation theory so far proposed.

Massachusetts Institute of Technology

REFERENCES

Chomsky, N. Review of Skinner's *Verbal behavior*. *Language*, 1959, xxxv, 26.

Katz, J. and Fodor, J. The structure of a semantic theory. *Language*, 1963, xxxix, 170–210.

Kimble, G. A. *Hilgard and Marquis' Conditioning and learning*. New York: Appleton-Century-Crofts, 1961.

Krasner, L. Studies of the conditioning of verbal behavior. *Psychol. Bull.*, 1958, lv, 148-170.

Mowrer, O. *Learning theory and the symbolic processes*. New York: Wiley, 1960.

Osgood, C. Motivational dynamics of language behavior. *Nebraska symposium on motivation*, 1957, 348–424.

Empirical methods in the study of semantics[a]

GEORGE A. MILLER

Methodology, the bread-and-butter of a scientist working in any given field, is usually spinach to those outside. I will try to keep my methodological remarks as brief as possible.

We now have some notion of what a theory of the interpretation of sentences might look like, and some glimmerings of the kind of lexical information about constituent elements that would be required. It is quite difficult, however, to launch directly into the compilation of a lexicon along the lines suggested by this theory; so many apparently arbitrary decisions are involved that it becomes advisable to try to verify them somehow as we go. One important kind of verification, of course, is given by a theorist's own intuitions as a native speaker of the language; such intuition is probably the ultimate court of appeal in any case. But, if possible, it would be highly desirable to have some more objective method for tapping the intuition of language users, especially if many people could pool their opinions and if information about many different words could be collected and analyzed rapidly. A number of efforts have been made to devise such methods.

As far as I am aware, however, no objective methods for the direct appraisal of semantic *contents* have yet been devised, either by linguists or psychologists. What has been done instead is to investigate the semantic *distances* that are implied by the kind of spatial and semispatial representations we have just reviewed. The hope is that from a measurement of the distances between concepts we can infer something about the coordinates of their universe. Distance can be related to similarity, and judgments of similarity of meaning are relatively easy to get and to analyze. Several methods are available.

The empirical problem is this. It is not difficult to devise ways to estimate semantic similarities among words. On the basis of such judgments of similarity we would like to construct, or at least test, theoretical descriptive schemes of the sort just reviewed. Any large-scale empirical attack on the problem would involve three steps:

(1) Devise an appropriate method to estimate semantic similarities and use it to obtain data from many judges about a large sample of lexical items. These data form a symmetric items-by-items matrix, where each cell a_{ij} is a measure of the semantic similarity between item i and j.

(2) Devise an appropriate method to explore the structure underlying the data

[a] This paper is reprinted from D. L. Arm (ed.), *Journeys in Science: Small Steps – Great Strides*, Albuquerque: University of New Mexico Press, 1967, pp. 51–73.

matrix. Among the analytic tools already available, factor analysis has been most frequently used, but various alternatives are available. In my research I have used a method of cluster analysis that seems to work rather well, but improvements would be possible if we had a solid theoretical basis for preferring one kind of representation over all others.

(3) Identify the factors or clusters in terms of the concepts in a semantic theory. In most cases, this is merely a matter of finding appropriate names for the factors or clusters. Given the backward state of semantic theory at present, however, this step is almost not worth taking. Eventually, of course, we would hope to have semantic descriptions, in terms of marker or list structures, say, from which we could not only construct sentence interpretations, but could also predict similarity data for any set of lexical items. At present, however, we have not reached that stage of sophistication.

Most of the energy that psychologists have invested in this problem so far has gone into step (1), the development of methods to measure semantic similarity. However, since there is no generally accepted method of analysis or established theory against which to validate such measures, it is not easy to see why one method of data collection should be preferred over the others. But, in spite of the sometimes vicious circularity of this situation, I think we are slowly making some headway toward meaningful methods.

There are four general methods that psychologists have used to investigate similarities among semantic atoms: (1) *scaling*, (2) *association*, (3) *substitution*, and (4) *classification*. I myself have worked primarily with classification, but I should mention briefly what alternative procedures are available. Where possible, examples of the methods will be cited, but no attempt at a thorough review is contemplated.

(1) The method of subjective *scaling* known as magnitude estimation, as described by S. S. Stevens in numerous publications, suggests itself as a simple and direct method to obtain a matrix of semantic similarity scores. So far as I know, this method has not been used in any systematic search for semantic features.

Scaling methods have been used in psychometric research. Mosier used ratings to scale evaluative adjectives along a favorable-unfavorable continuum, and Cliff used them to argue that adverbs of degree act as multipliers for the adjectives they modify. However, these studies did not attempt to construct a general matrix of similarity measures for a large sample of the lexicon, or to discover new semantic features.

One example of the use of scaling is a study reported by Rubenstein and Goodenough. They asked people to rate pairs of nouns for their 'similarity of meaning'. They used a five-point scale, where zero indicated the lowest degree of synonymy and four the highest. Their averaged results for 65 pairs of words included the following:

cord	smile	0·02
cushion	jewel	0·45
forest	graveyard	1·00
hill	woodland	1·48
magician	oracle	1·82
sage	wizard	2·46
asylum	madhouse	3·04
serf	slave	3·46
midday	noon	3·94

Although Rubenstein and Goodenough did not obtain a complete matrix of all comparisons among the 48 words they used, their results indicate that meaningful estimates can be obtained by this technique.

A difficulty that any procedure must face is that a truly enormous amount of data is required. If a lexicon is to contain, say, 10^6 word senses, then the similarity matrix with have 10^{12} cells to be filled. It is obvious immediately that any empirical approach must settle for judgments on strategic groups of word items selected from the total lexicon. But even with that necessary restriction, the problem is difficult. If, for example, we decide to work with 100 items in some particular investigation, there are still 4950 pairs that have to be judged. If we want several judges to do the task, and each judge to replicate his data several times, the magnitude of the data collection process becomes truly imposing. It is doubtful that judges could maintain their interest in the task for the necessary period of time. If a multidimensional scaling procedure is used in which judges decide which two of three items are most similar, the number of judgments required is even greater: 100 items give 161,700 triplets to be judged.

For this reason, scaling procedures do not seem feasible for any large survey of semantic items; preference must be given to methods that confront a judge with the items one at a time, where similarity is estimated on the basis of the similarities of his responses to the individual items – thus avoiding the data explosion that occurs when he must judge all possible pairs, or all possible triplets. Scaling methods should probably be reserved for those cases where we want a particularly accurate study of a relatively small number of items.

(2) Because of the historical importance of *association* in philosophical theories of psychology, more work on semantic similarity has been done with associative methods than with any other. This work has been reviewed by Creelman and also by Deese, who made it the starting point for a general investigation of what he calls 'associative meaning'.

In the most familiar form of the associative method, people are asked to say (or write) the first word they think of when they hear (or read) a particular stimulus word. When given to a large group of people, the results can be tabulated in the form of a frequency distribution, starting with the most common response and proceeding down to those idiosyncratic responses given by only a single person. Then the similarity of two stimulus words is estimated by observing the degree to which their response distributions coincide.

The procedure for estimating the degree of similarity from two response distributions is a very general one that has been used in one form or another by many workers. The logic behind it is to express the measure of similarity as a ratio of some measure of the *intersection* to some measure of the *union* of the two distributions. In the case of word associations, the responses to one word constitute one set, the responses to another word another. The intersection of these two sets consists of all responses that are common to the two; the union is generally interpreted to be the maximum number of common responses that could have occurred. The resulting ratio is thus a number between zero and one.

The argument is sufficiently general that intersection-union ratios can be used in many situations other than word association tests; their use in studies of information retrieval, where synonymy must be exploited to retrieve all documents relevant to a given request, has been reviewed by Giuliano and Jones and by Kuhns. The ratio has been invented independently by various workers, for in one version or another it is the natural thing to do when faced with data of this type. Consequently, it has been given different names in different context – intersection coefficient, coefficient of association, overlap measure, etc. – and minor details of definition and calculation have occasionally been explored, though rather inconclusively.

The utility of an intersection-union ratio is that estimates of similarity can be obtained without actually presenting all possible pairs to the judges. The assumption underlying it, of course, is that similarity of response reflects similarity of meaning. If, as some psychologists have argued, the meaning of any stimulus is all the responses it evokes, this argument is plausible. But the notion that the meaning of a word is all the other words it makes you think of should not be accepted without some reservations.

The principal recommendation for a word association technique is its convenience of administration; it is generally given in written form to large groups of subjects simultaneously. The method gives some information about semantic features, since an associated word frequently shares several semantic features of the presented word, but it is also sensitive to syntactic and phonological association. Attempts have been made to classify associates as either syntagmatic or paradigmatic, but the results have been equivocal, e.g., if *storm* elicits *cloud*, or *flower* elicits *garden*, is the response to be attributed to paradigmatic semantic similarity or to a familiar sequential construction? The method is sensitive only to high degrees of similarity in meaning; most pairs of the words elicit no shared responses at all. And no account is taken of the different senses that a word can have; when, for example, *fly* is associated with *bird* and also with *bug*, we suspect that *fly* has been given in different senses by different people, but the data provide no way to separate them.

A variation on the association technique that combines it with the scaling methods has been developed and extensively used by Osgood, who constrains a judge's response to one or the other of two antonymous adjectives. Several pairs of adjectives are used and people are allowed to scale the strength of their response. By constraining the judge's responses to one of two alternatives Osgood obtains for all his stimulus words distributions of responses that are sufficiently similar that he can correlate them, even for words quite unrelated in meaning. When these correlations are subjected to factor analysis, a three-dimensional space is generally obtained. The position of all the stimulus words can be plotted in a three-dimensional space defined by the antonymous adjectives. It is not clear that these three dimensions bear any simple relation to semantic markers – nor does Osgood claim they should. It is true that words near one another in the space often share certain semantic features, but the method gives little hint as to what those shared features might be. Osgood's method is most useful for analyzing attitudinal factors associated with a word.

(3) Another approach uses *substitution* as the test for semantic similarity. In linguistics a technique has been developed called 'distributional analysis'. The distributional method has been most highly developed in the work of Zelig Harris.

Consider all the words that can be substituted in a given context, and all the contexts in which a given word can be substituted. A linguist defines the distribution of a word as the list of contexts into which the word can be substituted; the distributional similarity of two words is thus the extent to which they can be substituted into the same contexts. One could equally well consider the distributional similarity of two contexts. Here again an intersection-union ratio of the two sets can provide a useful measure of similarity.

Closely related to distributional similarity is a measure based on co-occurrence. Co-occurrence means that the words appear together in some corpus, where 'appear together' may be defined in various ways, e.g., both words occur in the same sentence. We can, if we prefer, think of one word as providing the context for the other, thus making the distributional aspect explicit. A union-intersection measure of similarity can be defined by taking as the intersection the number of times the words actually co-occurred, and as the union the maximum number of times they could have co-occurred, i.e., the number of times the less frequent word was used in the corpus. Co-occurrence measures have the advantage that they can be carried out automatically by a properly programmed computer. Distributional measures can in general be made automatic if a very large corpus is available – large enough that the two words will recur many times.

Several psychologists have invented or adapted variations on this distributional theme as an empirical method for investigating semantic similarities. It is, of course, the basis for various sentence completion procedures insofar as these are used for semantic analysis. The basic assumption on which the method rests is that words with similar meanings will enjoy the same privileges of occurrence, i.e., will be substitutable in a great variety of contexts.

For example, *couch* and *sofa* can be substituted interchangeably in a great variety of contexts, and they are obviously closely related in meaning. In terms of a theory of semantic markers, some such relation would be expected, since the semantic features of the words in any meaningful sentence are interdependent. The predicate *is upholstered* imposes certain restrictions on the semantic markers of its subject, and only words that have those required features can be substituted as subject without violently altering the acceptability of the sentence. *Couch, sofa, chair*, etc., are substitutable in the context *The...is upholstered*, and are similar in meaning, whereas *sugar, hate, learn, delicate, rapidly*, etc., are not. If the method is used blindly, of course, it can lead to absurd results, e.g., *no* and *elementary* are not similar in meaning just because they can both be substituted into the frame *John has studied... psychology*.

If judges are asked to say whether or not two items are substitutable in a given context, they must be instructed as to what is to remain invariant under the substitution. Various criteria can be applied: grammaticality, truth, plausibility, etc. The results can be very different with different criteria. If meaning is to be preserved, for example, only rather close synonyms will be acceptable, whereas if grammaticality is to be preserved, a large set of words belonging to the same syntactic category will usually be acceptable.

Stefflre has used distributional techniques to obtain measures of semantic similarity. He takes a particular word and asks people to generate a large number of sentences using it. Then he asks them to substitute another word for the original one in each of the sentences they have written. Taking the sentences as contexts and the whole set of substitutions as his set of words to be scaled, he creates a context-by-word matrix, and has his subjects judge whether every context-word pair in the matrix is a plausible sentence. Then he can apply an intersection-union ratio to either the sets of contexts shared by two words, or to the sets of words admissible in two contexts.

In England Jones has proceeded along different lines toward a similar goal. She uses the Oxford English Dictionary to create a list of synnoyms, or near-synonyms, for every word sense, then computes an intersection-union ratio for the number of shared synonyms.

(4) A number of workers have resorted to *classification* methods for particular purposes, but until recently there appears to have been no systematic use of these methods to explore semantic features. At present several variations of the general

DATA

S₁: (COW, TIGER)(CHAIR, ROCK)(FEAR, VIRTUE)
        (TREE)(MOTHER)

S₂: (COW, MOTHER, TIGER, TREE)(CHAIR, ROCK)
        (FEAR, VIRTUE)

S₃: (COW, MOTHER, TIGER)(CHAIR, ROCK, TREE)
        (FEAR, VIRTUE)

MATRIX

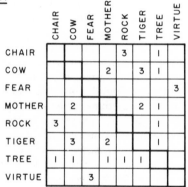

	CHAIR	COW	FEAR	MOTHER	ROCK	TIGER	TREE	VIRTUE
CHAIR				3		1		
COW				2		3	1	
FEAR								3
MOTHER	2					2	1	
ROCK	3						1	
TIGER		3		2			1	
TREE	1	1		1	1	1		
VIRTUE			3					

FIGURE I. *Illustration of how the classifications given by three judges would be tabulated in matrix form for subsequent analysis*

method are in use, but almost nothing of this work had appeared in print at the time this paper was written. In order to illustrate the classification method, therefore, I will describe a version of it that Herbert Rubenstein, Virginia Teller, and I have been using at Harvard.

In our method, the items to be classified are typed on file cards and a judge is asked to sort them into piles on the basis of similarity and meaning. He can form as many classes as he wants, and any number of items can be placed in each class. His classification is then recorded and summarized in a matrix, as indicated in Fig. 1, where data for three judges classifying eight words are given for illustrative purposes. A judge's classification is tabulated in the matrix as if he had considered every pair independently and judged them to be either similar (tabulate 1) or dissimilar (tabulate 0). For example, the first judge, S_1, uses five classes to sort these eight words; he puts 'cow' and 'tiger' together, 'chair' and 'rock' together, and 'fear' and 'virtue' together, but leaves 'tree' and 'mother' as isolates. In the data matrix, therefore, this judge's data contribute one tally in the cow–tiger cell, one in the chair–rock cell, and one in the fear–virtue cell. The data for the second and third judge are similarly scored, and the number of similarities indicated by their classifications are similarly tabulated in the matrix. Thus, in this example, all three subjects group the inanimate objects 'chair' and 'rock' together, so '3' appears in that cell; two subjects group the animate organisms 'cow' and 'mother' together; etc. After the classifications of several judges are pooled in this manner, we obtain a data matrix that can be interpreted as a matrix of measures of semantic similarity: our assumption is that the more similar two items are, the more often people will agree in classifying them together. In our experience, judges can classify as many as 100 items at a time, and as few as 20 judges will generally suffice to give at least a rough indication of the pattern of similarities.

The data matrix is then analyzed by a procedure described and programmed for a computer by S. C. Johnson of the Bell Telephone Laboratories. The general principle of this cluster analysis is suggested in Fig. 2. If we look at the data matrix of Fig. 1 for those items that all three subjects put together, then we have the five

CLUSTER ANALYSIS

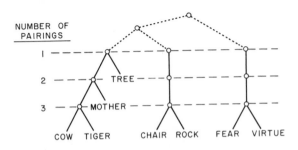

FIGURE 2. *Illustration of a cluster analysis of the data tabulated in the matrix of Fig. 1*

classes shown at the tips of the tree in Fig. 2: (cow, tiger) (mother) (tree) (chair, rock) (fear, virtue). If we relax our definition of a cluster to mean that two or more judges agreed, we have only four classes; 'mother' joins with 'cow' and 'tiger' to form a single class. And when we further relax our definition of a cluster to include the judgment of only one person, we have only three classes: 'tree' joins 'mother', 'cow', and 'tiger'. As Johnson points out, the level of the node connecting two branches can be interpreted as a measure of their similarity. The dotted lines at the top of Fig. 2 are meant to suggest that an object–nonobject dichotomy might have emerged with more data, but on the basis of the data collected, that must remain conjectural.

Our hope, of course, is that clusters obtained by this routine procedure will bear some resemblance to the kinds of taxonomic structures various theorists have proposed, and that the clusters and their branches can be labeled in such a way as to reflect the semantic markers or dimensions involved. Whether this hope is justified can be decided only by examining the results.

SEMANTIC CLUSTERS

In order to illustrate the kind of results obtained with this classification procedure, consider a study whose results are summarized in Fig. 3. Forty-eight nouns were selected rather arbitrarily to cover a broad range of concepts, subject to the constraint that half of them should be names of objects and the other half should be names of nonobjects (concepts). This important semantic marker was introduced deliberately

NOUNS

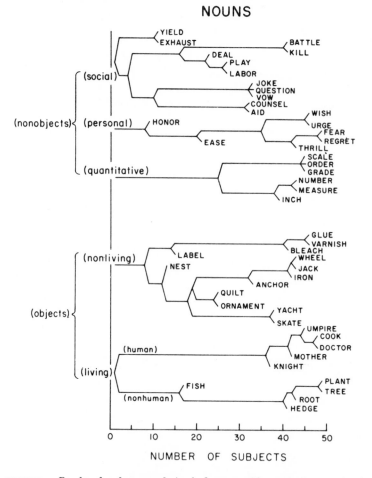

NUMBER   OF   SUBJECTS

FIGURE 3. *Results of a cluster analysis of 48 nouns, with suggested names for the clusters indicated in parentheses*

in order to see whether it would be detected by the clustering procedure; if so important a semantic marker would not come through clearly, then nothing would.

Each of the 48 nouns was typed on a 3″ × 5″ card, along with a dictionary definition of the particular sense of the word that was intended and a simple sentence using the word in that sense. The cards were classified by 50 judges, their results were tabulated in a data matrix, and cluster analysis was performed on the data matrix in order to determine what Johnson calls the optimally 'compact' set of clusters. The five major clusters that were obtained are shown in Fig. 3, where they are named, quite intuitively, 'living objects', and 'nonliving objects', 'quantitative concepts', 'personal concepts', and 'social concepts'. The finer structure within each of these clusters is also diagrammed in Fig. 3. For example, the tree graph shows that 48 of the 50 judges put 'plant' and 'tree' into the same class; that 42 or more judges put 'plant', 'tree', and 'root' into the same class; and that 40 or more judges put 'plant', tree', 'root', and 'hedge' into the same class.

Did the semantic marker that was deliberately introduced into the set of words reappear in the analysis? Yes and no. The clusters obtained did not contradict the hypothesis that our judges were sorting with this semantic distinction in mind, yet their data indicate a finer analysis into at least five, rather than only two clusters, so the object marker is not completely verified by these data. Nonetheless, the results were sufficiently encouraging that we continued to study the method.

The 48 nouns listed in Fig. 3 were also chosen to have the characteristic that each of them could also be used as a verb. In another study, therefore, the verb senses of these words were defined and illustrated on the cards that the judges were asked to classify. When they are thought of as verbs, of course, the object-concept distinction that is so obvious for these words in their noun usages is no longer relevant; the object marker would not be expected to appear in the results of the verb classifications, and in truth it did not. The results of the verb study are not presented here, however, because I do not yet understand them. The object marker did not appear, but neither did anything else that I could recognize. It is my impression that judges were too much influenced by other words in the particular sentences in which the verbs were illustrated. Perhaps the semantic analysis of predicates is basically different from the analysis of subjects; perhaps verbs signify rather special formulae – complex functions into which particular nouns can be substituted as arguments – and the classification of these functions is more complex, more contingent, more difficult than the classification of their arguments. This is a deep question which I am not prepared to discuss here.

There is, of course, an important syntactic basis for classifying English words, i.e., the classification into parts of speech. The data in Fig. 3 were obtained for a single part of speech – nouns – and so do not give us any indication of the relative importance of syntactic categories. Fig. 4, however, shows some results obtained by Jeremy M. Anglin with a set of 36 common words consisting of twelve nouns, twelve verbs, six adjectives, and six adverbs. Twenty judges classified these concepts and the analysis of their data reveals five major clusters. Four of these clusters reflect the syntactic classification, but one – 'sadly', 'suffer', and 'weep' – combines an adverb with two verbs. With this one exception, however, adult judges seem to work by sorting the items on syntactic grounds before sorting them on semantic grounds.

It is important to notice, however, that the results summarized in Fig. 4 were obtained with adult judges. Anglin also gave the same test to 20 subjects in the 3rd and 4th grades, to 20 in the 7th grade, and to 20 more in the 11th grade (average ages about 8·5, 12, and 16 years, respectively). The clusters obtained from the youngest group of judges are shown in Fig. 5. It is obvious that children interpret the task quite differently. When asked to put things together that are similar in meaning, children tend to put together words that might be used in talking about the same thing – which cuts right across the tidy syntactic boundaries so important to adults. Thus all 20 of the children agree in putting the verb 'eat' with the noun 'apple'; for many of them 'air' is 'cold'; the 'foot' is used to 'jump'; you 'live' in a 'house'; 'sugar' is 'sweet'; and the cluster of 'doctor', 'needle', 'suffer', 'weep', and 'sadly' is a small vignette in itself.

These qualitative differences observed in Anglin's study serve to confirm develop-

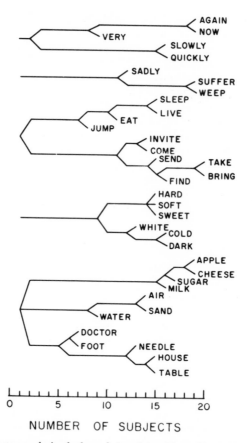

FIGURE 4. *Cluster analysis of 36 words by adult subjects. Note that syntactic categories are faithfully respected. Data from J. M. Anglin*

mental trends previously established on the basis of word association tests with children – an excellent discussion of this work has been given by Doris R. Entwisle – where it is found that children give more word association responses from different syntactic classes than do adults. This trend also appears in the classification data. In Fig. 6 some particular word pairs have been selected as illustrating most clearly the changes Anglin observed as a function of age. The thematic combination of words from different parts of speech, which is generally called a 'syntagmatic' response, can be seen to decline progressively with age, and the putting together of words in the same syntactic category, generally called a 'paradigmatic' response, increases during the same period.

Although there is no basic contradiction between results obtained with word association methods and with word classification methods, some aspects of the sub-

GRADES 3 AND 4

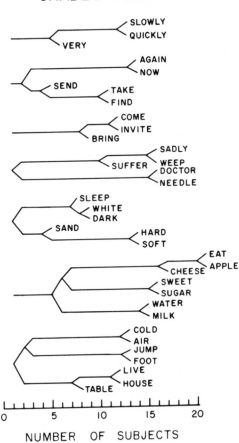

FIGURE 5. *The same 36 words of Fig. 4 were classified by children in the 3rd and 4th grades. Note violation of syntactic categories. Data from J. M. Anglin*

jective lexicon seem to be displayed more clearly with the classification procedures. In order to make some comparison between the two methods, we took word association data collected by Deese and used Johnson's cluster analysis on them. In this particular study, Deese used the word 'butterfly' as a stimulus and obtained 18 different word associations from 50 undergraduates at Johns Hopkins. Then he used these 18 responses as stimuli for another group of 50 subjects. He then had response distributions for 19 closely related words, so that he was able to compute intersection coefficients (a particular version of an intersection-union ratio) between all of the 171 different pairs that can be formed from these 19 words. Deese published the intersection coefficients in a matrix whose entries could be interpreted as measures of associative similarity between words. When Johnson's cluster analysis was carried out on this data matrix, the results showing in Fig. 7 were obtained.

The clusters obtained with the word association data are a bit difficult to interpret. If we ask whether these clusters preserve syntactic classes, the answer depends on whether we consider certain words to be nouns or adjectives. For example, 'blue' can be used either as an adjective (as in the phrase 'blue sky') or as a noun (as in the phrase 'sky blue'); 'flower' would normally be considered as either a noun or

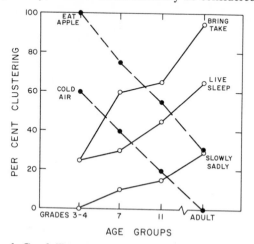

FIGURE 6. *Graph illustrating some developmental trends in the classification data. Data from J. M. Anglin*

a verb, but in 'flower garden' it functions as an adjective; 'fly' in the cluster with 'bird' and 'wing' is probably a verb; but it might be a noun; etc. One could also argue that many of the words have multiple meanings; for example, some of the subjects who associated 'spring' and 'sunshine' might have been thinking of 'spring' as a season of the year and others might have meant it as a source of water. In short, data of this sort are useful when words are to be dealt with in isolation, as they often are in verbal learning experiments, but they do not contribute the information we need in order to understand how word meanings work together in the interpretation of sentences.

For purposes of comparison, therefore, we appealed to a lexicographer: the 19 words in Deese's study were looked up in a child's dictionary, where a total of 72 different definitions were found. Each definition was typed onto a separate card, along with the word defined and a sentence illustrating its use. J. M. Anglin and Paul Bogrow tested 20 judges with these 72 items. Their results are shown in Fig. 8. Anglin and Bogrow found nine major clusters, which are quite different from Deese's associative clusters and much closer to the requirements of a semantic theory. For example, there were twelve senses of 'spring'. Ten of these comprise a single cluster, and the similarity measure suggests how a lexical entry for these ten might be organized. 'Spring' in the sense of a source of water did not fit into this cluster, nor did 'spring' in the seasonal sense; those two senses would have to have separate entries. Whereas the preceding studies illustrated the use of the classification for widely different concepts, this one indicates that the method might also be useful for investigating the finer details of closely related meanings.

One final example of this method may be of interest. As mentioned previously, Osgood and his coworkers have made extensive use of rating scales defined by antonymous adjectives in order to define a coordinate system in which meanings can be characterized by their spatial position. We decided, therefore, to use antonymous pairs of adjectives in a classification study. One hundred of the adjective

CLUSTER ANALYSIS
OF WORD ASSOCIATION DATA
(Deese, 1965)

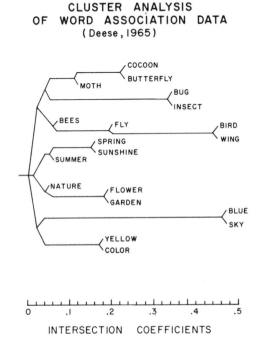

INTERSECTION   COEFFICIENTS

FIGURE 7. *The cluster analysis procedure developed by Johnson was applied to Deese's data on the intersection coefficients for word associations*

pairs Osgood had used were selected and typed on cards – this time without definitions or examples, since the antonymous relations left little room for ambiguity – and 20 judges were asked to classify them. The results for 41 of these 100 pairs are shown in Fig. 9 to illustrate what happened. (Data for the other 59 pairs is analogous, but limitations of space dictate their omission.)

Osgood finds rather consistently that the most important dimension in his semantic differential is the good–bad, or evaluative dimension. Most of the antonymous pairs that were heavily loaded on Osgood's evaluative dimension turned up in our cluster analysis in the three clusters shown in the lower half of Fig. 9. Inspection of these three clusters suggests to me that our judges were distinguishing three different varieties of evaluation which, for lack of better terms might be called moral, intellectual, and esthetic. To the extent that Osgood's method fails to distinguish among these varieties of evaluation, it must be lacking in differential sensitivity.

Fig. 9 also presents a large cluster of adjectives that, according to the introspective reports of some judges, might be considered as going together because 'they can all be used to describe people'. It is not easy to know what this characterization

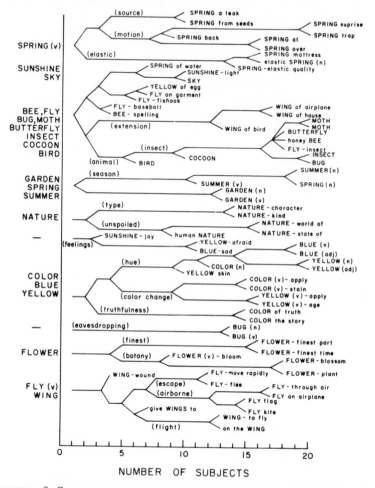

FIGURE 8. *Seventy-two senses of the nineteen words studied by Deese (see Fig. 7) were classified and a cluster analysis of the classification data was performed*

means, since almost any adjective can be used to describe someone, but perhaps it points in a suggestive direction. It should be noted, however, that this characterization is not given in terms of similarities of meanings among the adjectives, but rather in terms of similarities among the words they can modify. Once again, therefore, we stumble over this notion that the nouns may have a relatively stable semantic character, but the words that go with them, the adjectives and verbs, are much more dependent on context for their classification.

There are still difficulties that must be overcome before the classification method can be generally useful. Some way must be found to work with more than 100 meanings at a time. Some way should be sought to locate generic words at branching points. Effects of context – both of the sentence in which the meaning is exemplified, and also of the context provided by the other words in the set to be classified – must be evaluated. Relations of cluster analysis to factor analysis need to be better

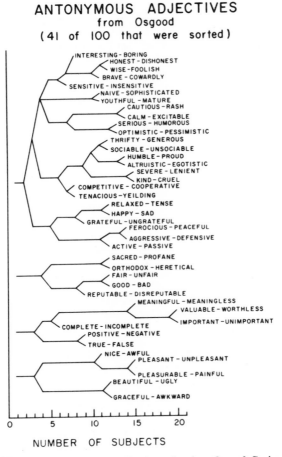

ANTONYMOUS ADJECTIVES
from Osgood
(41 of 100 that were sorted)

NUMBER OF SUBJECTS

FIGURE 9. *One hundred pairs of antonymous adjectives taken from Osgood, Suci, and Tannenbaum were sorted by 20 judges; 41 pairs are shown here as they clustered under the classification procedure. Note that the good–bad dimension so important in the semantic differential of Osgood is here analyzed into three separate clusters*

understood, and so on and on down a catalogue of chores. But the general impression we have formed after using the classification method is that, while it is certainly not perfect, it seems to offer more promise for semantic theory than any of the other techniques psychologists have used to probe the structure of the subjective lexicon.

Summary

A too short summary of this paper might be that language is what it is because we use it to say things. The capacity to say something – to affirm or deny some comment about some topic – may not be uniquely limited to man, but certainly he is better at it than any other animal. Saying something requires that we have certain grammatical machinery in our languages, ways to combine topics and comments, ways to make one sentence the topic for another comment. Saying something also requires that we have certain semantic machinery, so that what we say can be interpreted by our listener on the basis of what he knows about the meanings of its constituent parts. These are problems that linguists and psychologists share, and that form the kernel of the young science called psycholinguistics. If and when we are able to achieve a deeper understanding of what men do when they say something, we should be able to use that understanding to improve the communication of meaning.

Rockefeller University

BIBLIOGRAPHY

Chomsky, N. *Syntactic structures*. The Hague: Mouton, 1957.
  Aspects of the theory of syntax. Cambridge, Mass.: M.I.T. Press, 1965.
Cliff, N. Adverbs as multipliers. *Psychol. Rev.*, 1959, LXVI, 27–44.
Creelman, Marjorie B. *The experimental investigation of meaning: A review of the literature*. New York: Springer, 1966.
Deese, J. *The structure of associations in language and thought*. Baltimore: The Johns Hopkins Press, 1965.
de Languna, Grace A. *Speech: Its function and development*. First published 1927; reissued by Indiana University Press, 1963.
Entwisle, Doris R. *Word associations of young children*. Baltimore, Md.: The Johns Hopkins Press, 1966.
Fodor, J. A. and Bever, T. The psychological reality of linguistic segments. *J. verbal Learning verbal Behavior*, 1965, IV, 414–20.
Frege, G. On the sense and nominatum. Trans. by H. Feigl from Über Sinn und Bedeutung, *Z. f. Philos. u. Philos. Kritig*, 1892, and reprinted in H. Feigl and W. Sellars (eds.), *Readings in Philosophical Analysis*. New York: Appleton-Century-Crofts, 1949, 85–102.
Garrett, M. Structure and sequence in judgments of auditory events. Unpublished paper. Institute of Communication Research, University of Illinois, 1964.
Garrett, M., Bever, T. and Fodor, J. A. The active use of grammar in speech perception. *Perception & Psychophysics*, 1966, I, 30–2.
Giuliano, V. E. and Jones, P. E. Studies for the design of an English command and control language system. ESD-TDR-63-673, Arthur D. Little, Inc., Cambridge, Mass., 1963.
Harris, Z. S. Distributional structure. *Word*, 1954, X, 146–62.
Johnson, S. C. Hierarchical clustering schemes. *Psychometrika*, 1967, 32.
Jones, Karen S. Experiments in semantic classification. *Mech. Trans. & Computational Linguistics*, 1965, VIII, 97–112.
Katz, J. J. and Fodor, J. A. The structure of a semantic theory. *Language*, 1963, XXXIX, 170–210.
Katz, J. J. *The philosophy of language*, New York: Harper and Row, 1966.

Kuhns, J. L. The continuum of coefficients of association. In M. E. Stevens, L. Heilprin and V. E. Giuliano (eds.), *Proceedings of the Symposium on Statistical Association Methods for Mechanized Documentation*. National Bureau of Standards, Washington, D.C., 1965.

Ladefoged, P. and Broadbent, D. E. Perception of sequence in auditory events. *Quart. J. exper. Psychol.*, 1960, XII, 162–70.

Mosier, C. I. A psychometric study of meaning. *J. soc. Psychol.*, 1941, XIII, 123–40.

Osgood, C. E., Suci, G. J. and Tannenbaum, P. H. *The measurement of meaning*. Urbana, Illinois: University of Illinois Press, 1957.

Rubenstein, H. and Goodenough, J. B. Contextual correlates of synonymy. *Comm. ACM*, 1965, VIII, 627–33.

Stefflre, V. *Language and behavior*. Reading, Mass.: Addison-Wesley, forthcoming.

Some lexical structures and their empirical validity[a]

THOMAS G. BEVER and
PETER S. ROSENBAUM

Discussions of a semantic theory of language can flounder on the determination of the data which that theory is intended to describe. In this paper we shall avoid questions of the legitimacy of particular goals for semantic theory. Nor are we concerned primarily with the nature of the computational device used in analyzing the meanings of whole sentences. Rather we report some recent investigations of several kinds of lexical devices which are necessary formal prerequisites for the description of semantic phenomena.

The empirical reality of any descriptive device is initially supported by the facts it describes. It is further supported by the validity of predictions which it justifies. In this paper we show that certain hierarchical structures proposed for an adequate description of semantic lexical structure also correctly predict the potential occurrence of certain types of words and interpretations and correctly reject the potential occurrence of other types of words and interpretations.

The empirical phenomena and the formal devices which describe them are reflections of problems and solutions which arise in other behavioral phenomena (e.g. in anthropology or psychology). The basic phenomenon we explore is the representation of classes of words by features, like 'animateness', 'plantness', and 'livingness'. These features are themselves organized hierarchically (e.g. anything that is 'living' must be a 'plant' or 'animate'). Finally, certain hierarchies express relations between particular words rather than between classes or words. These relations are maintained in the metaphorical extensions of the literal interpretations of these words. At each point we show how the linguistic facts justify certain formal constraints, and we demonstrate the empirical extensions of each of these descriptive devices.

Binary features

Consider first the use of features in semantic description. It is not radical to propose that words are categorized in terms of general classes. Nor is it novel to consider the

[a] This work was supported by the MITRE corporation, Bedford, Mass., Harvard Society of Fellows, NDEA, A.F. 19(68)-5705 and Grant #SD-187, Rockefeller University and IBM. A preliminary version of this paper was originally written as part of a series of investigations at MITRE corporation, summer 1963, and delivered to the Linguistic Society of America, December 1964. We are particularly grateful to Dr D. Walker for his support of this research and to P. Carey and G. A. Miller for advice on this manuscript.

The paper is reprinted from R. A. Jacobs and P. S. Rosenbaum (eds.), *Readings in English Transformational Grammar*, Blaisdell, 1970.

fact that these classes bear particular relations to each other. Consider, for example, the English words in the matrix presented in ⟨1⟩:

⟨1⟩	boy	tree	sheep	carcass	log	water	blood	*'og'	*'triffid'
human	+	−	−	−	−	−	−	−	−
animate	+	−	+	+	−	−	−	−	+
living	+	+	+	−	−	−	+	−	+
plant	−	+	−	−	+	−	−	+	+

In this chart words are represented in terms of particular binary dimensions. In each case the particular semantic dimension, or 'semantic feature', is the marking which specifies particular aspects of the semantic patterning of the lexical item. In ⟨2⟩ the positive value for each feature is associated with some sample sentence frames which indicate the kind of systematic restrictions each feature imposes. Thus in ⟨2a⟩ any noun marked '+human' can, among other things, appear in the sentence 'The *noun* thought the problem over', but any noun marked '−animate' cannot appear in the frame 'the *noun* sensed the danger'; any noun marked '+plant' can appear in the frame 'the *noun* has damaged clorophyll'. The privileges of occurrence in these kinds of frames of the different nouns in the figure are indicated by the placement of the '+' and '−' markings on the separate features.

⟨2⟩ (a) +human: The ——————— thought the problem over
                               answered slowly
                               was a beautiful actor

    (b) +animate: The ——————— sensed the danger
                               ate the food
                               ran away

    (c) +living: The ——————— died

    (d) +plant: The ——————— has damaged clorophyll
                               needed to be watered
                               grew from the seed

It has rarely been questioned that such classifications play a role in natural language.[a] But the problem has been to decide which of the many possible aspects of classification are to be treated as systematically pertinent to a semantic theory. We cannot claim to have discovered all and only the semantic features of natural language. However, it is absolutely necessary to assume that there is some universal set from which particular languages draw their individual stock.

Furthermore, if the semantic analysis of natural language is to achieve explanatory adequacy, there must be a principled and precise manner of deciding among competing semantic analyses. In this way the semantic analysis which is ultimately chosen can be said to be the result of a formal device, and not due to the luck of a formal linguist. If the particular analysis is chosen on the basis of precise, formal criteria then the reality of the predictions made by the device constitutes a confirmation of the general linguistic theory itself.

Lexical analysis of single words includes a specification in terms of a set of semantic features. We propose, furthermore, that the most highly valued semantic analysis

---

[a] Modern discussions of such features and their integration within grammatical theory can be found in Katz and Fodor, 1963; Chomsky, 1965; Miller, 1967. In this article we will assume that the reader has a basic familiarity with the role of semantic analysis in current transformational linguistic theory.

which meets these constraints be the one which utilizes the smallest number of symbols in a particular form of semantic analysis. For a given natural language this will direct the choice of which features are drawn from the universal set as well as the assignment of predictable features in the lexicon itself.

A hierarchy of semantic features

For instance, the features in the matrix in ⟨1⟩ are specified with some redundancy. We can see from inspection of the frames in ⟨2a⟩ and ⟨2b⟩ that if a noun can act as '+human' then it can occur as '+animate' although the reverse is not necessarily true. For example, if we can say 'the man thought it over' we can also say 'the man sensed the danger', but if we can say 'the ant ran away' we cannot necessarily say 'the ant answered slowly'.

Consequently, any word marked '+human' can be predicted as being '+animate'. This prediction is represented in rule ⟨3a⟩, which obviates the necessity of lexical entries for 'animateness' in nouns which are marked as 'human'.

⟨3⟩ (*a*) +human ⟶ +animate
    (*b*) +animate ⟶ −plant

Similarly, it is predictable that if a noun is '+animate' it is '−plant' in English, although the reverse is not true. Then, rule ⟨3b⟩ allows the features '−plant' to be left out of the lexical entry for 'ant'. Such redundancy rules simply reflect a class-inclusion hierarchy among the semantic features (e.g. ⟨3c⟩).

⟨3⟩ (*c*)

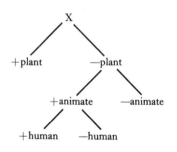

The evaluation criterion requires the simplest acceptable solution to be used. Accordingly, the solution which includes the semantic redundancy rules in ⟨3⟩ is clearly indicated over a solution without these rules, given that the theory allows such rules at all. (Notice that rule ⟨3b⟩ is *not* universal but is a rule peculiar to English, since there are languages in which at least some plants can be considered animate.)

We can now ask if the predictions made by this analysis about new lexical items are valid or not. Consider the two potential English words, 'og' and 'triffid', on the right hand side of figure ⟨1⟩. 'og' is intended to be a word which signifies the dead remains of an entire plant, in the way which 'carcass' indicates the dead remains of an entire animal. 'triffid', on the other hand, indicates a kind of pine tree that moves and communicates.

Neither of these words exists in everyday English, and they may both be classified as semantic lexical gaps – that is, they are combinations of semantic features which are not combined within a single lexical entry in English. Yet the reasons for their non-occurrence are radically different according to the analysis in ⟨3⟩. Rule ⟨3 b⟩ in principle blocks the appearance of any word like 'triffid' which is both plant and animal, but the word 'og' is not blocked by anything in the lexical rules. That is, 'triffid' is classified by our analysis as a *systematic gap*, while the non-occurrence of 'og' is entirely *accidental*.

This distinction, made as an incidental part of our analysis, appears to be supported by the intuitions of speakers of English. Nobody would be surprised to learn that professional loggers and foresters have some word in their vocabulary which indicates the whole of a dead tree. But everybody would be surprised to learn that foresters have a single word for trees which they think are animals. [It is that sense of surprise for English speakers which made possible *The Day of the Triffids*, a science-fiction novel in which the discovery of this concept is a primary vehicle for the plot.]

This result represents an independent corroboration of the general form of the semantic lexicon and the evaluation criteria which we proposed above. The rules in ⟨3 a, b⟩ were not introduced to block systematic lexical gaps, but to achieve the highest valued semantic analysis. The fact that the systematic gaps they predict correspond to the intuitive ones is a distinct empirical corroboration of the precise formal characteristics of this form of semantic theory.

We have shown that an explanatory lexical semantic theory is possible, and have indicated some of its characteristics and empirical validations. In the next sections we briefly indicate some other semantic lexical devices, the lexical distinctions which they predict, and the validity of those predictions.

Hierarchies among lexical items

Not all semantic phenomena can be handled by binary features, even if those features are related in a class-inclusion hierarchy. Consider the sentences in ⟨4⟩. Why is it that the sentences on the right are grammatically deviant? It is not simply the case that they are counterfactual, since ⟨4 e i⟩ is counterfactual, but does not involve the same kind of violation as ⟨4 e ii⟩. The first doesn't happen to be true; the second couldn't possibly be true.

⟨4⟩ (a) (i) the arm has an elbow   *(ii) the elbow has an arm
   (b) (i) the elbow is a joint   *(ii) the joint is an elbow
   (c) (i) the finger has a knuckle   *(ii) the knuckle has a finger
   (d) (i) the thumb is a finger   *(ii) the finger is a thumb
   (e) (i) the ant has an arm   *(ii) the arm has an ant

These facts justify the assumption that lexical entries are themselves arranged in two simultaneous hierarchies: one represents the inalienable inclusion of *Have*, and the other represents the class membership of *Be*. In ⟨5⟩ the solid vertical lines represent the *Have* hierarchy and the dotted horizontal lines represent the *Be* hierarchy. [We use the convention that the subject of an inalienable *Have* sentence must dominate the object in this tree, and that the subject of a generic *Be* must be dominated by the predicate object. Thus we can say, 'the body has an arm', 'the arm is a limb', but not the reverse.]

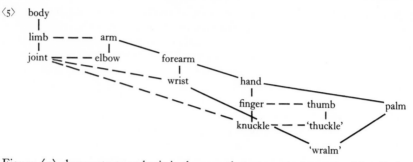

⟨5⟩

Figure ⟨5⟩ demonstrates why it is the case that semantic features with a limited number of states cannot handle these relations and block the kinds of incorrect sentences on the right in ⟨4⟩. Figure ⟨5⟩ allows the sentences in ⟨6⟩:

⟨6⟩ a body has an arm   an arm is a limb
    an arm has a forearm   an elbow is a joint
    a forearm has a hand   a thumb is a finger
    a hand has a finger   a wrist is a joint
    a finger has a knuckle

Any reversal of the nouns in these sentences would produce a deviant sentence, so that just to handle these cases a binary feature analysis would require six states.

It is clear that we must distinguish between lexically permanent hierarchies as in ⟨5⟩ and the productive use of 'have' and 'be' in actual sentences. Thus the fact that we can say 'the tree *is* nice' and 'a tree *has* importance' does not necessarily require a corresponding lexical hierarchy. Such sentences with 'Be' and 'Have' are 'productive' (or alienable) uses of *Have* and *Be* rather than the 'lexical' (inalienable) use exemplified in ⟨4⟩ and ⟨6⟩.

The implications of lexical hierarchies for some 'syntactic' phenomena

There are several 'syntactic' consequences of the inclusion of hierarchies as part of lexical structure. (As could be expected for grammatical structures dependent on the lexicon, the distinction between 'semantic' phenomena and the corresponding 'syntactic' phenomena is extremely difficult; and beside our point.)

SELECTIONAL RESTRICTIONS

Lexical items are marked as to the binary classes which they share. As demonstrated in ⟨2⟩ these class-markings restrict the kinds of noun-verb-noun combinations that may appear within a clause. Such constraints are referred to as 'selectional restrictions'.

The assumption that lexical entries are organized in several hierarchies makes possible some simplifications of the representation of selectional restrictions. Consider these sentences:

⟨7⟩ the firearm shoots bullets   *the sword shoots bullets
    the pistol shoots bullets   *the knife shoots bullets
    the rifle shoots bullets   *the dog shoots bullets
    the revolver shoots bullets
    the derringer shoots bullets
    the six-shooter shoots bullets

In general it is the case that if a lexical item can be selected in a construction then everything dominated by that lexical item in the *Be* hierarchy (directly or indirectly) can also fit into that construction (e.g. the hierarchy firearm $<$ pistol rifle etc. predicts that everything generically true of 'firearm' is specifically true of objects subordinate to 'firearm'). Notice that if a binary feature solution were sought for the above cases, that corresponding to every level in the hierarchy which has a unique set of possible constructions, there would be a separate feature (e.g. '±shoots-bullets').[a]

A similar simplification is achieved by use of the inalienable *Have* hierarchy.

⟨8⟩ car	the venturi mixes gas and air
engine	the carburator mixes gas and air
carburator	the engine mixes gas and air
venturi	the car mixes gas and air

That is, if an item in the *Have* hierarchy is a particular active construction then the items which dominate it in the *Have* hierarchy can also fit into that construction.[b]

FEATURE ASSIMILATION

There are independent syntactic reasons for postulating the existence of the *Have* and *Be* hierarchies: consider the following sentences.

⟨9⟩ (a) (i) the boy's knee aches
    (ii) *the statue's knee aches
    (iii) the statue's knee broke
  (b) (i) the man's neck itches
    (ii) *the man's book itches
    (iii) the man's book convinced me

The sentences in ⟨9a⟩ are both well-formed with respect to the *Have* hierarchy since both 'boy' and 'statue' stand in the same relation to 'knee'. But one observes that ⟨9ai⟩ has a conceptual status which differs significantly from sentence ⟨9aii⟩. The difference is a reflection of the fact that a 'statue' is normally both inanimate and not living and, hence, cannot be subject to the set of events in which nouns that are both animate and living can participate. The descriptive problem concerns the fact that the anomalous interpretation of the phrase 'knee aches' is dependent upon information which is not included in the lexical entry for the noun 'knee'. The information that a 'knee' belongs to a noun marked either '+animate, +living'

  [a] We are not claiming that there are *no* features which are pertinent to particular restricted sets of lexical items – in fact, if one argued that the feature '± shoots-bullets' should be used, then our argument simply is that the feature +*shoots bullets* is predictable for everything below its first occurrence in the *Be* hierarchy. In other words, if other aspects of the formal treatment of the above problem require features for uniformity of notation, this can be accommodated easily. Nevertheless, it remains the case that the hierarchy can be utilized to reduce intuitively the duplication of lexical information.

  [b] Note, however, that there is something odd about 'the car mixes gas and air', although it is technically correct. There are other cases like this: 'the electric lamp has tungsten', 'the body has fingernails'. There are several potential explanations for the oddness of these sentences. (1) Certain words (e.g. carburator) are designated as referring to 'the whole' of an object and feature assimilation cannot pass through them. (2) We must distinguish between various senses of 'have'; 'have *in* it'; 'have *as part of* it'; 'have *adjacent* to it'. Then a car might be said to have a 'carburator' *in* it but not as *part of* it, while the 'venturi' is *part of* the carburator. It would not be the case that a car and a 'carburator' 'have' a venturi in the same sense of 'have'. (3) There might be a principle of linguistic performance: the more nodes an assimilation of features passes through, the lower the acceptability of the sentence.

or '−animate, −living' is necessarily dependent upon some mechanism which assigns these features to 'knee' on the basis of the underlying phrase structure configurations. This mechanism provides a formal object which can be systematically interpreted by the semantic component.

The mechanism which assimilates the relevant features in sentences ⟨9a⟩ explicitly refers only to the lexical *Have* hierarchy rather than to any sentence with 'have'. This becomes clear on consideration of the fact that such assimilation does not apply in the event that the head noun does not stand in an inalienable *Have* relation to its modifier, as in ⟨9bii⟩. Underlying ⟨9bii⟩ is the string 'the man has a book' in which the interpretation of 'have' is one of 'possession'. As was assumed earlier, there are independent reasons for assuming that the non-lexical 'have' of 'possession' is distinct in certain ways from the lexical 'have' of 'inalienability'. This hypothesis is confirmed with respect to feature assimilation. The hypothesis that feature assimilation involves the lexical *Have* hierarchy is supported further by the fact that sentences in which the *Have* hierarchy is used to its fullest extension have properties identical to those in ⟨9a⟩, as demonstrated in ⟨10⟩:

⟨10⟩ (a) the toe of the foot of the leg of the boy aches
     (b) *the toe of the foot of the leg of the statue aches

There is a similar additional motivation for the *Be* hierarchy as a necessary component of the grammar: consider ⟨11⟩.

⟨11⟩ (a) a gun which is a cannon fires cannonballs
     (b) *a gun which is a six-shooter fires cannonballs

The descriptive problem posed by the sentences in ⟨11⟩ exactly parallels that posed by the sentences in ⟨9a⟩. The peculiarity of ⟨11b⟩ is reflected by the fact only a gun which has the properties of a 'cannon' can 'fire cannonballs'. The difference between the two sentences is formally characterized by an assimilation rule which assimilates the features of the modifying noun to the 'head' noun: in ⟨11⟩ the features of 'cannon' and 'six-shooter' are assimilated to 'gun'. Such a feature assimilation provides the information requisite to the semantic interpretation of the sentences in ⟨11⟩ which is not systematically given by the underlying structure.[a] In particular, it is necessary to know what kind of 'gun' is involved in order to determine the semantic well-formedness of the predicate.

[a] It is conceivable that the assimilation rules are to be incorporated into the interpretative semantic cycle itself rather than to provide a derived semantic structure upon which the interpretative component operates. Factors bearing on this decision will not be explored in this paper. At the time when we first proposed the problem raised by sentences like those in ⟨9⟩ several of our friends and teachers argued that a syntactic solution would be forthcoming. This has been incorporated as part of the analysis of adverbs proposed by G. Lakoff. Briefly ⟨9a⟩ would be analyzed as 'the boy aches in his leg', and 'the statue aches in his leg'. The anomaly of the second would be explained as a function of the fact that 'the statue aches' is anomalous. ⟨9bii⟩ would be anomalous because there is no sentence 'the man itched in his book', while there is a sentence corresponding to ⟨9ai⟩ 'the man itched in his neck'. Whether or not one is convinced by this analysis (which we are not), the fact remains that those nouns which can act as agents with verbs, like 'ache', 'itch', 'hurt', etc. are just those which are represented in the lexical Have hierarchy as inalienably part of the head noun. For example, there is a sentence 'I hurt' and 'I hurt in my arm', but not 'I hurt in my carburetor'. Similarly words like 'book', 'idea', or 'thought' are not entered lexically. Thus, either form of syntactic analysis of these examples presupposes a lexical *Have* hierarchy. (Note that this does not preclude a 'creative' grammatical component which would allow for new *ad hoc* Have and Be relations to be produced as needed. As we continually emphasize, the ontogenesis of lexical structures is not treated in this paper.)

COMPARATIVES

There is a set of restrictions on comparative constructions which is characterized exactly in terms of the *Be* hierarchy. Consider, by way of illustration, the following sentences:

⟨12⟩ (a) (i) a cannon is more deadly than a pistol
　　　　(ii) a pistol is more deadly than a cannon
　　　　(iii) *a cannon is more deadly than a gun
　　　　(iv) *a pistol is more deadly than a gun
　　　(b) (i) *a gun is more deadly than a cannon
　　　　(ii) *a gun is more deadly than a pistol

An examination of the compared nouns in the sentences above in terms of the *Be* hierarchy which involves weapons, ⟨13⟩, reveals that the restrictions observed in these sentences can be expressed in terms of the dominance relation obtaining in this hierarchy:

⟨13⟩

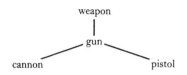

The comparative constructions in ⟨12⟩ are grammatical just in case a comparing noun neither dominates nor is dominated by a compared noun in the *Be* hierarchy. Thus, 'pistols' and 'cannons' may be compared. Furthermore, any item which 'pistol' dominates (e.g. 'derringer') can be compared either with 'cannon' or any other item dominated by 'gun' which neither dominates nor is dominated by 'pistol' (e.g. 'rifle') as in the following examples:

⟨14⟩ (a) a cannon is more deadly than a derringer
　　　(b) a rifle is more deadly than both a derringer and a cannon

Although the derivation of comparative constructions in general has not been fully resolved, it is nonetheless clear that certain restrictions on these constructions require reference to a lexical *Be* hierarchy.

Hierarchies are non-convergent

The examples we have discussed do not involve convergence within one hierarchy. That is, there are no instances of hierarchies like ⟨15⟩ in which the hierarchy branches

⟨15⟩

at X and converges at P. There are several reasons why convergent hierarchies are not tolerable. First, there are many instances in which a single lexical item is dominated by different words, e.g. ⟨16⟩:

⟨16⟩

If convergences were an acceptable part of the formalism, it would be possible to simplify ⟨16⟩ to ⟨17⟩:

⟨17⟩

But this would transfer to 'hands' all the selectional features of both 'humans' and 'clocks', many of which are mutually incompatible (notably '± animate'). Furthermore, if the senses of 'hands' are not distinguished we could not account for the oddity of sentences like:

⟨17^1⟩ The clock's face and hands are large but mine aren't.

Maintaining distinct senses of lexical items in distinct lexical hierarchies is motivated independently of the consequences for convergent hierarchies. There are also motivations against convergent hierarchies within the same 'lexical tree'. Consider the subsection of figure ⟨5⟩ in ⟨18a⟩:

⟨18⟩ (a)

```
limb — — — — — arm
  |               |
joint — — — — — elbow
        - -
            - - -
                - - knee
```

'Arm' could dominate 'joint' directly on the *Have* dimension as in ⟨18b⟩ (as well

⟨18⟩ (b)

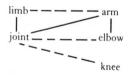

as via 'limb') since there is a sentence 'an arm has a joint'. However, this would transfer to 'joint' all the properties pertaining to 'arm'. This would in turn transfer all the features of 'arm' to 'knee', which is clearly incorrect.

The requirement that there are no convergences in the lexical hierarchy has the result that there are many distinct matrices in the lexicon which utilize the *Have* and *Be* hierarchies.[a] There are many problems about these matrices which require further investigation. We have merely outlined the essential motivations for the hierarchies as lexical structures, and shown that convergent hierarchies cannot be included.

Systematic and accidental lexical gaps explained by lexical hierarchies

The evidence above attests to the general need for lexical hierarchies as one of the structures of grammar. The hierarchies provide mechanisms for the blocking of certain obvious anomalies (e.g. $\langle 4a\text{ii}\rangle$) and for the simplification of general selectional processes. The hierarchies also make predictions about the kinds of lexical items which can and cannot occur. That is, they provide a further basis for distinguishing lexical gaps which are systematically explained by the lexical structure of the language from those gaps which have no such explanation and are therefore 'accidental'.

For example, the combination of the hierarchies as in $\langle 5\rangle$ makes several claims about English lexical structure. The placement of *finger, thumb* and *knuckle* in the lower right of $\langle 5\rangle$ strongly implies that the single lexical item indicating a 'thumb-knuckle' would be a regular extension of the present lexical system, since it would fit in the regular slot defined by the intersection of the existing *Have* and *Be* hierarchies (labeled 'thuckle'). An instance of an irregular extension would be the word which is a unique part of both the wrist and the palm (labeled 'wralm' in figure 5). The irregularity of 'wralm' is explained by the fact that it would require a convergence of the hierarchy, which is not allowed in our formulation, for independent reasons. Thus 'wralm' would require two separate hierarchies and involve added complexity.

Surely this result is an intuitive one – the concept 'thumb-knuckle' seems quite natural, but the concept of wrist-palm does not. It should be re-emphasized here that it is irrelevant to us whether or not these linguistic properties are reflections of something in the so-called 'real world'. Our present quest does not include the source for linguistic structure but seeks to define the nature of the structure itself.

The hierarchy may also predict instances in which certain uses of words are

---

[a] In fact, it might appear that an incorrect result follows from this restriction just in case a particular property or object is shared by many different words. For example, consider the word 'electron'. Convergent hierarchies would allow this word to appear only once in the lexicon:

The restriction against convergent hierarchies would appear to force such a word to have multiple representation:

However, 'electron' does not appear many times, once under each object, rather only once under the word 'object' itself. Any other word that *is* an object automatically acquires the property of *having* electrons, by the assimilation rules discussed above.

dropped from a language. Consider the treatment of the word 'sidearm' in the *Be* hierarchy introduced above:

a $\begin{Bmatrix} \text{pistol} \\ \text{sword} \end{Bmatrix}$ is a sidearm

a sidearm is a weapon

These sentences indicate that it would have to be represented with a convergence on 'pistol' as in ⟨19⟩, or with two lexical instances of pistol as in ⟨20⟩:

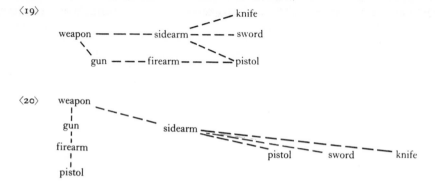

⟨19⟩

⟨20⟩

If it were not for the fact that swords (and knives, technically) are sidearms it would be possible to formulate the hierarchy without any double dominance relations:

⟨21⟩ weapon —— gun —— firearm —— sidearm —— pistol

In fact although we must allow for the possibility of the facts represented in ⟨19⟩ we have set up the formalism such that that type of structure involves a relatively large amount of complexity (as in ⟨20⟩). ⟨19⟩ is represented as two separate hierarchies to avoid the double dominance of 'pistol'. This offers an explanation of the tendency for the term *sidearm* (meaning both 'sword' *and* 'knife') to drop out of modern English, since its disappearance would reduce the complexity of ⟨20⟩ to ⟨21⟩.

As above we cannot rule out any particular aggregate of features (or possible constructions) on the basis of lexical structure rules since anything *can* occur as a lexical item. But the theory which we set up to represent lexical structure provides some formal basis for distinctions between lexical items which do occur. For instance, 'shotgun' is in an anomalous position in the hierarchy because of these sentences:

⟨22⟩ (*a*) a shotgun is a firearm
　　　(*b*) a firearm shoots bullets
　　　(*c*) *a shotgun shoots bullets[a]

This requires us to mark 'shotgun' as an exception to the hierarchy (or to the rule which predicts features on the basis of the hierarchy.) That is, sentence ⟨22c⟩ is marked as anomalous only itself as a process of an exceptional relation.

　[a] These sentences seem to suggest that sentences like 'a gun is a cannon' must be generated in the grammar in order to account for the relative clause formation. In other words, the original motivation for the lexical *Be* hierarchy would seem to disappear. This argument is untenable, however, since it is necessary to postulate that a relativized noun is definite, referring to a *specific* noun: e.g. 'a gun fires cannonballs if it (that gun) is a cannon'.

In certain cases the restriction on convergence in the hierarchy is supported directly by intuitions. For example, consider the word 'neck':

⟨23⟩ (*a*) John's head has a thick neck
John's head was frozen, including his neck
(*b*) John's torso (body) has a thick neck
John's torso (body) was frozen, including his neck

For some, sentences ⟨23 *a*⟩ are correct and ⟨23 *b*⟩ are incorrect, for others the reverse is true. For many, both ⟨23 *a*⟩ and ⟨23 *b*⟩ are acceptable at different times, *but not simultaneously*: that is, there is no single lexical hierarchy which includes a structure like ⟨24⟩

⟨24⟩

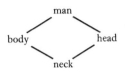

but rather the hierarchy is like ⟨25⟩.

⟨25⟩

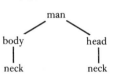

It is tempting to 'explain' these facts as an 'obvious reflection of the fact that the neck resides between the body and the head'. But there is nothing incompatible with the physical facts in an analysis which would delegate the neck as part of the 'head' and simultaneously part of the 'body'. Thus, it is not 'reality' which disallows convergences of lexical hierarchies, but the non-convergent property of the lexical system itself.

Metaphor and the form of lexical entry

The final case we shall consider involves 'historical metaphor'. For example, in modern English there is a regular rule (e.g. ⟨26⟩) which extends surface quality adjectives which are not drawn from a restricted set of abstract qualities. Thus in ⟨27 *b*⟩ the adjective 'colorful' is used with the noun 'ball' or 'idea'. Certain adjectives drawn from a restricted set are not affected by rule ⟨26⟩. For example, shapes and colors are not regularly applied to ideas, as can be seen in ⟨27 *d*⟩. Whenever

⟨26⟩ $\begin{bmatrix} \text{adjective concrete} \\ \text{surface quality} \end{bmatrix} \longrightarrow$ adjective abstract

shapes are used to modify abstract nouns, as in ⟨27 *e*⟩ they are defined by our analysis as isolated expressions and not the result of the general metaphor rule in ⟨26⟩.

⟨27⟩ (*a*) The ball is colorful
(*b*) The idea is colorful
(*c*) The ball is ovoid

  (*d*) *The idea is ovoid
  (*e*) The idea is square
  (*f*) The idea is variegated

This too seems intuitively correct. In 'the idea is square' we know that an abstract interpretation of 'square' is possible but unique – it is not the result of a general productive rule. In this case we happen to know the unique abstract interpretation of 'square' but we do not know how to interpret 'ovoid' in 'the idea is ovoid'. However, prior knowledge of the abstract interpretation of surface quality adjectives is not necessary for their metaphorical interpretation: for example, we may never have heard the sentence 'the idea is variegated' before, but we can immediately recognize it as a regular lexical extension of the 'literal' meaning of 'variegated' (i.e. 'the idea has many apparent components').

  The representation of lexical items in hierarchies facilitates the automatic interpretation of the metaphorical extensions of certain words from their original lexical structure. Consider the lexical analysis of the word 'colorful' using the *Be* and *Have* hierarchy simultaneously.

⟨28⟩

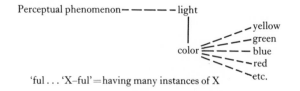

'ful . . . 'X–ful' = having many instances of X

Suppose we use the following metaphorical principle:
  a metaphorical extension of X includes the hierarchical relations of the literal
  interpretation without the specific labels:
That is, the literal meaning of 'color' is extended metaphorically as shown in ⟨29⟩:

⟨29⟩

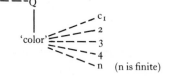

In this way the word 'colorful' as applied to 'idea' would be automatically interpreted as 'having simultaneously a large set of potentially distinguishing characteristics'. Similarly, the lexical hierarchical analysis predicts the metaphorical interpretation of 'shape' in 'the idea took shape'.

⟨30⟩

    Perceptual phenomenon — — — — object

                square
              round
       shape — — — ovoid
             rectangular
             etc. (n is finite)

In the extended usage 'shape' is interpreted as 'a particular character out of many (but finite) characters'. Note that although there is an infinite variety of shapes possible, there is a finite number listed in the lexicon: what 'the idea has shape' means is that the idea has one clear recognizable characteristic, as opposed to an unnamable characteristic.

Consider now the interpretation of 'the idea is square'. The lexical metaphor principle applies to ⟨28⟩ to give the interpretation 'the idea has a specific characteristic out of a finite list of them'. But there is no special indication as to which characteristic it is, or how it fits into the abstract realm of ideas. Thus it takes special knowledge to interpret the metaphorical extensions of lexical items at the most subordinate part of the hierarchy, if unique knowledge is also required in the original literal interpretation (e.g. that 'square' is a specific shape).

For these kinds of cases this principle provides a more general rationale for metaphorical interpretation than does the specific rule in ⟨26⟩. It distinguishes cases of metaphor which require no special knowledge ('shape, color') from those which do ('square, green'). It also offers some insight into the nature of the metaphorical extension itself.

There are many metaphors for which this sort of lexical account will not do. Thus, 'the secretary of defense was an eagle gripping the arrows of war, but refusing to loose them' is not interpretable in terms of any lexical hierarchy. What we have shown is that the interpretation of 'lexical metaphors' can be defined in terms of the original literal form of the lexical entry, and that this can distinguish between 'regular' metaphorical extensions and isolated cases of metaphor.

Conclusion

The force of the empirical validity of lexical distinctions and the corresponding descriptive mechanisms depends on the extent to which the analyses are arrived at following a formal evaluation criterion. The exact form of the distinctive features in ⟨1⟩, the feature hierarchy in ⟨3c⟩, the lexical hierarchy in ⟨5⟩ and the metaphor extension rule will ultimately be determined by the form of the universal semantic structures and the complexity of a particular description.

The descriptive devices we have presented do not exhaust what is necessary for semantic theory. What we have shown is that an explanatory semantic theory is possible, and have indicated some of its characteristics. Primarily, we have outlined the kinds of empirical considerations which must be included in the evaluation of different candidates for any explanatory semantic theory.

Rockefeller University
IBM, Yorktown Heights

Index of proper names